ITEM NO: 2025352

AIR POLLUTION

THIRD EDITION

VOLUME II

The Effects of Air Pollution

ENVIRONMENTAL SCIENCES

An Interdisciplinary Monograph Series

Editors: DOUGLAS H. K. LEE, E. WENDELL HEWSON, and DANIEL OKUN

A complete list of titles in this series appears at the end of this volume.

AIR POLLUTION

THIRD EDITION

VOLUME II

The Effects of Air Pollution

Edited by

Arthur C. Stern
Department of Environmental Sciences and Engineering
School of Public Health
University of North Carolina at Chapel Hill
Chapel Hill, North Carolina

ACADEMIC PRESS New York San Francisco London 1977

A Subsidiary of Harcourt Brace Jovanovich, Publishers

ACADEMIC PRESS, INC.
111 Fifth Avenue, New York, New York 10003

United Kingdom Edition published by
ACADEMIC PRESS, INC. (LONDON) LTD.
24/28 Oval Road, London NW1

Library of Congress Cataloging in Publication Data

Stern, Arthur Cecil.
The effects of air pollution.

(His Air pollution, third edition ; v. 2) (Environmental sciences)
Bibliography: p.
Includes index.
1. Air–Pollution. 2. Air–Pollution–Physiological effect. I. Title. II. Series: Environmental sciences.
TD883.S83 1976 vol. 2 363.6 77-7189
ISBN 0–12–666602–4

PRINTED IN THE UNITED STATES OF AMERICA

To Dick and Judy

Contents

3. Effects on Indoor Air Quality

John E. Yocom, William A. Coté, and Ferris B. Benson

PART B EFFECTS ON BIOLOGICAL SYSTEMS

4. Effects on Vegetation: Native, Crops, Forests

Walter W. Heck and C. Stafford Brandt

5. Biological Effects of Air Pollutants

David L. Coffin and Herbert E. Stokinger

6. Organic Particulate Pollutants—Chemical Analysis and Bioassays for Carcinogenicity

Dietrich Hoffmann and Ernst L. Wynder

List of Contributors

Numbers in parentheses indicate the pages on which the authors' contributions begin.

Ferris B. Benson (117), National Environmental Research Center, U.S. Environmental Protection Agency, Research Triangle Park, North Carolina

C. Stafford Brandt (157), Bellevue, Washington

David L. Coffin (231), Health Effects Research Laboratory, U.S. Environmental Protection Agency, National Environmental Research Center, Research Triangle Park, North Carolina

William A. Coté (117), The Research Corporation of New England, Wethersfield, Connecticut

Lars T. Friberg (457), Department of Environmental Hygiene, Karolinska Institute and National Environmental Protection Agency, Stockholm, Sweden

John R. Goldsmith* (457), National Cancer Institute, National Institutes of Health, Bethesda, Maryland

Peter Halpin (611), Air Pollution Technical Information Center, Office of Air Quality Planning and Standards, U.S. Environmental Protection Agency, Research Triangle Park, North Carolina

Walter W. Heck (157), Agricultural Research Service, U.S. Department of Agriculture, North Carolina State University, Raleigh, North Carolina

* Present address: Epidemiological Studies Laboratory, State Department of Health, Berkeley, California.

Dietrich Hoffmann (361), Naylor Dana Institute for Disease Prevention, American Health Foundation, Valhalla, New York

Elmer Robinson (1), Department of Chemical Engineering, College of Engineering, Washington State University, Pullman, Washington

Herbert E. Stokinger (231), Toxicology Branch, Division of Laboratories and Criteria Development, National Institute for Occupational Safety and Health, Cincinnati, Ohio

James B. Upham (65), Environmental Sciences Research Laboratory, U.S. Environmental Protection Agency, Research Triangle Park, North Carolina

Ernst L. Wynder (361), American Health Foundation, New York, New York

John E. Yocom (65, 117), The Research Corporation of New England, Wethersfield, Connecticut

Preface

This third edition is addressed to the same audience as the previous ones: engineers, chemists, physicists, physicians, meteorologists, lawyers, economists, sociologists, agronomists, and toxicologists. It is concerned, as were the first two editions, with the cause, effect, transport, measurement, and control of air pollution.

So much new material has become available since the completion of the three-volume second edition, that it has been necessary to use five volumes for this one. Volumes I through V were prepared simultaneously, and the total work was divided into five volumes to make it easier for the reader to use. Individual volumes can be used independently of the other volumes as a text or reference on the aspects of the subject covered therein.

Volume I covers two major areas: the nature of air pollution and the mechanism of its dispersal by meteorological factors and from stacks. Volume II covers the effect of air pollution upon plants, animals, humans, materials, and the atmosphere. Volume III covers the sampling, analysis, measurement, and monitoring of air pollution. Volume IV covers two major areas: the emissions to the atmosphere from the principal air pollution sources and the control techniques and equipment used to minimize these emissions. Volume V covers the applicable laws, regulations, and standards; the administrative and organizational strategies and procedures used to administer them; and the energy and economic ramifications of air pollution control. The concluding chapter of Volume II discusses air pollution literature sources and gives guidance in locating information not to be found in these volumes.

To improve subject area coverage, the number of chapters was increased from the 51 of the second edition (and 42 of the first edition) to 72. The scope of some of the chapters, whose subject areas were carried over from the second edition, has been changed. Every contributor to the second edition was offered the opportunity to prepare for this edition either a revision of his chapter in the second edition or a new chapter if the scope of his work had changed. Since 8 authors declined this offer and one was deceased, this edition includes 53 of the contributors to the second edition and 48 new ones.

The new chapters in this edition are concerned chiefly with aspects of air quality management, such as, data handling, emission inventory, mathematical modeling and control strategy analysis; global pollution and its monitoring; and more detailed attention to pollution from automobiles and incinerators. The second edition chapter on Air Pollution Standards has been split into separate chapters on Air Quality Standards, Emission Standards for Stationary Sources, and Emission Standards for Mobile Sources. Even with the inclusion in this edition of the air pollution problems of additional industrial processes, many are still not covered in detail. It is hoped that the general principles discussed in Volume IV will help the reader faced with problems in industries not specifically covered.

Because I planned and edited these volumes, the gap areas and instances of repetition are my responsibility and not the authors'. As in the two previous editions, the contributors were asked to write for a scientifically advanced reader, and all were given the opportunity of last minute updating of their material.

As the editor of a multiauthor treatise, I thank each author for both his contribution and his patience, and each author's family, including my own, for their forbearance and help. Special thanks are due my secretary, Susan Bigham, and her predecessors, who carried a hundred and one times the burden of the other authors' secretaries combined, and Eleanor G. Rollins for preparing the subject index for this volume. I should also like to thank the University of North Carolina for permitting my participation.

Arthur C. Stern

Contents of Other Volumes

VOLUME I AIR POLLUTANTS, THEIR TRANSFORMATION AND TRANSPORT

VOLUME III MEASURING, MONITORING, AND SURVEILLANCE OF AIR POLLUTION

Part A Sampling and Analysis

Part B Ambient Air Surveillance

Part C Source Surveillance

VOLUME IV ENGINEERING CONTROL OF AIR POLLUTION

Part A Control Concepts

Part B Control Devices

Part C Process Emissions and Their Control

VOLUME V AIR QUALITY MANAGEMENT

Part A Local, State, and Regional

Part B National and Worldwide

Part A

EFFECTS ON PHYSICAL AND ECONOMIC SYSTEMS

1

Effects on the Physical Properties of the Atmosphere

Elmer Robinson

I. Introduction

Pollutants in the atmosphere can bring about changes in atmospheric properties from the obvious observation of heavy smoke to subtle effects on urban temperature or regional precipitation. These active air pollutants may be either gases or particles. It is only with an understanding of

(a)

(b)

some of the basic physics of the polluted atmosphere that adequate programs of monitoring and impact assessment can be made.

This chapter will describe the various ways air pollutants may change atmospheric properties, but it will not catalog in detail all the results that have been noted in studies of air pollutants and the atmosphere. These are covered in other chapters.

Air pollutants can affect more than the visibility, although because it can be readily observed by the affected citizens, visible air pollution must be regarded as one of the more objectionable pollutant effects. Another important aspect of the atmosphere that can be affected by air pollutants is the mechanism of precipitation formation, because in this way air pollutants may play a role in altering weather patterns. An examination of the ways in which pollutants impact on these atmospheric mechanisms may provide an improved basis for judging the reality of a weather impact in a given area. In addition to precipitation changes, air pollutants have also been linked to fog frequency and persistence, especially in urban areas.

The presence of air pollutants can affect ambient temperatures through the absorption of radiant energy. The possible effects of increasing global carbon dioxide concentrations have been publicized frequently in this regard, but nitrogen dioxide and aerosols may also be factors in some local urban areas.

The most important impact on atmospheric properties, however, is still in terms of the visibility, where pollutants can seriously degrade the atmospheric transparency. Figure 1, showing two views of a portion of Seattle, Washington, one of America's more attractive urban areas, is an obvious example of the impact of air pollutants on urban visibility (*1*). There will probably be continuing strong pressure by the public for accelerated progress in air pollution control until scenes such as this haze-obscured view cease to be frequent occurrences. Thus, considerable detail on the basis for atmospheric visibility is considered justified, because a problem cannot really be tackled until it can be explained.

II. Mathematical Concepts of Visibility

A. Scattering and Absorption of Light

The effects of air pollution illustrated in Figure 1 occur because liquid and solid airborne materials both absorb and scatter light and cause a

Figure 1. Two views of Seattle, Washington, near the campus of the University of Washington. (a) March 11, 1969, at 2:30 p.m. (b) March 14, 1969, at 11 a.m. under conditions of heavy urban haze. (Photos courtesy of Dr. A. P. Waggoner, University of Washington.)

reduction in the visibility over the affected area. This reduced visibility is related to the size, concentration, and physical characteristics of the particulate pollutants present. The basis for atmospheric visibility or visual range theory has been presented in detail by a number of authors, and only some of the features generally pertinent to air pollution problems are presented here. In general, the notation and development of Middleton (*2*) are used.

Visibility is dependent upon the transmission of light through the atmosphere and the ability of the eye to distinguish an object because it contrasts with the background. A change in contrast with viewing distance occurs for both dark and light or bright objects. With dark objects, the atmosphere introduces light, called "air light," into the sight path, and the dark object appears lighter at increasing distances until it blends into the background at the horizon. In the case of light objects, light is lost from the line of sight with increasing distance. In both cases the result is the same—the contrast between the object and the background approaches zero, since the light coming along the line of sight from the target to the observer approaches the intensity of the light from the background at the horizon. When the eye can no longer distinguish a difference or contrast between the object and the background, the object cannot be seen, and it is said to be beyond the limit of visibility.

This alteration of contrast or light intensity is due to the absorption and scattering of light by the atmosphere. If a light path from an object to an observer of length x is illuminated by a light beam of intensity I, this light in passing through the incremental distance dx is reduced by I. This may be written as

$$dI = -\sigma I \, dx \tag{1}$$

where σ is the extinction or attenuation coefficient and the minus sign indicates that intensity is being reduced by the effect of the atmosphere. By integration over the path length Equation (1) becomes

$$I = I_0 \exp(-\sigma x) \tag{2}$$

where I is the intensity reaching the observer, I_0 is the original intensity, x is the length of the light path, and $\exp(A)$ indicates the base of the natural logarithm, e, to the power (A).

It can also be shown that apparent contrast C at a distance x can be expressed in a similar manner as

$$C = C_0 \exp(-\sigma x) \tag{3}$$

where C_0 is the actual contrast of the object relative to its background.

The extinction coefficient σ includes the effects of both scattering and absorption. It is sometimes useful to consider these two processes separately, in which case Equation (2) becomes

$$I = I_0 \exp[-(b + k)x] \tag{4}$$

where b is the scattering coefficient and k is the absorption coefficient.

1. Rayleigh Scattering

The extinction coefficient is generally determined by the particulate materials in the atmosphere, even though the air molecules are responsible for some scattering and absorption of incident light. Molecular scattering is the familiar Rayleigh scattering and is of little importance in air pollution work. If Rayleigh scattering were the only contributor to the extinction of light, visibility of more than 150 miles (240 km) could be calculated. Rayleigh scattering is discussed in detail by Middleton (*2*) and other authors. It is predominant for particles in the atmosphere which are much smaller in size than the wavelength of the incident light. Porch *et al.* (*3*) have investigated Rayleigh scattering as a limiting value to the observed scattering in an atmospheric sample and have found some instances where the atmosphere in remote areas did approach the Rayleigh scattering limit.

2. Mie Scattering

The scattering due to particulate air pollutants in the atmosphere is usually attributed to particles of a size comparable to the wavelength of the incident light, i.e., in the submicron size range. Mie developed the theory of scattering under these conditions, and the phenomenon is usually referred to as Mie scattering (*2*). The expression for the scattering coefficient for Mie scattering is given by

$$b = NK\pi r^2 \tag{5}$$

where N is the number of particles of radius r, and K is the scattering area ratio for a particle of radius r. This scattering area ratio K is the ratio of the area of the wave front acted on by the particle to the area of the particle. The value of K is dependent upon the radius of the particle, the wavelength of the incident light, and the refractive index m of the particulate material.

When particulate material is not homogeneous, the value of b is the sum of the n individual values and is expressed as

$$b = \sum_{i=1}^{n} N_i K_i \pi r_i^2 \tag{6}$$

Values of K are shown in Figure 2 for aerosols of sulfur, water, and oil. These data, given by Sinclair (*4*), are for light of wavelength $\lambda = 5240$ Å, which is a useful average value for daylight. The value of K will frequently be found expressed as a function of $2\pi r/\lambda$, for which α is the usual symbol.

Values of K are not available for all of the possible aerosol materials

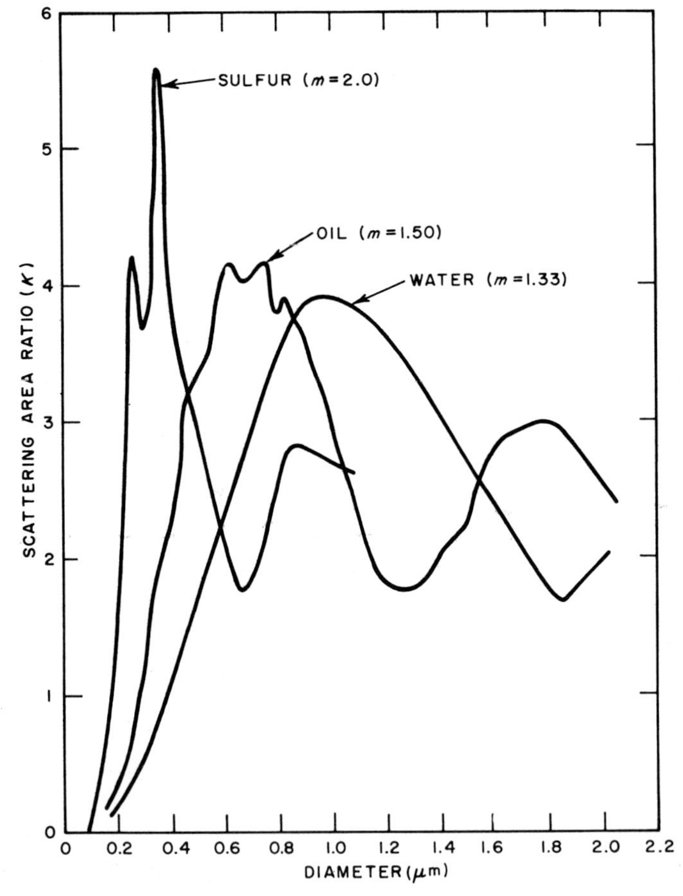

Figure 2. Scattering area ratio K as a function of diameter for spherical particles, $\lambda = 5240$ Å (*4*).

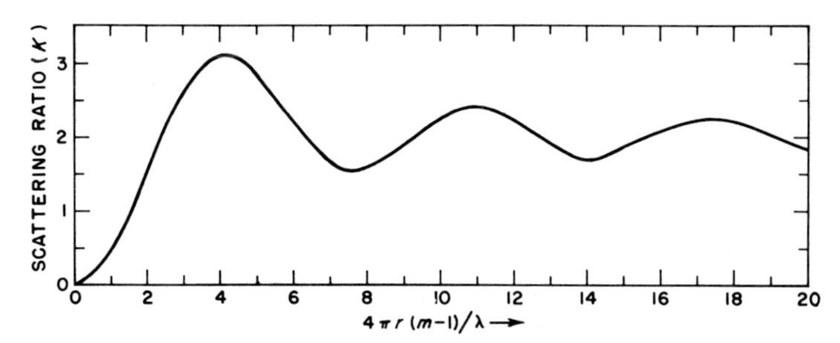

Figure 3. Approximate values of the scattering area ratio K for nonabsorbing spheres as a function of $4\pi r(m - 1)/\lambda$ (5).

of interest to air pollution investigators. However, van de Hulst (*5*) shows that for spheres that do not absorb but only scatter the incident radiation, K can be adequately approximated by the curve of Figure 3. The abscissa is in terms of $4\pi r(m - 1)/\lambda$, where m is the refractive index of the particulate material, r is its radius, and λ is the wavelength of the incident light.

The data of Figure 3 are of more than theoretical interest to the air pollution investigator, since K values for most condensed smokes can be approximated by these data (*5, 6*).

3. Absorption

In Equation (4), the extinction coefficient was separated into scattering and absorption factors. Although many common air pollutants, such as submicron-sized wood smoke (*6*) and probably oil mist and incineration smoke, produce light extinction primarily by scattering, for smokes of obvious color, such as open-hearth emissions and soot, absorption becomes an important factor in the light extinction calculation.

Middleton (*2*) summarizes a number of measurements of industrial "haze" in England during the 1940s showing that the absorption co efficient was more or less equivalent to the scattering coefficient. The absorption factor was particularly important close to the ground and then decreased with altitude; however, layers with strong absorption were sometimes observed aloft also.

Some calculated data on extinction, scattering, and absorption area ratios, or K values, as given by van de Hulst (*5*), are shown in Figure 4. These data are for iron spheres with a refractive index of $m = 1.27 - 1.37i$, where i signifies $\sqrt{-1}$. Note that the abscissa is in units of $2\pi r/\lambda$, or α. The significant feature of these calculations is the fact that the coefficients for absorption and scattering are comparable for a wide range of particle size.

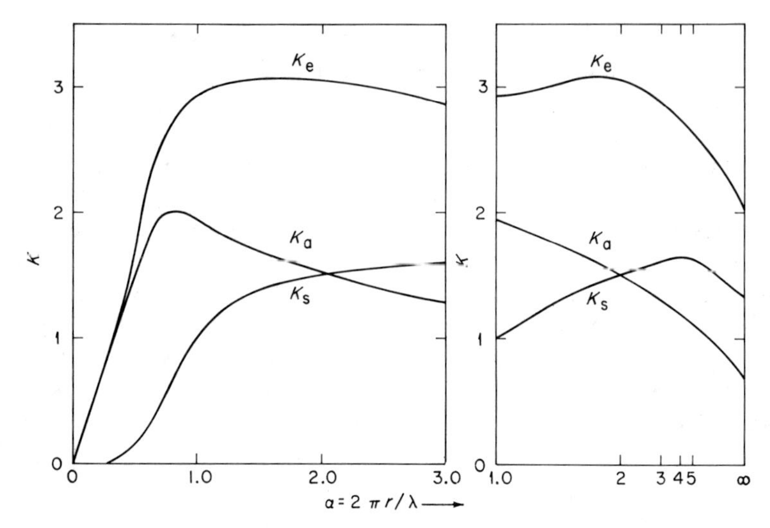

Figure 4. Area ratio values for extinction K_e, absorption K_a, and scattering K_s for iron spheres (*5*).

On the basis of calculations of the optical absorption coefficient for generalized atmospheric aerosols, Waggoner *et al.* (*7*) state that values of k, the absorption coefficient in Equation (4), over the distance x depends only on the total mass of the particles and an average value of the imaginary part of the refractive index. Waggoner argues that size and shape distribution measurements are not justified in a detailed assessment of particle absorption, because these procedures cannot be carried out within the accuracy needed to improve on the simpler mass and refractive index expression, namely, a factor of 2 or 3.

Limited information on the scattering and absorption characteristics of carbon for α less than 1.57 can be obtained from some of the calculations of Stull and Plass (*8*). Their calculations of extinction, scattering, and absorption cross sections for 0.2 μm diameter carbon particles at $\lambda = 4000$ Å, $\alpha = 1.57$, give $K_e = 3$, $K_s = 1.3$, and $K_a = 1.7$. These are very close to the values shown for iron in Figure 4. With decreasing values of α, K_s for carbon seems to follow closely the curve shown for iron, while K_e and K_a for carbon appear to decrease more rapidly than for iron.

An approximation of the radiative properties of atmospheric carbon aerosols was made by Twitty and Weinman (*8a*) based on the complex index of refraction data for a number of different carbonaceous materials including graphite, coals, and soots. From Mie theory and using two urban aerosol models, Twitty and Weinman (*8a*) estimated that for visible light the calculated extinction coefficient for carbonaceous aerosols was about twice the magnitude of the scattering coefficient.

The data on extinction coefficients as a function of particle size shows that submicron particles are more effective, per unit mass, in reducing visibility than are larger particles. For water droplets the most effective size is about 0.8 μm, where the scattering effectiveness per gram is about four times greater than for 2 μm particles. An oil aerosol is most effective when the particles are about 0.6 μm in diameter; i.e., the scattering effectiveness is about four times greater than that for 1 to 5 μm particles.

The applicability of this information should be readily apparent to the air pollution investigator who is concerned with the alleviation of visibility restrictions. If he is to bring about significant improvement in visibility, he must solve the problem of controlling the emission or atmospheric generation of submicron particles. His problem is compounded by the fact that emission control for small particles is generally much more expensive than is the control of larger-sized particles.

The investigator should not overemphasize the importance of small or submicron particles to the exclusion of larger particles. Reference to Equation (4) will show that scattering, and thus a major visibility impact, is dependent on the particle area with the incorporation of an efficiency factor K. Even though K is larger for particles in the submicron size range, on a particle-by-particle basis, a large particle can be expected to be more important in the visibility problem. It is only when numbers of particles constituting a given mass are considered and the fact that the number of particles increases very rapidly for a given unit mass when the size decreases that the small particle assumes a dominant role in the visibility problem.

4. Atmospheric Color due to Nitrogen Dioxide

Although aerosols are the most important factor in determining the loss of visibility in polluted atmospheres, there has also been considerable interest in sky color effects, especially the brown discoloration that might occur as a result of excessive nitrogen dioxide concentrations. First, it should be noted that nitrogen dioxide concentrations cause the atmosphere to have a different color because this gas is strongly absorbent over the blue-green area of the visible spectrum. This produces illumination that is overbalanced toward the yellow-red end of the spectrum and gives nitrogen dioxide mixtures in air their characteristic yellow-brown coloration in proportion to the nitrogen dioxide concentration.

Nitrogen dioxide extinction due to absorption as a function of wavelength in angstroms is shown in Table I, which was taken from data given

Table I Extinction Coefficient of Nitrogen Dioxide (9)

Wavelength (Å)	k' (ppm^{-1} $mile^{-1}$)
4000	2.60
4500	2.07
5000	1.05
5500	0.47
6000	0.18
6500	0.062
7000	0.026

by Leighton (9). In order to determine the attenuation effect of nitrogen dioxide at a given wavelength λ, the gaseous extinction coefficient k' from Table I is multiplied by the atmospheric concentration in parts per million. Thus, Equation (2) rewritten for the attenuation of a light beam by nitrogen dioxide of concentration c becomes

$$I = I_0 \exp(-k'cx) \tag{7}$$

For an urban atmosphere containing both aerosols and nitrogen dioxide Equation (4) becomes

$$I = I_0 \exp[-(b + k + k'c)x] \tag{8}$$

where b and k refer to aerosol scattering and absorption and $k'c$ to nitrogen dioxide attenuation.

The simplest nitrogen dioxide aerosol situation to consider is that of Equation (8) where illumination of I_0 is coming from a target at a distance x from the observer. In considering the effect of nitrogen dioxide attenuation on intensity, the ratio of the intensity with and without nitrogen dioxide can be calculated using the expression

$$I/I_{C=0} = \exp(-k'cx) \tag{9}$$

Figure 5 shows $I/I_{C=0}$ as a function of wavelength for a target at $x = 2$ miles for nitrogen dioxide concentrations of 0.1, 0.25, 0.5, and 1.0 ppm. This relative change in transmitted intensity due to nitrogen dioxide is not dependent on the aerosol conditions present in the atmosphere. Under the conditions used for Figure 5, it is unlikely that an observer would be aware of a target coloration when $c = 0.1$ ppm. However, at 1.0 ppm, color effects would probably be quite marked, and even at 0.25 ppm, they would most likely be noticed if the target were white. Target coloration is subject to many variations of illumination—shadow, basic color, etc.—which tend to reduce the observable effect of nitrogen dioxide absorption.

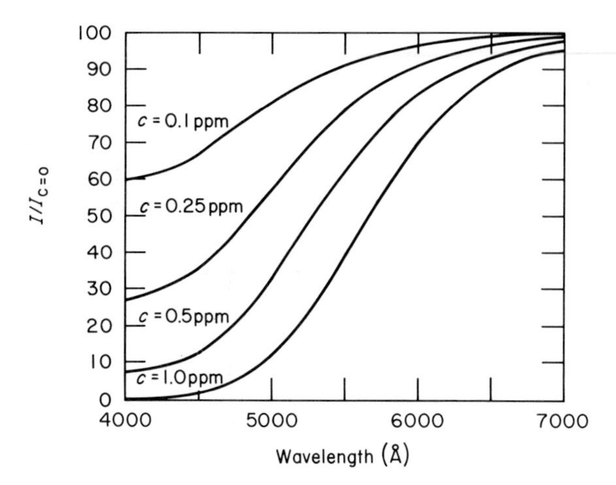

Figure 5. Intensity of illumination from a target as a function of nitrogen dioxide concentration relative to the intensity when nitrogen dioxide concentration is zero. Target distance, 2 miles (3 km).

In typical urban air pollution situations, the color of the horizon sky is probably a more important feature for the general public than is the discoloration of distant objects. Excessive concentrations of nitrogen dioxide in the urban atmosphere, probably above 0.25 ppm, can be significant in altering the color of the urban horizon. For the horizon sky, the target distance x is infinite and the target luminance B_0 is zero; thus, all of the horizon luminance is due to scattering. This scattering light, the air light, will be modified by nitrogen dioxide absorption. The luminance in an atmosphere with both scattering (air light) and absorption by nitrogen dioxide can be derived by using Koschmieder's approach as shown either by Middleton (*2*) or by Hodkinson (*10*). This results in an expression for the scattered luminance of

$$B_s = A(b/k'c + b)\{1 - \exp[-(k'c + b)x]\} \tag{10}$$

where B_s is the scattered luminance and A is a volume scattering function related to the luminance of the horizon B_h. Considering the ratio of scattered luminance with and without nitrogen dioxide gives the following expression:

$$\frac{B_s}{B_{s(C=0)}} = \frac{b}{k'c + b} \frac{\{1 - \exp[-(k'c + b)x]\}}{1 - \exp(-bx)} \tag{11}$$

For the horizon sky, x = infinity, and thus

$$\frac{B_s}{B_{s(C=0)}} = \frac{b}{k'c + b} \tag{12}$$

Rearranging this expression shows that B_s, the scattered luminance reaching a viewer from the horizon through a scattering and absorbing atmosphere, is equal to $B_{s(C=0)}$, the scattered luminance in the absence of absorption, multiplied by the ratio of the scattering coefficient b and the total extinction $k'c + b$. Figure 6 shows this alteration of luminance of the horizon sky as a function of wavelength and nitrogen dioxide concentration for an atmosphere of visibility 10 miles ($b = 0.39$ miles^{-1}) (16 km, $b = 0.63$ km^{-1}) due to aerosol scattering. The interpretation of Figure 6 is similar to that for Figure 5 in that 0.1 ppm nitrogen dioxide would probably not be noticed in this urban atmosphere, while 0.5 ppm and 1.0 ppm would produce proportionately more discoloration. Figure 6 shows where the nitrogen dioxide effect would be most noticeable. As aerosol concentrations increase and visibility due to scattering decreases, the relative effect of nitrogen dioxide decreases. Figure 7 shows the relative effect of a nitrogen dioxide concentration of 0.25 ppm in atmospheres of 10, 5, 3, and 1 mile (16, 8, 4.8 and 1.6 km) visibility [$b = 0.39$, 0.78, 1.30, and 3.9 miles^{-1} (0.63, 1.23, 2.09, and 6.3 km)]. It seems likely that under visibility conditions of about 3 miles (5 km) or less a nitrogen dioxide concentration of 0.25 ppm would cause only a minor discoloration of the horizon sky.

5. Atmospheric Color due to Aerosol Scattering

The preceding discussion of sky color due to nitrogen dioxide concentrations should not be interpreted to mean that all observed sky or haze

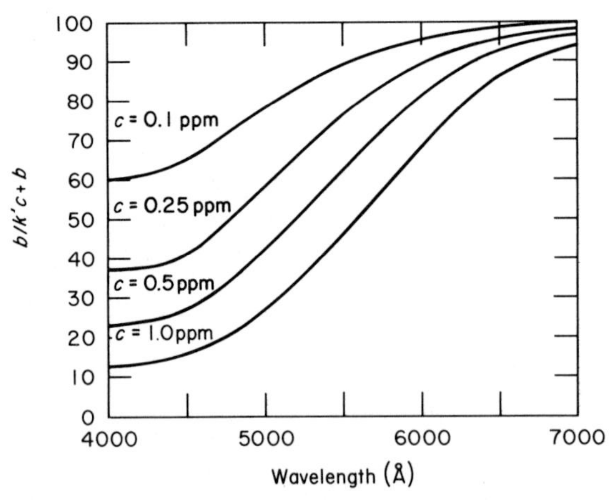

Figure 6. Intensity of illumination from the horizon sky as a function of nitrogen dioxide concentration, relative to conditions when nitrogen dioxide concentration is zero and when visibility is 10 miles (16 km).

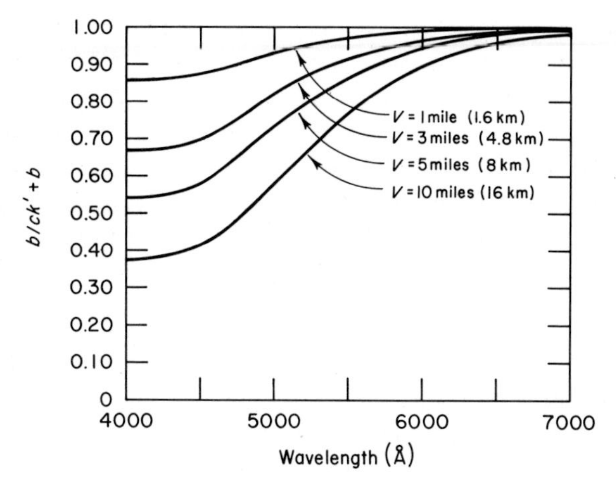

Figure 7. Relative intensity of illumination from the horizon sky when nitrogen dioxide concentration is 0.25 ppm as a function of visibility.

color is due to nitrogen dioxide. Charlson and Ahlquist (*11*) show that atmospheric extinction due to aerosol scattering is nearly proportional to the reciprocal of the square of the wavelength, and thus the atmospheric aerosol can act as a minus-blue filter to produce a yellow-brown haze color. The size dependence of aerosol scattering has already been described in preceding sections. Measurements of nitrogen dioxide and aerosol scattering in the Los Angeles, California, basin in August and September 1969 by Charlson *et al.* (*12*) indicated that nitrogen dioxide could be a significant contributor to the atmospheric extinction at 5500 Å about 20–30% of the time. These authors argue that only when the ratio of particle scattering to nitrogen dioxide absorption is less than 6 does nitrogen dioxide become a dominant factor in determining the haze color. Thus, in the Los Angeles, California, basin, specifically Pasadena, California, in the fall of 1969, the experimental data indicate that during 70–80% of the time, appearance in terms of the brown color of the air pollution haze cloud is determined by the aerosol particle scattering (*12*).

B. Calculation of Meteorological Visibility

The concepts of extinction, scattering, absorption, particle size, etc. that have just been described are all present in an observation of the visibility, because all these factors have a bearing on how far away a given object can be distinguished from its background. Visibility can be calculated from the extinction coefficient. This follows from the expression for the apparent contrast, Equation (3). Since, for visibility determinations, the target can be assumed to be black, $C_0 = -1$, and thus Equation (3) becomes

$$-C = \exp(-\sigma x) \tag{13}$$

In this expression, x is the unit of length, and σ is in the same units but to the -1 power. The minus sign occurs in Equation (13) because the target is assumed to be darker than the background. In order to obtain an expression for an observed visiblity in terms of the extinction coefficient from Equation (13), it is necessary to determine what lower limit of contrast can be distinguished by the eye, since the eye is the sensor specified for the determination of visibility. It is generally assumed that the limiting contrast for daytime visual determinations is 0.02, although experimental data show that the average observer has a contrast detection threshold closer to 0.05 (*2*). When a limiting visual contrast value is substituted for C in Equation (13), the distance x becomes identical with the definition of visibility V. Thus, for the theoretical limit, Equation (13) can be written

$$0.02 = \exp(-\sigma V) \tag{14}$$

which, when natural logarithms are taken, becomes

$$V = 3.9/\sigma \tag{15}$$

Here both V and σ must be in compatible units, i.e., meters and meters^{-1}. Calculated values of V based on the contrast limit of 0.2 are often referred to as the "meteorological range," with the specific definition being "that distance for which the contrast transmission of the atmosphere is 2%."

These derivations depend on several assumptions, which, fortunately, are reasonably applicable to the atmosphere. A homogeneous atmosphere is required, with uniform illumination between the observer and the horizon. Thus, the extinction coefficient σ is constant along the path of sight. Nonuniformity in the atmosphere is a more significant factor if it occurs close to the observer than if it is at the far end of the line of sight. Figure 8 shows the percentage of extinction that is derived from various sectors between the observer and the limit of visibility for a homogeneous atmosphere. It is important to note that 33% of the finally observed extinction occurs in the closest 10% of the path and 90% in the closest 50%. This feature of the extinction is due to the fact that air light introduced into the line of sight is also attenuated by the atmosphere closer to the observer, and this attenuation of the air light reduces the final effect of the remote portions of the sight path.

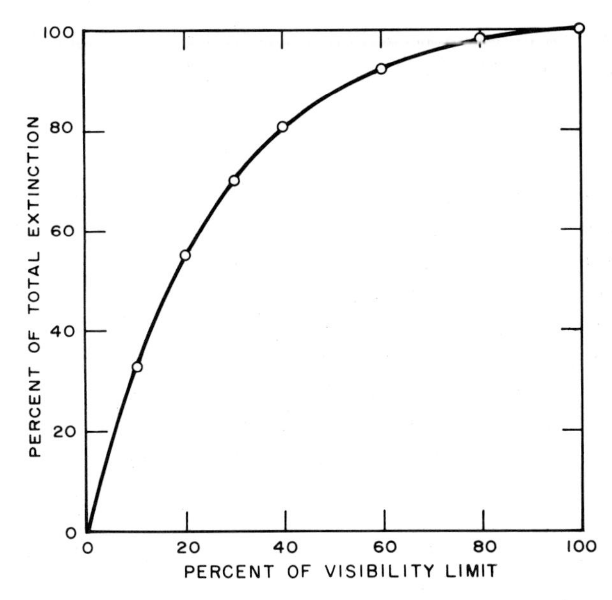

Figure 8. Net contributions to final extinction value due to various portions of the sight path.

This situation is obviously important also when instruments are used to obtain a visibility measurement, since an instrument reading will apply essentially to a single point. Good correlation between instrument data and visual observations may be expected in many situations because of the greater importance of the close-in air mass extinction in the visual observation.

C. Air Pollution Applications of Attenuation Concepts

1. Calculation of Atmospheric Particle Concentrations

Measurements of visibility can be used to obtain information on the particulate material in the atmosphere. Although the following arguments require several assumptions and are obvious simplifications, the results can still serve useful purposes when more detailed sampling data are not available. If the assumption is made that the particulate material in the atmosphere is of uniform size and that scattering alone accounts for the extinction, then the concentration of particulate material coinciding with a given visibility can be determined. This derivation uses Equation (15) for visibility and Equation (5) for the extinction coefficient. After substitution for σ and the introduction of the volume of a particle, it follows that visibility in meters can be expressed as a function of the concentration c (gm/m^3) of the material contained in uniform spherical particles of radius r (μm) and density ρ (gm/cm^3). This expression is

$$V = 5.2\rho r/Kc \tag{16}$$

Values of K as a function of r can be obtained from Figures 2, 3, or 4. For oil droplets of 0.6 μm diameter, where $\lambda = 5240$ Å, $\rho = 0.9$, and $K = 4.1$, the concentration c for a visibility of 1 mile (1.6 km) is 2.1×10^{-4} gm/m^3. The mass concentration for any particular visibility increases markedly with an increase in oil droplet particle size from 0.6 μm diameter.

If Equation (16) is rewritten as

$$Vc = 5.2\rho r/K \tag{17}$$

it is apparent that, for example, a doubling of the concentration of a given particulate material of radius r will reduce the visibility to one-half its former value. The units of Vc are mass per unit area, and the specific units of grams per square meter are used in the present discussion. Thus the product of the visibility and the mass concentration is an indication of the total mass that is between the observer and the limit of visibility per unit of cross-sectional area. The product Vc for the oil aerosol example just given is 0.34 gm/m^2. This means that the visibility is determined by the length of a column of 1 m^2 cross section that contains 0.34 gm of oil aerosol of 0.6 μm diameter. For a water aerosol of 1.0 μm diameter particles, where $K = 3.9$, the visibility limit is determined by 0.67 gm/m^2. These calculations substantiate a rule of thumb that is often useful: that half a gram of suspended material of the optimum particle size per square meter will determine the limit of visibility.

Hagen and Woodruff (13) have compared Equation (16) with measurements of dust storm characteristics including visibility, dust concentration, and particle characteristics. From the visibility and dust concentration data they derived the relationship

$$c = (56.0/V^{1.25}) \quad \text{mg m}^{-3} \tag{18}$$

where V is in kilometers. On the basis of an average dust particle density of 2 gm cm^{-1} and a particle radius of 11 μm at the geometric mean with $K = 2$ for large particles Equation (16) gives an expression

$$c = (57.2/V) \quad \text{mg m}^{-3} \tag{19}$$

A comparison of Equations 18 and 19 by Hagen and Woodruff (13) shows intersecting curves at about $V = 1$ km, with Equation 19 indicating higher dust concentrations than the observations of Equation 18 for comparable visibilities greater than 1 km.

Relationships between the total mass concentrations of atmospheric particulate material and visibility as indicated by the aerosol scattering coefficient have been carried out by Charlson *et al.* (*14*) using data from filter samples for particle mass and an integrating nephelometer for the scattering coefficient (15). Samuels *et al.* (*16*) have also compared nephelometer data, visibility observations, and high volume sampler collections of total suspended particles. Data for Charlson's study (*14*) were available from three urban areas, New York City, New York; San Jose, California; and Seattle, Washington. The results of this study showed that

$$Vc = 1.2 \text{ gm/m}^2 \tag{20}$$

where V is visibility in meters, and c is total suspended particle concentration in grams per cubic meter. A total of 238 cases were available for this derivation as indicated above; the modal value for the product Vc was 1.2 gm/m^2. The data showed a skewed distribution, but 90% of the Vc values were found between 0.7 gm/m^2 and 2.6 gm/m^2 in comparison to the ideal case, where the product of Vc was about 0.5 gm/m^2. When it is recognized that the atmospheric aerosol is not an ideal scattering medium, the range shown by Charlson's data from about -50 to $+100\%$ is not unrealistic when the magnitude of the various factors affecting the properties of the atmospheric aerosol are considered. In many respects, it is surprising that such a close argument was obtained.

The study by Samuels *et al.* (*16*) was a year-long program in three California cities—Los Angeles, Oakland, and Sacramento—and provided for the statistical comparison of three observations related to the atmospheric aerosol—total scattering by nephelometer, prevailing visibility from visual observations of the horizon, and the concentration of total suspended particles in the atmosphere through standard high-volume samples. In addition, the standard high-volume samples were supplemented by an additional unit preceded by a cutoff cyclone that passed only respirable-sized particles, i.e., generally smaller than 3 μm. These comparisons provide a detailed analysis of visibility and particle relationships in urban areas affected by varying concentrations of photochemical air pollution.

The comparison of nephelometer data in terms of b, the scattering attenuation coefficient in Equation (4), and prevailing visibility observations showed that an average value for all data was

$$V = 2.83/b \tag{21}$$

which can be compared with Equation (15) the expression for visual range, $V = 3.9/\sigma$. The reduction of the coefficient from 3.9 for optimum viewing of ideal black targets is in line with the practical contrast limit of 0.05 mentioned previously, and Steffens' (*17*) empirical value of 3.0 for the coefficient linking visibility and attenuation under conditions of less-than-ideal target contrast. Thus, the general applicability of using total scattering observations as an estimate of prevailing urban visibility was substantiated.

Statistical correlations between simultaneous observations of light scattering and prevailing visibility were very high. The lowest was —0.91 at Sacramento, California. Analyses of the total suspended particle mass as an indicator of either total scattering or visibility were unsuccessful. The general scatter of the data was so large, especially when visibility was calculated on the basis of the particle concentration, that a procedure for calculating visibility from high-volume data cannot be recommended. Using concentrations of only the smaller particles that passed the respirable cyclone did not improve the correlation.

One result of the Samuels *et al.* study (*16*) is the strong indication that nephelometer atmospheric scattering data could be translated to visibility data through the use of an expression such as Equation (21) rather than Equation (15), which is based on the theoretical visual limit.

For Charlson's (*14*) expression, Equation (20), to be meaningful, the relative humidity should be below 70% to avoid excessive water condensation and the hygroscopic effects, which will be described in a subsequent section. In addition, the observations should not be made in an air mass where recently emitted aerosol plumes can cause large variation in time and space, although as shown by Figure 8, complete air mass homogeneity is not necessary. These factors do not place overburdening restrictions on the use of Equation (20), which in conventional units with V in kilometers and c in micrograms per cubic meter may be expressed as

$$V = (1.2 \times 10^3)/c \tag{22}$$

or, using -50% and $+100\%$ as a likely range for Vc, the result is

$$(0.6 \times 10^3) < Vc < (2.4 \times 10^3) \tag{23}$$

This range probably reflects the typical variability of aerosol sources in an urban area and should be fully recognized when an expression such as Equation (22) is used. In addition, as shown by Samuels *et al.* (*18*), Equation (22) is probably more useful for an estimation of total particle mass concentration from a measurement of visibility than vice versa.

Experimental measurements of the aerosol mass versus scattering coefficient ratio, i.e., M/b, for several aerosols have been made by Elder *et al.* (*18*) and compared with Charlson's (*14*) expression, Equation (23). Various size distributions of coal dust, fly ash, and silica dust were used. Compared to the values shown in Equation (23), the chamber tests were all higher by factors ranging from 1.5 to 7. Although the M/b ratio was consistent for a given particle-size distribution and material, there were significant effects when the size distribution was changed. There were also differences between materials, but because size distributions of the materials could not be duplicated exactly, the effects of materials could not be quantified. Elder *et al.* (*18*) concluded that coal dust and fly ash scattered significantly less light than does a typical urban aerosol or silica dust. They also conclude that a variation of $\pm 100\%$ can be present in an estimate of total particle mass when based on particle scattering estimates. This is, of course, in line with Charlson's (*14*) results and Equation (23).

2. Plume Opacity Calculations for Open Burning

The principles of plume opacity and visibility have also been used to estimate the effect of scheduled open burning to dispose of debris such as that from land clearing. By first estimating this downwind effect, Duckworth (*19*) has shown how selection of optimum meteorological conditions can be used to limit the downwind effects of a scheduled open burn.

Using data from a scheduled open burning of brush from land clearing, Duckworth gives an example based on the following assumptions:

1. Twenty-five conical piles of brush 15 ft (4.5m) in diameter, 5 ft (1.5 m) high, containing 16.5 yd³, (12 m³), will completely burn in 1 hour.
2. Density of compacted wood brush is 200 lb/yd³ (134 kg/m³).
3. Twenty-four pounds of 0.7–7 μm diameter particulate material will be emitted per ton of wood burned (12 kg per metric ton).
4. Plume diffusion follows Sutton's continuous ground-level point-source equation (see Strom, Chapter 9, Vol. I):

$$c = \frac{2Q}{\pi C' \bar{u} x^{2-n}} \exp\left(\frac{-y^2}{C' x^{2-n}}\right) \tag{24}$$

where the coefficient $C' = 0.09$, $n = 0.2$, $\bar{u} = 4$ m/second and emission rate $Q = 82.4$ gm/second.

5. Plume width is the distance y normal to the plume axis at which the concentration c reaches 10% of the centerline value.

Successive calculations for one pile at downwind distances x of 100 and 500 m show plume widths of 60 and 240 m, respectively. At these distances, the amount of particulate material in the plume cross section can

Table II Theoretical Calculation[a] of Airborne Particulate Material in Half-Plume Cross Section at Two Downwind Distances (*15*)

x (m)	y (m)	c (gm/m^3)	Δyc (gm/m^2)
100	0–10	1.28×10^{-3}	0.013
	10–20	0.79×10^{-3}	0.008
	20–30	0.30×10^{-3}	0.003
	½ plume		0.024
500	0–40	0.72×10^{-4}	0.0029
	40–80	0.46×10^{-4}	0.0018
	80–120	0.19×10^{-4}	0.0007
	½ plume		0.0054

[a] Based on assumptions in text.

be estimated by multiplying the mean concentration $\bar{c}$ by the plume width interval Δy. Values of $\bar{c}$ for intervals of y and Δy from the plume axis to one edge are shown in Table II. The values in Table II are for one-half the plume from one brush pile. The full-plume value would be twice this value, and on the basis of 25 brush piles burning simultaneously, the amount of material in a section through the combined plume would be 50 times the half-plume values indicated. Figure 9 shows the mass of

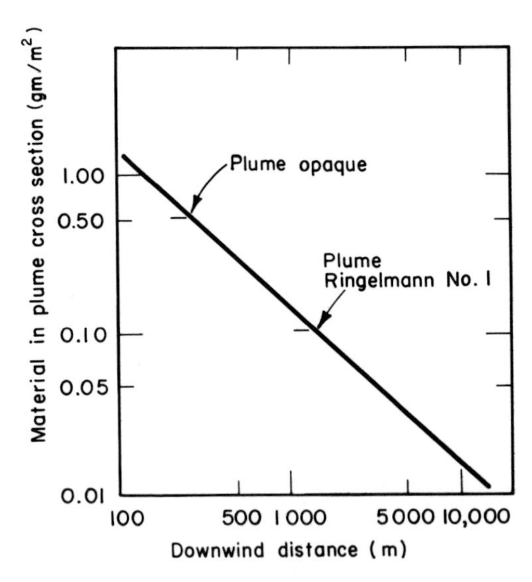

Figure 9. Mathematical estimate of visibility reduction resulting from the open burning of 28 tons (31 metric tons) of brush (*19*). Theoretical calculation using Sutton's equation and $Q = 82.4$ gm/sec.

particulate material in the total plume cross section as a function of distance.

It has previously been estimated that a plume cross-section density of about 0.5 gm/m^2 of ideal particles would produce an opaque cloud. In Figure 9, the cross-section density becomes less than this at a distance of about 300 m. When a plume is reduced in opacity to 20% of its opaque value, it should meet the definition of Ringelmann No. 1. In the context of this example, this would be the plume cross-section density at about 1100 m downwind from this multipile open burn. Using Charlson's (*14*) expression for Vc (Eq. 22) the predicated range of a visible plume would be considerably reduced, and thus, Duckworth's procedure meets the needs of a control agency for a "worst case" estimate.

Consistent with what would be expected from Charlson's data (*14*), Duckworth points out that where visual estimates have been made downwind of scheduled open burns and compared with calculated data, the calculated opacity values are higher than the observed values.

Further adjustment of the various assumptions in this estimation procedure is probably not justified because of the lack of precision with which the various parameters such as burning rate can be predicted. Furthermore, an estimate of the most probable maximum downwind effect is usually a more critical factor for establishing air pollution control than is some average value.

3. Calculation of Plume Opacity

Plume opacity in either Ringelmann numbers or equivalent opacity can be expressed mathematically in terms of the particulate concentration and other parameters. This follows from a restatement of Equation (2):

$$I/I_0 = \exp(-\sigma x) \tag{25}$$

In this situation, I_0 would be the intensity of the light entering the plume, I the intensity seen by the observer, σ the extinction coefficient, and x the distance through the plume. From Equation (5) it can be shown, since

$$c = N\rho \cdot \tfrac{4}{3}\pi r^3 \tag{26}$$

and thus

$$N\pi r^2 = 3c/4\rho r \tag{27}$$

that

$$\sigma = 3Kc/4\rho r \tag{28}$$

where K, c, ρ, and r have the same meaning as in Equation (16). Thus it follows from Equations (6) and (28), through substitution and taking logarithms, that

$$\log \frac{I}{I_0} = -\frac{3x}{4\rho} \sum_{i=1}^{n} \frac{K_{iCi}}{r_i} \tag{29}$$

In making an opacity calculation, log I/I_0 should equal 80% for Ringelmann No. 1, 60% for No. 2, etc.

The appearance of a smoke plume relative to its illumination and background has been examined using the basic relationships of contrast and transmittance by Jarman and de Turville (*20*). They were able to explain the appearance of a smoke plume under different conditions of illumination and background. Attenuation of the background by a nonblack plume, i.e., plume opacity, is shown to be a function of the brightness of the plume due to scattering of incident light and to the loss of light transmitted through the plume. When scattering predominates in a plume, it will appear bright against its background, while in other cases for the same plume when scattering is unimportant and plume attenuation of background illumination is dominant, the plume will appear to be dark. Experimental work on the obscuration caused by nonblack plumes has been reported by Crider and Tash (*21*).

4. Urban Turbidity Applications

The attenuation of solar radiation by a polluted atmosphere has been used by McCormick and Baulch (*22*) and McCormick and Kurfis (*23*) to determine the vertical extent of aerosols and to estimate diurnal changes in the pollution layer (see also Chapter 12, Vol. I). Some general relationships between total particulate mass and number concentrations have been formulated (*22*). The methods used in this procedure are relatively simple and the information obtained may provide useful descriptive features of polluted atmospheres.

If the transmission equation is converted into base-10 logarithms and is made to include terms permitting corrections for variations in the relative optical path, the equation for incident radiation intensity can be written as

$$I = (I_0/F)10^{-[ak(p/p_0)+B]d} \tag{30}$$

In this equation

I = incident solar radiation intensity
I_0 = extraterrestrial solar radiation intensity
F = reduction factor for mean solar distance

a = 0.0674 = atmospheric extinction coefficient of pure air including ozone per unit optical path length

k = factor to allow for the differences between the optical path length of the atmosphere for air, haze, and ozone. $k = 1.0$ for $d \leq 3$; $k = 0.97$ for $d = 6$; and $k = 0.924$ for $d = 10$

p, p_0 = ambient and sea level pressure, respectively

B = turbidity coefficient for haze, dust, etc.

d = relative optical path length = $1/\sin h$, when h is the apparent elevation angle of the sun

The experiments conducted by McCormick and Baulch have used the Voltz sun photometer (*22*). The measurements were made at a wavelength of 5000 Å. The coefficient B in the above expression is defined as the turbidity of the sample atmosphere. Values of the turbidity B can be determined from direct observations of incident solar radiation I since the other terms in the expression can be either determined independently or adequately estimated.

Observations with the sun photometer may be reported in terms of the incident solar transmissivity I although the ratio of I to I_0, the extraterrestrial solar intensity, is probably more meaningful. Figure 10 shows profiles of I/I_0 made in polluted and "clean" air near Cincinnati, Ohio (*22*). A helicopter was used to obtain these data. These observations clearly indicate the magnitude of the energy loss that can occur in polluted atmospheres. In clean atmospheres, the I/I_0 value is about 0.5, while in polluted areas on the same day the ratio of I/I_0 at ground level was between 0.2 and 0.3, indicating a 50% loss of intensity compared to rural surface conditions.

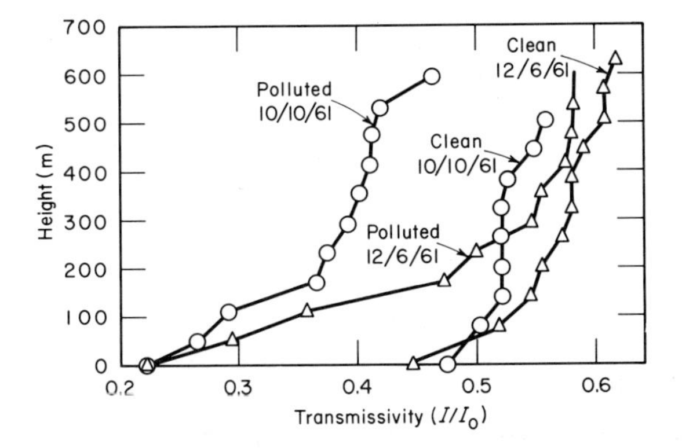

Figure 10. Transmissivity profiles for clean and polluted atmospheres, Cincinnati, Ohio (*22*).

Measurements of I may also be used to determine values of the turbidity B of the atmosphere (*22*). Vertical profiles of the turbidity may be expressed as the ratio of turbidity at a given height B_z, to ground-level turbidity B_0, and are shown in Figure 11 for the same days as in Figure 10 for I/I_0. As one moves downward into polluted surface air, the turbidity of the atmosphere increases; this is illustrated by Figure 11. Here, the ratio of turbidity values between upper clean air and surface air is about 0.2, and it is quite clear that similar conclusions about the loss of solar energy in polluted atmospheres would be obtained from either intensity or turbidity measurements.

McCormick and Baulch (*22*) conclude that values of the turbidity B are, in general, more useful than the transmissivity values, because they have developed a means of relating turbidity values to quantitative measures of the particulate pollutants. This development is based on an assumed size distribution of the type postulated by Junge (*24*), a particle size range between 0.1 and 1.0 μm, and solid, spherical particles of density 2 gm/cm^3. They then show that the number of particles per cubic meter at ground level in the 0.1 to 1.0 μm range can be expressed as

$$N = (17.3 \times 10^9 B_0) \tag{31}$$

Using the assumed nature of the particles stated above, the total mass concentration c in milligrams per cubic meter of the N particles is shown to be

$$c = 969B_0 \tag{32}$$

McCormick's measurements indicate a B_0 value for moderately polluted air of the order of 0.2, which from Equation (32) above would be

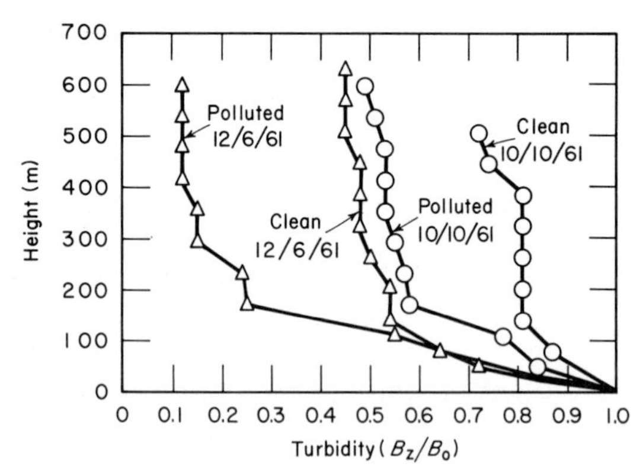

Figure 11. Turbidity profiles for clean and polluted atmospheres, Cincinnati, Ohio (22).

indicative of a particle concentration of about 3460×10^6 particles/m^3 and a mass concentration of about 194 μg/m^3. Both of these values are "reasonable" for a polluted area. However, they may not be overly precise in any given instance.

The concept of turbidity put forth by McCormick and his associates (*22, 23*) is very useful in air pollution surveys and monitoring as one indicator of air pollutant impact. It has been included in a number of monitoring programs including that of the World Meteorological Organization (WMO). However, before beginning a detailed application of these turbidity concepts, especially to particle mass concentrations, a comprehensive study of the assumptions set down in the original paper is strongly recommended. In the following section are described some of the problems that have been encountered in relating visibility to particle measurements. It would not be surprising to find that turbidity correlations with particulate levels are also complex and will need careful study. One apparent problem has been described by Randerson (*25*) when during turbidity measurements in Las Vegas, Nevada, a very apparent cycle was observed. He concluded that this cycle with maximum turbidity at local noon was due to incomplete correction of the sun photometer for changes in the solar angle. In a comment on Randerson's paper, Bullrich (*25*) considered Randerson's results to indicate most likely a nonlinear instrument response. They could also be due to an inability to distinguish between the intensity of the solar beam and the intensity of forward-scattered light (*26*).

Turbidity changes on a global and long-term basis have also been discussed as one indication of global atmospheric effects. Spectral pyrheliometric measurements for Jerusalem, Israel, are available for the periods 1930–1934 and 1961–1968, and these have been analyzed by Joseph and Manes (*27*). The data indicate a change in mean turbidity of between 16 and 30%, depending on the specific definition of turbidity that is used for the calculations. These values thus show an increase of between 5 and 10% per decade. The conclusion of Joseph and Manes (*27*) was that this increase in turbidity over more than 30 years was not due to a growth of local pollutant emissions, but that it was a change in global or at least northern hemispheric turbidity. This might be related to changes in global pollutant emissions but there are other conditions that must also be considered. One of these is the impact of volcanic emissions, because it is well known that major volcanic eruptions were minor contributors of atmospheric aerosol particles in the 1930s and important contributors in the 1950s and 1960s (*28*). Since Joseph and Manes (*27*) recorded a strong relationship between turbidity and synoptic weather, climatic change and resultant air mass frequencies would also have to be considered when a cause of long-term turbidity change is proposed. Long-term humidity changes and effects on visibility have been determined for Sydney, Australia, by Patterson (*29*). Solar radiation studies at Mauna Loa, Hawaii, by Pueschell *et al.* (*30*) show that atmospheric water vapor above at that station, elevation 3400 m, may attenuate the direct solar radiation intensity by as much as 10%. These measurements were made with an Eppley pyrheliometer, which has a broad spectral response from 0.35 to 2.5 m and thus covers many of the broad water vapor bands. A major value of the work at Jerusalem is the use of sophisticated actiometric instruments (*27*) and thus the ability to avoid some of the problem inherent in the use of the very simple sun photometer (*26*).

5. Visibility Observations for Urban Air Pollution Studies

Visibility data can often be used to obtain a more complete understanding of air pollution conditions in an area, particularly the assessment of local sources and determination of long-term trends. To determine how seriously a given area might be affected by air pollution, the suspect area can be compared with an unaffected area nearby or through a study of long-term data for the suspect area. But to make such a comparison, it is extremely important to determine that air pollution sources are the only significant variables in the situation. It should always be emphasized that wind conditions, humidity, and temperature inversions can all affect a visibility observation, and these effects must be factored out of the analysis before any conclusions can be drawn. The necessary visibility data for this type of study can often be obtained from the National Climatic Center, National Oceanic and Atmospheric Administration, Asheville, North Carolina.

Two-station comparisons can be made either by using average values from a large mass of climatological data or by a more detailed study of a selected group of days or events. In a study of this latter type, since it is concerned with fewer situations, more details of the weather conditions can be investigated. Such a study was carried out for Los Angeles, California, for July-September, 1949 (*31*). In this particular study, two control stations were used: Goleta, California, 100 miles north near Santa Barbara, California, and El Toro, California, Marine Air Station, 45 miles south. Humidity data were studied to rule out fog situations. It was concluded that 90% of the visibility reduction in downtown Los Angeles, on smog days was due to man-made pollution and only 10% to natural haze. This seems to be a more definitive and satisfactory result than what might be obtained from the comparison of long-term average data.

An analysis of visibility data for three airports east of the Mississippi

River–Akron, Ohio; Lexington, Kentucky; and Memphis, Tennessee—was carried out by Miller *et al.* (*32*) for the 8-year period 1962–1969. They found that the frequency of poor visibility, 0–6 miles (10 km), was markedly greater in the latter half of the period compared to the first 4 years. The data were classified to remove effects of high relative humidity, high winds, and precipitation, and still the frequency of 0– to 6–mile visibilities increased from an average of 16.7% in the first 4 years of the study period to 27.6% in the second 4 years. These authors concluded that the change was probably due to the effects of man-made pollutants, since they believed they had factored out the known meteorological variables, and changes in other natural phenomena could not be identified.

The fact that the visibility is affected by both the prevailing weather conditions and the previous history of the air mass makes it very difficult to use visibility as a measure of short-term changes in air pollution emissions. Short-term comparisons are apt to be attempted if, for example, a major air pollution source in a community ceases to operate for a period of time. In such situations there is often a strong tendency to try to show the degree to which the source pollutes the local atmosphere by displaying visibility records before and after shutdown. Depending upon the prevailing weather patterns that occur during the periods used, one could expect to show that the shutdown either reduces pollution, makes no change, or perhaps in some way increases the pollution. In most situations, a short-term, before-and-after type of visibility study is not worth the effort.

III. Meteorological Aspects of Visibility

A. Introduction

The transparency of the atmosphere, or more commonly, the visibility, is strongly dependent upon a variety of meteorological parameters that act in addition to any pollutant effects. These meteorological impacts, especially relative humidity, must be recognized and accounted for before visibility observations can be related to pollutant problems. Uses of visibility data from the official records of the United States National Weather Service must also consider the definition of "visibility" (*33*), the method by which it is observed, and the limitations that may be present in any observation program.

B. Evaluation of Visibility Data

The definition of visibility and the methods and techniques by which visibility observations are made by an observer are given by Hewson,

Chapter 11, Vol. I. It should be emphasized, however, that official records of the various national weather services specify an observation made with the unaided eye viewing known targets or, at night, lights. Furthermore, after visibilities have been determined around the entire horizon circle, the individual observations are resolved into a single value of "prevailing visibility" for reporting purposes (*33*).

In the United States, instructions for determining prevailing visibility are given in the Manual of Surface Observations used by the United States National Weather Service and the military services (*34*). Prevailing visibility is defined as the greatest visibility that is attained or surpassed around at least half of the horizon circle, but not necessarily in continuous sectors. To determine prevailing visibility under nonuniform conditions, the horizon is divided into several sectors of equal size in which the visibility is substantially uniform. The value reported as the prevailing visibility is the highest sector visibility that is equal to or less than the visibility of sectors that account for at least half of the horizon. Therefore, the visibility around at least half the horizon is equal to or better than the prevailing visibility reported. For example, if the horizon were divided into four sectors and the respective visibilites were 3, 4, 5, and 8 miles (4.8, 6.4, 8, and 12.8 km), the prevailing visibility would be 5 miles. The weather record will carry additional data on especially low or variable visibility situations under the notation of "Remarks."

The reporting of prevailing visibility has been the standard procedure in the weather services in the United States since January 1, 1939 (*35*). Prior to that date, the reported visibility for a station was the maximum visibility that could be observed around the horizon from the station (*36*). This change in definition is important in the evaluation of long-term visibility records and trends, because it might not be possible to compare directly pre-1939 data with later observational records.

In the use of tabulated visibility records, it is important to realize that the visibility reading as reported in a given station weather record is a composite value for the whole horizon and is not the visibility in just a single direction, nor is it a simple average value. Whenever a transmissometer or instrumental measurement is reported in a station weather record, it does not replace the regular visual observation but is listed under "Remarks and Supplemental Coded Data." It is also important to remember that meteorological visibility data are taken primarily for aircraft operations.

In some meteorological literature the term "visual range" is used instead of the more common term "visibility." It is argued that "visual range" is a more logical term to specify a distance at which something can be seen, while "visibility," at least outside of meteorological science, seems more indicative of the clarity with which objects stand out from their surroundings. However, an attempted switch to the term "visual range" has made little headway, and there is every indication that "visibility" will continue to be the commonly used term to denote how far one can see. It is used in this manner in this chapter.

As the definition indicates, visibility is determined by an observer looking around the horizon and noting, whether he can identify known landmarks or lights. It is readily admitted that subjective factors can influence the visibility record. However, considering the problem involved in reducing such a variable as the visibility around the horizon to a usable and reportable observation, the accepted system must be considered a satisfactory compromise.

A large number of the reporting weather stations observe and record the visibility at hourly intervals. In the United States, the visibility is reported in statute miles at land stations. There is no fixed upper limit for reportable visibility. However, many stations do not have suitable markers at distances beyond 15 miles (24 km), and under such conditions the visibility over 15 miles is recorded at 15+. Such a procedure could cause confusion in preparing summaries of two stations with varying target horizons if data were carelessly lumped together.

Since most airports seem to be located on the fringes of our urban areas, it is entirely possible for the prevailing visibility measurement to be obtained without any consideration being given to the visibility over the more congested urban and industrial areas. When there are important differences between the reported visibility for specific directions they are often entered in the "Remarks" column, and these frequently may describe smoke over adjacent urban areas. Such notations do not find their way into any climatological records, however, and can be obtained only by studying copies of the actual station records.

In addition to the prevailing visibility, a station weather record will contain information on the cause of "reduced visibility," i.e., a visibility less than 7 miles (11 km). Whenever this occurs, the observer must report the reason, which can be either precipitation—rain, snow, etc.—or an "obstruction to vision"—haze, smoke, fog, etc. These notations can often be used as a valuable source of information for air pollution studies. It should be noted, however, that such explanations are required only when the visibility is less than 7 miles (11 km).

There are obvious difficulties in separating haze and smoke in urban areas, and it is usual to find haze and smoke listed together as a cause of reduced visibility or obstruction to vision in the observation record. Sometimes fog and smoke are listed together. Normally, the experienced weather observer does not find it particularly difficult to discriminate between fog and haze.

Visibility observations in polluted areas frequently show strong directional variation that is dependent upon the angle of the sun. Lower visibilities will normally be observed in the direction of the sun rather than away from it, even in a uniformly mixed atmosphere. This is due to one aspect of light scattering by small particles in the polluted air, namely, that small particles scatter more light in the forward direction toward an observer looking toward the sun than they scatter backward toward the illuminating source (*2, 5*). In previous sections it was shown that the greater the amount of scattered light, the larger the extinction coefficient and the poorer the visibility. In situations where visibility variations are dependent upon the direction of illumination, it obviously would be impossible to relate the observed variations to different concentrations of pollutants.

C. Meteorological Factors Affecting Urban Visibility

1. Relative Humidity and Hygroscopic Particles

In addition to those factors that affect visibility through changes in the dispersion of particulate materials, there are those that alter visibility by affecting the size of hygroscopic particulate materials in the atmosphere (*37*). Under high but still not saturated humidity conditions, hygroscopic particles pick up water and increase in size. As they increase in size they become more effective in reducing visibility.

Direct determinations of sea salt nuclei have been made by Twomey by viewing the transition of particles under a microscope (*38*). He found the transition of sea salt from a crystal to a droplet to occur at 75% relative humidity. Table III shows the predicted humidities for the transition from a crystal to a droplet for a number of other materials.

An example of a relation found between visibility and relative humidity

Table III Predicted Crystal-to-Droplet Transition Humidities for a Number of Salts (*38*)

Compound	*Relative humidity* (%)
Potassium carbonate	43–45
Sodium bromide	58
Cobalt chloride	67
Ammonium sulfate	80–82
Potassium chloride	84–86
Potassium sulfate	97

at the Los Angeles International Airport in California is shown in Figure 12 (*39*). The sharp steady decrease in visibility above 67% relative humidity is obvious in these data. In view of the location of the airport within about 2 miles (3 km) of the Pacific Ocean, a source of sea salt particles, this close agreement was not unexpected.

An analysis of visibility, relative humidity, and air mass source has been made by Buma (*40*) for the city of Leeuwarden on the North Sea coast of the Netherlands. Daytime data were used and classified according to whether the wind was from the continental or maritime directions. Table IV shows some of the results of this study. Low visibilities at Leeuwarden were more frequent at all relative humidities for continental air than for maritime air. In the relative humidity range from 69 to 76%, low visibilities were ten times more probable for continental air than for maritime air. The maritime air data fit the previous discussion of the hygroscopic effects of sea salt. The continental air data probably show

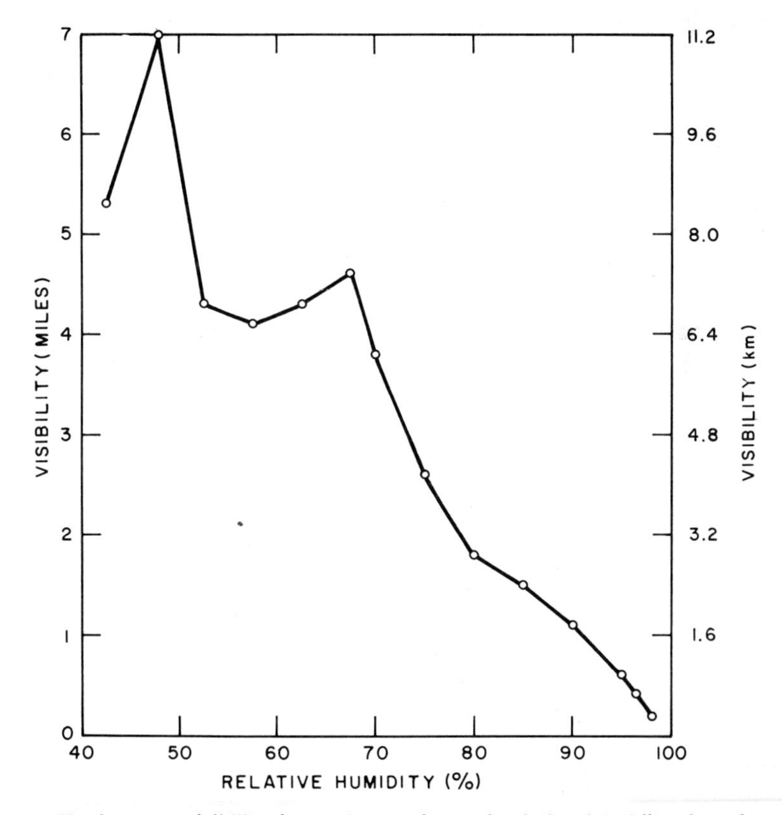

Figure 12. Average visibility for various values of relative humidity, Los Angeles International Airport, California (*39*).

Table IV Frequency of Visibility Less than 7 km (4 miles) as a Function of Relative Humidity, Leeuwarden, the Netherlands (*40*)

Relative humidity (%)	*Continental air* (% <7 *km*)	*Maritime air* (% <7 *km*)
100–99	100	95
98–97	97	89
96–93	88	62
92–89	72	32
88–83	57	19
82–77	47	11
76–69	45	4
68–61	35	1
60–50	20	0

the influence of air pollutants from various continental sources. The persistence of visibility effects at low relative humidities is one of the properties of sulfuric acid aerosols. There are no confirmatory sampling data, however.

Patterson (*29*) has analyzed 30 years of humidity and visibility data from Sydney, Australia, and found the expected result, namely, that periods of lower humidity were characterized by better visibility. However, he also was able to show that in the past 30 years, there was a consistant trend toward lower average relative humidity and thus toward increased visibility over the city. Variations related to the day of the week were observed as a result of secular pollutant sources, with Sunday having the best visibility. The long-term trend in humidity was large enough, however, that determinations of the effectiveness of smoke control ordinances was masked, and most, if not all, of the apparent improvement in visibility was attributed to the decrease in relative humidity (*29*).

Laboratory and field studies on aerosol optical effects, aerosol chemistry, and relative humidity have been carried out by Covert *et al.* (*41*). Their data show that chemically pure aerosols show two different characteristic types of behavior in an environment with a changing humidity. A variety of salts, e.g., sodium chloride, sodium sulfate, and ammonium sulfate, show deliquescent behavior with a sharp change in size that is identifiable with a narrow humidity range. This is similar to the results of Twomey (*38*) shown in Table III. Other materials, most importantly sulfuric acid, show a continuous increase in size with increasing humidity. Figure 13 shows the laboratory data for sea salt, mostly sodium chloride, and for sulfuric acid. The two different humidity relationships are

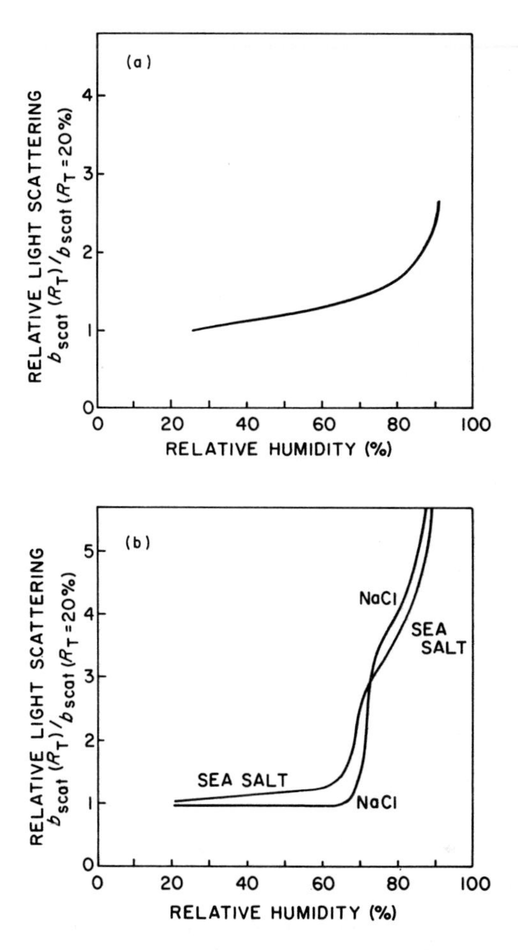

Figure 13. Laboratory measurements of light scattering ratio versus relative humidity; (a) sulfuric acid aerosol and (b) sea salt and sodium chloride aerosol (*41*).

obvious. In addition to their laboratory data, Covert *et al.* (*41*) also give examples of the humidity dependence of ambient urban aerosol scattering. Figure 14 shows examples for Altadena, California, and Denver, Colorado. The authors argue that the Altadena data show, as expected, an aerosol of mixed and varied composition without clear effects of deliquescent salts such as sodium chloride. The rapid change above 90% relative humidity is attributed to a limited number of hygroscopic particles. The lack of any humidity relationship below 80% for the Denver data is considered indicative of the fact that the aerosol particles were not strongly hygroscopic.

Winkler and Junge (*42*), in analyses of European urban aerosols in

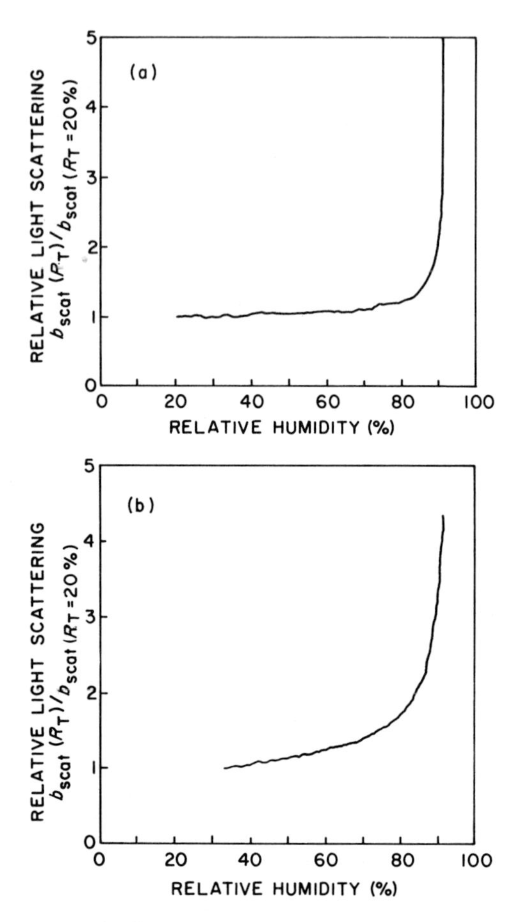

Figure 14. Urban atmospheric measurements of light scattering ratio versus relative humidity; (a) Denver, Colorado, and (b) Altadena, California (*41*).

terms of relative particle mass as a function of humidity, also found a hygroscopic relationship such as Covert (*41*) shows for Altadena, California, and for sulfuric acid. Aerosol particles in several size ranges were separated by Winkler and Junge (*42*), and this process showed that when only the giant particles—larger than 1 μm radius—in advected maritime air masses were considered was there a characteristic deliquescent pattern to the humidity–particle mass relationship. In this case, the pattern was similar to that found experimentally for sea salt.

Measurements of the free liquid water content of airborne pollutant particles have been made by Ho *et al.* (*43*) in southern California. For maritime and urban aerosol particles, a significant fraction of the particle mass is free water in the relative humidity range between 40 and 75% This is generally below the hygroscopic threshold indicated for urban aerosols as shown in Figures 12–14. The free-water data were compared by Ho *et al.* (*43*) with scattering coefficient values—a reasonable approximation of visibility—and they found a linear relationship between aerosol water content and the atmospheric scattering coefficient.

In situ measurements of atmospheric aerosol size distributions as a function of relative humidity have been made by Sinclair *et al.* (*44*) using diffusion battery techniques on humidified urban aerosol streams. These results show that atmospheric particles of all sizes increase in diameter by a factor of 2 or more, depending on the amount of hygroscopic material, as the humidity increases from 0 to 100%.

Thus, there is a demonstrated effect of relative humidity and the aerosol liquid water content on the optical properties of the atmospheric aerosol and of the ambient atmosphere. Furthermore, this importance apparently extends over a wide range of relative humidities and is not confined just to the hygroscopic range above 70% relative humidity.

2. Wind and Mixing Depth

Visibility is dependent on the concentration of the particles and, for a given pollutant emission condition, on atmospheric dilution processes. Dilution can be directly related to wind speed and to the depth through which the materials are mixed, i.e., the mixing depth. It would normally be expected that the visibility will improve as wind speed and mixing depth increase, all other factors being equal. Figure 15 shows the improvement in visibility statistics as wind speed increases from calm to 15 mph at Bakersfield, California, averaged over a 10-year period (*45*). However, there is a clear trend for poorer visibility at higher wind speeds. Since this site is in an area of extensive agricultural activity and is adjacent to open desert areas, this change in tendencies with the higher wind speeds represents the effect of blowing dust lifted by the stronger winds.

3. Air Quality Criteria

The effect of high relative humidity on visibility reduction resulting from typical air pollutants was recognized by the United States Environmental Protection Agency (USEPA) when the air quality criteria for particulate materials were presented (*46*). In this case, a criterion for adverse visibility as a function of particle concentration was cited as

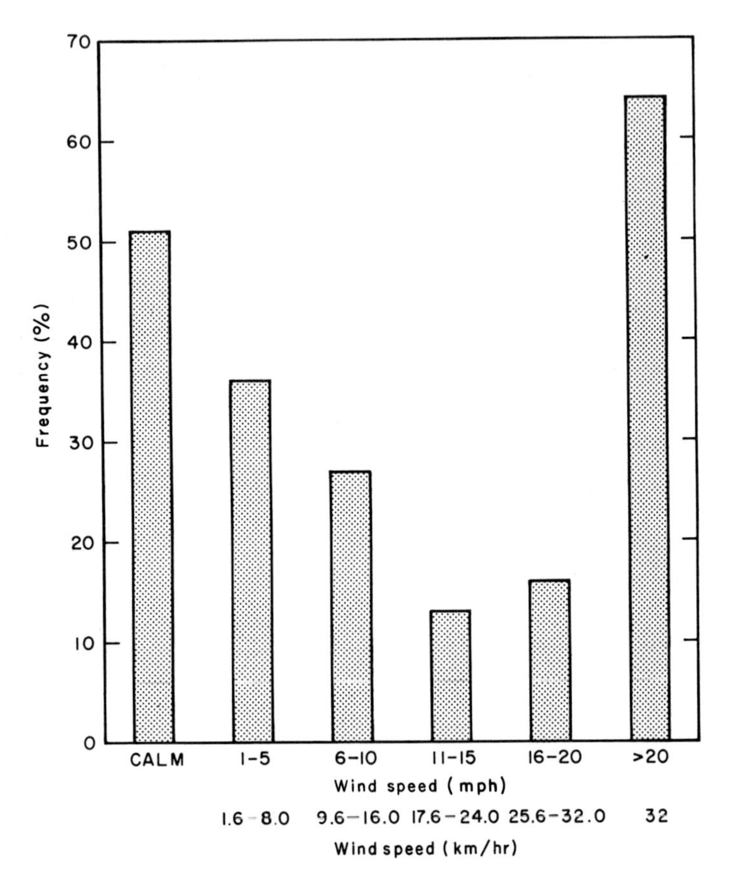

Figure 15. Percent frequencies of 0–10 mile (0–16 km) visibility by wind speed classes, Bakersfield, California, 7:30 a.m.–5:30 p.m., May, July, September, November, 1948–1957 (*45*).

applicable only when the relative humidity was less than 70%. The introduction of the humidity qualification in the visibility limit in these criteria is an attempt to restrict the influence of water droplets and hygroscopic pollutants and in this manner to obtain a more accurate measure of the man-made air pollutants. This seems generally reasonable, as indicated by the data shown in Figure 14, but, as shown by Ho *et al.* (*43*), aerosol particle water content is a factor in determining visibility at lower humidities. The limit of 70% relative humidity is a stricter discriminant than is the 4°F (2°C) temperature–dew point difference that is used by the United States National Weather Service to differentiate between haze and fog.

IV. Air Pollution Effects on Weather and Atmospheric Conditions

A. Introduction

Air pollutants have effects on the atmosphere that are more subtle and in some ways more important than just a reduction in visibility. These include effects on cloud and precipitation formation and on processes that may cause changes in the radiant energy received at the earth's surface or changes in the electrical properties of the atmosphere. A summary of various possible effects of pollutants on the atmosphere has been made by Hobbs *et al.* (*47*). The results of their assessment in terms of a tabulation of effects is shown in Table V. This table also includes an estimate of the pollutant concentration that might be related to a detectable change in background amospheric parameters affected by pollutants. Thus a change in ice nuclei concentration might be detectable at extremely low concentrations because of the low ambient levels of ice nuclei and the very small-sized particles that can act in this manner.

B. Air Pollutants and Precipitation Mechanisms

1. Condensation Mechanisms

In the atmosphere, condensation of water vapor occurs around nucleating particles when the relative humidity reaches a sufficiently high

Table V Approximate Minimum Concentration Thresholds of Atmospheric Effects due to Minor Constituents (*47*)

Species	*Effect*	*Threshold*: *Volume per unit volume of air*	*Threshold*: *kg/km³ of air*
Ice nuclei	Cloud structure and precipitation	10^{-18}	10^{-6}
CCN	Cloud structure and precipitation	10^{-15}	10^{-3}
Aerosols	Visibility and heating rates	10^{-12}	10^{0}
HCl, H_2SO_4	pH of rain	10^{-11}	10^{-2}
Aerosols	pH of rain	10^{-10}	10^{2}
NH_3	pH of rain	10^{-10}	10^{-1}
SO_2	pH of rain	10^{-8}	10^{1}
NO_2	Visibility and heating rates	10^{-7}	10^{2}
O_3	Heating rates	10^{-6}	10^{3}

value to "activate" available nuclei. In the laboratory, it is possible to cause condensation to occur on essentially all the available particles in an air sample by producing very high values of supersaturation. This is generally the case in the Aitken "nuclei" counter and other devices where large expansions are used to cool rapidly a volume of moist air and in this way to increase very rapidly the humidity in the chamber. Droplets grow rapidly around all nuclei in such conditions, but the humidities attained in such devices are highly supersaturated and very much beyond humidities occurring in the atmosphere. Since large supersaturation values do not occur in the atmosphere, it has been found that natural condensation is dependent on the presence of a limited number of more effective nuclei—cloud condensation nuclei (CCN). These nuclei generally make up only a small fraction of the total atmospheric particle load, but they are able to initiate condensation at supersaturations below 1%, a probable maximum for atmospheric processes. The reasons whereby particles may act as CCN have not been completely defined, although larger hygroscopic or, at least, soluble, particles seem to be generally the most effective. In the preceding discussion on visibility and in Figures 12–14 and Table III, it was shown that at relative humidities of greater than about 70%, condensation began on hygroscopic sea salt particles and, as shown by the decrease in visibility with increasing humidity, the particles increased in size at higher humidity values.

Once condensation begins in a cloud, the nature of the development is strongly dependent on the number of active nuclei present. The larger the number of active nuclei among which the available water must be divided, the smaller will be the size of the droplets. Following condensation, precipitation-sized drops are formed in liquid water clouds at above freezing temperatures by coalescence of the small cloud droplets. Since larger cloud droplets agglomerate more effectively, clouds with larger average particle sizes will more readily develop precipitation. This has been shown by comparative observations of marine and continental cumulus clouds; over the ocean, where CCN are fewer, cloud droplets are larger, and precipitation is formed at an earlier stage than is the case with similar clouds over continental areas (*48*). The expected effect of air pollutants in a continental situation, if the pollutants acted as CCN, would be to inhibit even further the development of precipitation because of the increased number of nuclei and perhaps to cause a reduction in the total observed rainfall.

Warner (*49*), in a study of the area around a sugar cane growing area in eastern Australia, attributed a long-term reduction in average rainfall of perhaps 25% in downwind areas to increases in smoke and smoke-generated nuclei (either freezing or condensation nuclei) from the preharvest burning of the cane fields. His thesis was that the increased smoke introduced additional nuclei, leading to more numerous and smaller cloud particles, thereby producing a cloud system that was a less-effective precipitation system.

Similar effects were observed in the Pacific northwest in measurements of cumulus cloud characteristics by Eagan *et al.* (*50*). In the forest, fire-induced clouds had a narrower droplet size distribution and a higher concentration of droplets relative to adjacent clouds unaffected by the smoke. Eagan *et al.* (*50*) estimated that the burning wood in the controlled slash fire produced 6×10^{10} CCN (active at 0.5% supersaturation) per gram of fuel. This compares with about 6×10^{12} CCN (active at 0.5% supersaturation) reported by Warner (*49*).

The relationship between the introduction of CCN and precipitation is not a simple one, however, nor is it possible to even predict the direction of the possible change. In contrast to the results of Warner (*49*) and Eagan *et al.* (*50*), Hobbs *et al.* (*51*) found apparent increases in precipitation of as much as 30% in areas downwind of paper mills and similar industries in the state of Washington. Hobbs *et al.* (*51*) attributed these effects to the addition of CCN of such a nature that a more effective precipitation system was generated. Subsequent measurements by Eagan *et al.* (*52*) downwind of kraft paper mills tend to substantiate this earlier hypothesis. The observations showed that CCN concentrations were about four times higher in the plume from such a mill than in the ambient air and that the droplet size spectrum was much broader. In the mill plumes, most clouds contained droplets greater than 30 μm diameter. These conditions could explain the increased tendency for precipitation observed in the earlier observations (*51*). The industries producing these positive effects were also sources of large emissions of water vapor and heat, both of which could be important factors in the generation of clouds and precipitation, and, in fact, convective cloud bands were observed to form on occasion in the downward plume from these industries (*51*).

If the effect on precipitation caused by the introduction of pollutant-generated CCN into a warm cloud system is to be predicted, considerable information on the CCN content both before and after the addition of the pollutant materials must be available. Apparently when a cloud system does not contain a significant number of giant particles, e.g., greater than a micrometer in diameter, the addition of large numbers of small nuclei into this system will reduce its potential to produce precipitation, because the coalescence process of rain formation is less effective when the cloud droplet size distribution is reduced and the number of nuclei increased. However, if the system, prior to the introduction of pollutant CCN, contained significant numbers of giant particles so that the coalescence process could be initiated, the effect of the small-sized CCN would probably be relatively unimportant. On the other hand, if the pollutant CCN were large hygroscopic materials, then it might be expected that coalescence processes in the cloud system could be enhanced as these pollutant CCN supplemented the natural concentration of giant nuclei. Then, an increase in precipitation might be expected. The variety of effects on cloud droplet size distributions from different aerosol compositions and concentrations has been modeled numerically by Fitzgerald (*53*). His calculations show droplet number related to the number concentration of soluble particles but droplet size relatively insensitive to aerosol particle type.

The different possible precipitation impacts have been postulated on the basis that the introduction of pollutants does not affect the air mass in any way other than through the introduction of CCN. This is, of course, not realistic. Pollutant sources, as in the case of Hobbs' study (*51*), contribute heat and water vapor to the air, both of which under common situations can be important in the initiation of shower-type precipitation. In addition, the common pollutant source areas in urban and industrial regions are areas of increased surface roughness. Such areas can also lead to shower initiation in certain air mass conditions by the promotion of vertical motions. Thus, while pollutants may be present in the atmosphere and altered concentrations of CCN may occur, the result is not necessarily predictable, and considerable care must be exercised in any attempted assignment of cause and effect, even if a change in precipitation has been observed.

2. Nucleation Mechanisms and Effects

When a liquid water cloud becomes supercooled (i.e., cooled to a temperature below freezing) through a major part of its volume, then mechanisms other than condensation and coalescence become important in the generation of precipitation-sized particles. When supercooled droplets and ice particles coexist, the saturation vapor pressure around the droplet exceeds that around the ice particle, and a vapor pressure gradient exists between the two. The result is that the ice particles will grow at the expense of the supercooled droplets. Since the usual situation is for there to be significantly fewer ice particles than water droplets, perhaps by factors of 10^5 or more, this process tends to form fewer but larger particles in the cloud, and the process can result in the growth of particles large enough to fall from the cloud as precipitation. This precipitation mechanism is frequently called the Bergeron–Findeisen process because of the two meteorologists who played major roles in its identification in the 1930s (*54*).

The effectiveness of the Bergeron–Findeisen mechanism in the production of rain is dependent on the presence of a sufficient number of ice nuclei to provide the necessary ice particles for the system. The atmosphere is frequently deficient in available ice nuclei. In the late 1940s, it was found that silver iodide vapor or dry ice could increase significantly the available ice nuclei in a cloud system and that the addition of these materials at the right time and place might increase the precipitation received from a given cloud system or, under other conditions, cause clouds or fog to dissipate. For many years this concept has been the basis for cloud seeding and weather modification programs.

Ice nuclei are classified in terms of the temperature at which they become active in initiating freezing in a cloud of supercooled droplets in the ambient atmosphere. On the average, in the ambient atmosphere natural ice nuclei becoming active at —20°C number about one per liter. Significantly larger numbers become active at lower temperatures. By contrast, silver iodide is active as an ice nucleus at —5°C, while dry ice may be useful at even higher temperatures. There are indications that certain pollutant emissions also can act as ice nuclei (*55, 56*). To the extent that this is true pollution might alter precipitation patterns. A summary of the chemistry of ice nuclei, both natural and artificial, has been made by Montefinale *et al.* (*57*). They also discuss briefly the problem of overseeding in weather modification operations and inadvertent weather modification due to air pollutant emissions.

The introduction of ice nuclei into a supercooled cloud in sufficient concentrations to "overseed" the cloud probably has the effect of reducing the effectiveness of precipitation mechanisms, as was mentioned for CCN in warm clouds. Overseeding has been used in research and operational efforts on hail suppression and thunderstorm modification, where the goal is to slow down or inhibit natural precipitation processes, and in the dissipation of supercooled fogs and clouds.

Modification of clouds by induced thermal effects may result from the introduction of ice nuclei into clouds because of the release of significant amounts of latent heat when large numbers of ice particles are formed (*58*).

Although large numbers of cloud-seeding experiments and operational programs have been carried out since the initial discovery of the mechanisms in the late 1940s, there is currently no indication in the literature that air pollutant emissions are involved to any significant degree in causing large-scale changes in ice nuclei concentrations in the atmosphere. However, some examples of possible local situations do exist.

Telford (*55*) and Langer (*56*) have discussed pollutant sources of ice nuclei; steel mills and ferrous metallurgical operations are particularly noted as being active nuclei sources. Ice nucleation has been shown by Murty and Murty (*59*) to be a property of ordinary Portland cement particles with an activity temperature as high as —5°C; however, the activity of Portland cement was less effective than silver iodide. Battan (*60*) reports that Portland cement particles are used in the USSR as a cloud-seeding agent, especially with the objective of cloud dissipation. Telford, making measurements of nuclei activity at —24°C, also found urban areas as well as steel mills to be sources of ice nuclei. Ogden (*61*) surveyed the area around the steel mill complex at Port Kembla, Australia, for ice nuclei and for rainfall increases. He found no detectable increase in precipitation, and at distances of 16 km in the plume could detect no ice nuclei active at either —11.5° or —15.8°C. Urban areas around Seattle, Washington, have been surveyed by Hobbs and Locatelli (*62*), and concentrations of ice nuclei about 6 times the local background level were recorded. In contrast to these data, Hidy *et al.* (*63*) reported ice nuclei measurements (—20°C activation) in the Los Angeles, California, basin that indicate no increase due to air pollutants, and there was some possibility that urban measurements were less than in the surrounding background areas. Measurements by Pueschel and Langer (*64*) in traffic areas of Hilo, Hawaii, also failed to find any difference between traffic and background areas. Clear indications of the deactivation of ice nulei as detected by the membrane filter technique are reported by Braham and Spyers-Duran (*65*). These experiments were carried out in St. Louis, Missouri, in 1973 using aircraft sampling. The source or type of deactivating agent could not be identified.

Direct analyses by Pueschel and Langer (*64*) of smoke from sugar cane burning in Hawaii have shown the presence of ice nuclei in the smoke by measurements close to the fire and with laboratory burns. However, the concentrations of ice nuclei found within 30–300 m of cane fires ranged between 5 and 10 per liter, active at —20°C. This level of ice nuclei generation may be significant compared to background levels of about 1–2 per liter in marine air, but it seems that natural dilution would rapidly bring about at least a ten-fold dilution of the cloud to near background levels. It was concluded that the ice nuclei were identified with the mineralized cane ash particles rather than with the soot particles from incomplete combustion (*64*).

3. General Urban Effects on Precipitation

When the question of possible air pollution effects on precipitation is mentioned, it is probably a safe bet that La Porte, Indiana, will be mentioned. La Porte is a small town in northwestern Indiana, about 11 miles (18 km) from Lake Michigan and 30 miles (48 km) east of the heavy industrial area of Chicago, Illinois. It has become noted in air pollution and meteorological circles because of claims and counterclaims that air pollution did or did not bring about a 31% annual increase in annual precipitation compared to surrounding areas during the period of 1951 to 1965.

The first publication of La Porte, Indiana, precipitation conditions was by Stout in 1962 (*66*) when he mentioned that the La Porte records of precipitation and cloud cover indicated a sharp rise in precipitation during prior decades. In 1968, a more detailed presentation of the "La Porte anomaly" was published by Changnon (*67*). His conclusion was that beginning in about 1925, the La Porte area, when compared to nearby stations such as South Bend, Indiana, showed increases in precipitation, moderate rain days, thunderstorm days, and hail days. In the latter part of the study period, taking averages for 1951–1965, the claimed increases were sizable at La Porte relative to the surrounding stations, as shown in Table VI. Changnon's conclusion was that the increased precipitation activity in the La Porte area was due to an increase in warm-season convective activity. Both Stout (*66*) and Changnon (*67*) pointed to the apparent temporal correlation between the La Porte situation and air pollution in the Chicago area, especially from the steel industry, as a possible cause of the precipitation anomaly through emissions of condensation and freezing nuclei, water vapor and heat. Since this paper was published, it has often been cited as proving that Chicago air pollution was the cause, although Changnon himself made no such claim.

Changnon's 1968 paper (*67*), but not Stout's earlier one (*66*), was challenged in 1970 by Holzman and Thom (*68*) of the United States Weather Bureau, National Weather Service on the basis that the La Porte record was itself faulty because of station rain gauge and observer changes. This thesis was in turn challenged by Changnon (*69*) in a detailed reply in which counterarguments were given to the points developed

Table VI Difference between Average Precipitation Data for La Porte, Indiana, and Surrounding Stations, as Indicated by the Period 1951–1965 (*67*)

Annual precipitation	+31%
Warm season precipitation	+28%
Annual number of days over 0.25 inches precipitation	+34%
Annual number thunderstorm days	+38%
Annual number hail days	+246%

by Holzman and Thom. Changnon presented additional statistical correlations of La Porte annual precipitation with Chicago smoke–haze frequencies. He also tabulated data on La Porte hail damage losses showing that they were generally higher than surrounding areas (*69*) and used this as a further substantiation of greater warm-season convective activity that was not dependent on the rain gauge record.

In 1971, Hidore (*70*) published a study of stream run-off in the Kankakee basin in which La Porte is located. Hidore's conclusion was that his results supported the "La Porte anomaly," especially calling attention to the fact that the Kankakee Basin annual run-off increased at a greater rate than precipitation had increased at any basin station except La Porte. The findings of Hidore (*70*) were also challenged by Holzman (*71*) with other but less detailed run-off data from the La Porte area. In a reply to Holzman, Hidore (*72*) disputed Holzman's conclusions and tried to show that Holzman's data could be used to support the validity of Hidore's original conclusion.

An attempt was made by Ashby and Fitts (*73*) to use tree-ring growth to reflect changes in precipitation claimed for La Porte. The results were inconclusive, and they stated that their particular analysis neither "proves nor disproves the precipitation anomaly at La Porte" (*73*). Subsequent discussion on the question of a tree-ring effect in the La Porte area by Charton and Harmon (*74*) came to the same conclusion, but these authors also pointed out that because of coarse-textured soils, trees in the La Porte area would be relatively insensitive indicators of moisture changes. Both papers dealing with tree-ring patterns (*73, 74*) also raised the question of whether there was also some toxic effect of Chicago area pollutants on trees in the La Porte area. Charton and Harmon (*74*) also point out that reported smoke–haze days at Chicago were normally accompanied by surface winds that were not directed toward the La Porte area. This latter remark seems to cloud the issue unnecessarily, since there is no necessity for haze and smoke observations in Chicago or any upwind area to be correlated on a day-to-day time basis with rain in La Porte. In fact, synoptic situations favoring haze buildup would generally not be equally favorable for the production of convective showers. The fact that major urban pollutant sources are essentially continuous means that the contribution of effluents can be made to downwind areas regardless of the weather conditions in the local region. Changnon (*67*) and Stout (*66*) both use the Chicago haze data as a general air pollution index for the region not as a day-to-day indicator. Holzman and Thom (*68*) made some very brief references to general air mass trajectories corresponding to rain situations at La Porte but gave little details as to the nature of their analysis.

In common with many problems in meteorology, there is no firm or conclusive resolution of the "La Porte anomaly" question—"fact or fiction" was Changnon's original question (*67*). The reader is strongly urged to decide for himself after he reviews the series of papers cited above. As far as the La Porte area is concerned the anomalous precipitation pattern seems to have been disappearing since the mid-1960s, and thus further on-site assessment is no longer possible. Mother nature has again covered her tracks.

Further research by Changnon (*75*) tends to show that La Porte has not been a unique situation and that city-center or downwind rainfall increases may be observed in Chicago and Champaign-Urbana, Illinois; St. Louis, Missouri; Long Island, New York; Tulsa, Oklahoma; and Washington, D.C. The situation around St. Louis, Missouri, has subsequently been analyzed in detail by Huff and Changnon (*76*), who showed that average summer rainfall increased from 6 to 15% as a result of urban effects when compared to adjacent control areas. For St. Louis, Missouri, the most important but not exclusive urban factor for precipitation increase was apparently added heat from the urban "heat island." The precipitation increases also tended to be greater on weekdays relative to weekends—11% weekdays increase compared to 4% on weekends. Thus, temporal factors such as pollutant or water vapor emissions may also be significant factors while increased roughness in the urban area leading to greater turbulence and possible convective cloud formation was apparently of less importance in the St. Louis area.

Schaefer (*77, 78*) has found that lead aerosols, such as contained in automobile exhausts when the fuel contains lead additives, can be made to form active ice nuclei by exposing the exhaust aerosol particles to high concentrations of iodine vapor. He has also suggested that this process might occur in the atmosphere and cause altered pollutant patterns downwind of major urban traffic areas. Urban area measurements of ice nuclei do not indicate that this is an active atmospheric process (*63, 64*), although there is some contradiction, as indicated by the results of Hobbs and Locatelli (*62*). Although lead may frequently be detected in both rain and snow, these concentrations probably result from precipitation scavenging of airborne lead aerosol particles rather than from the nucleation of the precipitation by the lead compounds.

Further analysis of rainfall variations as related to urban areas is reported by Schickedanz (*79*) as a result of detailed analysis of convective storm patterns in and around St. Louis, Missouri. The analysis was of the size and intensity of rain cells within the general larger area convective storms. From this study, Schickedanz concluded that rain cells occurring in the urban–industrial region of the city produced an average rainfall volume 176% greater than control sample cells. For rain cells occurring in the industrial area of Wood River, near St. Louis, the average rainfall volume was 262% greater than in control raincells. The urban-induced rainfall increases have not been identified with any single cause. Some possible reasons include the urban heat-island effect, increased nuclei concentrations, and anomalous moisture sources (*79*).

4. Fog-Producing Mechanisms

An increased frequency of fog in urban areas has often been mentioned as being characteristic of urban areas and a result of air pollutant emissions. Landsberg indicates that urban areas may experience 100% more winter fog and 30% more summer fog than adjacent rural areas (*80*). More persistent and long-lasting fogs in urban areas are attributed to air pollution effects for about the same reasons as given for cloud modification, namely, large numbers of additional nuclei, it is argued, will produce a fog cloud with increased numbers of smaller drops and thus a situation that is more stable than would otherwise occur. The smaller particles will also cause a higher degree of obscuration with no increase in liquid water content.

To examine the relationship between pollutant air and fog formation, Neuberger and Gutnick (*81*) carried out a series of laboratory experiments in which fog characteristics were studied as a function of condensation nuclei concentration by measurement of light transmission. The laboratory fogs were formed using an expansion-cloud chamber. A constant expansion ratio of 20% was used in all experiments. Condensation nuclei concentrations, but not CCN as would be most applicable to actual urban situations, were determined using an Aitken counter. Nuclei concentrations of up to 50,000/ml could be obtained by using either laboratory or outside air in the experiments. Higher concentrations were obtained in the room by lighting a Bunsen burner adjusted for good combustion, i.e., blue flame. Low concentrations of less than 1000/ml were obtained by filling the chamber with laboratory air and leaving it over night. A total of 113 individual tests for seven groups of nuclei concentration was carried out during the study.

Figures 16 and 17 are taken from Neuberger and Gutnick's paper (*81*). Figure 16 shows the minimum transmission and the persistence in terms of both total duration of the fog and the half-time, i.e., the time required to recover 50% of the maximum transmission decrease, all as functions of Aitken nuclei concentration. The duration curves show that for very clear air, less than 25 nuclei per cubic millimeter, the change in fog duration is relatively small as compared to more rapid increase in

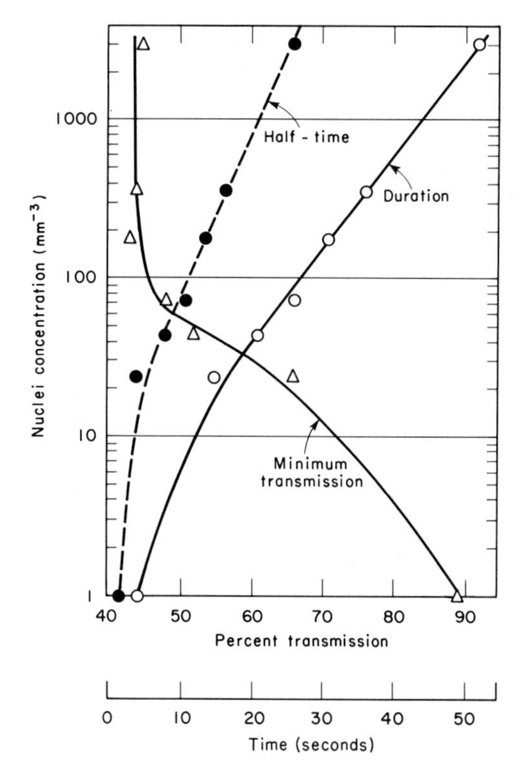

Figure 16. Fog persistence and density for various nuclei concentrations *(81)*.

fog duration for much higher nuclei concentrations such as would be characteristic of urban areas. Minimum transmission reached in the fog experiments shows a decrease with increasing nuclei until a concentration of about 50 mm^{-3} is reached and then more or less constant transmission. Figure 17 shows that calculated visual range decreases rapidly as the nuclei concentration increases to about 50 mm^{-3} and then remains constant for higher concentrations of nuclei. This is due to the fact that although fog droplet numbers increase with increasing nuclei concentrations, the changes are not at all proportional, and the size of the droplets decrease as their concentration increases. The result is the constant visual range plot at nuclei concentrations greater than 50 mm^{-3}, as shown by Figure 17.

Field verification of increased fogs in air pollution areas has been obtained in a number of studies, and there does not seem to be much doubt that the proposed effect is real. Much of the field study data come from England. For example, Eggleton *(82)* examined the chemical aerosols and visibility problems in the Tees River area of northeast England. More persistent and extensive fogs and mists were attributed to inter-

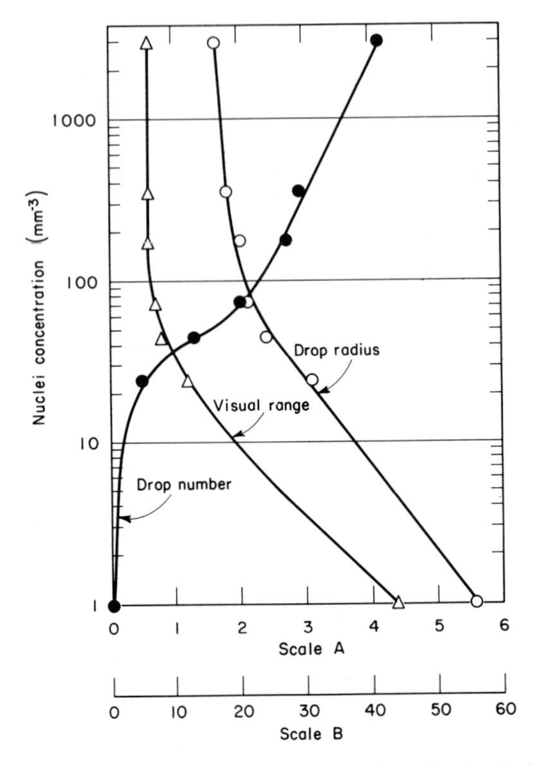

Figure 17. Radius (scale A in microns) and number (scale B in mm^{-3}) of fog droplets and visual range (scale B in meters) as functions of nuclei concentrations (*81*).

actions between the natural North Sea fogs and massive emissions from a variety of heavy industrial sources. Also in England, Collier (*83*) has examined the effects of urban area pollution and fog frequency in Manchester, where the city may have experienced more frequent fog than the airport. In the London area, Chandler (*84*) has complied data showing increased fog frequencies in the urban area.

Heat-island effects resulting from urban heating have, at least in Manchester, England, been credited with less frequent heavy fogs (*83*). The adoption of smokeless zones in London, England, has according to Lawther (*85*) caused a very significant decrease in the intensity of the legendary London fog. Atkins (*86*) also indicates that a change in English fog characteristics may have occurred as a result of the smokeless zones resulting from the Clean Air Act.

Another aspect of fog and pollutant interactions has been examined by Berlyand and Kanchan (*87*), namely, the dispersion of air pollutants during fog conditions and the uptake of pollutants by the fog particles.

On the basis of model calculations of both fog formation and dispersion, they showed that after about a 0.5 km travel in a fog, a gaseous contaminant would be almost all dissolved in the fog droplets. Measurements of aerosol size distribution changes in fog in the Elburs area of the Soviet Union by Ivlev *et al.* (*88*) showed, in general, a decrease in submicron particle concentration and an increase in larger particles. Coagulation is given as the explanation.

C. Pollutant Interaction with Radiation

Studies of gross comparative solar radiation climatology have been made for a number of areas, including Boston, Massachusetts (*89*); Seattle, Washington (*90*); and Los Angeles, California (*91*). In these studies, the techniques were similar: total radiation data taken in downtown and outlying locations were compared. When clouds or other weather phenomena did not complicate the picture, the amount of radiation received at the two sites was compared, and it was argued that air pollution in the downtown area was responsible for less radiation being received in the urban or downtown area. Holzworth gives more details on air pollution radiation climatology in Chapter 12, Vol. I.

These studies (*89–91*) on radiation diminution have been concerned with changes in total solar radiation. However, radiation loss due to air pollution is strongly wavelength dependent, with the shorter wavelengths being much more seriously affected than the long wavelengths. In fact, one of the effects of air pollution that was an early concern of investigators in the era of heavy coal smoke was the excessive loss of ultraviolet radiation from the sunlight when it penetrated an urban smoke pall. Solar ultraviolet radiation is a major factor in the body's generation of natural vitamin D. The excessive loss of the shorter wavelength ultraviolet compared to losses of visible light is readily explained by the inverse relationship between the wavelength of the light and extinction caused by small particles. In the case of very small particles, Rayleigh scattering is applicable, and the extinction is dependent upon λ^{-4}.

Losses of ultraviolet and visible solar radiation due to Los Angeles, California, smog clouds were measured by Stair (*92, 93*) in Pasadena, California, in the fall of 1954. Measurements of conditions during the morning, prior to the midday arrival of heavy smog clouds, permitted the effects of the smog to be evaluated. He reported that while there were large drops in solar intensities at all wavelengths when the smog reached the observation location, the greater percentagewise reductions occurred with the shorter wavelengths. In some cases the smog reduced the intensities for shorter wavelengths by 90% or more (*93*).

Stair was able, under smog conditions, to detect losses in transmittance both from general scattering and from absorption (*92*). The scattering was attributed to the small particles in the smog, while the absorption was attributed to gaseous pollutants such as nitrogen dioxide. The absorption occurred in two general parts of the shortwave spectrum, with increased absorption occurring at wavelengths shorter than 3200 Å and in a broad band centered around 3600–4000 Å. The absorption in the region below 3200 Å was attributed primarily to ozone, although calculations seemed to indicate that the expected ozone concentrations could not account for all the absorption that was observed in this region. The broad absorption band around 3600–4000 Å was attributed to nitrogen dioxide.

A detailed field study of ultraviolet radiation in the Los Angeles, California, area involving simultaneous urban ground-level, aircraft, and remote mountain-top measurements was conducted in 1965 in the Los Angeles Basin on 5 days of varying photochemical smog intensity (*94*). Comparing the attenuation through the lower 5000 ft (1500 m) for incoming solar radiation on smoggy and clear days showed that days of heavy smog had up to four times greater attenuation than clear days. The maximum attenuation occurred between about 9 a.m. and noon at times when ozone, nitrogen dioxide, and particle concentrations showed high values and visibility was only about $\frac{1}{2}$ mile (0.8 km). These were relatively broad-band spectral measurements, and significant wavelength dependence could not be observed.

While pollutants may scatter incoming radiation and reduce the net amount that reaches the earth's surface, if the pollutant layer absorbs significant radiation, this energy may be transferred to the atmosphere as heat and result in an increase in the air temperature in the polluted layer. Absorption by aerosol particles may be one source of atmospheric heating. The absorption coefficient, the total particle cross section, and the particle concentration are significant parameters in this problem. The only gaseous pollutant that probably has significant absorption at visible wavelengths and sufficient concentration to affect urban heating rates is nitrogen dioxide.

Theoretical estimates of lower atmospheric heating rates for both gaseous pollutants and aerosol particles have been made by Try (*95*) and by Atwater (*96, 97*). For nitrogen dioxide, which has strong absorption in the visible spectrum resulting in the atmospheric color effects already noted in the previous discussion, a typical background concentration of 3 parts per billion (ppb) produces a calculated heating rate of 0.06°C/17-hour day at the earth's surface according to Try (*95*). Urban concentrations in Los Angeles, California, have exceeded 1 ppm on occasion and concentrations above 0.25 ppm are expected regularly. For a daily average concentration of 1 ppm Try's estimated heating rate is 12°C/12-hour day. This could be a very significant contribution to observed temperature contrasts between urban and rural areas.

Heating rates for atmospheric particles as a result of incoming visible solar radiation have also been calculated by Try (*95*). For background tropospheric conditions with a particle concentration of about 10 $\mu g/m^3$, he calculated a heating rate of about 0.5°C/12-hour day using a simple scattering model. For urban atmospheres with higher particle concentrations and using a model that accounted for multiple scattering, Try (*95*) obtained heating rates of up to 15°C/12-hour day for particle concentrations up to 1000 $\mu g/m^3$. He found that the estimated heating rate was dependent on the particle concentration and vertical distribution, real and imaginary indices of refraction, particle size distribution, albedo of the earth's surface, and the optical depth of the atmosphere.

The question of whether atmospheric aerosols under realistic distribution conditions might produce net atmospheric heating through radiation absorption, or cooling through the effect of an increased albedo has been examined by Mitchell (*98*) and by Ensor *et al.* (*99*), as well as by other authors. Both discussions (*98, 99*) point out that resultant heating or cooling by an aerosol layer depends on whether the ratio of aerosol absorption to backscatter for incoming solar radiation is greater or less than some critical value. Factors in determining the critical value of this ratio include surface albedo, surface heating by solar radiation, convective contact between the atmospheric aerosol and the earth's surface, and physical aerosol parameters. Ensor *et al.* (*99*) show that with regard to the aerosol parameters, if the absorption coefficient for the global aerosol, as described by the imaginary part of the refractive index $n_2 i$, has a n_2 value less than 10^{-3}, the qualitative effect of an increase in aerosol concentration would be a cooling of the earth. If n_2 is greater than 10^{-1}, the effect would be global heating. For values between the two limits the resultant effect is ambiguous. Mitchell (*98*) points out that while refractive index and other data on tropospheric aerosols are seriously lacking, the information that is available indicates that the most likely effect of tropospheric aerosols is heating rather than cooling, except for arid areas such as deserts and perhaps some local urban areas.

Interference with the terrestrial radiation balance through anomalous absorption of long-wave radiation from the ground with reradiation to the ground rather than completely into space can also effect the heat balance of the atmosphere. In fact, normal ambient concentrations of carbon dioxide and water vapor significantly absorb heat radiated by the ground. There is some concern that the atmospheric concentration of infrared absorbing pollutants such as carbon dioxide will increase significantly, with the result that average surface air temperatures will gradually increase. This is the popularly identified "greenhouse effect," although the physical analogy with the atmosphere is not really an accurate one. Concern over atmospheric affects from the gaseous pollutants is almost exclusively related to the observed gradual increase in carbon dioxide in the atmosphere e.g., see Mitchell (*100*). On the basis of atmospheric radiation models, an increase of 10 ppm in atmospheric carbon dioxide concentration would result in an increase of about 0.1°C in global mean temperature. The estimated carbon dioxide average global concentration was about 320 ppm in 1970. Very large particles in the atmosphere could affect the infrared radiation balance in a manner similar to carbon dioxide. However, concentrations would have to be very great (*5*), and thus such an effect would be localized in some limited industrial areas or perhaps dust storms.

Atwater (*96, 97*) has carried out model calculations showing that increasing concentrations of absorbing and radiating pollutants, such as nitrogen dioxide, sulfur dioxide, and aerosol particles such as would be found in the atmospheric boundary layer over an urban area, can lead to the development of an inversion above the surface pollution layer. These model results indicate that the presence of absorbing and radiating pollutants in the surface layer increases the downward radiative flux and reduces the solar flux at the ground. At night, this would increase the surface temperature and accentuate the heat-island effect. When a pollutant layer cools by radiation, the upper part of the layer can cool more rapidly than the clear air above it leading to the occurrence of an elevated inversion surface between the polluted and clear layers. In many respects the process is similar to the development described in 1952 by Fleagle *et al.* (*101*) of a stable layer above and instability within a ground fog layer. Observations of temperature lapse rate conditions during fog formation show clearly the sequence of the development of an elevated inversion above the turbid layer.

Actual measurements of comparative urban and rural radiation and temperature conditions at Hamilton, Ontario, Canada, were made by Rouse *et al.* (*102*). Their results generally paralleled the results calculated by Atwater from his model including the formation of an elevated inversion, the attenuation of solar radiation, and an increase of approximately 12% for total in-coming long-wave radiation at the ground.

The possible impact on the global atmosphere of increased radiational heating or cooling leads to more than a simple temperature change, as pointed out by Newell *et al.* (*103*). They show through model calculations that an increase in atmospheric carbon dioxide leads to a warmer troposphere and a cooler stratosphere. However, this change in the thermal structure of the atmosphere can result in complex changes in available potential energy in the atmospheric system. The net impact on the total atmospheric system cannot be predicted until further advances in three-dimensional atmospheric modeling are made.

Observations showing apparent parallel cycles in local temperatures and air pollutant levels were reported by Schuck *et al.* (*104*). These cycles in the Los Angeles, California, area were apparently related to changes in traffic patterns between weekdays and weekends. This study covered data from nine oxidant air monitoring stations for the period from July 1962 through June 1964. Average data at inland stations showed weekday maxima and weekend minima and the opposite cycle at coastal stations. This was interpreted as resulting from changes in traffic patterns, from business and industrial travel during the week, to recreational travel toward coastal areas on weekends. Weekend temperatures were found to be lower by 0.5°F (0.3°C) at stations having minimum weekend oxidant, and higher by 0.5°F at the coastal stations on weekends where oxidant values also tended to be at a maximum on weekends. As an example, at Pasadena. California, an inland Los Angeles, California basin station characterized by a weekday oxidant maximum, the average daily mean temperature during the period July–December 1962 on weekdays was 0.6°F (0.33°C) higher than on weekends and 0.9°F (0.5°C) higher on weekdays for the period January–June 1963. Although no definitive conclusions were drawn as to the reasons for these temperature changes, it was postulated by Schuck (*104*) that they could be due in part to thermal emissions from traffic. However, since there is a strong positive correlation between nitrogen dioxide and oxidant levels and the local generation of significant aerosol concentrations in photochemical smog, it is not unreasonable to argue that nitrogen dioxide and particle absorptions could be a more significant factor, as has been suggested by Try (*95*) and Atwater (*96, 97*).

D. Fine-Particle Concentrations and Atmospheric Conductivity

An increase in atmospheric fine-particle concentrations over the period from 1910 to 1967 is described by Gunn (*105*) and Cobb and Wells (*106*) on the basis of electrical conductivity measurements over the oceans (Fig. 18). Conductivity measurements of this type are most closely related to particles in the Aitken range, 2×10^{-7} cm to 10^{-4} cm radii, which act as a sink to remove from the atmosphere the highly mobile small ions that are responsible for the electrical conductivity of the atmosphere. The greater the concentration of Aitken nuclei, the lower the conductivity, because the concentration of small ions is reduced.

Except for the two earliest years, the North Atlantic measurements of

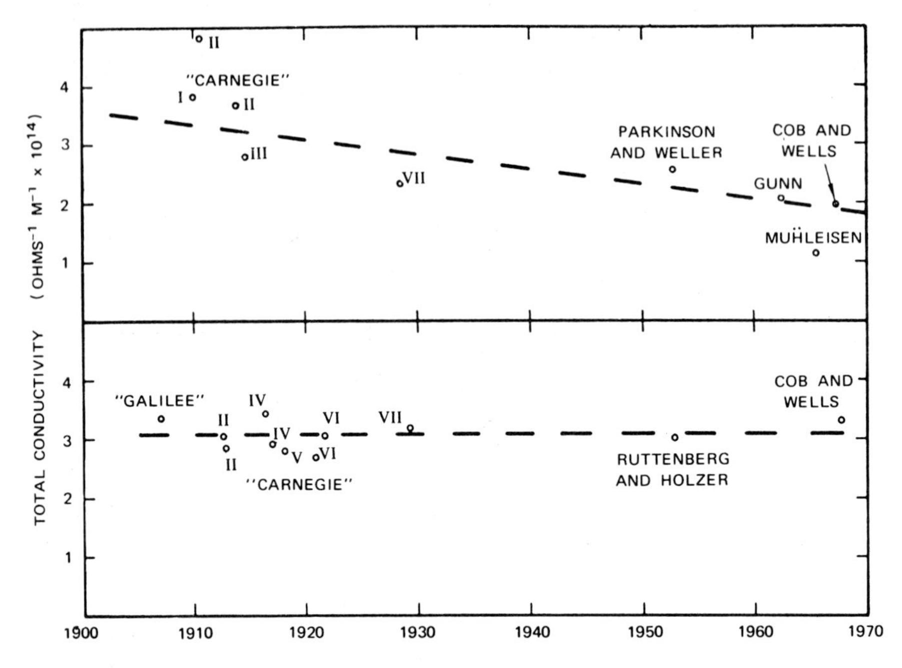

Figure 18. Change in atmospheric conductivity in (*top*) the North Atlantic (latitude, 20°N–50°N; longitude, 20°W–60°W) and (*bottom*) the South Pacific (latitude, 10°S–50°S; longitude, 80°W–180°W) between 1907 and 1967 (*106*).

Figure 18 all show lower conductivity than do the South Pacific measurements. It can probably be inferred from this that Atlantic air masses carry a greater load of fine particles than do the southern hemispheric Pacific air masses. As of 1970, North Atlantic conductivity appears to have decreased by about 45% since about 1910 and by about 35% since 1929. No comparable changes occurred in the southern hemisphere, based on the South Pacific data in Figure 18. The fact that pollutant emissions in the southern hemisphere are 10% or less than those north of the equator is well known.

Additional global data on marine atmospheric conductivity by Cobb (*107*) is given in Figure 19 and shows that the North Atlantic pattern of decreasing conductivity in more recent years has also been observed in the North Pacific east of Japan and in the Indian Ocean south of India. East of Japan and the China mainland conductivities decreased by about 20% between the period 1915–1929 and 1966, while south of India the change was more than 50%, larger than in the North Atlantic, between the 1911–1920 period and 1967. The pattern in the North Pacific seems to follow the general westerly wind flow off the Asian continent, while the area in the Indian Ocean doubtless represents the influence of the Indian monsoon.

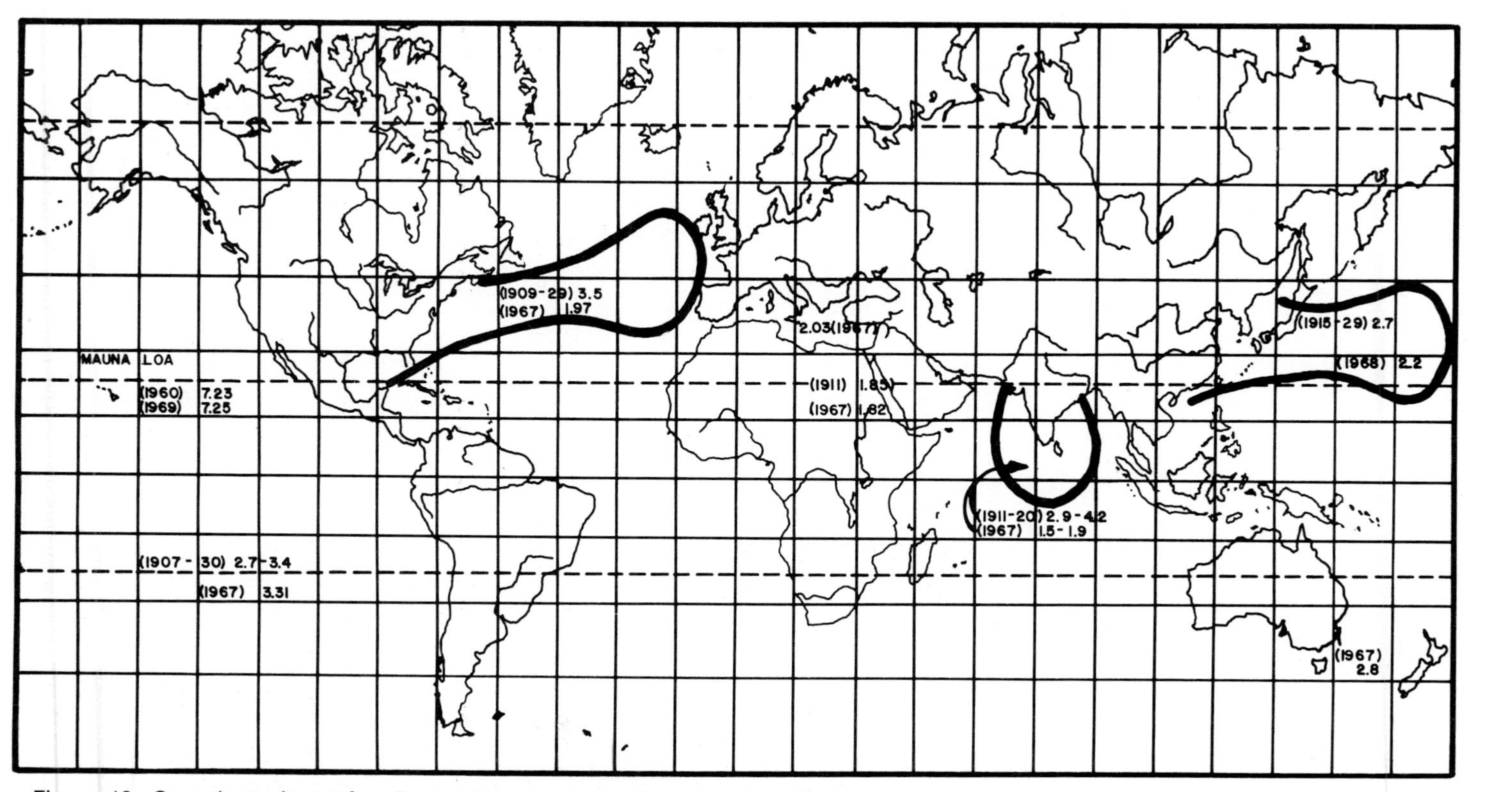

Figure 19. Oceanic regions of anthropogenic aerosol pollution as indicated by a secular decrease in atmospheric conductivity. Conductivity units are $ohms^{-1}m^{-1} \times 10^{14}$ (*107*).

The change in conductivity over the North Atlantic shown by Figure 18 could result from a doubling of the fine-particle concentration (*106*). There seems to be little doubt that long-distance transport of fine particles from pollution sources in North America could be responsible for these North Atlantic changes in electrical conductivity. This is not an unreasonable conclusion considering the fact that the Atlantic is both smaller than the Pacific and generally dominated by heavily industrialized North America, while the South Pacific is primarily influenced by much less industralized areas.

That fine particles from pollutant sources can be found regularly over the North Atlantic has been shown by condensation nuclei data collected by Hogan *et al.* (*108*) during the period 1966–1971. Their data showed continental influences in terms of relatively high concentrations, e.g., 1000–4000/ml, north of 40°N and in north–south bands several hundred miles wide along the North American east coast and the European west coast. The general correspondence between the conductivity data (*107*) and the condensation nuclei data (*108*) seems reasonably good, although parallel measurements in these oceans are not available.

The transport of larger-sized pollutant materials over wide stretches of the North Atlantic is shown by the work of Parkin *et al.* (*109*). These investigators were able to identify fly ash spherule particles in collections of particles greater than 20 μm diameter made over the open ocean between Newfoundland and Ireland. Typically, the spherule fraction in the 1969 summer data was between 5 and 10% of the total number of particles collected.

The collection of pollutant materials in areas remote from sources may be an indication of long-distance transport but may not be an indication of a global change in the atmosphere. Thus, Figures 18 and 19 cannot be interpreted as indicating more than long-distance transport in terms of atmospheric impact. Ellsaesser (*110*) has made a detailed assessment of the problem of global particle concentration change and was able to make a strong case for the absence of any global trend. He points out in some detail the problems encountered by various investigators of background-level pollutants, including both conductivity and turbidity observations. Ellsaesser also makes a valid point that in major urban areas, demonstrable progress is being made in attaining lower levels of particulate pollutants, and thus he argues that perhaps it should be expected that the global atmosphere should be getting cleaner (*110*).

E. Precipitation Chemistry

One of the major mechanisms for the removal of air pollutants from the atmosphere, as described by Junge (*24*), is through their incorporation first into clouds and then precipitation. Precipitation scavenging may remove both particulate materials such as insoluble lead compounds and highly soluble gases such as sulfur dioxide. The frequent finding of significant amounts of ammonium sulfate in atmospheric aerosol particles is used as a basis for arguing that a reaction has taken place between ammonia and sulfur dioxide (*24*). It is generally believed that this reaction occurs when both gases are in solution in rain or cloud droplets (*24*) although other processes may also be involved.

There is no doubt that precipitation chemistry represents the chemical content of the atmosphere both where the cloud was formed and the air through which it had fallen before reaching the earth. In remote areas, the chemical content of the precipitation is indicative of the low chemical background concentrations of these air masses, while in urban and industrial regions a higher chemical content would be expected in the precipitation as a reflection of the higher atmospheric pollutant burden in these areas.

Selezneva (*111*) has used the Soviet Union precipitation chemistry network to estimate the relative impacts of various sources as contributors of atmospheric particles. It was determined that 30–40% was background and 60–70% was local sources. Of this latter portion, 20–30% was the natural contribution, such as soil dust, and the rest were air pollutants. Selezneva (*111*) concluded that in the Soviet Union pollutant sources contributed 20–30% of the air chemistry levels even in remote clean areas.

Modeling of the chemical scavenging process in the atmosphere is characteristically expressed as the ratios of the concentrations of the pollutant and water vapor in the precipitation particles to those in the air, e.g., see Junge (*24*), Engelmann (*112*), and others. Engelmann (*112*) presents the following introduction to his cloud scavenging model. "We visualize the rain cloud as a machine which pumps or processes air through it, removing pollutants and water as precipitation with efficiencies E_1 and E_2, respectively, before the air exits. This is a very different concept from that concept of the cloud as a box wherein the polutants are removed to the cloud elements at some fractional rate, ..." Englemann (*112*) combines both in-cloud and below-cloud processes into the ground-level washout ratio for a pollutant as follows:

$$\left(\frac{k}{C}\right)_v = \frac{n}{qE_2} + \frac{(1-n)\rho x}{q} + \frac{H\Lambda}{R} \qquad (33)$$

where

k = pollutant concentration on volume basis in rain
C = pollutant concentration on volume basis in air
ρ = density of water

q = absolute humidity
E_2 = the efficiency of the cloud in converting water vapor to precipitation, value range: 0.05–0.65
n = fraction of the pollutant that nucleates and is scavenged
x = a reactivity factor for gases with water, value is near 1.0
H = height of the cloud base
Λ = the washout coefficient for the given precipitation rate
R = the rate of precipitation

In this expression, the first two terms on the right-hand side relate to in-cloud processes and the third term to below-cloud washout. For sulfur dioxide in air and rain, the k/C ratios on a mass basis are in the range of 19–500, while for particulate material, the ratio usually is in the range of 290–2700 (*112*).

Precipitation washout effects on the atmospheric aerosol size distribution have been examined both theoretically and experimentally by Ivlev *et al.* (*88*) in the Soviet Union. The experiments made in Leningrad in 1969 included particle-size sampling prior to, during, and after periods of rain, snow, drizzle, and fog. The data for precipitation cases showed a reduction in aerosol particle concentration and an optimum washout generally in the size range with the radius greater than 1 μm (*88*). However, there was also significant loss of submicron particles and considerable scatter in all the data. Considerable scatter is not unexpected considering the diverse nature of both precipitation and urban aerosol particles. Selezneva and Petrenchuk (*113*) have concluded that clouds forming below an anticyclonic inversion are effective scavengers of air pollutants present in this layer. These authors (*113*) also show that in precipitation situations in regions of low pollutant concentrations, more than 50% of the scavenging occurs in the clouds, while in areas with heavy pollution, 70-80% of the contaminant scavenging occurs as washout below the clouds. A significant vertical gradient of pollutant concentration in high emission areas could be one explanation for such a result.

The results of measurements of average annual sulfate deposition in rainfall over the United States is shown in Figure 20 (*114*). West of the Mississippi River, relatively low values are indicated, about 0.2 gm/m^2; and these are probably indicative of rural areas with little pollutant sulfur emission. East of the Mississippi River, and in particular in the northeastern states, sulfate deposition rates increase dramatically, doubtlessly in response to the increased atmospheric sulfur dioxide burden from combustion and industrial sources. In Europe, longer-distant travel of pollutants is shown by the fact that during the years 1968–1971 snow in southern Norway was observed to have a banded, or laminated, appear-

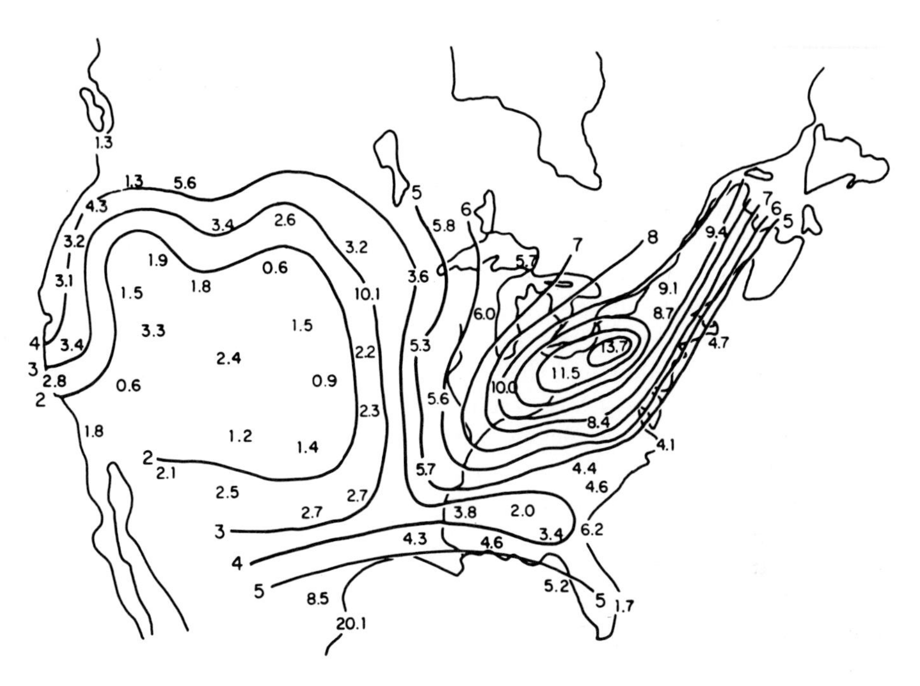

Figure 20. Excess sulfur in precipitation over the United States in units of kilograms per 10^4 m^2 (0.1 gm/m^{-2}) per year (*114*).

ance, with the bands being the result of heavy air pollutant contamination (*115*). The gray-colored bands were highly acid and higher in concentration of sulfur and the other elements than white bands. It was concluded that the sources of the pollutants was outside Norway and probably in western and central Europe.

In the atmosphere, changes in precipitation chemistry are probably of little consequence unless the condensation and precipitation processes themselves are affected; this was discussed in a prior section. The effects on the soil or vegetation of precipitation with a significant chemical content is a possible long-term factor in air pollution chemistry, as shown by studies in the English midlands (*116*), where effects on plant nutrition were traced to fly ash and other pollutant contents of precipitation. Analysis of plants affected by airborne lead aerosols showed that significant amounts of lead remained in washed leaves, but some had also been translocated to the plant roots (*117*). Significant amounts of nitrate and sulfate compounds in soil and water could result from the scavenging of reaction products of nitrogen dioxide and sulfur dioxide. In fact, Prince and Ross (*118*) argue that there are many areas of sulfur-deficient soils and that in these areas the deposition of pollutant emissions could be beneficial as long as biologically damaging levels were not attained. Hanley and Tierney (*119*) show that sulfur in precipitation is one of the important sources of this element for agriculture in Ireland, where coastal areas receive about twice as much in precipitation as do inland stations. Much of this sulfur presumedly has a natural source. A similar line of reasoning might also be used with regard to oxides of nitrogen or ammonia emissions and the usefulness of additional nitrogen for vegetation growth. Needless to say, there is considerable room for controversy in the subject of "beneficial" air pollutant concentrations.

Another factor that has received increased attention in recent years is the pH of precipitation, and there is some evidence that precipitation pH can affect the pH of surface waters and of soils (*120*). If either pH or chemical changes are significant, they can cause a variety of effects in the biosphere (*120*). Some fish are especially sensitive to pH, while in some soils nutrient availability is a function of pH.

V. Conclusions

The list of atmospheric properties that can be affected by the presence of air pollutants includes a wide variety from the aesthetic factors in the visibility problem and possible short-term changes in urban precipitation patterns to long-range global climate changes that could result if, for example, a widespread introduction of freezing nuclei were to occur over a prolonged period of time. Even with this wide variety, there is one aspect that is common to all pollutant impacts on atmospheric properties; the common feature is that pollutant impacts are not unique effects but rather are changes in the magnitude of processes that are otherwise natural aspects of the atmosphere. Pollutants, for example, can reduce the visibility over a given region, but reduced visibility by itself does not necessarily indicate air pollution unless fog, marine haze, blowing dust, and other natural phenomena can be ruled out. The example of La Porte, Indiana, and the question of whether there were anomalous changes in precipitation is an excellent example of the complexities that may be encountered when considering possible atmospheric impacts.

This lack of uniqueness of pollutant impacts becomes an important consideration when the identification of background perturbations or threshold changes is attempted. In order to separate pollutant effects from natural atmospheric variability a careful sampling over an extended period of time is required. It will probably be necessary to identify true background conditions and then to compare the suspected pollutant impact to the background levels. Such programs will not be simple or inexpensive, but efforts should be made to determine whether air pollutants are causing subtle changes in the atmospheric environment. The alternate course of action, to wait until a suspected effect is evident, can obviously lead to an unsatisfactory situation of having to respond with corrective actions much too late in a cause-and-effect cycle.

References

1. A. P. Waggoner, University of Washington, Seattle, Washington, private communication, 1973.
2. W. E. K. Middleton, "Vision Through the Atmosphere," Univ. of Toronto Press, Toronto, Ontario, 1952.
3. W. M. Porch, R. J. Charlson, and L. F. Radke, *Science* **170,** 315–317 (1970).
4. D. Sinclair, *in* "Handbook on Aerosols," Chapter 7. U.S. At. Energy Comm., Washington, D.C., 1950.
5. H. C. van de Hulst, "Light Scattering by Small Particles." Wiley, New York, New York, 1957.
6. W. W. Foster, *Brit. J. Appl. Phys.* **10,** 416 (1959).
7. A. P. Waggoner, M. B. Baker, and R. J. Charlson, *Appl. Opt.* **12,** 896 (1973).
8. V. R. Stull and G. N. Plass, *J. Opt. Soc. Amer.* **50,** 121 (1960).

8a. J. T. Twitty and J. A. Weinman, *J. Appl. Meteorol.* **10,** 725–731 (1971).

9. P. A. Leighton, "Photochemistry of Air Pollution." Academic Press, New York, New York, 1961.
10. J. R. Hodkinson, *Int. J. Air Water Pollut.* **10,** 137 (1966).
11. R. J. Charlson and N. C. Ahlquist, *Atmos. Environ.* **3,** 653–656 (1969).
12. R. J. Charlson, D. S. Covert, Y. Tokima, and P. K. Mueller, *J. Colloid Interface Sci.* **39,** 260–265 (1972).
13. L. J. Hagen and N. P. Woodruff, *Atmos. Environ.* **7,** 323–332 (1973).
14. R. J. Charlson, N. C. Ahlquist, and H. Horvath, *Atmos. Environ.* **2,** 455 (1968).
15. N. C. Ahlquist and R. J. Charlson, *Environ. Sci. Technol.* **2,** 363 (1968).
16. H. J. Samuels, S. Twiss, and E. W. Wong, "Visibility, Light Scattering and Mass Concentration of Particulate Matter." Air Resources Board, State of California, Sacramento, California, 1973.
17. C. Steffens, *in* "Air Pollution Handbook" (P. L. Magill, F. Holden, and C. Ackley, eds.), Section 6, pp. 6-1–6-43. McGraw-Hill, New York, New York, 1956.
18. J. C. Elder, H. J. Ettinger, and R. Y. Nelson, *Atmos. Environ.* **8,** 1035–1048 (1974).
19. F. S. Duckworth, *J. Air Pollut. Contr. Ass.* **15,** 274 (1964).
20. R. T. Jarman and C. M. de Turville, *Int. J. Air Water Pollut.* **10,** 465 (1966).
21. W. L. Crider and J. A. Tash, *J. Air Pollut. Contr. Ass.* **14,** 161 (1964).
22. R. A. McCormick and D. M. Baulch, *J. Air Pollut. Contr. Ass.* **12,** 492 (1962).
23. R. A. McCormick and K. R. Kurfis, *Quart. J. Roy. Meteorol. Soc.* **92,** 392 (1966).
24. C. E. Junge, "Air Chemistry and Radioactivity." Academic Press, New York, New York, 1963.

25. D. Randerson, *Atmos. Environ.* **7,** 271–279 (1973).
26. K. Bullrich, *Atmos. Environ.* **7,** 665 (1973).
27. J. H. Joseph and A. Manes, *J. Appl. Meteorol.* **10,** 453–462 (1971).
28. H. H. Lamb, *Phil. Trans. Roy. Soc. London, Ser. A* **266,** 425–533 (1970).
29. M. P. Paterson, *Atmos. Environ.* **7,** 281–290 (1973).
30. R. F. Pueschel, C. J. Garcia, and R. T. Hansen, *J. Appl. Meteorol.* **13,** 397–401 (1974).
31. Stanford Research Institute, "The Smog Problem in Los Angeles County," 3rd Interim Rep. Western Oil and Gas Ass., Los Angeles, California, 1950.
32. M. E. Miller, N. L. Canfield, T. A. Ritter, and C. R. Weaver, *Mon. Weather Rev.* **100,** 67–71 (1972).
33. R. E. Huschke, ed., "Glossary of Meteorology." Amer. Meteorol. Soc., Boston, Massachusetts, 1959.
34. "Manual of Surface Observations," Circ. N, 7th ed. pp. 2–3. U.S. Govt. Printing Office, Washington, D.C., 1966.
35. U.S. Weather Bureau, "Instructions for Airway Meteorological Service," Circ. N, 4th ed., pp. 31–32. U.S. Govt. Printing Office, Washington, D.C., 1939.
36. U.S. Weather Bureau, "Instructions for Airway Meteorological Service," Circ. N, 3rd ed., p. 6. U.S. Govt. Printing Office, Washington, D.C., 1935.
37. C. Junge, *in* "Compendium of Meteorology" (T. F. Malone, ed.), pp. 182–191. Amer. Meteorol. Soc., Boston, Massachusetts, 1951.
38. S. Twomey, *in* "Atmospheric Chemistry of Chlorine and Sulfur Compounds" (J. P. Lodge, ed.), Geophys. Monogr. No. 3, p. 4. Amer. Geophys. Union, Washington, D.C., 1959.
39. M. Neiburger and M. G. Wurtele, *Chem. Rev.* **44,** 321 (1949).
40. T. J. Buma, *Bull. Amer. Meteorol. Soc.* **41,** 357 (1960).
41. D. S. Covert, R. J. Charlson, and N. C. Ahlquist, *J. Appl. Meteorol.* **11,** 968–976 (1972).
42. P. Winkler and C. E. Junge, *J. Rech. Atmos.* **6,** 617–638 (1972).
43. W. Ho, G. M. Hidy, and R. M. Govan, *J. Appl. Meteorol.* **13,** 871–879 (1974).
44. D. Sinclair, R. J. Countess, and G. S. Hooper, *Atmos. Environ.* **8,** 1111–1117 (1974).
45. G. C. Holzworth and J. A. Maga, *J. Air Pollut. Contr. Ass.* **10,** 430 (1960).
46. National Air Pollution Control Administration (Now: Environmental Protection Agency), "Air Quality Criteria for Particulate Matter," No. AP-49. Natl. Air Pollut. Contr. Admin., Washington, D.C., 1969.
47. P. V. Hobbs, H. Harrison, and E. Robinson, *Science* **183,** 909–915 (1974).
48. S. Twomey and T. A. Wojciechowski, *J. Atmos. Sci.* **26,** 684–688 (1969).
49. J. Warner, *J. Appl. Meteorol.* **7,** 247–251 (1968).
50. R. C. Eagan, P. V. Hobbs, and L. F. Radke, *J. Appl. Meteorol.* **13,** 553–557 (1974).
51. P. V. Hobbs, L. F. Radke, and S. E. Shumway, *J. Atmos. Sci.* **27,** 81–89 (1970).
52. R. C. Eagan, P. V. Hobbs, and L. F. Radke, *J. Appl. Meteorol.* **13,** 535–552 (1974).
53. J. W. Fitzgerald, *J. Atmos. Sci.* **31,** 1358–1367 (1974).
54. B. J. Mason, "The Physics of Clouds," 2nd ed. Oxford Univ. Press (Clarendon), London, England and New York, New York, 1971.
55. J. W. Telford, *J. Meteorol.* **17,** 676–679 (1960).
56. G. Langer, *Proc. Nat. Conf. Weather Modification, 1st, 1968* p. 220–227. Atmospheric Sciences Program, S.U.N.Y., Albany, New York, (1968).

57. A. C. Montefinale, T. Montefinale, and H. M. Papee, *Pure Appl. Geophys.* **91,** 171–210 (1971).
58. A. I. Weinstein and P. B. MacCready, Jr., *J. Appl. Meteorol.* **8,** 936–947 (1969).
59. A. S. R. Murty and B. V. R. Murty, *Tellus* **24,** 581–585 (1972).
60. L. J. Battan, *Bull. Am. Meteorol. Soc.* **50,** 924–925 (1969).
61. T. L. Odgen, *J. Appl. Meteorol.* **8,** 585–591 (1969).
62. P. V. Hobbs and J. D. Locatelli, *J. Atmos. Sci.* **27,** 90–100 (1970).
63. G. M. Hidy, W. Green, and A. Alkezweeny, *J. Colloid Interface Sci.* **39,** 266–271 (1972).
64. R. F. Pueschel and G. Langer, *J. Appl. Meteorol.* **12,** 549–551 (1973).
65. R. R. Braham, Jr. and P. Spyers-Duran, *J. Appl. Meteorol.* **13,** 940–945 (1974).
66. G. E. Stout, *in* "Air Over Cities," SEC Tech. Rep. A62-5. Taft Sanit. Eng. Cent., Pub. Health Serv., U.S. Dept. of Health, Education, and Welfare, Cincinnati, Ohio, 1962.
67. S. A. Changnon, Jr., *Bull. Amer. Meteorol. Soc.* **49,** 4–11 (1968).
68. B. G. Holzman and H. C. S. Thom, *Bull. Amer. Meteorol. Soc.* **51,** 335–337 (1970).
69. S. A. Changnon, Jr., *Bull. Amer. Meteorol. Soc.* **51,** 337–342 (1970).
70. J. J. Hidore, *Bull. Amer. Meteorol. Soc.* **52,** 99–103 (1971).
71. B. G. Holzman, *Bull. Amer. Meteorol. Soc.* **52,** 573–574 (1971).
72. J. J. Hidore, *Bull. Amer. Meteorol. Soc.* **52,** 573–574 (1971).
73. W. C. Ashby and H. C. Fitts, *Bull. Amer. Meteorol. Soc.* **53,** 246–251 (1972).
74. F. L. Charton and J. R. Harmon, *Bull. Amer. Meteorol. Soc.* **54,** 26 (1973).
75. S. A. Changnon, Jr., *Bull. Amer. Meteorol. Soc.* **50,** 411–421 (1969).
76. F. A. Huff and S. A. Changnon, Jr., *J. Appl. Meteorol.* **11,** 823–842 (1972).
77. V. J. Schaefer, *Science* **154,** 1555 (1966).
78. V. J. Schaefer, *J. Appl. Meteorol.* **7,** 113 (1968).
79. P. T. Schickedanz, *J. Appl. Meteorol.* **13,** 891–900 (1974).
80. H. Landsberg, "Physical Climatology," 2nd ed. Gray Printing Co., Dubois, Pennsylvania, 1964.
81. H. Neuberger and M. Gutnick *in* "Proceedings of the 1st National Air Pollution Symposium (A. M. Zarem, ed.), pp. 90–96. Stanford Res. Inst., Menlo Park, California, 1949.
82. A. E. J. Eggleton, *Atmos. Environ.* **3,** 355–372 (1969).
83. C. G. Collier, *Weather* **25,** 25–29 (1970).
84. T. J. Chandler, "The Climate of London." Hutchinson, London, England, 1965.
85. P. J. Lawther and J. A. Bennell, *in* "Proceedings 2nd International Clean Air Congress (H. M. Englund and W. T. Berry, eds.), pp. 213–216. Academic Press, New York, New York, 1971.
86. J. E. Atkins, *Meteorol. Mag.* **97,** 172 (1968).
87. M. E. Berlyand and Ya. S. Kanchan, *in* "Atmosfernaya Diffuziya i Zagryaznenie Vozdukha" (M. E. Berlyand, ed.), pp. 1–19. Gidrometeoizdat, Leningrad, U.S.S.R., 1973 (in Russian). (Available in English translation as "Air Pollution and Atmospheric Diffusion-2." Wiley, New York, New York, 1974.)
88. L. S. Ivlev, V. A. Ionin, A. Yu. Semova, and N. K. Spazhakina, *in* "Atmosfernaya Diffuziya i Zagryaznenie Vozdukha" (M. E. Berlyand, ed.), pp. 156–167. Gidrometeoizdat, Leningrad, U.S.S.R., 1973. (Available in English translation as "Air Pollution and Atmospheric Diffusion-2." Wiley, New York, New York, 1974.)
89. I. F. Hand, *Bull. Amer. Meteorol. Soc.* **30,** 242 (1949).

90. R. G. Taylor, "Report on an Air Pollution Study for the City of Seattle." Environ. Res. Lab., University of Washington, Seattle, Washington, 1952.
91. N. A. Renzetti, ed., "An Aerometric Survey of the Los Angeles Basin," Rep. No. 9, pp. 193–200. Air Pollut. Found., Los Angeles, California, 1955.
92. R. Stair, *in* "Proceedings of the National Air Pollution Symposium" (A. M. Zarem, ed.), pp. 48–55. Stanford Res. Inst., Menlo Park, California, 1955.
93. R. Stair, *Int. J. Air Water Pollut.* **10**, 665 (1966).
94. J. S. Nader, ed., "Pilot Study of Ultraviolet Radiation in Los Angeles, October 1965," Publ. No. 999-AP-38. Pub. Health Serv., U.S. Dept. of Health, Education, and Welfare, Cincinnati, Ohio, 1967.
95. P. D. Try, Ph.D. dissertation, University of Washington, Seattle, Washington, 1972.
96. M. A. Atwater, *J. Appl. Meteorol.* **10**, 205–214 (1971).
97. M. A. Atwater, *J. Atmos. Sci.* **28**, 1367–1373 (1971).
98. J. M. Mitchell, Jr., *J. Appl. Meteorol.* **10**, 703–714 (1971).
99. D. S. Ensor, W. M. Porch, M. J. Pilat, and R. J. Charlson, *J. Appl. Meteorol.* **10**, 1303–1306 (1971).
100. J. M. Mitchell, Jr., *in* "Global Effects of Environmental Pollution" (S. F. Singer, ed.), pp. 139–155. Reidel Publ., Dordrecht, Netherlands, 1970.
101. R. G. Fleagle, W. H. Parrott, and M. L. Barad, *J. Meteorol.* **9**, 53–60 (1952).
102. W. R. Rouse, D. Noad, and J. McCutchean, *J. Appl. Meteorol.* **12**, 798–807 (1973).
103. R. E. Newell, G. F. Herman, T. G. Dopplick, and G. J. Boer, *J. Appl. Meteorol.* **11**, 864–867 (1972).
104. E. A. Schuck, J. N. Pitts, and J. K. S. Wan, *Int. J. Air Water Pollut.* **10**, 689–711 (1966).
105. R. Gunn, *J. Atmos. Sci.* **21**, 168 (1964).
106. W. E. Cobb and H. J. Wells, *J. Atmos. Sci.* **27**, 814–819 (1970).
107. W. E. Cobb, *J. Atmos. Sci.* **30**, 101–106 (1973).
108. A. W. Hogan, V. A. Mohnen, and V. J. Schaefer, *J. Atmos. Sci.* **30**, 1455–1460 (1973).
109. D. W. Parkin, D. R. Phillips, and R. A. L. Sullivan, *J. Geophys. Res.* **75**, pp. 1782–1793 (1970).
110. H. W. Ellsaesser, "The Upward Trend in Airborne Particulates that Isn't," UCRL-7370, Rev. 1. Lawrence Livermore Lab., University of California, Livermore, California, 1972.
111. E. S. Selezneva, *Tellus* **24**, 122–127 (1972).
112. R. J. Engelman, *J. Appl. Meteorol.* **10**, 493–497 (1971).
113. E. S. Selezneva and O. P. Petrenchuk, *in* "Meteorological Aspects of Air Pollution" (M. E. Berlyand, ed.), pp. 253–259. Hydrometeorological Publishing House, Leningrad, U.S.S.R., 1971 (in Russian, English abstracts).
114. E. Eriksson, *Tellus* **12**, 63–109 (1960).
115. K. Elgwork, A. Hagen, and A. Langeland, *Environ. Pollut.* **4**, 41–52 (1973).
116. E. G. Hallsworth and W. A. Adams, *Environ. Pollut.* **4**, 231–235 (1973).
117. M. Rabinowitz, *Chemosphere* **4**, 175–180 (1972).
118. R. Prince and F. F. Ross, *Air, Water, Soil Pollut.* **1**, 286–302 (1972).
119. P. K. Hanley and S. L. Tierney, *Ir. J. Agr. Res.* **8**, 19–27 (1969).
120. A. Hagen and A. Langeland, *Environ. Pollut.* **5**, 45–57 (1973).

2

Effects on Economic Materials and Structures

John E. Yocom and James B. Upham

I. Introduction

Air pollution has long been a significant source of economic loss in urban areas. Damage to nonliving materials may be exhibited in many ways, such as corrosion of metal, rubber cracking, soiling and eroding of building surfaces, deterioration of works of art, and fading of dyed materials. This chapter examines some of the damage mechanisms, attempts to determine the importance of air pollution in relation to other variables in producing damage, and considers the problems involved in translating observable air pollution damage into terms of economic impact.

II. Mechanisms of Deterioration in Polluted Atmospheres

Air pollutants damage materials by five mechanisms.

1. *Abrasion*—Solid particles of sufficient size and traveling at high velocities can cause destructive abrasion. Large, sharp-edged particles imbedded in fabrics can accelerate wear.

2. *Deposition and removal*—Solid and liquid particles deposited on a surface may not damage or change the material itself except, perhaps, to spoil its appearance. However, the removal of these particles may cause some deterioration. Although a single washing or cleaning may not cause noticeable deterioration, frequent cleaning ultimately does.

3. *Direct chemical attack*—Some air pollutants react irreversibly and directly with materials to cause deterioration; for example, the tarnishing of silver by hydrogen sulfide and the etching of a metallic surface by an acid mist.

4. *Indirect chemical attack*—Certain materials absorb pollutants and are damaged when the pollutants undergo chemical changes. Sulfur dioxide absorbed by leather, for instance, is converted to sulfuric acid, which deteriorates the leather.

5. *Electrochemical corrosion*—Much of the atmospheric deterioration of ferrous metals is by an electrochemical process (*1, 2*). Numerous small electrochemical cells form on ferrous metal surfaces exposed to the atmosphere. Anodes and cathodes result from the local chemical or physical differences on the metal surfaces. The distance between the anodes and cathodes is usually small. The difference in potential between the anode and cathode is the driving force for the corrosive action. If the metal is clean and dry, no current flows and no corrosion occurs. If water is present, even as a molecular layer on a surface that appears to be dry, current flows. If the water is then contaminated with air pollutants, it is very likely to have more electrical conductivity, and corrosion will proceed faster. Larrabee (*2*) reports that when iron remains in dry air for an appreciable length of time, it develops a protective film of oxygen. This film insulates the iron from atmospheric moisture, but when sulfur

dioxide is present, the protective oxygen layer is broken down, thus exposing the surface to corrosive action.

III. Factors That Influence Atmospheric Deterioration

The more important factors that influence the attack rate of damaging pollutants include moisture, temperature, sunlight, and air movement.

1. *Moisture*—Without moisture in the atmosphere, there would be little, if any, atmospheric corrosion even in the most severely polluted environments. Visible wetting of surfaces is not required for corrosion to take place. For several metals, there seems to be a critical atmospheric humidity, which when exceeded, produces a sharp rise in the rate of corrosion. Sanyal and Bhadwar (*3*) report that, for atmospheres containing sulfur dioxide, aluminum has a critical humidity of 80% and mild steel has two at 60 and 75%. Aziz and Godard (*4*) state the critical humidities for nickel and copper in the presence of sulfur dioxide as 70 and 63%, respectively, whereas zinc and magnesium show critical humidities in unpolluted air of 70 and 90%, respectively.

Vernon's experiments (*5*) to disclose the relative roles of humidity, sulfur dioxide, particulates, and time in producing metallic corrosion showed that, even for high pollutant levels, corrosion was minimal at relative humidities below 60%. As the humidity was increased toward 80%, the protective oxygen film on the metal surface apparently broke down and corrosion began. Above 80% relative humidity, corrosion proceeded rapidly under the test conditions of 100 ppm sulfur dioxide and various combinations of particulates. In pure air, corrosion was minimal even with high humidity.

Air polluted with sulfur dioxide and particles of charcoal produced a much more rapid corrosion rate than air polluted with sulfur dioxide alone. From this observation, Vernon reasoned that the action of charcoal particles was primarily physical in that they increase the concentration of sulfur dioxide by sorption and thus create "hot spots" at the point of contact between the metal surface and charcoal particles. He concluded that in England, where the relative humidity is generally high, the controlling factor in atmospheric corrosion of metals is provided entirely by the pollutants present in the atmosphere.

Moisture in the form of rain often reduces corrosion rates of metals in polluted atmospheres. The effect is probably the result of the dilution and washing away of corrosive materials, although under some conditions the corrosives can accumulate at the lower edge of a metallic object and cause accelerated corrosion.

2. *Temperature*—The most obvious influence of temperature is on the rate of the chemical reaction resulting in deterioration. Lowered surface temperature, however, may increase the chance for damage, because objects exposed during a radiation temperature inversion lose heat rapidly and usually cool to temperatures below that of the ambient air. If their surface temperature falls below the dew point, the surface becomes moist and, in the presence of corrosive pollutants whose concentrations are increasing under the stable influence of the temperature inversion, conducive to certain types of damage to materials.

3. *Sunlight*—In addition to producing damaging agents such as ozone through a series of complex photochemical reactions in the atmosphere (Chapter 6, Vol. I), sunlight can cause direct deterioration of certain materials. In cases of rubber cracking or fading of certain dyes, direct sunlight damage cannot always be distinguished from that caused by ozone.

4. *Air movement*—Wind speed is significant in determining whether solid and liquid agents impact on vertical surfaces, settle on horizontal surfaces, or produce abrasion. Where deterioration is caused by pollutants released nearby, wind direction is a most important variable. Air movement where leatherbound books are stored is a critical factor when the air is contaminated by sulfur dioxide. The air movement continually supplies sulfur dioxide-contaminated atmosphere to the leather surface of the books.

5. *Other factors*—Position in space is an important variable in the testing of metals for atmospheric corrosion. Corrosion test samples are usually mounted 45° from the horizontal. The under surfaces often are corroded more rapidly than the upper surfaces because corrosive agents are not as well washed off them by the rain.

The order in which substances contact a surface can also be highly significant. Aziz and Godard (*4*) state that when copper is first exposed to unpolluted air, a thin oxide film develops, which protects it against attack by hydrogen sulfide.

Freezing and thawing in the presence of moisture accelerates the deterioration of porous materials such as concrete and building stone. Biological action such as mildew and bacteria can affect materials of natural origin such as cotton and wool fabrics.

In addition to these natural variables, the degree of atmospheric corrosion is dependent on protective measures. Protective paint coatings are commonly applied to metals. Additions of small quantities of chromium, nickel, and manganese to mild steel also add to the corrosion resistance (*6*).

IV. Methods of Measuring Atmospheric Deterioration*

Deterioration of materials manifests itself and is measured in many ways. Corrosion of metals is frequently detected by weight change. In tests of short duration, a metal sample will gain weight for some weeks after it is exposed because of the formation of corrosion products. In steel samples, however, the weight gain due to rust formation will gradually be offset as the rust begins to slough off the sample. Consequently, it is better to use weight-loss measurements in steel for test periods of long duration. For this measurement a clean, weighed sample is exposed for periods of months or years; the sample is then recovered, cleaned, and reweighed. The difference in weight represents the corrosion that has taken place and is usually reported as milligrams of weight loss per square decimeter of surface per day or for the period of test. These results can also be converted to loss of metal thickness in units such as mils per year. Change in thickness of a sample can also be measured directly by a micrometer. Irregular changes in thickness, however, reduce the validity of this technique.

Mechanical tests of metal samples also offer possibilities. Bending, tension, fatigue, and impact tests may be employed to reveal internal weaknesses in metal not apparent from weight-change measurements. Mechanical tests reveal changes such as intergranular corrosion, stress cracking, or dezincification of brass.

Electrical resistance measurements can also be used to detect corrosion and subsequent loss of thickness in the test specimen. This technique has merits in industrial monitoring.

An interesting approach to measurement of metal corrosion in a polluted atmosphere is found in the work of Lodge and Havlik (*7*), in which metal films placed on glass slides by vacuum deposition serve as indicators of the relative corrosiveness of atmospheres. If different metals are used, some specificity for individual pollutants is obtained. Light transmission, electrical conductivity, microchemical analysis, X ray, and electron diffraction patterns offer possible means of reporting significant changes to the surfaces.

Since atmospheric corrosion of metals is a relatively slow process, accelerated tests have been developed so that materials, protective coatings, and the interplay between the various causes of corrosion can be evaluated in short-term experiments.

As Vernon (*8*) has pointed out ". . . the criterion of an accelerated test is that it shall accurately reproduce on a shortened time scale the relative

* See also Volume III, Chapter 10.

behavior of materials in service. . . ." In order to achieve this similarity of action, the most suitable arrangement is to accentuate one of the factors at play in the environment. Preston and Straud (*9*) have described laboratory accelerated tests in which metals with various experimental coatings are immersed in beakers containing aqueous sulfur dioxide solutions. The solutions are replenished twice a day for a period of test, which runs from 1 to 9 weeks. Evaluations are made by means of the weight-loss method for cleaned samples and by determining the time for an average of one rust spot per square inch of surface to appear through the protective coating. Humid conditions alone are not sufficient to produce corrosion except in the most poorly protected samples. Tests using the sulfur dioxide-enriched solution give 1-week corrosion values that are similar to 12-week values for sheltered outdoor field samples exposed to an urban environment.

Preston (*10*) employs another method for atmospheric corrosion testing of metal samples. The bottom of a large beaker containing a small quantity of water or sulfur dioxide–water solution is heated, and a water jacket around the top of the beaker is cooled to provide a saturated atmosphere in the beaker. Metal samples are suspended in this saturated atmosphere. The highest corrosive rates for mild steel samples occur at temperatures around 50°C with sulfur dioxide concentrations in the underlying solution of 0.02%. The corrosion rate under these conditions is three times that under test conditions without sulfur dioxide and approximately seven times that of steel exposed outdoors in winter in an urban atmosphere.

Building materials such as stone and mortar may be discolored or leached away by pollutants, but these effects are not usually measured quantitatively. Dramatic demonstrations, however, can occasionally be made by comparing photographs of buildings before and after cleaning, as shown in Figure 1, or by comparing photographs of the same artwork taken at different times. If sufficient time elapses between photographs, substantial damage may be illustrated.

The measurement of damage to lead-based paint from hydrogen sulfide by means of color change or darkening is difficult because of the tendency of hydrogen sulfide to produce a mottled effect in actual outdoor exposures. Quantitatively relating the degree of discoloration with hydrogen sulfide concentrations together with other environmental factors can best be done in exposure chambers, using synthetic polluted atmospheres.

The weakening of leather exposed to sulfur dioxide has been identified qualitatively, and this effect could probably be quantitated by means of tensile or tearing tests. Since leather is a natural product of variable thickness and density, there would be certain inherent errors in making

Figure 1. Old post office building being cleaned in St. Louis, Missouri, 1963. Photo by H. Neff Jenkins.

such tests. The embrittlement of paper caused by sulfur dioxide has not been measured quantitatively, but tests could conceivably be devised to measure resistance to folding.

The weakening of cloth by acidic pollutants such as sulfur dioxide and sulfur trioxide can be easily measured by means of loss in tensile strength; however, the effects of biological processes (e.g., mildew) in conjunction with moisture can also weaken cotton cloth. The United States Department of Agriculture (*11*) has devised methods for differentiating between the two types of damage. Samples of exposed and unexposed cotton are dissolved, and the viscosities of the resulting solutions are measured. Since sulfur dioxide and sulfur trioxide depolymerize cellulose, cotton damaged in this way will contain cellulose molecules of lower average molecular weight, and the viscosity of this solution will be lower than that from a sample of unexposed cloth. Although biological damage weakens the cotton, the attack is physical rather than chemical, and the average molecular weight of the cellulose is unchanged by this process. Thus, the resulting solution has a viscosity approximating that of an unexposed sample.

Damage to nylon hose by sulfur dioxide or sulfur trioxide has thus far been measured only visually by counting breaks in the stocking fibers. In a survey of air pollution in Staten Island, New York (*12*), exposed stockings were stretched over inverted Mason jars and were microscopically examined for fiber breaks. Since nylon hose are knit, a test of tensile strength of a swatch from such material would not subject fibers to straight-line tension.

Dyes are affected by pollutants such as nitrogen dioxide and ozone. The extent of such effects on exposed samples of cloth can be measured by means of specially designed colorimeters that can detect small changes of color within narrow ranges of the visible spectrum.

The cracking of rubber by ozone has been quantitated to a considerable extent. The measurement of crack depths in bent samples of specially formulated natural rubber is considered to be one of the most specific measures of ambient ozone. The Los Angeles County Air Pollution Control District in California for many years considered this method to be a standard for ozone measurement. More recently, greater emphasis has been placed on other measurement methods because of their ease of adaptation to continuous analyzers, even though some of these methods measure other ozidizing materials in addition to ozone.

The surface deterioration of glass and ceramic materials caused by air pollution is thus far a little-recognized problem. The quantitation of this effect should be straightforward and should involve surface reflectance or, in the case of clear glass, light transmission measurements.

The effect of air pollutants on electrical contacts usually involves increased resistance resulting from buildup of particulate materials, pitting and accompanying loss of contact surface, or formation of reaction products on the contact surface. Thus, resistance versus exposure time or number of contacts is the most logical measure of this effect.

As a summary of the foregoing, and as a means of introducing the detailed material in the next section, Table I presents a generalized

Table I   Air Pollution Damage to Various Materials

Materials	*Typical manifestation*	*Measurement*	*Principal air pollutants*	*Other environmental factors*
Metals	Spoilage of surface, loss of metal, tarnishing	Weight gain of corrosion products, weight loss after removal of corrosion products, reduced physical strength, changed reflectivity or conductivity	SO_2, acid gases	Moisture, temperature
Building materials	Discoloration, leaching	Not usually measured quantitatively	SO_2, acid gases, sticky particulates	Moisture, freezing
Paint	Discoloration, softened finish	Not usually measured quantitatively	SO_2, H_2S, sticky particulates	Moisture, fungu
Leather	Powdered surface, weakening	Observation, loss of tensile strength	SO_2, acid gases	Physical wear
Paper	Embrittlement	Decreased folding resistance	SO_2, acid gases	Sunlight
Textiles	Reduced tensile strength, spotting	Reduced tensile strength, altered fluidity	SO_2, acid gases	Moisture, sunlight, fungus
Dyes	Fading	Fading by reflectance measurements	NO_2, oxidants, SO_2	Sunlight, moisture
Rubber	Cracking, weakening	Loss in elasticity, increase in depth of cracks when under tension	Oxidants, O_3	Sunlight
Ceramics	Changed surface appearance	Changed reflectance measurements	Acid gases	Moisture

statement on the nature of air pollution damage to materials in all of the categories covered in this chapter.

V. Materials Damage

A. Metals

1. Ferrous Metals

As pointed out earlier, a number of variables affect metal corrosion rates—principally, moisture, air pollutants, and temperature. Corrosion rates for steel are most rapid at first and then slow down as a partially protective film develops. Hudson (*13*) described a study in which steel specimens were exposed for 1 year at twenty locations throughout the world. Samples exposed at dry or cold locations showed the least corrosion, tropical and marine exposures produced intermediate values, and industrial exposures produced the greatest. Corrosion values for Frodingham, England, were one hundred times greater than those found in Khartoum, Sudan.

Committee A-5 of the American Society for Testing and Materials (ASTM) (*14*) has since 1926 been conducting extensive studies of protected and unprotected iron and steel shapes, including sheets, wires, and various structural forms. Various galvanized, lead, and aluminum coatings have also been tested. Exposure sites have included rural, seacoast, and industrial areas. In a given location, the life of the specimen is proportional to the thickness of the protective coating, and the most rapid corrosion occurs in the most highly industrialized area.

In an air pollution survey in the Tulsa, Oklahoma, area (*15*), wrought iron disks were exposed, and their change in weight was used as a measure of air pollution-induced corrosion. The results showed high corrosion rates in a fairly clear-cut area adjacent to a group of industrial plants that included oil refinery, fertilizer, and sulfuric acid manufacturing operations.

Although the studies mentioned above described the materials damage accurately, they did not evaluate the air pollution parameters in any detail. Recently, a number of investigators have recorded air pollution levels and weather observations in addition to the corrosion measurements and have attempted to describe the relationships between these variables.

Figure 2 summarizes the work of Sereda (*16*), which represents an early attempt to develop a model to describe the corrosion of steel in

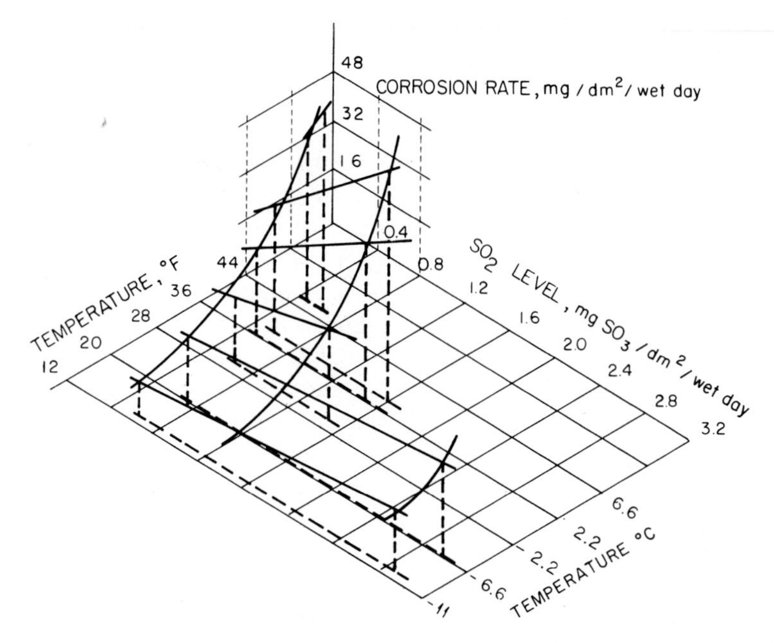

Figure 2. Corrosion rate of steel versus temperature and sulfur dioxide level, based on a day of wetness (*16*).

terms of several environmental factors. Samples of steel were exposed over an 8-month period. The time of wetness was determined for the exposure period by measuring surface moisture on both sides of the samples, the temperature of the exposed surfaces was recorded continuously, and sulfur dioxide concentrations were estimated by the lead peroxide method. A concentration of approximately 0.025 ppm sulfur dioxide resulted in deposition in lead peroxide of 1 mg sulfur trioxide per square decimeter per day. By means of the method of least squares, the best equation for the plane in Figure 2 was found to be

$$Y = 0.131X + 0.018Z + 0.787 \quad (1)$$

where

Y = logarithm of the corrosion rate (mg/dm^2/wet day)
X = sulfur dioxide pollution rate (mg SO_3/dm^2/day)
Z = monthly average temperature during the time of wetness (°F)

Starting in April, 1963, Upham (*17*) exposed steel plates [4 in. × 6 in. × 0.035 in. (10.25 cm × 15.36 cm × 0.09 cm)] at thirty-five sites in metropolitan St. Louis, Missouri. Two plates were removed from each site after exposures of 2, 4, 8, and 16 months. The average weight loss for each pair of plates was determined. Corrosion rates were 30–80%

greater in urban and industrial than those in suburban and rural areas. No relationship between dustfall and corrosion could be detected. A relationship was found, however, between corrosion and sulfation rates, the sulfation being determined by use of lead peroxide candles. This relationship was strongest during early stages of exposure and became weaker with time.

Upham (*18*) also performed a metal corrosion survey in Chicago, Illinois. Steel plates were exposed at twenty stations, and corrosion rates after 3-, 6-, and 12-month exposure periods were determined by weight-loss methods. Downtown values were about 70% greater than suburban values. Sulfur dioxide levels were measured at seven of the corrosion sites, and a direct relationship was observed as shown in Figure 3.

Haynie and Upham (*19*) evaluated in detail corrosion data for three different types of steel at locations within eight United States cities where continuous air monitoring data were available for carbon monoxide, nitric oxide, nitrogen dioxide, sulfur dioxide, total hydrocarbons, and total oxidants. In addition, data were collected for total suspended particulate matter. The three types of steel represented (1) plain carbon steel, (2) copper-bearing steel, and (3) low-alloy weathering steel. The samples were exposed 4, 8, 16, 32, and 64 months. A statistical analysis of the data showed that over 90% of the variability in corrosion behavior could be accounted for in the variability of two pollutants—sulfur dioxide increased corrosion and oxidant decreased corrosion. While the corrosive

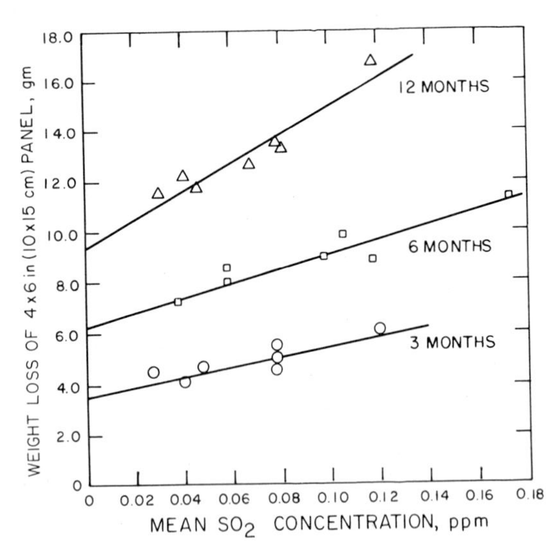

Figure 3. Relationship between corrosion of mild steel and corresponding mean sulfur dioxide concentration at seven Chicago, Illinois, sites (*17*).

effect of sulfur dioxide on ferrous metals is well known, the effect of oxidant is a new observation. The authors postulated that the ozone inhibits the effect of sulfur dioxide and tends to produce a more highly oxidized and more resistant corrosion product (ferric oxide as opposed to ferrous oxide and magnetic iron oxide or magnetite).

Weight-loss data were translated to corrosion depth, and it was found that corrosion was related to exposure time and environmental factors by the following relationship:

$$\ln y = \ln a + b \ln t \tag{2}$$

or

$$y = at^b \tag{3}$$

where

y = depth of corrosion (μm)
t = exposure time (years)
a and b = constants related to environmental conditions and type of steel.

Figure 4 is an extrapolation of the 5-year data from this study to 10 years, showing the predicted effect of sulfur dioxide and ozone on carbon steel. The authors point out that this prediction underestimates corrosion to typical carbon steels, since the samples used contained unusually high concentrations of copper, which contribute to corrosion resistance.

Harding and Kelley (*20*) measured corrosion and sulfation rates throughout an area of Jacksonville, Florida, which included seacoast as

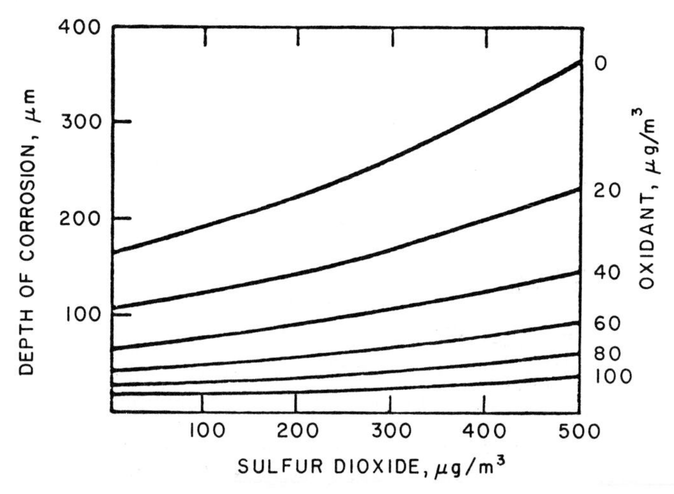

Figure 4. Effects of sulfur dioxide and oxidant concentrations on depth of corrosion of carbon steel exposed for 10 years.

well as industrial test sites. Corrosion rates did not correlate well with sulfation rates, but the corrosion pattern showed clearly the independent influence of sea salt and air pollution emissions. Steel corrosion panels mounted at various heights along a 1000 ft (300 m) television tower showed a pattern of corrosion with maxima occurring below the 800 ft (240 m) level. Nighttime radiation inversions in Jacksonville vary in depth between 200 and 600 ft (60 and 180 m), and since most of the stacks in the area are 150 ft (45 m) or less, corrosive pollutants would normally accumulate within the inversion. The data clearly showed this effect and substantiated a statement by the owner of a television station claiming excessive maintenance requirements to the transmitting tower.

Controlled laboratory experiments indicate that particulate matter is an important factor in the corrosion of metals, especially in the presence of gaseous pollutants of an acidic nature. Field work at the Chemical Research Laboratory in Teddington, England, (*21*) revealed that the rusting of iron in a moist atmosphere containing sulfur dioxide is greatly accelerated by the presence of particulate matter. In one experiment, a sample was exposed to a moist atmosphere containing traces of sulfur dioxide; another sample was exposed to the same atmosphere but protected from particulate matter by means of a muslin cage, which permitted only the gaseous constituents to contact the sample. The rusting of the protected sample was negligible compared with the sample that was unprotected. How much of the moisture and sulfur dioxide were absorbed by the cage fabric or whether equilibrium was established were not determined.

The same effect was confirmed by Preston and Sanyal (*22*), but in a laboratory environment. They "inoculated" various metallic surfaces (coated and uncoated) with fine powders or "nuclei," using such materials as sodium chloride, ammonium sulfate, ammonium chloride, sodium nitrate, and flue dust. The samples were then exposed to atmospheres held at various humidities, and the resulting corrosion was measured. The corrosion observed was of the filiform type, so named because of its filamental configuration. Corrosion of all the uncoated and all coated surfaces except the ammonium chloride-coated surface, increased with humidity.

Barton (*23*) determined the corrosive effects of a variety of artificial dusts on metals. He concluded that the quantity of water-soluble components in the dust, the pH of the resulting solution, and the concentration of chloride and sulfate ions were important factors in this type of corrosion. He stated that, contrary to previously expressed opinions, corrosive action was not affected by dusts with high adsorptive capacities for water and sulfur dioxide.

Corrosion studies in the outdoor air of communities have not confirmed the influence of particulate matter in accelerating corrosion rates, although it doubtless occurs. The presence of sulfur dioxide along with the particulate matter in those areas where such studies have taken place undoubtedly dominates the corrosion rate to such an extent that the independent effect of particulate matter is not evident.

A number of corrosion surveys have been conducted in Japan. Some of these illustrate the difference in corrosion rates between low- and high-pollution areas, and some have involved the role of temperature and moisture in accelerating corrosion rates (*24–27*).

2. Nonferrous Metals

The corrosion rates of commercially important nonferrous metals in polluted atmospheres, although generally less than those for steel, cover a wide range. The relative corrodability of some of these nonferrous metals can be compared in Table II. This summary of 20 years of investigation by the ASTM Committee on Atmospheric Corrosion compares corrosion rates for different metal formulations and under different exposure conditions.

Uhlig (*28*), Evans (*29*), and Speller (*30*) have assembled in considerable detail information on corrosion of metals. Muffley (*31*) performed a rather extensive review of the literature that pertains to the corrosion of metals under differing atmospheric regimes, and Greathouse and Wessel (*32*) have brought together much diverse information with respect to the deterioration of materials other than metals.

a. ZINC. Although zinc is widely used in galvanizing to protect ferrous metals from atmospheric corrosion, zinc itself is subject to corrosion. Anderson has reported on 20 years of investigations to evaluate atmospheric corrosion of zinc (*33*). He found that industrial air pollution is the greatest factor in zinc corrosion, because it destroys the basic carbonate protective coating that normally forms on zinc.

Tice (*34*), in summarizing the work of Hudson (*13*) and Larrabee and Ellis (*35*), compares data on zinc corrosion in Pittsburgh, Pennsylvania, with rural State College, Pennsylvania, in 1926 and 1960. Figure 5 shows that zinc corrosion was reduced by a factor of 3.5 between those two dates. This was associated with a decrease in average ambient sulfur dioxide concentrations in Pittsburgh by a factor of 3 over the same period. Note that corrosion rates at State College were essentially unchanged.

Based on an ambitious long-term zinc exposure program conducted at a single site where various atmospheric factors were simultaneously mea-

Table II Weight Loss of Metal Panels[a] after 20 Years Exposure in Various Atmospheres (about 1930–1954)[b]

City	*Expoeure classification*	*Average loss in weight, %*					
		Commercial copper (99.9% + Cu)	*Commercial aluminum (99% + Al)*	*Brass (85% Cu, 15% Zn)*	*Nickel (99% + Ni)*	*Commercial lead (99.92% Pb, 0.06% Cu)*	*Commercial zinc (99% Zn, 0.85% Pb)*
Altoona, Pennsylvania	Industrial	6.1	—	8.5	25.2	1.8	30.7
New York, New York	Industrial	6.4	3.4	8.7	16.6	—	25.1
La Jolla, California	Seacoast	5.4	2.6	1.3	0.6	2.1	6.9
Key West, Florida	Seacoast	2.4	—	2.5	0.5	—	2.9
State College, Pennsylvania	Rural	1.9	0.4	2.0	1.0	1.4	5.0
Phoenix, Arizona	Rural	0.6	0.3	0.5	0.2	0.4	0.8

[a] Panels—9 × 12 × 0.035 in (22.86 × 30.48 × 0.089 cm).
[b] Data from H. R. Copson Report of ASTM Subcommittee VI, of Committee B-3 on Atmospheric Corrosion, *Amer. Soc. Test. Mater., Spec. Tech. Publ.* **175** (1955). Used by permission of American Society for Testing and Materials, Philadelphia, Pennsylvania.

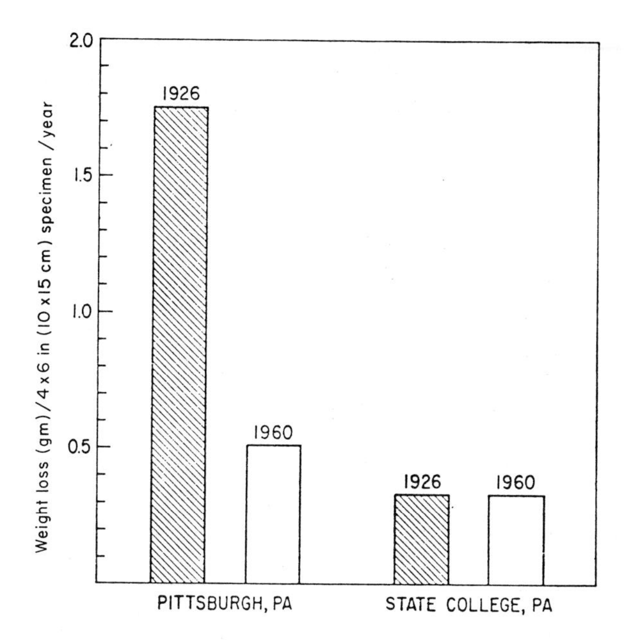

Figure 5. Reduction of corrosion rate of zinc at Pittsburgh and State College, Pennsylvania, 1926 versus 1960 (*34*). 1926 data are from losses for wire specimens (ASTM A-5 Committee) converted to 4 × 6 in (10 × 15 cm) panel; 1960 data are for losses of 4 × 6 in specimens.

sured, Guttman (*36*) confirmed the sensitivity of zinc to environmental factors. Time of panel wetness and atmospheric sulfur dioxide content during the time that panels are wet exert critical effects on the corrosion rate of zinc; temperature, however, was insignificant. Guttman has developed the following empirical equation relating corrosion to these factors:

$$Y = 0.00546\mathrm{A}^{0.8152}(B + 0.02889) \tag{4}$$

where

Y = corrosion loss (mg per 3 × 5 in panel)
A = time of wetness (hours)
B = atmospheric sulfur dioxide content during the time panels are wet (ppm)

This equation accounts for most of the observed variation in corrosion losses and is valid for exposure periods of up to 256 weeks at this site. Figure 6 presents this equation graphically and shows the corrosion loss versus time of wetness and the effect of different levels of sulfur dioxide.

As part of their eight-city study, Haynie and Upham (*37*) exposed panels of zinc for a 5-year period and found that zinc corrosion varied

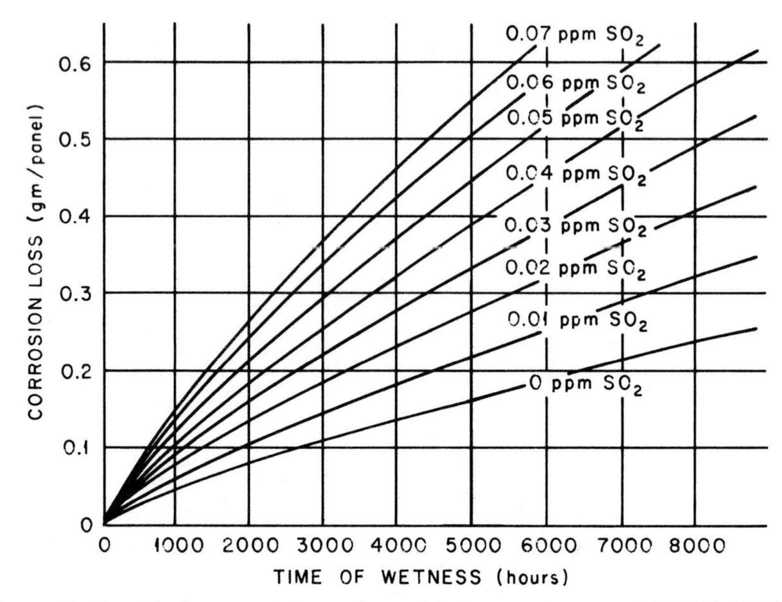

Figure 6. Graphical presentation of empirical equation $y = 0.005461A^{0.8152}(B + 0.02889)$ showing the effects of time of panel wetness and atmospheric sulfur dioxide on the corrosion of zinc (*36*).

linearly with sulfur dioxide concentration and relative humidity according to the equation

$$y = 0.001028(H - 48.8)S \tag{5}$$

where

y = zinc corrosion rate (μm/year)
H = average relative humidity (%)
S = average sulfur dioxide concentration ($\mu g/m^3$)

The constant 48.8 is the percent relative humidity below which zinc wetting did not occur. This equation accounted for 92% of the variation of expected corrosion rates.

These results were used to predict expected useful life of galvanized steel products and compare them with observed life for several types of areas. Table III shows such a comparison based on the assumption that galvanized steel is coated with 53 μm of zinc.

b. ALUMINUM. Aluminum is resistant to concentrations of sulfur oxides normally found in polluted atmospheres (*38*). Aziz and Godard (*4*) studied the corrosion of aluminum under laboratory conditions at various humidities and at high concentrations of sulfur dioxide (280 ppm). The

Table III Predicted Useful Life of Galvanized Sheet Steel with a 53 μm Coating at an Average Relative Humidity of 65% (*37*)

		Useful life (yrs)		
SO_2 Concentration, μg/m³	*Type of environment*	*Predicted best estimate*	*Predicted range*	*Observed range*
13	Rural	244	41.0	30–35
130	Urban	24	16.0–49	
260	Semiindustrial	12	10.0–16	15–20
520	Industrial	6	5.5–7	
1040	Heavy industrial	3	2.9–3.5	3–5

authors believed that their experiments, despite the high concentration, give insight into the mechanism of atmospheric corrosion of aluminum by sulfur dioxide.

At relatively low relative humidity (52%), samples of aluminum showed corrosion rates (weight increase) that reached a limiting value with time, and the shape of the corrosion curve did not differ appreciably with that for the direct oxidation of aluminum in the absence of sulfur dioxide. Thus, at low relative humidities, sulfur dioxide does not influence the atmospheric corrosion of aluminum, and there is no apparent visual change to the metal surface.

At the higher humidities (72 and 85%), both kinds of aluminum samples corroded much faster than at the lower humidity. The white powdery deposit on the surface of the samples under these conditions was found to be aluminum sulfate, an indication that sulfur dioxide plays an essential role in the corrosion of aluminum. Hydrogen sulfide detected at the end of each run was probably produced by the reduction of sulfur dioxide according to the reaction

$$2Al + SO_2 + H_2O \rightarrow Al_2O_3 + H_2S \quad (6)$$

The American Society for Testing and Materials has reported some of the most comprehensive studies on atmospheric corrosion of nonferrous metals (*39*). Twenty-four nonferrous metals and alloys were exposed at seven locations for periods up to 20 years. The investigations started in 1930 and included about 9000 individual test specimens. Tests included weight loss, differences in tensile strength, and depth of pitting. The test locations included seacoast, industrial, and rural atmospheres.

Table IV illustrates the relative effects of the different atmospheric conditions on aluminum alloys (*40*). The La Jolla, California, test site was several hundred feet from the ocean and was continually subjected

Table IV Percent Loss in Tensile Strength of Five Aluminum Alloys Exposed 20 Years (*40*)

Specimen, [0.035 in (0.90 cm) thick]	*Loss in tensile strength, %*
Seacoast	
La Jolla, California	30
Sandy Hook, New Jersey	8
Key West, Florida	4
Industrial	
Altoona, Pennsylvania	17
New York City	14
Rural	
State College, Pennsylvania	1
Phoenix, Arizona	<1

to salt spray or mist but did not have the benefit of rainfall to rinse the surfaces clean. Although the panels were within a few hundred yards of the ocean, the Sandy Hook, New Jersey, and Key West, Florida, station values were lower than the La Jolla values. These panels benefited from more frequent cleansing rains and were exposed to less salt mist because of the direction of prevailing winds and less surf action. The corrosive salt spray from the ocean is markedly diminished at distances a few hundred yards inland, and at distances greater than about $\frac{1}{4}$ mile (0.4 km) the atmospheric corrosivity for aluminum alloys may be similar to that of inland areas.

At first glance, it would appear strange that the corrosivity of the atmosphere in Altoona, Pennsylvania exceeded that of New York City. The Altoona samples were placed on the roof of a building in a railroad yard and were exposed to large amounts of smoke and gas from shops and locomotives, whereas the New York City sample was on the roof of the Bell Telephone Laboratories and reflected more nearly the average urban air pollution. The two rural stations showed minimum changes. Slightly lower values for Phoenix, Arizona, might be accounted for by the drier weather there than at State College, Pennsylvania.

Aluminum alloys have much better weathering characteristics than ferrous materials because of what has been called a self-limiting effect, an effect probably due to the gradual buildup of a stable, tenacious, protective film.

c. Copper and Silver. Copper and copper alloys in most atmospheres develop a thin, stable surface film, which inhibits further corrosion. Initial atmospheric corrosion is a brown tarnish of mostly copper oxides and sulfides, which can thicken to a black film. Then in a few years the familiar green patina forms. Analyses by Freeman and Kirby (*41*) and Vernon and Whitby (*42*) indicate this to be either basic copper sulfate or, in marine atmospheres, basic copper chloride, both of which are extremely resistant to further atmospheric attack. However desirable this effect may be from the standpoint of aesthetics and further attack, these same reaction products are unwanted when they form on electrical contacts made of copper and thereby change the electrical resistance of the contacts.

In the presence of hydrogen sulfide, copper and silver tarnish rapidly. Copper that has first been exposed to unpolluted air for a significant period resists the effects of hydrogen sulfide. In the case of silver, both moisture and oxygen must be present for hydrogen sulfide to cause tarnishing (*43*). The sulfide coating formed on open copper and silver electrical contacts can increase greatly the resistance across these contacts when they are closed and may also result in welding the contacts together in the closed position.

In the ASTM studies, Tracy (*44*) showed that copper and copper alloy corrosion rates were essentially negligible in rural atmospheres during the 20 years of study [averaging less than 0.025 mil (0.63 μm) per year loss in thickness]. Corrosion rates in industrial atmospheres [less than 0.1 mil (2.5 μm) per year] were greater than those in marine atmospheres. A number of formulations were tested, and with the exception of a high-tensile brass, they all appear suitable for use in industrial atmospheres.

d. Nickel. The fogging of nickel surfaces is due to the simultaneous presence of sulfur dioxide and water vapor (*45*). The surface film formed is basic nickel sulfate. Sulfur dioxide in the presence of water vapor is catalytically oxidized by the nickel surface to sulfuric acid. However, in general, nickel and nickel alloys, including stainless steels, are quite resistant to corrosion even in highly polluted atmospheres.

e. Stress Corrosion. In 1959, the Pacific Telephone and Telegraph Company noticed considerable breakage of their nickel–brass (12Ni–65Cu–23Zn) wire springs in some relays located in Los Angeles, California, area central offices (*46–48*). Failures often occurred within 2 years after installation. These failures were totally unexpected, since the wire springs had been used with excellent results for years throughout the nation. Investigators found that breakage occurred on wires that were under moderate stress and a positive electrical potential, and it was concluded that the failure mechanism was a form of stress–corrosion cracking. Bell Laboratories subsequently showed that high-nitrate concentrations in airborne dust, which had accumulated on surfaces adjacent to cracked areas, produced the failures. The nitrate content of dust from Los Angeles was from 5 to 15 times greater than from most eastern and midwestern cities. Furthermore, Los Angeles dust consisted of light-colored, very fine, claylike materials and considerable organic matter in a highly oxidized, polar condition. This combination results in dust that is more sensitive and reactive to moisture than the carbonaceous, oily, siliceous, high-sulfate content of most eastern dusts. Nitrates are also more hygroscopic than sulfate salts. Additional tests have shown that failures take place only when surface nitrate concentrations are above 2.4 $\mu g/cm^2$ and when the relative humidity, a very important controlling factor, is above 50%. Other salts will cause stress corrosion, but only when the relative humidity is greater than 75%; for sulfate salts it must exceed 95%.

Some time after this stress–corrosion problem was first observed in Los Angeles, scattered failures were reported, not only in wire springs, but in other nickel–brass components elsewhere in California, and in New York City, New York, and in Philadelphia, Pennsylvania. Westinghouse (*47*) has reported failures in Texas and New Jersey. The telephone company took several measures to correct the stress–corrosion problem. Researchers found that when zinc is left out of the nickel–brass alloy, stress corrosion no longer occurs. To prevent future problems, a copper–nickel material was specified for subsequently manufactured wire-spring relays. In high-nitrate areas, local central offices protected existing nickel–brass relays by installing high-efficiency filters in outside-air intakes of ventilating systems and by redesigning their cooling systems to keep relative humidity below 50%.

B. Building Materials

Building materials are corroded and disfigured by air pollution in a number of ways in addition to normal weathering processes. Smoke and tarry, sticky aerosols adhere to stone, brick, and other building surfaces to produce unsightly coatings. Rain may remove some of these materials, but streaking at window sills, for example, shows that rain removal is only a partial process.

Under conditions of high wind speed, larger particulates can be reentrained in the wind stream and actually produce a slow erosion of building surfaces similar to sandblasting. This is a much more subtle damage than that caused by reactive chemical pollutants. Acid gases such as sulfur dioxide and sulfur trioxide in the presence of moisture can react with limestone ($CaCO_3$) to form calcium sulfate ($CaSO_4$) and gypsum ($CaSO_4 \cdot 2H_2O$), both of which are rather soluble in water. Carbon dioxide in the presence of moisture produces carbonic acid; the acid (*49*) converts the limestone into a water-soluble bicarbonate, which is then leached away. This type of damage is notable in works of art.

Speaking as a geologist, Winkler (*50*) points out that corrosive atmospheric substances inflict damage to stone in urban areas. The rate of decay is doubled or tripled in areas with adverse atmospheric conditions. He points out that the damage can be reduced by the selection of stone that is more resistant to these atmospheric stresses. Stones such as granite and certain sandstones in which the grains are cemented together with materials containing no carbonate are relatively unaffected by sulfur dioxide in the atmosphere (*51*).

When a downtown office building is cleaned, it undergoes a dramatic change in appearance. The dirt, grime, and bird droppings can be cleaned off in a number of ways. One of the most common is to use water and air under high pressure [1200 psi (84 kg/cm^2)]. Occasionally, sand or dilute acids may be added to help on especially tenacious soiling. Some buildings are cleaned by either wet or dry sandblasting. Although a clean building is the most visible sign of this type of maintenance, contractors also have to repair cracks, replace mortar and broken or missing bricks. The cost for such cleaning may run 15 to 20 cents per square foot, and for a 15-story office building occupying one-quarter of a city block, would be approximately $10,000. A 1964 estimate to clean Trinity Church in New York was given at $15,000–$20,000 (*52*).

In a recent study by Beloin and Haynie (*53*), six different types of building materials were exposed at five sites in Birmingham, Alabama, to determine soiling rates in comparison to different levels of suspended particulate matter. The sampling sites had suspended particulate concentrations ranging from 60 to 250 $\mu g/m^3$. The materials exposed were (1) cedar siding coated with three different types of white paint,* (2) concrete block, (3) brick, (4) limestone, (5) white asphalt shingles, and (6) window glass. Degrees of soiling were determined throughout the 24-month exposure period by reflectance measurements for the opaque materials and haze measurements through the glass. A portion of the surfaces of the concrete block, brick, and limestone structures were coated with a silicone sealant to determine its effectiveness in reducing atmospheric soiling.

The degree of soiling of the painted siding was directly proportional to the square root of the suspended particulate dose (concentration

* Alkyd house paint, tint-base house paint (both oil base), and a water-base acrylic emulsion paint.

times exposure time). Asbestos shingle soiling was directly proportional to suspended particulate dose. The regression equations describing this soiling could account for 74–92% of the variability in reflectance measurements. On the other hand, similar regressions accounted for 34–50% of soiling variability for brick. Poor correlations were obtained for concrete, limestone, and window glass. The silicone surface sealant was ineffective in reducing soiling of brick. Coated limestone showed increases in reflectance with time, probably because of the increase in opacity of an initially clear coating.

Ozone affects coating-grade asphalts, as indicated by oxidative changes noted through use of infrared spectroscopy and by cracks observed in thin asphalt films. The accelerated tests employed by Wright and Campbell (*54*) used ozone concentrations as high as 52 ppm. Whether these oxidative changes are significant to road asphalts at normal air pollution concentrations has not as yet been determined.

C. Protective Coatings

Paint contains both pigment and vehicle. Pigments, such as white lead, titanium dioxide, and zinc oxide, provide color, hiding power, and durability. The vehicle, consisting of binder and additives, holds the pigment to the surface; together they enhance the attractiveness of the surface and protect the underlying material from corrosion or weathering. Air pollutants may limit both of these functions by damaging the protective coating and by exposing the underlying surface to attack (*55*). Some of the more common pollutants that can cause this damage are sulfur dioxide, ozone, hydrogen sulfide, tarry and greasy aerosols, and metal salts.

Holbrow (*56*) has reported a number of experiments to determine effects of sulfur dioxide on newly applied paints. Drying time for various oil-base paints exposed to 1–2 ppm of sulfur dioxide was increased 50–100%. Films thus dried had either softer finish or in some cases a more brittle finish, and, as a result of either change, the paints were likely to provide less than optimum durability. Discoloration of Brunswick green paints was noted if the fresh paints were exposed to sulfur dioxide, warmth, and moisture. Under these conditions, the lead chromate in the pigment was bleached and became blue. Sulfur dioxide in the presence of moisture and ammonia can result in crystalline bloom on paint and varnish surfaces. Small (0.1–1.0 μm) crystals of ammonium sulfate are efficient in light scattering and, therefore, become noticeable on the surface as a bloom.

Thus far, no studies have been carried out on the effects of sulfur dioxide on the drying of water-based paints, which now dominate the exterior paint market. It is conceivable that sulfur dioxide and other water-soluble acid gases could interfere with the stability of the polymer–pigment–water emulsion and produce an incomplete paint film.

Mell (*57*) pointed out that sulfur oxides have little effect on dry and hard paint films, even though they do damage incompletely dried films. He has suggested for high pollution areas use of rapid-drying paints with alkyd resins that afford hard protective finishes to minimize this exposure hazard. In addition to gas damage during painting, the anticorrosive properties of varnish and paint films are seriously impaired by the presence of included dust particles (*58*). The particles can act as wicks in a moist environment to transfer corrosive agents to the underlying metal surface.

House paint containing lead compounds is rapidly darkened in the presence of even low concentrations of hydrogen sulfide by the formation of black lead sulfide. Figure **7** is a photograph of a building at the

Figure 7. Paint damage from hydrogen sulfide emitted from polluted bay waters. Photo by J. E. Yocom.

southern end of San Francisco Bay in California. The paint on this building has been discolored by hydrogen sulfide emitted from polluted bay waters. The work of Hess (*59*) points out, however, that the dark lead sulfide is in turn oxidized to lead sulfate, thus the paint film eventually turns white again if the atmosphere remains free of hydrogen sulfide. Apparently, the severity of discoloration is related to the amount of lead present in the paint, amount of hydrogen sulfide in the air, duration of exposure, and moisture available at time of exposure. Very little damage occurs if both surface and air are dry. This may offer an explanation as to why most residential paint discoloration is observed by home owners when they first go outside in the morning. Relative humidity is greatest during nighttime hours, and the moisture formation that is most apt to occur during such time enhances the pollutant effect. Wohlers and Feldstein (*60*) have concluded that old lead-base paints are more susceptible to hydrogen sulfide damage than are new paints and that blackening can be brought about by several hours' exposure to hydrogen sulfide at concentrations averaging as little as about 0.05 ppm.

Since highly polluted atmospheres often contain appreciable quantities of dark particulate matter, dirtying of a painted surface by this kind of material can occur in addition to lead sulfide darkening. One of the principal reasons for repainting houses in urban areas is the accumulation of discoloring particulate matter on the surfaces. The particles are placed there by thermal, electrostatic, and mechanical forces. Newly painted houses are especially susceptible, because the surface is sticky and even large particles can attach themselves easily and permanently.

LeClerc (*61*) collected figures from building trade experts in France and found that on the average both outside paint and inside papering were required once every 4 years in urban areas, but only every 6 years in rural areas. It was not determined which portion of the difference is due to air pollution damage and which is due to possibly different maintenance patterns between urban and rural areas. Michelson and Tourin (*62*) determined the frequency of house painting in two cities in the Upper Ohio River Valley and in three suburbs of Washington, D.C., in comparison with levels of suspended particulate matter. They found that painting frequency (reciprocal of interval between paintings) was linearly related to particulate levels as shown in Figure 8. On the other hand, the results of a similar study by Booz, Allen and Hamilton, Inc. (*63*) carried out in the Philadelphia, Pennsylvania, area showed that there was no statistically significant difference in painting frequency as a function of atmospheric particulate levels. The authors believed that this was caused by a tendency for higher particulate levels to exist in those areas populated by lower income families.

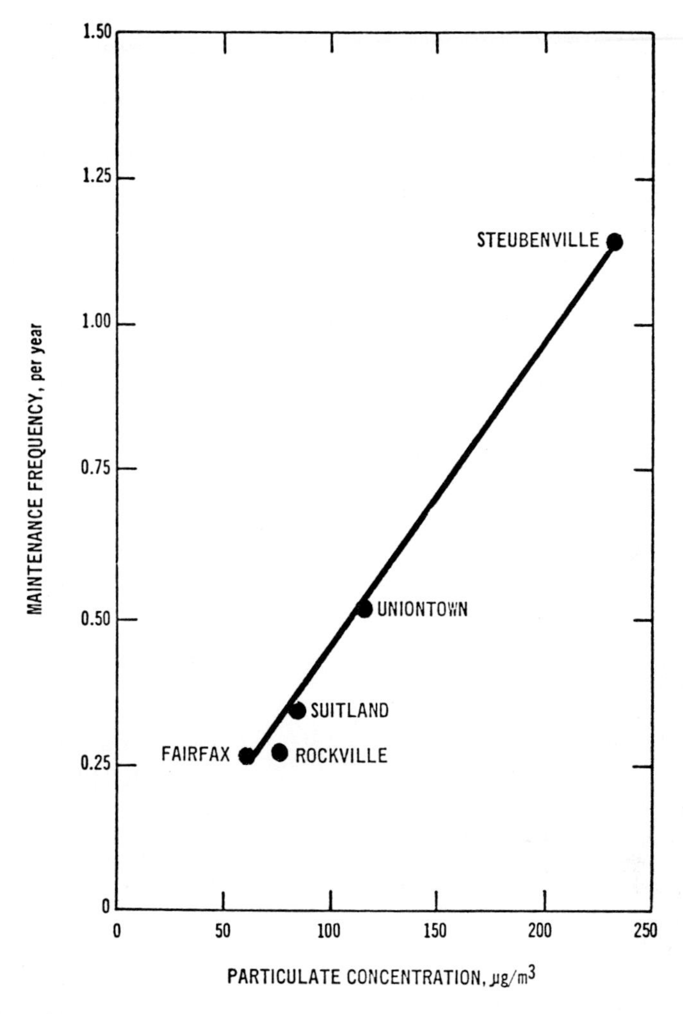

Figure 8. Relationship of maintenance frequency for exterior repainting to particulate concentration in five cities (*62*).

As Schwegmann (*64*) points out, a number of biological conditions can and do produce paint damage. Blistering due to moisture below the paint is readily recognized; however, spotting from lichen growth, droppings from bees, and pollen strains require close examination to distinguish them from chemical damage. A paint surface darkening that has been confused with hydrogen sulfide damage is caused by certain types of fungus. This kind of damage occurs most frequently where the humidity remains above 60% (*65*). It is most likely to occur with paints containing a high percentage of raw linseed oil and with some of the newer latex-emulsion-based exterior paints.

While specialists conducted air pollution research in a Public Health Service building in Cincinnati, Ohio, their cars in the parking lot were covered with brown spots that washing failed to remove (Fig. 9). A nearby industry accidently released vapors from one of its chemical processes, which sprayed droplets on cars and dwellings for about a mile downwind. Although one likes to think of this as an isolated event, such accidents are not uncommon. On another occasion, iron particles from a grinding operation damaged cars in a parking lot (*66*). Brown particles attached themselves tenaciously to the paint of automobiles. A brown stain surrounded each particle and many cars had to be repainted. The particles at the center of each stain were iron grindings or cuttings, identified microscopically by their peculiar shape. The investigators postulated that the deposit resulted from the formation of ferrous hydroxide in the presence of moisture. Being in collodial form, the substance diffused into the paint film and, upon drying and oxidizing, left the brown stain of ferric oxide.

Figure 9. Car soiling in Cincinnati, Ohio. Photo by S. D. Moran.

Recent laboratory research and field studies by Sherwin-Williams (*67*) indicate that air pollutants damage paint; furthermore, the degree of damage depends on the type of paint. The researchers evaluated four important types of paint: (1) house paints (acrylic latex and oil-based); (2) a urea-alkyd coil coating; (3) a nitrocellulose–acrylic automotive refinishing paint; and (4) an alkyd industrial maintenance coating. Although a number of techniques were used to assess damage, erosion measurements (weight loss converted to mils loss) proved to be the most meaningful. Erosion, which is a gradual weathering of the surface of a paint film, is the normal mechanism of failure that limits the service life of a well-formulated and properly applied exterior coating.

Panels coated with different paints were exposed to five separate controlled environment conditions; (1) clean air; (2) sulfur dioxide at 262 $\mu g/m^3$ (0.1 ppm); (3) sulfur dioxide at 2620 $\mu g/m^3$ (1.0 ppm); (4) ozone at 196 $\mu g/m^3$ (0.1 ppm); and (5) ozone at 1960 $\mu g/m^3$ (1.0 ppm). The exposure chamber operated on a 2-hour light–dew cycle consisting of 1 hour of xenon light at 70% relative humidity and a black panel temperature of 66°C, followed by 1 hour of darkness at 100% relative humidity and 49°C during which period moisture condensed on the coated panels. To assess the influence of simulated sunlight, one group of panels was exposed to the light cycle while a like group of panels was exposed shaded from light during the entire exposure period. Erosion measurements were made after exposure periods of 400, 700, and 1000 hours; erosion rates were then calculated.

Generally, exposures to 1 ppm sulfur dioxide or 1 ppm ozone produced statistically significant erosion rate increases compared to clean air (zero pollution) conditions. Erosion rate increases, however, varied considerably among paint types. Oil-based house paint experienced the largest erosion rate increases, latex and coil coatings moderate increases, and the industrial maintenance coating and automotive refinish the smallest increases. Exposures to sulfur dioxide produced higher erosion rates than ozone for three of the paints; however, of more importance was that sulfur dioxide produced much larger erosion rate increases than ozone. Unshaded panels eroded more than shaded panels. Table V presents erosion rate data for the unshaded exposures. Exposures to 0.1 ppm pollutants did not produce significant erosion rate increases over clean air exposures.

The researchers point out that in judging erosion-rate increases a distinction should be made between statistical significance and practicality. For example, unshaded panels of automotive refinish developed an erosion rate of 1.8×10^{-5} mil (45×10^{-5} μm) loss per hour in clean air and 3.1×10^{-5} mil (79×10^{-5} μm) loss per hour in 1 ppm sulfur dioxide. Both values are small and, although the increases is statistically signifi-

Table V Paint Erosion Rates and T-Test Probability Data for Controlled Environmental Laboratory Exposures (67)

	Mean erosion rate (mil loss $\times 10^{-5}$/hour with 95 % confidence limits) for unshaded panels and probability that differences exist		
Type of paint	*Clean air control*	*SO_2 (1.0 ppm)*	*O_3, (1.0 ppm)*
House paint			
oil	20.1 ± 7.2	141.0 ± 19.0 99 %	44.7 ± 10.5 99 %
latex	3.5 ± 1.5	11.1 ± 1.0 99 %	8.5 ± 5.9 93 %
Coil coating	11.9 ± 2.3	34.1 ± 4.7 99 %	14.9 ± 2.5 94 %
Automotive refinish	1.8 ± 0.8	3.1 ± 2.6 75 %	5.1 ± 1.3 99 %
Industrial maintenance	18.6 ± 5.1	22.4 ± 7.0 66 %	28.1 ± 14.0 85 %

cant, for all practical considerations the increase has no significant implication, especially in view of the severe exposure conditions.

Field exposures were conducted at four locations with different environments: (1) rural—clean air, (2) surburban, (3) urban—sulfur dioxide-dominant (annual mean level 60 μg/m³), (4) and urban—oxidant-dominant (annual mean ozone level 40 μg/m³). Panels were exposed facing both north and south, and were evaluated after 3, 7, and 14 months exposure. Table VI shows the erosion-rate results for the various coatings exposed facing south. In most cases, southern exposures produced somewhat larger erosion rates; this agrees with the unshaded versus shaded results of the laboratory study. Oil-based house paint again experienced by far the largest erosion rate increases, followed in order by the coil coating, latex house paint, industrial maintenance paint, and automotive refinish. Generally, the oxidant-dominant environment was more damaging than the sulfur dioxide dominant environment. It is noteworthy that the oil-based house paint and coil coating experienced the largest erosion rate increases in both the field and laboratory sulfur dioxide exposures. These coatings were the only ones that contained a calcium carbonate extender—a substance that is sensitive to attack by acids.

D. Leather

Sulfur dioxide causes leather to lose much of its strength and ultimately to disintegrate. As early as 1843, Faraday came to the conclusion that

Table VI Paint Erosion Rates and T-Test Probability Data for Field Exposures (*67*)

	Mean erosion rate (mil loss × 10^{-3}/month with 95% confidence limits) for panels facing south and % probability that differences exist			
Type of paint	*Rural (clean air)*	*Suburban*	*Urban (SO_2 dominant)*	*Urban (oxidant dominant)*
House paint				
oil	4.3 ± 7.5	14.8 ± 2.6 99.3%	14.2 ± 4.9 98.1%	21.0 ± 6.2 99.2%
latex	1.8 ± 0.5	3.0 ± 0.7 99.2%	3.8 ± 0.3 97.8%	6.5 ± 5.6 94.3%
Coil coating	2.1 ± 0.8	10.0 ± 1.9 99.9%	9.5 ± 0.8 99.9%	8.8 ± 1.7 99.9%
Automotive refinish	0.9 ± 1.1	2.3 ± 0.7 97.6%	1.6 ± 0.4 86.2%	1.7 ± 0.4 91.6%
Industrial maintenance	3.6 ± 1.6	8.2 ± 4.2 97.3%	6.6 ± 3.9 91.2%	7.8 ± 2.4 99.7%

the rotting of leather upholstery on chairs in a London club was the direct result of sulfur compounds in the air (*68*). The storage of leather-bound books in libraries can pose a serious problem. The bindings of books stored in the open in rooms with polluted air were found to deteriorate much more rapidly than those stored in confined spaces or inside glass cases. Chemical decay of bookbindings can be observed initially by the cracking that takes place on the top inside hinges of the book. The cracks gradually spread apart, and as further material is exposed, the leather loses its resiliency and disintegrates to a reddish brown powder. After some time, the entire back may become detached.

Plenderleith (*69*) and Innes (*70*) have described in some detail the steps in chemical destruction of leather and the attempts made to protect it. At one time it was thought that a leather initially free of sulfuric acid would not decay, but leather originally free of sulfuric acid was found to accumulate as much as 7 wt % acid if it were exposed to an atmosphere containing sulfur dioxide. It has been suggested that leather absorbs the sulfur dioxide, and that minute quantities of iron in the leather serve as a catalyst capable of oxidizing the sulfur dioxide to its acid form.

In 1932, the British Leather Manufacturing Research Association initiated an experiment to assess air pollution damage to various leathers used in bookbinding. Duplicate sets of books were bound with a variety of leathers. One set of books was placed in the British Museum Library in London, and the other in the National Library of Wales in Aberystwyth, which at that time was relatively free of air pollution. After 15 years, initial signs of chemical damage were observed in some volumes in London, but not in any of the books at the National Library. None of the artificially protected leather in either location showed chemical damage.

E. Paper

Paper made prior to about 1750 is not seriously damaged by sulfur dioxide (*71*). This date is about the point in history when chemical methods for papermaking were introduced. Apparently, the small amounts of metallic impurities in "modern" paper accelerate the conversion, in the presence of moisture, of absorbed sulfur dioxide to sulfuric acid. The sulfuric acid content of some papers has been found to be as high as 1%, which makes the paper extremely brittle. Kimberly (*72*) found that exposure of books and writing paper to sulfur dioxide in concentrations of 2–9 ppm for 10 days caused embrittlement and decreased their folding resistance.

More recent research (*73*) shows that relative humidity does not significantly affect the long-term pickup of atmospheric sulfur dioxide by paper. Relative humidity only affects the quantity of sulfur dioxide taken up for about the first 48 hours. Afterward, the rate remains essentially constant for prolonged exposures and appears to be proportional to the square root of the gas-phase sulfur dioxide concentration.

Wallpapers form an important part of the total indoor surface area available for sulfur dioxide sorption. Researchers (*74*) measured the sulfur dioxide sorption characteristics of wallpapers on exposure to maximum initial sulfur dioxide concentrations of 100–125 $\mu g/m^3$. Sorption depended largely on surface finish and design pattern. Conventional wallpapers showed a greater sulfur dioxide uptake than plastic-coated wall coverings. The researchers suggested that sulfur dioxide sorption accelerated the deterioration of wallpapers.

F. Textiles

Sulfur oxides are capable of causing deterioration of natural and some synthetic textile fibers. Cotton is a cellulosic fiber and, like paper, is weakened by sulfur dioxide. The work of Race (*75*) in England showed that the breaking point of cotton fabrics in winter was considerably less than that in summer. Without atmospheric pollutants, one would expect the reverse to be the case because of the higher sunlight intensity in summer, which in itself can cause deterioration of the fabric. Howard (*76*) reasoned that this seasonal difference can be accounted for by the higher winter pollution levels. When acid aerosols are deposited on exposed fabrics, they attack and weaken the cellulose chain at the glucosidic linkage.

Damage to nylon hose by air pollution (presumably sulfur dioxide or sulfur trioxide) has made newspaper headlines in many areas. The exact mechanism of attack has not yet been determined, but even dilute solutions of sulfuric acid reduce the strength of nylon (a polymeric amide) significantly. It has been postulated that extremely small atmospheric particles containing adsorbed sulfur dioxide (*77*) or tiny droplets of sulfuric acid, which have formed around particles (*78*), become attached to the very thin nylon fibers and these, being under some tension, fail. Nylon fabrics used in clothing such as shirts and dresses are woven from fibers of much larger diameter than those of nylon hose and furthermore are not under the same tension; hence, they are not so easily damaged by sulfur oxides.

The dirtying of fabrics by particulate matter is not in itself damaging unless the dust is highly abrasive and the fabric is frequently flexed. The deterioration arises mainly from repeated attempts to clean the fabric. Rees (*79*) discusses the mechanical, thermal, and electrostatic mechanisms by which cloth is soiled. He used a light reflectance method to assess dirtiness of cloth samples exposed to moving air containing finely divided carbon and found that the most tightly woven cloth was most resistant to soiling. Soiling by thermal precipitation, as one might expect, was found to be directly related to the degree the surface temperature was cooled below the air temperature. Thicker samples of cloth had higher surface temperatures and, therefore, collected less dust. Morris (*80*) found that airborne soil slightly increased the photochemical degradation of nylon yarn but had no measurable effect on breaking strength of soiled nylon yarns that had not been exposed to sunlight.

The soiling of certain types of cloth, such as acetate rayon, by electrostatic attraction of particulate matter has been a manufacturing problem. The cloth becomes electrostatically charged by friction with the metallic parts of the loom. Cotton and viscose rayon are relatively good conductors, and their charges do not build up. In the experiments described by Rees (*79*), soiling of cotton fabric samples exposed to laboratory air at both positive and negative electric potentials rose with increased potential. The samples exposed at positive potentials showed greater soiling than those exposed at equivalent negative potentials because, perhaps, of the preponderance of negatively charged particles in the atmosphere.

Petrie (*81*) summarized the problems that atmospheric pollutants make for fabrics. Cellulose fibers such as linen, hemp, cotton, and rayon are especially susceptible to acid damage. Sulfuric acid reacts with cellulose fibers to produce a water-soluble product that has very little tensile strength.

Curtains suffer badly in polluted areas because they hang at open windows and to some extent serve as filters for dust, soot, and acid droplets. Curtains, thus weakened, often split in a characteristic manner, in parallel lines matching the folds in the hanging fabric. The stress in these folds is accentuated, since the fabric structure is relatively more open at these points, and greater opportunity is afforded for impinged acidic materials to reach the inner fibers.

Animal fibers of wool, fur, and hair are more resistant than synthetics to atmospheric pollutants since they already contain nitrogenous and sulfur compounds and apparently are more resistant to acid aerosols.

A study was made of cotton fabrics exposed at seven locations in the St. Louis, Missouri, area to determine the relationship between air pollution and fabric degradation (*11*). Two types of fabric were exposed on racks from June 1963 to June 1964, and individual samples of each fabric were removed from the exposure rack each month. Samples were evaluated with respect to their tensile strengths, and a significant relationship was observed between air pollution levels and degradation of the exposed fabrics. Figure 10 shows the effects of exposure and time on the breaking strength of a cotton duck material. The authors demonstrated

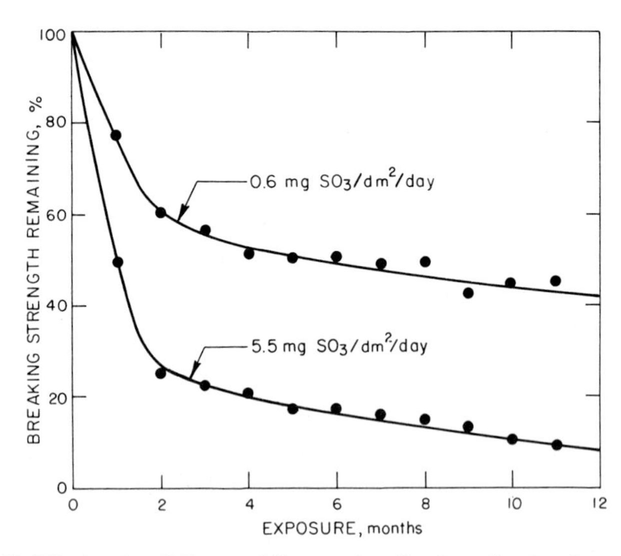

Figure 10. Effects of sulfation and time on tensile strength of cotton duck (*68*).

that heavy air pollution can substantially reduce the service life of exposed fabrics.

Trommer (*82*) states that nylon polymers are subject to oxidation by nitrogen oxides and other oxidants. If oxidation occurs, nylon fibers would have less affinity for certain types of dyes. Trommer cautions that nylon yarns should not be exposed for prolonged periods to factory environments in which fork trucks emit exhaust fumes.

Zeronian (*83*) carried out laboratory exposures in which he exposed cotton and rayon fabrics for 7 days to clean air with and without 250 $\mu g/m^3$ (0.1 ppm) sulfur dioxide. Both controlled environments included continuous exposure to artificial light (xenon arc) and a water spray turned on for 18 minutes every 2 hours. Loss in strength for all fabrics exposed to clean air averaged 13%, while the fabrics exposed to air containing sulfur dioxide averaged 21%. Zeronian *et al.* (*84*) also exposed fabrics made from manmade fibers—nylon, polyester, and modacrylic—to controlled environmental conditions similar to the cotton exposures, except that the sulfur dioxide level was 486 $\mu g/m^3$ (0.2 ppm). He found that only the nylon fabrics were affected, losing 80% of their strength when exposed to sulfur dioxide and only 40% when exposed in clean air. Zeronian (*85*) also examined the surfaces of exposed nylon fibers under a scanning electron microscope. Depending on the exposure conditions, he observed varying numbers and sizes of surface pits and cavities.

G. Textile Dyes

The fading of textile dyes by air pollutants, primarily nitrogen dioxide and ozone, has been a particularly vexing problem for the textile industry during much of the twentieth century. The problem first surfaced just prior to World War I, when a German dye manufacturer investigated some unusual cases of fading of stored woolen goods (*86*). The cause was traced to nitrogen oxides in the air produced by open electric-arc lamps and incandescent gas mantles. All the susceptible dyes contained free or substituted amino groups.

In subsequent years, increased replacement of older forms of lighting with electric filament lamps led to a general decline of the wool-fading problem. In the mid-1920s, however, a newly developed manmade fiber, cellulose acetate rayon, was introduced. Since traditional dyes were of little use on this fiber, chemists developed an entirely new line of dyestuffs called disperse dyes. Many are derived from anthraquinone and therefore contain amino groups. Shortly thereafter, a puzzling type of fading began to show up on blue, green, and violet shades of dyed acetate goods. This mysterious fading was called "gas fading" because it was frequently observed in rooms heated by gas heaters. Acetate fading became a serious problem during the 1930s, and in 1937, Rowe and Chamberlain (*87*) independently reached the same conclusion, indicting nitrogen oxides as did the earlier German researchers. Since then, much research has been carried out in an effort to understand gas fading and to develop ways to prevent it.

Gas fading of sensitive dyes on acetate is marked by a definite and characteristic reddening of the fabric. Previous research (*86, 88*) supported by recent controlled environmental studies (*89*) shows nitrogen dioxide to be the nitrogen oxide responsible for gas fading. During the mid-1950s, researchers found that nitrogen dioxide also causes certain dyed cotton textiles to fade. This discovery came about as a result of investigating a series of complaints that some colored cotton fabrics were fading during the drying cycle in home gas-fired dryers (*90, 91*). The investigators traced the fading to nitrogen oxides formed during the combustion of natural gas used to heat the dryers. Furthermore, fading only occurred while the cotton goods were moist.

In subsequent field exposures (*92, 93*) conducted in the absence of sunlight, additional evidence of cotton fading was observed. Certain dyed (blue and green shades) cotton fabrics faded after 2–3 months at several exposure sites. The location of the sites indicated that relative humidity might be an important controlling factor. Laboratory investigations showed that the combination of nitrogen oxides and relative humidity greater than 70% was necessary to produce fading in sensitive dyed cotton textiles (*94, 95*).

A more recent gas-fading problem confronting the textile industry concerns the yellow discoloration of undyed white or pastel-colored fabrics woven from any number of common fibers (*91, 95*). Since discoloration occurs on whites, researchers studied various additives that are applied to fibers and fabrics to enhance certain properties. These include optical brighteners, antistatic and soil-release finishes, softeners, and resinous processing agents. Laboratory studies showed that many of these additives yellowed on exposure to nitrogen oxides. Today, the problem is best handled by selecting resistant, but more expensive, additives. In the mid-1950s, Salvin and Walker (*96*) synthesized new blue disperse dyes that exhibited a high resistance to gas fading (NO_x) when tested in the laboratory. However, during subsequent field testing some of these dyes showed considerable fading. Furthermore, the observed fading was characterized by a bleached, washed-out appearance rather than the familiar reddening that NO_x-sensitive dyes develop. A thorough investigation showed that atmospheric ozone was responsible. To describe this fading phenomenon, Salvin and Walker coined a new term, "O-fading."

The discovery that ozone, in addition to nitrogen oxides, is a prime cause of fading was useful in helping to explain much of the anomalous fading of certain dyed fabrics observed during subsequent field trials (*97–100*). These trials also alerted the textile industry to the possibility of potential consumer complaints. Such complaints began to appear in the early 1960s, and mainly concerned two distinctly different textile materials—polyester–cotton permanent press fabrics and nylon carpets. Fading complaints on permanent press fabrics first showed up on the folds and edges of slacks stored in warehouses or on the stock shelves of retail outlets (*95, 97*). Incidents occurred in California, Texas, and Tennessee. Because of the volume of garments involved, some producers suffered heavy economic losses. Researchers found ozone, and to a lesser extent nitrogen oxides, to be responsible for this fading problem. The fading mechanism is unique and complex. It takes place as a result of the curing operation and involves the disperse dyes used to color the polyester fibers rather than the vat dyes on the cotton. During curing, some disperse dyes partially migrate to the permanent press finish. The finish is a combination of reactant resins, catalysts, softeners, and nonionic wetting agents. Dyes migrate to the solubilizing agents (nonionic surfactants and softeners) in the finish and are then in a medium where fading by air contaminants can easily occur. Softeners are especially good media for absorbing gases. Carefully selecting the materials that make up the permanent press finish or replacing vulnerable dyes with dyes that resist migration will essentially eliminate the fading problem.

Fading complaints on nylon carpets originated mainly in the warm humid areas from Texas to Florida, and because of this, became known as "Gulf Coast fading" (*95, 101*). In one incident, nylon carpeting in a Texas apartment complex faded 30 days after installation. Avocado was a particularly sensitive color, and fading was characterized primarily by loss in blue as the green color gradually turned to a dull orange shade. Subsequent research showed that the combination of atmospheric ozone and high relative humidity (75% or more) caused pronounced fading of this particular blue dye. The problem may be prevented by using more ozone-resistant dyes and by using nylon fibers that have been modified and textured to decrease the accessibility and diffusion rate of ozone.

Recent research has shown that the combination of ozone and high humidity causes pronounced fading of certain dyes on cotton and rayon fabrics. Consumers have complained about faded drapery and upholstery fabrics (*101*). The United States Environmental Protection Agency (USEPA), in cooperation with the American Association of Textile Chemists and Colorists (AATCC), has investigated the effects of air pollution on various dyed textiles by conducting both field and laboratory exposures.

The field study consisted of exposing a wide range of dyed fabrics in light-tight cabinets located at four urban and corresponding rural (control) sites (*102*). These sites represented a cross section of various types of pollution and climates. The study was carried out over a 2-year period consisting of eight consecutive 3-month season exposures.

Color change data along with air pollution and weather measurements were statistically analyzed in an effort to identify factors causing fading. The researchers found that about two-thirds of the fabrics showed appreciable fading. Most of these fabrics faded significantly more at urban sites than at corresponding rural sites, and the amount of fading varied between metropolitan areas and among seasons. Air pollution was assumed to account for most of the environmental differences between urban and corresponding rural sites. Analysis showed that of the pollutants measured—nitrogen dioxide, ozone, and sulfur dioxide—appeared to be those most responsible for causing fading. Generally, it was impossible to separate out the effects of individual pollutants because they were confounded with the effects of other pollutants.

The USEPA laboratory studies were designed to assess directly the effect of individual pollutants on twenty selected dyed fabrics (*89*). The fabrics were a cross section of those that showed the greatest tendency to fade during the field study. Samples were exposed for twelve consecutive weeks in the absence of light to clean air (controls) and to two levels of individual pollutants (dynamic system) at four combinations of temperature (13°C, 32°C) and relative humidity (50%, 90%). The pollutants and concentrations were: sulfur dioxide (0.1 ppm, 1.0 ppm), nitric oxide (0.1 ppm, 1.0 ppm), nitrogen dioxide (0.05 ppm, 0.5 ppm), and ozone (0.05 ppm, 0.50 ppm).

Results showed that nitrogen dioxide and ozone produced appreciable dye fading in a number of fabrics. Sulfur dioxide caused visible fading on wool fabrics; the magnitude of fading, however, was considerably less than the magnitude of fading that many of the other fabrics developed during exposure to nitrogen dioxide or ozone. Nitric oxide had little or no fading effect on any of the fabrics. In all cases, high pollutant levels produced significant fading in more fabric samples than low levels under similar exposure conditions; furthermore, the color changes were more pronounced. Low nitrogen dioxide and ozone levels nevertheless produced visible fading in a number of fabrics. This is important because the low levels were similar to levels frequently existing in urban areas. The study also clearly demonstrated that high relative humidity, and to a lesser extent high temperature, are significant factors in promoting and accelerating fading by pollutants.

Over the years, the textile industry has taken important steps to prevent or mitigate the fading effects that air pollutants may produce. They have accomplished this by developing resistant dyes and chemical inhibitors. Resistant dyes, however, are more expensive than the sensitive dyes. Furthermore, many have poor dyeing properties that result in slower processing rates and require greater care during application. These disadvantages have increased the use of fading inhibitors, which are added to dyed fabrics and selectively react with pollutants. Inhibitors, unfortunately, are only effective temporarily, since they are eventually consumed and the dye again becomes vulnerable to fading. Other means to circumvent the fading problem include using alternate dye–fiber systems or fibers with colored pigments incorporated in them. In the final analysis, each producer and end user must balance the pros and cons in deciding on the degree of protection textile goods will possess.

H. Elastomers

One of the early indications of the smog problem in Los Angeles, California, was the excessive cracking of rubber products. It had been known that atmospheric oxidants, especially ozone, could cause rubber cracking. Crabtree and Kemp (*103*, *104*) carried out extensive studies as early as 1946 on the attack of natural rubber by light-energized oxygen and ozone. The cracking of rubber in an ozone atmosphere even of high concentration (1–5%) does not begin until the rubber is stretched. Even the slightest stretching under these conditions brings about rapid deterioration.

The Los Angeles County Air Pollution Control District has used as one of its measurements of ozone concentration in the atmosphere a method developed by Bradley and Haagen-Smit (*105*) in which bent strips of rubber of precise formulation are exposed to the air to be sampled. The depths of cracks at the end of definite exposures are measured to estimate the ozone content of the atmosphere.

The present theory is that elastomers of the unsaturated type are attacked by ozone at the double bond in the carbon-to-carbon chain (*106*). If the chain is under stress, it breaks and leaves its neighbor under additional strain. Unsaturated natural and synthetic rubbers, such as butadiene–styrene or butadiene–acrylonitrile are affected in this same manner. Neoprene, although unsaturated, resists ozone attack, presumably because of the presence of the chlorine atom adjacent to the double bond. Butyl, thiokol, and silicone polymers, being saturated, are highly resistant to attack by ozone. This property is especially important to tires and rubber-insulated electrical wires.

Figure 11 shows the effects of ozone on various types of synthetic

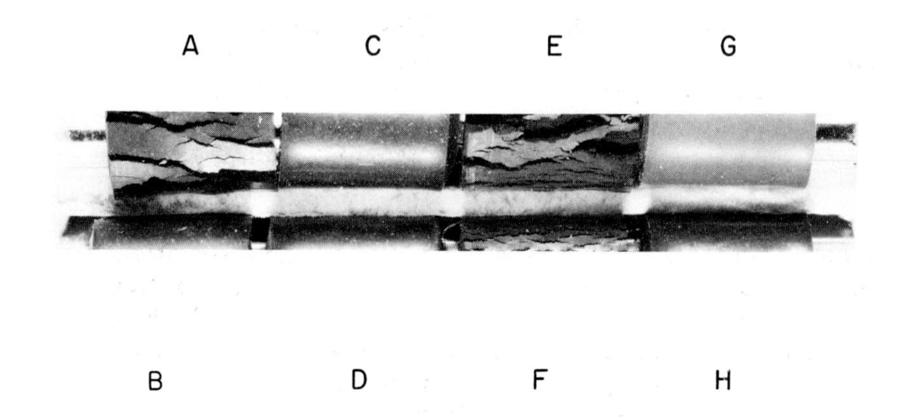

Figure 11. Effect of ozone exposure on samples of various rubber components: (A) GR-S; (B) butyl; (C, D) neoprene; (E) Buna-N; (F) natural rubber; (G) silicone; (H) Hypalon. Photo courtesy of F. H. Winslow, Bell Telephone Laboratories.

rubber and natural rubber. In addition to the rubber formulation and the ozone concentration, the degree of strain experienced by the elastomer has bearing on the extent of damage. A high degree of strain produces many shallow cracks, whereas low strain produces fewer but deeper cracks. Cracking reaches a maximum at elongations of 5–20%.

Overcoming the damaging effect of ozone on stockpiled material can be a substantial factor in storage logistics. Techniques used to protect rubber in storage (*107*) include coating with new wax and plastic and incorporating chemical substances in the rubber to inhibit the ozone action. If adequately protected, tires can be safely stored under severe conditions up to 3 years.

Rubber cracking normally attributed to ozone occurred at a military storage area in one of the most unlikely spots on earth, northern Greenland (*108*). Although low, the observed ozone concentrations were equivalent to those shown to have produced rubber cracking in more temperate environments. Other environmental factors were suspected of contributing to the damage.

Observed cracking of some rubber products in less than 2 years of service led a power company to develop a materials specification (*109*). Failed products included gate seals, cable jackets, conveyor belt covers, and pressure hose covers. Accelerated exposures, in which test specimens cut from a variety of products were subjected to 980 $\mu g/m^3$ (50 parts per hundred million) ozone at 38°C, showed that specimens cut from products that failed in less than 2 years of service cracked in 20 hours or less. Specimens from products that did not fail prematurely in service lasted from 20 to 1000 hours in the accelerated test. As a result of the laboratory observations, the power company developed a specification requiring all mechanical goods to pass 500 hours in the accelerated test.

I. Glass and Ceramics

Although enamels and glasses are especially resistant to the chemical action of air pollutants, a 3-year exposure test of porcelain enamels was conducted at seven cities in the United States (*110*). Although the enamels protected the base metal for the 3-year period, the surface appearance changed. Moisture and atmospheric pollution by acidic substances appeared to play a major role in surface degradation.

Fluorides, especially hydrogen flouride, are capable of attacking a wide range of ceramic materials and glass through their ability to react with silicon compounds. In the past, window glass in areas near synthetic fertilizer and enamel frit plants has been rendered opaque through the action of fluorides. However, the concentrations required to produce this type of damage are far in excess of those necessary to kill sensitive vegetation. Restrictive legislation on emissions of fluorides has essentially eliminated this type of effect except in occasional localized situations. Thus, fluoride damage to materials appears to be an insignificant economic effect.

J. Works of Art

In many parts of the world, air pollution is silently eating away irreplaceable works of art (*111*). Some of the finest monuments of antiquity and thousands of pieces of sculpture and carvings on historic buildings and cathedrals show vivid evidence of the insidious effects of polluted atmospheres. The frieze of the Parthenon in Athens, Greece, is a good example (*112*). A plaster cast made in 1802 shows the relatively minor damage occurring to the marble during the first 2240 years. A photograph of the same marble frieze taken in 1938 is almost unrecognizable because of rapid deterioration during the intervening 136 years of increasing industrialization.

Similar cases have been reported elsewhere. The Coliseum and Arch of Titus in Rome, Italy, and the San Marco Basilica in Venice, Italy, all show accelerated decay attributable to air pollution. The situation in Florence, Italy, is described as disastrous. In France, conservation specialists have removed statues from the exterior of cathedrals and replaced them with copies. A team of experts has been fighting decay and corrosion destroying the massive, twin-spired Cologne Cathedral, the most magnificent church building of the German high gothic era of around 1200 A.D. Polluted atmospheres are threatening centuries-old shrines, temples, and buildings in highly industrial areas of Japan. Cleopatra's needle, the large stone obelisk moved from Alexandria, Egypt, to London, England, has suffered more deterioration in the damp, smoky, acid atmosphere of London in 80 years than in the earlier 3000 or more years of its history. Another ancient Egyptian obelisk standing in Central Park in New York City, New York, shows similar decay.

Conservation scientists recognize that a number of environmental agents hasten decay. These include weathering (wind, rain, heat, snow, and frost), traffic vibration, plant growths such as molds, lichens, and pigeon droppings. However, air pollution, the unwanted by-product of the industrial age, is considered the major villain. The most important single pollutant is sulfur dioxide because of its potential conversion in the presence of moisture to sulfuric acid. Carbonate-containing stones, such as limestone, marble, and some sandstones, are particularly vulnerable to acid attack. Sulfuric acid reacts to produce soluble sulfates that rains easily leach away. In winter, moisture penetrates the resulting network of small cracks and crevices and subsequent freezing forces small pieces of stone to flake off. The air inside museums, churches, and other buildings can also contain air pollutants. Paintings, drawings, books, fabrics, antique costumes and other art objects have suffered deterioration.

It is impossible to make adequate restitution for the irreversible losses sustained by works of art. Once damage has occurred, the loss can never be recovered. Although these losses cannot be reversed, conservation experts have scientific means of preserving and restoring them. For outside exposures, preservation normally involves the application of specially formulated coatings. Such coatings, however, have not always been successful. Sensitive art objects displayed inside buildings have been placed in hermetically sealed containers. Central air conditioning and purification using activated charcoal filters are frequently used as protective measures.

K. Other Materials Effects

Gaseous pollutants and particulate matter have been constant sources of trouble to the electronics industry (*113, 114*). Many kinds of electronic components and equipment have been damaged. Low-power (millivolts, microamperes) electrical contacts, which are used in numerous electrical devices, are particularly sensitive components. They find wide application in computers, communications equipment, and electronic instrumentation.

Contacts are made from a variety of metals, either in the pure state, or more frequently as combinations. Air pollutants can cause thin insulating films to develop on contacts resulting in open circuits and malfunctioning equipment. Sulfur gases, sulfur dioxide and hydrogen sulfide, tarnish copper and silver contacts by producing sulfide films. These pollutants even cause gold–silver bonded contacts to fail; the silver sulfide creeps through pores in the gold and across boundaries onto the gold surfaces. Atmospheric particles can settle on contacts and prevent intimate surface contact when closed. If the particles contain corrosive components, direct chemical attacks may occur.

Bell Laboratories (*48*) also reported contact corrosion problems that have been observed in such widely scattered locations as Cincinnati, Ohio; Cleveland, Ohio; Detroit, Michigan; Los Angeles, California; New York, New York; and Philadelphia, Pennsylvania. The nickel bases of palladium-capped contacts of crossbar switches corroded forming bright greenish corrosion products that gradually crept up over the palladium cap of the contacts, resulting in electrically open circuits. Investigators concluded that the "creeping green" corrosion was promoted by the presence of anions, principally nitrates, in accumulated dust.

Contact problems have alerted engineers to make a special effort to specify appropriate contact materials for various applications. For example, a computer that controls the operations of a chemical plant could cease to function because of an inexpensive tarnished electrical contact. The resulting down time could cost thousands of dollars. Therefore, because of the critical nature of contact materials, frequently the only solution is to specify precious metals such as gold, platinum, and palladium, an expensive alternative.

Power companies have a unique problem associated with air pollution (*115*). Airborne particulate matter is deposited on high-voltage transmission line insulators. During conditions of high humidity, fog, or rain the deposited material serves as a conductor and results in insulator flashover. This has been a severe problem in those areas downwind from industrial sources emitting large amounts of particulate matter. The most successful method of minimizing contamination flashover without interrupting service has been a preventive schedule of washing the insulators with very high pressure streams of water.

Sulfur dioxide, ammonia, and hydrogen sulfide have produced spots on microfilm under laboratory conditions. Henn and Wiest (*116*), investigating small spots on stored microfilm, have eliminated bacteria, fungi, and radioactivity as possible causes. Air pollutants may cause oxidation of silver grains on microfilm that has been stripped of its protective gelatin layer. These small spots have been observed on film from a number of manufacturers. They are more pronounced on older films and on film stored at high temperatures, high humidity, and in contaminated atmospheres.

VI. Economics of Air Pollution Effects

A. Early Estimates

Air pollution corrodes metal, weakens fabric, and soils clothing. It causes building stone to crumble and paint to discolor. It destroys leather, fades dyed fabric, and cracks rubber. It damages vegetation and kills livestock. It reduces visibility, and it is harmful to human health. There is no question that these adverse effects are caused directly by air pollution and that they are extremely costly to the inhabitants of urban and industrialized areas throughout the world. Determining accurately the true cost of these effects has not yet been accomplished nor is it likely that these costs can be developed during the next few years. Costs of vegetation damage in areas where agricultural crops and ornamental flowers are grown can be developed relatively easily. On the other hand, damage to vegetation where a crop of no direct economic value is involved (e.g., city parks and home flower gardens) is difficult to appraise. Most of the air pollution damage to nonviable materials is in this latter class, and assessment of damage is further complicated by the many other environmental factors involved.

Let us examine some of the problems involved in translating a qualitative observation of damage into a quantitative estimate of cost. If the mechanism of damage is used as a starting point, it should suggest the most pertinent variables, such as pollutants, moisture, temperature, and sunlight. The next step is to evaluate these variables and determine the relative contribution of a particular pollutant or combination of pollutants. Then a survey is required to determine the extent of damage. Next, significance of the damage, and the response to it, must be considered. The cost of damage might be represented either by a shortened useful life or by protective measures such as over design, protective coating, substitution of materials, or reduced market value. These responses need to be translated into dollar values, and, finally, a proper basis for extrapolation must be developed before the various components can be summed up on a national scale.

From time to time, studies have dealt with limited aspects of the problem. Some of these, for example, report that on the average, paint lasts 6 years in rural areas as compared with 4 years in polluted cities (*55*); zinc-work that would normally last 30 years has had its useful life reduced to 5 years in cases of severe pollution (*117*); and in France the city dweller requires 50% more laundry service than the rural dweller (*118*).

These partial assessments lead to estimates of the total cost of pollution. One of the most extensive assessments was made in Pittsburgh, Pennsylvania, in 1913. A cost of $20 per capita per year in tangible losses was estimated. This did not include aesthetics or damage to health. Many additional estimates have been made in the United States and in England. In some cases, the estimates have merely updated previous figures by adjusting for population growth and change in the purchasing power of the monetary unit. In other cases new effects and cost data have been considered. The various United States estimates range from about $2 billion to $12 billion a year, or about $10–$60 per person per year; the difference depends on what is included in the estimate. The health aspect, which should be our primary concern, is rarely included in these estimates (*119–121*).

Considerable effort has been made in England to determine the cost of the effects of air pollution. The Beaver Report (*122*) states that the total economic loss from air pollution is of the order of £250 million per year. Of this about £150 million are direct costs, and the remaining 100 million result from loss in efficiency, for example, in the operation of transportation facilities. The distribution of direct costs as listed in the report is tabulated below.

Cost item	*Millions of £ sterling per year*
Laundry	25
Painting and decorating	30
Cleaning and depreciation of buildings other than houses	20
Corrosion of metals	25
Damage to textiles and other goods	52.5
	152.5

Most of the deterioration of materials by air pollution goes unnoticed, because it cannot be distinguished from what might be called normal or natural deterioration. One source for the Beaver Report estimated, however, that one-third of the cost of replacing steel railroad rails in England was attributable to corrosion induced by air pollution. The report further states that, as a direct result of the 1952 London smog, the owner of one chain store had to reduce prices of damaged goods by £90,000 in order to sell them.

One might reason that air pollution would depress land values and that the resulting loss could be estimated (*123*). The few studies made thus far indicate that the cost-depressing effect of air pollution is masked by the many other variables at play in determining land value.

The book edited by Wolozin (*124*) assembled considerable information on the economics of air pollution. Although this book provides little in the way of data on the actual costs of air pollution, it presents a wealth of background information on approaches to the problem of gathering such data, and brings together the opinions of economists, engineers, and scientists. The committee on Pollution of the National Academy of Sciences, in their report "Waste Management and Control" (*125*), presents some approaches to determining the costs of various types of waste management programs, but the report is not limited to the area of air pollution.

One study (*126*) of two cities in the Upper Ohio River Valley compared costs of air pollution effects in Steubenville, Ohio, and Uniontown, Pennsylvania. Steubenville experiences heavy air pollution, and, by comparison, Uniontown is relatively clean. The study showed that the per capita annual cost for outside and inside maintenance of houses, laundry, and dry cleaning, and hair and facial care were $84 higher in Steubenville than in Uniontown. The authors attribute the bulk of this difference to air pollution.

Ridker (*127*) recognized the complexities of developing meaningful costs and presented a number of strategies for approaching various aspects of the problem. Useful estimates are presented for the costs of health impairment related to air pollution, an especially difficult type of estimate to make. The results of detailed studies of two specific localized air pollution problems are of special interest:

1. In March 1965, a firing failure in a large pulverized coal-fired institutional heating plant in Syracuse, New York, released, over a 20-minute period, a large amount of soot and partially burned coal that was deposited over a fairly well-defined area downwind of the plant. Shortly after the episode, householders and others who were affected were interviewed. From an analysis of these interviews, estimates were made of the cleaning costs and insurance claims resulting from the episode. The total cost was about $38,000—the result of the emission of approximately 225 lb (100 kg) of particulate matter above and beyond that which would normally have been released from the heating plant during the same 20-minute period.

2. During 1962, a metal-fabricating firm started operation in a residential neighborhood in St. Louis, Missouri. The malodorous and irritating gases emanating from the plant created in the surrounding area

a nuisance that up until mid-1966 had not been abated. A study was made of the effect of the pollution from this plant on property values. Property value indexes over the period 1956–1965 for the census tract encompassing the affected area were compared with an adjoining tract similar in all respects to the affected tract except for the significant pollution levels. This analysis showed that the average householder in the affected tract who sold his home after 1962 incurred a loss of about $1000 compared to that of the householder in the control area.

B. More Recent Estimates

The United States federal government air pollution program has sponsored several contracts to estimate nationwide costs of air pollution damage to selected classes of materials. Stanford Research Institute (SRI) investigated the cost impact of air pollution on electrical contacts and found it to consist mainly of the cost of precious metals for contact surfaces and protective measures, such as air conditioning to lower relative humidity and air purification to remove particulate and gaseous pollutants (*113*). Annual estimated costs are shown in the following tabulation.

	Millions of dollars
Precious metals	20
Protective measures	25
Loss due to failures	10
Research	5
	60

SRI also estimated an annual cost of $4 million for cleaning high-voltage transmission line insulators.

In a similar survey (*114*) of electrical components (semiconductor devices, integrated circuits, television picture tubes, connectors, transformers, relays, etc.), annual cost estimates associated with air pollution were about $2 million for protective measures and $13 million for maintenance costs—cleaning, repairing, and replacing defective equipment. The costs for electrical components did not include the above costs related to electrical contacts.

Battelle Memorial Institute (*128, 129*) surveyed the effects of air pollution on elastomers and estimated the total cost at the consumer level to be about $500 million annually. Of this total, about $170 million was associated with costs of protection (use of resistant polymers, antiozonants, waxes, protective finishes, etc.); $225 million was for reduced service life (premature replacement of damaged products); and the remainder for labor costs connected with early replacement of damaged products.

Battelle (*130*) estimated the cost of corrosion damage to metallic systems and structures attributable to air pollution to be $145 billion annually. This includes early replacement costs and added maintenance costs; the latter derived mainly from painting frequencies in clean and polluted air. The investigators concluded that accelerated corrosion of zinc-galvanized products by sulfur dioxide accounts for about 90% of increased atmospheric corrosion damage.

Spence and Haynie (*131*) studied the effect of air pollution (direct chemical attack and the indirect effects of soiling) on the service life of four classes of paint—household, automotive refinishing, coil coating, and maintenance. Total cost at the consumer level was estimated to be about $700 million annually. Household paint damage represents over 75% of this total.

Barrett and Waddell (*132*) estimated the cost of materials damage by air pollution in the United States to be $4.8 billion in 1968. Sulfur dioxide was responsible for $2.2 billion, closely followed by the combined effects of nitrogen oxides and oxidants ($1.9 billion), and particulate matter ($0.7 billion). In an updating of the above study, Waddell (*133*) arrived at a figure of $3.8 billion for 1970. This lower estimate results from a reassessment of information on materials damage. Waddell states that this figure is a best estimate and probably falls within the range of $2.2–$5.4 billion.

References

1. U. R. Evans, *in* "Corrosion Handbook" (H. H. Uhlig, ed.), Sect. 1, p. 3. Wiley New York, New York, 1948.
2. C. P. Larrabee, *Corrosion* **15**, 36 (1959).
3. B. Sanyal and D. V. Bhadwar, *J. Sci. Ind. Res., Sect. A* **18**, 69 (1959).
4. P. M. Aziz and H. P. Godard, *Corrosion* **15**, 39 (1959).
5. W. H. J. Vernon, *Trans. Faraday Soc.* **31**, 1668 (1935).
6. H. T. Shirley and J. E. Truman, *J. Iron Steel Inst., London* **160**, 367 (1948).
7. J. P. Lodge, Jr. and B. R. Havlik, *Int. J. Air Pollut.* **3**, 249 (1960).
8. W. H. J. Vernon, *J. Roy. Soc. Arts* **97**, 570 (1949).
9. R. St. J. Preston and E. G. Straud, *J. Inst. Petrol.* **35**, 457 (1950).
10. R. St. J. Preston, *J. Iron Steel Inst., London* **160**, 286 (1948).
11. R. J. Brysson, B. J. Trask, J. B. Upham, and S. G. Booras, *J. Air Pollut. Contr. Ass.* **17**, 294 (1967).

12. G. A. Jutze, R. L. Harris, Jr., and M. Georgevich, *J. Air Pollut. Contr. Ass.* **17,** 291 (1967).
13. J. D. Hudson, *J. Iron Steel Inst., London* **148,** 161 (1943).
14. Reports of Committee A-5 on Corrosion of Iron and Steel, *Amer. Soc. Test. Mater., Proc.* **52,** 106 (1952).
15. W. C. Galegar and R. O. McCaldin, *Amer. Ind. Hyg. Ass., J.* **22,** 187 (1961).
16. P. J. Sereda, *Ind. Eng. Chem.* **52,** 157 (1960).
17. J. B. Upham, *J. Air Pollut. Contr. Ass.* **17,** 398 (1967).
18. J. B. Upham, *J. Air Pollut. Contr. Ass.* **17,** 400 (1967).
19. F. H. Haynie and J. B. Upham, *Mater. Prot. Performance* **10,** No. 12, 18 (1971).
20. C. I. Harding and T. R. Kelley, Paper No. 66–116, *Annu. Meet.* Air Pollut. Contr. Ass., Pittsburgh, Pennsylvania, 1966.
21. A. R. Meetham, "Atmospheric Pollution, its Origin and Prevention," 2nd ed. Pergamon, Oxford, England, 1956.
22. R. St. J. Preston and B. Sanyal, *J. Appl. Chem.* **6,** 26 (1956).
23. K. Barton, *Werkst. Korros.* **9,** 547 (1958).
24. "Air Pollution in Yokohama-Kawasaki Industrial Area (1957–1962)," Tech. Rep. Kanagawa Prefectural Govt., Yokohama, Japan, 1963.
25. Y. Nose and T. Shimizu, *Bull. Sch. Technol., Yamaguchi Univ.* **3** (1952).
26. Y. Nose, *Annu. Rep. Res. Inst. Ind. Med.* [6] (1962).
27. Y. Kitagawa, Y. Hayashi, and S. Mashino, "Experimental Studies on Effects of Exposure of Metal Plates to Air Pollutants in Ube City." Dep. Hyg. Pub. Health, Yamaguchi Med. School, Ube, Japan, 1954.
28. H. H. Uhlig, "Corrosion and Corrosion Control." Wiley, New York, New York, 1963.
29. U. R. Evans, "The Corrosion and Oxidation of Metals: Scientific Principles and Practical Applications." Arnold, London, England, 1960.
30. F. N. Speller, "Corrosion Causes and Prevention." McGraw-Hill, New York, New York, 1951.
31. H. C. Muffley, "Influence of Atmospheric Contaminants on Corrosion—Literature Report." Defense Documentation Center, Alexandria, Virginia, 1963.
32. G. A. Greathouse and C. J. Wessel, eds., "Deterioration of Materials, Causes and Preventive Techniques." Van Nostrand-Reinhold, Princeton, New Jersey, 1954.
33. E. A. Anderson, *Spec. Tech. Publ.* **175,** 126. Amer. Soc. Test. Mater., Philadelphia, Pennsylvania (1956).
34. E. A. Tice, *J. Air Pollut. Contr. Ass.* **12,** 553 (1962).
35. C. P. Larrabee and O. B. Ellis, *Amer. Soc. Test. Mater., Proc.* **59,** 183 (1959).
36. H. Guttman, *Spec. Tech. Pub.* **435,** 223–239. Amer. Soc. Test. Mater., Philadelphia, Pennsylvania (1968).
37. F. H. Haynie and J. B. Upham, *Mater. Prot. Performance* **9,** No. 8, 35–40 (1970).
38. W. W. Binger, R. H. Wagner, and R. H. Brown, *Corrosion* **9,** 440 (1953).
39. Symposium on Atmospheric Corrosion of Non-Ferrous Metals, *Spec. Tech. Publ.* **175,** 21, Amer. Soc. Test. Mater., Philadelphia, Pennsylvania (1956).
40. C. J. Walton and W. King, *Spec. Tech. Publ.* **175,** 21–46, Amer. Soc. Test. Mater. Philadelphia, Pennsylvania (1956).
41. J. D. Freeman and P. H. Kirby, *Metals Alloys* **3,** 190 (1932).
42. W. H. J. Vernon and L. Whitby, *J. Inst. Metals* **42,** No. 2, 181 (1929).
43. R. H. Leach, *in* "Corrosion Handbook" (H. H. Uhlig, ed.), Sect. 2, p. 319. Wiley, New York, New York, 1948.

44. A. W. Tracy, *Amer. Soc. Test. Mater., Spec. Tech. Publ.* **175**, 67–76 (1956).
45. W. H. J. Vernon, *J. Inst. Metals* **48**, No. 1, 121 (1932).
46. H. W. Hermance, "Combatting the Effects of Smog on Wire-Spring Relays," Bell Lab. Rec. 48–52. Bell Telephone Laboratories, Murray Hill, New Jersey, 1966.
47. N. McKinney and H. W. Hermance, *Spec. Tech. Publ.* **452**, 274–291, Amer. Soc. Test. Mater. Philadelphia, Pennsylvania (1967).
48. H. W. Hermance, C. A. Russell, E. J. Bauer, T. F. Egan, and H. V. Wadlow, *Environ. Sci. Technol.* **5**, 781–785 (1971).
49. L. Whitby, *Trans. Faraday Soc.* **29**, 844 (1933).
50. E. M. Winkler, *Science* **147**, 459 (1965).
51. H. B. Meller, *Trans. Amer. Soc. Heat. Vent. Eng.* **37**, 217 (1931).
52. J. J. Rorimer, *N. Y. Times* (April 13, 1964).
53. N. J. Beloin and F. H. Haynie, "Soiling of Building Materials." United States Environmental Protection Agency, Research Triangle Park, North Carolina, 1973.
54. J. R. Wright and P. G. Campbell, *J. Res. Nat. Bur. Stand., Sect. C* **68**, No. 4, 115 (1964).
55. R. I. Larsen, *Amer. Paint J.* **42**, 94 (1957).
56. G. L. Holbrow, *J. Oil Colour Chem. Ass.* **45**, 701 (1962).
57. C. Mell, *Paint Technol.* **20**, 135 (1956).
58. C. Graff-Baker, *J. Appl. Chem.* **8**, 590 (1958).
59. M. Hess, "Paint Film Defects: Their Causes and Cure." Van Nostrand-Reinhold, Princeton, New Jersey, 1951.
60. H. C. Wohlers and M. Feldstein, *J. Air Pollut. Contr. Ass.* **16**, 19 (1966).
61. E. LeClere, *Monogr. Ser.* **46**, 279. World Health Organ Geneva (1961).
62. I. Michelson and B. Tourin, "Report on Study of Validity of Extension of Economic Effects of Air Pollution Damage from Upper Ohio River Valley to Washington, D.C. Area," Contract No. PH-27-68-22. Environmental Health and Safety Research Association, New Rochelle, New York, 1967.
63. Booz, Allen and Hamilton, Inc., "Study to Determine Residential Soiling Costs of Particulate Air Pollution," Final Rep., Contract No. CPA 22-69-103. United States Environmental Protection Agency, Research Triangle Park, North Carolina, 1970.
64. J. C. Schwegmann, *J. Air Pollut. Contr. Ass.* **14**, 48 (1964).
65. P. F. Klens and C. F. Koda, *J. Air Pollut. Contr. Ass.* **6**, 243 (1957).
66. E. G. Fochtman and G. Langer, *J. Air Pollut. Contr. Ass.* **6**, 243 (1957).
67. G. G. Campbell, G. G. Schurr, D. E. Slawikowski, and J. W. Spence, *J. Paint Technology* **46** (No. 593), 59–71 (1974).
68. A. Parker, "The Destructive Effects of Air Pollution on Materials." Nat. Smoke Abatement Soc., London, England, 1955.
69. H. J. Plenderleith, "The Preservation of Leather Bookbindings." British Museum, London, England, 1946.
70. F. R. Innes, *Smokeless Air* No. 68, p. 23 (1948).
71. W. H. Langwell, *Proc. Roy. Inst. Gt. Brit.* **37**, Part II, No. 166, 210 (1958).
72. A. E. Kimberly, *J. Res. Nat. Bur. Stand.* **8**, 159 (1932).
73. C. J. Edwards, F. L. Hudson, and J. A. Hockey, *J. Appl. Chem.* **18**, 146–148 (1968).
74. D. J. Spedding and R. P. Rowlands, *J. Appl. Chem.* **20**, 143–146 (1970).
75. E. J. Race, *J. Soc. Dyers Colour.* **65**, 56 (1949).

76. J. W. Howard and F. A. McCord, *Text. Res. J.* **30,** 75 (1960).
77. H. A. Belyea, personal communication (1958).
78. L. Greenburg and M. Jacobs, *Amer. Paint J.* **39,** 64 (1955).
79. W. H. Rees, *Brit. J. Appl. Phys.* **9,** 301 (1958).
80. M. A. Morris and B. W. Mitchell, *Text. Res. J.* **31,** 488 (1961).
81. T. C. Petrie, *Smokeless Air* No. 67, p. 62 (1948).
82. K. H. Trommer, *Can. Text. J.* **85,** 31–34 (1968).
83. S. H. Zeronian, *Text. Res. J.* **40,** 695–698 (1970).
84. S. H. Zeronian, K. W. Alger, and S. T. Omaye, *in* "Proceedings of the Second International Clean Air Congress" (H. M. England and W. T. Beery, eds.), pp. 468–476. Academic Press, New York, New York, 1971.
85. S. H. Zeronian, *Text. Res. J.* **41,** 184–185 (1971).
86. C. H. Giles, *J. Appl. Chem.* **15,** 541–550 (1965).
87. F. M. Rowe and K. A. J. Chamberlain, *J. Soc. Dyers Colour.* **53,** 268–278 (1937).
88. V. S. Salvin, W. D. Paist, and W. J. Myles, *Amer. Dyest. Rep.* **14,** 297–304 (1952).
89. N. J. Beloin, *Text. Chem. Color.* **5,** 128–133 (1973).
90. "A Study of the Destructive Action of Home Gas-Fired Dryers on Certain Dyestuffs," *Amer. Dyest. Rep.* **45,** 471 (1956).
91. V. McLendon and F. Richardson, *Amer. Dyest. Rep.* **54,** 305–311 (1965).
92. V. S. Salvin, *Amer. Dyest. Rep.* **47,** 450–451 (1958).
93. V. S. Salvin, *Amer. Dyest. Rep.* **53,** 12–20 (1964).
94. V. S. Salvin, *Amer. Dyest. Rep.* **58,** 28–29 (1969).
95. V. S. Salvin, *Text. Qual. Contr. Pap.* **16,** 56–64 (1969).
96. V. S. Salvin and R. A. Walker, *Text. Res. J.* **25,** 571–585 (1955).
97. C. H. A. Schmitt, *Amer. Dyest. Rep.* **51,** 664–675 (1962).
98. C. H. A. Schmitt, *Amer. Dyest. Rep.* **49,** 974–980 (1960).
99. V. S. Salvin, *J. Air Pollut. Contr. Ass.* **13,** 416–422 (1963).
100. V. S. Salvin, *Amer. Dyest. Rep.* **53,** 33–41 (1964).
101. V. S. Salvin, *Text. Chem. Color.* **1,** 245–251 (1969).
102. N. J. Beloin, *Text. Chem. Color.* **4,** 43–48 (1972).
103. J. Crabtree and A. R. Kemp, *Ind. Eng. Chem.* **38,** 278 (1946).
104. J. Crabtree and A. R. Kemp, *Anal. Chem.* **18,** 769 (1946).
105. C. E. Bradley and A. J. Haagen-Smit, *Rubber Chem. Technol.* **24,** 750 (1951).
106. J. E. Gaughan, *Rubber World* **133,** 803 (1956).
107. W. D. England, *Proc. Metropol. Conf. Air Pollut. Contr., Cincinnati Ohio, 1962.*
108. H. C. McKee, *J. Air Pollut. Contr. Ass.* **11,** 562 (1961).
109. D. C. Cordingley, *Ont. Hydro Res. Quart.* **17,** 17 (1945).
110. D. G. Moore and A. Potter, *Nat. Bur. Stand. (U.S.), Monogr.* **44,** 13 (1962).
111. M. Esterow, *N. Y. Times,* New York, New York. (April 13, 1964).
112. H. J. Plenderleith, "The Conservation of Antiquities and Works of Art." Oxford Univ. Press, London, England and New York, New York, 1957.
113. R. C. Robbins, "Inquiry into the Economic Effects of Air Pollution on Electrical Contacts," Contract PH-22-68-35. Stanford Res. Inst., Menlo Park, California, 1970.
114. ITT Electro-Physics Laboratories Inc., "A Survey and Economic Assessment of the Effects of Air Pollutants on Electrical Components," Contract CPA 70–72. ITT Electro-Phys. Lab. Inc., Columbia, Maryland, 1971.
115. T. W. Steading, *J. Air Pollut. Contr. Ass.* **15,** No. 3, 99 (1965).
116. R. W. Henn and D. G. Wiest, *Photogr. Sci. Eng.* **7,** No. 5 (1963).

117. A. R. Meetham, "Atmospheric Pollution—Its Origin and Prevention." Pergamon, Oxford, England, 1956.
118. W. L. Faith, *J. Occup. Med.* **2**, No. 9, 439 (1960).
119. J. J. O'Connor, *Bull.* **4** Mellon Inst, Pittsburgh, Pennsylvania (1913).
120. "The Economic Cost of Air Pollution," Appendix II from Committee on Air Pollution Rep. HM Stationery Office, London, England, 1954.
121. R. G. Gustavson, *Publ.* **654** Pub. Health Serv. U.S. Department of Health Education and Welfare, Washington, D.C. (1959).
122. H. Beaver, chairman, "Committee on Air Pollution Report." HM Stationery Office, London, England, 1954.
123. American Industrial Hygiene Association, "Air Pollution Manual," Chapter 7. Amer. Ind. Hyg. Ass., Detroit, Michigan, 1960.
124. H. Wolozin, ed., "The Economics of Air Pollution." Norton, New York, New York, 1966.
125. Waste Management and Control—A Report to the Federal Council for Science and Technology by the Committee on Pollution, *Publ.* **1400** Nat. Acad. Sci.—Nat. Res. Counc. Washington, D.C. (1966).
126. I. Michelson and B. Tourin, *Publ.* **81,** No. 6, 505, Pub. Health Serv. U.S. Department of Health Education and Welfare, Washington, D.C. (1966).
127. R. G. Ridker, "Economic Costs of Air Pollution—Studies in Measurement." Praeger, New York, New York, 1967.
128. W. J. Mueller and P. B. Stickney, "A Survey and Economic Assessment of the Effects of Air Pollution on Elastomers," Contract No. CPA 22-69-146. United States Environmental Protection Agency, Research Triangle Park, North Carolina, 1970.
129. P. B. Stickney, W. J. Mueller, and J. W. Spence, *Rubber Age* **103,** 45–51, (1971).
130. F. W. Fink, F. H. Buttner, and W. K. Boyd, "Technical-Economic Evaluation of Air-Pollution Corrosion Costs on Metals in the U.S.," Contract No. CPA 70–86. Battelle Memorial Inst., Columbus, Ohio, 1971.
131. J. W. Spence and F. H. Haynie, "Paint Technology and Air Pollution: A Survey and Economic Assessment," Publ. No. AP-103. United States Environmental Protection Agency, Research Triangle Park, North Carolina, 1972.
132. L. B. Barrett and T. E. Waddell, "Cost of Air Pollution Damage: A Status Report," Publ. No. AP-85. United States Environmental Protection Agency, Research Triangle Park, North Carolina, 1973.
133. T. E. Waddell, "The Economic Damages of Air Pollution," EPA-60015-74-012. United States Environmental Protection Agency, Washington, D.C., 1974.

3

Effects on Indoor Air Quality

John E. Yocom, William A. Coté, and Ferris B. Benson

I. Introduction

The primary thrust in defining air quality and its relationship to human health effects has been directed at the outdoor ambient atmosphere. In fact, air pollution is defined in many air pollution control laws as the presence in the outdoor atmosphere of one or more contaminants, or combination thereof, in such quantities and of such duration as may be or tend to be injurious to human, plant, or animal life or property.

This emphasis on the outdoor atmosphere is surprising in view of the large percentage of time that the average person spends indoors where the quality of air can differ considerably from that outdoors. There are no accurate estimates of the percentage of time that people spend indoors depending on where they live, their occupation, mobility, and other factors. Forest rangers and construction workers may spend one-third to two-thirds of their time indoors, depending on the season, while city dwellers, the old and infirm, the very young, and young mothers may spend almost 100% of their time indoors, especially in the winter.

Therefore, it is clear that indoor air quality in terms of the duration of exposure is far more important than outdoor air quality in determining health and welfare. The indoor exposures to air pollutants are made up of those discrete periods (and associated concentrations) spent by individuals primarily in the home and in office buildings, but also in transportation vehicles, public buildings, stores, schools, restaurants, barrooms, theaters, hotels, and other enclosed inhabited spaces *(1)*. Indoor air quality as discussed in this chapter excludes occupational exposure. The air quality of the interior of office buildings is included to the extent that it is determined by nonoccupational factors.

II. Factors Influencing Indoor Air Quality

The quality of the indoor atmosphere varies widely and can be influenced by the following factors:

a. *Outdoor Air Pollution*

Pollutants present in the outdoor ambient air surrounding a building can penetrate indoor spaces through natural or controlled ventilation, leakage, and diffusion. The type of pollutant will influence the character and degree of penetration; e.g., particulates are in part removed in the penetration process, while carbon monoxide, an unreactive gas, readily penetrates.

b. *Indoor Generation of Pollutants*

Indoor pollutants are generated by many activities, e.g., cooking, cleaning, smoking, painting, cosmetic application, and the simple act of moving around and stirring up particulate matter. Pollutants produced inside a building will diffuse throughout the closed space and build up to equilibrium concentrations based on leakage out of the space, leakage into the space from the same pollutant present in the outdoor atmosphere, and pollutant removal processes such as adsorption and absorption on interior surfaces, and atmospheric reactions.

c. *Building Permeability*

The ability of outdoor pollutants to leak in and indoor pollutants to leak out as a result of building permeability will have a strong influence on indoor air quality. The principal factors are numbers of windows and doors and general tightness of building construction.

d. *Ventilation and Air Conditioning Systems*

A positive ventilation system in a building can have a profound effect upon indoor air quality. One that provides outdoor ventilation will produce interior concentrations approaching those outdoors, while one that recirculates indoor air will produce indoor air quality dominated by pollutants generated indoors. An air conditioning system that includes filtration, air washing, humidification or dehumidification, and activated carbon adsorption can be expected to reduce indoor concentrations of particulates and reactive gaseous pollutants (e.g., sulfur dioxide).

e. *Meteorological and Geographic Factors*

Weather conditions influence indoor air quality in a number of ways.

1. Weather conditions for a given location can determine outdoor air quality. (e.g., temperature inversions can cause outdoor concentrations to become elevated). Under these conditions, outdoor air penetrating the building will degrade indoor air quality.

2. Indoor–outdoor temperature relationships can determine tendency for penetration of outdoor pollutants. In cold weather, a heated building exhibits a considerable stack effect, which produces significant pressure differences between inside and outside, tending to draw in outdoor air at the lower levels and exhaust indoor air near the top of the structure.

3. Wind can produce pressure differences between inside and outside and enhance the tendency for outdoor pollutants to penetrate or indoor pollutants to dilute more rapidly.

f. *Location with Respect to Outdoor Sources*

The location of a building will determine the general ambient outdoor concentrations, but beyond this, nearby sources whose impact on the generalized ambient air may not be well defined, can strongly influence the quality of the building's indoor air. Nearby stacks, recirculation of combustion products from the building's heating system, and a truck loading dock near the building's air intake are examples of this factor.

g. *Energy Conservation Measures*

A number of energy conservation measures can influence indoor air quality.

1. Heavy insulation and tight construction of homes can prevent penetration of outdoor pollutants, but will permit build-up of pollutants

generated indoors plus carbon dioxide and moisture from human occupancy.

2. Maximum recirculation of heated or conditioned air in winter and summer will reduce the impact of outdoor air quality and will maximize the influence of indoor pollutants.

3. Use of outdoor ventilation when outdoor temperatures and humidities permit substitution for recirculated heated or conditioned air will produce indoor air quality approaching that of outdoors.

It is clear from the above list that indoor air quality is a complex function of many factors. Attempts have been made to model indoor air quality, notably the work of Shair and Heitner (*2*), and Holcombe and Kalika (*3*).

III. Indoor–Outdoor Air Quality Relationships

The indoor environment is basically an extension of the outdoor environment. Depending on the type of structure, outdoor temperature and humidity, and type of building and its ventilation system, the indoor environment may be similar to or quite different from the outdoor environment. Nevertheless, except for totally enclosed spaces such as submarines and space vehicles, indoor air quality depends on outdoor air quality and a review of what is known about indoor–outdoor air quality relationships provides considerable insight into the nature of indoor air quality.

A. Gaseous Pollutants

Gaseous pollutants in the outdoor atmosphere can be expected to move inside a building as easily as the air. Except as the pollutant gases are reactive and disappear through various removal processes or are produced inside, the concentrations indoors will depend upon the degree of penetration of outdoor air.

1. Carbon Monoxide

Carbon monoxide is an extremely unreactive gas and can be expected to penetrate the indoor atmosphere without loss. In a comprehensive study of indoor–outdoor air quality relationships for six buildings of various types in Hartford, Connecticut, Yocom *et al.* (*4, 5*) showed that indoor–outdoor ratios for carbon monoxide were about 1.0 or slightly above. Figure 1 from this study shows indoor–outdoor profiles for three of the

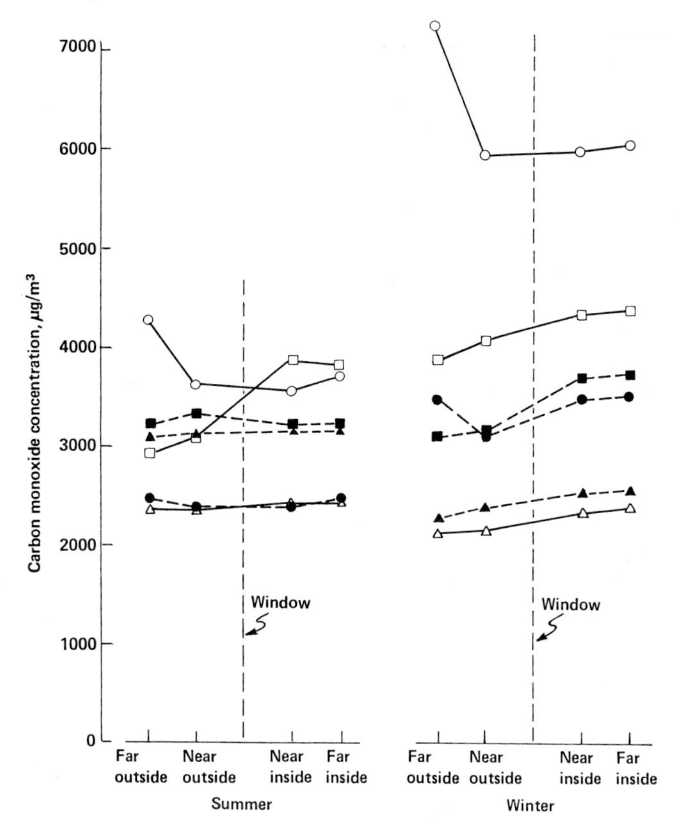

Figure 1. Indoor/outdoor profiles for carbon monoxide in several types of buildings—summer and winter. Non-air conditioned city library: ○ = day, ● = night; air conditioned office building: □ = day, ■ = night; non-air conditioned private home: △ = day, ▲ = night (*4*).

buildings studied both summer and winter. The spacing of sampling points in the diagram is arbitrary. The "far outside" sample was generally taken about 20 feet (6 m) from the building, the "near outside" and "near inside" sampling stations were taken on each side of a window in the building, and the "far inside" sample was taken in an interior room of the building. These data show that indoor levels of carbon monoxide are slightly higher than those immediately outside, probably as a result of indoor sources such as smoking. The appreciably higher levels of carbon monoxide at the "far outside" station at the city library is the result of its being an air-rights structure, built over a four-lane freeway connector.

General Electric Company measured indoor–outdoor concentrations of carbon monoxide at two types of high rise buildings in New York, New York (*6*). One was a modern apartment house built over a multi-lane highway, with central heating and optional window-mounted air conditioners. The other building was an older "canyon structure" with central heating and minimum air conditioning by a few window-mounted units. Carbon monoxide concentrations were measured inside and outside at a number of elevations above the street. The results showed that during the winter the stack effect of the building strongly influenced indoor air quality. Contaminated air from the streets and roadways was brought into the building at the lower levels and distributed throughout the building. Since outdoor concentrations fall off exponentially with height above the roadway, indoor concentrations during the winter exceeded those outdoors at the upper floors of the buidings. During the summer, wind direction and speed were the principal factors influencing indoor carbon monoxide levels.

Some limited studies were made of carbon monoxide levels in Russian houses close to industrial operations capable of emitting carbon monoxide. (Table I) (*7*). They seem to contradict the data from the Hartford, Connecticut, study. Except for the one home where natural gas was used, indoor concentrations were less than those outdoors. While the exact methodology of this study is not clear, the results represented 24-hour samples, and perhaps an equilibrium relation between indoor and outdoor concentrations had not been reached. In the Hartford study, sampling was carried out simultaneously and continuously at each of the sampling locations over 2-week periods using a special sampling system that utilized a single calibrated carbon monoxide monitor. In this way, small differences between sampling points could be measured accurately.

Table I Indoor–Outdoor Relationships and Carbon Monoxide in Russian Homes Located near Industrial Sources (*7*)

Distance from source (m)	*Carbon monoxide levels (ppm)*		*Indoor–outdoor ratio*
	Indoors	*Outdoors*	
50	11.6	17.8	0.65
100[a]	16.3	16.5	0.99
250	9.0	14.9	0.60
500	7.6	12.8	0.59
300	39.9	48.8	0.82
800–1000	22.4	34.1	0.66

[a] Home was natural gas equipped.

2. Sulfur Dioxide

Sulfur dioxide is a reactive gas that is susceptible to oxidation and, in the presence of moisture, can be absorbed or adsorbed on surfaces and react with them. Therefore, it is not surprising to find that indoor–outdoor ratios for sulfur dioxide are less than 1.0 with considerable scatter in the data. Figure 2 shows results of several studies (*8–11*) that indicate that typically indoor sulfur dioxide levels are about half those outdoors. However, in a study conducted in Boston and Cambridge, Massachusetts, during a period when low to moderate sulfur dioxide concentrations were measured (90–130 $\mu g/m^3$), indoor concentrations were roughly equal to those outdoors (*12, 13*). But when outdoor concentrations were above 130 $\mu g/m^3$, indoor concentrations remained below 130 $\mu g/m^3$. In this same study, indoor–outdoor relationships were determined for a building that housed both offices and laboratories. During one period while this building was being sampled, indoor concentrations exceeded those outdoors.

On several occasions during the studies conducted in Hartford, Connecticut (*4, 5*), indoor concentrations of sulfur dioxide appeared to exceed those outdoors. On close examination of the record, it was deter-

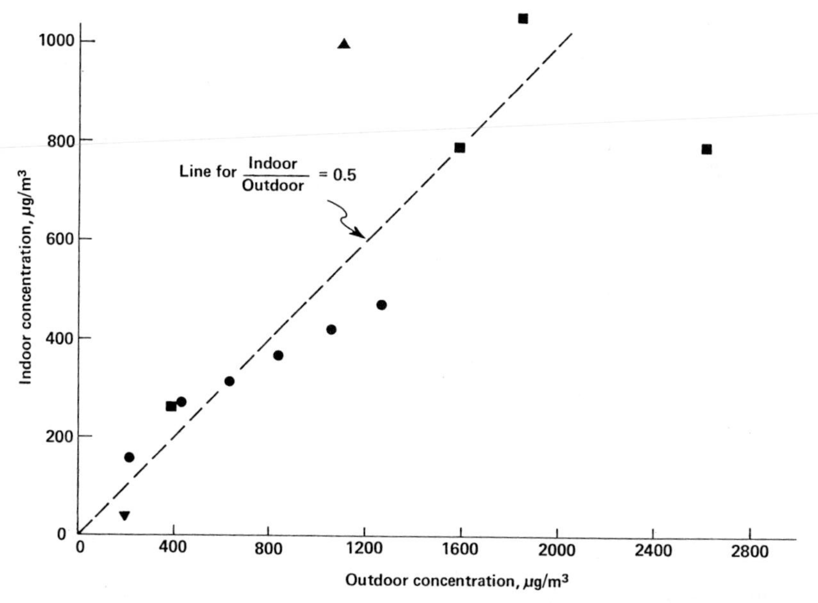

Figure 2. Indoor–outdoor relationships for sulfur dioxide; ● = hospital, Cincinnati, Ohio (*5*); ■ = houses, Moscow, Soviet Union (*6*); ▲ = houses, near viscose plant, Soviet Union (*7*); ▼ = house, Rotterdam, the Netherlands (*8*).

mined that the conductometric instrument being used was probably responding to indoor contaminants (other than sulfur dioxide) such as ammonia from use of household cleaners.

An extensive program of indoor–outdoor sulfur dioxide sampling in Germany indicated that indoor concentrations could be expected to range from 4 to 28% of outdoor levels for outdoor concentrations greater than 400 $\mu g/m^3$, but that they might be as high as 80–100% if windows were open and a high wind was blowing (*14*).

3. Carbon Dioxide

Since carbon dioxide in the outdoor atmosphere remains relatively constant and in the indoor atmosphere is produced by human activity (breathing, cooking, and smoking), indoor concentrations of occupied spaces will invariably be higher than those outdoors. Assuming that outdoor concentrations are normally around 0.03%, concentrations in several types of office buildings were found by Ishido to range from 1 to over 10 times outdoor levels as shown in Table II (*15*). According to Ishido, a space of 10 m^3 per person and a recirculation rate of 30 m^3/hour are required to maintain carbon dioxide concentrations below 0.1% in rooms where people are doing office work.

B. Nonviable Particulate Matter

Data from a number of sources on indoor–outdoor relationships for particulate matter in terms of weight and number concentration are presented graphically in Figure 3 (*4, 11, 16–33*), from which it appears that indoor–outdoor ratio decreases with outdoor concentration. This is a reasonable conclusion, since one would expect that when outdoor particulate concentrations are high, relatively larger particles would predominate

Table II Indoor Concentrations of Carbon Dioxide for Several Buildings in Osaka, Japan (*15*)

Type of building	*Season*	*Indoor concentration range (%)*
Office building	Not specified	0.06–0.32
Old office building	Winter	0.08–0.28
Old office building	Summer	0.04–0.09
New air conditioned office building	Winter	0.06–0.23
New air conditioned office building	Summer	0.04–0.13
Newer air conditioned building	Not specified	0.03–0.14

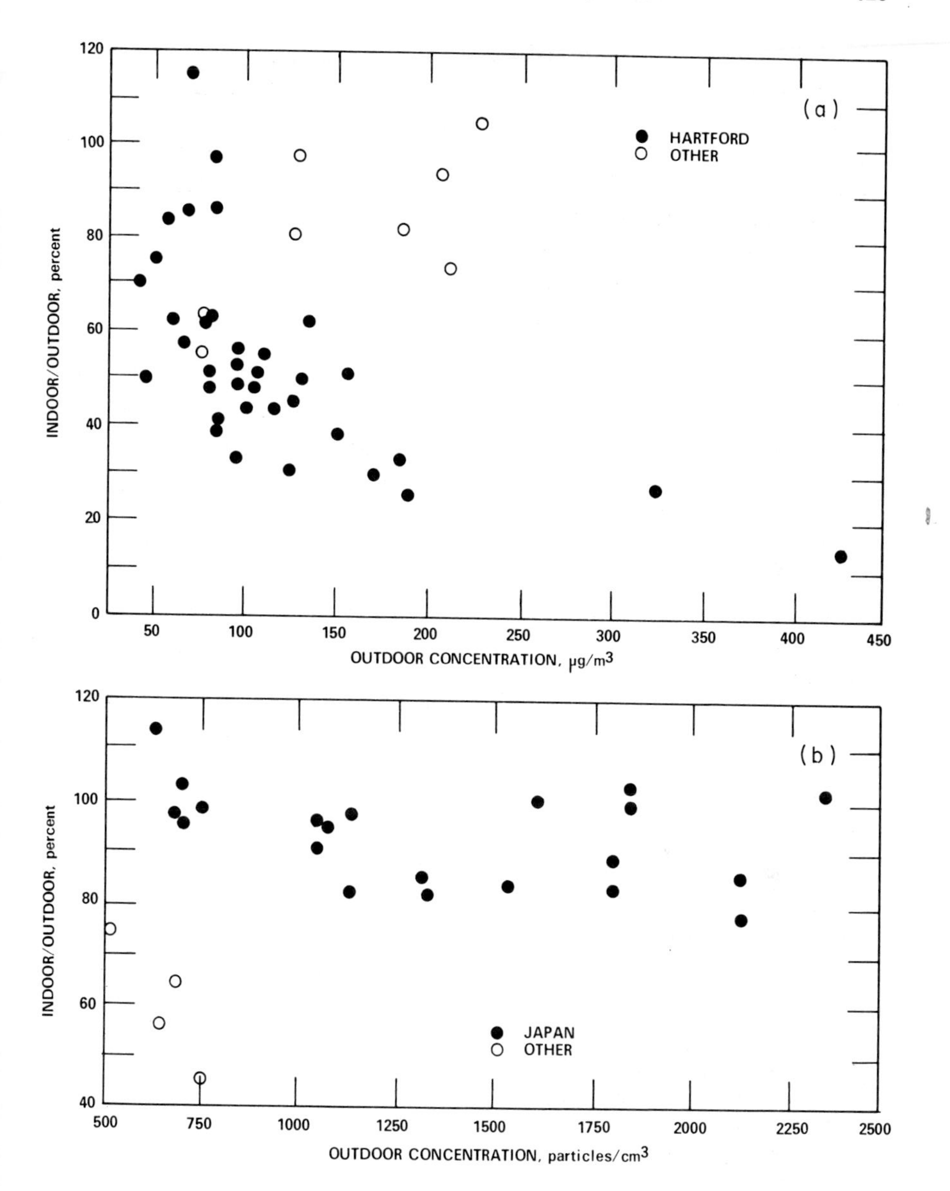

Figure 3. Indoor suspended particulate matter concentrations as a function of outdoor concentrations. (*a*) By weight. (*b*) By particle count (*4, 11, 16–33*).

and that these larger particles would tend to be removed in penetrating a building. This conclusion is borne out by some limited results reported by Yocom *et al.* (*5*) in the Hartford study. Particle size distributions were determined simultaneously indoors and outdoors by means of cascade impactors. In general, outdoor samples contained relatively more large particles when total suspended particulate concentration was high. In addition, indoor samples exhibited a relatively higher percentage of small particles, except in some of the samples from private homes in which elevated concentrations of larger particles appeared to be related to indoor activities such as cleaning or simply moving about the house and stirring up dust.

Three possible trends in the relationship between indoor and outdoor particulate concentrations have been identified.

a. Ishido and his colleagues concluded, as a result of their studies in Japan, that, even in relatively air-tight buildings (*34*) and in schools and hospitals, as well as in small rooms (*16*), indoor suspended particulate levels are completely under the influence of outdoor changes, and that the generation of dust by daily activities may have some effect, but that it is of relatively short duration and is not directly reflected in daily variations in indoor dust concentrations (*17, 19*). Although change in indoor concentration lags behind outdoor change and the range of concentrations is smaller indoors, indoor levels are nearly equal to outdoor levels if mean values over 24-hour periods are considered (*34*). These conclusions are supported by statistical analyses of the results of two studies (*24, 26*), which indicated that the differences in indoor and outdoor concentrations were not significant at the 5% level. A study in Cincinnati, Ohio, indicated that "under normal atmospheric conditions, the main component of suspended matter in the home was drawn from outside air, while during 'smog' periods the correspondence of the two measurements was even closer (*28*)."

b. A study in a London, England, office (*21, 22*) lends some support to the relationship indicated by Figure 3a, but not at the same concentrations or percentages. Indoor and outdoor concentrations were found to be about equal up to concentrations of 300 $\mu g/m^3$. It has also been noted that indoor and outdoor levels showed fair agreement when windows were kept open, but that indoor levels were sometimes less than half of outdoor levels when windows were closed, particularly at night (*30*).

c. Romagnoli concluded that indoor dust content does not seem to reflect the outside dust levels (*31*). Kanitz attached equal importance to outside concentrations and to the presence and activities of people inside (*35*). In the crowded classrooms where Romagnoli (*31*) obtained his data, the presence and activities of people inside may be of greater importance than outdoor concentrations.

The data in Figure 3b on indoor–outdoor relationships for number concentration of particulates are based primarily on several studies carried out in Japan (*16, 19*). These data indicate that particle counts indoors are only slightly less inside than outdoors, and this relationship appears to be independent of outdoor concentrations. The limited data from other sources do not confirm these Japanese data. One can only surmize that the one or more of the many variables that influence indoor–outdoor concentrations produced this apparent anomaly.

In the Hartford, Connecticut, study (*4, 5*) some data on relative sampling position, day and night sampling periods, different building types, and summer and winter seasonal trends were obtained for both total suspended particulate matter and soiling index (Figs. 4–6). Note in Figure 4 the relatively lower outdoor concentrations of total suspended particulate during the summer as compared with winter. Also, note the steeper gradient from indoors to outdoors during winter as opposed to summer. This is a function of both higher outdoor concentrations and the tighter closure of buildings (except for the air conditioned building) during the winter.

The data in Figure 5 show a similar profile for soiling particulates but not nearly as pronounced as for particulate weight concentrations. In general, the indoor–outdoor ratios for soiling particulates were higher than for total suspended particulate matter. This finding is logical, since soil-

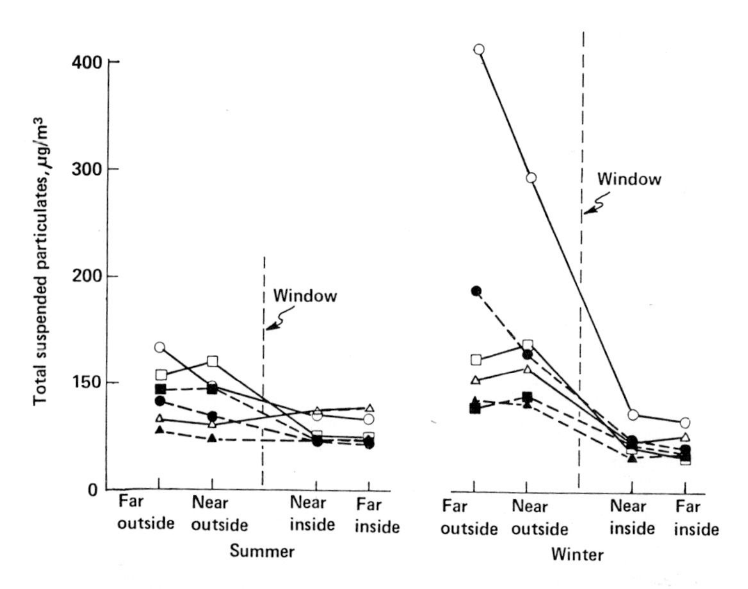

Figure 4. Indoor–outdoor profiles for total suspended particulate matter in several types of buildings—summer and winter. Non-air conditioned city library; ○ = day, ● = night; air conditioned office building: □ = day, ■ = night; non-air conditioned private home: △ = day; ▲ = night (*4*).

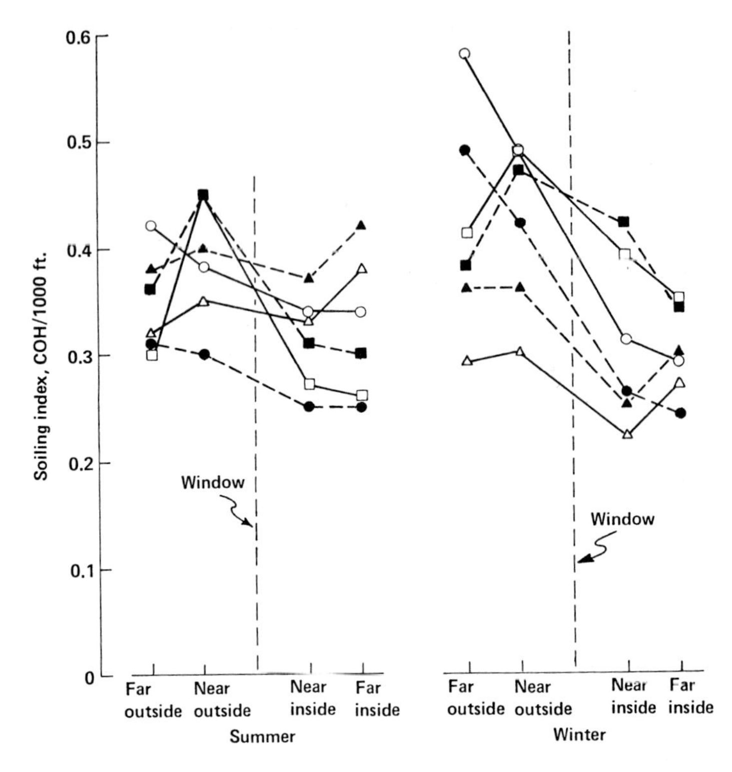

Figure 5. Indoor–outdoor profiles for soiling particulate matter in several types of buildings—summer and winter. Non-air conditioned city library: ○ = day, ● = night; air conditioned office building: □ = day, ■ = night; non-air conditioned private home: △ = day; ▲ = night (*4*).

ing effects of particulate matter are created largely by small particles, which penetrate more readily than large particles.

There are indications that the composition of indoor particles differ from that of outdoor particles. The Hartford study showed larger particles outdoors than those indoors and a difference in penetration of soiling and total particulates (*4, 5*). In one study (*28*), median particle diameter inside was found to be 0.36 μm, compared with 0.46 μm outside. In an air conditioned office building, 99% of the particles were smaller than 0.7 μm, while 89% of outdoor particulate mass were smaller than 0.7 μm (*36*). In another study, 85% of indoor particles were found to be 1 μm or smaller, while only 74% of those outside were 1 μm or smaller (*26*).

In still another study (*24*), the ash content of indoor samples ranged from 1.5 to 38.0%, with a mean of 13.3%. Ash in outdoor samples ranged

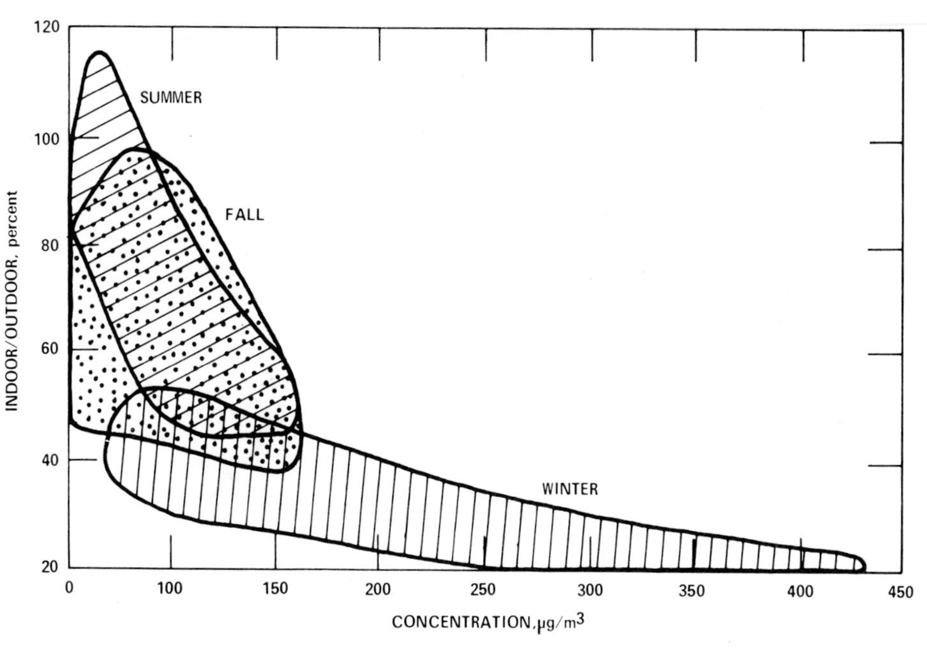

Figure 6. Seasonal variation of suspended particulate matter concentrations and indoor/outdoor ratios in Hartford, Connecticut (*2*).

from 2.1 to 80%, with a mean of 29.3%. This difference, which was highly significant at the 1% level, indicates that indoor air contains more organic material than outdoor air. In the Hartford, Connecticut, study, indoor air particulates were found to contain significantly higher percentages of benzene-soluble organic material and marginally higher percentages of lead in relation to outdoor particulates (*4, 5*).

C. Viable Particulate Matter

The atmosphere both indoors and outdoors contains viable particulate matter, e.g., bacteria, viruses, fungal and plant spores, and pollen. Pollens and fungal spores usually have their origin out of doors, but some fungal spores are produced on surfaces indoors, e.g., air conditioning components (air washers, humidifiers, and dehumidifiers). Although bacteria and viruses are present outdoors and can penetrate the indoor atmosphere, of principal concern is their transmission from person to person within the indoor atmosphere.

1. Spores

Many studies have been carried out on the relative indoor–outdoor concentrations of mold spores (Table III) (*37–52*). In spite of some results showing indoor concentrations to be extremely high, presumably because of poor hygienic conditions, the results predominantly show indoor–outdoor ratios less than 100%, with an average around 40%. There are also indications that spore populations are different indoors and outdoors. Spores of the *Penicillium genus* are by far the most common type in indoor air.

2. Pollen

As is the case with spores, the data on indoor levels of pollen are extensive (*37, 51, 53–59*). While much of the data have been collected in evaluation of air conditioners, Benson *et al* (*1*) have summarized the results for non-air conditioned buildings (Fig. 7). Four data points for which outdoor concentrations were greater than 100 pollen grains/m³

Table III Indoor–Outdoor Concentration Ratios for Spores of the Ten Most Commonly Occurring Fungi[a] *(37–52)*

Fungus	*Range of indoor–outdoor ratios (%)*	*Studies in which ratio reported was:*		*Total studies*
		<100%	*>100%*	
Penicillium	29 – 56	4	4	8
Cladosporium	0.3– 26	7	0	7
Aspergillus	24 –138[b]	4	2	6
Hormodendron	18 – 20	2	0	2
Mycelia sterilia[c]	24 – 30	2	0	2
Mucor	90 –300[d]	1	4	5
Pullularia	4 – 50	4	0	4
Yeasts	27	1	0	1
Alternaria	0 – 44	6	0	6
Phoma	3 – 75	4	0	4

[a] Data from Spain (*45*) excluded, since indoor–outdoor ratios were much higher than general data trend.

[b] Range does not include an instance in which *Aspergillus* was found indoors but not outdoors; ratio would approach infinity.

[c] The majority of these organisms are in the family *Deutromycetes*.

[d] Range does not include two instances in which *Mucor* was found indoors but not outdoors; ratio would approach infinity.

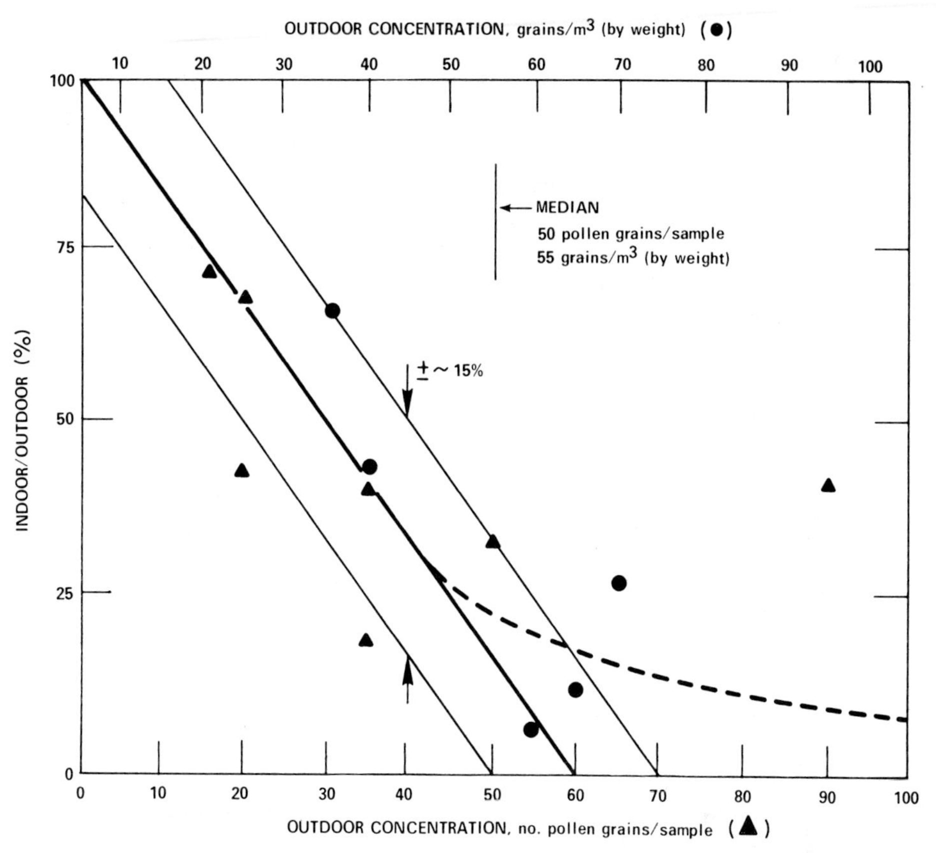

Figure 7. Indoor pollen concentrations as a function of outdoor concentrations—non-air conditioned buildings (*1*).

were excluded to allow plotting on a more convenient scale. To facilitate comparison between concentrations in pollen grains per cubic meter and number per sample, the medians are plotted coincident with each other. A pattern of decreasing indoor–outdoor ratios with increasing outdoor concentration is indicated by the data bands in the figure. Consideration of data above 50 grains/m^3 in the figure and above 100 pollen grains/m^3 shows that the relation is probably not linear above about 50 pollen grains/m^3, but is asymptotic, approaching a limit between 1 and 5% for outdoor concentrations above 100 pollen grains/m^3. Thus, it appears that indoor concentrations will vary from 85 to 100% of outdoor concentrations for low outdoor concentrations to 1–20% at high outdoor concentrations.

3. Bacteria

Data on indoor–outdoor concentration ratios for bacteria are available from a number of workers (*17, 19, 37, 60*) (Table IV). Indoor–outdoor ratios obtained for the house in Osaka, Japan, are exceptionally high compared to the other data, and these values have been excluded in the following analysis. The remaining indoor–outdoor ratios in the table range from 62 to 273%, and half of the ratios are greater than 100%. A great disparity is noted between data obtained in Japan and in the United States. However, perhaps because the Japanese data are for total bacteria while most of the United States data are for streptococci and total "microbes" (which includes spores, as well as bacteria), indoor–outdoor ratios based on the Japanese data range from 62 to 225%, with only 38% of the values greater than 100%, whereas the range for the United States data is 75–273%, with 67% of the values greater than 100%.

The disparities among measuring and reporting procedures preclude analysis of indoor–outdoor ratios as a function of varying outdoor concentrations. The consensus of the investigators, however, is that bacterial counts indoors do not reflect fluctuations in the outdoor air (*17, 19, 37*). Dust density and bacterial counts indoors reportedly show different tendencies, but the data are insufficient for proving them unrelated (*19*). The influence of living conditions and daily activities on changes in indoor bacterial count is considered relatively great (*17*).

IV. Effects of Indoor Generation of Pollutants and Other Factors on Indoor Air Quality

A. Indoor Generation of Pollutants

Some of the most important indoor activities that result in the indoor generation of pollutants are smoking, use of personal products, cleaning, cooking, heating, maintenance, hobbies, and electrical appliances.

1. Gaseous Pollutants

There are many gaseous pollutants that may be generated within a structure. Most are generated for only short periods at infrequent intervals. A few are generated for longer periods and more frequently. The gaseous pollutants generated indoors vary widely in their reactivity and capability of being adsorbed or absorbed.

a. CARBON MONOXIDE. Carbon monoxide is generated indoors by combustion (smoking, heating, cooking) (*5*). The effect of combustion can be

Table IV Indoor and Outdoor Concentrations of Bacteria

Type	*Location*	*Building type*	*Concentration measurement*	*Range*		*Mean*		*Indoor–outdoor (%)*	*Remarks*
				Indoor	*Outdoor*	*Indoor*	*Outdoor*		
Total bacteria	Osaka, Japan (*17*)	Apartment	Colonies/sample, 5-minute exposure	—	—	27	16	169	October–November (48-hour culture)
				—	—	40	43	93	May
				—	—	16	21	76	June
				—	—	18	8	225	October–November (24-hour culture)
		House		—	—	57	4	1425	October–November (24-hour culture)
				—	—	71	6	1183	October–November (48-hour culture)
	Toyonaka, Japan (*19*)	Apartment	Bacteria/sample, 5-minute exposure	5–126	7–147	35.0	43.0	82	Living room, May
				8–134	7–147	44.0	43.0	102	Bedroom, May
				2– 68	1–118	13.0	21.0	62	Living room, June
				4– 78	1–118	18.0	21.0	86	Bedroom, June
	Philadelphia, Pennsylvania (*37*)	Houses	Colonies/sample, 15-minute exposure	0– 45	0– 60	—	—	75[a]	With air conditioner
				0– 45	0– 60	—	—	75[a]	Without air conditioner
Streptococci	New York, New York (*60*)	Offices	Number/100 ft^3	—	—	22	11	200	—
		Schools		—	—	30	11	273	—
Microbes (bacteria and spores)	New York, New York (*60*)	Offices	Number/ft^3	—	—	87	52	167	Average for cultures at 20° and 37°C
		Schools		—	—	96	72	133	Cultures at 20°C

[a] Based on maximum values.

Table V Carbon Monoxide Concentrations near a Plant with an Open Hearth Furnace (7)

Distance from plant (m)	*Concentration (ppm)* Indoor	Outdoor	*Indoor–outdoor (%)*
50	11.6	17.8	65
100	16.3	16.5	99
250	9.0	14.9	60
500	7.6	12.8	59

seen in the data from Russia (Table V). Indoor concentrations in the natural-gas-equipped home 100 m from a steel plant were higher than those in a home without natural gas located closer to the plant (7).

According to Yocom *et al* (4), well-vented gas heating systems do not affect indoor carbon monoxide concentrations, but gas stoves and attached garages do. The effects of stoves and garages on indoor carbon monoxide concentrations can be seen in Figure 8, which shows the carbon monoxide concentrations in a house in Hartford, Connecticut, having a gas range and an attached garage. The family room is between the

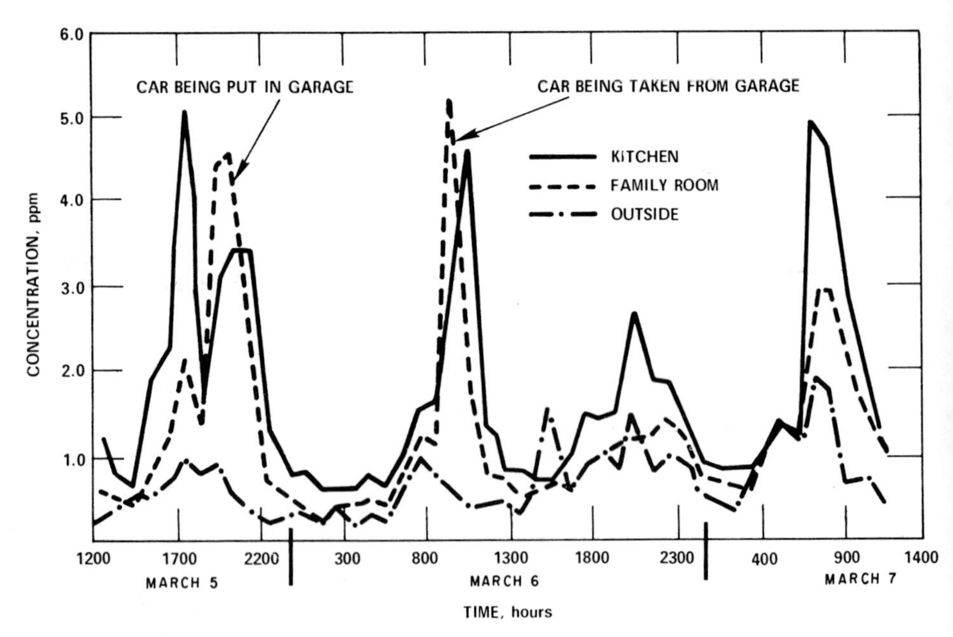

Figure 8. Carbon monoxide concentrations in house with gas range and furnace and with attached garage (5).

kitchen and garage. For this house, carbon monoxide concentrations are generally much higher than and unrelated to outdoor levels. Peak concentrations in the kitchen correspond to the periods when meals are being cooked, and concentrations in the family room generally follow those in the kitchen rather than those outside. For two periods in the record, when the car was being put into or taken out of the attached garage, the emissions from the garage are the controlling influence on both the family room and kitchen concentrations (*5*).

A laboratory study of gas stove and unvented space heater emissions showed the carbon monoxide emission quantities, depending upon the heat input to the appliance (*61*), listed in the following tabulation.

	Carbon monoxide emissions	
Appliance	*(μg/kcal)*	*(mg/hour)*
Stove		
Burners	300– 500	1000–3000
Oven	500–1500	1100–3500
Space heater	300– 600	1500–2000

b. Oxides of Nitrogen. In the laboratory study by Coté *et al* (*61*), oxides of nitrogen emissions were measured from gas stoves and unvented space heaters. Results are shown in the following tabulation.

	Pollutant emissions			
	Nitric oxide		*Nitrogen dioxide*	
Appliance	*(μg/m³)*	*(mg/hour)*	*(μg/m³)*	*(mg/hour)*
Stove				
Burners	100–150	250–1500	50–80	150–700
Oven	75–100	150– 200	50–75	100–300
Space heater	75–150	200– 800	40–50	150–300

As part of the same study, extensive indoor and outdoor air quality measurements were made using several residential structures. Figure 9 is a graph of 2-hour average nitrogen dioxide levels for a typical 2-day period at one of the structures used in the study (*61*). This graph shows the rapid response of nitrogen dioxide levels in the kitchen to stove use, followed quickly by response at the other two indoor locations. The oven,

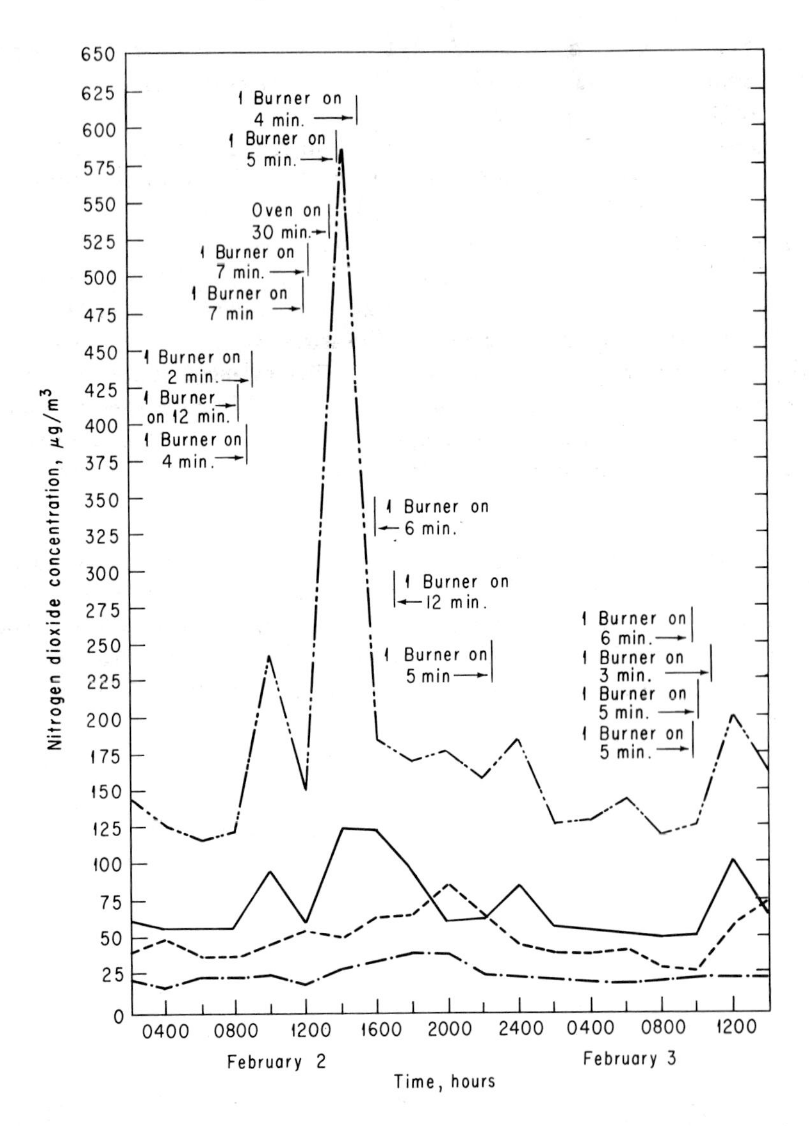

Figure 9. House No. 4. A time history of nitrogen dioxide concentrations, 2-hour averages (Winter 1974). (··—··) Kitchen, over stove; (———) kitchen, 1 m from stove; (– – –) living room; (— · —) outside (*61*).

in particular, seems to have the greatest influence on indoor nitrogen dioxide concentrations. Examination of the raw data showed that the generation of nitrogen dioxide by the oven is greatest during initial oven start-up. Average nitrogen dioxide concentrations in the kitchen during inactive periods were slightly greater than nitrogen dioxide levels at the other two indoor locations, which closely approximated the outdoor nitrogen dioxide concentrations. The higher kitchen concentrations are undoubtedly caused by the stove's pilot lights.

Figure 10 shows the diurnal pattern of nitrogen dioxide data for another of the houses monitored. These curves were based on a "composite day." Note that the levels of nitrogen dioxide in the rooms away from the kitchen are considerably below the kitchen curve *(61)*.

Table VI summarizes the indoor–outdoor air quality and stove use data for one house over a 2-week monitoring period *(61)*. The higher concentrations of pollutant gases in the kitchen as compared with the other rooms show that the stove is a significant source of the pollutants measured. In addition, there is some indication that stove use correlates with kitchen concentrations, although outdoor concentrations and season, with its effect on the permeability of the house, complicate this relationship.

c. SULFUR DIOXIDE. Interior generation of sulfur dioxide is probably limited to faulty heating systems burning oil or coal *(5)*. Bicrsteker *et al.* reported that indoor sulfur dioxide concentrations were not generally affected to a significant extent by the heating method used. However, in one 30-year-old home presumed to have a faulty heater, indoor concen-

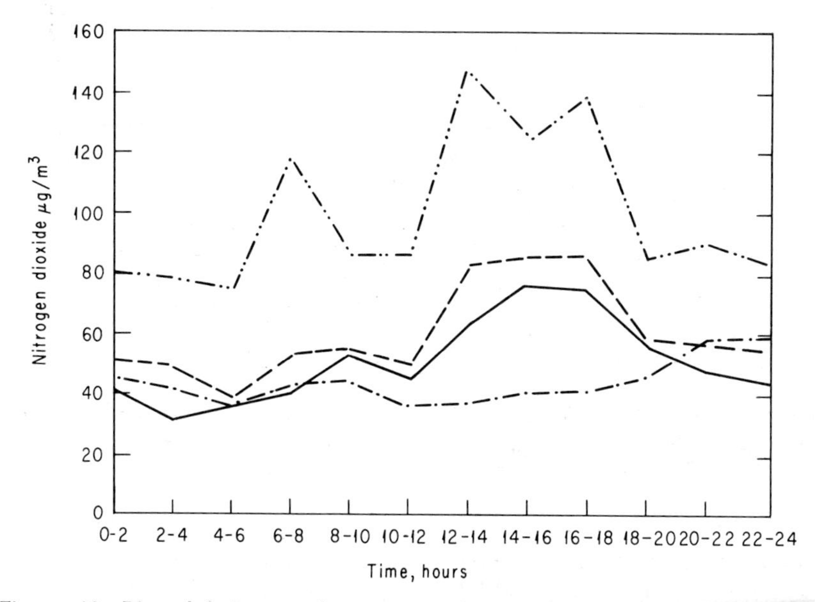

Figure 10. Diurnal indoor–outdoor pattern for nitrogen dioxide—House No. 1, Spring–Summer 1973, composite day based on 6 days of data. (·· — ··) Kitchen, 1 m from stove; (– – –) living room; (—— —) bedroom; (— · —) outside *(61)*.

Table VI House No. 4, Summary of Nitrogen Dioxide and Nitric Oxide Concentrations–Winter (January–February) 1974 (*61*)

Data category	*Kitchen—above stove*	*Kitchen—1 m from stove*	*Living room*	*Out-doors*	*Average stove use (minutes/day)*		
					Oven	*Burners*	*Total*
Nitrogen dioxide concentrations ($\mu g/m^3$)[a]	213	120	71	3.9	66	73	139
Nitrogen dioxide indoor–outdoor ratios	5.46	3.08	1.82	—			
Nitric oxide concentrations ($\mu g/m^3$)[a]	305	229	156	19	66	73	139
Nitric oxide indoor–outdoor ratios	16.1	12.1	8.2	—			

[a] Average concentrations are based on daily averages for those days in which 12 valid 2-hour averages were obtained.

trations averaged 3.8 times the outdoor levels (*11*). Table VII shows a comparison of sulfur dioxide concentrations for new and old coal-heated houses in Hartford, Connecticut (*5*). The exceptionally high indoor concentrations for the old coal-heated house are presumed to be caused by a faulty heating system. Indoor concentrations at this house were found to be unrelated to outdoor concentrations; peak values were related instead to the stoking periods of the furnace. Indoor concentrations at the new coal-heated house were much lower than at the old house, even though outdoor concentrations were slightly higher at the new house (*5*).

Indoor sulfur dioxide concentrations are reduced by adsorption or absorption. According to Chamberlain, walls and ceilings should provide a perfect sink for sulfur dioxide. Thus, the rate of adsorption should be

Table VII Sulfur Dioxide Concentrations for Two Coal-Heated Houses (*5*)

Type of building	*Concentration (ppm)*		*Indoor–outdoor (%)*
	Indoor	*Outdoor*	
New house	5	14	36
Old house	78	10	780

controlled by the rate of diffusion across the boundary layer to the surface, and vigorous circulation, which would decrease boundary-layer resistance, should cause increased reductions in sulfur dioxide concentration (*62*). Wilson (*63*) found that removal of sulfur dioxide from indoor air was limited by the properties of interior surfaces and only slightly by transport to the surfaces. The ceiling (fiberboard with eggshell paint) was found to be effective in removing sulfur dioxide. The floor (lacquered cork), walls (painted with emulsion paint), and treated wood surfaces were not. "Stirring" the air was found to reduce concentrations by 10–40%, with the most reduction effected at higher concentrations (*63*).

d. Aerosol Propellants. No data from field or laboratory measurement programs were found in the literature, but one study (*61*) cited emission estimates and projected indoor air quality related to specific aerosol product usage. Table VIII, based on a user survey, shows emission estimates for propellants (primarily fluorinated hydrocarbons) associated with use of various household products (*61*). These emission factors, together with use patterns established by the survey, provide a basis for estimating indoor propellant concentrations in rooms of various sizes and at various times after product use.

2. Nonviable Particulate Pollutants

Particulates are generated by combustion (heating, cooking, smoking) (*5*). Smoking has been found to significantly increase particulate concentrations indoors (*11, 64*). According to Lefcoe and Inculet, smoking just one cigar raises particle counts by a factor of 10–100. Elevated counts persisted for a period of 1–3 hours (*64*). Yocom *et al* (*4*) noted that the higher concentration of organic particles indoors may result in part from interior generation of pollutants from cooking or smoking, al-

Table VIII Emission Estimates of Propellants for Aerosol Products (*61*)

Product category	*Propellant emission estimates (gm/use)*
Deodorant sprays	1.0–1.2
Hair sprays	4.9–6.5
Shaving foam	0.3–0.4
Air freshener spray	5.6–11.2
Disinfectant spray	7.5
Furniture polish spray	8.4
Dust spray	4.2–8.4
Oven cleaners	20–25

Table IX Indoor Particulate Distribution by Height above Floor for Waking and Sleeping Periods (*19*)

Height above floor (cm)	*Concentration (particles/cm³)*	
	Waking hours	*Sleeping hours*
40	676	664
100	629	640
150	636	587
210	669	538

though the greater penetration of smaller organic particles, as compared to the larger inorganic particles, is also partly responsible for the difference.

Particles that are present inside are resuspended and/or kept in suspension by the activities of the people inside. Seisaburo *et al* (*19*) reported particulate concentrations at different heights during waking and sleeping periods (Table IX). These measurements indicate that particles are distributed rather uniformly from floor to ceiling because of activities of people during the day, but that during the night they tend to settle and become concentrated near the floor. Table X (*31*) shows particle counts in Italian schools before, during, and after classes. Counts were much higher during classes than before in two of the four cases, presumably because of the presence and activities of the students.* There was also an increase in particle size from a mean of 0.5 μm before class to 1.2 μm during class. The measurements made after class indicate that con-

Table X Particle Counts before, during, and after Classes in Schools (*31*)

Location of school	*Particle counts (particles/cm³)*		
	Before	*During*	*After*
Urban	348	347	410
Suburban			
Residential	100	360	315
Industrial	180	421	420
Rural	449	392	490

* Measurements were made at set times and do not necessarily indicate peak concentrations. Thus, higher values measured after class in two instances do not indicate a continuing increase in concentration.

centrations do not drop rapidly after activities have ceased (*31*). This conclusion is supported by the data of Lefcoe and Inculet (*64*), which indicate that high particle counts resulting from cleaning and dusting persist for a period of at least several hours.

A number of investigators have concluded that indoor generation and entrainment of particles have a significant effect on indoor–outdoor relationships. In an office with an air filtration system that reduced interior concentrations to 24% of outside levels, the amounts of dust generated in a room was found to be proportional to the number of people in the room (*36*). An equation was developed relating the number of people in the room to the amount of dust they generated, expressed as the room's ventilation rate times its dust content, measured by the optical density of the standarddiameter filter-paper disk, through which a standard amount of room air is passed.

Based on limited measurements for air conditioned office buildings in Hartford, Connecticut, it was concluded that internal generation of suspended and soiling particulates was a significant factor in the estimation of interior concentrations. For these buildings, the ratio of internal generation to exterior concentration was estimated to range from 0 to 0.6 for indoor–outdoor ratios of 30–116% (*3*). Internal generation may also contribute to the varying indoor–outdoor ratios and to the indoor–outdoor ratios greater than 100% (*3, 4*).

3. Viable Particulate Pollutants

As pointed out previously, indoor bacterial concentrations appear to be more closely related to indoor living conditions and activities than to outdoor concentrations (*17–19*). Pollen, in contrast, is almost completely dependent on outdoor concentrations, as would be expected. The importance of internal generation of spores is not clearly established. Maunsell found, however, that activities such as cleaning and dusting cause spores to be entrained in the air. The resulting increase in entrained spores was mainly in *Penicillium, Cladosporium, Pullularia,* and yeasts. Spores of larger sizes, which were absent in undisturbed air, were found to be present after dust was raised (*42*).

B. Influence of Time on Indoor Air Quality

Indoor air quality varies with time as those factors listed in Section II vary with time. Figures 1, 4, 5, and 6 have already shown how indoor air quality can vary by season. This section discusses shorter-term variability.

1. Gases

In Figure 1, the relationship between day and night indoor concentrations of carbon monoxide was shown for several buildings in the Hartford, Connecticut, study (*4*). Table XI shows day/night ratios for carbon monoxide for each of the six buildings studied. These day/night ratios show that concentrations both indoors and outdoors are higher during the day except at the Blinn Street home for all three seasons and at the Carroll Road home during summer and winter. The lower day/night ratios in and around the homes were ascribed to their location relatively near a busy freeway with significant nighttime traffic whose carbon monoxide contributions were amplified by nighttime inversions as compared to the public and office buildings in downtown Hartford with little nighttime traffic. With few exceptions, day/night ratios were lower indoors than out; that is, there is generally less difference between day and night concentrations indoors than outdoors. A seasonal effect on diurnal patterns can also be inferred from the data in Table XI. In most cases, there is less difference between day and night concentrations in the summer than in the winter.

As was shown in Figures 9 and 10, short-term fluctuations and the diurnal patterns of nitrogen dioxide concentrations responded to gas stove use. This work also presented these time-variable relationships for nitric oxide and carbon monoxide. As part of this same study, a diffusion experiment was carried out in a home unoccupied at the time. All stove burners were operated until one or more of the sensing instruments read maximum values, and then the stove was turned off and nitrogen dioxide, nitric oxide, and carbon monoxide were allowed to diffuse through the

Table XI Day/Night Ratios of Carbon Monoxide Concentrations—Hartford, Connecticut (*4*)

	Day/night ratio					
	Summer		*Fall*		*Winter*	
Building	*Indoor*	*Outdoor*	*Indoor*	*Outdoor*	*Indoor*	*Outdoor*
Library	1.49	1.72	1.05	1.30	1.72	2.09
City Hall	1.52	1.72	1.14	1.30	1.74	2.02
Office, 100 CP	1.18	0.91	1.44	1.36	1.18	1.26
Office, 250 CP	1.07	1.05	1.25	1.38	0.97	1.76
House, Blinn St.	0.76	0.80	0.91	0.95	0.98	0.98
House, Carroll Rd.	0.78	0.76	1.09	1.23	0.93	0.92

rest of the house. Carbon monoxide, an unreactive gas, showed a half-life of 2.1 hours. This dilution rate is considered to be the natural ventilation air change rate for this house under the test conditions. Nitric oxide showed a half life of 1.8 hours, indicating little loss through processes other than simple dilution. On the other hand, nitrogen dioxide had a half life of only 0.6 hour, indicating a considerable amount of loss by processes such as adsorption, absorption, and atmospheric reaction (*61*). Figure 11, based on this experiment, shows concentration-time relationships for both nitrogen dioxide and nitric oxide at three locations in the house in comparison to outside concentrations.

2. Particulates

Figure 12 shows the diurnal pattern obtained during the summer for a Japanese apartment (*19*). The pattern should be fairly typical for the Japanese studies, because similar patterns were found throughout the year, and indoor and outdoor patterns were generally found to be almost identical (*17–19*). The pattern may also be grossly applicable to the United States. It has been noted that daytime levels are higher than night levels (*65*), and major peak at around 8 a.m. has been identified (*20, 30*).

A slight lag time can be seen for the indoor concentrations in Figure 12, and it is reported that the lag time at night is even more apparent during the winter (*19*). The effect of the lag time in the example illustrated is relatively minor, but it does result in indoor levels higher than outdoor levels twice during the period covered—"at about 1800 hours and from 2300 to 0100 hours." Lag times, sometimes amounting to an hour or more, have been reported in other instances, and indoor curves may show fewer sharp peaks than outdoor curves (*16, 30*).

Measurements of daytime and nighttime particulate concentrations similar to those presented above for carbon monoxide, were also taken as part of the Hartford study, and the resulting day/night ratios are listed in Table XII (*4*). Day/night ratios are greater than 1 except for two values that are nearly equal to 1. These values indicate that daytime concentrations of particulates are higher than nighttime levels by as much as 100%. For the offices and public buildings, indoor day/night ratios are lower than outdoor day/night ratios in the summer and winter, indicating that there is less difference between day and night concentrations inside than out. For the houses, there is a greater difference between daytime and nighttime concentration inside than outside, except, perhaps, during the fall. The ratios generally increase from summer to winter, indicating that there is more variation in concentrations, both inside

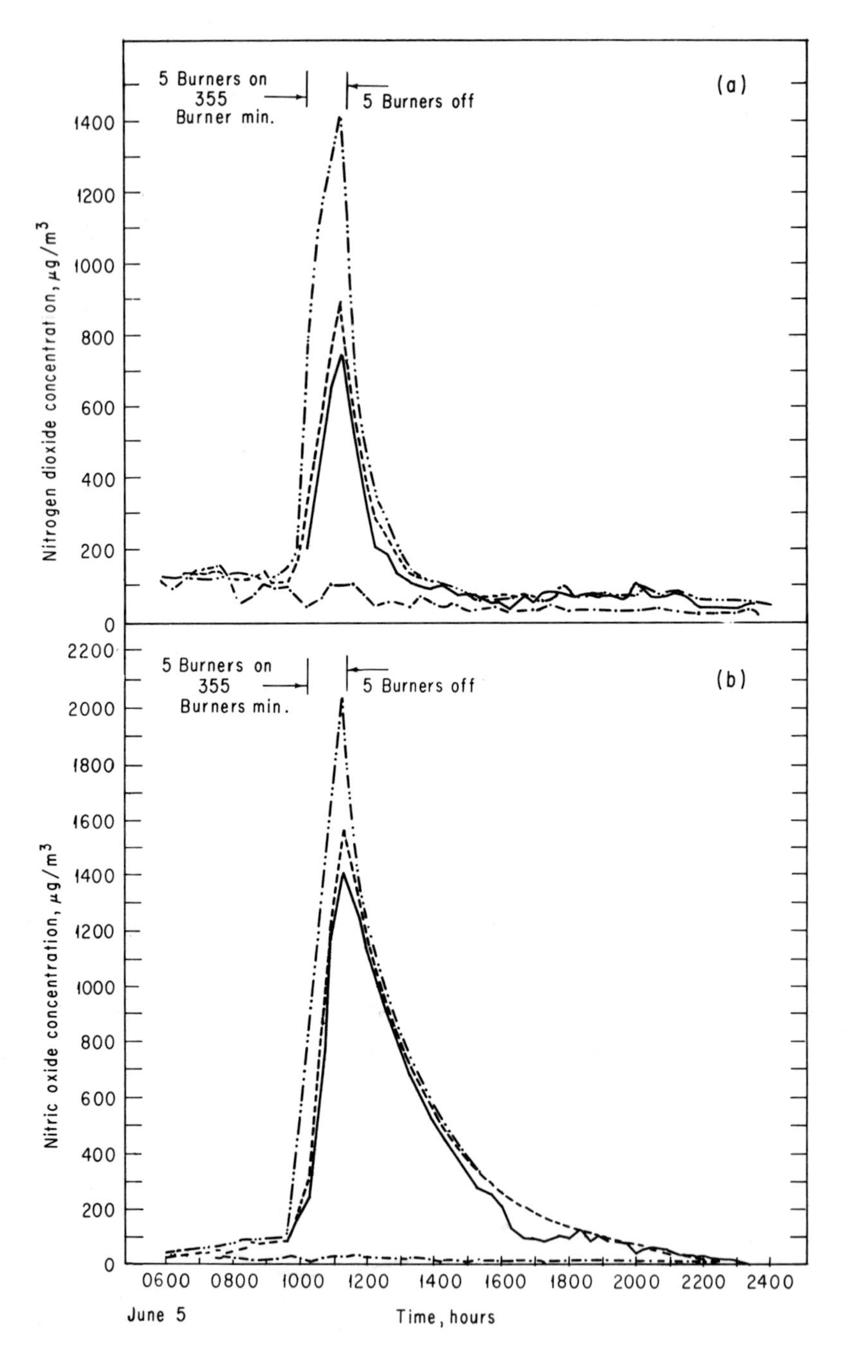

Figure 11. Dilution and loss with time for nitrogen dioxide and nitric oxide diffusing through a house; (a) nitrogen dioxide concentrations, (b) nitric oxide concentrations. (·· — ··) Kitchen; (– – –) living room; (———) bedroom, upper level; (— · —) outside (*61*).

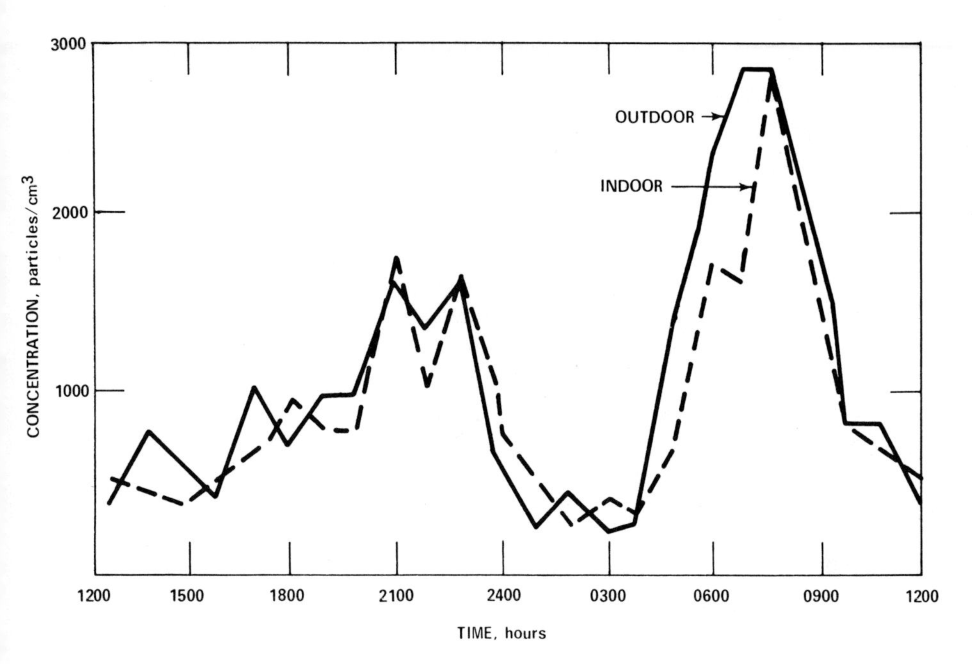

Figure 12. Concentration of particles in an apartment in Toyonaka City, Japan May 21–22, 1956 (*19*).

and outside, in the winter than in the fall and more variation in the fall than in the summer.

Day and night concentrations of soiling particulate matter were found to be about the same in the Hartford study in contrast to total suspended particulate matter, which generally showed higher levels during the day. This difference in behavior between total suspended and soiling

Table XII Day/Night Ratios of Suspended Particulate Matter Concentrations—Hartford, Connecticut (*4*)

	Day/night ratio					
	Summer		*Fall*		*Winter*	
Building	*Indoor*	*Outdoor*	*Indoor*	*Outdoor*	*Indoor*	*Outdoor*
Library	1.53	1.61	1.30	1.50	1.49	2.24
City Hall	1.59	1.96	1.64	1.41	1.70	1.94
Office, 100 CP	1.09	1.12	1.33	1.26	0.98	1.53
Office, 250 CP	0.94	1.14	1.65	1.43	1.87	1.89
House, Blinn St.	1.25	1.21	1.20	1.29	1.40	1.33
House, Carroll Rd.	1.62	1.13	2.00	1.28	1.60	1.21

particulates is probably the result of size difference; the smaller soiling particulates tended to stay suspended at night, whereas the larger particles contributing to the weight concentration of total suspended particulate matter tended to settle out (*4*).

C. Type of Building

One would logically expect the type of building, its permeability, indoor uses, and other characteristics to influence indoor air quality. However, little comparable data are available to evaluate the effects of building type (*4, 66*). Figures 1, 4 and 5 show indoor air quality patterns in relation to outdoor concentrations for several types of buildings in Hartford, Connecticut, for carbon monoxide, suspended particulate matter, and soiling particulate matter, respectively (*4*).

1. Gases

Carbon monoxide concentrations were measured in pairs of houses, office buildings, and public buildings in Hartford, Connecticut (*4*). As discussed in the next section, abnormally high indoor–outdoor ratios were measured at an office at 100 Constitutional Plaza (CP) because of the way in which the air conditioner was operated. Discounting these values, average indoor–outdoor ratios for the houses were about 105%; for the remaining office, about 95%; and for the public buildings, about 90% (Table XIII). However, outdoor concentrations were generally lower in the vicinity of the homes than at the office and the public buildings. Thus, it is difficult to determine if the differences in indoor–outdoor ratios are related to building type or to differences in outdoor pollution level.

In the study of Coté *et al.* (*61*), indoor concentrations of nitrogen

Table XIII Average Carbon Monoxide Concentrations for Several Types of Buildings—Hartford, Connecticut (*4*)

Building	*Mean concentration (ppm)*		*Indoor–outdoor (%)*
	Indoor	*Outdoor*	
Library	3.84	4.35	88
City Hall	3.78	4.21	90
Office, 100 CP	3.21	2.69	119
Office, 250 CP	3.18	3.33	96
House, Blinn St.	2.84	2.68	106
House, Carroll Rd.	2.56	2.44	105

dioxide, nitric oxide, and carbon monoxide were measured in four different private homes with gas stoves, included single-story, split-level, and two-story houses and a two-story apartment. In general, it was found that the smaller the home and the tighter the construction, or more closed up the mode of the house (e.g., winter), the higher the indoor concentrations of the pollutant gases.

2. Particulates

As shown in Table XIV, average indoor–outdoor particulate ratios for the houses in the Hartford, Connecticut, study were around 65%; for the office, around 45%; and for the public buildings, around 35% (*4*). However, outdoor pollution levels were lower in the vicinity of the houses.

Figure 13 is a plot of the data for the range of common outdoor concentrations. A limited amount of data from Whitby *et al.* (*32*), which can be plotted in the same form, is included for comparison. Outdoor concentrations for the homes fall between about 50 and 125 $\mu g/m^3$. Data for the offices and public buildings for this level of pollution generally fall within the data scatter band for the houses but are concentrated in the lower portion of the band.

D. Weather and Natural Ventilation

The importance of atmospheric conditions and natural ventilation was recognized in a study in Cincinnati, Ohio, that revealed large differences in domestic concentrations over short distances in the city, depending on window ventilation, the proximity of buildings to pollution sources, wind direction, and temperature inversions (*29*). The way in which these factors can interact to influence indoor–outdoor pollution relation-

Table XIV Average Particulate Concentrations for Several Types of Buildings—Hartford, Connecticut (*4*)

Building	*Mean concentration* ($\mu g/m^3$) *Indoor*	*Outdoor*	*Indoor–outdoor* (%)
Library	45	189	26
City Hall	66	159	42
Office, 100 CP	39	81	48
Office, 250 CP	45	104	43
House, Blinn St.	52	86	60
House, Carroll Rd.	54	75	72

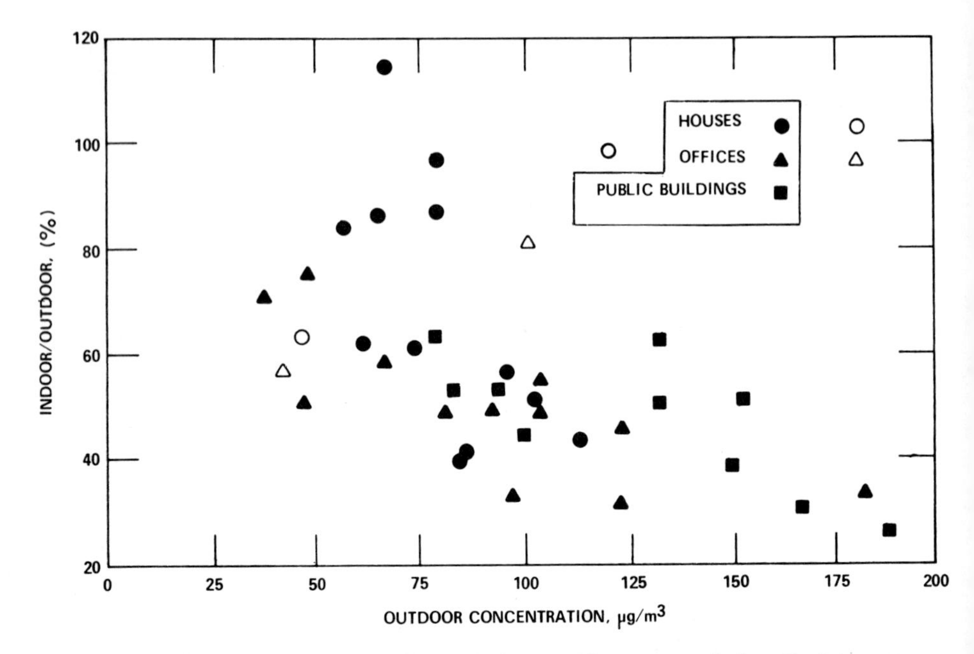

Figure 13. Effect of type of building on indoor–outdoor suspended particulate matter concentrations; solid symbols indicate data from Yocom *et al.* (*4*), open symbols indicate data from Whitby *et al.* (*32*).

ships can be seen in the following example from the study in Hartford, Connecticut (*4, 5*).

Simultaneous carbon monoxide samples were taken inside the dining room and on the outside of a house in Hartford. Sampling for 1 day is shown in Figure 14. On the evening illustrated, outside concentrations increased rather rapidly to about 12 ppm because of a light wind from a nearby interstate highway. Indoor concentration remained around 5 ppm, about equal to the outdoor concentration before it increased. Because the windows and doors of the house were closed and there was relatively little influx of air, interior concentrations reacted much more slowly to the

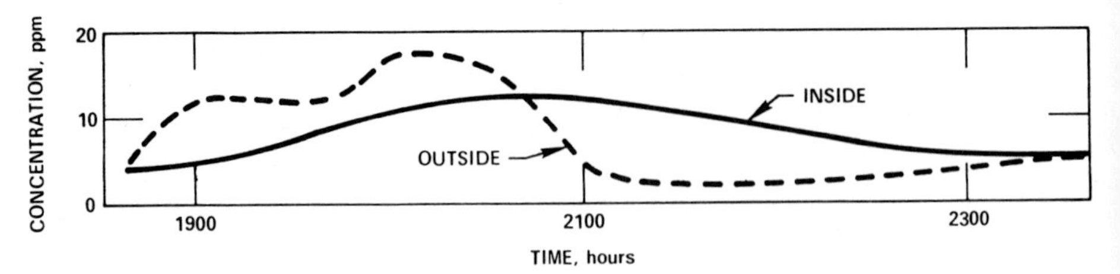

Figure 14. Carbon monoxide concentrations for a house in East Hartford, Connecticut, September 22, 1969 (*4, 5*).

change in wind direction than did outdoor concentrations. After 2 hours, outdoor concentrations had increased to about 16 ppm, but indoor concentrations were still significantly lower at 10 ppm. At this time, the wind direction changed, causing outside concentrations to drop rapidly to about 3 ppm. Inside concentrations remained high, however, and required 2.5 hours to return to the low outside ambient level. Thus, over a 5-hour period, indoor concentrations ranged from much lower to much higher than outdoor levels (indoor–outdoor ratios ranged from about 40 to more than 300%) (*4, 5*). If either windows or doors had been even partially open or if a stronger wind had been blowing directly at the windows, resulting in greater natural ventilation, interior concentrations would probably have more nearly reflected those outdoors (*14*).

As mentioned, several studies conducted in Japan have led to the conclusion that indoor particulate concentrations are not affected by natural ventilation but are controlled entirely by outside concentrations (*15–19, 34*). Other investigators, however, have noted a significant effect from natural ventilation. Studies in Cincinnati, Ohio, indicated that indoor and outdoor levels were in fair agreement when windows were open but that indoor concentrations were sometimes less than half of outdoor concentrations when windows were closed. Average indoor concentrations were found to be roughly 15% higher with windows open than with windows closed (*29*). Results of the Hartford, Connecticut, study mentioned above also support the conclusion that natural ventilation affects indoor particulate levels. A seasonal decrease was noted in the indoor–outdoor percentage from summer to winter, and it was hypothesized to be the result of shutting up buildings for the winter (*4*). It was also noted that particulate levels were lower in public buildings than in homes, which can be explained by lower air infiltration per unit building volume for public buildings (*5*).

Indoor pollen concentrations were found to be closely associated with both wind speed and window opening (*56*). When windows were closed, indoor–outdoor ratios remained relatively constant at approximately 20% for wind speeds up to 8 miles/hour (13 km/hour). For higher wind speeds, there was a nearly linear increase in indoor–outdoor ratios up to 97% at 15 miles/hour. When windows were open, penetration of pollen was quite different, but the amount of opening apparently made little difference.

E. Air Conditioning and Filtration

Air conditioning engineers are confident that air conditioning systems can be designed, built, and operated to remove air pollutants so that indoor air in buildings and vehicles will be continuously comfortable and free from the stress effects of air pollution (*67*). It has been alleged, however, that "in contrast to what most people comfortably assume, much of the pollution of the outdoor air *enters* our buildings directly through the air conditioning equipment as supplied and installed today," and the data with which to refute this charge have yet to be gathered (*12*). It has been noted instead that the current employment of air conditioning is largely dictated by the economics of heating and cooling with little regard for changes in indoor air quality and how it is affected by outside pollutant levels, air-conditioning system parameters, and internal pollutant generation (*68*).

1. Gases

A study in Boston, Massachusetts, indicated that sulfur dioxide concentrations were reduced to 60% of outside levels simply by bringing the air inside, and that further reductions were not affected by air conditioning systems unless the systems included water sprays on the cooling coils. This study also revealed that ozone concentrations indoors were not generally affected by air conditioning. Air conditioning systems with electrostatic precipitators actually caused a slight increase in ozone concentrations, but never enough to be of concern (*12*).

Carbon monoxide, being unreactive, is not removed by air conditioning systems. The indoor–outdoor profiles for carbon monoxide in Figure 1 from the Hartford, Connecticut, study (*4*) show that concentrations inside the air conditioned building are essentially the same as outdoors. In fact, indoor concentrations during the day are consistently higher than those outdoors, probably as a result of smoking or the addition to the indoor air of polluted outdoor make-up air taken in during the morning rush hours. In an experiment (*4*) "stale" air trapped in a building during the overnight shutdown was purged each morning with "fresh" air drawn from the outside. This "fresh" air, however, was drawn from near street level during the time of the morning peak traffic period. The dilution of this initial charge of carbon monoxide provided by the 10% make-up air used during the remainder of the day was not sufficient to reduce the indoor concentration to the outdoor level.

2. Particulates

Air conditioning systems can significantly reduce indoor particulate concentrations when efficient filters are employed. An air conditioning system that maintained a positive interior pressure was found to reduce indoor concentrations to 24% of outdoor levels (*36*). This system employed two filters with high dust-removal efficiency. Electrostatic particulate collectors have also been noted to be highly effective in eliminating indoor suspended particulate matter (*34*). In a Boston, Massachusetts, study, significant reductions were noted for a building with a central air conditioning system having, in succession, an electrostatic precipitator, a roll screen backing filter, water spray, and cooling coils (*12*).

Yocom *et al.* (*4*) concluded that the roughing filters normally used in air conditioning systems are moderately effective in removing particulates. This conclusion is based on data from two air conditioned offices sampled in the Hartford, Connecticut, study showing that indoor–outdoor percentages averaged less than 50%. However, when the data are examined for the range of common outdoor concentrations as in Figure 13, indoor–outdoor ratios are not found to be reduced when compared with the non-air conditioned public buildings, nor even consistently reduced when compared with the houses. Thus, it is not clear whether the apparent reduction at the offices was a result of the air conditioning system or was, in fact, a result of higher outdoor pollution levels. Significant reductions in indoor particulate levels for the Boston, Massachusetts, study were found only for the air conditioning system described above. In five other air conditioned buildings, indoor–outdoor relationships were about what one would expect for non-air conditioned buildings (*12, 13*).

3. Viable Particles

Pollen appears to be the only pollutant that is unequivocally reduced by air conditioning. Comparative concentrations for air conditioned and non-air conditioned buildings are presented in Table XV. These data indicate that air conditioning significantly reduces indoor pollen concentrations. Indoor–outdoor ratios for air conditioned rooms or buildings range from 0.2 to 2%, whereas those for non-air conditioned rooms or buildings in comparison tests range from 6 to 68%. Some types of air filtration and purification devices were found to be effective in reducing pollen (*55, 57, 59*). Spiegelman and Friedam (*53, 54*) found little difference in pollen concentration between air conditioned rooms with or without filters in place. The authors suggest that the coils and other parts of the air conditioning mechanism act as a baffle to obstruct the course of the pollen.

Concentrations of bacteria and spores may also be lower in air conditioned buildings, but the data are highly limited and inconclusive. In one study, mold and bacteria in an air conditioned room were found to be

Table XV Effect of Air Conditioners, Filters, and Purifiers on Indoor Pollen Concentrations and on Indoor–Outdoor Ratios

Location and building type	*Concentration measurement*	*Range*		*Mean*		*Indoor–outdoor (%)*	*Remarks*		*Reference*
		Indoor	*Outdoor*	*Indoor*	*Outdoor*				
Pittsburgh, Pennsylvania Hospital	Pollen count	—	—	144	1539	9.4	Without air filter		(57)
		—	—	0	1539	0	With air filter		
Richmond, Virginia Hospital	Grains/day	7 –407	71 –1188	—	—	23	Non-air conditioned room		(58)
		0 – 2	71 –1188	—	—	0.2	Air conditioned room		
Chicago, Illinois Hospital	Grains/cm²	0 – 23	10 – 350	6	133	4.5	With air filter		(59)
Philadelphia, Pennsylvania Houses	Grains/m³	0 – 74	0 –1100	—	—	6	Without air conditioner		(37)
		0 – 28	0 –1100	—	—	2	With air conditioner		
Houses	Grains/m³	—	2 –1100	11	205	5	Non-air conditioned house		(53)
		—	2 –1100	11	205	5	Air conditioner and air filter off	Test house	
		—	2 –1100	2	205	1	Air filter off	Test house	
		—	2 –1100	2	205	1	Air conditioner and air filter on	Test house	
Hospital	Grains/m³	12.7– 98.2	31.9– 110	42.2	61.8	68	Windows open	No filter in air conditioner	(54)
		12.4–141	31.9– 110	53.5	61.8	86	Windows open with air purifier	No filter in air conditioner	
		0.4– 2.8	31.9– 110	1.6	61.8	2	Air conditioned	No filter in air conditioner	
		0.4– 2.8	31.9– 110	1.0	61.8	2	Air conditioned with air purifier	No filter in air conditioner	
		11.9– 68.3	60.0– 272	33.8	119	28	Windows open	Standard filter in air conditioner	
		26.8– 82.0	60.0– 272	57.9	119	49	Windows open with air purifier	Standard filter in air conditioner	
		0.7– 2.6	60.0– 272	1.6	119	1	Air conditioned	Standard filter in air conditioner	
		0.7– 6.6	60.0– 272	2.6	119	2	Air conditioned with air purifier	Standard filter in air conditioner	
Chicago, Illinois Hospital	Grains/m³	6.6–392	14.4– 914	92.8	262	36	Without air filter		(55)
		1.3– 23.6	14.4– 919	7.9	262	3	With air filter		

only 9% of those in a non-air conditioned room with windows open (*54*). But in another study by the same authors, mold counts ranged from 0 to 20 colonies/dish in a non-air conditioned house and from 0 to 25 colonies/dish in an air conditioned house, while bacteria counts in both houses ranged from 0 to 45 colonies/dish (*37*).

REFERENCES

1. F. B. Benson, J. J. Henderson, and D. E. Caldwell, "Indoor-Outdoor Air Pollution Relationships," Vols. I and II, Publ. Nos. AP-112 and AP-112b. United States Environmental Protection Agency, Research Triangle Park, North Carolina, 1972.
2. F. H. Shair and K. L. Heitner, *Environ. Sci. Technol.* **8**, No. 5, 444–451 (1974).
3. J. K. Holcombe and P. W. Kalika, "The Effect of Air Conditioning System Components on Pollution in Intake Air, "Res. Proj. TRP-93. American Society of Heating, Refrigerating and Air Conditioning Engineers, New York, New York, 1970.
4. J. E. Yocom, W. L. Clink, and W. A. Coté, *J. Air Pollut. Contr. Ass.* **21**, 251–259 (1971).
5. J. E. Yocom, W. A. Coté, and W. L. Clink, "Summary Report of A Study of Indoor Outdoor Air Pollution Relationships to the National Air Pollution Control Administration, "Contract No. CPA-22-69-14, Vols. I and II. TRC—The Research Corporation of New England, Hartford, Connecticut, 1970.
6. "Indoor-Outdoor Carbon Monoxide Pollution Study," Final Report to the United States Environmental Protection Agency, Contract No. CPA-70-77. General Electric Co., Philadelphia, Pennsylvania, 1972.
7. N. N. Skvortsova, *Gig. Sanit.* **22**, 3–9 (1957).
8. J. J. Phair, R. J. Shepard, G. C. R. Carey, and M. L. Thompson, *Brit. J. Ind. Med.* **15**, 283–292 (1958).
9. Ts. P. Kruglikova and V. K. Efimova, *Gig. Sanit.* **23**, 75–78 (1958).
10. N. M. Tomson, Z. V. Dubrovina, and M. I. Grigor'eva, *In* "U.S.S.R. Literature on Air Pollution and Related Occupational Diseases" (B. S. Levine, transl.), Vol. 8, pp. 140–144. Office of Technical Services, U.S. Department of Commerce, Washington, D.C., 1963.
11. K. Biersteker, H. de Graaf, and C. A. G. Nass, *Int. J. Air Water Pollut.* **9**, 343–350 (1965).
12. "Field Study of Air Quality in Air Conditioned Spaces, Second Season (1969–1970), RP-86. American Society of Heating, Refrigerating, and Air Conditioning Engineers, New York, New York, 1970.
13. "Field Study of Air Quality in Air Conditioned Spaces," RP-86. American Society of Heating, Refrigerating and Air Conditioning Engineers, New York, New York, 1969.
14. K. Grafe, *in* "Proceedings of the First International Clean Air Conference," Part I, pp. 256–258. Nat. Soc. Clean Air, London, England, 1966.
15. S. Ishido, *Air Clean.* (*Tokyo*) **3**, 11–15 (1965).
16. S. Ishido, *Bull. Dep. Home Econ., Osaka City Univ.,* (*Osaka, Jap.*) **6**, 53–59 (1959).
17. S. Ishido, K. Kamada, and T. Nakagawa, *Bull. Dep. Home Econ., Osaka City Univ.* (*Osaka, Jap.*) **4**, 31–37 (1956).

18. S. Ishido, T. Tanaka, and T. Nakagawa, *Bull. Dep. Home Econ., Osaka City Univ.* (*Osaka, Jap.*) **3**, 35 (1955).
19. S. Seisaburo, K. Kiyoko, and N. Tatsuko, *Bull. Dep. Home Econ., Osaka City Univ.* (*Osaka, Jap.*) **4**, 31–37 (1959).
20. G. C. R. Carey, J. J. Phair, R. J. Shephard, and M. L. Thompson, *Amer. Ind. Hyg. Ass., J.* **19**, 363–370 (1958).
21. M. L. Weatherly, "Air Pollution Inside the Home." Warren Spring Laboratory Investigation of Atmospheric Pollution, Standing Conference of Cooperating Bodies. Proceedings of the 1966 Blackpool (England) Meeting of the Royal Society of Health, London, England, 1966.
22. M. L. Weatherly, *Int. J. Air Water Pollut.*, **10**, 404–409 (1966).
23. Kh. B. Berdyev, N. V. Pavlovich, and A. A. Tuzhilina, *Gig. Sanit.* **32**, 424–426 (1967).
24. L. J. Goldwater, A. Manoharan, and M. B. Jacobs, *Arch. Environ. Health* **2**, 511–515 (1961).
25. M. B. Jacobs, L. J. Goldwater, and A. Fergany, *Int. J. Air Water Pollut.* **6**, 377–380 (1962).
26. M. B. Jacobs, A. Manoharan, and L. J. Goldwater, *Int. J. Air Water Pollut.* **6**, 205–213 (1962).
27. A. Manaharan, M. B. Jacobs, and L. J. Goldwater, *Ann. Meet.*, Air Pollut. Contr. Assoc., Pittsburgh, Pennsylvania, 1961.
28. R. J. Shephard, G. C. R. Carey, and J. J. Phair, *J. Appl. Physiol.* **15**, 70–76 (1970); *Health* **17**, 236–252 (1958).
29. R. J. Shepard, *AMA Arch. Ind. Health* **19**, 44–54 (1959).
30. R. J. Shepard, M. E. Turner, G. C. R. Carey, and J. J. Phair, *J. Appl. Physiol.* **15**, 70–76 (1970).
31. G. Romagnoli, *Ital. Rev. Hyg.* **21**, 410–419 (1961).
32. K. T. Whitby, A. B. Algren, R. C. Jordan, and J. C. Annis, *Heat., Piping Air Cond.* **29**, 185–192 (1957).
33. K. T. Whitby, R. C. Jordan, and A. B. Algren, *Amer. Soc. Heat. Refrig. Air-Cond. Eng. J.* **4**, 79–88 (1962).
34. S. Ishido, "Air Pollution in Osaka City and Inside Buildings," APTIC No. 11653. Department of Home Economics, Osaka City University, Osaka, Japan.
35. S. Kanitz, *J. Hyg. Prevent. Med.* (*Italy*) **1**, 57–68 (1960).
36. Air Filtering System Design Committee, *Air Clean.* (*Tokyo*) **4**(5), 1–31 (1967).
37. J. Spiegelman, H. Friedman, and G. I. Blumstein, *J. Allergy* **34**, 426–431 (1963).
38. M. A. Swaebly and C. M. Christensen, *J. Allergy* **23**, 370–374 (1952).
39. H. E. Prince and M. B. Morrow, *S. Med. J.* **30**, 754–762 (1937).
40. K. Maunsell, *Proc. Int. Congr. Allergy, 1st, 1952* pp. 306–314 (1952).
41. K. Maunsell, *Int. Arch. Allergy* **5**, 373–376 (1954).
42. K. Maunsell, *Int. Arch. Allergy* **3**, 93–102 (1952).
43. I. Nilsby, *Acta Allergol.* **2**, 57–90 (1949).
44. E. W. Flensborg and T. Samsoe-Jensen, *Acta Allergol.* **3**, 49–65 (1949).
45. C. Jimenez-Diaz, J. M. Alex, F. Ortiz, F. Lahoz, L. M. Garcia, and G. Canto, *Acta Allergol. Suppl.* **7**, 139–149 (1960).
46. M. E. Wallace, R. H. Weaver, and M. Scherago, *Ann. Allergy* **8**, 202–211 (1950).
47. M. Dowrin, *Ann. Allergy* **24**, 31–36 (1966).
48. M. Richards, *J. Allergy* **25**, 429–439 (1954).
49. E. Ripe, *Acta Allergol.* **17**, 130–159 (1962).
50. O. Rostrup, *Bot. Tidsskr.* **29**, 32–41 (1908).

51. K. F. Adams and H. A. Hyde, *J. Palynol.* pp. 67–69 (1965).
52. E. Rennerfelt, *Sv. Bot. Tidskr.* **21**, 283–294 (1947).
53. J. Spiegelman and H. Friedman, *J. Allergy* **42,** 193–202 (1968).
54. J. Spiegelman, G. I. Blumstein, and H. Friedman, *Ann. Allergy* **19,** 613–618 (1961).
55. T. Nelson, B. Z. Rappaport, and W. H. Welker, *J. Amer. Med. Ass.* **100,** 1385–1392 (1933).
56. A. N. Dingle and E. W. Hewson, *J. Air Pollut. Contr. Ass.* **8,** 16–22 (1958).
57. L. H. Creip and M. A. Green, *J. Allergy* **7,** 120–131 (1936).
58. W. T. Vaughan and L. E. Cooley, *J. Allergy* **5,** 37–44 (1933).
59. B. Z. Rappaport, T. Nelson, and W. H. Welker, *J. Amer. Med. Ass.* **98,** 1861–1864 (1932).
60. C. E. A. Winslow and W. W. Browne, *Mon. Weather Rev.* **42,** 452–453 (1914).
61. W. A. Coté, W. A. Wade, III, and J. E. Yocom, "A Study of Indoor Air Quality," Final Rep., Contract 68-02-0745, Publ. No. EPA 650/4-74-042. United States Environmental Protection Agency, Research Triangle Park, North Carolina, 1974.
62. A. C. Chamberlain, *Int. J. Air Water Pollut.* **10,** 403–409 (1966).
63. M. J. G. Wilson, *Proc. Roy. Soc., Ser. A* **300,** 215–222 (1968).
64. N. M. Lefcoe and I. I. Inculet, *Arch. Environ. Health* **22,** 230–238 (1971).
65. J. E. Yocom and W. A. Coté, "Indoor/Outdoor Air Pollutant Relationships for Air-Conditioned Buildings." American Society of Heating, Refrigerating and Air Conditioning Engineers, New York, New York, 1971.
66. E. De Fraja Frangipane, C. F. Saccani, and V. Turolla, *New Ann. Hyg. Microbiol. (Rome)* **14**(6), 403–421 (1963).
67. A. F. Bush and M. Segall, Personal communication, University of California, Los Angeles, California, 1970.
68. P. W. Kalika, J. K. Holcombe, and W. A. Coté, *Amer. Soc. Heat. Refrig. Air-Cond. Eng. J.* **12,** 44–48 (1970).

Part B

EFFECTS ON BIOLOGICAL SYSTEMS

4

Effects on Vegetation: Native, Crops, Forests

Walter W. Heck and C. Stafford Brandt

I. Introduction

Injury to vegetation has been one of the earliest manifestations of air pollution. The significant and sometimes devastating effects of sulfur dioxide and fluoride gases were first investigated near the middle of the nineteenth century in Europe, primarily in Germany. The most dramatic effects were those from sulfur dioxide emitted from smelters. The intensity of interest and thus the number of experimental investigations at any time have been directly related to the periodic concern for localized or general air pollution problems.

A. History of Air Pollution Hazards to Vegetation and Their Current Importance

The earliest recorded air pollution problems to vegetation were related to the sulfur oxides. Sulfur dioxide is the major agent among the sulfur oxides causing injury to vegetation, although plants are injured to some extent by other sulfur compounds such as sulfuric acid aerosols. Haselhoff and Lindau had several decades of experience and observations to draw upon by the time they issued their handbook (*1*) on the effects of sulfur dioxide on vegetation. Even with the evidence found in this publication and other reports from German investigations, the effects of sulfur dioxide were often ignored by industry, as evidenced by such major episodes as occurred at San Francisco, California—The Selby Report (*2*); Trail, British Columbia, Canada (*3*); Sudbury, Ontario, Canada (*4*); and Ducktown (Copper Hill), Tennessee (*5*). Copper sulfide at Ducktown was roasted in piles by huge bonfires until about 1900. The sulfur was released as sulfur dioxide, and the copper pyrite was converted to copper metal. The trees that were not cut for fuel were killed by the sulfur dioxide, and the land was denuded for several miles around the source.

Today, complete destruction of vegetation near point sources rarely occurs because of a reduction in sulfur dioxide concentrations. However, vegetation is still severely injured around such point sources as smelters and power plants. Indeed, the sulfur dioxide problem is of increasing concern because of the many sources in urban areas and the current widespread development of high-capacity power plants using sulfur-containing fossil fuels. These new sources of sulfur dioxide emissions plus the reported interactions between sulfur dioxide and ozone, nitrogen dioxide, and fluoride make sulfur dioxide pollution a high-priority concern. The additional focus on acidic precipitation (*6*), with sulfur dioxide as a major contributor, adds urgency for continued studies on sulfur dioxide as an air pollutant.

Injury to vegetation by fluorides from industrial sources was well described by the early 1900s (*7*), but fluoride did not become a serious threat to vegetation until industrial expansion began about 1940 (*8*). By this time, characteristic fluoride symptoms were known, the concept of a spectrum of sensitivity among species had developed, fluoride was known to be an accumulative poison, and foliar analysis for fluoride content was an accepted diagnostic tool. Because fluoride is widely distributed in nature, evaluation of fluoride as an air pollutant to plant and animal life is complex. The problem has been important in agriculture, because vegetation appearing perfectly normal, but with an elevated fluoride content, can seriously harm grazing animals.

Although reduction in visibility was probably the earliest manifestation of photochemical air pollution in the Los Angeles, California, area, injury to vegetation was the earliest recognized biological effect. Middleton *et al.* (*9*) first characterized this injury in 1944 as glazing, silvering, and bronzing of lower leaf surfaces of broadleaf plants. This type of injury was recognized over a large segment of southern California and in the San Francisco, California, bay area by 1950 (*10*). Components of photochemical pollution injure plants in most, if not all, of the major metropolitan areas of the United States, Canada, and Mexico. Plants are affected by photochemical oxidants in major metropolitan areas throughout the world, but scientists were slow to recognize these effects (*11*). Three phytotoxic oxidants have been isolated from the photochemical complex, and research in several laboratories suggests the presence of additional phytotoxicants.

Ozone, the most important phytotoxicant in the complex (*12*), was first identified by Richards *et al.* (*13*) in 1958 as the cause of a stipple on the upper surface of grape leaves.

Peroxyacyl nitrates (PANs) were first identified in 1961 by Stephens *et al.* (*14*) as the primary cause of the undersurface glazing and bronzing of certain sensitive leaves.

Nitrogen dioxide has generally been considered of lesser importance, because its threshold of injury to vegetation appears to be severalfold higher than for sulfur dioxide. However, increased emissions of nitric oxides, plus knowledge of their long-term growth effects and interactions with sulfur dioxide and possibly other pollutants, suggest that the importance of this phytotoxicant should be reassessed.

Ethylene is one of the major petrochemicals and a major product of auto exhaust. It was first identified as a toxic component of illuminating gas in commercial greenhouses near the turn of the century. Later research suggests that ethylene is more than fifty times as phytotoxic as other hydrocarbon gases and that it contributes to the formation of photochemical oxidants. Concentrations in metropolitan areas, primarily from auto exhaust, are sufficiently high to cause early senescence and possibly yield reductions (*15*).

Other phytotoxic air pollutants considered to be important include airborne pesticides, chlorine, heavy metals, acid aerosols, ammonia, aldehydes, hydrogen chloride, hydrogen sulfide, and particulates, such as cement dust (*12*). These pollutants are released primarily from industrial sources or as agricultural applications, but they are either less widespread or less concentrated than the major pollutants. Thus they have received less study and their effects are less well understood.

It is extremely difficult to firmly assess the economic, ecologic, or aesthetic costs of air pollution damage to vegetation. Estimates have ranged up to $500 million for economic losses just to crop plants in the United States. The United States estimate of $135 million loss for 1969 from effects causing visible injury to crops and ornamentals is lower than previous estimates, but probably more realistic (*16*). None of the estimates includes possible growth and yield reductions in the absence of visible injury, effects on native ecosystems, decreased aesthetic values when ornamentals are injured, reduction in vigor, predisposition to invasion by pests, the continued cost of erosion from the "Ducktown desert," or other important concerns. If we could include these diverse factors, the total annual cost estimate would undoubtedly exceed a billion dollars.

Research efforts in many countries have greatly expanded since around 1965. Research strength has developed in many Eastern and Western European countries, although Great Britain long felt that they were relatively free of any major air pollution problems. Germany is still the leader in sulfur dioxide research. The Germans have recognized oxidant problems and are doing research with other pollutants. Japan, where severe air pollution problems exist because of a lack of past concern for the environment, has a strong research program. Japan's interests, as well as those of most countries, has been focused more on industrial rather than general urban pollution. North America was the first to accept the overall importance of photochemical air pollution.

B. Citation and Evaluation of Key Reviews

A plethora of from poor to excellent review articles covering general or specialized areas of air pollution effects on vegetation are available. Rather than detail all of these reviews we have developed a fairly comprehensive listing of review articles by subject matter (Table I) *(1, 3, 16–66)*. The criteria document reviews developed by the United States Environmental Protection Agency (USEPA) and the National Research Council of Canada *(32–36)* contain the best summaries of dose–response relations and present an overall understanding of pollutant effects.

II. Air Pollution Effects on Vegetation

The effects of air pollutants on vegetation are generally discussed in terms of effects on individual plants. Such effects can be viewed as either visible or subtle. Visible effects are identifiable morphological, pigmented, chlorotic, and/or necrotic foliar patterns that result from major physiological disturbances in plant cells. Subtle effects are those that do not result

Table I Relevant Review Articles—A Topical Listing

Subject	*Reference*
Historic (literature prior to 1950)	*(1, 3, 17–22)*
General information	*(23–31)*
Criteria documents	*(32–36)*
Foreign investigations	*(26, 37–42)*
Visible symptoms and listing of sensitive species	*(30, 43, 44)*
Photochemical oxidants	*(24, 33a, 36, 44–50)*
Sulfur oxides	*(3, 18–21, 26, 33, 39, 40, 42–44, 51)*
Fluoride	*(44, 52–55)*
Nitrogen oxides	*(32, 44, 44a)*
Hydrocarbons	*(19, 35, 44)*
Particulate matter	*(34, 44)*
Physiological and biochemical factors	*(29, 45, 49, 54, 55)*
Effects of environmental factors	*(19, 46, 56, 57)*
Plants as monitors (indicators)	*(58–64)*
Problems of recognition and evaluation	*(16, 65, 66)*

in visible injury but are measurable when growth or physiological changes occur in the plant. Both visible and subtle effects are induced by physiological and biochemical changes within the plant systems. Some subtle changes may affect the reproductive and/or genetic systems of plants. The cumulative changes within these plant systems subjected to air pollution stress may result in changes in plant populations and communities. Indeed, over several years, the ecosystem may lose its ability to support higher plants.

A. Visual Symptoms

Visible injury is usually classified as acute or chronic. Acute injury always causes cell death, although it may affect only a small part of a given leaf. It is caused by disruption of cell membranes with a resultant loss of cell contents, followed by cell death. Subsequent development of necrotic patterns tend to be characteristic of a given pollutant. When not characteristic, acute symptoms demonstrate the presence of a chemical toxicant. Acute symptoms are associated with short exposures (measured in hours) to varying concentrations of pollutant and usually appear within 24 hours after exposure.

Chronic injury may be mild or severe. Initial disruption of normal cellular activity may be followed by chlorosis and/or other color or pigment changes that may eventually cause cell death. Chronic injury may appear as early leaf senescence with or without leaf abscission. Chronic injury patterns are generally not characteristic for a given pollutant and are easily confused with symptoms caused by parasitic diseases, insects, nutritional factors, or other environmental stresses. Chronic symptoms are normally associated with long-term or intermittent exposures to low concentrations of a gas. However, repeated short-term exposures to varying pollutant concentrations will cause physiological changes responsible for chronic symptoms. Sometimes such exposures produce small but additive amounts of acute injury that are mistakenly referred to as chronic injury.

1. General Types of Injury

A general understanding of the structure of a leaf will help the physical scientist understand the vegetation effects of air pollutants. In this context a generalized diagram of the structure of a "normal" leaf is given in Figure 1. If you hold a leaf up to the light, you will normally see a network of denser structures, the veins, all interconnecting back to the base or stem of the leaf, the petiole. Specialized cells within these veins serve as

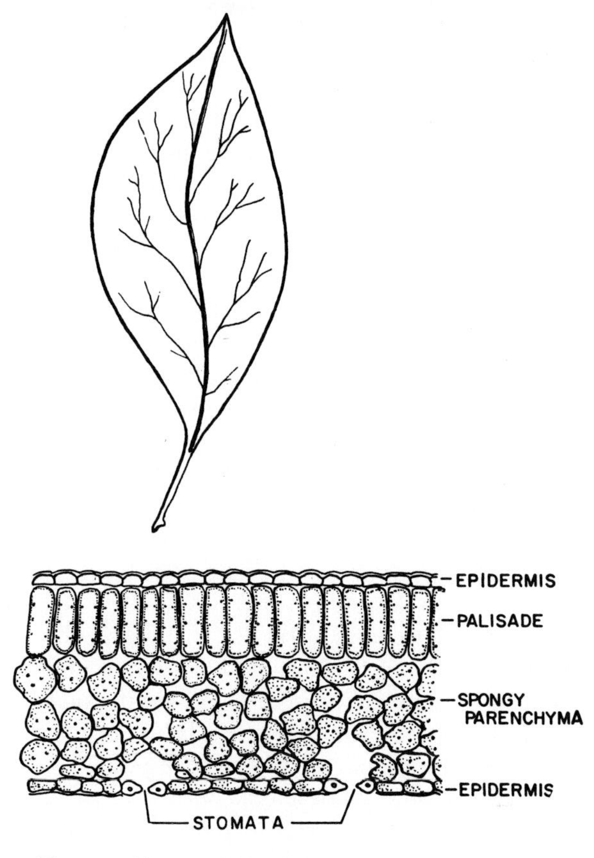

Figure 1. Normal dicotyledon leaf and cross section.

the transport systems of the leaf. In the interveinal areas, there are normally three specialized layers. On the upper and lower surface there is a single layer of thick-walled cells that comprise the epidermis, or "skin," of the leaf. The epidermis is covered with a layer of cutin, a waxy material of varied thickness. Directly under the upper epidermis is the palisade layer—one or more rows of rather uniform, elongated cells standing on end and packed closely together. Between the lower edge of the palisade layer and the lower epidermis, you will find a loosely packed area of cells that are somewhat irregularly shaped, the spongy parenchyma. Some plants, such as most grasses, have no palisade layer; other plants have more than one palisade layer. On more careful examination of the epidermal layers, you will find pairs of specialized guard cells edging an opening, or stoma (plural, stomata). These openings provide the necessary mechanism for gas exchange between the plant tissues and the atmosphere. The guard cells can change shape, thus regulating the opening or closing of the stoma.

a. TISSUE COLLAPSE AND NECROTIC PATTERNS. Acute injury from air pollutants results in the plasmolysis of cells and subsequent collapse of the tissue. The injury may occur in the spongy cells from exposure to peroxyacetyl nitrate (PAN), in the palisade cells from exposure to ozone, or in both cell types from exposure to fluorides or sulfur dioxide. Following plasmolysis and loss of cell structure, the adjacent segments of the leaf may be affected, depending on the nature of the toxicant, its concentration, the species of plant, and many other factors. In most cases, the first visible symptom on the intact leaf is a slightly water-soaked or bruised appearance. The affected areas generally dry out, producing the necrotic pattern characteristic of the toxicant.

Tissue collapse and necrotic patterns may be associated with chronic injury. Perhaps the most distinctive necrotic patterns from chronic exposure result from fluoride phytotoxicity, where visual injury is normally confined to leaf margins or leaf tips.

b. CHLOROSIS AND OTHER COLOR PATTERNS. Chlorosis, the loss or reduction of the green plant pigment, chlorophyll, is a very common and nonspecific symptom in plants. The loss of chlorophyll usually results in a pale green or yellow pattern. Chlorosis generally indicates a deficiency of some nutrient required by the plant. In many respects, it is analogous to anemia in animals. Sometimes other colors appear from pigments normally masked by the green of the chlorophyll or from new pigments that develop as a result of the stress that caused a reduction in the chlorophyll. Over the years, we have learned to recognize certain patterns of chlorosis as being characteristic of certain nutrient deficiencies. There are other color changes in the leaf that we associate with maturity or senescence, such as the color change in leaves during autumn. Similar changes may take place in plant leaves during the summer months in the presence of air pollutants.

Tissue severely injured by air pollutants often has a characteristic color; we associate bleaching with sulfur dioxide, a yellowing with ammonia, and a browning with fluoride. A dark band of color marking the edge of necrotic tissues often indicates fluoride injury. The pigmentation (stipple) of small necrotic areas of the palisade tissues, caused by cellular polymerization of *o*-quinones with amino acids and proteins (*67*), seems to be characteristic of ozone injury in some plants. A silvering or bronzing of the under surface of some leaves is associated with PAN injury. Chlorosis may appear in association with necrotic tissue after exposure to sulfur dioxide or oxidants.

Chronic injury from sulfur dioxide may be expressed as a diffuse chlorosis of the older leaves. This may be a saltlike toxicity resulting from excessive sulfate accumulation and often resembles the chlorosis associated with senescence. In at least two species, corn and citrus, fluorides produce a characteristic chlorosis before the typical edge and tip burn occurs. In corn, scattered chlorotic flecks near the leaf tip may extend downward with higher doses, especially along the margin where an initially narrow chlorotic band may increase in width, length, and intensity (*68*). In citrus, the fluoride chlorosis produces a spotty yellow interveinal development, somewhat similar to manganese deficiency, but the fluoride pattern usually develops more toward the leaf tip (*69*). In some cases the oxidant type of smog also produces a chlorosis. The pattern, however, is not distinctive, usually appearing as an early leaf senescence.

c. Growth Abnormalities. Although reduction in growth from air pollutants is well established, few distinctive alterations of growth can be attributed to air pollutants. Herbicides (such as 2, 4-D) that drift out of desired spray patterns produce marked abnormalities in growth of sensitive species. These abnormalities include twisting and/or elongation of leaves and stems.

Ethylene induces epinasty and abscission of plant parts. These responses resulted in an almost prostrate growth of cotton in a field near a polyethylene plant in Texas (*70*) with almost complete loss of yield. Similar effects have been reported for laboratory exposures. Ethylene is a normal constituent of plant tissues and, as such, acts as a plant hormone. This physiological activity accounts for the ability of external sources of ethylene to alter normal growth patterns.

An investigation into an alleged air pollution problem in Scotch pine Christmas tree plantations suggests that sulfuric acid aerosols in conjunction with particulate matter may cause growth abnormalities and restrict lateral bud development (*71*). However, insects may be the causative factor.

2. Injury Patterns from Specific Pollutants

Descriptions of injury to vegetation by air pollutants are found in most reviews and in many original papers. This chapter will not attempt to be all-inclusive but will be definitive in summarizing the significant points in the symptomatology of the most important air pollutants on vegetation. The most complete descriptions are found in the 1970 pictorial atlas (*44*), where specialists on specific pollutants have described and illustrated symptoms. Descriptions are found also in other review articles and in the

United States criteria documents (*32–36*). The most complete discussion of sulfur dioxide effects is found in the German atlas compiled by van Haut and Stratmann (*43*) (with text in German, French, and English). A good presentation for nitrogen dioxide is found in van Haut and Stratmann (*72*). In the following outline, the significant points in the symptomatology of air pollution effects on vegetation are summarized. Pollutants are listed in order of worldwide importance.

a. Ozone

i. Leaf markings are divided into three categories by plant type: (1) *Broadleaf*—upper surface, pigmented, red-brown spot (stipple); bleached tan to white areas (fleck); small irregular bifacial collapsed (necrotic) areas that may coalesce to form irregular necrotic blotches; chlorosis and premature senescence may occur. (2) *Grasses*—scattered bifacial necrotic areas (fleck); sometimes larger lesions or necrotic streaking may occur. (3) *Conifers*—brown-tan necrotic needle tips with no separation between dead and healthy tissue.

ii. Similar Markings. Red spider mite and certain insects may cause an upper-surface fleck. Some leaf spot fungi may give similar patterns. Severe ozone-induced lesions may resemble those caused by sulfur dioxide. Chronic injury may resemble normal leaf senescence. Pathogen-induced tip burn in conifers may resemble ozone injury, but the color shading from ozone is fairly distinctive.

iii. Sensitive Plants. Oat, petunia, pinto bean, potato, radish, soybean, sycamore, tobacco, tomato, white ash, white pine.

iv. Resistant Plants. Beet, geranium, gladiolus, maple, mint, pepper, rice.

b. Sulfur Dioxide

i. Leaf Markings are divided into three categories by plant type: (1) *Broadleaf*—irregular, bifacial, marginal, and interveinal necrotic areas bleached white to tan or brown; chlorosis may be associated with necrotic areas, or a general chlorosis of older leaves may develop; diffuse to stippled colors ranging from white to reddish-brown have been observed. (2) *Grasses*—irregular, bifacial, necrotic streaking between larger veins that is bleached light tan to white; chlorosis usually is not pronounced. (3) *Conifers*—brown necrotic tips of needles often with a banded appearance; generally chlorosis of adjacent tissue; needles of the same age are uniformly affected.

ii. Similar Markings. White spot of alfalfa, leafhopper injury, rose chafer injury, various mosaic viruses, cherry leaf spot, and other fungal diseases producing blotchy markings; high-temperature scorch on maple and horse chestnut; Victoria blight on oats, bacterial blight of barley and other grains; terminal bleaches in cereals; winter, drought, and red spider mite injury in conifers.

iii. Sensitive Plants. Alfalfa, apple, barley, cotton, giant ragweed, pine, squash, wheat.

iv. Resistant Plants. Cantaloupe, celery, corn, oak, rhododendron.

c. NITROGEN DIOXIDE.

Symptoms are similar to those from sulfur dioxide, but much higher concentrations are required to produce acute injury. Lower concentrations may increase chlorophyll levels, but long exposures may cause early leaf senescence and abscission.

d. FLUORIDE

i. Leaf Markings are divided into three categories by plant type: (1) *Broadleaf*—necrotic tip and/or leaf margins with occasional interveinal blotches; area between dead and living tissue is sharp and usually accentuated by a narrow, darker, brown-red band; a narrow chlorotic band may be found adjacent to the necrotic area; in some species, the necrotic area may fall off, leaving a "chewed" edge on a seemingly healthy leaf; citrus, sweet cherry and certain other plants show a mottled interveinal chlorosis. (2) *Grasses*—brown necrotic tip burn extending in irregular streaks down the leaf, sharp demarcation between dead and healthy tissue; some plants develop a chlorotic mottle. (3) *Conifers*—brown to red-brown necrotic needle tips; necrosis may affect entire leaf; all of same age needles not uniformly affected.

ii. Growth and Yield Effects. Alteration of growth patterns ranging from increased needle length (Douglas fir), smaller leaf size (citrus), reduced tree growth (several species), to increased twig development (citrus) have been reported. Fluoride also produces a malformation in peach fruit (soft suture), and may induce poor fruit set or excessive early fruit drop in cherry.

iii. Similar Markings. Various fungal leaf spots that normally do not show the characteristic edge and tip injury caused by fluoride. Wind, high temperature, salt spray, and drought can cause mimicking symptoms. Bacteria-induced scalds, blotches, and streaks on grasses could be confused with fluorides, as could *Botrytis* blight and *Verticillum* on gladiolus. Winter injury or sulfur dioxide injury on conifers may resemble fluoride burn.

iv. Sensitive Plants. Chinese apricot, gladiolus (light-colored varieties are more sensitive than darker ones), grape, Italian prune, pine.

v. Resistant Plants. Alfalfa, cotton, elm, pear, tobacco, tomato.

e. Peroxyacetyl Nitrate (PAN)

i. Leaf Markings are divided into three categories by plant type: (1) *Broadleaf*—collapse of spongy mesophyll giving a glazed, silvered, or bronzed appearance to the underside of the leaf; some leaves show a bifacial collapse, usually in a banded pattern; plants often show an early senescence with leaf abscission. (2) *Grasses*—irregular collapsed banding bleached yellow to tan, sometimes appearing more as a chlorotic or bleached band than necrotic. (3) *Conifers*—not specific, needle blight with some chlorosis or bleaching.

ii. Similar Markings. "Sun scald"; various fungus, virus, and bacterial diseases that produce blotchy patterns on broadleaf or streaked patterns on grasses. The typical silvering is seldom duplicated by other agents. Red spider mites, drought, and excess salts cause confusing symptoms on conifers.

iii. Sensitive Plants. Annual bluegrass, petunia, pinto bean, romaine lettuce.

iv. Resistant Plants. Broccoli, chrysanthemum, corn, cotton, sorghum.

f. Ethylene

i. Leaf Markings are divided into three categories by plant type: (1) *Broadleaf*—epinasty and/or abscission of leaves without markings; general chlorosis of older leaves, sometimes resulting in necrosis and abscission; stimulation of lateral buds and adventitious roots; abscission of young flower buds and/or failure of floral blooms to open; growth reduction and loss of apical dominance on more resistant plants. (2) *Grasses*—retardation of growth with increased tillering, no visible injury even at high concentrations. (3) *Conifers*—abscission of needles; retarded elongation of new needles; abscission of young cones or poor cone development.

ii. Similar Markings. Water stress, bacterial wilts, root stresses, nematode injury, early senescence produced by other pollutants.

iii. Sensitive Plants. Cotton, cowpea, orchid blossom, tomato.

iv. Resistant Plants. Grasses, lettuce.

g. Other Pollutants

i. Irradiated Auto Exhaust. Exposure to simulated photochemical oxidant atmospheres produce both PAN and ozone injury symptoms. In addition, several types of glaze that might be different from PAN injury and a fleck that is differentiated from ozone injury do occur.

ii. Chlorine. Acute injury may resemble sulfur dioxide or ozone effects. Defoliation with no leaf symptoms may occur.

iii. Hydrogen Sulfide. Acute exposure causes necrosis of immature tissue before older tissue is injured. This sensitivity of immature tissue differs from its sensitivity to other pollutants. The growing tip may be injured.

iv. Hydrochloric Acid. Injury resembles acute sulfur dioxide effects. Acid burn may occur at higher concentrations.

v. Ammonia. Acute injury is similar to symptoms from sulfur dioxide, but necrotic areas are yellowish. High concentrations may cause tissue death without complete breakdown of chlorophyll.

vi. Mercury. Chlorosis and abscission of older leaves, growth reduction and general decreased vigor.

vii. Herbicides. Growth malformations, necrosis, and chlorosis on sensitive plants.

3. Problems of Diagnosis

A plant is a product of its genetics and its environment, responding in many ways to both the stresses and the support provided by the environment. Air pollution is simply an abnormal component of the natural environment, which includes climate, soil, insects, and diseases, as well as care or abuse by man. Other environmental factors may modify or obscure the injury created by an air pollutant. Environmental extremes may cause patterns of injury that are difficult or impossible to distinguish from air pollution effects. The modifying effects of genetic makeup must also be considered. Different species and varieties within a species usually respond differently to air pollution as well as to other environmental factors. One variety of gladiolus may have 50% of the leaf destroyed by fluoride, whereas another variety growing next to it may have only the tip of the blade injured. The susceptibility of tobacco to ozone-induced fleck is markedly affected by genetic makeup.

The modifying effects of rainfall, temperature, and wind are difficult to assess. In alfalfa, white spot disease, which is related to water stress, is difficult to distinguish from sulfur dioxide injury. High temperatures can produce a scorch on maple and some other trees that will mimic sulfur dioxide injury. Temperature extremes and wind, under certain conditions of soil moisture, can cause a marginal burn on the leaves of many plants that is similar to fluoride injury. In any of these cases, the climatic stress may be of short duration, and its occurrence therefore difficult to establish after the symptoms have developed.

Edaphic factors—soil, nutrition, and management—can duplicate chlorosis, senescence, and growth effects of air pollutants. It would be foolish to diagnose a chlorotic pattern on citrus as an atmospheric fluoride effect without carefully considering soil fertility and other management practices. Dieback on apricot could be caused by fluoride, by nutritional deficiency, by winter injury, by a combination of these, or by a host of other factors. Most of our ornamentals and many of our crops are foreign to the area in which we desire them to grow. To get such material to thrive, man must modify the natural environment of the area through management practices. Abnormal symptoms or poor growth of urban ornamentals could be due to air pollution or to poor soil or care. Even in areas such as the Los Angeles, California, basin, where there is ample evidence that air pollution causes early senescence and limited growth, management practices must be considered.

When faced with the list of bacteria, fungi, viruses, nematodes, and insects that can destroy plants, one is inclined to adopt the concept of a vengeful nature set upon destroying the plant world. Many of these agents produce injuries that are similar to those produced by air pollutants. Leafhoppers on alfalfa produce a blotchy white interveinal necrosis that is similar to that caused by sulfur dioxide. Rose chafer and sulfur dioxide injuries are somewhat similar. *Botrytis* infection of gladiolus is often difficult to distinguish from fluoride injury. Many of the bacterial and fungal diseases of grasses can produce the banding, tip burn, and striping similar to the effects of photochemical oxidants, fluoride, or sulfur dioxide. In the summaries of air pollution symptoms (Section II,A,2), only a few of the possible mimicking symptoms caused by biotic organisms and insects are noted. To properly diagnose air pollution effects on vegetation, the problem must be seen in the field. The diagnosis should preferably be supported by measurements of ambient air concentrations of the suspected pollutant(s). Furthermore, the observer must have a thorough knowledge of local cultural conditions.

B. Subtle Effects

The presence of subtle effects suggests that air pollutants interfere with physiological and biochemical processes, alter plant growth, and affect both reproduction and yield without visible symptoms developing.

1. Growth and Yield

That air pollution can reduce growth and yield, with or without leaf symptoms, has been known for some time. Studies using chambers and greenhouses provided with either ambient or filtered air show that growth and/or fruiting reductions caused by air pollution are a widespread problem. Field identification of subtle effects is impossible without control plantings that are protected from the air pollutants.

Low levels of fluoride, sulfur dioxide and ozone stimulate growth in certain plants. Stimulation by sulfur dioxide is associated with a low soil sulfur level.

Treshow and Harner (*73, 74*) reported decreased needle growth of Douglas fir and decreased fresh and dry weights of bean from exposure to fluoride concentrations that did not cause visible injury. Although such reports are not common, most growth reductions from exposure to fluoride are accompanied by visible injury.

Classic research by Thomas and associates (*20, 21*) clearly demonstrated that acute exposure to sulfur dioxide did not reduce growth unless there also was visible injury. However, two greenhouse studies demonstrated that chronic exposure to sulfur dioxide or ozone could decrease growth of radish roots (*75*) and alfalfa (*76*) without visible injury being significant. A mixture of the two gases reduced growth more than either gas by itself, but in general, the effects were less than additive. Ethylene exposure reduced growth of grasses and several other plants without attendant symptoms (*15, 77*). However, the plants recovered rapidly after the gas was removed (*77*).

Bleasdale (*78*) reported that growth of ryegrass in ambient air was reduced when the maximum average sulfur dioxide concentrations for 24 hours were between 0.1 and 0.2 ppm for 2–3 days during two different experiments. However, possible interactions with other pollutants, such as ozone, or the possible effects of short-term high concentrations of sulfur dioxide were not considered. Either may have contributed to the reduced growth.

In general, growth reductions have been related to visual injury symptoms. Quantitative determinations of growth reductions are discussed in Section IV.

2. Physiological and Biochemical Processes

A plethora of reports on the effects of fluoride, ozone, and PAN on physiological processes (such as net photosynthesis, stomatal response, and changes in water relations) and on metabolic activity (including *in vivo* and/or *in vitro* studies of individual enzymes, enzyme systems, metabolic pathways, metabolic pool relationships, cell organelles, and plant tissue studies) have appeared since 1964 (*29, 45, 49, 54, 55, 78a, 78b*). Little work on the effects of sulfur dioxide was reported from the mid-1950s to the mid-1970s, but early reports are available. Ethylene, as a plant hormone, has been extensively studied since 1961, and its effects are summarized by Abeles (*79*). The remaining pollutants, other than the pesticides, have not been extensively investigated.

Physiological and biochemical effects obviously are a prerequisite to measurable visual and growth effects. Thus they must be understood before we can hope to evaluate our pollution problem. Early studies by Thomas and associates (*20, 21*) showed that sulfur dioxide decreased net photosynthesis without visible injury but that plants recovered rapidly when exposure ended. Results have been similar for ozone, PAN, nitric oxide, and nitrogen dioxide (*80–82*). White *et al.* (*83*) reported a synergistic inhibition of net photosynthesis in alfalfa from a mixture of sulfur dioxide and nitrogen dioxide when the gas concentrations were 15–25 pphm but not at 50 pphm. All these studies show a decrease in net photosynthesis, with a return to normal after removal of the pollutant. Hill and Bennett (*81*) associated these changes with both stomatal opening and rates of transpiration. The report that sulfur dioxide is reduced to hydrogen sulfide by several plant species and that the process may be related to photosynthesis is of interest, but its importance is unclear (*84*).

Stomata are the principal avenue of pollutant uptake by plant leaves. Stomatal closure will effectively protect plants from pollutant injury, except in the unusual case of nitrogen dioxide (dark exposures produce more injury than light exposures) (*72, 84a*). Thus, conditions that favor stomatal opening result in increased gas entry, plant assimilation, and resultant injury. Early studies suggested that oxidants cause stomatal closure; some work substantiates those early reports (*81, 85*). Although the work of Dugger *et al.* (*86*) did not show stomatal closure in response to ozone, their research was not specifically designed to show the effects of ozone on stomatal response. Stomatal closure in onion is related to a genetic factor; the stomata of the resistant onion plants close in response to ozone, whereas the stomata of sensitive plants do not close (*87*). Stomatal opening in bean plants was stimulated after exposure to sulfur dioxide at relative humidities above 40% but was suppressed at 32% relative humidity (*88*). Unsworth *et al.* (*89*) found that sulfur dioxide concentrations from 5 to 50 pphm stimulated stomatal opening in both bean and corn when relative humidity was 50–60%. The effect was noted in well-watered and water-stressed plants. Thus, the effect of pollutant and pollutant concentration on stomatal opening, although not clearly understood, is dependent upon many interacting factors; those representing water stress should be the most important.

Isolated enzymes and enzyme systems are affected by exposure to PAN and ozone, and to fluoride (normally added as sodium fluoride). High pollutant concentrations were used in many of these studies without knowing the concentration at the reaction sites (*89a*). However, it is recognized that strong oxidants should interfere with oxidation reactions within plant systems. Metabolic pools have been studied in terms of nitrogen (amino acids and proteins) and carbohydrate levels, and the sensitivity of plants to specific pollutants has been related to these pool concentrations. There are many suggested mechanisms of action for air pollutants. Sulfhydryl groups appear to be a key to understanding the mechanism of action of oxidant pollutants and perhaps of sulfur dioxide. Unsaturated lipid components of cell membranes are an early site of action for ozone and possibly PAN.

Although we will not attempt to explain the mechanism of action, we will present a reasonable mechanistic approach that might explain some of the classic results of acute dose–response symptoms and chronic growth–yield reductions. The knowledge that all biological systems must have repair capability is inherent in the following discussions.

The acute response of plants to air pollutants must result from a massive dose (massive in terms of its ability to saturate sensitive sites and to overcome normal repair mechanisms) that causes disruption of the cell membrane (through an attack on sulfhydryl linkages and/or unsaturated lipid linkages) and a loss of the differential permeability of the membrane. Cellular water and solutes may be lost and the cell plasmolyzed, or water may enter and cause cell rupture. Either disruption may cause cell death. However, if environmental conditions are not severe and if exposure dose is minimal, some recovery may take place. The extent of recovery depends upon the severity of the external stresses and the ability of the cells to initiate repair mechanisms. Repair mechanisms are inherent within plants but are dependent upon the age and overall physiological health of the exposed tissue. If we assume that a given tissue is growing under conditions that maximize its sensitivity to a particular pollutant, then the severity of the injury is a function of the inherent resistance of the plant. This inherent resistance is essentially the ability of the plant to both "inactivate" the pollutant and complete tissue repair. Thus, mechanistically, we should expect a sigmoidal response surface to acute exposures as concentration is increased. This sigmoidal response surface has been observed for many plants exposed to different pollutants (Section V). We would expect some cells to be more sensitive than others, but most cells should have a similar response time. Thus, a slight shift in concentration could involve a major increase in response by sensitive cells. A given dose over a longer time would result in a modified sigmoidal curve due to the repair mechanism of the plant. The physical resistances to gas movement within the tissue could account for the variation in apparent cell sensitivity. Exposures conducted over time under variable environmental stress could result in a near-linear function. Thus, the dose–response effects noted for acute injury can be explained as a massive attack on cellular membranes.

Chronic injury and subtle effects result primarily from secondary reactions, either as the result of acute effects on cellular membranes or from continuous or intermittent exposures to lower concentrations of air pollutants. Any membrane injury will induce secondary reactions (at least repair reactions) that could disrupt cellular organelles or shift metabolic pathways, which in turn may induce additional adverse effects in cells. Sulfur dioxide is rapidly changed to sulfite, which may be used as a sulfur source, as a precursor in the production of hydrogen sulfide, or in the initiation of other metabolic effects. Nitrogen dioxide could form nitrite in the cell, which would cause harmful secondary reactions. Ozone, PAN, and other oxidants could cause the formation of free radicals and/or other more stable oxidants (such as hydrogen peroxide), which in turn would cause secondary reactions. Any of these secondary reactions could, over time, cause senescence via the increased production of cellular ethylene. The formation of ethylene as a possible byproduct of air pollutant stresses on plants has been suggested by several workers, and work by Cracker (*90*) has shown that ethylene is produced by tissue exposed to air pollutants. These secondary effects of pollutant stress probably play only a minor role in the development of acute injury; however, they may predispose plants to greater injury from subsequent acute exposures by limiting the plant's repair capability. Predisposition has been noted in several reports, the most specific of which considers tobacco exposed to ambient oxidants (*91*).

The above discussion, although not comprehensive, offers a rational explanation of dose–response functions for pollutants studied to date.

3. Reproduction

Several researchers report that fluoride, ozone, and ambient oxidants affect reproductive structures and thus may influence fruit set and yield in the absence of visible injury or any obvious effect on total biomass. Ethylene has long been known to stimulate floral abscission, although it is usually accompanied by either growth alterations and/or abscission of other plant parts (*77*).

Exposure of sorghum to fluoride during tasseling and anthesis decreased the seed yield and top growth, while exposures before and after this relatively short developmental period had no effect on yield (*92*). Fluoride inhibited pollen germination and pollen tube length in tomato (*93*) and sweet cherry (*94*) and decreased citrus yield when exposures occurred during the bloom period (*95*). Fluoride also caused chromosomal aberrations in tomato, corn, and onion roots and phenotypic abnormalities in the second generation of tomato (*96, 97*). Thus, flouride can affect reproductive structures and initiate genetic abnormalities. Feder (*98*) reported that ozone decreased tobacco pollen germination and pollen tube growth, and decreased flower production in carnations (*99*). Germination in petunia pollen was also lowered (*99a*) and was associated with a loss of organelles in the peripheral layer of cytoplasm. Ozone can induce formation of free radicals and thus may also cause genetic abnormalities. Suzuki *et al.* (*100*) found that 0.3 ppm of sulfur dioxide for an hour inhibited pollen germination and growth of the pollen tube in pear. Shkarlet (*101*) related the size and development of Scots pine pollen to distance from industrial pollution where sulfur dioxide was the major component. It is apparent that the major pollutants can affect plant pollen and thus affect yield without a corresponding effect on plant foliage.

We know that conditions under which seeds are formed may influence the susceptibility of the next generation to various stresses. We suspect that certain of the air pollutants may thus influence the next plant generation.

C. Effects on Populations and Communities

The "ecological crisis" emphasizes a major weakness in our scientific knowledge relating to effects of pollutants on either plant populations or communities. Perhaps the best community studies are those early reports on sulfur dioxide from point sources. The classic Trail smelter case was a major ecological study of the effects of sulfur dioxide, even though it was not so described (*3*). The work at Ducktown, Tennessee, (Copper Hill) showed that sulfur dioxide caused the disappearance of vegetation and the resulting problems with soil erosion (*5*). The Ducktown case is an extreme example of what a noxious pollutant can do if not controlled. The United States has been derelict in not developing a major study around this site, as was done at Sudbury in Canada (*28*). Guderian (*102*) has reported changes in the composition of plant communities after exposure to sulfur dioxide and Knabe (*103*) has correlated the distribution of Scots pine in the Ruhr region with sulfur dioxide concentrations in the air. An extensive study of the effects of oxidants upon the loss of Ponderosa pine and on other responses in a major forest ecosystem in the San Bernardino mountains east of the Los Angeles, California, basin is receiving major emphasis from the United States Environmental Protection Agency, the United States Forest Service, and the State of California (*104–108*).

Treshow covered community concepts in his review (*109*), and several community studies have been conducted at the University of Utah (*110, 111*). Several publications have conceptualized the need for community studies in relation to air pollution effects on plants (*12, 111a*). It is obvious that these studies are badly needed but will be expensive.

III. Factors Affecting the Response of Vegetation to Air Pollutants

The sensitivity of plants to air pollutants is conditioned by many factors. Our understanding of the importance of any given factor on a specific variety or species is fragmentary or preliminary at best. We know that the response of a given plant to a specific pollutant cannot be predetermined by the response of related plants nor by the response to similar doses of different pollutants. Before we can predict how a plant variety will respond to a specific pollutant, we must understand many interrelated factors. These include but are not limited to an understanding of genetic variability (both between and within species), the influence of climatic and edaphic factors, interaction with other pollutants, interaction with biotic pathogens and insects, and the growth or physiological age of susceptible plant tissue. The overall conceptualization of relationships between pollution exposure and ultimate effects is shown simplistically in an adaption from van Haut and Stratmann (*72*) in Figure 2. This section considers factors other than dose, which is discussed in Section V.

A. Genetic Factors

Knowledge of the influence of genetic variability on plant response to pollutants has been obtained from field observations and from chamber experiments with controlled additions of pollutants. Plant response varies between species of a given genus and between varieties within a given species. Such variation is simply a function of genetic variability as it affects plant morphological, physiological, and biochemical characteristics. Thus, pollutants may act as a selective pressure mechanism in native populations and in breeding experiments (whether planned or accidental). Plants do not necessarily show similar susceptibility to different pollutants. For instance, some plants are sensitive to fluoride but resistant to sulfur dioxide.

Major variations in the response of different species to air pollutants have been documented in several review articles (*43, 44, 51, 53, 72, 111b*) and in the United States and Canadian criteria documents (*32–36*). Variations in varietal responses were reported for oat, potato, petunia, spinach, tobacco, white pine, and onion (*46*). Reports since 1968 include ornamentals (*112–117a*), vegetable crops (*117b–121*), potato

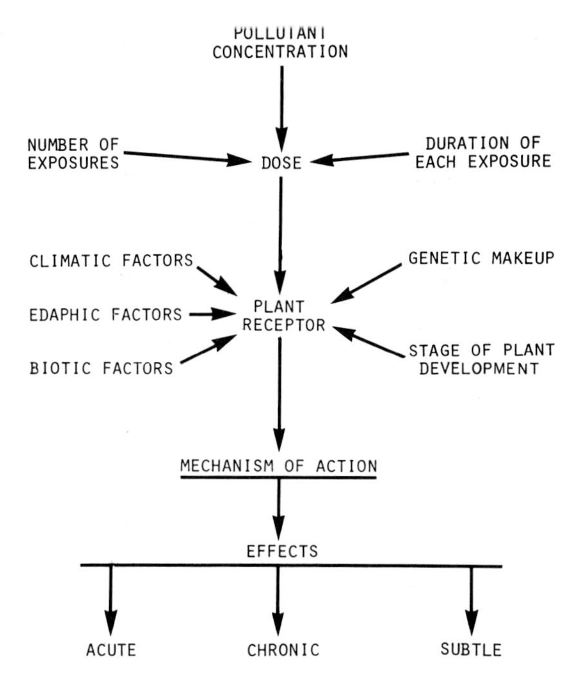

Figure 2. Conceptual model of factors involved in air pollution effects on vegetation. Adapted from van Haut and Stratmann (*72*).

(*117a, 122*), tomato (*123–125*), sweet corn (*126, 127*), forage legumes (*128, 129*), cotton (*129a*), soybean (*129b, 130*), sorghum (*131*), tobacco (*132–137*), turfgrass (*138, 139*), wheat (*139a*), and pine species (*140, 141*). Varietal variations in fluoride response in gladiolus have long been recognized, and several reports suggest that white pine selections show differential susceptibility to sulfur dioxide, ozone, and fluoride.

Varietal variations in tobacco have been most extensively studied (*132–137, 142–144*). Damage to susceptible varieties in the Connecticut River Valley tobacco fields and elsewhere has led to the development of more resistant varieties for commercial use. The very sensitive variety Bel W_3 and other cigar wrapper tobacco varieties are used widely as biological indicators of pollution.

The sensitivity to ozone of two onion cultivars is controlled by a single gene pair. The resistant gene is dominant (Section II,B,2) (*87*). The stomata of the resistant cultivar closes after exposure to ozone and injury does not occur, whereas the stomata of the sensitive cultivar remains open and injury occurs. Resistance mechanisms such as this must be explored in other species.

B. Climatic Factors

The effects of climate on the response of sensitive plants to air pollutants have been studied primarily under laboratory conditions, but field observations have often substantiated these results. Most of the studies have dealt with individual climatic factors and only one or two response measures (usually injury). A better understanding of how climatic factors affect the response of plants to specific pollutants is necessary before factors affecting field injury can be adequately interpreted.

Environmental conditions before, during, and after the exposure of plants to air pollutants may alter their response. Plants may change in sensitivity to a given pollutant after being grown for 1 to 5 days under a given set of conditions. Sensitivity appears to change most within about 3 days. Conditions during exposure can be extremely critical. Conditions after exposure are important, but probably less so than those before and during exposure. Several studies suggest that interactions between various climatic factors may play an important role in plant response to air pollutants.

1. Duration of Light (Photoperiod)

A given light duration within a 24-hour period controls certain aspects of plant growth and development. Annual bluegrass, pinto bean, and certain tobacco varieties were much more sensitive to ambient oxidants or to ozone when grown under an 8-hour photoperiod than under a 16-hour photoperiod (*145–148*). This response of annual bluegrass was independent of temperature variations during growth when exposed to ambient oxidants (*146*). Plant age did not affect this photoperiodic response of tobacco to ozone (*147*).

2. Light Intensity

Light intensity during growth affects the sensitivity of pinto bean and tobacco to a subsequent ozone exposure. Sensitivity increased with decreasing light intensities within the range of 4000 to 900 foot-candles (ft-c) (*145, 149, 150*). However, the sensitivity of pinto bean to PAN increased with increasing light intensity (*149*). Buckwheat was more susceptible to sulfur dioxide when grown at 3000 ft-c than at 7000 or 10,000 ft-c (*19*).

Early work with sulfur dioxide (*19*) and later work with products of photochemical smog (*146*) and ozone (*57*) showed that increased light intensity up to 3000 ft-c during exposure increased sensitivity. However, other work (*150*) reports no difference in response to tobacco Bel W_3 to ozone with light intensities between 2000 and 4000 ft-c during exposure, but a 40% increase in injury to pinto bean. An interaction between light and relative humidity during exposure of pinto bean to ozone was also reported (*150*). High light intensity increased sensitivity at lower humidities (60%) but was less important at higher humidities (>80%). These results emphasize the difficulty of drawing generalizations based on a few reports.

Plants exposed to pollutants in the dark (except for nitrogen dioxide) are generally not sensitive. At low light intensities, plant response is closely correlated with stomatal opening. However, since full stomatal opening occurs at about 1000 ft-c, light intensity must have an effect on plant response in addition to its effect on stomatal opening. Nitrogen dioxide may be an exception, since plants appear to be more sensitive during night than during day exposures (*72, 84a*). In addition, night exposures of sensitive plants to low concentrations of sulfur dioxide may predispose these plants to chronic sulfur dioxide injury (*151, 152*). Under summertime stresses of soil moisture and with high light intensity, tobacco is a more sensitive monitoring plant under 50% shade than in full sunlight (*153*).

3. Light Quality

The quality of light (wavelength) affects plant growth and development and may play a role in determining plant response to pollutants. In pinto bean, injury from PAN is maximal at 420 and 480 nm; injury at 640 nm is less than half as much (*149*). Apparently, the response is associated with changes in the carotenoid rather than in the chlorophyll pigments. No similar studies have been reported for other pollutants or plants. However, numerous reports suggest a differential sensitivity between plants grown outdoors, in greenhouses, and in growth chambers. Some of these differences may be due to variations in light quality. If such a relationship exists, it probably relates more to the effect of light quality on the physiology of the plant, which in turn affects sensitivity, rather than to a direct photochemical response such as reported for PAN.

4. Temperature

Plants grown at temperatures below 5°C lose sensitivity to air pollutants. Generally, sensitivity to oxidants increases with increasing temperature to about 30°C. However, in *Poa annua* (*146*), the effect of growth temperature on sensitivity varied with plant age. As the growth temperature was reduced, plants were most sensitive as they started to flower. Sensitivity of *Poa annua* growing under a given temperature regime is reversed within 3 days when changed to a different temperature regime. Soybeans are more sensitive to ozone when grown at 28°C than at 20°C, regardless of exposure temperature, potassium nutrition, or ozone doses (*154*). This same report showed that the response of pinto bean to a 20° and 28°C growth temperature was dependent upon both exposure temperature and ozone dose, but not on potassium nutrition. MacDowall (*147*) found that the sensitivity of tobacco to ozone was increased by low day and high night temperatures. Adedipe and Ormrod (*154a*) reported growth suppression in radish grown at 20°C but not at 30°C.

It is almost impossible to separate the effect of temperature at the time of exposure from the effect of light intensity, since a positive correlation exists between the two. Thus, early reports suggest a positive correlation between plant response to air pollutants and increased exposure temperature. However, when pinto bean or tobacco plants were exposed to ozone under controlled lighting and temperature conditions, sensitivity decreased with increasing exposure temperature between about 17° and 30°C (*57, 155*). Similar results were obtained for Virginia pine and white ash when exposed at temperatures between about 10° and 32°C (*156, 157*). Soybean response to ozone was the same when exposed at 20° or 28°C, although the plants were more sensitive to certain doses at the lower temperature (*154*). However, pinto bean was more sensitive at the higher exposure temperature (28°C) when grown at 28°C, and sensitivity was not affected by exposure temperature when grown at 20°C. It was more sensitive at the lower exposure temperature (20°C) when exposed to a fairly high ozone concentration for 2 or 4 hours. It is apparent from this and other work that the sensitivity of pinto bean to ozone is greatly dependent on many environmental conditions. This great variability raises a question as to the validity of in-depth studies with pinto bean until the causes for the variability are better defined.

Hull and Went (*158*) found a positive correlation between post exposure temperature and severity of injury to five plant species using temperature variations between 3° and 36°C.

5. Humidity

Field observations generally indicate that plants are more sensitive to air pollutants at higher humidities. However, these observations did not separate effects during growth from those during exposure. Dunning and Heck (*150*) showed that tobacco was slightly more sensitive to ozone when grown at 60% relative humidity instead of 80%. In this same study, the sensitivity of pinto bean to ozone was not affected by growth relative humidity when plants were grown at 4000 ft-c. However, sensitivity was increased as growth relative humidity was raised from 60 to 80% when plants were grown at 2000 ft-c. The relative humidity range (60 and 80%) was narrow and should be extended in future work.

Research supports the thesis that sensitivity to air pollutants increases as exposure relative humidity increases. However, the relative humidity differentials may have to be greater than 20% before differences are shown. Thomas and Hendricks (*21*) reported a 90% loss in sensitivity from sulfur dioxide exposures when the relative humidity was dropped from 100% to essentially zero. Davis and Wood (*156*) reported greater sensitivity in Virginia pine to ozone when exposed to ozone at 85% relative humidity rather than 60%. MacLean *et al.* (*159*) found gladiolus plants were more sensitive to fluoride as relative humidity increased from about 50 to 80%.

6. Carbon Dioxide

The carbon dioxide concentration within the leaf influences stomatal action and thus should influence plant sensitivity to pollutants. The only study on this subject (*145*) reported that tobacco, but not pinto bean, was less sensitive to ozone when 500 ppm of carbon dioxide was added immediately before and during exposure to ozone. The variability in ambient carbon dioxide concentrations in various areas suggests a need to further investigate this variable. If carbon dioxide protects plants from pollutant injury, it could be used in greenhouse management.

7. Pollutant Mixtures

One of the obstacles to understanding the effects of pollutants on plants is the lack of information about possible plant–pollutant interactions (additive, antagonistic, or synergistic) when mixtures of pollutants occur in the atmosphere (*159a*). Early research suggested that no interactions were present, or if present were antagonistic. Menser and Heggestad (*160*) first reported a positive interaction when they injured Bel W_3 tobacco by exposure to mixtures of ozone and sulfur dioxide at concentrations that were not independently injurious.

These results have since been substantiated by several investigators. Some, but not all, species appear to show a greater than additive interaction to mixtures of ozone and sulfur dioxide in certain ratios (*137, 161, 162*). Decreased growth and possible reductions in yield and productivity from mixtures of ozone and sulfur dioxide have been reported (*75, 76, 163, 164*).

Tingey *et al.* (*165*) reported that mixtures of nitrogen dioxide and sulfur dioxide injured six crop plants at concentrations that independently did not injure the plants. Traces of injury were observed at concentrations of 5 pphm of either gas when used in combination with from 5 to 25 pphm of the other gas. Exposure to the mixture caused chlorotic or necrotic upper surface fleck or pigmented lesions resembling ozone injury, instead of the interveinal bifacial necrosis commonly found after exposure to either gas separately. The general order of decreasing sensitivity to the pollutant mixture was soybean, radish, tobacco, pinto bean, oat, and tomato. Hill and associates reported a greater than additive reduction in net photosynthesis at sulfur dioxide and nitrogen dioxide mixtures between 15 and 25 pphm (*83*). They found no interaction between the two gases when seven native desert plants were exposed to gas mixtures (*166*).

Preliminary results with pollutant mixtures suggest that interactions do not occur with all gas ratios. Results with pollutant mixtures may partially explain the inconsistencies between results obtained in laboratory studies, where only single pollutants were used, and the results obtained in the natural environment.

8. Meterology

Several attempts to relate meteorological variables to plant injury have been reported. The most thorough studies were from Canada, where correlations between plant injury, ambient oxidant dose, and meteorological conditions were made (*167, 168*). A correlation was found when an empirical relationship involving evapotranspiration (the coefficient of evaporation) was developed and used to modify the dose information. By monitoring meteorological conditions, the Canadian workers have used this empirical relationship with fair success to predict oxidant occurrences that would cause ozone injury (fleck) to tobacco and other crops.

The relative importance of other factors, such as wind speed and barometric pressure, has seemed insignificant in relation to the primary factors already discussed. Air movement would be expected to play a role under ambient conditions because of its known effect on leaf boundary layers. It appears to have little effect in most chamber work (*169*) but might be important when wind velocity is greater than 1 mph (1.6 km/hour) (*170*). Under ambient conditions, it may be impossible to separate these other meteorological factors.

9. Diurnal Effects

Plants are generally more sensitive in mid to late morning and early afternoon. Under some conditions, depending on leaf maturity, there may be a midday loss of sensitivity. This time-of-day effect is related to the overall effects of the environment on plant physiological processes that then directly affect the sensitivity of the plant to air pollutants.

C. Edaphic Factors

Soil, water, and fertility affect the sensitivity of plants to air pollutants; however, this whole area of edaphic effects is critically in need of work. Hoffman *et al.* (*171*) and Maas *et al.* (*172*) studied the effect of soil water stress (—0.4, —2.4, and —4.4 bars) on the growth and injury response of pinto bean to ozone. They found that ozone caused less injury as soil water stress was increased, irrespective of ozone dose. Further studies of this type are needed to better elucidate other edaphic effects.

Soil water stress during plant growth and during exposure to air pollutants is probably the most significant environmental factor affecting the response of plants to pollutants. Khatamian *et al.* (*173*) found a reduction in injury and an increase in dry matter production in tomato exposed at the three-leaf stage to 50 or 100 pphm of ozone for 1 hour, when plants were grown in a water-stressed condition that did not itself cause a reduction in growth. Humidity and available soil water are the two primary factors that control overall water stress in plants. The relative importance of these two factors is difficult to separate under field conditions. Drought during growth causes physiological changes that increase the resistance of plants to pollution stress, and drought during exposure causes a decrease in stomatal opening, with a resultant reduction in the amount of pollutant in leaves. Ozone may cause stomatal closure in some plants only when they are under water stress (*174*). Several workers have recommended withholding water from greenhouse and irrigated crops during times of high pollution potential. Ting and Dugger (*175*) suggested that the relative sensitivity of Bel B and Bel W_3 tobacco is related to the small root system of Bel B that results in slower rates of water transport to the leaves.

There is a general lack of understanding as to the importance of soil fertility in affecting plant sensitivity to air pollutants, but ion availability must play a role. However, plants may be more sensitive when grown under low fertility. Nitrogen nutrition has received the most attention, but evidence on its effects is conflicting. Numerous studies suggest that plants growing in the presence of minimal nitrogen are more sensitive.

However three studies found the greatest sensitivity at optimum nitrogen concentrations and a reduced sensitivity at either low or high nitrogen nutrition (*176–178*). Brewer *et al.* (*179*) reported an interaction between potassium and phosphorus. Dunning *et al.* (*154*) found that pinto bean and soybean are more sensitive to ozone when the nutrient solution contains about a sixth the normal potassium level. Leone and Brennan (*180*) showed an increase in the sensitivity of tomato to ozone with increasing phosphorus in the nutrient solution.

Plants are not as sensitive when grown in heavy-textured soils, possibly due to decreased soil oxygen tension. The effects of soil temperature, aeration, texture, compaction, and composition have not been studied.

D. Other Factors

The interaction between pollutants and various biological agents as they affect plant sensitivity is receiving considerable effort, but the subject is still not well understood. The review by Heagle (*56*) points up the concept that most pathogenic organisms induce protection from ozone injury to the host plant. In general, pollution inhibits diseases caused by fungi but greatly increases the parasitism of *Botryti cinerea* and *Armillarea mellea* on several plants. Several papers should be reviewed for these effects (*181–186*). Reinert (*187*) found that the systemic presence of three tobacco viruses enhances the sensitivity of tobacco to ozone even when virus symptoms are not pronounced.

Cobb (*104*) reviewed the problem of Ponderosa pine decline in relation to oxidants and several species of bark beetles. It is generally considered that the beetles invade trees injured by oxidants and can increase the rate of decline. Pollution stress on sensitive tree species, such as sycamore, might result in increased feeding by insects such as lace bugs and mites, thereby causing early leaf senescence and abscission.

The development of chemicals to protect plants from oxidant pollutants has been stimulated by the increasing prevalence of the oxidant syndrome throughout agricultural areas, especially in the vicinities of large metropolitan centers. Antioxidants, various fungicides and other chemicals can decrease oxidant injury to a variety of plants, but no economically practical means of application have been found. Results of experiments utilizing chemicals as protectants can be found in many research papers. We have listed several of the more useful publications (*188–196*).

The sensitivity of plants is also conditioned by leaf maturity, as discussed under injury symptoms in Section II,A. Ting and Dugger (*85*) found cotton leaves were most sensitive to ozone at about two-thirds full expansion, which is representative of all work reported for both ozone and sulfur dioxide. Generally, studies show that young tissues are more sensitive to PAN and hydrogen sulfide, and maturing leaves are most sensitive to the other pollutants. Linzon (*197*) observed the greatest chronic injury to second-year needles of white pine.

E. Discussion

A full understanding of air pollution effects on vegetation is difficult to gain because of the many factors that determine plant response. This section has discussed the most important factors except for the pollutant concentration–exposure time (dose) relationships. The inherent genetic resistance is probably the most important single factor affecting plant response to a given pollutant dose. Environmental factors, however, also play major roles.

Any genetic or environmental factor that causes stomatal closure will markedly reduce plant sensitivity. Factors that affect the sensitivity of the plant to pollutants during exposure are probably those that affect stomatal aperture. Factors that occur during growth periods before exposure will affect the response of the plant through different mechanisms. Environmental stresses, such as extremes in moisture or temperature, if present for long, will cause physiological changes in the plant that make it more resistant to the added stress of air pollution. Many of these environmental stresses alter membrane physiology and make the membranes less sensitive to pollution stress. These changes include both the sulfhydryl linkages of membrane proteins and the unsaturated bonding of membrane lipids. These membrane changes explain the acute responses (Section II,B,2). The many changes in metabolic pool relationships that result from environmental stress during growth probably have a relatively minor effect on plant sensitivity to acute exposures but may decrease their sensitivity to chronic exposure.

IV. Effects of Air Pollutants on Biomass and Yield

Air pollutants may reduce growth without injury being visible. However, the literature generally supports the thesis that yield and/or growth reductions are associated with visible injury. Since there are many difficulties in correlating plant growth and productivity with plant injury, this section has been divided into discussions of each of the major pollutants.

A. Sulfur Dioxide

Katz and Ledingham (*197a*) found that sulfur dioxide did not affect alfalfa growth until at least 5% of the foliage was visibly injured, and Hill and Thomas (*198*) reported that yield reductions from acute sulfur dioxide injury were roughly equivalent to those occurring when the same amount of leaf tissue was removed. However, Guderian (*102*) found that alfalfa was highly sensitive when leaf necrosis was used as the sensitivity measure but was resistant when ranked in regard to yield. Light-to-moderate defoliation of cotton by sulfur dioxide did not affect fiber grade, staple length, or ginning percentage (*199*). Guderian and Stratmann (*200, 201*) reported a decrease in growth and yield of potato with increased pollution intensity. During the second year, plants started from tubers taken from the heavily polluted areas did not yield as well as plants started from tubers of similar size obtained in control areas.

For alfalfa the relationship between decreased yield and the percentage of leaf area destroyed is well expressed by a simple regression equation:

$$y = a + bx \tag{1}$$

where y = yield, expressed as percent of control; a = a constant of approximately 100%; b = slope of the yield–leaf-destruction curve; x = percentage of leaf area destroyed. Hill and Thomas (*198*) exposed alfalfa to sulfur dioxide one, two, or three times during the growing season and found the following regressions:

$$\text{single exposure} \quad y = 99.5 - 0.30x \tag{2}$$
$$\text{double exposure} \quad y = 95.5 - 0.49x \tag{3}$$
$$\text{triple exposure} \quad y = 96.6 - 0.75x \tag{4}$$

These results were obtained using exposures of 1–5 ppm for 1–2 hours. A similar equation was developed (*21*) for alfalfa using exposures of 0.1–3 ppm for from 1 to 600 hours:

$$y = 99 - 0.37x \tag{5}$$

Results of barley, wheat, and cotton differ from those for alfalfa, because yield of the first three is not always directly proportional to vegetative growth. With these types of plants, pollutants have a greater effect during the blossom and fruit development stages than at other growth stages. Examples for barley (*201a*) show this well:

$$\text{early vegetative stage} \quad y = 98 - 0.06x \tag{6}$$
$$\text{heading out (flowering) stage} \quad y = 98 - 0.40x \tag{7}$$

Similar equations have been developed for other crop species.

The comprehensive growth–yield experiments (*200, 202*) near a sulfur dioxide source at Biersdorf, Germany, have not been adequately studied in the United States. During the 7-month growing seasons of 1959 and 1960, a variety of plants, including cereals, vegetables, trees, forage, and fruit crops, were studied. Sulfur dioxide concentrations at five test stations varied with distance from the source and two other stations were developed as controls (chosen by distance and direction from the smelter). Plants were grown in large containers, and comprehensive injury, growth and yield data were obtained. The investigators spent considerable time attempting to develop a mathematical expression to relate pollution concentration, time of exposure, and time between exposures to effects (*65, 203*). This concept is briefly discussed in Section V. Plants were injured and growth and yield were decreased at all five of the pollution sites (Table II). The maximum 30-minute average concentrations reported and the percentage of time when measurable concentrations existed implies that numerous acute episodes caused the injury, primarily acute type, and decreased growth. The average growing season concentrations are probably indicative of the possible occurrence of higher concentrations but are not responsible, per se, for the effects found. The average concentrations during times sulfur dioxide was actually present are probably more indicative of effects and are used in Table II. No attempt was made to

Table II Sulfur Dioxide Concentrations, Duration of Exposures, and Effects on Biomass from the Biersdorf Studies (*200*)

Site	*Distance from source (m)*	*Maximum half hour concentration*[a] *(ppm)*	*Average concentration*[b] *(ppm)*	*Pollution time*[c] *(%)*	*Plant response*[d]
I	325	6.0	0.45	30	5
II	600	5.1	0.34	26	4
III	725	2.3	0.24	20	3
IV	1350	1.6	0.18	12	2
V	1900	1.3	0.15	8	1
VI	6000	(Control plot—reference station)			0
VII	Control	(Control plot—nursery station)			0

[a] Average of maximum values for 1959 and 1960.
[b] For the 7-month growing seasons of 1959 and 1960 (April 1 through October 31). These are the average concentrations during the time that measurable sulfur dioxide concentrations were recorded.
[c] The percentage of time that measurable pollution was present over the two 7-month growing seasons.
[d] Use of rating of 0–5: 0 = no effect, 1 = slight to no effect, 2 = light to no effect, 3 = moderate effect, 4 = moderate to severe, 5 = severe to plant death.

relate injury to growth effects by means of the regression equations reported earlier.

Sulfur dioxide concentrations and injury and growth suppression in white pine were reported for the 6-month growing season over a 10-year period around a smelter complex near Sudbury, Ontario, Canada (*4, 197, 204*). A summary of results (Table III) indicates that short-term averages around 0.25 ppm were probably required to reduce growth. It was not possible to separate the effects of acute exposures from probable effects of chronic exposures. Here again, the averages shown represent the times that sulfur dioxide fumigations actually occurred and may be indicative of expected effects. The injury reported is descriptive of that associated with both chronic and acute effects. No attempt was made in this study to determine whether injury was required to decrease growth.

Decreased growth attributable to sulfur dioxide is well presented in numerous publications, but very little information is available to suggest growth reductions unless visible injury occurs. Bell and Clough (*205*) found a 46% depression in final yield of ryegrass exposed to 0.12 ppm of sulfur dioxide for 9 weeks and a 52% depression when plants were exposed for 26 weeks to 0.067 ppm. Exposed plants showed chronic injury but no acute injury. Guderian (*102*) has reported reduced growth in a number of species, when grown singly or in combination with each other, after exposure for 8 or 12 hours to about 1 ppm of sulfur dioxide and in a mixture of three species after a 48-hour exposure to about 0.4 ppm. Plants were injured in these exposures. Adedipe *et al.* (*112*) showed a reduction

Table III Sulfur Dioxide Concentrations, Duration of Exposure, and Effects on Net Tree Volume of Eastern White Pine, Sudbury, Ontario ***(197, 204)***

Distance from Sudbury		*Maximum half-hour concentration (ppm)*	*Average concentration*[a] *(ppm)*	*Pollution time*[b] *(%)*	*Hours per 6 month growing season above stated concentration (ppm)*			*Average net annual gain or loss in tree volume (%)*
Miles	*Kilometers*				*0.25*	*0.50*	*1.0*	
16–19	26–31	3.64	0.216	20.8	240	95	15	−1.3
25–27	40–43	1.24	0.105	16.2	31	3.5	0.5	−0.6
40–43	64–69	0.63	0.078	10.3	12	0.5	0.0	+1.6[c]

[a] For the 6-month growing season (May 1 through October 31) over a 10-year period (1954–1963). These are the average concentrations during the time that measurable sulfur dioxide concentrations were recorded.

[b] The percentage of time that measurable pollution was present over the ten 6-month growing seasons.

[c] This plot was not out of the fumigation zone, but the net increase in tree volume was the same as from plots located well out of the fumigation zone.

in shoot weight (15%), flower weight (20%) and flower number (30%) of petunia (cultivar Capri) when exposed to 0.50 ppm of sulfur dioxide for 2 hours with from slight to severe visible injury. Cultivars of coleus and snapdragon showed reductions in these three parameters without development of injury symptoms. Radish root weight was reduced (*75*) after exposure to 5 pphm of sulfur dioxide for 8 hours per day, 5 days per week, for 5 weeks, with only a trace of injury. Several long-term studies in which oxone–sulfur dioxide mixtures were used with sulfur dioxide controls are summarized in Table V.

A major project was initiated in July of 1971 (*206*), when sulfur dioxide from a power plant was observed to cause extensive injury in a number of soybean fields. The study involved many fields of soybean and four varieties. The fumigation occurred in the mid-vegetative stage, before bloom. By the time of fruit set, the sulfur dioxide injury was not noticeable and no yield reductions were found. It is probable that severe yield reductions would have occurred if the fumigation has occurred around pod filling time.

B. Fluoride

Fluoride can decrease the growth and yield of many plant species. This topic is well presented in reviews dealing with criteria considerations for the possible setting of standards (*52, 53, 207, 208*). McCune (*207*) has devised a method to show effects on a log–log graph using mean concentrations of atmospheric fluoride versus duration of exposure. These are simple representations but are probably the best possible with the limited data available.

Growth reductions have normally been associated with leaf injury, but no attempts have been made to relate injury to yield reductions. However, some relationships must exist based on physiological considerations. Growth reductions have been quantified in the field (around fluoride sources) and in controlled exposures. Growth effects reported include decreased radial growth of trees, decreased dry weights of many plants (including rose, alfalfa, grass, lettuce and sorghum), loss of flower quality and gladiolus corm size, and reduction in flower or fruit numbers of tomato, citrus, sorghum, and bean. Decreased root growth is reported. Published reports show that the effects occur under a broad range of atmospheric fluoride concentrations, durations of exposures, times of exposure in relation to plant development, and both climatic and edaphic factors. Effects of fluoride on fruit quality and number have been reported for peach (soft suture), apricot, pear, cherry, and citrus.

There has been considerable interest in attempting to relate effects of fluoride with concentrations of fluoride in plant tissue. Leonard and Graves (*209*) developed linear regression equations for orange and grapefruit yield against fluoride content of the leaves using Equation (1), where y was yield of citrus and x was leaf fluoride in parts per million. They found significant yield reductions at leaf fluoride concentrations below 50 ppm. Growth of pinto bean was decreased when the fluoride concentrations in the leaves exceeded 300 ppm; growth of alfalfa decreased at leaf concentrations above 200 ppm, and growth of citrus and Douglas fir may decrease at leaf concentrations near 100 ppm (*74*). With respect to injury, some highly sensitive plants may have a threshold leaf level around 15–25 ppm and certainly under 150 ppm; resistant plants can probably tolerate tissue levels above 200 ppm.

C. Oxidants

Data are not available to develop specific equations for growth and yield effects of ozone or ambient oxidants on plants. Graphs similar to those produced by McCune for fluoride, however, may be of help in estimating oxidant dose–response relations over time. Several long-term studies involving ambient oxidants under field conditions are shown in Table IV (*126, 210–213*). Severe injury was reported in all these studies.

Table IV Effects of Ambient Oxidants on Plant Growth[a]

Plant species	*Location*	*Concentration*[b] *(pphm)*	*Duration of test*	*Effects (% reduction from control)*	*Reference*
Lemon	California	>10	Growing season	42% Yield	*213*
Orange	California	>10	Growing season	54% Yield	*213*
Grape	California	≥25	Often (May–Sept.)	12% Yield, year one	*212*
				61% Yield, year two	*211*
				47% Yield, year three	*211a*
Tobacco, Bel W_3	North Carolina	5–10	Often during growing season	30% Leaf weight	*210*
Corn, sweet	California	20–35	Often during growing season	5% of 34 Cultivars were unmarketable	*126*

[a] Plants were exposed continuously to ambient air. All plant species except sweet corn were compared with plants growing in charcoal-filtered air.

[b] The concentration was equal to or above the levels shown for some period of time during each day. Sufficient data were not available to calculate total dose.

Studies using controlled pollutant additions permit accurate determination of total dose for long-term studies of chronic injury. Although these studies are suggestive of "real world" effects, cumulative acute exposures are probably responsible for chronic effects under most field conditions. A summary of these long-term controlled chronic studies is shown in Table V (*75, 127, 163, 164, 171, 172, 214–216*). Plants in two studies (*127, 163*) were grown in the field and enclosed in field chambers to ensure controlled pollutant dosages. Three of these studies (*127, 163, 172*) show that a correlation between injury and certain growth parameters may occur and may then be useful in predicting reduced yield in field situations (Table VI). However, two studies (*75, 164*) showed growth reductions with little or no associated injury.

The data from three studies (*75, 163, 164*) examined possible growth effects from ozone–sulfur dioxide mixtures (Table V). The mixture produced additive or less-than-additive yield reductions for radish (*75*). In laboratory studies of soybeans (*164*), the mixture produced greater-than-additive effects, but the effects were the same as or less than a doubling of ozone alone. In field studies of soybeans (*163*), the mixture produced only additive effects.

Several investigators measured growth responses to acute ozone exposures after a given period of regrowth (Table VII). The data showed that growth measurements gave good response separations for acute exposures and thus may be useful in the development of dose–response equations. Although injury percentages are shown, they were obtained in a subjective fashion from the papers referenced and thus are indicative only.

Field observations suggest that growth and yield have been reduced for many years in many plants, but little specific data are available to quantify these observations. When a field of spinach is badly injured by oxidant air pollution, it may be a complete loss to the farmer. Thus, although the presence of pollution effects are known, they are not always well documented.

D. Other Pollutants

Some growth and yield information is available for nitrogen dioxide and ethylene effects on plants. Thompson *et al.* (*217*) reported leaf drop and decreased yields of navel oranges exposed to 0.25 ppm of nitrogen dioxide for 8 months. Taylor and Eaton (*80*) found a decrease in dry weight and an increase in chlorophyll content of pinto bean exposed to 0.3 ppm of nitrogen dioxide for 10–19 days. They found a decrease in dry weight of tomatoes exposed to from 0.15 to 0.26 ppm of nitrogen dioxide for a similar time.

Table V Effects of Long-Term Controlled Ozone Exposures on Plant Growth[a]

Plant species	*Exposure type*	*Dose [concentration (pphm) × time (hours)]*[b]	*Effects (% reduction from control)*	*Reference*
Lemna	Laboratory	700 (10; 5/day, 14 days)	100% Flowering	*214*
		700 (as above)	36% Flowering (after 1 week)	
Carnation	Laboratory	15,210 (5 to 9; 2160)	50% Flowering	*215*
Geranium	Laboratory	7,268 (7 to 10; 9.5/day, 90 day)	50% Flowering	*215*
Radish	Laboratory	1,000 (5; 8/day, 5 days/week, 5 weeks)	54% Root fresh weight; 20% leaf fresh weight	*75*
		1,000 + 1000 SO_2 (as above, for each gas)	63% Root fresh weight; 22% leaf fresh weight	
		1,000 SO_2 (as above)	30% Root fresh weight; 7% leaf fresh weight	
Soybean	Laboratory	1,200 (10; 8/day, 5 days/week, 3 weeks)	24% Root fresh weight; 21% top fresh weight	*164*
		600 (5; same time frame)	3% Root fresh weight; 2% top fresh weight	
		600 + 600 SO_2 (as above, for each gas)	24% Root fresh weight; 12% top fresh weight	
		600 SO_2 (as above)	0% Root fresh weight; 5% (increase) top fresh weight	
		2,400 SO_2 (20; same time frame)	2% Root fresh weight; 3% top fresh weight	
Pinto bean	Laboratory	210 (30; 0.5/day, 14 days)	11% Dry weight of leaves	*172*
		420 (30; 1/day, 14 days)	40% Dry weight of leaves	
		840 (30; 2/day, 14 days)	70% Dry weight of leaves	
		1,260 (30; 3/day, 14 days)	76% Dry weight of leaves)	
Pinto bean	Laboratory	1,890 (15; 2/day, 63 days)	33% Plant dry weight; 46% pod fresh weight	*171*
		3,150 (25; 2/day, 63 days)	95% Plant dry weight; 99% pod fresh weight	
		4,410 (35; 2/day, 63 days)	97% Plant dry weight; 100% pod fresh weight	
Pinto bean	Laboratory	2,912 (13; 8/day, 28 days)	79% Top fresh weight; 73% root fresh weight	*216*
Sweet corn	Field	1,920 (5; 6/day, 64 days)	9% Kernel dry weight; 12% other yield measures	*127*
Sweet corn		3,840 (10; 6/day, 64 days)	45% Kernel dry weight; 35% other yield measures	

Table V *(Continued)*

Plant species	*Exposure type*	*Dose [concentration (pphm) × time (hours)]*[b]	*Effects (% reduction from control)*	*Reference*
Soybean	Field	3,990 (5; 6/day, 133 days)	3% Seed weight; 22% plant fresh weight	*163*
		7,980 (10; 6/day, 133 days)	55% Seed weight; 65% plant fresh weight	
		7,980 + 7,980 of SO_2 (as above for each gas)	63% Seed weight; 72% plant fresh weight	
		7,980 SO_2 (as above)	0% Seed weight; 10% (increase) plant fresh weight	

[a] Three studies show the effects of ozone–sulfur dioxide mixtures along with the effects on the sulfur dioxide controls.
[b] This column contains considerable information and reads as follows for Lemna, the first entry: 700 pphm-hours (10 pphm; 5 hours/day, 14 days). The information in parentheses gives sufficient information to calculate the dose given. Concentrations are always given first followed by a semicolon, the remaining information shows the time periods that the plants were exposed to the given concentration.

Ethylene is a growth hormone in plants and is naturally produced. Thus, it is not surprising that growth reductions and yield losses have been reported as a direct effect of ethylene in cotton (*218*) and in numerous other plants (*77*).

Accidental release and spills of gaseous pollutants causing severe injury to vegetation within a limited area can cause slight to severe decreases in yield and productivity. One has but to see a field of soybean completely

Table VI Injury versus Growth Reductions in Plants Exposed to Ozone

Plant species	*Exposure type*	*Dose (pphm-hours)*	*Plant growth measures (% reduction from control)*	*Foliar injury (% increase over control)*	*Reference*
Pinto bean	Laboratory	210	11	9	*172*
		420	40	46	
		840	70	69	
		1260	76	78	
Sweet corn	Field	1920	12	14	*127*
		3840	35	25	
Soybean	Field	3990	22	19	*163*
		7980	65	37	
		7980 + 7980 SO_2	72	46	
		7980 SO_2	10 (increase)	7	

Table VII Effects of Short-Term Acute Ozone Exposures on Plant Growth

Plant species	Concentration (pphm)	Time (hours)	Effects: Percentage growth reduction from control	Effects: Percentage increase in injury from control[a]	Reference
Cucumber (Ohio mosaic)	100	1	19% Top dry weight	1	*120*
	100	4	37% Top dry weight	18	
Onion (spartan era)	20	24	0	0	*120*
	100	1	21% Plant dry weight	2	
	100	4	48% Plant dry weight	6	
Potato (Norland)	100	4	0	0	*120*
	100	4 (2 times)	30% Tuber dry weight	Some	
Begonia (white Tausendschon)	10	2	5% Growth[b]	0	*112*
	20	2	10% Growth	0	
	40	2	19% Growth	5	
	80	2	37% Growth	65	
Petunia (Capri)	10	2	9% Growth	0	*112*
	20	2	11% Growth	0	
	40	2	21% Growth	5	
	80	2	31% Growth	—	
Coleus (scarlet rainbow)	10	2	2% Growth	0	*112*
	20	2	17% Growth	0	
	40	2	24% Growth	0	
	80	2	39% Growth	0	

[a] These are estimated values taken from data presented or from the discussion of the data.
[b] The growth effects in the rest of this column are the averages of reductions from control values for three parameters—shoot weight, flower weight, flower number.

burned from an accidental release of ammonia to understand that the field would not produce soybean that year. However, there are no studies concerning the frequency and extent of such accidental releases.

Brandt and Rhoades (*219*) found a reduction of lateral growth (about 18%) in three trees and an increase of lateral growth (76%) in one species in an area fumigated with limestone dust since 1946. They suggest the increase in the tulip poplar was a result of competitive advantage of this species over the other three due to its tolerance of limestone dust.

V. Dose–Response of Vegetation to Air Pollutants

An understanding of the effects of a given dose (concentration of pollutant times duration of exposure) on plants is essential for the development of intelligent air quality standards and as a basis for understanding the effects of pollutants. The ideal criteria would be a set of standard equations that would relate response to concentration and duration of exposure (time), and would reflect the effects of all other factors that control the response of the plant. In theory, such a model should be practicable with a different set of constants for each plant species under a multiplicity of experimental conditions. However, such a multivariate model could become so complex as to be of little practical importance. If the model were to assess only the most critical resistance factors, if these factors could be easily measured, and if average values for these factors were to reflect a given number of plant susceptibility groupings, then such a model would be of tremendous importance to both the air pollution control official and the agricultural specialist.

Earlier discussions (Section II,B,2) suggest that two types of models are required if we are to define "what is happening." These two models would not be mutually exclusive. One should explain the acute membrane response, and the second should explain the chronic and subtle responses. When yield and growth effects are measured, the acute exposures will include some chronic responses. Under ambient exposures, long-term chronic responses will undoubtedly reflect one to many low-level acute responses. The basic models should account for these varied responses.

A. Acute Response

The first attempt to develop an acute dose–response model was an empirical relationship developed by O'Gara for sulfur dioxide and later modified by Thomas and Hill (*220*). This equation was developed from acute exposures of alfalfa to sulfur dioxide when grown under conditions of maximum sensitivity. Visible injury was the response measure used. The equation is usually written in a form that emphasizes the concept of a threshold concentration and dose:

$$t(c - a) = b \tag{8}$$

where t is exposure time in hours at a concentration c parts per million, a is a concentration threshold for a specific degree of injury, and b is a constant reflecting the other components of the model, such as inherent plant resistance and the external factors affecting resistance.

To facilitate comparison with equations developed by others, it is convenient to arrange the O'Gara equation in the form:

$$t = b/(c - a) \tag{9}$$

Thomas and Hill (*220*) reported a and b values for alfalfa at three plant injury levels when c was measured in parts per million of sulfur dioxide and t in hours:

	a	b
For incipient (threshold) injury	0.24	0.94
For 50% leaf destruction	1.4	2.1
For 100% leaf destruction	2.6	3.2

The leaf injury occurred on leaf tissue showing maximum sensitivity to the sulfur dioxide.

Guderian *et al.* (*65*) developed an exponential equation that they felt gave a better fit to their experimental data than the O'Gara equation. This equation is written in a form to make it consistent with the O'Gara equation above:

$$t = K \exp[-b(c - a)] \tag{10}$$

where K is given as the vegetation lifetime in hours. This constant probably makes the equation somewhat more applicable for long-term exposures. However, the equation was developed from data obtained from plants that had received several acute exposures.

Zahn (*152*) proposed a function that he suggested would fit experimental data over longer time periods than either the O'Gara or Guderian equations. The equation is basically of the same form as the O'Gara equation.

$$t = b\frac{1 + 0.5c}{c(c - a)} \tag{11}$$

In his original equation, Zahn used KP for the b term, where K is a resistance factor expressing varying environmental conditions (external conditions) and P is an inherent plant resistance factor. In fact, the b term in all three equations is an attempt to incorporate known and unknown resistance factors. Unless we know these resistance factors or have some means of comparing their relative importance, the b factor is just as useful as the KP of Zahn.

All three of the above equations were developed from sulfur dioxide exposures and reflect the fact that concentration plays a greater role in the response of plants to a pollutant than does time. It has long been known, but often forgotten, that equal doses do not give equal responses. A given dose applied for a short time, produces a much greater plant response than does the same dose applied over a longer time frame. In fact, the very concept of a concentration threshold assures this basic dose concept. Mechanistically, this concept must hold, since biological organisms have some inherent mechanism for detoxifying toxic materials. Zahn contends (*152*) that the O'Gara and Guderian equations do not adequately reflect this relationship. For the O'Gara equation, this contention has some validity, since the O'Gara equation promises an equal dose effect after subtracting the threshold concentration, and this is not always valid. The Zahn expression does say that for a given effect to occur in a short time period, a lower concentration than would be predicted from the O'Gara equation is needed. For long-term exposures, it suggests that the time needed to reach an effects threshold is much longer than would be predicted by O'Gara. However, within the accuracy of present observations, all three equations appear to predict acceptably the acute vegetation response. Thus, it appears that the O'Gara equation might be a useful and acceptable model to follow.

None of these equations adequately describes the variation in response as either time or concentration, or both, vary. A suitable description for the acute exposure of pinto bean and tobacco to ozone was developed using a three-dimensional surface (*221*). The graph for pinto bean is reproduced in Figure 3. Graphical surfaces such as these highlight the steep slope in both the response–concentration and the response–time planes. The steepness of these slopes gives a relative measure of the degree of variability to be expected in data collected under practical experimental conditions.

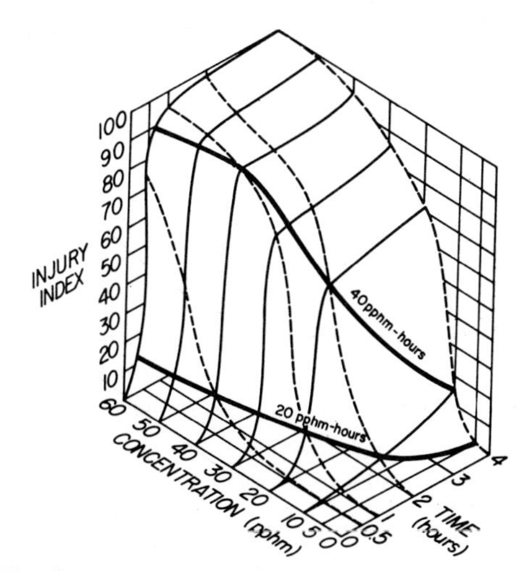

Figure 3. Interrelations of time and concentration on the sensitivity of pinto bean plants to ozone. From Heck *et al.*, (*221*).

To better relate concentration, response, and time, a group of plants was exposed to one (4 hours) or two ozone concentrations at each of five exposure times (0.5, 1.0, 2.0, 4.0, and 7.0 hours) and the acute response was determined. The data were used to develop an equation following the basic O'Gara equation (*222*):

$$c = a + b_1 I + b_2/t \tag{12}$$

where I is the percentage of plant response, b_1 is the inherent resistance within the plant, b_2 reflects the external resistance factors, and c is measured in parts per hundred million. The equation permits the development of a response surface, but it forces the concentration–response two-dimensional curves into linear relationships that are not representative of the sigmoidal curves normally generated from actual exposures. These equations do depict an accurate conceptualization of concentration–time graphs for specific response levels. Thus, they can be used to generate the O'Gara type of curve represented by the equation:

$$c = a + b/t \tag{13}$$

where a c versus t plot permits an easy interpretation.

We have critically reviewed the time–concentration–response studies reported in the literature for controlled ozone exposures for inclusion in this chapter. From these studies, we developed two-dimensional curves for both the concentration–response and the time–response of over one hundred species and varieties of plants. From these curves, we generated the times and concentrations required for 5 and 33% plant responses. The values generated were used collectively to develop the c versus t equations (Eq. 13) for each of the response percentages. The equations developed were:

For 5% response $\quad c = 15.93 + 11.53/t$
For 33% response $\quad c = 20.60 + 21.10/t$

Plots of the 95% confidence limits for these two curves are shown in Figure 4. These two curves (5 and 33%) represent response measures to ozone from many species and varieties of plants. It is suggested that the confidence limits reflect the expected response of plants of intermediate sensitivity to ozone, since the data are representative of plants of mixed inherent resistances exposed under varied external conditions. The lower confidence limit, then, may reflect the threshold for plants of intermediate sensitivity.

The response information developed for Figure 4 could be developed for sulfur dioxide and for nitrogen dioxide, although much less information is available for nitrogen dioxide. We suggest this as an effective approach to the development of criteria for use in establishing ambient air quality standards, whenever sufficient information is available.

The information presented in this section is pertinent only to those concentrations of a pollutant that will produce an acute response within a given pollution episode—no longer than 10 to 12 hours. None of the data should be extrapolated beyond this time period. The four equations (Eq. 9–12) have been developed to explain the acute response. The Guderian and Zahn equations (Eqs. 10 and 11) attempt to relate to repeated acute responses over the growth period and appear to have some empirical validity. However, they suffer from the lack of experimental data to develop numbers for their constants. Experimental designs developed specifically for that purpose are necessary before a simplified multivariant model can be developed.

Although a response threshold for sensitive plants cannot be developed from Figure 4, the figure was developed from over one hundred experimental points. When these points are inserted into the figures as a scatter diagram, they permit the development of projected ozone concentrations

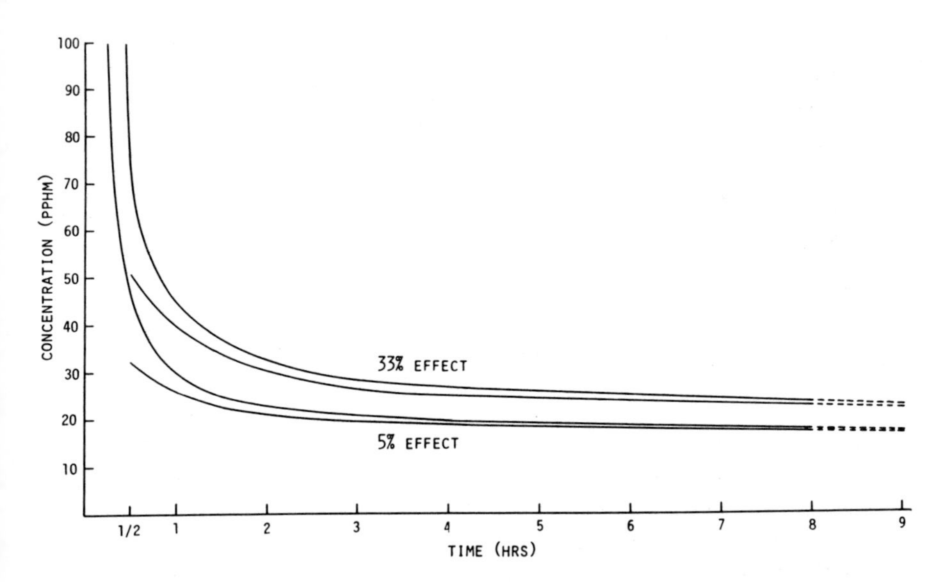

Figure 4. Ozone concentration versus time ($c = a + b/t$) from Equation (13) needed to produce 5% and 33% acute injury. The upper and lower 95% confidence limits curves are shown for each percent injury. Data were obtained from exposures of over one hundred plant species and cultivars. The 5% curves fit the intermediate susceptibility grouping in Table VIII. The 33% curves fit the resistant susceptibility grouping in Table VIII.

Table VIII Projected Pollutant Concentrations for Short-Term Exposures That Will Produce about 5% Injury[a] to Vegetation Grown under Sensitive Conditions

Pollutant	Time (hours)	Concentrations (ppm) producing 5% injury		
		Sensitive	*Intermediate*	*Resistant*
Ozone	0.5	0.20–0.35	0.30–0.55	≥0.50
	1.0	0.10–0.25	0.20–0.35	≥0.30
	2.0	0.07–0.20	0.15–0.30	≥0.25
	4.0	0.05–0.15	0.12–0.26	≥0.23
	8.0	0.03–0.12	0.10–0.22	≥0.20
Sulfur dioxide	0.5	1.0–4.0	3.5–10	≥9.0
	1.0	0.50–2.5	2.0–7.5	≥7.0
	2.0	0.30–2.0	1.5–5.0	≥4.5
	4.0	0.15–1.25	1.0–3.5	≥3.0
	8.0	0.10–0.75	0.50–2.0	≥1.5
Nitrogen dioxide	0.5	6.0–12	10–25	≥20
	1.0	3.0–10	9.0–20	≥18
	2.0	2.5–7.5	7.0–15	≥13
	4.0	2.0–6.0	5.0–12	≥10
	8.0	1.5–5.0	4.0–9.0	≥8

[a] These values were taken after some revision from Heggestad and Heck (*27*), the United States Environmental Protection Agency (*32*, *33*), and the National Air Pollution Control Administration (*36*).

that would produce a threshold response in three susceptibility groups of plants (Table VIII). The values projected are similar to those generated earlier (*27*). Similar values are shown (Table VIII) for sulfur dioxide and nitrogen dioxide. Such tables can be used to list plant species in three general susceptibility groups. Such subjective delineations of species and varietal susceptibilities are found in several reviews (*32, 33, 36, 43, 44, 53*). Similar listings are found in many earlier reviews and in original research papers.

B. Chronic Response

Although the equations by Guderian *et al.* (Eq. 10) and Zahn (Eq. 11) were developed to apply to long-term chronic responses, data have not been developed to adequately test the equations. The research required for the development of such equations is time consuming, tedious, and marked with pitfalls (*65, 152, 203*). Stratmann (*203*) covers, in some detail, problems related to atmospheric measurements and how they can be used to develop averaging times that are useful in understanding plant effects.

The best attempt to depict graphically the chronic time and concentration effects of pollutants on vegetation was developed by McCune (*207, 208*) for fluoride. The relationship between atmospheric fluoride concentrations and plant response has been studied extensively in controlled exposures using both greenhouse and field exposure chambers. Since gaseous fluorides constitute the most significant phytotoxic form of fluoride, most research has dealt with exposures to hydrogen fluoride; some utilized fluorine, silicon tetrafluoride, fluorosilicic acid, or a particulate form. The effects of specific gaseous fluoride concentrations are generally determined from continuous exposures at constant fluoride concentrations. Therefore, problems arise in the interpretation of controlled experiments in relation to nonuniform field conditions where the frequency, sequence, concentration, and duration of exposures vary in the ambient air. Although these ambient variables influence the plant response, controlled pollutant additions have been the only acceptable method of relating plant responses to the concentration and duration of fluoride exposures.

Since the nature of the relationship between plant response, length of exposure, and atmospheric concentration is not known, the simplest assumption is that the degree of injury and fluoride accumulation have the same relation to concentration and time. Some mathematical relationship is needed to develop threshold concentrations over an extended time period.

McCune and Weinstein (*54, 207, 222a*) plotted the available data for eight plants or groups of plants—tomato, alfalfa, sorghum, corn, gladiolus, tree fruits (other than citrus), citrus, and conifers. The data plots consisted of a notation of response (leaf symptoms, no symptoms, yield effect, etc.) on a graph defined by the log of the mean atmospheric concentration (micrograms of fluoride per cubic meter) and the log of exposure duration (days). From these plots, threshold curves were developed for various effect parameters of the plants listed. Although these are not calculated curves and may be subject to some error, they satisfy the concept of threshold value, below which no injury should occur. Figure 5 (*207*, Fig. 10) has been included to show threshold effect curves for tomato, corn, sorghum, gladiolus, tree fruits, and conifers. The suggested threshold lines for the three most sensitive groups (gladiolus, sorghum, and conifers) terminate at 0.5 μg F/m^3 over the exposure period used (2–4 weeks). The data for citrus (not shown in the figure) suggest 0.5 μg F/m^3 or less as a long-time exposure limit (18 months) for an effect to be seen.

Projected concentration ranges of gaseous fluoride that will injure

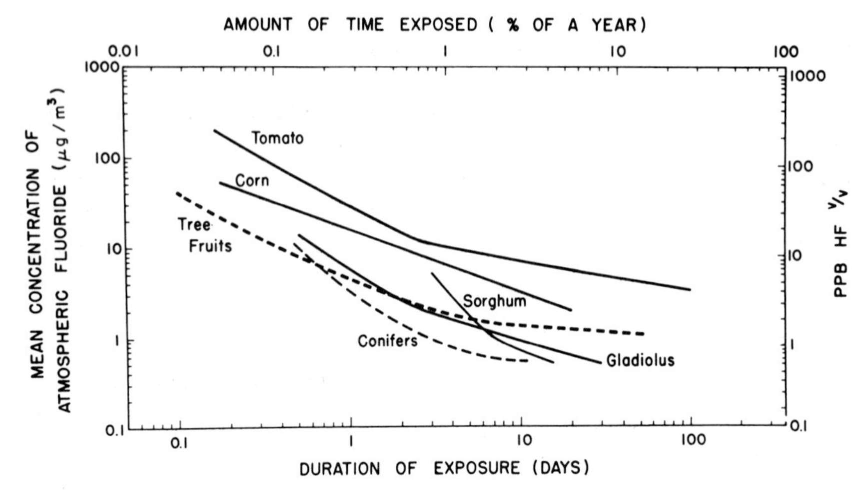

Figure 5. Possible injury threshold for atmospheric fluoride, with reference to different plant species. From McCune (*207*).

three different susceptibility groups of plants are shown in Table IX. The susceptibility groupings are the same as those used in references (*44*) and (*53*). It may be possible to treat chronic responses to sulfur dioxide and oxidant (ozone) in a similar manner.

Table IX Projected Gaseous Fluoride Concentration Ranges That Will Produce about 5% Injury[a] to Vegetation Grown Under Sensitive Conditions

	Concentration ($\mu g/m^3$) *producing 5% injury*		
Time	*Sensitive*	*Intermediate*	*Resistant*
8 Hours	2.0–6.0	5.0–30	≥25
12 Hours	1.5–5.0	4.0–27	≥22
24 Hours	1.0–4.0	3.0–20	≥15
1 Week	0.75–2.0	1.5–8	≥7
1 Month	0.5–1.0	1.0–5	≥3
Growing season	0.3–0.7	0.5–2	≥1
Year[b]	—	0.2–0.5	—

[a] Injury is defined as any measurable effect on the plant as a result of exposure to fluoride. These values can be related to sensitivity groupings in Jacobson and Hill (*44*) and the Committee on the Biological Effects of Atmospheric Pollution Weinstein and McCune (*222a*).

[b] Important only for a few southern tree crops and ornamentals (specifically for citrus).

VI. Vegetation as a Biological Monitor for Air Pollutants

The use of plants to monitor air pollution problems has often permitted the early detection of these problems, since plants are often the most sensitive biological receptor. Plants have been used in field surveys to determine the accumulation of certain pollutants and to determine the phytotoxicity of pollutants generated in special irradiation chambers.

A. Historical Use in Field Surveys

Since plants may develop characteristic symptoms when exposed to acute or chronic concentrations of certain air pollutants, they can serve as a useful tool in field surveys. However, even characteristic symptoms are not necessarily definitive of cause, since disease, insects, cultural conditions, etc., singly or in combination, may produce similar, if not identical, injury patterns. Thus, considerable caution must be exercised in defining an air pollution problem.

Field surveys have a significant place in the assessment of air pollution problems (*222b*). Basic techniques originated in the nineteenth century in work involving sulfur dioxide. The field survey attempts to answer the question: Does this area have a significant air pollution problem? Often this is somewhat in the category of confirming the obvious, since a survey would not be initiated unless there were subjective reasons for assuming that the answer would be affirmative. In such cases, the surveys are designed to produce objective data.

Baseline data on air pollution effects on vegetation cannot be established after an air pollution source has been built. Baseline data in urban areas are very difficult to obtain, because other environmental variables interact with air pollution. The effects of a given pollutant from a single source can be determined if some history of the source is available. Thus, baseline studies are needed before a new pollutant source begins operation, but the investigator must realize that such baseline studies still will include background pollution levels.

If both the pollutant and its effect on vegetation are known, we can examine a field area for a relatively few characteristic symptoms. Even identifying the visible injury symptoms will not permit an investigator to deduce with a high degree of confidence the amount, duration, or number of exposures to an injurious pollutant. A competent observer must make a series of inferences. Any individual judgment is subject to a high degree of uncertainty arising from the tissue age, the site, the weather conditions, and the history of the monitoring plant. But through a series of these inferences or judgments concerning different plants in different stages of growth, a competent observer is often able to say with a rather high degree of credibility, "This area has been fumigated with this pollutant."

The limitations of this approach are obvious. First, by observation of vegetation we can identify with any certainty, specific necrotic symptoms caused by only a relatively few pollutants—sulfur dioxide, fluoride, ozone, and PAN. The symptoms caused by ethylene, chlorine, and other pollutants are more difficult to identify. Second, we are dependent upon the presence of susceptible varieties at a susceptible stage of growth when the exposures occur. Third, a competent observer must be present to observe the symptoms after they have developed but before they have become obscured by other changes. The interpretation of observations collected in this way, although dependent upon the frequency or the extent of the observed field fumigation(s), must rest finally with the observer. The observer must always keep in mind that the vegetation responds to all environmental factors. Thus, the observation that the appearance of the vegetation around a pollution source is comparable to that of the vegetation in an adjoining "clean" area by no means denies the hypothesis of a significant air pollution problem.

Field surveys involving sulfur dioxide go back to the nineteenth century (*20*). A comprehensive survey was made at the Trail smelter (*3, 223*) to interrelate several meteorological parameters with topography, species susceptibility, and levels of sulfur dioxide. With the increase in power consumption and, therefore, the number of large power stations burning coal, the surveillance program initiated by the Tennessee Valley Authority (TVA) has interesting implications (*224, 225*). Using sensitive natural indicator species, the pattern of injury around the power station was found to be dependent upon the pollutant dispersion pattern for the power station. This in turn was conditioned by topography, ground cover, local weather conditions, and power station design. Thus, a knowledge of the dispersion pattern proved a useful tool in determining where and how much vegetation injury to expect under given conditions.

B. Pollutant Accumulation in Plants

In some cases, such as some salts, sulfur oxides, and fluorides, accumulation of the pollutant in foliar tissue may identify the cause for the observed symptoms. Salt injury along roads can often be confirmed by chloride analysis. For most situations involving occasional acute fumigations with sulfur dioxide, tissue analysis for sulfate will not satisfactorily confirm the causative agent (*226*), because the normal sulfate content of leaf tissue varies over a fairly wide range of values depending upon species, soil fertilizer practices, and other factors. However, for chronic air pollution exposures (*226*) or for acute exposures imposed upon a chronic exposure, such as reported by McKee and Bieberdorf (*227*), sulfate levels in the leaf tissue may confirm sulfur dioxide as the causative agent for the observed field symptoms.

Fluoride accumulation in plant tissues has special significance, because even small amounts can indicate fluoride problems. Plants grown in areas free from fluoride air polution problems rarely accumulate as much as 20 ppm fluoride on a dry weight basis. Fluoride accumulation has been used to help confirm the effects of fluoride injury to plants and of fluoride levels in forage. The visible leaf symptoms produced by fluorides are generally quite distinctive. Because the injury develops as a result of fluoride accumulation in the tissue, analyses for fluoride will provide confirmation of fluoride as the causative agent. Because excess fluoride in animal diets can be seriously detrimental to teeth and bones, the fluoride content of forage is an important consideration in animal management (*52*, *227a*). In several areas, special monitoring programs are used to provide information on the fluoride content of the vegetation.

Many workers have recognized the importance of relating atmospheric fluoride concentration to flouride uptake by plants over time but have felt that such relationships were not possible considering all the important variables in pollutant uptake (*44*, *55*, *208*). Brandt (*52*) fitted the data from several investigators (*228–230*) to the simple dose rate relation

$$F = Kct \tag{14}$$

where F is tissue fluoride concentration (parts per million) accumulated over background, c is the average atmospheric fluoride concentration (micrograms per cubic meter) over the exposure time t (days) and K is an accumulation coefficient. He noted that this relation assumed that the plant acted as a passive accumulator and that, if this were the case, K would be a constant dependent only upon the variety of the plant. Obviously, except for short periods of time under constant environmental conditions (temperature, light, windspeed, etc.), plants do not act as passive accumulators. Stomata open and close, leaf areas and surface to mass ratios change, and windspeed varies. All of these factors will affect the coefficient K.

In summarizing their extensive data, the group at Boyce Thompson Institute (*231*, *232*) used a similar relation

$$F = act + b \tag{15}$$

where a and b are regression coefficients, and the other terms are the same as noted in Equation (14). The coefficient a has the same dimensions and interpretation as the K (Eq. 14). These workers also studied the loss in fluoride over a postexposure "weathering" time period under field conditions. For alfalfa, the loss is quite significant, with an estimated half-life of 8–22 days. The loss in orchard grass was not significant. The "accumulation coefficients" from these studies are 1.89 for alfalfa and 1.13 for orchard grass. The average K values derived by Brandt (*52*) from the data available at the time were 2.4 for alfalfa and 1.4 for orchard grass. A more detailed review of this approach including the pertinent tables from Brandt's review are more readily available in the National Academy of Science document on fluorides (*222a*).

An accumulation coefficient K is derived for use as a tool for developing air quality standards by relating exposure time and concentration to accumulation. The derived K values are not strictly constant. They vary somewhat with concentration and time. However, agreement in the values derived from data obtained under widely different environmental conditions (California, New York, Germany) and different cultural techniques suggest that the uncertainty in the value may be within a factor of 2. Work by Israel (*233*) adds support to the concept and its applicability to the field problem.

Israel (*233*) used data from an extensive monitoring program around an aluminum plant in a dairy region of Maryland. The twenty-two monitoring sites were distributed around the plant at various distances up to about 4 miles (6 km) from the plant center. The program provided 3 years of data on fluoride content of forages (hay and pasture), accumulation by lime paper, and the measurement of atmospheric fluoride using dual bicarbonate tube samplers at several sites. Israel's analysis of the data demonstrates good correlation between the fluoride content of forage samples and the gaseous fluoride concentration measured by the bicarbonate tube sampler. The data do not permit calculation of individual K values for the various species in the forage, alfalfa, orchard grass, bluegrass, and clover. The K for the mixed forage is 3.8 ± 0.8 based upon free gaseous fluoride in the atmosphere.

This value is somewhat higher than we would expect from the K values derived from fumigation experiments, but is within the suggested uncertainty of a factor of 2 (*52, 222a*). However, there are some unique aspects to the atmospheric sampling program used in this study. The data reported by Israel (*233*) suggest that, of the total fluoride in the atmosphere around this plant, 13% occurs as free gaseous fluoride, 23% occurs as gaseous fluoride adsorbed on particulate, and 64% occurs as particulate. If the total gaseous fluoride (free and adsorbed on particulate) is used to determine K, then K for the mixed forage is 2.0 ± 0.3.

The concept of a simple accumulation coefficient K was initially suggested (*52*) as a value that could be derived from existing data that might serve as a basis for a first approximation of an air quality criterion for fluoride in relation to forage. The additional data from Boyce Thompson (*231, 232*) added support to the idea, and the data from Israel (*233*) added the needed first confirmation from the field. If the management practices of the area are such that an accumulation of 30 ppm fluorine over 30 days can be tolerated, average gaseous fluoride levels should not exceed 0.5 $\mu g/m^3$ (for a K of 2) or 0.3 $\mu g/m^3$ (for a K of 3.5). If field management practices permit, a longer exposure time or a different accumulation tolerance for safe air levels are readily estimated. These are only estimates of the limiting air concentration needed to limit tissue accumulation at a safe level for grazing animals. The uncertainty in the estimate is no greater than the uncertainty in the translation of an emission rate to the ambient level at any given location.

C. Native Plants as Monitors

Benedict and Breen (*234*) were interested in identifying native weed species of widespread occurrence that were sensitive to different air pollutants. Thus, they determined the sensitivity of ten selected weed species to six phytotoxic air pollutants. The experimental concentrations were unrealistically high in comparison with natural pollution levels. Thus, although acute symptoms were described, no information was available on chronic symptoms for use in field investigations.

The presence and abundance of lichen and moss species have been mapped for large urban or industrial areas for a number of years (*58a*). Many of these investigations associated both decreased presence and abundance of these plants with increased air pollution, primarily sulfur dioxide. Some lichenologists have felt that the decrease of lichens in polluted areas was due primarily to other environmental factors (such as increased temperature and decreased humidity) rather than air pollution (*235*). The lack of air quality data in early studies contributed to the controversy, but several studies since 1967 have correlated sulfur dioxide concentration with either the absence of, or a decrease in the presence and/or abundance of lichen or moss species. LeBlanc and Rao (*63*) have presented an excellent review of the pollution and the drought hypotheses for the presence and absence of lichen and moss species in urban environments. However, even from this review, it is not possible to separate the various factors that influence the survival of lichen and moss communities. There is a growing consensus among research workers that lichen and moss species respond to the general sulfur dioxide–dust–dirt–grime complex of an urban–industrial area (*58*). There is no evidence that they respond to other specific pollutants. Therefore, lichens and mosses serve as general long-term indicators of the total environment and may identify historical air pollution problems and other environmental imbalances, but they give no indication of current problems. Literature from Sweden reports that substrate pH is important for lichen growth. Thus, the deterioration of lichens may be more closely related to the effect of sulfur dioxide on bark pH than to a direct effect on the lichen itself. Schönbeck (*236*) and LeBlanc and Rao (*237, 238*) have made extensive use of "lichen boards" to monitor sulfur dioxide air pollution problems. Before attempting to use lichen or moss species as monitors for air pollution, workers should carefully review several publications (*236–247*).

D. Bioassay for Field Monitoring

Plants may be used for fiield monitoring after general field surveys have established the presence of specific pollutants or groups of pollutants. Some of the uncertainties of the field survey are reduced by exposing plants under controlled culture in field situations where surveys are conducted. Species or varieties are selected for pollutant specificity, sensitivity, and symptom characteristics. Plants may be cultured under controlled conditions, exposed to ambient air, and returned to a controlled environment for symptom development and evaluation. In other cases, plants are germinated and grown to a given stage under known cultural conditions, transplanted to the field, and "read" periodically in the field. Various modifications of these basic techniques have been used. This technique improves the specificity and sensitivity of field monitoring by exposing more uniform plant material over the entire growing season. Generally, our ability to quantitate the field problem is improved by use of this type of plant material. Since the response measured is an estimation of the extent of a specific symptom on a single plant variety, training of competent observers is easier.

Annual bluegrass and petunia were used in this type of study in Los Angeles, California (*248*) in an attempt to correlate plant response with oxidant concentrations. Plants were grown under controlled conditions, transported to holding chambers throughout Los Angeles County, exposed for 24 hours, and returned to a filtered-air chamber where symptoms developed (*146*). The correlation between oxidant concentrations and plant injury was relatively poor.

A similar study (*249*) used the primary leaves of pinto beans grown under greenhouse conditions to monitor oxidant or other pollution in the Los Angeles area. Five monitoring stations were established and daily readings taken for 4 months during the fall of 1954. Maximum and mean oxidant values were recorded at all stations. Aldehyde, carbon monoxide, hydrocarbon, and nitrogen dioxide concentrations were recorded at two stations. Results were recorded as the percentage of days that plant injury occurred and the average injury index for each station. Although significant plant injury–oxidant correlations were found at only two stations, a general relationship did exist between injury and oxidant. However, the association between the two measurements was obscure.

Several reasons may be given for the apparent lack of correlation between plant response and total oxidant concentration in the two foregoing studies. First, some ambient phytotoxicants may not be oxidants. Second, some phytotoxic oxidants (such as PAN) may not react or may not completely react with standard oxidant measurement methods. Environmental conditions may control the proportion of various phytotoxic oxidants present in the ambient air at any given time. Interactions between pollutants could help account for the lack of correlation between plant injury and pollution levels measured as total oxidants. MacDowall *et al.* (*167*) correlated tobacco injury with oxidant concentrations by considering ozone flux into the plant leaves (Section III,B,8).

The report (*250*) that injury to petunia grown under ambient conditions correlated more closely with total aldehyde than with total oxidant was probably an artifact, since the injury reported was similar to that from PAN, and there are no reports of aldehydes injuring plants at the concentrations reported.

The use of tobacco to monitor photochemical oxidants was reported by Heck *et al.* (*153*) and Heck and Heagle (*61*). Tobacco is an excellent monitor because (i) it produces new leaves continuously over the 6- to 8-week growth period, (ii) leaves of different maturity differ in sensitivity but are uniformly sensitive at a given growth stage, (iii) new injury is visually separated from old injury, and (iv) a highly susceptible tobacco variety with quite specific symptoms of oxidant injury is available. The response of Bel W_3 tobacco was used to determine the distribution, frequency of occurrence, and the approximate average levels of photochemical oxidants in and around Cincinnati, Ohio, over a period of four growing seasons. Injury results from monitoring sites set up as far as 75 miles (120 km) east of Cincinnati showed that average concentrations of photochemical oxidants were as high in rural as in urban areas. Although oxidant concentrations (obtained from the Cincinnati monitoring site) did not correlate well with plant injury at weekly intervals, the seasonal ratio of injury to oxidant was similar over a 3-year period. The authors suggested that all areas east of the Mississippi River have sufficient photochemical pollution to injure sensitive plants at certain times during their development. This type of monitoring system could provide communities with estimates of the frequency of occurrence of phytotoxic levels of oxidants, of the relative severity of each episode, and of the regional distribution of oxidant air pollutants, but would not give reliable estimates of oxidant concentrations.

Tobacco has been used as part of a regional study (*251*) to determine the importance of oxidant pollutants to sensitive crops in the northeastern part of the United States (Maryland, Delaware, New Jersey, New York, Massachusetts, Connecticut, and Maine). Tobacco has also been used as an oxidant indicator in Ohio, North Carolina, Washington, Arizona, South Dakota (*252*), and in other locations. The technique was used to demonstrate for the first time that photochemical oxidants are widespread in Germany (*11*).

Sensitive white pine clones show a high degree of uniformity in the amount of injury from exposure to ozone (*253*). Some studies suggest that it may be possible to develop white pine clones that are sensitive just to ozone, fluoride, or sulfur dioxide. These clones could be used to identify the specific pollutant and might be valuable in areas where the relative importance of the three pollutants were in question.

Several varieties of gladiolus show leaf tip burn after a few weeks of exposure to less than 1 $\mu g/m^3$ of fluoride. They are therefore useful field monitors for relatively low atmospheric concentrations of fluoride. A measure of the length or area of tip burn has been used to indicate relative atmospheric fluoride concentrations.

The technique of transplanted lichens seems to provide a usable bioassay index for the sulfur dioxide–dust complex of the Ruhr Valley in Germany (*241*).

E. Bioassay of Controlled Exposures

The isolation and identification of toxic components of the photochemical smog complex have been complicated by the lack of adequate chemical techniques. In certain instances, plants have been used to help elucidate the chemical nature of the phytotoxic components. Plants have been used to monitor the products from six chemical reaction systems that are important in the photochemical complex: ozone plus hydrocarbon, ozone plus hydrocarbon (irradiated), nitric oxide plus hydrocarbon (irradiated), irradiated automobile exhaust, irradiated aldehydes, and ambient photochemical oxidants.

Seven types of phytotoxicants are suggested from an analysis of the experimental results using the response of certain sensitive plant tissues. First, ozone causes a fleck or stipple on the leaves of pinto bean. Second, an ozone-like injury is distinguished from ozone injury by the differential response of mature tobacco plants. Third, PAN causes an undersurface glaze on young pinto bean primary leaves. Fourth, a PAN-like injury to the undersurface of young pinto bean primary leaves occurs in response to a nitrogen-free reaction system [ozone plus hydrocarbon (irradiated)]. Fifth, a PAN-like injury is found on the undersurface of older pinto bean primary leaves in response to a nitrogen-free reaction system (ozone plus hydrocarbon). Sixth, an upper surface glaze occurs on young tobacco leaves exposed to irradiated auto exhaust. Seventh, a nonoxidant that passes a reducing filter produces a smog-type glaze on tomato.

These results imply additional phytotoxicants in the photochemical complex. Literature has largely ignored the unidentified phytotoxicants and has concentrated on PAN and ozone, which are relatively stable. The difference in half-lives between phytotoxicants produced by ozone reactions with two different olefins (*254*) and the isolation of several higher homologs of PAN, which are more phytotoxic (*255*), suggest that series of chemically similar phytotoxicants are produced under ambient conditions. Higher molecular weight homologs in other series of phytotoxicants may also be more reactive. The potential importance of nonoxidant types of phytotoxicants should not be ignored in light of the poor correlation between plant injury and oxidant concentrations.

F. Discussion

Plants have long been recognized as sensitive indicators of air pollutants. Their use in field surveys, although limited to studies after the fact, is necessary, along with chemical monitoring and meterological data, before we can understand air pollution conditions that are conducive to plant injury. The collection of chemical data at any given site will reduce the need for extensive field surveys that involve plants as monitors. Vegetation surveys should, however, be a routine part of any surveillance program. Studies that relate meteorological conditions to pollution episodes and plant injury are necessary before forecasting pollution episodes capable of causing plant injury can become routine. No one should consider surveys as any more than an indication of the obvious. There is a need, however, for the obvious to be shown.

A major limitation in the use of plants as monitors is the lack of studies on correlations between specific effects on a sensitive monitoring species and effects on important plant species. At present, we only evaluate the effect on the monitor itself. Thus, we have not identified the growth and reproductive effects that can have a major impact on plant communities.

The annual bluegrass and pinto bean bioassay were helpful in delineating the extent and frequency of Los Angeles, California, photochemical pollution, but they did not measure the growth reductions reported by Noble (*248*). Tobacco served as an excellent index of photochemical oxidants but gave no measure of the growth effects found in filtered versus unfiltered air studies (*210, 212, 213*). A similar fault applies to the lichen index in Germany. Perhaps, as suggested by Brandt (*58*), the information from these bioindicators can be extended. We must determine if the response of transplanted lichens, tobacco, pinto beans or other indicators correlates with general growth responses of important plant species in filtered versus unfiltered air studies. The readable response of the indicator may serve as an index of growth response. Thus, the system may give us at least some first approximation (rough scale measure) of loss in productivity over large areas.

VII. Economic Evaluation of Response

While the markings on the leaves of a plant may be identified with an air pollutant, it is often difficult to evaluate these markings in terms of the question: Has the plant been so altered by the air pollutant as to significantly affect its growth, survival, yield, or use? For example, 2,4-D may markedly and adversely alter the growth habit of grape. A severe fumigation with sulfur dioxide may so weaken a pine tree that it will not survive the winter. If ethylene causes the young bolls of cotton to drop off, the yield will certainly be affected. Oxidant pollution that bronzes the leaves of lettuce just before harvest may not greatly affect the yield but will affect its use.

In evaluating some of the leaf injury effects on economic loss, the problem, although difficult, is relatively straightforward when the appearance of the plant is paramount or when there is a loss of a desired product or the failure of a product to develop. In other cases, some assumption is required as to how leaf injury affects quality and reproductive capability. Finally we must determine how subtle injury from air pollutants affects the growth, development, survival, or use of a plant.

A. *Injury versus Damage*

Guderian *et al.* (*65*) presented the German approach to the general problem of pollutant effects on vegetation. They made a distinction between "injury" and "damage" that is much in keeping with our views. They define injury as any identifiable and measurable response of a plant to air pollution. They define damage as any identifiable and measurable adverse effect upon the desired or intended use or derived product of the plant that results from an air pollution injury. Thus, leaf necrosis on soybean or lettuce is injury. To be classified as damage, the injury must affect yield or use. Metabolic changes that accompany low pollutant concentrations (injury) might not cause visible injury but may cause damage (economic loss).

B. General Problems of Evaluation

1. Visible Leaf Injury

The value of romaine lettuce, like many truck crops, depends primarily on its size and appearance. If the outer leaves are injured, the crop may not be marketable. The cost of stripping the outer leaves and the resulting change in shape may not be justified. The number of crates per acre may be unaffected, but the crop may still be a total loss. In this case, injury may be minimal but damage is complete.

Gladioli sold as flowers are priced according to grades based on bloom size and spike length. Even though fluorides may not have altered the grade distribution, foliar injury decreases the net income because the buyer also expects healthy foliage. Therefore, damage has occurred because injured foliage must be removed by the grower, thereby increasing labor costs.

Where a crop, such as alfalfa, is marked by sulfur dioxide, evaluation is more complicated. Hill and Thomas (*198*) showed that the yield of such crops was reduced in proportion to the area of leaf destroyed, whether the reduction in leaf area was caused by sulfur dioxide or by clipping (Eq. 1). This offers a method of evaluating the damage sustained by such crops. Since yield reduction varies somewhat with the stage of growth at the time of injury, severe injury may affect subsequent cuttings in a similar way to that of cutting at too early a stage. This type of injury can, however, be evaluated as damage within reasonable limits by a competent observer.

The effect of leaf injury on damage to fruit trees is more difficult to evaluate and no data are available. Correlations between percentage leaf destruction and total yield or distribution of fruit grades are not available. We know that severe injury can decrease tree growth but what about subsequent crops? The problem is similar in forest trees where there is evidence that leaf injury from air pollution may reduce the diameter of growth rings. We do not know how to evaluate this effect on a crop to be harvested in 10–20 years.

The assessment of economic damage to ornamentals is also difficult.

Severe leaf injury in the nursery may make the stock unsalable. If injury is less severe, pruning may make the stock salable at a reduced price. The damage (economic loss) in these cases is usually estimatable. If the stock is held over for subsequent seasons, it may fully recover and be perfect the following season. In some cases, if its growth was somewhat stunted, this could be a benefit to the grower. In these cases assessment is more difficult. After ornamentals are planted in a yard, park, or cemetery, the concept of damage loses its meaning. That blue spruce in the yard may be scrubby, showing red spider mite and winter injury as well as sulfur dioxide injury. It could easily be replaced for fifty dollars except that it was planted on Johnny's first birthday and is, therefore, irreplaceable! In such cases, the damage is not so much to the leaves of the plant as to the emotions of the persons concerned.

2. Growth and Physiological Effects

The question of damage without visible injury in the strict sense of the term may be purely academic. The important point is whether the air pollutant has affected the growth, yield, or survival of the plant. When sulfur dioxide reduces net photosynthesis, injury has occurred, but the effect may not be sufficient to significantly alter the growth of the plant and produce damage (economic loss). Hill and his co-workers (*256*) carefully measured tomato growth under controlled conditions and reported that damage from fluorides did not occur without visible injury. Fluorides reduced the yield of sorghum with no visible leaf symptoms. Some of the apparent effects of fluoride on citrus growth (*95, 257*) and sweet cherry yield (*94*) are not consistent with the degree of leaf injury. Certainly the reduced growth observed in filtered and unfiltered air studies represents a real loss in productivity and must be considered damage.

C. Estimates of Economic Loss

Attempts to assess air pollution damage to agriculture require major judgments by the investigator who collects the data. Landau and Brandt (*258*) believe that the success of surveys on crop yields conducted by the Statistical Reporting Service, United States Department of Agriculture, are based on the collection of data that require a minimum of subjective decisions. We must develop similar evaluation methods that will require few subjective decisions, if the best estimates of loss from air pollution are to be obtained.

The first survey of economic losses was made in California (*64*) for several important crops in the state. Trained observers reported all observable air pollution effects on a field-by-field basis. They made no attempt to judge reduced growth or yield. The survey provided an excellent record of the distribution of plant injury from air pollutants.

A similar survey was begun during the 1969 growing season in Pennsylvania after an intensive training program. First-year reports from this survey gave an estimate of $11.5 million loss for the State (*259*). Similar training sessions and surveillance programs have been developed in Massachusetts (*260*), New Jersey (*261*), California, Michigan, Georgia and West Virginia to develop statewide estimates of economic losses. These surveys considered only injury and made no direct assessments of growth and yield effects. Nevertheless, subjective estimates of damage can be obtained from accurate assessment of visible effects.

A study was initiated in 1969 to develop a model for assessing damage to vegetation from which dollar values could be calculated (*16*). The model used effects of pollutant doses obtained from laboratory and field exposures of various crops. It also used data obtained from photochemical reaction chambers to predict the occurrence and concentration of secondary pollutants produced from given concentrations of primary pollutants (primarily hydrocarbons). The researchers then used hydrocarbon values from over one hundred statistical reporting areas within the United States to estimate expected oxidant values (related oxidant data were used where possible). Injury and damage estimates for specific crops were then calculated for each of the statistical reporting areas. The approach, by its very nature, uses many subjective assumptions and relates primarily to visible injury symptoms. However, such approaches should place damage functions within some useful perspective.

We lack the information necessary to accurately assess the overall impact of air pollution on the agricultural economy. The survey by Benedict *et al.* (*16*) projected about $135 million losses to vegetation in the United States. Gross estimates in the past have run between $500 and $1000 million annually for the United States. Worldwide estimates would at least double this figure. When we project preliminary greenhouse and field data that relate to reduced growth and yield, and take into account effects on ornamentals, wildlife, and esthetic values, we believe that these past estimates may be low, possibly by a factor of 2.

VIII. Vegetation as a Sink for Air Pollutants—Greenbelts

The potential value of vegetation in controlling air pollution has been discussed for many years. The concept has received encouragement from the proponents of greenbelts, as an aid in protecting urban areas from industrial pollution (*262–266*). Plant scientists favor the development of greenbelts for use as parks within large urban–industrial complexes, but have been skeptical of the greenbelt as a means of controlling pollution.

Most city planners look to the greenbelt as a park system to provide isolation of areas of differing function, to provide areas for recreation, and simply to break the monotony of the urban complex. Some have stressed the potential screening effect of greenbelts to remove air pollutants and thus serve to benefit human health. As often happens, emotionalism over improving health has been used to promote the greenbelt concept beyond available evidence. Others, claiming the greenbelt has no potential for filtering air pollutants, fail to see the real significance of the greenbelt.

There are many reasons for considering greenbelts in urban planning; their value as a filter and sink for air pollutants is just one. The potential effectiveness of these two air pollution aspects of the greenbelt concept can not be properly evaluated unless research is undertaken to determine the effectiveness of vegetation as a pollutant sink.

It has long been recognized that trees filter particulates. Abundant observations appear in the literature (*34, 266*), and considerable quantitative data are available. It is reasonable to assume that trees can filter particulates, at least under certain conditions. The larger the particulate and the more stable the atmospheric conditions, the better the tree screen will operate. With small particulates and increased turbulence, the effectiveness of the screen will decrease.

The sink concept has become of interest to atmospheric chemists and meteorologists who are attempting to develop air pollutant budgets. What are the ultimate sinks of air pollutants released by the activities of man? Reports indicate that plants, soils, and soil organisms are effective sinks for gaseous air pollutants under certain conditions (*82, 170, 262, 265, 267–270*). Soils are apparently minor sinks per se, but the microbial constituents of the soils are effective via metabolic activity. Hill and associates (*170, 268*), using plants grown and exposed under controlled conditions, attempted to quantify pollutant uptake. They developed pollutant uptake values for eight gaseous pollutants from which they calculated the total uptake of each pollutant per acre of alfalfa under given pollutant loads. This work, though preliminary, strongly suggests that vegetation is a significant pollutant sink. Bennett *et al.* (*271*) have presented a model that simulates pollutant exchange with isolated leaves. The model is able to integrate those factors that are known to affect pollutant uptake rates. Additional techniques and plant species must be considered as we explore the sink capacity of vegetation.

It is one thing to find that plants are major pollutant sinks and another to extrapolate this information to the development of greenbelts primarily for the purpose of decreasing pollutant concentration. Pollutant uptake by plants involves many factors, including the inherent variation in potential for uptake by different plant species, the direct effect of the pollutant on the uptake potential of the plant, the effect of other environmental stresses on the uptake potential of the plant, and meteorological factors affecting pollutant distribution. Do plants normally resistant to pollutants act as sinks or do only sensitive plants act as sinks? If only sensitive plants are major sinks, then we must raise a serious question as to their use in greenbelts. What is the mechanism of resistance? If some plants are resistant because the pollutant initiates stomatal closure, then pollutant uptake will be stopped. Stomata of plants under water stress will be closed, and even sensitive plants will not take up gases under these conditions. How effectively will plants absorb pollutants under other stress conditions? How do meteorological conditions affect pollutant distribution and the availability of pollutants for plant uptake?

It is probable that plants, over both distance and time, are major pollutant sinks. The atmospheric chemist knows that much pollutant loss is due to washout and to atmospheric reactions followed by washout. Any atmospheric reaction that supports aerosol (particulate) formation will aid the washout phenomena. These processes can occur over a relatively short time, but in arid areas or under certain atmospheric conditions, these processes can be very slow. This atmospheric cleansing process is not necessarily related to the greenbelt concept, since the greenbelt is in a fixed location within an urban area. Its air pollution aspect is to reduce pollution from the urban–industrial complex so as to protect human health. Plants may act as effective filters of pollutants from single sources but should not be expected to act as effective filters for gaseous air pollutants from the general urban–industrial pollution complex.

We do not visualize greenbelts as an effective barrier against gaseous pollutants, although they are effective in removing larger-sized particulates. This should not reduce the overall effectiveness and need for greenbelts in city planning, and they are of supplemental benefit for air pollution hazards. A major point to remember is that plants do not respond in a predictable way over time and are not active over the greater part of the year. Thus, they should not be expected to act as reliable pollutant filters. Greenbelts should not be pushed as an air pollution control strategy. Vegetation, on a worldwide basis, should prove an effective sink for air pollutants. The total capacity of this sink and its capacity without harm to the receptors or to ecological communities are not known.

IX. Reduction of Air Pollution Injury to Vegetation

The major thrust in attempting to reduce injury to vegetation from polluted air has been toward the promulgation of ambient air quality standards. Air pollution controls will not immediately reduce pollution. Thus, some level of pollution is inevitable for the foreseeable future. It is time we started to seriously consider other means by which injury from pollutants can be reduced. At this time, the means most widely discussed are breeding programs, cultural and management practices, and land use practices.

A. Breeding for Resistance

The genetic studies (Section III,A) clearly show considerable genetic variability in pollutant sensitivity within varieties of any given species. Some selection of resistant varieties has probably occurred, since plant breeders normally select plants with the highest yield and least injury regardless of cause. Natural selection pressures have, no doubt, increased the tolerance of populations of some native species located near urban–industrial areas. An intensified effort is needed to identify resistant plant material and to develop more resistant varieties. This need is especially critical for sensitive species such as tobacco, soybean, and truck crops commonly grown in more heavily polluted areas.

Breeding studies were initiated in the late 1950s to develop varieties of wrapper tobacco that were resistant to ozone. As resistant varieties were developed, the sensitive varieties were dropped from production. This research as it relates to several types of tobacco is continuing, primarily in Connecticut, Maryland, and Florida in the United States, and in Ontario, Canada. A program has been initiated to increase the resistance of loblolly pine to ozone (*272*).

Except for tobacco, there are no definitive breeding programs to increase plant resistance to air pollutants. However, researchers concerned with other crops are beginning to discuss the development of such programs. Varietal studies (Section III,A) have begun to identify possible sources of resistant genetic material. It is time for the plant scientist to accept air pollution as simply another stress phenomenon that must be incorporated into breeding programs.

Although breeding programs appear to be a necessity, they must take into account the desirable characteristics bred into present varieties (such as flavor, texture, growth habit, and insect and disease resistance, to name a few). Although air pollution stress should be added to present breeding programs, such programs will be costly, and resistant varieties thus developed may be injured if the concentrations of ambient pollution increase. Thus, breeding programs are not in themselves sufficient.

B. Cultural and Management Practices

An understanding of environmental effects on plant sensitivity will enable the recommendation of certain cultural practices that could be instituted during air pollution alerts. Recommendations have already been made with regard to soil moisture for greenhouse- and irrigation-grown plants, since we know that infrequent watering during times of high air pollution potential will make sensitive crops more resistant.

Some chemical sprays have been used experimentally to protect plants from air pollution (Section III,D). Reducing substances sprayed on plants react with oxidants in the air and reduce their effective concentration. Ascorbic acid has been used in California experiments, as have several reducing compounds sprayed on shade cloth over tobacco wrapper varieties in the Connecticut Valley. However, nothing yet tried is economically feasible for field or greenhouse use. Four basic weaknesses are present in any program thus far suggested: the frequency of application needed for continued resistance, the cost of chemicals, the possibility of undesirable residues, and the inability to accurately predict high oxidant days. However, lime sprays have long been recommended as a protectant against soft suture in peach induced by fluoride. This practice needs to be reviewed, but it does not seem to offer a general approach to plant protection from air pollutants.

If added carbon dioxide increases plant resistance to air pollutants (Section III,B,6), it might prove of supplemental benefit to those greenhouse operations that increase their carbon dioxide concentrations to improve plant growth.

C. Land Use

There is a growing recognition that land use must undergo some type of regulation. This issue and the substantive efforts toward innovative land use laws in many states of the United States is well presented in a report to the United States Council on Environmental Quality (*273*). The report makes the observation that we must start considering land both a resource and a commodity. There are serious efforts in most countries to develop an intelligent land use policy.

The air quality needed for optimum growth in a given area depends upon the type of vegetation. Since land use defines the type of vegetation cultivated, land use planning and air quality standards must be viewed together. Certain speciality crops can no longer be grown in some areas because of the adverse effects of air pollution. If the best land use plans suggest that a given agricultural pursuit be located in a given area, then air pollution must be controlled to protect that use. On the other hand, the best land use plans may dictate that specific crops or agricultural pursuits be discontinued in the vicinity of large specific pollutant sources. There are no universal answers. There will always be certain pollution sources and certain agricultural pursuits that are incompatible under any reasonable control regime.

Air pollution must be considered an important factor in the development of effective land use. We must use our land effectively for the living space and services we demand. We must also protect the land we require for food and fiber.

REFERENCES

1. E. Haselhoff and G. Lindau, "Die Beschädigung der Vegetation durch Rauch." Bornträger, Berlin, Germany, 1903.
2. J. A. Holmes, E. C. Franklin, and R. A. Gould, eds., "Report of the Selby Smelter Commission," Bull. No. 98. U.S. Bur. Mines, Washington, D.C., 1915.
3. National Research Council of Canada, "Effects of Sulfur Dioxide on Vegetation," NRC 815. Ottawa, Ontario, 1939.
4. S. N. Linzon, "The Influence of Smelter Fumes on the Growth of White Pine in the Sudbury Region," Contrib. No. 439. Forest Biol. Div., Dep. Agr., Ottawa, Ontario, 1958.
5. K. J. Seigworth, *Amer. Forests* **49,** 521 and 558 (1943).
6. S. Oden, *Swed. Natur. Sci. Res. Counc., Bull.* No. 1, pp. 1–86 (1968).
7. H. Ost. *Z. Angew. Chem.* **20,** 1689 (1907).
8. G. Bredeman, "Biochemie und Physiologie des Fluors." Akad. Verlagsges., Berlin, Federal Republic of Germany, 1951.
9. J. T. Middleton, J. B. Kendrick, Jr., and H. W. Schwalm, *Plant Dis. Rep.* **34,** 245 (1950).
10. J. T. Middleton, E. F. Darley, and R. F. Brewer, *J. Air Pollut. Cont. Ass.* **8,** 9 (1958).
11. W. Knabe, C. S. Brandt, H. van Haut, and C. J. Brandt, *in* "Proceedings of Third International Clean Air Conference," p. A110. Int. Union Air Prev. Ass., Dusseldorf, Federal Republic of Germany, 1973.
12. W. W. Heck, O. C. Taylor, and H. E. Heggestad, *J. Air Pollut. Contr. Ass.* **23,** 257 (1973).
13. B. L. Richards, J. T. Middleton, and W. B. Hewitt, *Agron. J.* **50,** 559 (1958).
14. E. R. Stephens, E. F. Darley, O. C. Taylor, and W. E. Scott, *Int. J. Air Water Pollut.* **4,** 79 (1961).
15. F. B. Abeles and H. E. Heggestad, *J. Air Pollut. Contr. Ass.* **23,** 517 (1973).
16. H. M. Benedict, C. J. Miller, and J. S. Smith, "Assessment of Economic Impact of Air Pollutants on Vegetation in the United States: 1969 and 1971," SRI Project LSU-1503. Stanford Res. Inst., Menlo Park, California, 1973.

17. W. Crocker, "Growth of Plants," Chapters 4 and 5. Van Nostrand-Reinhold, Princeton, New Jersey, 1948.
18. M. Katz, *Ind. Eng. Chem.* **41,** 4250 (1949).
19. C. Setterstrom and P. W. Zimmerman, *Contrib. Boyce Thompson Inst.* **10,** 155 (1939).
20. M. D. Thomas, *Annu. Rev. Plant Physiol.* **2,** 293 (1951).
21. M. D. Thomas and R. H. Hendricks, *in* "Air Pollution Handbook" (P. L. Magill, F. R. Holden, and C. Ackley, eds.), Chapter 9. McGraw-Hill, New York, New York, 1956.
22. H. Wislicenus, "Sammlung von Abhandlungen über Abgase und Rauchschäden." Parey, Berlin, Germany 1914.
23. D. F. Adams, *AMA Arch. Ind. Health* **14,** 229 (1956).
24. E. F. Darley, W. M. Dugger, Jr., J. B. Mudd, L. Ordin, O. C. Taylor, and E. R. Stephens, *Arch. Environ. Health* **6,** 75 (1963).
25. E. F. Darley and J. T. Middleton, *Annu. Rev. Phytopathol.* **4,** 103 (1966).
26. K. Garber, *Angew. Bot.* **36,** 127 (1962).
27. H. E. Heggestad and W. W. Heck, *Advan. Agron.* **23,** 111 (1971).
28. S. N. Linzon, "The Effects of Air Pollution on Forests" (presented at the 4th Chem. Eng. Conf.). Can. Soc. Chem. Eng., Vancouver, British Columbia, Canada, 1973.
28a. J. B. Mudd and T. E. Kozlowski, eds., "Response of Plants to Air Pollutants," Academic Press, New York, New York, 1975.
29. J. A. Naegele, ed., "Air Pollution Damage to Vegetation," Advan. Chem., Ser. No. 122, Amer. Chem. Soc., Washington, D.C., 1973.
30. M. D. Thomas, *World Health Organ., Monogr. Ser.* **46,** 223 (1961).
31. M. Treshow, "Environment and Plant Response." McGraw-Hill, New York, New York, 1970.
32. Environmental Protection Agency, "Air Quality Criteria for Nitrogen Oxides," Chapter 8, AP-84. United States Environmental Protection Agency, Washington, D.C., 1971.
33. Environmental Protection Agency, "Effects of Sulfur Oxides in the Atmosphere on Vegetation," EPA-R3-73-030. United States Environmental Protection Agency, Research Triangle Park, North Carolina, 1973.
33a. S. N. Linzon, W. W. Heck, and F. D. H. Macdowall, *in* "Photochemical Air Pollution: Formation, Transport and Effects," p. 89, NRC Associate Committee on Scientific Criteria for Environmental Quality, Report No. 12, Publ. No. NRCC 14096, National Research Council of Canada, Ottawa, Ontario (1975).
34. National Air Pollution Control Administration, "Air Quality Criteria for Particulate Matter," Chapter 6, AP-49. U.S. Department of Health, Education, and Welfare, Washington, D.C., 1969.
35. National Air Pollution Control Administration, "Air Quality Criteria for Hydrocarbons," Chapter 6, AP-64. U.S. Department of Health, Education, and Welfare, Washington, D.C., 1970.
36. National Air Pollution Control Administration, "Air Quality Criteria for Photochemical Oxidants," Chapter 6, AP-63. U.S. Department of Health, Education, and Welfare, Washington, D.C., 1970.
37. K. Garber, "Luftverunleinigungen, eine Literoturabersicht," Bericht No. 102 der Eidg., Anstalt fur des forstliche Versuchswesen, CH-8903. Birmensdorf, Federal Republic of Germany, 1973.
38. S. Godzik and P. Zdzislaw, *Wiad. Bot.* **13,** 239 (1969).

39. R. Guderian, *Z. Pflanzenkr. (Pflanzenpathol.) Pflanzenschutz* **73,** 241 (1966).
40. W. Knabe, *Forstarchiv* **37,** 109 (1966).
41. W. Knabe, *Ber. Landwirtschaft, Berlin* **50,** 169 (1972).
42. C. O. Tamm and A. Aronsson, Department of Forest Ecology and Soils, Research Notes, No. 12. Royal College of Forestry, Stockholm, Sweden, 1972.
43. H. van Haut and H. Stratmann, "Farbtafelatlas über Schwefeldioxid-Wirkungen an Pflanzen." Verlag W. Girardet, Essen, Federal Republic of Germany, 1970.
44. J. S. Jacobson and A. C. Hill, eds., "Recognition of Air Pollution Injury to Vegetation: A Pictorial Atlas." Air Pollut. Contr. Ass., Pittsburgh, Pennsylvania, 1970.
44a. D. C. MacLean, Staub-Reinhalt. Luft **35,** 205 (1975).
45. W. M. Dugger and I. P. Ting, *Annu. Rev. Plant Physiol.* **21,** 215 (1970).
46. W. W. Heck, *Annu. Rev. Phytopathol.* **6,** 165 (1968).
47. H. E. Heggestad, *J. Air Pollut. Contr. Ass.* **19,** 424 (1969).
48. J. T. Middleton, *Annu. Rev. Plant Physiol.* **12,** 431 (1961).
49. S. Rich, *Annu. Rev. Phytopathol.* **2,** 253 (1964).
50. O. C. Taylor, *J. Occup. Med.* **10,** 485 (1968).
51. R. H. Daines, *J. Occup. Med.* **10,** 516 (1968).
52. C. S. Brandt, "Report on Fluorides," mimeo. Prepared for the National Air Pollution Control Administration by the U.S. Dept. of Agriculture, Beltsville, Maryland, 1970.
53. Committee on Biologic Effects of Atmospheric Pollutants "Fluorides." Natl. Acad. Sci., Washington, D.C., 1971.
54. D. C. McCune and L. H. Weinstein, *Environ. Pollut.* **1,** 169 (1971).
55. M. Treshow, *Annu. Rev. Phytopathol.* **9,** 21 (1971).
56. A. S. Heagle, *Annu. Rev. Phytopathol.* **11,** 365 (1973).
57. W. W. Heck, J. A. Dunning, and I. J. Hindawi, *J. Air Pollut. Contr. Ass.* **15,** 511 (1965).
58. C. S. Brandt, *in* "Indicators of Environmental Quality" (W. A. Thomas, ed.), p. 101. Plenum, New York, New York, 1972.
58a. B. W. Ferry, M. S. Baddley, and D. L. Hawksworth, eds., "Air Pollution and Lichens." Oxford Univ. Press (Athlone), London, England, and New York, New York, 1973.
59. W. W. Heck, *Int. J. Air Water Pollut.* **10,** 99 (1966).
60. W. W. Heck, *Int. Symp. Ident. Meas. Environ. Pollut.* [*Proc.*], *1971* p. 320. National Research Council of Canada, Ottawa, Ontario (1972).
61. W. W. Heck and A. S. Heagle, *J. Air Pollut. Contr. Ass.* **20,** 97 (1970).
62. H. E. Heggestad and E. F. Darley, *in* "Proceedings of the First European Congress on the Influence of Air Pollution on Plants and Animals," p. 329. Centre for Agricultural Publishing and Documentation, Wageningen, The Netherlands, 1969.
63. F. LeBlanc and D. N. Rao, *Bryologist* **76,** 1 (1973).
64. J. T. Middleton and A. O. Paulus, *AMA Arch. Ind. Health* **14,** 526 (1956).
65. R. Guderian, H. van Haut, and H. Stratmann, *Z. Pflanzenkr.* (*Pflanzenpathol.*) *Pflanzenschutz* **67,** 257 (1960).
66. M. Treshow, *J. Air Pollut. Contr. Ass.* **15,** 266 (1965).
67. R. K. Howell and D. F. Kremer, *J. Environ. Qual.* **2,** 434 (1973).
68. A. E. Hitchcock, L. H. Weinstein, D. C. McCune, and J. S. Jacobson, *J. Air Pollut. Contr. Ass.* **14,** 503 (1964).

69. I. W. Wander and J. J. McBride, Jr., *Science* **123,** 933 (1956).
70. W. C. Hall, G. B. Truchelut, C. L. Leinweber, and F. H. Herrero, *Physiol. Plant.* **10,** 306 (1957).
71. Environmental Protection Agency, "Mount Storm, West Virginia-Maryland, Air Pollution Abatement Activity," APTD-0656. United States Environmental Protection Agency, Research Triangle Park, North Carolina, 1971.
72. H. van Haut and H. Stratmann, "Schriftenreihe der Landesanstalt für Immissions und Bodennutzungsschutz des Landes Nordrhein-Westfalen," No. 7, p. 50. Essen, Federal Republic of Germany, 1967.
73. M. Treshow, F. K. Anderson, and F. M. Harner, *Forest Sci.* **13,** 114 (1967).
74. M. Treshow and F. M. Harner, *Can. J. Bot.* **46,** 1207 (1968).
75. D. T. Tingey, W. W. Heck, and R. A. Reinert, *J. Amer. Soc. Hort. Sci.* **96,** 369 (1971).
76. D. T. Tingey and R. A. Reinert, *Environ. Pollut.* **9,** 117 (1975).
77. W. W. Heck and E. G. Pires, *Tex. Agr. Exp. Sta., Misc. Publ.* **MP-613** (1962).
78. J. K. A. Bleasdale, *Nature* (*London*) **169,** 376 (1952).
78a. C. Coulson and R. L. Heath, *Plant Physiol.* **53,** 32 (1974).
78b. L. S. Evans and I. P. Ting, *Amer. J. Bot.* **61,** 592 (1974).
79. F. B. Abeles, "Ethylene in Plant Biology." Academic Press, New York, New York, 1973.
80. O. C. Taylor and F. M. Eaton, *Plant Physiol.* **41,** 132 (1966).
81. A. C. Hill and J. H. Bennett, *Atmos. Environ.* **4,** 341 (1970).
82. D. T. Tingey, M. A. Thesis, Botany Department, University of Utah, Salt Lake City, Utah, 1968.
83. K. L. White, A. C. Hill, and J. H. Bennett, *Environ. Sci. Technol.* **8,** 574 (1974).
84. L. DeCormis and J. Bonte, *C. R. Acad. Sci., Ser. D* **270,** 2078 (1970).
84a. O. C. Taylor and D. C. MacLean, *in* "Recognition of Air Pollution Injury to Vegetation: A Pictorial Atlas" (J. S. Jacobson and A. C. Hill, eds.), p. E1. Air Pollut. Contr. Ass., Pittsburgh, Pennsylvania, 1970.
85. I. P. Ting and W. M. Dugger, Jr., *J. Air Pollut. Contr. Ass.* **18,** 810 (1968).
86. W. M. Dugger, O. C. Taylor, C. R. Thompson, and E. Cardiff, *J. Air Pollut. Contr. Ass.* **13,** 423 (1963).
87. R. L. Engle and W. H. Gableman, *Proc. Amer. Soc. Hort. Sci.* **89,** 423 (1966).
88. O. Majernik and T. A. Mansfield, *Nature* (*London*) **227,** 377 (1970).
89. M. H. Unsworth, P. V. Biscoe, and H. R. Pinckney, *Nature* (*London*) **239,** 458 (1972).
89a. J. B. Mudd, *Advan. Chem. Ser.* **122,** 31 (1973).
90. L. E. Craker, *Environ. Pollut.* **1,** 299 (1971).
91. A. S. Heagle and W. W. Heck, *Environ. Pollut.* **7,** 247 (1974).
92. A. E. Hitchcock, P. W. Zimmerman, and R. R. Coe, *Contrib. Boyce Thompson Inst.* **28,** 175 (1963).
93. C. W. Sulzbach and M. R. Pack, *Phytopathology* **61,** 1247 (1972).
94. T. J. Facteau, S. Y. Wang, and K. E. Rowe, *J. Amer. Soc. Hort. Sci.* **98,** 234 (1973).
95. C. D. Leonard and H. B. Graves, Jr., *Proc. Fla. State Hort. Soc.* **83,** 34 (1970).
96. A. H. Mohamed, *J. Air Pollut. Contr. Ass.* **18,** 395 (1968).
97. A. H. Mohamed, *Can. J. Genet. Cytol.* **12,** 614 (1970).
98. W. A. Feder, *Science* **160,** 1122 (1968).

99. W. A. Feder and F. J. Campbell, *Phytopathology* **58,** 1038 (1968).
99a. B. H. Harrison and W. A. Feder, *Phytopathology* **64,** 257 (1974).
100. S. Suzuki, Y. Matsuoka, T. Kawamura, and I. Mitani, *in* "Proceedings of the First International Citrus Symposium," Vol. 2, p. 747. University of California, Riverside, California, 1969.
101. O. D. Shkarlet, *Sov. J. Ecol.* **3,** 38 (1972).
102. R. Guderian, "Schriftenreihe," No. 4, p. 80. Landesanstalt fur Immissions und Bodennutzungsschutz des Landes Nordrhein-Westfalen, Essen, Federal Republic of Germany, 1967.
103. W. Knabe, *Staub* **30,** 32 (1970).
104. F. W. Cobb, Jr. and R. W. Stark, *J. Forest.* **68,** 147 (1970).
105. J. G. Edinger, M. H. McCutchan, P. R. Miller, B. C. Ryan, M. J. Schroeder, and J. V. Behar, *J. Air Pollut. Contr. Ass.* **22,** 882 (1972).
106. P. R. Miller, Ph.D. Thesis, University of California, Berkeley, California, 1965.
107. P. R. Miller, M. H. McCutchan, and H. P. Milligan, *Atmos. Environ.* **6,** 623 (1972).
108. R. N. Larsh, P. R. Miller, and S. L. Wert, *J. Air Pollut. Contr. Ass.* **20,** 289 (1970).
109. M. Treshow, *Phytopathology* **58,** 1108 (1968).
110. R. J. Eastland, Ph.D. Thesis, University of Utah, Salt Lake City, Utah, 1971.
111. H. Price and M. Treshow, *in* "Proceedings of the International Air Pollution Conference," p. 275. Clean Air Society of Australia and New Zealand, Melbourne, Australia, 1970.
111a. W. W. Heck, *Advan. Chem. Ser.* **122,** 118 (1973).
111b. E. J. Ryder, *Advan. Chem. Ser.* **122,** 75 (1973).
112. N. O. Adedipe, R. E. Barrett, and D. P. Ormrod, *J. Amer. Soc. Hort. Sci.* **97,** 341 (1972).
113. E. Brennan and I. A. Leone, *Plant Dis. Rep.* **56,** 85 (1972).
114. H. M. Cathey and H. E. Heggestad, *J. Amer. Soc. Hort. Sci.* **97,** 695 (1972).
115. H. M. Cathey and H. E. Heggestad, *J. Amer. Soc. Hort. Sci.* **98,** 3 (1973).
116. W. A. Feder, J. Donoghue, and I. Perkins, *Florogram* **5,** 13 (1972).
117. H. E. Heggestad, K. L. Tuthill, and R. N. Stewart, *HortScience* **8,** 337 (1973).
117a. I. A. Leone and D. Green, *Plant Dis. Rep.* **58,** 683 (1974).
117b. D. D. Davis and L. Kress, *Plant Dis. Rep.* **58,** 14 (1974).
118. R. K. Howell and C. A. Thomas, *Plant Dis. Rep.* **56,** 195 (1972).
119. W. J. Manning, W. A. Feder, and I. Perkins, *Plant Dis. Rep.* **56,** 832 (1972).
120. D. P. Ormrod, N. O. Adedipe, and G. Hofstra, *Can. J. Plant Sci.* **51,** 283 (1971).
121. R. A. Reinert, D. T. Tingey, and H. B. Carter, *J. Amer. Soc. Hort. Sci.* **97,** 711 (1972).
122. E. P. Brasher, D. J. Fieldhouse, and M. Sasser, *Plant Dis. Rep.* **57,** 542 (1973).
123. C. D. Clayberg, *HortScience* **6,** 396 (1971).
124. A. G. Gentile, W. A. Feder, R. E. Young, and Z. Santner, *J. Amer. Soc. Hort. Sci.* **96,** 94 (1971).
125. R. A. Reinert, D. T. Tingey, and H. B. Carter, *J. Amer. Soc. Hort. Sci.* **97,** 149 (1972).
126. J. W. Cameron, H. Johnson, Jr., O. C. Taylor, and H. W. Otto, *HortScience* **5,** 217 (1970).
127. A. S. Heagle, D. E. Body, and E. K. Pounds, *Phytopathology* **62,** 683 (1972).
128. E. Brennan, I. A. Leone, and P. M. Halisky, *Phytopathology* **59,** 1458 (1969).
129. R. K. Howell, T. E. Devine, and C. H. Hanson, *Crop Sci.* **11,** 114 (1971).

129a. A. S. Heagle, Agricultural Research Service, Raleigh, North Carolina (personal communication).
129b. V. L. Miller, R. K. Howell, and B. E. Caldwell, *J. Environ. Qual.* **3,** 35 (1974).
130. D. T. Tingey, R. A. Reinert, and H. B. Carter, *Crop Sci.* **12,** 268 (1972).
131. R. E. Schneider and D. C. MacLean, *Contrib. Boyce Thompson Inst.* **24,** 241 (1970).
132. M. K. Aycock, Jr., *Crop Sci.* **12,** 672 (1972).
133. J. J. Grosso, H. A. Menser, Jr., G. H. Hodges, and H. H. McKinney, *Phytopathology* **61,** 945 (1971).
134. G. H. Hodges, H. A. Menser, Jr., and W. B. Ogden, *Agron. J.* **63,** 107 (1971).
135. H. A. Menser, Jr. and G. H. Hodges, *Agron. J.* **62,** 265 (1970).
136. H. A. Menser, Jr. and G. H. Hodges, *Agron. J.* **64,** 189 (1972).
137. H. A. Menser, Jr., G. H. Hodges, and C. G. McKee, *J. Environ. Qual.* **2,** 253 (1973).
138. E. Brennan and P. M. Halisky, *Phytopathology* **60,** 1544 (1970).
139. A. C. Wilton, J. J. Murray, H. E. Heggestad, and F. V. Juska, *J. Environ. Qual.* **1,** 112 (1972).
139a. J. G. Shannon and C. L. Mulchi, *Crop Sci.* **14,** 335 (1974).
140. C. R. Berry, *Phytopathology* **61,** 231 (1971).
141. D. D. Davis and F. A. Wood, *Phytopathology* **62,** 14 (1972).
142. H. E. Heggestad, F. R. Burleson, J. T. Middleton, and E. F. Darley, *Int. J. Air Water Pollut.* **8,** 1 (1964).
143. H. A. Menser, *Tobacco* **162,** 32 (1966).
144. F. D. H. MacDowall, L. S. Vickery, V. C. Runeckles, and Z. A. Patrick, *Can. Plant Dis. Surv.* **43,** 131 (1963).
145. W. W. Heck and J. A. Dunning, *J. Air Pollut. Contr. Ass.* **17,** 112 (1967).
146. M. Juhren, W. M. Noble, and F. W. Went, *Plant Physiol.* **32,** 576 (1957).
147. F. D. H. MacDowall, *Can. J. Plant Sci.* **45,** 1 (1965).
148. H. A. Menser, Ph.D. Thesis, University of Maryland, College Park, Maryland, 1962.
149. W. M. Dugger, Jr., O. C. Taylor, C. R. Thompson, and E. Cardiff, *J. Air Pollut. Contr. Ass.* **13,** 423 (1963).
150. J. A. Dunning and W. W. Heck, *Environ. Sci. Technol.* **7,** 824 (1973).
151. H. van Haut, *Staub* **21,** 52 (1961).
152. R. Zahn, *Staub* **23,** 343 (1963).
153. W. W. Heck, F. L. Fox, C. S. Brandt, and J. A. Dunning, "Tobacco, A Sensitive Monitor for Photochemical Air Pollution," AP-55. U.S. Department of Health, Education, and Welfare, Cincinnati, Ohio, 1969.
154. J. A. Dunning, W. W. Heck, and D. T. Tingey, *Water, Air, Soil Pollut.* **3,** 305 (1974).
154a. N. O. Adedipe and D. P. Ormrod, *Z. Pflanzenphysiol.* **71,** 281 (1974).
155. A. M. Cantwell, *Plant Dis. Rep.* **52,** 957 (1968).
156. D. D. Davis and F. A. Wood, *Phytopathology* **63,** 371 (1973).
157. R. G. Wilhour, Ph.D. Thesis, Pennsylvania State University, University Park, Pennsylvania, 1970.
158. H. M. Hull and F. W. Went, in "Proceedings of The Second National Air Pollution Symposium," p. 122. Stanford Res. Inst., Pasadena, California, 1952.
159. D. C. MacLean, R. E. Schneider, and D. C. McCune, *in* "Proceedings of The Third International Clean Air Congress," pp. A143–145. Int. Union Air Prev. Ass., Dusseldorf, Federal Republic of Germany, 1970.

159a. R. A. Reinert, W. W. Heck, and A. S. Heagle, *in* "Response of Plants to Air Pollutants" (J. B. Mudd and T. E. Kozlowski, eds.), p. 159. Academic Press, New York, New York, 1975.
160. H. A. Menser and H. E. Heggestad, *Science* **153**, 424 (1966).
161. L. S. Dochinger, F. W. Bender, F. L. Fox, and W. W. Heck, *Nature (London)* **225**, 476 (1970).
162. D. T. Tingey, R. A. Reinert, J. A. Dunning, and W. W. Heck, *Atmos. Environ.* **7**, 201 (1973).
163. A. S. Heagle, D. E. Body, and G. E. Neely, *Phytopathology* **64**, 132 (1974).
164. D. T. Tingey, R. A. Reinert, C. Wickliff, and W. W. Heck, *Can. J. Plant Sci.* **53**, 875 (1973).
165. D. T. Tingey, R. A. Reinert, J. A. Dunning, and W. W. Heck, *Phytopathology* **61**, 1506 (1971).
166. A. C. Hill, S. Hill, C. Lamb, and T. W. Barrett, *J. Air Pollut. Cont. Ass.* **24**, 153 (1974).
167. F. D. H. MacDowall, E. I. Mukammal, and A. F. W. Cole, *Can. J. Plant Sci.* **44**, 410 (1964).
168. E. I. Mukammal, *Agr. Meteorol.* **2**, 145 (1965).
169. A. S. Heagle, W. W. Heck, and D. Body, *Phytopathology* **61**, 1209 (1971).
170. J. H. Bennett and A. C. Hill, *J. Air Pollut. Contr. Ass.* **23**, 203 (1973).
171. G. J. Hoffman, E. V. Maas, and S. L. Rawlins, *J. Environ. Qual.* **2**, 148 (1973).
172. E V. Maas, G. J. Hoffman, S. L. Rawlins, and G. Ogata, *J. Environ. Qual.* **2**, 400 (1973).
173. H. Khatamian, N. O. Adedipe, and D. P. Ormrod, *Plant Soil* **38**, 531 (1973).
174. S. Rich and N. C. Turner, *J. Air Pollut. Contr. Ass.* **22**, 718 (1972).
175. I. P. Ting and W. M. Dugger, *Atmos. Environ.* **5**, 147 (1971).
176. I. A. Leone, E. Brennan, and R. H. Daines, *J. Air Pollut. Contr. Ass.* **16**, 191 (1966).
177. H. A. Menser and G. H. Hodges, *Agron. J.* **60**, 349 (1968).
178. R. Zahn, *Z. Pflanzenkr. (Pflanzenpathol.) Pflanzenschutz* **70**, 81 (1963).
179. R. F. Brewer, F. B. Guillemet, and R. K. Creveling, *Soil Sci.* **92**, 298 (1961).
180. I. A. Leone and E. Brennan, *Phytopathology* **60**, 1521 (1970).
181. A. S. Heagle, *Phytopathology* **60**, 252 (1970).
182. A. S. Heagle and L. F. Key, *Phytopathology* **63**, 397 (1973).
183. A. S. Heagle and A. Strickland, *Phytopathology* **62**, 1144 (1972).
184. W. J. Manning, W. A. Feder, and I. Perkins, *Phytopathology* **60**, 669 (1970).
185. H. M. Resh and V. C. Runeckles, *Can. J. Bot.* **51**, 725 (1973).
186. M. Treshow, G. Dean, and F. M. Harner, *Phytopathology* **57**, 756 (1967).
187. R. A. Reinert, Agricultural Research Service, Raleigh, North Carolina (personal communication).
188. N. O. Adedipe, H. Khatamian, and D. P. Ormrod, *Z. Pflanzenphysiol.* **68**, 323 (1973).
189. N. O. Adedipe and D. P. Ormrod, *Z. Pflanzenphysiol.* **68**, 254 (1972).
190. H. C. Dass and G. M. Weaver, *Can. J. Plant Sci.* **48**, 569 (1968).
190a. A. Koiwai, H. Kitano, M. Fukuda, and T. Kisaki, *Agr. Biol. Chem.* **38**, 301 (1974).
190b. W. J. Manning, W. A. Feder, and P. M. Vardaro, *J. Environ. Qual.* **3**, 1 (1974).
191. P. M. Miller and G. S. Taylor, *Plant Dis. Rep.* **54**, 672 (1970).

191a. J. W. Moyer, H. Cole, Jr., and N. L. Lacasse, *Plant Dis. Rep.* **58**, 136 (1974).
192. L. Ordin, O. C. Taylor, B. E. Propst, and E. A. Cardiff, *Int. J. Air Water Pollut.* **6**, 223 (1962).
193. R. A. Reinert and H. W. Spurr, Jr., *J. Environ. Qual.* **1**, 450 (1972).
194. S. Rich and G. S. Taylor, *Science* **132**, 150 (1960).
195. R. C. Seem, H. Cole, Jr., and N. L. Lacasse, *J. Environ. Qual.* **2**, 266 (1973).
196. E. K. Walker, *Can. J. Plant Sci.* **47**, 99 (1967).
197. S. N. Linzon, *J. Air Pollut. Contr. Ass.* **21**, 81 (1971).
197a. M. Katz and G. A. Ledingham, *in* "Effects of Sulfur Dioxide on Vegetation," NRC 815, p. 332. Natl. Res. Counc. Can., Ottawa, Ontario, 1939.
198. G. R. Hill and M. D. Thomas, *Plant Physiol.* **8**, 223 (1933).
199. C. R. Davis, G. W. Morgan, and D. R. Howell, *Agron. J.* **57**, 250 (1965).
200. R. Guderian and H. Stratmann, *Forschungsber. Landes Nordrhein-Westfalen* **1118**, 1–102 (1962).
201. R. Guderian and H. Stratmann, *Forschungsber. Landes Nordrhein-Westfalen* **1920**, 1–114 (1968).
201a. A. E. Wells, *U.S., Bur. Mines, Bull.* **98**, 213 (1914).
202. R. Guderian, *Staub* **20**, 334 (1960).
203. H. Stratmann, *Forschungsber. Landes Nordrhein-Westfalen* **1184**, 1–69 (1963).
204. B. R. Dreisinger, *58th Annu. Meet.*, Paper No. 65–121, pp. 1-21. Air Pollut. Contr. Ass., Pittsburgh, Pennsylvania, 1965.
205. J. N. B. Bell and W. S. Clough, *Nature* (*London*) **241**, 47 (1973).
206. H. C. Jones, J. R. Cunningham, S. B. McLaughlin, N. T. Lee, and S. S. Ray, "Investigation of Alleged Air Pollution Effects on Yield of Soybeans in the Vicinity of the Shawnee Steam Plant," E-EB-73-3. Tennessee Valley Authority, Chattanooga, Tennessee, 1973.
207. D. C. McCune, "Air Quality Monograph," No. 69-3. Amer. Petrol. Inst., New York, New York, 1969.
208. D. C. McCune, *Amer. Ind. Hyg. Ass., J.* **32**, 697 (1971).
209. C. D. Leonard and H. B. Graves, Jr., *Fluoride* **5**, 145 (1972).
210. A. S. Heagle, D. E. Body, and W. W. Heck, *J. Environ. Qual.* **2**, 365 (1973).
211. C. R. Thompson and G. Katz, *Calif. Agr.* **24**, 12 (1970).
211a. C. R. Thompson, University of California, Riverside, California (personal communication.
212. C. R. Thompson, E. Hensel, and G. Katz, *HortScience* **4**, 222 (1969).
213. C. R. Thompson and O. C. Taylor, *Environ. Sci. Technol.* **3**, 934 (1969).
214. W. A. Feder, *Science* **165**, 1373 (1969).
215. W. A. Feder, *Environ. Pollut.* **1**, 73, (1970).
216. W. J. Manning, W. A. Feder, P. M. Papia, and I. Perkins, *Environ. Pollut.* **1**, 305 (1971).
217. C. R. Thompson, G. Katz, and E. G. Hensel, *Environ. Sci. Technol.* **5**, 1017 (1971).
218. W. W. Heck, E. G. Pires, and W. C. Hall, *J. Air Pollut. Contr. Ass.* **11**, 549 (1961).
219. C. J. Brandt and R. W. Rhoades, *Environ. Pollut.* **4**, 207 (1973).
220. M. D. Thomas and G. R. Hill, Jr., *Plant Physiol.* **10**, 291 (1935).
221. W. W. Heck, J. A. Dunning, and I. J. Hindawi, *Science* **151**, 577 (1966).
222. W. W. Heck and D. T. Tingey, *in* "Proceedings of the Second International

Clean Air Congress" (H. M. Englund and W. T. Berry, eds.), p. 249. Academic Press, New York, New York, 1971.
222a. D. C. McCune and L. H. Weinstein, *in* "Fluorides," p. 77. Natl. Acad. Sci., Washington, D.C., 1971.
222b. L. H. Weinstein and D. C. McCune, *in* "Recognition of Air Pollution Injury to Vegetation: A Pictorial Atlas" (J. S. Jacobson and A. C. Hill, eds.), p. G1. Air Pollut. Contr. Ass., Pittsburgh, Pennsylvania, 1970.
223. T. C. Scheffer and G. G. Hedgcock, *U.S., Dep. Agr., Tech. Bull.* **1117,** 1–49 (1955).
224. G. A. Cole, *Agron. J.* **50,** 553 (1958).
225. B. W. Ellertsen and H. C. Jones, "Report on Study of a Disease of White Pine in East Tennessee." Tennessee Valley Authority, Chattanooga, Tennessee, 1969.
226. R. Guderian, *Z. Pflanzenkr. (Pflanzenpathol.) Pflanzenschutz* **77,** 200, 289, and 387 (in 3 parts) (1970).
227. H. C. McKee and F. W. Bieberdorf, *J. Air Pollut. Contr. Ass.* **10,** 222 (1960).
227a. J. W. Suttie, *in* "Fluorides," p. 133. Natl. Acad. Sci., Washington, D.C., 1971.
228. H. M. Benedict, J. M. Ross, and R. H. Wade, *Int. J. Air Water Pollut.* **8,** 279 (1964).
229. H. M. Benedict, J. M. Ross, and R. H. Wade, *J. Air Pollut. Contr. Ass.* **15,** 253 (1965).
230. R. Guderian, H. van Haut, and H. Stratmann, *Forschungsber. Landes Nordrhein-Westfalen* **2017,** 1–54 (1969).
231. A. E. Hitchcock, D. C. McCune, L. H. Weinstein, D. C. MacLean, J. S. Jacobson, and R. H. Mandel, *Contrib. Boyce Thompson Inst.* **24,** 363 (1971).
232. D. C. McCune and A. E. Hitchcock, *in* "Proceedings of the Second International Clean Air Congress" (H. M. Englund and W. T. Berry, eds.), p. 289. Academic Press, New York, New York, 1971.
233. G. Israel, *Atmos. Environ.* **8,** 167 (1974).
234. H. M. Benedict and W. H. Breen, *in* "Proceedings of the Third National Air Pollution Symposium," p. 177. Stanford Res. Inst., Pasadena, California, 1955.
235. J. Rydzak, *Ann. Univ. Mariae Curie-Sklodowska, Sect. C* **23,** 131 (1968).
236. H. Schoenbeck, *Staub* **29,** 14 (1969).
237. F. LeBlanc and D. N. Rao, *Bryologist* **69,** 338 (1966).
238. F. LeBlanc and D. N. Rao, *Ecology* **54,** 612 (1973).
239. O. L. Gilbert, *New Phytol.* **67,** 15 (1968).
240. O. L. Gilbert, *New Phytol.* **69,** 629 (1970).
241. R. Guderian and H. Schoenbeck, *in* "Proceedings of the Second International Clean Air Congress" (H. W. Englund and W. T. Berry, eds.), p. 266. Academic Press, New York, New York, 1971.
242. F. LeBlanc and J. DeSloover, *Can. J. Bot.* **48,** 1485 (1970).
243. J. Rydzak, *Ann. Univ. Mariae Curie-Sklodowska, Sect. C* **25,** 149 (1970).
244. E. Skye, *Acta Phytogeogr. Suecia* **52,** 1 (1969).
245. K. R. Sundström and J. E. Hällgren, *Ambio* **2,** 13 (1973).
246. H. Taoda, *Jap. J. Ecol.* **22,** 125 (1972).
247. H. Taoda, *Hikobia* **6,** 237 (1973).
248. W. M. Noble and L. A. Wright, *Agron. J.* **50,** 551 (1958).
249. J. T. Middleton, J. B. Kendrick, Jr., and E. F. Darley, *in* "Proceedings of the Third National Air Pollution Symposium," p. 191. Stanford Res. Inst., Pasadena, California, 1955.
250. E. G. Brennan, I. A. Leone, and R. H. Daines, *Science* **143,** 818 (1964).

251. J. S. Jacobson and W. A. Feder, "A Regional Network for Environmental Monitoring: Atmospheric Oxidant Concentrations and Foliar Injury to Tobacco Indicator Plants in the Eastern United States," Exp. Sta. Bull. No. 604, University of Massachusetts, Amherst, Massachusetts, 1974.
252. W. S. Gardner, *Plant Dis. Rep.* **57,** 106 (1973).
253. C. R. Berry and G. H. Hepting, *Forest Sci.* **10,** 2 (1964).
254. W. N. Arnold, *Int. J. Air Pollut.* **2,** 167 (1959).
255. E. F. Darley, K. A. Kettner, and E. R. Stephens, *Anal. Chem.* **35,** 589 (1963).
256. A. C. Hill, L. G. Transtrum, M. R. Pack, and W. S. Winters, *Agron. J.* **50,** 562 (1958).
257. R. F. Brewer, R. K. Creveling, F. B. Guillemet, and F. H. Sutherland, *Proc. Amer. Soc. Hort. Sci.* **75,** 236 (1960).
258. E. Landau and C. S. Brandt, *Environ. Res.* **3,** 54 (1970).
259. N. L. Lacasse and T. C. Weidensaul, *in* "Proceedings of the Second International Clean Air Congress" (H. M. Englund and W. T. Berry, eds.), p. 132. Academic Press, New York, New York, 1971.
260. J. A. Naegele, W. A. Feder, and C. J. Brandt, "Assessment of Air Pollution Damage to Vegetation in New England," Suburban Exp. Sta. University of Massachusetts, Amherst, Massachusetts, 1972.
261. A. Feliciano, "1971 Survey and Assessment of Air Pollution Damage to Vegetation in New Jersey," Coop. Ext. Serv. Rutgers University, New Brunswick, New Jersey, 1972.
262. G. P. Hanson and L. Thorne, *Lasca Leaves* **22,** 60 (1972).
263. M. Kadota, *Rodo No Kagaku* **27,** 77 (1972).
264. W. Knabe, *Forstarchiv* **44,** 21 (1973).
265. L. Thorne and G. P. Hanson, *Environ. Pollut.* **3,** 303 (1972).
265a. A. M. Townsend, *J. Amer. Soc. Hort. Sci.* **99,** 206 (1974).
266. J. L. Warren, "Green Space for Air Pollution Control," Tech. Rep. No. 50, pp. 1–118. School of Forest Resources, North Carolina State University, Raleigh, North Carolina, 1973.
267. F. B. Abeles, L. E. Craker, L. E. Forrence, and G. R. Leather, *Science* **173,** 914 (1971).
268. A. C. Hill, *J. Air Pollut. Contr. Ass.* **21,** 341 (1971).
269. R. E. Inman, R. B. Ingersoll, and E. A. Levy, *Science* **172,** 1229 (1971).
270. D. J. Spedding, *Nature (London)* **224,** 1229 (1969).
271. J. H. Bennett, A. C. Hill, and D. M. Gates, *J. Air Pollut. Contr. Ass.* **23,** 957 (1973).
272. R. Weir, North Carolina State University, Raleigh, North Carolina (personal communication).
273. F. Bosselman and D. Callies, "The Quiet Revolution in Land Use Control." Council on Environmental Quality, Washington, D.C., 1971.

5

Biological Effects of Air Pollutants

David L. Coffin and Herbert E. Stokinger

I. Photochemical Oxidants

The photochemical process by which secondary pollutants are formed by the interactions of primary pollutants energized by sunlight results in a number of irritating or toxic chemicals. Photochemical oxidants, such as ozone and peroxyacyl nitrates, are gases that exert their toxic influence via inhalation or by surface contact. If present in sufficient concentrations, these gases are capable of causing death. At sublethal concentrations, they produce more occult, but nonetheless health-impairing physiological malfunctions or anatomic lesions.

Data from exposure at several orders of magnitude above the highest ambient concentrations, while of great basic toxicological importance, are not relevant for determining air quality criteria in ambient air. Nevertheless, such information is included in tabular form to provide historical background and to develop a framework of relative toxicity where clear-cut effects such as mortality, pulmonary edema, and the like can serve as endpoints. It is obviously impossible to detect such effects in community health surveys. It is likewise unwise to draw inferences directly applicable to air quality criteria from animal exposure data at such concentrations, since the mode of action of any given pollutant may be different at high and low levels of exposure. Where appropriate, however, experiments conducted at high levels will be cited for purposes of elucidating mechanisms or sites of action or where new information is available that might serve as a basis for future studies. The reader is also referred to the United States air quality criteria document for photochemical oxidants (*1*), review articles by Stokinger (*2*) and Jaffe (*3*), and the previous edition (1968) of this book (*4*).

II. Ozone

Investigations of the acute toxicity of ozone, begun as early as 1904, determined that ozone was a remarkably toxic gas. Early work was concerned mainly with mortality, gross examination of the lung for evidence of inflammation, wet weight–dry weight ratios for evidence of edema, histopathological examinations, and the like (Table I) (*2, 5–41*).

It should be recognized in what follows that the experimentally determined biological effects of ozone may not have been determined on precisely the same ozone and other related oxidant species that exist in the ambient air. Experimentally produced ozone is not 100% pure triatomic oxygen, but contains other oxygen species, probably oxygen free radicals, possibly ozone homologs, O_4, O_5, etc., in varying amounts, depending upon the form of energy used to produce it, and, if prepared from air also contains a few percent of nitrogen dioxide.

Our attention was first drawn to the varying quality of laboratory-generated ozone when ozone generated by two different dielectric-type room air ozonizers that differed in current density repeatedly and consistently yielded LC_{50} values for small rodents that differed by a factor of almost two (*42*). Although attempts were made to identify the oxygen species involved, by means of a mass spectrometer specially constructed to measure reactive substances and free radicals, none were found because of lack of sensitivity of the instrument. Indirect evidence for oxygen species with short half-life was found when ozone generated from scrubbed air and "aged" for 20–30 minutes was less toxic for rodents than similarly prepared ozone that had not been aged (rat LC_{50}, aged O_3, 8.2 ppm versus 6.2 ppm) (*42*).

Table I Summary of Toxic Effects of Ozone in Animals Arranged according to Concentration

Ozone[a] *(ppm)*	*Length of exposure*	*Observed effect(s)*	*Species*	*Reference*
0.08	3 hours	Increased susceptibility to streptococcus (group C)	Mice	(*5*)
0.13–1.5	90-day continuous	Severe toxic signs with deaths in all species except dog at concentrations >0.3 ppm	Rats, dogs, mice, monkeys, guinea pigs	(*6*)
0.20	30 minutes	Increased sphering of red blood cells when irradiated	Rabbits, rats, mice	(*7*)
0.20	6 hours	Decreased voluntary running activity	Mice	(*2*)
0.20	5 hours/day/ 3 weeks	Structural changes in heart myocardial fibers	Mice	(*2*)
0.26, 0.5, 1.0	4.6 hours	Changes in dynamic compliance and pulmonary resistance at 1.0 and 0.5 ppm	Cats	(*8*)
0.3–4.4	4 hours	Tolerance to acute lung edema begins at 0.3 ppm	Mice	(*9*)
0.34	2 hours	30% increase in frequency of breathing; 20% decrease in tidal volume	Guinea pigs	(*10*)
0.4	6 hours/day, 5 days/week/ 10 months	Pulmonary arterial wall thickening	Rabbits	(*11*)
0.5	2–6 hours	Swelling and sloughing of alveolar type I cells	Rats	(*12*)
0.5	6–10 hours	Damage to type I with proliferation type II cells	Rats	(*13*)
0.5, 3.0, 22.0	2 days	Tolerance to pulmonary edema but not against cytotoxic effects	Rabbits	(*14*)
0.5, 1.2, 2.5, 3.5	6 hours	Reduction in DNA synthesis	Mice	(*15*)
0.5, 0.75, 3.0	3 hours	Decreased cell viability; increase in lung polymorphonuclear leukocytes	Rabbits	(*16*)
0.5	8 hours/day/ 7 days	Changes in lung macrophage osmotic fragility	Rabbits	(*17*)
0.6, 1.3	6–7 hours, 1 or 2 days	Disruption of endothelial lining of alveolar capillaries	Mice	(*18*)
0.68	2 hours	No significant increase in flow resistance	Guinea pigs	(*10*)
0.7, 0.8	Continuous 5 and 7 days	Increased lysosomal activity in lungs from inflammation	Rats	(*19*)

Table I *(Continued)*

Ozone[a] *(ppm)*	*Length of exposure*	*Observed effect(s)*	*Species*	*Reference*
0.75, 3.0, 10.0	3 hours	Increasing inhibition of benzpyrene hydroxylase with increasing concentrations	Hamsters	*(20)*
0.84	4 hours/5 days/2 weeks	Increased susceptibility to *Klebsiella pneumoniae*	Mice, hamsters	*(21)*
1.0	1 hour	Chemical changes in ground substance and lung protein	Rabbits	*(22)*
1.0	4 hours	Engorged blood vessels and excess leukocytes in lung capillaries	Mice	*(23)*
1.0–7.0	3 hours	Decrease in numbers of pulmonary alveolar macrophages	Rabbits	*(24)*
1.0	6–7 hours daily for 2 years	Bronchitis, bronchiolitis, emphysematous and fibrotic changes; acceleration of lung tumor development	Mice	*(25)*
1.0	6 hours	60% increase in mortality as a result of exercise for 15 minutes/hour	Rats	*(2)*
1.0	8.2–18.5 days continuous	Acceleration of vitamin E deficiency	Rats	*(26)*
1.08	2 hours	Increased flow resistance	Guinea pigs	*(10)*
1.3	3 hours	Increased susceptibility to *Klebsiella pneumoniae*	Mice, hamsters	*(21)*
2.0	3 hours	Increased lung weight, decreased tidal volume, decreased minute ventilation	Rats	*(23)*
2.3	8 hours/day, 5 days/week, 6 weeks	Mildly toxic signs and no deaths	Monkeys, rats, dogs, mice, guinea pigs	*(6)*
3.0	8 hours daily, 18 months	Squamous metaplasia of bronchiolar epithelium	Dogs	*(27)*
3.0	2 hours	Change in neutrophil–lymphocyte ratio	Rats	*(28)*
3.1	20 hours	Increased liver weight; increased liver alkaline phosphatase	Rats	*(29)*
3.2	4 hours	Gross pulmonary edema	Mice	*(23)*
4.0	4 hours	Decreased mortality with age; young 50% mortality, old 10% mortality	Mice	*(2)*
4.5	2 hours every 3rd day, 75 days	Severe epithelial changes in lung in tumor resistant strain	Micc	*(30)*

Table I *(Continued)*

Ozone[a] *(ppm)*	*Length of exposure*	*Observed effect(s)*	*Species*	*Reference*
5.0	2 hours	Increased lung compliance, increased susceptibility to histamine	Guinea pigs	(*31*)
5.0	3 hours	Decreased activity of bacteriocidal lysozyme	Mice, rabbits	(*32*)
5.0–6.0	90 minutes	Production of hydrogen peroxide in erythrocytes	Rats, mice	(*33*)
6.0	4 hours	Gross pulmonary edema, increased lung serotonin	Rats	(*34*)
6.0	4 hours	Decreased brain serotonin	Rats	(*35*)
8.0	4 hours	Decrease in erythrocyte acetylcholinesterase activity	Mice	(*36*)
8.0[b]	3–5 hours	Protection versus lethal effects by various enzyme-inducing agents	Rats	(*37*)
8.0–45.0	1 hour/week up to 49 weeks	Damage to epithelium of the lower trachea and bronchioles; fibrosis	Rabbits	(*23*)
10.4	4 hours	Vitamin E deficiency increases lethality of ozone	Rats	(*38*)
10.0, 30.0	8 hrs/day/ 18 mos	Decrease in brain catecholamine metabolism	Dogs	(*39*)
15.0	30 minutes	Decreased tidal volume, decreased oxygen consumption	Rabbits	(*23*)
15.0	426 minutes	Protection from acute lethal effects by *p*-aminobenzoic acid	Rats	(*40*)
21.0	3 hours	50% mortality	Mice	(*41*)
21.8	3 hours	50% mortality	Rats	(*41*)
34.5	3 hours	50% mortality	Cats	(*41*)
36.0	3 hours	50% mortality	Rabbits	(*41*)
51.7	3 hours	50% mortality	Guinea pigs	(*41*)

[a] The concentrations of ozone listed are the lowest for which the observed effects have been recorded.
[b] Average.

A. Effects of Short-Term Exposure to Ozone

Brief exposure to ozone at higher than ambient concentrations elicits edema and an acute inflammatory response. The edemagenic response, while striking at higher levels of exposure, diminishes with descending concentrations of the gas. The exact threshold for this effect is dependent on the species of animal and the method employed to determine excess pulmonary water. No evidence of edema was noted in mice exposed to 1 ppm for 4 hours, while exposure to this concentration for 20 hours resulted in very slight edema when gravimetric measurement of lung water was the method employed. Scheel *et al.* (*23*) and Stokinger (*2*) detected evidence of edema in rats after exposure to 2 ppm ozone for 3 hours. Exposures at higher levels, 3.2 ppm, quickly produces gross evidence of edema in mice (*2, 4*) (Table I).

While it is generally believed that a gross response by edema is probably not elicited in any species at ambient concentrations, more refined methods for the detection of edema based on the recovery of radiolabeled serum albumin by means of pulmonary lavage conducted by Alpert *et al.* (*16*) suggests that the ozone threshold for the preliminary leakage of plasma in rats lies between 0.25 and 0.50 ppm for 6 hours.

Acute pulmonary inflammatory response is a regular feature of ozone exposure of experimental animals. Excess polymorphonuclear leukocytes were observed in pulmonary capillaries after exposure of mice to 1 ppm for 4 hours, and inflammatory exudate was noted in the alveolar spaces after exposure at 1 ppm for 20 hours (*23*). The chemotactic effect of exposure to ozone has been shown at considerably lower concentrations and exposure times by methods employing pulmonary lavage and differential counts of the lavaged cells as studied by Gardner (*43*) and Coffin and Gardner (*44*). Thus, in rabbits a statistically significant increase in polymorphonuclear leukocytes recoverable from the bronchial tree through lavage has been observed after exposure to 1 ppm and above for 3 hours. This response was noted immediately after exposure to the gas, peaked at about 6 hours after cessation of the exposure, then gradually declined but was still evident 24 hours after exposure. In 1959, Scheel *et al.* (*23*), by means of paraffin sections, showed that exposure to from 1 ppm to 3.2 ppm ozone for 4 hours in mice produced engorged capillaries containing an excess of polymorphonuclear leukocytes and desquamation of the superficial epithelial cells. Exposure at the higher level produced edema. Subsequent pathological studies employing the technique of plastic-embedded 1 μm sections, as well as paraffin sections, for light microscopy and by means of transmission and scanning electron microscopy have further elucidated the pulmonary lesions of ozone. Thus, when Watanabe *et al.* (*45*) exposed cats to 0.26, 0.5, and 1 ppm ozone for 4.6 hours, lesions consisting of loss of ciliated cells throughout the airway were noted at all concentrations employed and appeared to be roughly dose related. The most consistent ultrastructural changes in the airway were cytoplasmic vacuolization of the ciliated cells and condensed mitochondria with abnormal cristae configuration. These changes appeared most pronounced in airways of 0.8–1.7 mm in diameter. These same authors reported lesions consisting of swelling and denudation of type I alveolar lining cells, swelling and breakage of the capillary endothelium, and red cell lysis within the interalveolar capillaries. The distribution of the airway lesions as judged by both light and scanning electron microscopy appeared scattered.

Boatman and Frank (*46*) have reported essentially similar findings in rabbits exposed to 0.5 and 1.0 ppm ozone for 3 hours in a model in which one lung was collapsed by means of unilateral (left) pneumothorax designed to elucidate the tolerance mechanisms. For the purpose of this discussion, however, only the lesions induced in the right, or nontolerant, lung will be described. These authors report that with exposure to the higher concentration (1 ppm), up to one-third of the periphery of the large airways was randomly denuded of epithelial cells, with lesions in the small airways (0.17–0.3 mm in diameter) consisting of loss of cilia and vacuolization of the cytoplasm. At 0.5 ppm, changes were observed mainly in the larger airways, though abnormal mitochondria, increased electron density, and smaller longitudinally aligned cristae were noted at both concentrations. Changes noted in the alveoli by light microscopy consisted of focal areas of intraalveolar hemorrhage and accumulation of leukocytes. Study by means of transmission electron microscopy showed changes consisting of either swelling or desquamation of type I cells and swollen or ruptured endothelial cells at the 1.0 ppm concentration, with minimal swelling being apparent at the 0.5 ppm level. Stephens *et al.* (*13*), studying the lesions induced by ozone exposure in rats, have indicated that degenerative changes in the type I alveolar cells occur after exposure to as low as 0.2 ppm ozone for 3 hours and that these cells are replaced by type II alveolar cells after 24 hours, as demonstrated for nitrogen dioxide by Evans *et al.* (*47*). This same group of researchers has studied this lesion more extensively for nitrogen dioxide exposure and has worked out the sequences of destruction, desquamation, and regeneration of alveolar cells (*47*). Presumably, the sequence is similar for ozone, except that the lesion appears with only about one-twentieth the dose of ozone because of the greater reactivity of ozone (*48–50*). Bils (*18*) has investigated by means of electron microscopy the effect of exposure to 0.6 and 1.3 ppm ozone for 4 hours, noting essentially similar lesions from both concentrations, namely, swelling of the epithelial alveolar lining cells but no evidence of fluid accumulation in the alveolar space. This was also associated with focal swelling of the endothelial cells and an occasional break in the basement membrane.

Electron microscopic studies of lungs of animals exposed to ozone at concentration several times ambient have been carried out by Werthamer *et al.* (*30*) and Plopper *et al.* (*51, 52*). Such exposures are known to produce gross evidence of edema and are thus not of relevance for immediate application to air quality criteria. However, since these data provide a continuum of effect at higher levels, they provide an upper dose–response point. These studies, by virtue of their stronger effect, characterize some of the lesions seen from exposures to ambient or near-ambient concentrations, thereby providing a link with earlier studies made by light microscopy after exposure to lower concentrations of ozone.

Thus, Werthamer *et al.* (*30*) found acute inflammatory bronchiolar lesions following ozone exposure of mice at 5 ppm for 3 hours. Plopper *et al.* (*51*) reported findings from exposure of rats to approximately 3 ppm ozone for 4 hours. Histopathological observations made by light microscopy on freeze-substituted sections indicated irregularly distributed focal or confluent lesions. Changes predominating were bronchiolar epithelial necrosis and sloughing and alveolar edema, particularly in the region of the terminal airway units. These changes are essentially the same as those described by other authors by standard histological methods after high level ozone exposure.

Ultrastructural studies on gluteraldehyde-fixed sections embedded in epoxy indicated that cilia and epithelial surface projections were diminished on the bronchiolar epithelial cells. In addition to edema and fibrinous exudative elements in the alveoli, specific cellular injury was observed. Swelling of the membranous pneumocyte (type I alveolar cells) was noted. Additionally, there was cytoplasmic swelling and internal blebing of the capillary endothelium. Interstitial edema was also present. The cellular lesions appeared to be confined to the proximal portion of the pulmonary acini and adjacent terminal bronchioles. However, endothelial lesions were observed peripherally as well as proximally in the acini.

B. Effects of Long-Term Exposure to Ozone

A number of investigators have either continuously or intermittently exposed various animals to ozone over considerable periods. Stokinger *et al.* (*25*) reported that chronic bronchitic, bronchiolitic, emphysematous, and fibrotic changes occur in the lungs of rats, mice, hamsters, and guinea pigs exposed 6 hours daily, 5 days a week for 14 months to 1 ppm ozone. Freeman *et al.* (*53*) and Stephens *et al.* (*27*) have reported on light and electron microscopic studies of the lungs of dogs exposed to doses ranging from 1 ppm to 3 ppm for 8–24 hours daily for a period of 18 months. The various changes observed appear roughly proportional to the time and dose of the ozone exposure. These consist of increase in the number of pulmonary alveolar macrophages beginning at exposures of 1 ppm for 8 hours daily, deposition of fibrous elements, thickening of the terminal airway and respiratory bronchiolar walls, with consequent reduction of the lumina of the small airways. There is also accumulation of lymphocytes, plasma cells, and fibroblasts in the peribronchiolar tissue, with an increase in the proportion of mucus-secreting to nonsecreting cells and squamous metaplasia of columnar and cuboidal cells.

Ultramicroscopic studies performed in dogs exposed to 3 ppm ozone for 8 hours per day for 18 months revealed abnormalities in the granular pneumocyte (type II alveolar epithelium) and in the endothelial cells of the interalveolar capillaries (*27*). The authors speculated that they represent metabolic alterations in the cells. These changes consisted of an accumulation of a material with a recurring periodicity of 754 Å within dilated cisternae of the rough endoplasmic reticulum in type II cells. Ribosomes and rough endoplasmic reticulum were also more abundant. The authors also describe accumulation of tubular membrane complexes forming lattice-worklike structures in the cytoplasm of capillary endothelial cells.

Freeman *et al.* (*54*) have studied rats exposed to 0.54 ppm and 0.88 ppm ozone for varying periods up to 6 months. They report the following pathological sequence from observations in lungs filled intratracheally with Zenkers fluid to a pressure of 25 cm of water. (Tissue blocks were embedded in paraffin and sectioned at 4 μm and 16 μm.) Observations made at 0.54 ppm on animals sacrificed after 12 hours exposure were essentially negative except for aggregations of alveolar macrophages. After 24 hours exposure, the respiratory bronchioles and alveolar ducts contained hypertrophic epithelium as well as the presence of amorphous material, suggesting necrosis of cells adjacent to the terminal bronchioles. After 48 hours exposure, the alveolar ducts showed hypertrophic alveolar epithelium that contrasted sharply with controls. Numerous mitotic figures were present in the terminal hypertrophic bronchiolar and proximal alveolar epithelium. In the proximal air spaces, macrophages were prominent, and there was indication that fibroblasts and connective tissue elements were increased in the subepithelial tissue of the respiratory bronchioles and ducts in some places amounting to replacement of the respiratory epithelium. The animals were also sacrificed at 6 days, 8 days, 3 weeks, and 6 months. Progressive changes in the epithelial cells were reported. The hyperplastic epithelium became cuboidal and the bronchioloductal tissue incorporated additional connective tissue. There was a gradual reversion of the connective tissue to a less cellular appearance in the terminal bronchioles. However, connective tissue deposits tended to disrupt the symmetry of the small airways and proximal alveoli and possibly restrict the lumina of the smallest airways.

When rats were exposed to 0.88 ppm ozone (*54*), evidence of epithelial injury was present at least at 4 hours after beginning of the exposure.

In general, the observations noted during the 0.54 ppm exposure were exaggerated at 0.88 ppm and occurred sooner. Also, after 2 days of exposure at the higher concentration, metaplastic-like activity was noted in the bronchiolar and ductal epithelium associated with fibroplastic proliferation. As the lesion progressed, there was noted replacement of normal alveolar epithelium by cuboidal cells (bronchiolization of the alveoli), the presence of adenoma-like structures, abundant macrophages, invasion by fibroblasts, and a flattening and smoothing of the bronchiolar epithelium (Fig. 1). Occasionally, the bronchiolar epithelium and its lamina propria were replaced by a bed of fibroblasts. Whereas rats exposed to the lower concentration were alive after 3 weeks of exposure, half of the animals exposed to 0.88 ppm had died. The survivors had voluminous air-filled lungs.

C. Biochemistry of Ozone

The biochemical mechanism by which exposure to ozone damages biological systems has been the subject of considerable speculation and investigation. Among the mechanistic postulations which have been advanced are the following: (a) more or less nonspecific stress, with the release of histamine; (b) oxidation of sulfur-bearing compounds (sulfhydryl groups) or their precursers; (c) oxidation of polyunsaturated lipid mainly contained in cell membranes; (d) formation of toxic compounds (ozonides, peroxides, etc.) through reaction with polyunsaturated lipid; (e) formation of free radicals; and (f) injury mediated by a pharmacological action, i.e., via a neurohumoral mechanism. Despite rather extensive investigation of the problem, there is still no generally accepted explanation of the biochemical mechanism or mechanisms through which ozone exerts its toxic action.

Dixon and Mountain (*55*) have shown that the release of histamine from the lung occurs immediately after exposure to ozone and persists for as long as 11 days in mice and 17 days in rats. Pretreatment with the antihistamine (promethazine) reduces the response following an edemagenic dose of ozone. The ability of aspirin to reduce the effects of ozone-induced edema may be related to its ability to block prostaglandin synthetase. Bradykinin appears to be intimately related with prostaglandins in the lung and its potentiation is possible.

While Mendenhall and Dixon (*56*) could find no role for serotonin in ozone treatment, Fairchild *et al.* (*57*) suggested that edema was inhibited by the prior administration of agents that block the histamine-sensitizing factor. Easton and Murphy (*31*) demonstrated that ozone-exposed animals react more severely to parenterally administered histamine.

Oxidation of sulfhydryl groups was indicated in 1959 by Fairchild

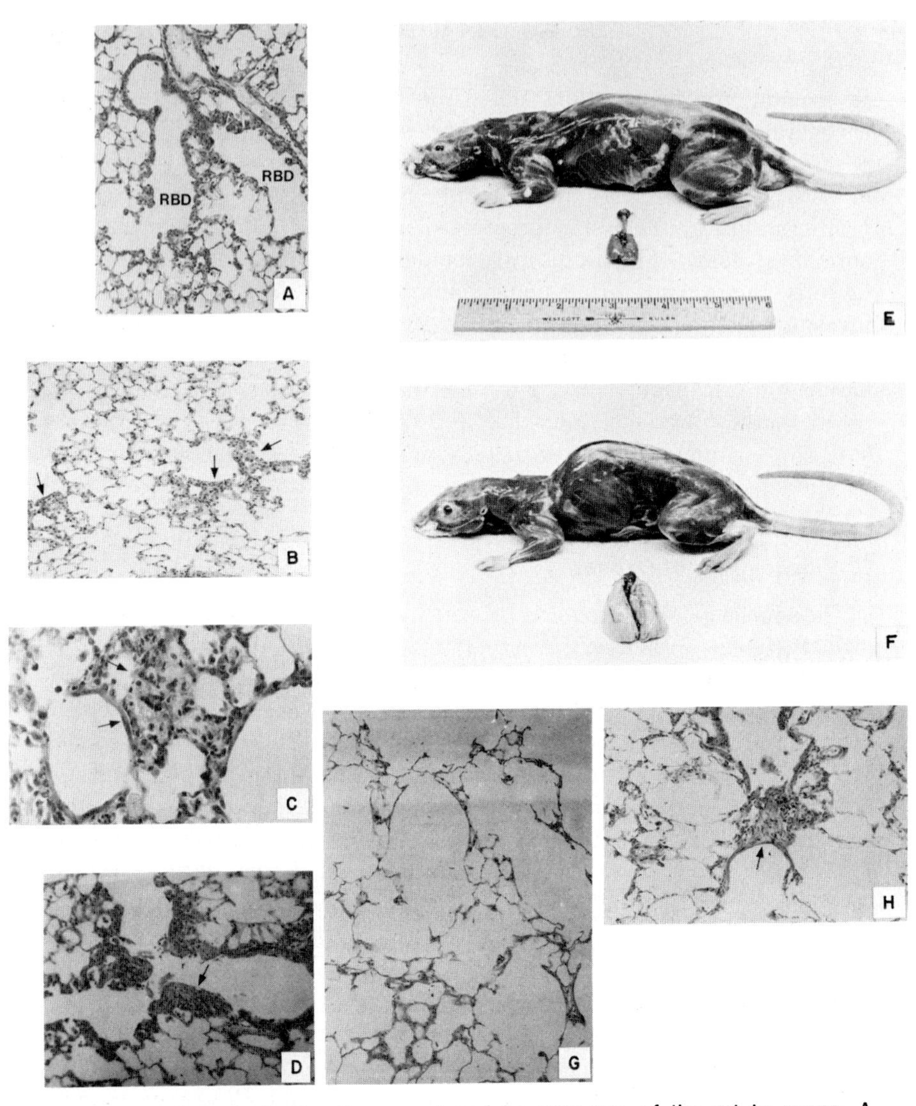

Figure 1. Pathological alterations produced by exposure of the rat to ozone. A, Lung of rat exposed to 0.54 ppm ozone for 24 hours, showing hypertrophic epithelium of respiratory bronchioles and alveolar ducts. B, Hypertrophy of ductal and adjacent alveolar epithelium after exposure to 0.54 ppm ozone for 4 hours. C, Spindle-shaped cells (arrow) lining alveolar duct after 6 days' exposure to 0.9 ppm ozone. D, Strata of fibroblast-like cells (arrow) lining the junction between the terminal and respiratory bronchioles after exposure to 0.9 ppm ozone. E, Normal rat and excised lungs under atmospheric pressure. F, Similar subject after exposure to 0.9 ppm ozone and 0.9 ppm nitrogen dioxide for 60 days. Note the strongly kyphotic spine due to aberration of the thorax from pressure of the enlarged lung. G, Attenuation and loss of alveolar walls suggesting emphysema in same lung as shown in F. H, Same lung as G showing deposition of connective tissue at the junction of the terminal bronchiole and alveolar duct. All photomicrographs 80× (Gustave Freeman) (*54*).

et al. (*58*), who showed that the addition of sulfhydryl compounds was protective. Subsequent studies by a number of investigators have confirmed these findings and extended the depth of study. Mudd *et al.* (*59*) studied ozone oxidation of glutathione in an *in vitro* model and showed that both fast and slow oxidation occur. Studies by Mountain (*60*) showed that *in vivo* ozone exposure causes oxidation of reduced glutathione (GSH) and depression of succinate dehydrogenase activity in the lung, while studies by King (*61*) demonstrated the loss of sulfhydryl content and loss of enzymatic function of partially purified glyceraldehyde-3-phosphate dehydrogenase in rat lungs following exposure of 1.2 ppm ozone for 4 weeks. Menzel (*62*) and Mudd *et al.* (*59*) demonstrated that the oxidation of reduced glutathione (GSH) proceeded to oxidized glutathione (GSSG), although some of the sulfhydryl groups formed higher oxidation products (i.e., sulfoxide and sulfone). DeLucia *et al.* (*63*) found that the concentration of protein sulfhydryls (PSH) and nonprotein sulfhydryls (NPSH) were reduced in the lungs of rats by exposure to 2 ppm ozone for 4–8 hours. Additionally, several lung enzymes —glucose-6-phosphate dehydrogenase, glutathione reductase, reduced nicotine adenine diphosphate (NADH), and succinate–cytochrome *c* reductase—were reduced by similar exposure. DeLucia *et al.* (*63*) have also reported that exposure to as little as 0.2 ppm of ozone for 4 hours results in a statistically significant depression of GSH and associated enzymes up to 24 hours after exposure followed by a subsequent elevation above normal, persisting for several days (Fig. 2). It is also well known that supplementing with glutathione or its precurser, methionine, will block this effect and will moderate various other manifestations of ozone exposure—as reported by Green (*64*) and Hurst and Coffin (*65*). It should be pointed out, however, that the cycle of depression and augmentation of lung glutathione mimics the cytological cycle induced by ozone exposure, namely, necrosis of type I alveolar epithelial cells with subsequent cellular hyperplasia, particularly of type II cells, said to be rich in these compounds. Thus, it is possible that the GSH reduction cycle is a result, rather than a cause, of ozone injury. Systemic reduction of reduced glutathione cannot be overlooked since blood GSH levels are suppressed throughout exposure to 1.0 ppm ozone for up to 60 days (*63*).

Oxidation (peroxidation) of polyunsaturated fatty acids has been advanced as a toxic mechanism in both ozone and nitrogen dioxide injury by Thomas *et al.* (*66*), Goldstein *et al.* (*67*) and Roehm *et al.* (*68*). This postulation is based on the proclivity of ozone to react with the ethylene groups of unsaturated fatty acids. This reaction can lead to the formation of free radicals, which, in the presence of molecular oxygen, proceeds to the peroxidation of the unsaturated fatty acid. In support of the

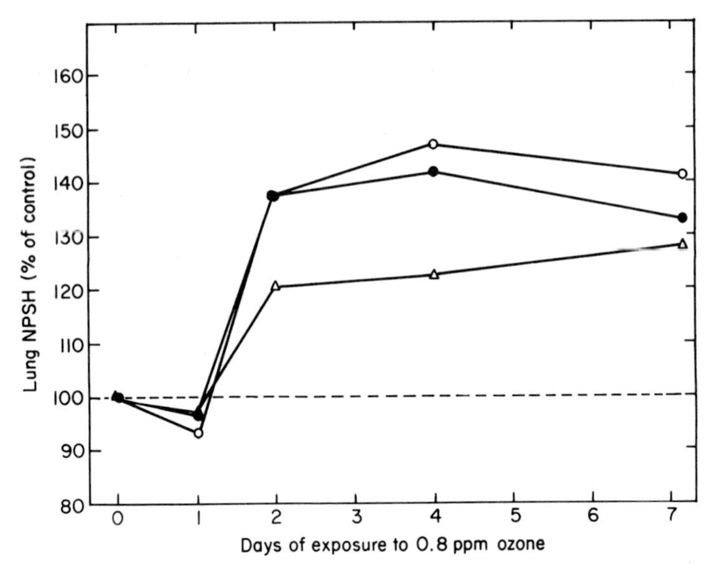

Figure 2. Decrement in various glutathione-related enzymes in the lung following ozone exposure and their subsequent recovery curves. Changes in the levels of nonprotein sulfhydryls (NPSH) and reduced glutathione (GSH) in lung tissue as a function of time during exposure of rats to 0.8 ppm ozone. The elevation in lung content of NPSH and GSH was significant from $p < 0.01$ to $p < 0.001$ for the exposure from day 2 to day 7. Data were obtained from three separate experiments involving 120 rats. ○ = nonprotein sulfhydryls, ● = nonprotein sulfhydryls after borohydride reduction, △ = reduced glutathione, --- = control, (*63*).

above hypothesis, Goldstein *et al.* (*38*) and Menzel *et al.* (*69*) have performed experiments that indicate that a deficiency of vitamin E, the principal cellular free radical trap, increases the toxicity of ozone for the rat. Goldstein *et al.* (*67*) speculate that this process could be affected either by direct ozonolysis or by free radicals resulting from unsaturated fatty acid breakdown. If such proves to be the case, a chain reaction beginning with direct ozonolysis and perpetuated by the resultant free radicals could be envisioned as a consequence of ozone exposure.

Another chemical pathway derived from unsaturated fatty acid oxidation by ozone has been demonstrated and might be considered as a mode of toxicity. Roehm *et al.* (*68*) and Menzel *et al.* (*69*) studied the oxidation of polyunsaturated lipid in an *in vitro* model in which samples of pure methyl esters of fatty acids were exposed to dynamically flowing atmospheres of ozone and nitrogen dioxide. There is a marked difference in the mode of oxidation between the two gases, ozone being the most reactive. Oxidation by ozone was not completely inhibited by phenolic antioxidants. Methyl linoleate was completely reacted by ozone to yield a number of compounds, including monozonides and diozonides. The authors speculate that ozone acts by direct addition of ozone across the fatty acid double bond, postulating that the mechanism proposed by Criegee (*70*) is the primary mechanism for ozonolysis. This reaction proceeds as follows:

$$\mathrm{RCH{=}CH{-}R} + \mathrm{O_3} \rightarrow \mathrm{R\overset{OOO}{\overset{|\quad|}{CH}}CHR} \rightarrow \mathrm{R\overset{OO^{\ominus}}{\overset{|}{C}}\oplus} + \mathrm{H\overset{O}{\overset{\|}{C}}R} \rightarrow \mathrm{R{-}CH\underset{O}{\overset{OO}{\diamond}}CHR}$$

The ozonide, once formed, appears stable. Menzel *et al.* (*69*) have studied the effects of fatty acid ozonides *in vivo* and with *in vitro* models. Intradermal injections of fatty acid ozonides of oleic, linoleic, linolenic, and arachnidonic acids in amounts ranging from 10 pg to 10 μg elicited increased vascular permeability of the capillaries as measured by pontamine-blue-bound serum proteins. The reaction appeared to be mediated by histamine, since it was blocked by prior treatment with compound 48/80 or by simultaneous injection by antihistamines. According to these same authors, fatty acid ozonides also rapidly initiated peroxidation of isolated microsomes and mitochondria. They also observed that erythrocytes incubated with fatty acid ozonides formed Heinz bodies and lysed red cells, indicating oxidation of the red cell membrane, and they suggest that fatty acid ozonides may be the toxic intermediaries in ozone toxicity. Formation of Heinz bodies by ozonide is inhibited by vitamin E supplementation (*71*).

So-called radiomimetic properties have been ascribed to ozone on the basis that certain pathological events are common to both ozone and ionizing radiation. Evidence for the release of free radicals by ozone rests on pathological alterations, and physical chemical theory and observation. The pathological alterations produced from ozone exposure that have been likened to those caused by ionizing radiation are "chromosome aberrations," reported by Fetner (*72*) and by Zelac *et al.* (*73*); premature aging, reported by Scheel *et al.* (*23*); retardation in the deoxygenation of hemoglobin in the capillary blood (common to radiant energy), reported by Brinkman and Lamberts (*74*); and avoidance behavior in mice, described by Peterson and Andrews (*75*). Theoretical support for the existence of free radicals are: (a) While ozone is predominantly an ion, it reacts to generate free radical species because of the vigorous nature of the oxidant and the formation of peroxy intermediaries. Free-radical-mediated chain reactions can thus occur on oxidation of biological material. (b) There is fairly good theoretical support that ozone reacts with sulfhydryl groups oxidizing them to thionyl free radical prior to oxidation to high oxidation states (e.g., sulfoxide and sulfone). (c) It also has been speculated that during the peroxidation of polyunsaturated lipid by ozone, free radicals are liberated via peroxy intermediaries and that exposure to ozone confers some protection against whole-body irradiation (*76*). (d) Other facts that may be indicative of free radicals from ozone are that protection is provided against both ozone and ionizing radiation by sulfur compounds that are known to be free radical scavengers, as pointed out by Fairchild *et al.* (*58*); that breakdown products of oxidized lipid include substances that may cause polymerization and cross-linking of protein, as suggested by Buell *et al.* (*22, 77*); and finally, that free radical signals have been demonstrated by electron paramagnetic resonance spectrometry during the ozone catalyzed oxidation of linoleic acid *in vitro* by Goldstein *et al.* (*67*).

Neurohumoral reaction to ozone exposure has been indicated in reports of Fairchild and co-workers (*78–80*). Evidence for the involvement of the pituitary–thyroid–adrenal axis is that hypophysectomy, thyroidectomy, and adrenalectomy will increase the resistance of rats to otherwise lethal doses of ozone by decreasing the elicitation of pulmonary edema (*79*). Ablation of the thyroid has the strongest effect, followed in decreasing importance by the pituitary and adrenals. The administration of antithyroid drugs such as methimozole and thiourea increase resistance to the lethal effects of ozone exposure (*79*), while conversely, hormonal preparations such as thyroxine increase the edemagenic response. Fairchild and Bobb (*80*) have further demonstrated that exposure to ozone inhibits the release of ^{131}I from the thyroid at levels as low as 0.5 ppm ozone. It has been suggested that the inhibitory role of the endocrine gland ablation rests on the elaboration of pressor amines and consequent hemodynamic shifts, rather than alteration of metabolic rate, since the production of a similar effect has been reported by lesions of the spinal cord at the sixth and seventh cervical and second thoracic segments (*81*).

D. Physiological Alteration from Ozone Exposure

A number of studies have demonstrated the physiological response of the lungs of several species of animals to ozone exposure. Most of these suffer from the difficulties in performing delicate pulmonary mechanical maneuvers on the lungs of small noncooperative animals—the necessity of employing anesthesia and the use of unrealistically high concentrations of the gas, all of which may introduce artifacts into the data. Possibly the most effective use of animals has been for the development of models for estimating gas absorption in the respiratory tract and evaluating the various mechanisms of ozone action as they are reflected in animal models in which various surgical or pharmacological interventions can be conveniently carried out. For determination of the influence of ozone on the pulmonary physiological parameters of man and establishing air quality criteria, the best model is man himself. Consequently, several investigators have carried on experimental exposures on volunteers with low, but physiologically reactive concentrations of ozone in which no persistent deleterious effects were reported. Only studies performed at 1 ppm or below will be cited here unless there is some mechanistic reason for doing otherwise.

The respiratory uptake of ozone has been measured in dogs by Vaughan *et al.* (*81*) and Yokoyama and Frank (*82*) by means of the surgically isolated upper airway. In the former study, the uptake in the upper airway alone was 40–60% of 9–12 ppm ozone, and 15–20% when concentrations of 7 ppm or less were employed. In the latter study, where exposures were to concentrations of 0.2–0.4 and 0.7–0.85 ppm, the rate of uptake was found to be greater at lower velocities and at lower concentrations. The rate of uptake was twice as high in the nose as in the mouth, but highest in the lower airway (80–87%) (Fig. 3). The authors estimated that approximately 90% of the ozone in the respired air was taken up during quiet respiration. It is speculated that mouth breathing during exposure may be a factor that heightens the effect of exercise during such exposure. Work is being carried on utilizing this area of study to derive other data for the construction of a mathematical uptake model at various segmental levels of the upper and lower respiratory tract. Preliminary computations indicate that uptake is a function of gas solubility. Thus, with ozone, which is slowly soluble, penetration is deep. The preliminary model predicts that the peak dosage for ozone occurs between the sixteenth and the twentieth segmental level of the human lung (*83*).

Physiological response of the lungs has been studied by Murphy *et al.* (*10*) in guinea pigs employing the forced pressure oscillatory method. Pulmonary flow resistance was not significantly altered by 0.34 and 0.68 ppm, but rose as much as 47% with 1.08 ppm and 1.35 ppm. Watanabe *et al.* (*8*) studied the effect of ozone exposure on pulmonary mechanics in cats exposed to 0.26, 0.5, and 1 ppm ozone for 2–6.5 hours. Their findings indicate that pulmonary air flow resistance was increased more or less proportionally to the ozone dose, while dynamic compliance fell less frequently and with less amplitude. There was no change in vital capacity, but the frequency of reduction of the carbon monoxide diffusing capacity of the lung among the various animals exposed, as measured, appeared to be a function of ozone concentration. There was an apparent

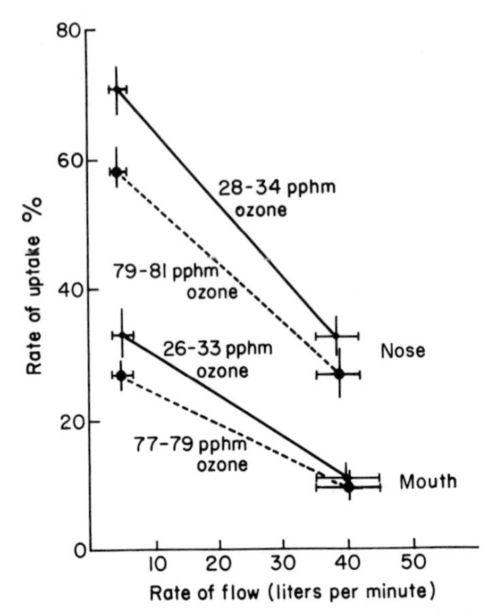

Figure 3. Rate of uptake of ozone in the upper respiratory tract of eleven anesthetized dogs ventilated by means of respirator at "low" and "high" concentrations at two flow rates and with nose and mouth breathing. Note that the absorption rate is inversely proportional to the flow rate and concentration and that the absorption rate is approximately twofold greater when breathing through the nose. Horizontal bars indicate range of flow, vertical bars indicate standard deviation (*82*).

association of reduced diffusion capacity and increased airway resistance in individual animals. The authors point out that the change in airway resistance had both reflex and nonreflex components. Concomitant pathological studies on similarly exposed animals show severe injury to the bronchial and bronchiolar epithelium, alveolar type I cells, and interalveolar capillaries, although no evidence of intraalveolar edema was discerned. A number of additional papers report pulmonary physiological studies at higher levels of ozone (Table I).

Several investigators have studied the influence of exposure to ozone on extrapulmonary physiological parameters. Exposure to ozone reduced the gross motor activity of guinea pigs, rats, and mice. Murphy *et al.* (*10*) found depression of spontaneous running activity of mice after exposure to 0.2 and 0.5 ppm ozone for 6 hours. Konigsberg and Bachman (*84*) also noted a diminution in gross activity of rats related to concentration of ozone when exposures varied from 0.1 to 1.0 ppm.

The effect of vagotomy has been studied by Vaughan *et al* (*85*) in exposures of dogs to 40–55 ppm ozone for 5 minutes. They produced an apneic phase of approximately 1 minute associated with a slight fall in blood pressure and slowing of the heart rate. These changes did not occur at the 40 ppm level, apparently having been blocked by vagotomy. Dynamic compliance was reduced and pulmonary air flow increased in both nonvagotomized and vagotomized animals.

The influence of unilateral exposure on pulmonary physiological parameters was studied by Frank (*86*) and Alpert and Lewis (*87*). In the former study (*86*), unilateral exposure was accomplished by means of one-sided pneumothorax. The animals were subjected to unilateral exposure of either 0.5 ppm or 1.0 ppm ozone for 3 hours. Following this exposure, the pneumothorax was relieved and the animals exposed bilaterally on three successive days to various combinations of exposure varying from 0.5 to 1.0 pm for 3 hours daily. Single exposures of 1.0 ppm altered the pressure volume relationship and no effect was noted at 0.5 ppm. Multiple exposures did not show this response. A second study (*87*) consisted of unilateral exposure via tracheotomy and unilateral bronchiolar catheterization in Flemish giant rabbits. The ozone dosage was 12 ppm for 6 hours. The respiratory rate increased for about $3\frac{1}{2}$ hours, then gradually declined. Tidal volume fell in both control and ozone treated lungs, with the terminal values for the treated below the control. No alteration in unilateral functional parameters occurred prior to the development of edema. The authors speculate that the physiological effect might be due to neural or chemical response, possibly secondary to edema.

Changes in the elasticity of the lung have been reported from exposure of various animal species to ozone. Reduction in compliance has been reported by Alpert and Lewis (*87*) in rabbits and Vaughan *et al.* (*85*) and Frank *et al.* (*88*) in dogs. These exposures were all above the edema-producing level and quite likely are the result of the influence of edema fluid increasing the rigidity of the lung. Watanabe *et al.* (*8*), however, reported a slight fall in compliance irregularly in cats exposed to concentrations of 1 ppm and less. Yokoyama (*89*) has also reported the reduction of variations in compliance in isolated dogs' lungs and a significant decrease in pressure–volume relationship after exposure to 5–10 ppm ozone. Perhaps the lowest exposure of the lung that caused significant alteration of the pressure–volume relationship was carried out by Bartlett (*90*), who exposed rats to 0.2 ppm ozone continuously for 30 days. Lung volumes were increased 16% and static pressure volume measurements indicated that the lungs were overdistended by high transpulmonary pressures. These changes were interpreted by the authors as indicating a decrease in lung elasticity and suggesting mediation by an effect on collagen and elastin.

It is probably relevant to the subject of lung elasticity to discuss the role of ozone in alteration of the pulmonary surface-active material, since it is known that this substance provides stability to the pulmonary parenchyma by virtue of its reduction of surface tension when compressed. Mendenhall and Stokinger (*91*) showed an attenuation of surface pressure when lung washings were exposed to ozone *in vitro*. This action could be blocked by the addition of sulfhydryl-containing substances. Yokoyama (*92*) noted a questionable increase in minimal surface tension after *in vitro* exposure of lung extracts to 0.5–1.2 ppm ozone. No change was evident when similar experiments were carried out with lung lavage fluid and dipalmitoyllethicin. However, no alterations in surface tension of either pulmonary lavage fluid preparations or of dipalmitoyllethicin were noted by Gardner *et al.* (*93*). It is questionable that ozone could attack the highly saturated molecule of dipalmitoyllethicin, generally concluded to be the principal component of the surface-active material. Thus, if changes in surface tension occur, some other mechanism must be postulated.

E. Exposure of Human Volunteers to Ozone

A number of studies of pulmonary physiological response employing experimental exposure of human beings have been reviewed by Bates and Hazucha (*94*). Griswold *et al.* (*95*) exposed one person to 2 ppm ozone for 2 hours and noted a transient fall in vital capacity and forced expiratory volume. Clamann and Bancroft (*96*), in an exposure of approximately 1 ppm ozone for 2 hours, saw reduction in vital capacity and maximal breathing capacity. Bennett (*97*) exposed subjects to 0.5 ppm ozone for 3 hours per day for 6 weeks and produced a progressive decrease in the 1-second forced expiratory volume. (In view of the production of persistent lesions in animals after similar exposure, some restraint might be suggested for exposures of this type in man.) Lagerwerff (*98*), however, noted no change in vital capacity after a single exposure to 0.5 ppm for 3 hours. More extensive exposures were conducted by Young *et al.* (*99*) on eleven volunteers via mouthpiece apparatus. Exposures were at 0.6–0.8 ppm ozone for 2 hours. The authors reported a significant reduction in steady-state pulmonary diffusing capacity and in forced respiratory volume. Hallett (*100*) exposed volunteers in a study that proved retrospectively to have been from 1 ppm to 3 ppm. As might be expected, the data showed a variable response in vital capacity, forced expiratory volume, and diffusion capacity. Experiments were conducted by Goldsmith and Nadel (*101*) on four human volunteers by exposure to 0.1–1.0 ppm ozone for 1 hour. The authors noted changes in two of the four subjects during exposure to 0.1 and 0.4 ppm.

A significant change in the regimen of exposure was undertaken by the group of investigators at McGill University who have evolved an experimental protocol containing a second variable (light exercise) in conjunction with exposure to ozone (*101a, 102*). The exercise is standardized so that the volunteers alternate 15 minutes of exercise on a bicycle ergomometer with 15 minutes of rest. Exercise employed was sufficient to double the minute ventilation and was said by the authors to be equivalent to walking on a level surface. Exposure took place in a Plexiglass-enclosed space by natural respiration and avoided the use of face masks, mouth pieces, nose clips, etc. In addition to the more routinely applied lung function tests, measurements of residual volume employing a helium dilution method and of closing volume using a bolus of 133 xenon (*94, 101a*) were performed.

Data from exposures employing intermittent exercise may be summarized as follows: significant decrement in maximal midexpiratory flow rate at 0.37 ppm and 0.75 ppm and in forced expiratory volume (1 second) at 0.37 ppm and 0.75 ppm. The authors calculated the probable effect of varying concentrations of ozone on forced vital capacity, forced expiratory volume, and the maximum midexpiratory flow rate (Fig. 4).

These same authors have also studied the effects of sulfur dioxide and combined sulfur dioxide and ozone in human subjects. These studies show that the combined effect of sulfur dioxide and ozone is greater than the effect from either gas alone. These data suggest that there is a biological interaction of the two agents. However, it is posible that the increase in effect is due to the conversion of sulfur dioxide to sulfuric acid, which can be facilitated by the presence of ozone (Fig. 5) (*94*).

Recent pathological evidence indicates that the small airways are a prime target for respiratory tract injury by ozone. Customarily used tests for ventilatory performance are insensitive to this region. Thus, tests to measure closing volume as a percentage of vital capacity, or closing capacity as used by Bates and Hazucha (*94*) are important, since they focus on this region. Their data indicate that there was a considerable increase in closing capacity, thus indicating that there were obvious effects in the small airways. The volunteers were categorized as smokers and nonsmokers for the above experiments. Results indicate that while differences between smokers and nonsmokers were questionable at the 0.37 ppm ozone concentration, exposure at 0.75 ppm produced more severe decrements in function parameters in smokers.

Human volunteers generally experience pain or discomfort while undergoing exposure. Subjective symptoms have included cough, substernal soreness, pharyngeal irritation, dyspnea, headache, and conjunctival discomfort. However, these symptoms do not persist after cessation of the exposure period.

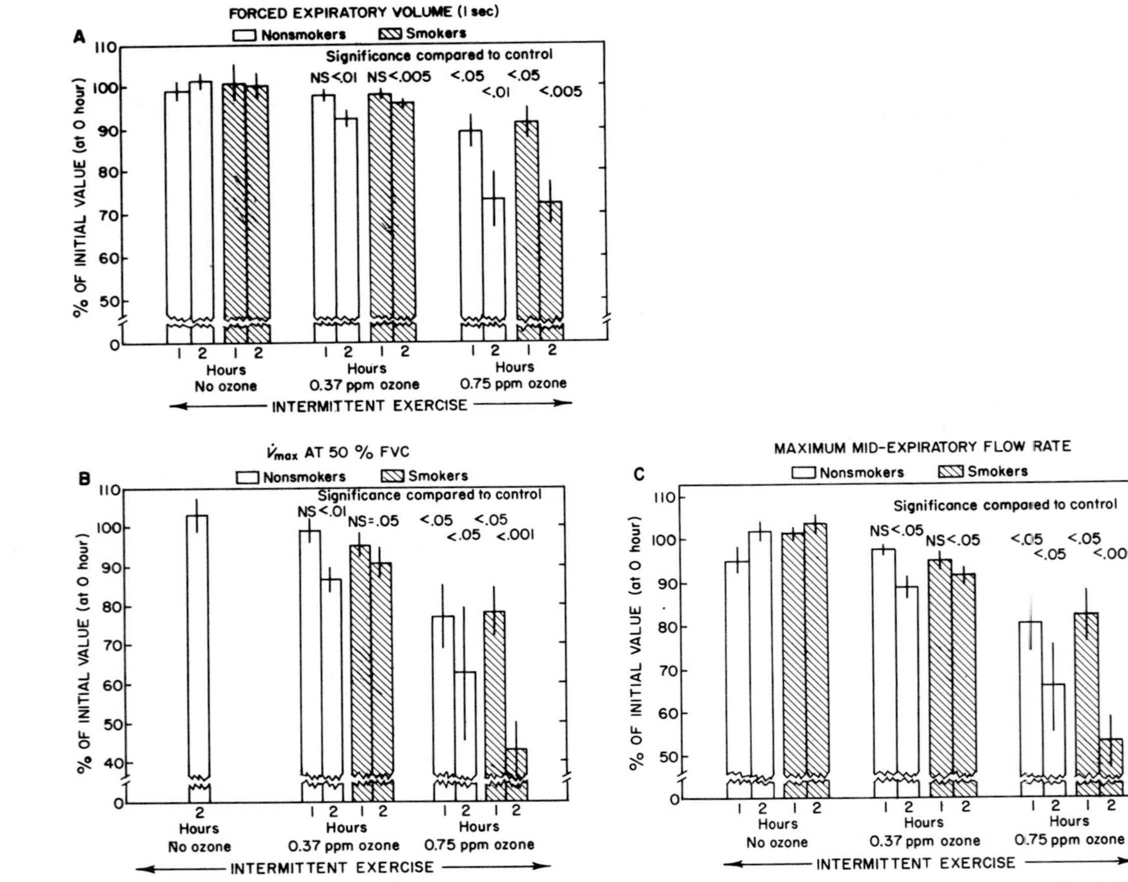

Figure 4. Decrement in pulmonary function resulting from short-term exposure to ozone in smoking (hatched bars) and non-smoking (plain bars) human volunteers while intermittently exercising. The mean values are expressed in percent of preexposure measurements. A, Forced expiratory volume at one second. B, Maximal expiratory flow rates of 50% of the average pre-

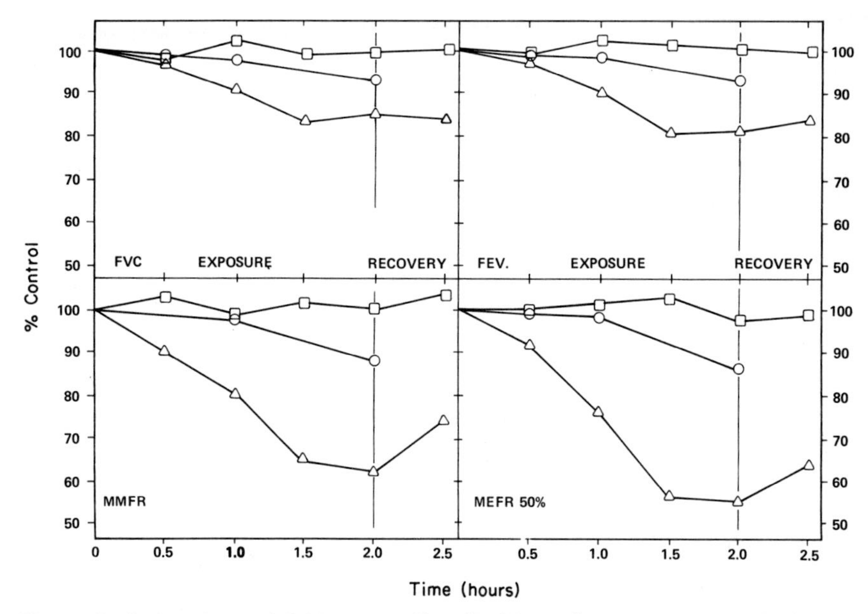

Figure 5. Comparison of 0.37 ppm sulfur dioxide and ozone exposure in human subjects as single agents and combined under conditions of light exercise. Note the striking increment in effect when these gases were combined. □ = 0.37 ppm sulfur dioxide; ○ = 0.37 ppm ozone, △ = 0.37 ppm sulfur dioxide and ozone (*94*). FVC = forced vital capacity; FEV = forced expiratory volume; MMFR = maximal midexpiratory flow rate; MEFR = midexpiratory flow rate.

F. Nonpulmonary Effects of Ozone Exposure

A number of so-called "systemic" effects have been ascribed to ozone exposure. These include changes in elements in the circulating blood—sphering of erythrocytes (*7*), chromosome breakage in lymphocytes (*104*), retardation of deoxygenation of hemoglobin (*7*), alterations in the cell membranes and nuclei of myocardial cells (*7*), decrease in brain serotonin (*35*), increase of the RNA/DNA ratio in the liver (*23*), transitory increase in liver weight and alkaline phosphatase (*29*), alteration of the contralateral lung when the lungs were exposed unilaterally (*88*), premature aging (*105*), morphologic alterations of the parathyroid gland (*106*), and increased sleeping time from sodium pentobarbital (*107*).

Some of these effects can be explained on the basis of local action of ozone within the lung, and true systemic action need not necessarily be postulated. In the case of chromosome breakage, for instance, ozone-generated free radicals or peroxidation products might react with circulating lymphocytes as they pass through the pulmonary capillaries. Therefore, only mature lymphocytes would be affected, and there would be limited mutagenic significance as compared to actual injury of cells capable of mitotic activity within the various nidi of lymphocytic production. This is an important point that needs further elucidation. A similar mechanism in the pulmonary capillary bed might explain the sphering changes in erythrocytes on the basis of ozone attacking the polyunsaturated lipids of the red cell envelope and stroma, and the slowed deoxygenation might result from abnormal oxidation of the hemoglobin. These hypotheses are in agreement with the formation of Heinz bodies, which characteristically result in a higher oxygen affinity curve and sphereocytosis when induced by other chemicals (primaquine, phenylhydrazine, etc). Changes in the RNA/DNA ratio, alkaline phosphatase, liver size, and the like might reflect stress and blood volume shift into the splanchnic area. It is possible that the effect on the contralateral lung as reported by Frank *et al.* (*88*) may be artifactual on the basis of two points. The authors achieved their unilateral exposure by producing one-sided pulmonary collapse by means of pneumothorax. Thus, there is a possibility that (a) there was some minimal ventilation due to imperfect collapse and consequent minimal ozone exposure or (b) that artifactual alteration of the lung was induced by collapse and reinflation. Alpert and Lewis (*87*) who created the same sort of model by unilateral exposure by means of differential catheterization did not observe edema in the contralateral lung. Several of the changes mentioned above, however, cannot, on any basis, be attributed to purely local chemical action within the lung or the production of a shock state by pulmonary irritation. Some circulating toxic factor must have been responsible if the changes in the heart muscle cells reported by Brinkman *et al.* (*7*) are valid. Two mechanisms appear theoretically tenable, (a) a chain reaction involving free radicals (postulated by Goldstein *et al.* (*38*) on the basis of *in vitro* studies) that, once started in the capillary bed of the lung, might continue long enough with respect to the blood polyunsaturated lipid to reach distant organs or (b) systemic reaction that results from the formation of fatty acid ozonides or peroxides in the lung. The latter appears an attractive alternative, since such radicals, once formed in the lung, might then gain admission to the blood and circulate for some time reaching all parts of the body, exerting highly toxic effects at any susceptible distant site.

The experiments by Gardner *et al.* (*107*) and Abudonia *et al.* (*108*) are interesting in that they appear to indicate the effect of ozone on a specific enzyme system in the liver of mice, which apparently requires conditioning by two ozone exposures prior to the induction of anesthesia by sodium pentobarbital. The phenomenon was noted only in females, suggesting attack on cytochrome *P448* if an analogy may be made from rats. While the exact mechanism is as yet not elucidated, it appears that some circulating product of ozone gains admission to the liver and inhibits a detoxifying mechanism. The sex difference in this effect suggests that the ozone product destroys cytochrome *P448*, which has been demonstrated *in vitro* with partially oxidized unsaturated fatty acids such as fatty acid ozonide (*38*) or reduction in cofactors such as NADH and NADPH, which are known to be oxidized by ozone (*62, 63, 95*).

The aging alterations reported for ozone exposure are of great interest, and, with what we now know of the chemistry of ozone, are to be expected. The possibility of a free radical effect has already been discussed. However, it is well known that ozone is highly active as a tissue oxidant (*62, 63*). It not only oxidizes unsaturated fatty acid, but has strong oxidant action against thiol. Glutathione is oxidized, and a significant portion cannot be again reduced by nicotine adenine triphosphate reduced form (NADPH)—glutathione reductase (*62, 63*). Tocopherol is protective against many oxidative processes, including oxidation of polyunsaturated lipid, although apparently not against reduction of thiols (*62*). Ozone exposure is destructive to tocopherol and artificially tocopherol-deficient animals are more susceptible to ozone exposure (*38*). Direct evidence for such effect has been produced by Harmon (*109*), who has shown that the maximum life span of at least two strains of mice can be significantly prolonged by means of high antioxidant supplement of the ration. Thus, it might be tenable to postulate that ozone contributes to aging on the basis of accelerated oxidation.

G. Tolerance to Ozone

Many irritant inhalants have the property of stimulating "tolerance," a state in which the toxicity of a second dose is blocked to a greater or lesser extent by a prior exposure to a stressing, but nonlethal, dose. So-called cross-tolerance develops between different irritants when administered in a tolerance-developing sequence, i.e., ozone against phosphene, or vice versa (*110, 111*).

This phenomenon has elicited a great deal of interest among toxicologists and consequently has been studied rather extensively. Ozone is a particularly strong tolerance-inducing pollutant, either against itself or through cross-tolerance with other agents (Table II). With ozone, tolerance begins to be initiated at 0.3 ppm, reaching its maximum protective effect at 3–4 ppm for the initiating dose (*110*). Tolerance begins to develop within 30 minutes, reaches its maximum in approximately 24 hours, and it is thought to persist for several months or longer (*110, 112*).

Several mechanisms have been proposed as modes of action for tolerance development: (a) mobilization of a protective chemical by the sensi-

Table II List of Agents for Which Cross Tolerance Has Been Tested (*4*)

Agent	*Adrenaline*	*ANTU*	*Alloxan*	*NH_4Cl*	*CdO*	*CCl_3NO_2*	*GeO_2*	*Histamine*	*H_2S*	*Ketene*	*NO_2*	*NOCl*	*O_2*	*O_3*	*Phenyl-thiourea*	*$COCl_2$*	*Thiourea*	*OF_2*
Adrenalinc	–										–			–				
ANTU		+							+			+		+	+	+		
Alloxan			?															
NH_4Cl				?														
CdO					+						+							
CCl_3NO_2						+					+			+		+		
GeO_2							±							+				
Histamine								–						–				
H_2S									+					+		+		
Ketene										+				+				
NO_2						+					+			+		+	+	
NOCl												+		+				
O_2													+					
O_3						+			+	+	+	+		+		+	+	
Phenyl-thiourea											+			+	+			
$CoCl_2$						+					+			+		+	+	
Thiourea		+									+			+				
OF_2																		+

tizing dose; in this category might be mentioned histamine antagonists, glutathione (GSH), acid hydrolases, or the induction of a specific enzyme; (b) physical thickening of the alveolar walls as the result of edema fluid; and (c) development of cellular resistance. None of the above mechanisms appears to satisfactorily explain tolerance in all its manifestations. Fairchild (*113*) has observed that lung GSH levels were maintained in ozone-tolerant animals but not in nontolerant animals. This situation could be brought about either through the enzymatic pathway, which would stimulate augmented production of GSH in response to a second exposure to ozone, or by directly blocking the oxidation of GSH by some means. It appears that the cycle of depression and augmentation of GSH following a single exposure to ozone does not necessarily result in augmented GSH levels at the point where maximum tolerance should occur, i.e., 24 hours after exposure (*63*) (Fig. 2). Another factor that might mitigate against the GSH protective mechanism or of other specific augmentation or induction of specific blocking agents on a molecular level is the extremely divergent chemical character of the various agents capable of eliciting cross-tolerence.

Also, in regard to GSH it is entirely possible that the maintenance of these thiols in the tolerant animals was merely secondary to their protected state, rather than the cause of it. While the development of a histamine release-blocking agent might be postulated as a consequence of pretreatment, the reappearance of histamine in the lung exposed to ozone does not parallel the duration of tolerance.

The physical swelling of the interalveolar septa on the basis of pulmonary edema producing a blockade of irritant effect as proposed by Henschler *et al.* (*114*) appears an attractive possibility, though it is difficult to justify by virtue of a report of persistence of tolerance for 14 weeks or longer (*4*). Information based on studies of tolerance induction through unilateral lung exposure suggest that there is little or no protection unless the particular lung has actually been previously exposed to ozone, i.e., that no significant crossover occurs from the contralateral lung (*87, 88*). This suggests that there is no basis for a circulating humoral agent and that the phenomenon is a purely locally inspired mechanism (Fig. 6).

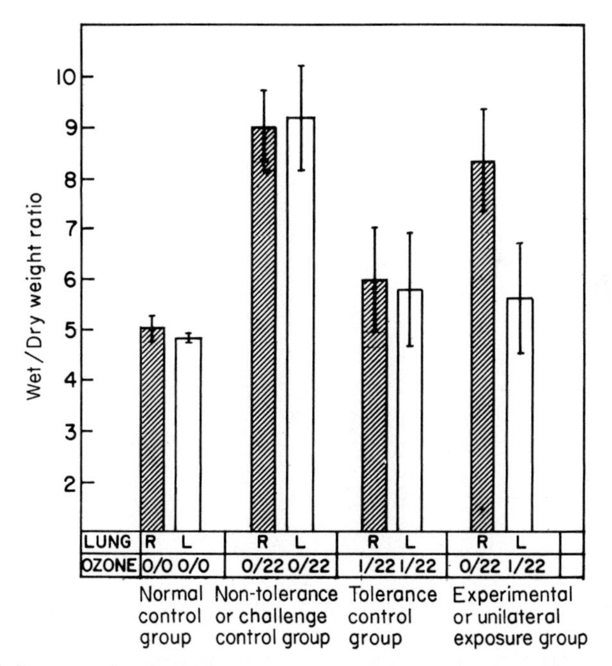

Figure 6. Influence of exposure to a 1 ppm prechallenge dose of ozone when followed by a challenge of 22 ppm ozone on wet–dry weight ratios. Note that no statistically significant difference exists between the tolerance induced by exposure to both lungs when exposure was bilateral and to one lung when exposure was unilateral and that no protection is evident in the contralateral lung (R, right; L, left) (*14*).

The significance of tolerance in the protection of the human population at risk in community air pollution may be questionable. In small rodents, there is a sharp reduction in the elicitation of tolerance when the initiating dose is below 2–3 ppm, and it appears to be extinguished at approximately 0.3 ppm (*110*). This is approximately the cut-off point for production of demonstrable edema in rats by means of recovery of labeled albumin via lung lavage, where the minimal effective dose lies between 0.25 and 0.5 ppm for 6 hours (*115*). Thus, it may be necessary to administer a minimally edemagenic dose before tolerance to subsequent edema formation develops. Therefore, the ozone injury that is blocked by tolerance may be one that specifically induces development of pulmonary edema. Such injury is unlikely to occur in human beings casually exposed to ambient air. Experiments conducted in mice by Coffin and Gardner (*44, 116*); in which the parameter of resistance to bacterial infection was the indicator of toxicity from ozone exposure, suggest that there is a biphasic response with partial protection above 0.3 ppm and that there is no protection at all below this concentration (Fig. 7) (*44*). Since it is known that the presence of edema fluid in the lung increases the risk of pulmonary infection, it seems reasonable to postulate that the portion of the reduction of mortality attributable to tolerance was due to blocking of the edemagenic response. Furthermore, the residual in-

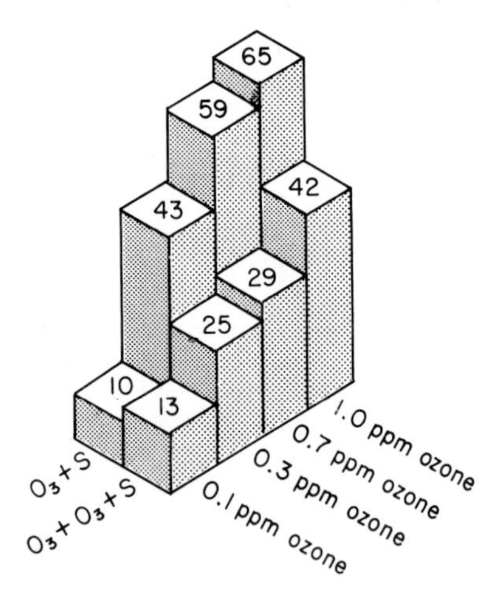

Figure 7. The percent of increase in mortality over controls after a preliminary dose of ozone to induce "tolerance" in an infective system is illustrated at four ozone concentrations. Note that tolerance inducing a 50% reduction is effected at the two higher doses, but is ineffective at the lower doses 0.1 ppm. $O_3 + S$ indicates that mice received streptococci immediately after ozone exposure. $O_3 + O_3 + S$ indicates that mice were exposed to a tolerance-inducing dose of ozone before a later exposure to ozone and streptococci (*44*).

crement in mortality (above 0.3 ppm) and the lack of any evidence for tolerance below this point might be interpreted as the result of ozone action on cellular and noncellular specific defense systems in which tolerance could not be induced. Experiments by Gardner *et al.* (*14*) indicate that while reduction in edema development could be induced by single or repetitive ozone exposures, no such protection was obtained for the chemotactic expression of ozone for the lung of the rabbit (Fig. 8). This finding might be interpreted as suggestive of a dual mechanism, one blocked by tolerance (edema formation), and one not affected (influx of inflammatory cells), thus separating edema and inflammation on the basis of tolerance.

In the latter study, it was also apparent that there was no influence of tolerance on the ozone injury to the macrophage system (as measured by numbers of recoverable cells in pulmonary lavage), in the phagocytosis index, or in alterations of the macrophage lysosomal enzymes (Fig. 8).

It is germane to discuss another phenomenon that may or may not be related to tolerance, but might be better termed an adaptation state. This was first noted for nitrogen dioxide exposure by Freeman *et al.* (*48*) in the bronchiolar epithelium of the rat but preliminary observations by the same group of investigators show that ozone exposure results in a similar effect. The sequences are similar when repetitive or continuous exposures are performed for either gas, although the minimal effective dose for ozone is 10–20 times lower than for nitrogen dioxide (*54*). Animals

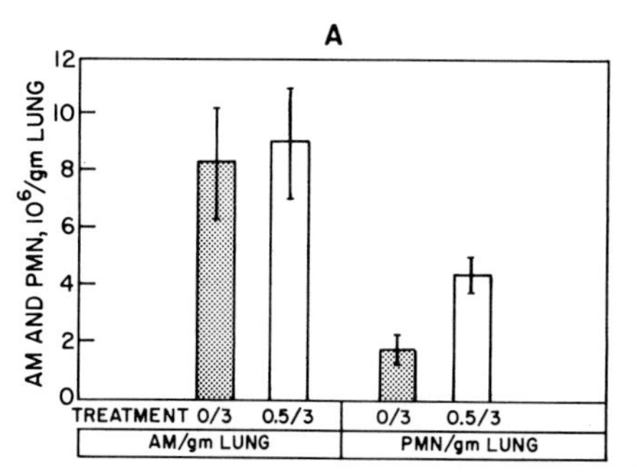

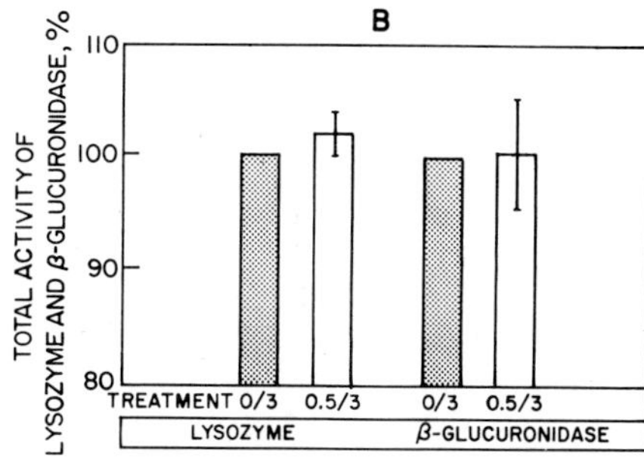

Figure 8. Influence of exposure to 0.5 ppm prechallenge dose of ozone on pulmonary cells (A) and macrophage lysosomal enzyme (B). Note that no tolerance was induced as indicated by no difference from the contralateral lung in the macrophage parameters and that the prechallenge increased the number of recoverable polymorphonuclear leukocytes. No change at all was noticed in the activity of the two lysosomal enzymes. (See Fig. 6). Treatments are expressed as prechallenge, tolerance-inducing dose/challenge dose. Shaded bars indicate right lung, unshaded bars indicate left lung; AM = alveolar macrophages, PMN = polymorphonuclear leukocytes (*14*).

undergoing either continuous or intermittent exposure to 0.25–0.9 ppm ozone show within a few hours necrotic and desquamative changes of some cells of the terminal portion of the bronchioles and proximally located alveoli (*48*). By 24 hours, most of these cells are replaced with new cells that can be distinguished morphologically from their predecessors. In the pulmonary alveolus, the squamous or type I epithelial lining cell is cast off and replaced by mitotic division of the granular pneumocyte type II cell. Characteristic morphological differences also occur in the terminal bronchiolar epithelia. The sequence of the change has been characterized for nitrogen dioxide by pulse labeling dividing cells with tritiated thymidine and measuring the rate of cell division among the several types in the newly formed cells (*116, 117*). Two points of interest may be mentioned in this connection: (a) in the face of continuous or intermittent exposure to the same concentration of the gas, the initially injured and desquamated cells are replaced by relatively resistant ones, which eventually revert to resemble the original cells; and (b) if the exposure is maintained long enough, alteration of the structure of the lung occurs with development of fibrosis and an emphysema-like state (*48*). Thus, in spite of the apparent adaptation of some cell types to the gaseous environment, the changes do not connote universal protection of the lung. At this point, it is uncertain whether the adaptation on the part of the lung cells requires substitution of a more resistant normal cell for a less resistant one or whether some adaptation may occur metabolically without replacement of cells. Pace *et al.* (*118*) has described an adaptive response among cells *in vitro* in which resistance to a subsequent exposure to ozone was induced by prior exposure. (It is possible that the Pace experiment may represent selection of a resistant cell type.)

H. Ozone as a Possible Mutagen

Reference has been made elsewhere in this chapter to chromosome breaks in lymphocytes in hamsters attributed to exposure to ozone, and the case for a possible nonpulmonary effect has also been discussed. Hence, interest has been generated in the possible potential of ozone attacking the genetic material of the body on a systemic level as a mutagen. It follows, then, that if the effect of ozone is mediated via a free radical, and if this action can be perpetuated distally from the lung, a strong possibility would exist that ozone could in fact act as a mutagen. The apparent similarity of certain effects of ozone to effects of ionizing radiation has also been alluded to as evidence for its being potentially a mutagen. However, all these things are merely circumstantial and not of great weight as far as actual proof of this is concerned.

In order to act as a mutagen, a given agent must damage genetic material so that the abnormality results in a succession of cells with new genetic potential resulting in anatomic or functional aberration in newly formed cells within an individual or in the transmission of a genetic defect to a succeeding generation.

In vitro studies with ozone have shown its ability to change the adsorption spectrum of nucleic acids (*119*) and to alter pyrimidine bases of nucleic acids from *Escherichia coli* (*120*). It also has been claimed that *E. coli* exposed in water through which ozone had been bubbled develop specific mutants (*121*).

In vitro exposure of chick embryo (*122*) and cultured human cells to ozone has produced chromosome aberrations that apparently are dose related (*123*). Chromosome aberrations were noted in roots of *Vicia faba* (*124*) and, as mentioned previously, changes were noted in nucleic acids and their derivatives following *in vitro* exposure to 0.2 ppm ozone for 5 hours (*119*). In the latter experiment, the ozone was generated by an ultraviolet light source situated within the exposure chamber but shielded so that the animals had no direct irradiation by ultraviolet light. Two other experiments both performed through *in vivo* exposure of mice are of interest. Brinkman *et al.* (*7*) exposed inbred gray mice and C57 blacks to 0.1 ppm and 0.2 ppm ozone for 7 hours a day 5 days a week for 3 weeks prior to parturition. They noted an approximate fourfold increase in neonate mortality among the offspring from the ozone-exposed mothers, which might logically have resulted from simple injury to the fetus by such exposure.

Hueter *et al.* (*125*) exposed both male and female white Swiss mice prior to breeding and continued exposure of the females in an experiment designed to determine the relative importance of the exposure on either sex. The following breeding schedule was used: exposed males to unexposed females, exposed females to unexposed males, exposed males to exposed females and unexposed males to unexposed females. The cages were examined for cervical plugs to ascertain the efficiency of breeding, and uterine scars were examined to ascertain the number of implantations. Their data indicate that a diminution of litter size attributable to fewer implantations occurred from matings in which only males had been exposed to ozone.

In view of the foregoing findings, it might be germane to cite somewhat similar exposure studies made with an oxidant-containing atmosphere (irradiated auto smog). Hueter *et al.* (*125*) found that female mice exposed to this atmosphere before, during, and after gestation showed a marked decrease in number of litters, total number of pups born, frequency of litters per mother, and survival of infant mice. Lewis *et al.*

(*126*) in a study aimed at elucidating the above findings, explored the subject when mice were exposed in a factorially designed experiment. These authors found that pretreatment of males with irradiated auto smog and switching of the females from one atmosphere to another appeared to reduce the number of births. This phenomenon is apparently not associated with absorption of fetuses *in utero* or by a diminution of copulation. The authors also claim that heightened infant mortality was also associated with pretreatment of the males. They speculate that their findings may have been due to alteration of the genetic composition or other cellular components of the sperm.

I. Discussion of the Effects of Ozone

Evidence is constantly accumulating that ozone is an extremely toxic gas. Data from early studies on laboratory animals indicated that death from pulmonary edema and inflammation occurred after exposure to concentrations of 15–30 ppm and above for several hours and that pulmonary edema as measured by wet weight–dry weight ratios could be produced by exposure of the order of 1–2 ppm for 3 hours in small rodents. Data reviewed here indicate that evidence of transudation of blood protein into the lung (as measured by recovery of labeled albumin in pulmonary lavage fluid) occurs after exposure of rats to between 0.25 and 0.5 ppm for 6 hours. Pathological studies have demonstrated that ozone attacks all levels of the bronchial tree with microscopic lesions already evident after exposures to as little as 0.25 ppm for a few hours. Cells are quickly damaged and cast off. Regeneration of the same or different cell types is apparent as early as 24 hours after exposure. This is probably not a specific lesion, but is indicative of the presence of a noxious influence at this anatomical level. There is a turnover of cells in the terminal portion of the bronchioles and proximal alveoli, which may produce a characteristic lesion transiently consisting of uniform epithelium in the bronchioles and alveolar epithelial cuboidal cells (hyperplasia of type II epithelial cells). The tentative mathematical model for ozone uptake in the respiratory tract of man predicts significant dosage at all levels of the airway. However, on the basis of pathological studies, the terminal portion of the airway appears more severely affected.

It is of interest in this connection that concomitant exposure to sulfur dioxide and ozone in man causes exaggerated decrements in various pulmonary flow rate responses (*102*). Such response might be predicted from a two-stage action, i.e., proximal airway reflex stimulated by sulfur dioxide (and probably ozone) plus a second, more distal airway direct tissue response by ozone. There appears to be some difference of opinion regarding the segmental level for the most severe airway response based on the pathological anatomy observed in laboratory animals. Some investigators claim that the most affected tissues are the terminal bronchioles, alveolar ducts, and proximal alveolar epithelium, whereas others observe them to be in the upper airway. It is now known, however, that there is fairly diffuse uptake throughout the pulmonary portion of the airway to the effect that the mathematical model mentioned above probably underestimates ozone uptake by conducting airways and overestimates that by respiratory airways. A model recently proposed by Miller (*102a*) that considers the reactivity of mucus provides predictions more congruent with empirical observations and with little relative differences in uptake in people, rabbits, and guinea pigs. Fibrosis and emphysema-like alterations are a feature of exposure to ozone of longer duration either continuously or intermittently. The role of the epithelial lesion in the genesis of structural modification of the lung is at this time unclear. For instance, it may only reflect a sequential cycle of epithelial injury and replacement, a feature common to most epithelial tissue recovering from relatively minor injuries. On the other hand, it may also be associated in some manner with fibroplastic and degenerative changes important in the genesis of emphysema. It is tempting to speculate that the emphysema-like state, characteristic of longer-term exposure to both ozone and nitrogen dioxide, may be secondary to at least two independent factors that may act in concert; (a) partial obstruction of the distal airway by the maturation and contraction of the fibroplastic lesions in this region and (b) alteration in the structural integrity of the lungs from loss of tone and elastic properties of the preexisting connective tissue framework due to chemical alteration of the collagen and elastin. Microscopic observations show fibroplastic proliferation in the distal airway. The role of ozone (and nitrogen dioxide) in the denaturation of collagen molecules has been referred to earlier in this chapter (*22*). Enlargement of collagen fibrils, simulating the effects of aging, has been reported for nitrogen dioxide. Evidence for both diminished dynamic compliance (increased rigidity) and recoil of lungs has been provided by physiological observations in animals. Increased rigidity occurs principally with edemagenic exposures of ozone and presumably is associated with increased pressure within tissue due to fluid and possibly with dilution of the surface active material by edema fluid. When nonedemagenic doses are administered, diminished recoil or elasticity (contractile force) demonstrated by pressure–volume curves employing both air and saline is noted, suggesting increased compliance and diminished contractile force of the lung (*90*). These changes are accompanied by marked pulmonary overdistention and a noticeable reduction in number of alveoli.

anges may herald the onset of pulmonary emphysema, which op much later than elevated expiratory flow resistance. Furthere is constriction of the lower airway in response to direct ozone injury in animals at this segmental level. Experiments with human volunteers in which closing capacities were determined suggest that this level in the bronchiolar tree contributes significantly to physiological alterations in human beings and that ozone acts reflexively by stimulation of centers in the nasopharynx causing constriction of the larger airways and nonreflexively by direct chemical action in the terminal portions of the airways. An interesting feature of the acute response of human beings to ozone exposure has been the apparent augmentation of effect when persons at risk to high ambient oxidant levels indulged in sports. This has shown up in *post hoc* analytic studies of performance records of high school athletes (*127*) and in systematically conducted clinical observations of school children in Japan (*127a*). A common observation has been that, while exercising, subjects are at higher risk. The observation that volunteers sustain a greater decrement in dynamic pulmonary efficiency and that animals suffer greater mortality due to pulmonary edema (*2*) and susceptibility to pulmonary infection (*128*) when exercised during exposure to ozone lend credence to these studies. It is reasonable to suppose that exercise produces an increased dose by virtue of increased minute volume. Another factor peculiar to human exposure is the effect of mouth breathing during violent exercise. Ozone uptake experiments conducted in anesthetized dogs suggest that a change from nose to mouth breathing doubles the amount of ozone delivered to the deeper airways. This mechanism, however, is not normally used by exercising rodents, in whom mouth breathing is abnormal. Nonetheless, Stokinger (*2*) reports a severalfold diminution in the concentration of ozone required to kill rats by means of pulmonary edema when the animals were exposed in a turning drum, and Gardner *et al.* (*128*) note a similar reduction in lethal dosage when interaction with bacterial infection is used as the end point and mice were exposed during continuous walking.

III. Nitrogen Dioxide

A number of oxides of nitrogen are theoretically present in ambient air. Nitric oxide (NO) is formed during the combustion of any sort of material at high temperature by the oxidation of atmospheric nitrogen. Nitric oxide is oxidized to nitrogen dioxide (NO_2) in the atmosphere. This oxidation of nitric oxide to nitrogen dioxide is particularly rapid in the atmospheric photochemical process. Nitrogen pentoxide is also thought to be present, but problems of analysis prevent its actual detection.

Nitrogen dioxide is the major toxicant of this group because of its relatively high toxicity and its ubiquity in ambient air. A number of occupational syndromes and pathological entities attributed to it have been described, such as silo fillers' disease and bronchiolitis obliterans. The high concentration of nitric oxide in tobacco smoke (up to 600 ppm) makes it tenable to consider its conversion to nitrogen dioxide in the airway as a possibility for the etiology of certain human diseases. The nitric oxide in tobacco smoke most likely results from conversion of nitrate during combustion, since the low temperature of the process precludes oxidation of atmospheric nitrogen. The alkaline nature of the smoke, due to the presence of ammonia, prevents any immediate conversion of the nitric oxide to nitrogen dioxide.

A. Animal Data

While nitrogen dioxide is a serious human health hazard with death-threatening potential from exposures of short duration (probably as short as 3 seconds at concentrations of 50 ppm and upward or of 10–40 ppm over a longer duration), such acute effects should be considered impossible at the relatively low concentrations present in ambient air. It has been difficult, therefore, to construct animal models that demonstrate measurable anatomical or functional alterations at concentrations of this gas commonly found in ambient air. It is still possible, however, that subtle effects may occur within ambient ranges from either short or long continued exposures. Therefore, models will be discussed here that explore the possibility that subtle acute effects, slowly progressive chronic lesions, or interaction with infectious disease agents or carcinogens may take place even though most of the striking effects are reported at concentrations well above the ambient (Tables III–V).

B. Pathological Effects on the Respiratory Tract

Nitrogen dioxide appears to exert its toxic action mainly on the deep lung and peripheral airway, although distal or systemic effects probably occur. Continuous exposure for 6 months or longer at 10–25 ppm have produced full-blown emphysema-like alterations of the lung with eventual fatal outcome in a variety of species, including rats, (*160*), rabbits, and dogs (*166*). Similar lesions have been reported for mice at much lower concentrations after exposure to various intermittent schedules for 3 months or longer (*135*). This process has been most intensively studied in the rat by Freeman *et al.*, who have suggested the lesion as a model for human emphysema (*168, 169*). The initial pathological lesions appear within 24 hours after the beginning of the exposure, and they consist of

Table III Summary of Toxic Effects of Nitrogen Dioxide Arranged according to Concentration

(Short-Term, Intermittent Exposure)

Nitrogen dioxide (ppm)	*Length of exposure*	*Observed effect(s)*	*Species*	*Reference*
0.5	4 hours	Degranulation of lung mast cells	Rats	(*129*)
1.0	1 hour	Degranulation of lung mast cells—reversible 24 hours postexposure	Rats	(*129*)
1.0	4 hours	Peroxidation of lung lipids	Rats	(*129*)
1.0	1 hour	Structural changes in lung collagen—partly reversible	Rabbits	(*77*)
1.5–3.5	2 hours	Increased susceptibility to infection with *Klebsiella pneumoniae*	Mice	(*130*)
1.6–5.0	~2 minutes	Increased airway resistance	Human subjects w/chronic respiratory disease	(*131*)
5.0	15 minutes	Decreased diffusion capacity	Healthy human subjects	(*131*)
5.0	2 hours	Reduced lung clearance of *Klebsiella pneumoniae* and increased mortality	Mice	(*130*)
5.0	2 hours	No increased mortality by *Klebsiella pneumoniae*	Hamster	(*130*)
5.2–13.0	2–4 hours	Reversible respiratory rate increase and decreased tidal volumes	Guinea pig	(*132*)
3.0–16.0	1 hour	3 ppm, alveolar endothelial damage; 5 ppm, decreased pulmonary lecithin; 7 ppm, intraalveolar edema	Dogs	(*133*)

bits of fibrin in the airway, increased numbers of macrophages, and altered appearance of the cells in the distal airway and adjacent pulmonary alveoli. When pulmonary lavage is performed at this stage, there is an elevation of protolytic enzyme, presumably derived from lysed epithelial cells and macrophages. As seen in plastic-embedded sections by light and electron microscopy, desquamation of type I alveolar cells (the squamous pneumocyte) occurs with replacement by the type II cells (the granular pneumocyte). Since the type II cell is larger in cross section, the microscopic appearance is quite distinctive and can resemble gland-like tissue (adenomatosis). This same lesion is described for ozone and also appears to be a feature of oxygen toxicity (*138, 170, 171*).

Table IV Summary of Toxic Effects of Nitrogen Dioxide Arranged according to Concentration

(Long-Term, Intermittent Exposure)

Nitrogen dioxide (ppm)	*Length of exposure*	*Observed effect(s)*	*Species*	*Reference*
0.25	4 hours/day/ 6 days	Irreversible structural changes in lung collagen	Rabbits	(*134*)
0.5	6 hours/day/ 3–12 months	Preemphysematous changes; pneumonitis; severity increasing to 6 months, receding at 9 months with thickening of alveolar septa; bronchiolar epithelial erosion at 12 months	Mice	(*135*)
0.5	6 hours/day, 7 days/week, 12 months	Increased susceptibility (mortality) to *Klebsiella pneumoniae* at 6 months, reduced capacity to clear bacteria from lung, 12 months	Mice	(*136*)
1.0	6 hours/day, 5 days/week, 18 months	No significant differences from controls in body weight, formed blood elements, biochemical indices, histology, or oxygen consumption (rabbits); moderate dilation alveolar ducts and sacs (dogs)	Dogs, rats, rabbits, guinea pigs, hamsters	(*137*)
0.4–2.7	180 days	Increase in blood lipids, lipoproteins and cholesterol, paralleling effects seen in guinea pigs	Human occupational exposure	(*149*)
2.0	40 days	Enlarged collagen fibrils	Rats	(*138*)
5.0	6 hours/day, 5 days/week, 18 months	Only differences from controls were mild dilation peripheral air spaces, small amount edema fluid (dogs); possible acceleration lung tumor in lung-tumor susceptible mouse strain; tolerance demonstrated in rats	Dogs, rats, rabbits, guinea pigs, hamsters, mice (3 strains)	(*137*)
5.0	4 or 7.5 hours/ day, 5 days/ week, 5.5 months	Formation of presumably lung antibody increasing with exposure; no change in expiratory flow resistance	Guinea pigs	(*139*)
15–20.0	2 hours/day, 5 days/week, 21 months	Inflammation of broncheolar epithelium with apparent recovery in 3 weeks	Guinea pigs	(*140*)

Table IV *(Continued)*

Nitrogen dioxide (ppm)	*Length of exposure*	*Observed effect(s)*	*Species*	*Reference*
24.0	6 hours/day, 5 days/week, 43 and 45 days	Transient methemoglobinemia in rats (13.6%), rabbits (2.8%)	Rats, rabbits	(*141*)
25.0	2 hours/day, 5 days/week, 21 months	Inflammation of broncheolar epithelium with apparent recovery in 3 weeks	Rabbits	(*140*)
25.0	6 hours/day, 5 days/week, 18 months	All exposed species: Increased incidence pulmonary congestion at 3 and 6 months; interstitial pneumonitis in most after 10 months, the same as controls. No evidence that nitrogen dioxide increased susceptibility to (natural) infection. Slight transitory increase in O_2 consumption (rabbits)	Rabbits, guinea pigs, rats, hamsters	(*137*)

The sequential replacement of type I by type II cells is probably the basic reaction to injury of this segment of the lung. Observations of these areas by means of radioautography in animals receiving tritium-labeled thymidine, carried out by Evans *et al.* (*117*), indicate that acceleration of generation of new type II cells begins within 24 hours, peaks at 2 days, and returns to preexposure levels by 6 days after onset of exposure. These latter changes were most pronounced at 17 ppm; although less marked, but statistically significant, labeling was noted after exposure to 2 ppm (Fig. 9). Similar observations have been made for exposure of mice to 90% oxygen by Adamson and Bowden (*170*) and rats to pure oxygen by Weibel (*171*). Though these authors also noted similar events, there was greater incorporation of thymidine in endothelial nuclei. These latter effects may be a special feature of oxygen toxicity. Although the influence of exposure of endothelial tissue to nitrogen dioxide and ozone is at this time unclear, it appears that greater concentrations are required to elicit the endothelial lesion than the epithelial one. Observation of these lesions in these sequences leads to the hypothesis that the actual events are that nitrogen dioxide-sensitive type I epithelial cells are destroyed and replaced by proliferating type II cells from which new type I epithelial cells eventually derive as the lesion reverts to normal.

Table V Summary of Toxic Effects of Nitrogen Dioxide Arranged according to Concentration

(Long-Term, Continuous Exposure)

Nitrogen dioxide ppm	*Length of exposure*	*Observed effect(s)*	*Species*	*Reference*
0.3–0.5	6 months	Infiltration of lymphocytes and plasma cells in lung resembling chronic bronchitis (no controls)	Mice	(*144*)
0.36	7 days	Increase in 2,3-diphosphoglycerate of red blood cells, indicating tissue deoxygenation	Guinea pigs	(*142*)
0.4	7 days	Increase in protein of lung washings, indicating leakage in lung capillaries	Guinea pigs	(*143*)
0.5	12 months	Enhanced susceptibility to respiratory infection (*Klebsiella pneumoniae*)	Mice	(*136*)
0.5–0.8	1 month	Degeneration of mucous epithelium in lungs and trachea	Mice	(*145*)
0.5–0.8	1–1.5 months	Edematous changes in alveolar epithelial cells of bronchus resembling hyperplastic foci of pulmonary adenoma	Mice	(*146*)
0.7–0.8	1 month	Alternating decrease and return to normal levels of reduced glutathione due to direct oxidation	Mice	(*147*)
0.8	730 days (lifetime)	Tachypnea and terminal bronchiolar hypertrophy	Rats	(*148*)
1.0	180 days	Increase in blood lipids, lipoproteins and cholesterol with decrease in liver cholesterol, indicating possible atherosclerotic effect	Guinea pigs	(*149*)
1.0	493 days	In virus challenge, increased amounts of serum neutralization antibodies over controls; slight emphysema and thickened bronchiolar epithelium	Monkeys	(*150*)
1.3	17 weeks	6% reduction in weight gain compared with controls; depression of phagocytic activity	Rabbits	(*151*)
2.0	425 days	Hypertrophied bronchial epithelium; bronchial epithelium changes equivocal	Monkeys, rats	(*152*)

Table V *(Continued)*

Nitrogen dioxide ppm	*Length of exposure*	*Observed effect(s)*	*Species*	*Reference*
2.0	Up to 365 days	Renewal of cells of alveoli and airways; terminal bronchial hyperplasia	Rats	(*117*)
2.0 ± 1	Up to 730 days	Increased amounts of collagen in terminal bronchioles; basement membranes thickened, with metabolic changes consistent with structural alterations, no edema or cell destruction	Rats	(*138*)
2.0	7–21 days	Alveolar type I cell replaced by type II cell	Guinea pigs	(*153*)
2.0	7–21 days	Hypertrophy and hyperplasia of type II pneumocyte	Guinea pigs	(*154*)
2.9 ± 0.7	9 months, 24 hours/day, 5 days/week	12.7% increase in lung weight; 13% decrease in lung compliance; reduced lipid content of lung and surface-active properties	Rats	(*155*)
3.0	15 weeks	Transitory leucocytosis in peripheral blood	Rabbits	(*151*)
5.0	30–60 days	Susceptibility to *Klebsiella pneumoniae* increased; influenza virus challenge caused deaths in 2 of 7	Monkeys	(*156*)
6.0	~45 days	Depressed mucociliary transport—reversible	Rats	(*157*)
10.0	10–49 days	Increase in alveolar macrophages and aggregations, indicating tissue injury	Guinea pig	(*158*)
10.0	1 month	Increased susceptibility to A/PR-8 virus caused death in all 6 exposed	Monkeys	(*156*)
10.0	90–120 days	About 50% animals died in 14 days; survivors showed airway obstruction, hyperinflated lungs, arterial oxygen desaturation	Rabbits	(*159*)
12.5	~30 weeks	Advanced hyperplasia and hypertrophy of epithelial cells; proliferation of connective tissue stroma; distension of alveolar ducts and alveoli; 20% decrease in rate of weight gain	Rats	(*160, 161*)

Table V *(Continued)*

Nitrogen *dioxide* *ppm*	*Length of exposure*	*Observed effect(s)*	*Species*	*Reference*
15.0	10 weeks	Increased oxygen consumption, spleen, kidney; increased lactic dehydrogenase activity, lung, liver, kidney; increased aldolase activity, lung, liver, spleen, kidney, serum	Guinea pigs	(*162–164*)
15.0	1 year	Production of serum antibodies to normal lung proteins	Guinea pigs	(*139*)
15.0	Lifetime	Hypertrophy of mucous cells and bronchial epithelium, with metaplasia, papillomatous changes; thickened alveolar septa in two-thirds of alveoli	Rats (very young)	(*165*)
16.0	2 days	Increased number of type II pneumocytes; increased turnover of type I cells	Rats	(*117*)
17.0	7 days	Increased turnover of bronchial and alveolar cells of terminal bronchioles and type II cells	Rats	(*117*)
17.0 ± 2	90 days	Increased collagenization of terminal bronchioles; thickened basement membranes (15 times normal)	Rats	(*166*)
25–26	6 months	Bullous emphysema with scattered small bullae; diffuse increase in collagen (ferric oxide inhalation decreased effects)	Hamsters	(*167*)

If exposure continues, other changes appear consisting of the loss of the normal contour of the cuboidal cells in the terminal airway at a higher level with replacement by a lower, more evenly contoured cell without the epithelial cell budlike projections of individual cells normally seen at this level of the rodent airways. Small bronchioles lose cilia. Ciliagenesis appears disturbed even though the basal bodies, from which these structures are derived, persist. In some instances, cilia appear to be growing within cystlike spaces enclosed by cytoplasm. Large crystaloid depositions are also observed in the cuboidal cells as exposure continued at the higher levels (above 10 ppm). Stephens *et al.* (*172*) exposed rats

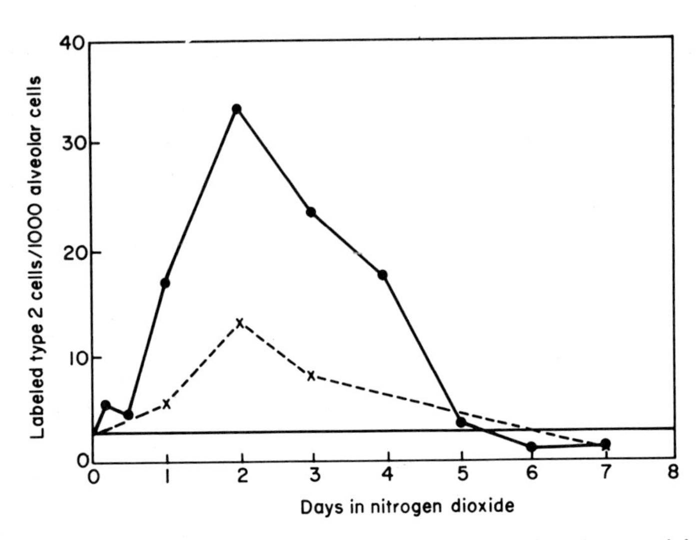

Figure 9. Mean index of incorporation of tritiated thymadine into nuclei of type II alveolar cells. Note that this peaks at 2 days and subsides to normal by 5 days after beginning of continuous exposure. Shaded area indicates standard deviation from the mean: ● = 15–17 ppm nitrogen dioxide, × = 2 ppm nitrogen dioxide, — = control (*117*).

to 17–20 ppm of nitrogen dioxide for up to 610 days, providing rest periods at intervals necessary to insure longevity of the animals for such lifetime exposures. In the rats intermittently exposed, swelling of the individual collagen fibrils and thickening of the basement membranes were apparent by electron microscopy after 90 days. These changes become more severe with time and are quite striking in animals surviving to the termination of exposure. These same animals develop sporadic narrowing and fibrosis of the terminal bronchioles which occasionally appear obstructed (*48*). These changes are synchronous with gross enlargement of the lung with dilation of the air spaces. Morphometric studies indicate that the increased volume of the lung was real. Exposure to 15 ppm nitrogen dioxide yields total enlargement accompanied by a 50% reduction in the number of air spaces and a 25% reduction in alveolar surface area. Thus, the lesion appears to be an actual destructive process rather than one of simple hyperinflation. Qualitative changes in the collagenous tissue observed by electron microscopy consist of swelling of the individual fibrils, which is quite evident on cross section. This may be an accelerated aging phenomenon (*172*). The hyperplastic alterations in the terminal bronchioles regress to normal when exposure is discontinued, but once structural alterations occur, no resolution of the emphysematous lesion is evident. Alterations of lung size are seen after continuous exposure to 2 ppm for the natural lifetime of the rats (*48, 169*). This alteration is associated with subtle changes in the bronchiolar epithelium and reduced exfoliative activity and surface blebbing and is also associated with cytoplasmic crystals. Bronchiolar changes similar to those described above were observed after lifetime exposures to 0.8 ppm, though no change in overall lung size has been reported (*48*).

Studies were performed on mice by Blair *et al.* (*135*), who exposed the animals to 0.5 ppm nitrogen dioxide for 5 days a week at daily schedules of 6, 18, and 24 hours and sacrificed them at 3, 6, 9, and 12 months. They report the presence of septal breaks and pulmonary enlargement at all sacrifice periods. However, their findings appear confounded by the presence of interstitial pneumonitis, and since the changes in the lungs appeared variable from place to place, it is difficult to interpret their findings from their photomicrographs. However, morphometric analysis of size of alveolar spaces suggests that emphysema was present (Fig. 10).

Another study of mice that appears to provide confirmation of the above has been reported by Aranyi *et al.* (*173*). Emphysematous changes occurred in the terminal airways and alveoli in animals exposed for 7 months continuously to 2 ppm or to a continuous base dose of 0.5 ppm with 1-hour peaks of 2.0 ppm for 5 days a week.

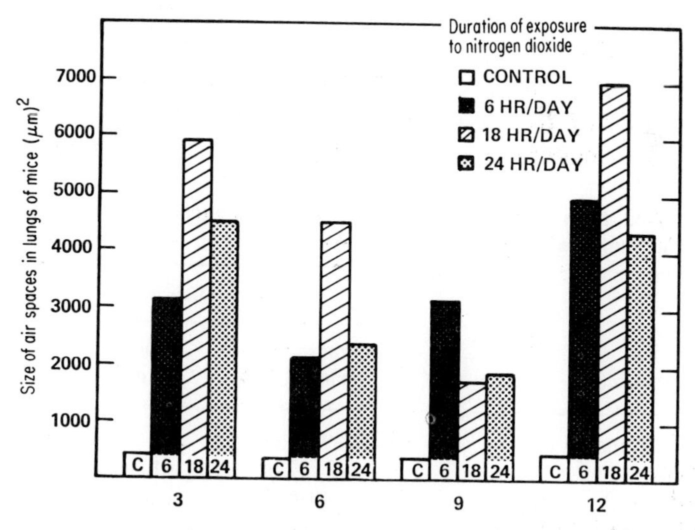

Figure 10. Comparison of size of air spaces in lungs of mice sacrificed at 3, 6, 9, and 12 months, respectively, after exposure of controls to filtered air and of exposed animals to 0.5 ppm of nitrogen dioxide for 6, 18, and 24 hours per day, respectively (*135*).

Enlarged lungs have been reported in dogs after exposure to 26 ppm nitrogen dioxide continuously for 6 months (*166*). The grossly enlarged organs do not collapse after removal from the thorax. Volumetric measurements made by water displacement indicate statistically significant alteration from controls in total size of the lung after removal from the thorax while the tracheas were ligated and in the collapsed lungs after removal of the ligature (Table VI). Possible partial healing and scarring of the nitrogen dioxide-altered lung has been reported by Freeman *et al.* (*174*), who describe experiments conducted for the purpose of determining the course of the nitrogen dioxide-induced lesion in the rat lung following cessation of exposure. Observations were made after exposure from 1 to 20 weeks to 15 ppm nitrogen dioxide, with recovery periods up to 52 weeks. The increased lung weight associated with nitrogen dioxide appears to subside after cessation of the exposure, being reduced to control weight by 20 weeks after termination of exposure. By that time recovery coincides with a rat age of 60 weeks or longer; however, the weight of the lungs again increases above controls. In three out of four groups, the changes were statistically significant. Regression also appeared in microscopic alterations. Epithelial hyperplasia subsided by 8 weeks and alveolar distortions also subsided, but some "inhomogeneity" of alveolar dimensions and abnormal staining of fibrous tissue elements persisted after a 52-week recovery period in air. The increased lung weight is attributed to hyperplasia of connective tissue elements. These changes associated with a so-called healing phase can be interpreted as being produced by recovery of the epithelial change and contracture of newly formed connecting tissue following cessation of exposure. Eventu-

Table VI Data from Dogs Exposed to 26 ppm Nitrogen Dioxide for 6 Months Compared to Controls Exposed to Filtered Air[a]

	Mean value		
Parameter	*Nitrogen dioxide exposed dogs*	*Control dogs*	*Probability*
Lung volume inflated (LV_I)	392.3 ml	291.5 ml	<0.10
Lung volume deflated (LV_D)	252.7 ml	164.8 ml	<0.05
Lung weight (LW)	139.8 gm	94.7 gm	<0.01
Body weight at autopsy (BW)	10.82 kg	11.18 kg	<0.5
LV_I/BW	3.597	2.685	<0.10
LV_D/BW	2.307	1.513	<0.02
LW/BW	1.29	0.864	<0.01

[a] Note hyperinflation, reduced deflation and increased lung weight apparently unrelated to pulmonary edema (*166*).

ally, there appears to be an increased predeliction of the previously exposed lungs to deposit collagen in response to aging.

Freeman *et al.* (*175*) have studied the influence of exposure to nitrogen dioxide on the development of the lung of the growing rat from shortly before birth to 75 days of age. In their study, exposure of pregnant rats to 15 ppm nitrogen dioxide began on the eighteenth day of gestation (mean gestation period is 22 days) and was continuous until sacrifice of the pups at the termination of the experiment, at which time various morphometric analyses of the lungs were made. This level of exposure results in considerable mortality in the dams and the newborn pups. Growth rate is slower in the exposed animals as expressed in both body weight and skeletal length. The formation of alveoli is delayed in the nitrogen dioxide-exposed animal for about the first 10 days, but later, alveoli are formed at a faster rate in exposed animals than in controls, exposed animals eventually ending with about the same number of alveoli at 75 days. Since the lungs of the exposed were larger, they contained fewer alveoli per unit volume. The initial lag in development of alveolar spaces is probably a nutritional factor associated with intrauterine disturbance. In the overall exposure, there does not appear to be a developmental failure of formation of pulmonary air spaces. This suggests that in their previous studies beginning with weanling rats (approximately 30 days of age), decreased alveolar number as determined in the whole lung was a function of septal breaking rather than failure of formation of new alveoli.

Bartlett (*90*) has also studied the growing rat for determination of the influence of ozone and nitrogen dioxide on maturation of the lung. While exposure to 0.2 ppm ozone results in significant increases in lung volume, mean cord length, and alveolar surface area, exposure to the same concentration of nitrogen dioxide causes no alterations. This observation for nitrogen dioxide appears entirely reasonable, since an equivalent concentration of nitrogen dioxide would be of the order of 2–5 ppm according to most comparative studies. His conclusion is that exposure to 0.2 ppm ozone does not influence the rate of development of alveoli and that increases in lung volume are attributable to decrease in lung tissue recoil.

Reference has been made to production of pulmonary emphysema-like lesions produced by exposure to both ozone and nitrogen dioxide. A comparative study of certain emphysema-like states has been described by Port and Coffin (*176*), who considered by means of light and scanning electron microscopy the lesions produced by nitrogen dioxide exposure in the rat and mouse, by enzyme instillation in the hamster, and spontaneous emphysema in the horse and man. In the rat exposed to nitrogen dioxide, the decrease of lung alveolar surface was readily apparent. The uniformity of the alveolar pattern was lost and the alveolar septa were flattened and appeared stretched. The alveoli contained numerous type II cells and the interalveolar pores of Kohn were infrequent and small. In the mouse, the emphysematous changes appeared confined principally to the respiratory bronchioles and alveolar ducts. There was some stretching and possibly breaking of alveolar septa. However, a most prominent feature of the normal mouse was the striking number and size of the alveolar pores. After nitrogen dioxide exposure, they appear somewhat enlarged. However, the enlargement of the pores in mice does not approach the condition in man or the remarkable fenestration and trabeculation occurring in spontaneous emphysema in the horse (Table VII). The enzyme-induced emphysema-like state in the hamster presents rather discrete "holes" in the area of alveolar ducts and central alveoli, with no effect in the peripheral alveoli. The alveolar septum appeared flaccid and pushed against the sides of the dilated areas. The flaccid appearance of the enzyme-induced lesion contrasted strikingly with the stretched aspect of nitrogen dioxide-induced and spontaneous lesions (Fig. 11).

It appears that the most likely sequence in the pathogenesis of structural alteration of the lung from exposure to both nitrogen dioxide and ozone is more or less as follows:

a. Loss of type I epithelial cells within a few hours after beginning exposure, their replacement by type II epithelia within 1 day with almost complete replication of the new epithelium by 6 days.

Table VII Number of Alveolar Pores in Various Animal Species[a]

Species	*Number of pores per 100 alveoli*	*Ratio pore/alveolus*
Rat	128	1.28
Mouse	308	3.08
Hamster	364	3.64
Human	614	6.14
Monkey	850	8.50
Horse	2713	27.13

[a] Table showing frequency of alveolar pores in normal lungs of various species as evaluated by scanning electron microscopy. Note the striking species difference in the number of these structures. From (*176*).

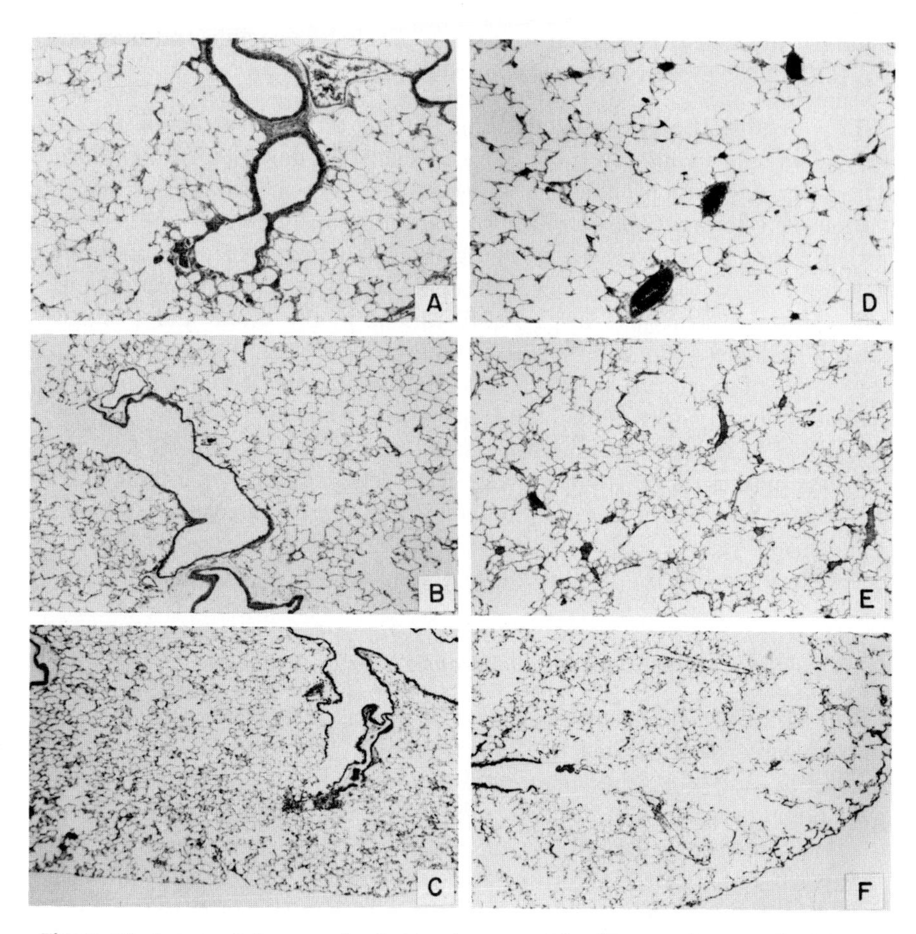

Figure 11. Lungs of A, normal rat; B, rat exposed to 15 ppm nitrogen dioxide for 118 days; C, normal hamster; D, hamster treated with intratracheal injections of 3 mg papain weekly for 3 weeks; E, normal mouse; and F, mouse exposed to 0.1 nitrogen dioxide with daily peaks of 1.0 ppm. Note the enlarged alveoli separated by numerous expanded air spaces in B, the enlarged alveolar ducts and central alveoli flaccidly "pushed" against the surrounding intact tissue in D, and the enlarged respiratory bronchioles and alveolar ducts in F (*176*).

b. A weakening of the connective tissue framework by chemical or physical degradation of the interalveolar collagen and elastin, resulting in lower lung recoil or contractility, resulting in an enlarged and relatively noncontractile lung. This process is reversible if exposure is discontinued; however, if exposure is continued, connective tissue becomes hyperplastic leading to scarring and some distortion, loss of alveoli and enlarged air spaces. These latter changes are not reversible and may ensue later in life even if exposure is discontinued.

Loss of alveoli and consequent lung surface area appears to be the result of breaks in interalveolar septa rather than developmental failure. The differences in morphological aspects of emphysema-like states among various species may be a result of anatomical differences in the lung rather than basic differences in their genesis. For instance, much attention has been given to the presence of septal bands in human emphysema, whereas these structures are notably absent in the ozone and nitrogen dioxide-induced lesion in the rat. If one postulates that enlargement of the pores of Kohn is a feature of emphysema, the difference in such fenestration is readily explainable, since this pulmonary feature is poorly developed in the rat. On the other hand, spontaneous emphysema of the horse is most striking for the presence of fenestration and interalveolar septation. The latter species has numerous large interalveolar pores (Table VI) (Fig. 12). It is possible, therefore, that a single mechanistic principle underlies the production of these several emphysemas.

C. Physiological Responses from Nitrogen Dioxide Exposure

The uptake in the respiratory tract of ozone, nitric oxide, and nitrogen dioxide have been studied by Vaughan *et al.* (*81*), who noted that the uptake in nose-to-trachea preparations for ozone, nitric oxide, and nitrogen dioxide was 40–60%, 73%, and 90%, respectively. These rates of uptake probably reflect the combined solubilities and chemical reactivities of these gases. According to Svorcova and Kant (*177*), nitrogen dioxide is rapidly converted to nitrite (NO_2^-) and nitrate (NO_3^-) ions in the lungs, and these ions appear immediately in the blood and are found within 15 minutes in the urine after exposure to 24 ppm nitrogen dioxide. They found that latency of uptake response was lengthened at lower concentrations.

Alterations in pulmonary mechanics have been observed from exposure to nitrogen dioxide by Murphy *et al.* (*10*), who reported an increased rate of respiration associated with diminished tidal volumes in guinea pigs after exposure to from 1.3 to 5.2 ppm for 2–4 hours. Increased respiratory rate was also seen by Henry *et al.* (*178*) after exposure of squirrel monkeys to 50 ppm for 2 hours. No evidence of pulmonary resistance was observed by Balchum *et al.* (*139*) after exposure of guinea pigs to 5 ppm nitrogen dioxide for various periods up to 18 months. Transient increased oxygen consumption was noted in rabbits after exposure to 25 ppm continuously for 18 months, but not for 1 and 5 ppm (*137*).

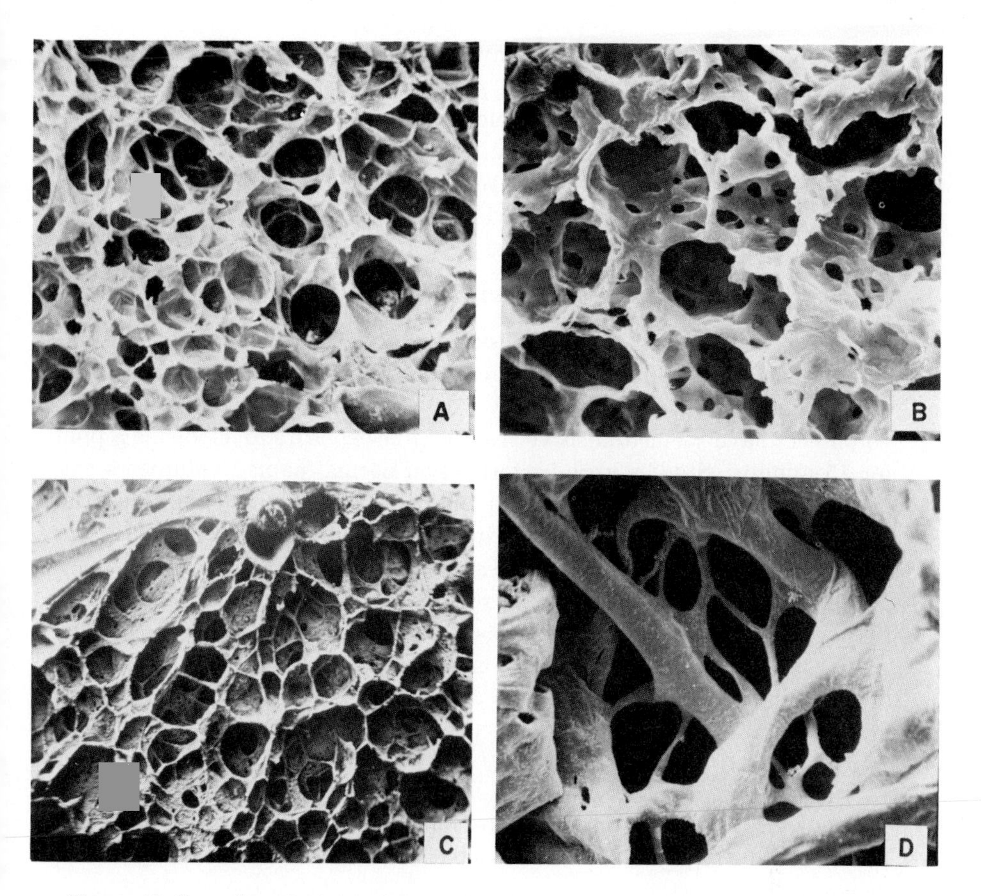

Figure 12. Scanning electron micrographs of induced and spontaneous emphysema A, Lung of rat exposed to 15 ppm nitrogen dioxide continuously for 118 days. Note inhomogeneity of alveoli separated by stretched septa (Magnifications 67×). B, Lung of mouse exposed to nitrogen dioxide 0.1 ppm continuously with daily peaks of 1.0 ppm. Note the variation in size of the size of the alveolar pores, some of which suggest fenestrae (680×). C, Lung of man with mild spontaneous pan acinar emphysema. Note variation in size of alveolar pores with development of fenestration (57×). D, Lung of horse with spontaneous emphysema. Note the remarkable enlargement of the alveolar pores with striking fenestration and trabeculation (700×) (*176*).

Vaughn *et al.* (*81*) could see no change in body weight, carbon monoxide diffusion capacity, compliance, or respiratory resistance after exposure of beagle dogs to varying schedules of nitrogen dioxide combined with nitric oxide at low concentrations (approximately 0.9–2 ppm nitrogen dioxide and 0.2 ppm nitric oxide) 16 hours a day for 18 months.

Williams *et al.* (*179*) described subtle changes in hysteresis curves after exposure to 15 ppm for 28 days. Changes in pulmonary compliance and surfactant activity have been reported only after sufficient nitrogen dioxide was administered to produce pulmonary edema (*133, 180*). Exposure of rabbits to 30–50 ppm nitrogen dioxide for 15 minutes is reported to have resulted in decreased perfusion–distribution in the lung *in vivo* as measured by radiolabeled isotopes (*181*).

D. Experimental Exposure of Human Volunteers

Despite the apparent lack of positive effect of nitrogen dioxide on pulmonary mechanics of animals (unless the edemagenic dose is exceeded), experimental studies conducted by von Nieding *et al.* (*182*) in human volunteers have yielded positive results after exposure to as low as 5 ppm for 15 minutes. Studies were performed on normal human volunteers and chronic bronchitics (*131, 183*). In both groups arterial oxygen partial pressure (PaO_2) decreased after exposure to 4 ppm nitrogen dioxide for 15 minutes but not at 2 ppm. These authors also report a slight increase in pulmonary arterial pressure after exposure to 5 ppm for the same time. Increased pulmonary resistance was described by these same authors at 1.6–2 ppm and above. This reaction was blocked by the administration of an antihistamine (Fig. 13).

E. Other Physiological Alterations

A slowing of weight gain has been reported in rats exposed continuously to 12 ppm nitrogen dioxide for 9 months (*174*) and rabbits exposed to

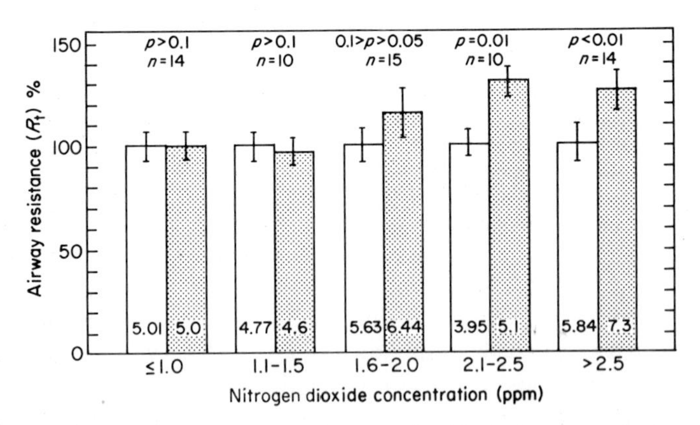

Figure 13. Influence of exposure to nitrogen dioxide in human volunteers before and immediately after inhalation of different concentrations of the gas. The initial response to inhalation of air represents 100% (*131*).

1.3 ppm for 17 weeks, whereas, intermittent exposure of rabbits, guinea pigs, rats, and hamsters exposed to as much as 25 ppm at 6 hours a day for 18 months showed no alteration in growth patterns. No affect was noted in dogs on the same exposure schedule at 5 ppm (*137*).

The influence of nitrogen dioxide exposure on physical activity has been examined in mice and rats. A transitory reduction in spontaneous running was noted in mice after approximately 8 ppm nitrogen dioxide and above (*10*). Decrements in swimming ability occurred in rats exposed to nitrogen dioxide for various periods. For instance, in rats exposed to 5 ppm, a 25% reduction occurred after 5 weeks of exposure (*184*).

Considerable alteration in the blood cellular constituents has been shown for nitrogen dioxide exposure. Freeman *et al.* (*161*) observed polycythemia in rats exposed continuously to 2 ppm or more of nitrogen dioxide. The number of erythrocytes rose to as much as 100% above controls. The blood picture was characterized by a reduction in mean corpuscular size and hemoglobin content, and the erythrocytes appeared thinner than normal; hematocrit and blood hemoglobin values were essentially normal.

Monkeys (*Macaca speciosa*) exposed to 2 and 9 ppm continuously also showed polycythemia. In rats exposed continuously, the polycythemia was transient, declining after approximately 3 months exposure. When the concentration of nitrogen dioxide was increased, however, the polycythemia reappeared (*152*). In contrast to the above, Wagner *et al.* (*137*) found no alteration in the erythrocytes in dogs exposed to 25 ppm nitrogen dioxide six hours a day for 18 months. Leukocytosis occurred in the peripheral blood of rabbits after exposure to both 1 and 3 ppm nitrogen dioxide for 15 weeks. It receded after cessation of the exposure. Phagocytic activity was also depressed by these exposures (*151*).

A great deal of interest has been expressed in the possibility that exposure to nitrogen dioxide or nitric oxide may induce methemoglobin formation. Since methemoglobin does not bind oxygen, its formation would impair oxygen-carrying capacity of the red cells to a degree commensurate with its concentration. Despite the strong methemoglobin-inducing effect of nitric oxide and nitrogen dioxide shown by *in vitro* exposures and the positive influence of ingested nitrite in this regard, there is no evidence of the induction of significant levels of methemoglobin by inhalation exposure to either nitrogen dioxide or nitric oxide at concentrations possible in ambient air.

The literature concerning the reaction of nitrogen dioxide (and nitric oxide) with hemoglobin is confusing. Methemoglobin and nitric oxide and nitrogen dioxide hemoglobin are readily formed during *in vitro* exposures of whole blood. No such effect from *in vivo* exposure in consistently reported. This appears paradoxical in face of the report by Svorcova and Kant (*177*) that nitrogen dioxide is rapidly converted to nitrite and nitrate after inhalation, these ions appearing in the blood almost immediately after exposure and being excreted into the urine within 5 minutes.

According to Wagner (*185*), no detectable change in methemoglobin occurred in rats exposed to 9.4 mg/m^3 (5 ppm) for as long as 10 months; whereas, 1 hour exposure to 18.8 mg/m^3 (10 ppm) causes an approximate doubling of methemoglobin over background. Other investigators have exposed animals to various high levels of nitrogen dioxide with little or no detectable alteration of hemoglobin pigment (*186*), but animals exposed to arc welding fumes developed appreciable amounts of methemoglobin. Elevated methemoglobins have been reported in rabbits, rats (*141*), and dogs (*187*) when exposed to high levels of nitrogen dioxide. Thus, while reaction with hemoglobin readily occurs *in vitro,* the formation *in vivo* has little or no significance, being reported only from exposures to excessive amounts of nitrogen dioxide and then usually amounting to low or insignificant levels of a transitory nature.

Oda *et al.* (*188*) studied the development of nitrosylhemoglobin (NOHb) in mice and rats exposed to nitric oxide by means of electron spin resonance (ESR) spectroscopy. When exposure to 10.6 ppm nitric oxide containing 0.8 ppm nitrogen dioxide, was carried out for 1 hour, NOHb rose to 0.13% of the total hemoglobin within 20 minutes then remained at this level for the remainder of the exposure. Concentrations returned to baseline values within 30 minutes.

The weight of the experiments seeking to develop abnormal hemoglobin with nitric oxide and nitrogen dioxide seem to indicate that while these compounds are readily induced *in vitro,* they are induced with difficulty in whole animal systems where the effect is minimal and transient; the pigment not persisting long enough for any appreciable concentration, as, for instance, is the case with carbon monoxide hemoglobin. The transient or reversible character of the NOHb development is probably the action of the enzyme methemoglobin dysmutase, which is a known component of adult human blood. Furthermore, it is germane to point out that the dose of nitrate or nitrite derived from absorbed nitrogen dioxide or nitric oxide combined with inhaled airborne nitrate present in ambient air amounts to only approximately one-five hundredth of that derived from water and food. Therefore, even in the human infant, where an appreciable amount of fetal hemoglobin having a stronger affinity for methemoglobin binding may be present and in whom methemoglobin reductase may be deficient, the amount of nitrate or nitrite derived from airborne sources should be considered trivial.

F. Biochemical Effects of Nitrogen Dioxide

The biochemical mechanisms by which nitrogen dioxide reacts with tissue to some degree resemble those by which ozone reacts by virtue of their common oxidizing properties. Thus, lipid peroxidation has been reported from both *in vivo* and *in vitro* exposure to nitrogen dioxide. Exposure of rats to 1 ppm nitrogen dioxide produces evidence of oxidation, as apparent from the appearance of diene conjugation of polyunsaturated fatty acids in extracts of lung lipids, as described by Thomas *et al.* (*129*). Peroxidation appears maximal 24 hours after exposure and persists for at least 48 hours after exposure. When exposure consists of six daily 4-hour exposures, the diene conjugates increase cumulatively. Similarly to ozone, animals receiving a tocopherol-deficient diet are more susceptible to lipid peroxidation (*68*). The reaction differs from ozone, at least in thin films of polyunsaturated fatty acids, in that phenolic antioxidants retard oxidation from nitrogen dioxide, whereas they have little or no effect on ozone-catalyzed oxidation. It is speculated that the nitrogen dioxide catalyzed reaction proceeds indirectly by the formulation of a nitroso free radical (*189*). The nitroso free radical in turn initiates autooxidation by the abstraction of methylene hydrogen. The alkyl free radical then reacts with molecular oxygen forming fatty acid hydroperoxides. Peroxidation continues in the same manner as with other free radical initiators. Antioxidants thus prevent nitrogen dioxide-catalyzed oxidation by reaction with the peroxy free radicals generated after reaction with nitrogen dioxide and not with nitrogen dioxide directly.

Ozone reacts rapidly and almost completely with ethylene groups. Ozone can produce free radicals either directly by rearrangement of the initial adduct of ozone and the ethylene group or indirectly through the decomposition of ozonides. Nitrogen dioxide, on the other hand, reacts more slowly and incompletely with ethylene groups. Peroxidation proceeds through a free radical-mediated chain reaction differing from spontaneous peroxidation only in the initiation step. Vitamin E and other phenolic antioxidants are effective inhibitors of nitrogen dioxide-catalyzed peroxidation but not of ozone, because phenolic antioxidants react most readily with peroxyl free radicals involved in the chain reaction. The free radical chain leading to peroxidation is thus most easily and efficiently intercepted by vitamin E and other peroxide decomposers. Perhaps nitrogen dioxide is less toxic than ozone because of the incomplete reaction of nitrogen dioxide with fatty acids and the ease of inhibition of the nitrogen dioxide catalyzed chain reaction by phenolic antioxidants and peroxide decomposers (such as heme proteins) and enzymes (such as glutathione peroxidase). The autocatalytic reaction induced by nitrogen dioxide might be expected to produce a slow development of the products of initiation followed by a rapid reaction once a critical concentration is achieved. Thus, normal cellular protective mechanisms, such as the antioxidants, may extend this period more or less indefinitely, unless an overwhelming concentration of nitrogen dioxide is employed. Whereas, for ozone, the reaction leads directly to peroxidation and membrane damage, and repair forces would be ineffective. The role of free radicals in the toxicity of ozone and nitrogen dioxide has been discussed in detail by Menzel (*190*).

Exposure to nitrogen dioxide may alter the molecular structure of lung collagen (*191*). When animals are subjected to one exposure, the reaction appears reversible, but animals exposed to 0.25 ppm repeatedly show persistent structural alterations when viewed by electron microscopy. These changes persist for at least 7 days after the final exposure (*191*). Considerable attention has been paid to the ability of nitrogen dioxide to cause alterations in blood or tissue enzymes. Buckley and Balchum (*162–164*) in a series of papers have described the influence of acute and chronic exposure to nitrogen dioxide on the total activity of lactic dehydrogenases (LDH) and on the proportion of its various isoenzymes. Exposure for 10 weeks or longer at 10 ppm or for 2 hours at 50 ppm cause changes both in tissue oxygen consumption and LDH and aldolase activity. A shift to the anaerobic isoenzyme occurs in the lung. This anaerobic shift is also noted in the heart muscle of exposed hamsters (*130*). Sherwin *et al.* (*192*) report that the hyperplastic epithelial cells formed in response to nitrogen dioxide exposure stained heavily for LDH and suggest that the demonstration of these cells by histochemical technique might be a useful index of exposure. This increase in number of LDH positive cells was also noted by Buckley and Loosli (*193*) after exposure of germ-free mice to nitrogen dioxide. LDHs could also reflect an alteration in cell populations, as has been demonstrated for both ozone and nitrogen dioxide (*117*).

The significance of the above findings are at this time unclear, since the high levels of exposure may have produced an artifact. However, recent unpublished and unconfirmed data suggest that a number of blood and tissue enzymes may be altered in guinea pigs by intermittent exposure for 4 months at levels as low as 0.5 ppm. These data urgently need confirmation, since their implications for biological effects of nitrogen dioxide are evident.

G. *Discussion of the Effects of Nitrogen Dioxide*

The toxicology of nitrogen dioxide is complex by virtue of a number of modifying factors, many of which are probably equally true for ozone: (a) skewing of concentration $\times$ time (CT) in favor of concentrations; (b) development of tolerance; (c) apparent healing of lesions even while exposure is underway; (d) multiple sites and possible mechanisms of action; (e) possible conditioning, by early exposure to later chronic change; and (f) synergism with other factors, e.g., sulfur dioxide, heat, exercise, bacterial infection.

The natural history of the exposure of man to nitrogen dioxide in ambient air is that of irregular intermittent peak concentrations of short duration (frequently superimposed on a much lower "base dose"). In photochemical atmospheres where nitrogen dioxide (and ozone) are prevalent in highest concentrations, the exposure norm is for an exposure of 3–5 hours at peak concentrations during optimum production. Due to variations of production of precursors and meteorological factors, peak periods may occur consecutively, daily, or at irregular intervals with considerable periods of much lower exposure intervening. Because of the short daily exposure time and irregularity of the exposure, an examination of the influences of concentration versus time in lesions induced by nitrogen dioxide exposure may be revealing. This factor was examined by Gray *et al.* (*194*) and LaTowsky *et al.* (*195*), both of whom observed that short exposures at high concentrations had considerably more weight than a much greater CT applied over a longer interval when direct lethal concentration for 50% of the animals (LC_{50}) was the criterion.

Coffin *et al.* (*196*) have examined this CT factor in an interacting system with bacteria where the end point is mortality increment from pulmonary infections. This system shows toxicity at a dose far lower than that of standard LC_{50}. In these experiments, in which CT was held constant and time and concentration varied, there was striking interaction with concentration (Fig. 14). In the same series of experiments when animals were continously exposed to various concentrations of nitrogen dioxide and sampled at intervals of increasing time and subjected to exposure to a bacterial aerosol, the mortality curves ascend according to the concentration. (Fig. 15). Significantly, however, intermittent exposures at the same concentration have strikingly similar effect on the basis of CT (Fig. 16). The same influence over a longer period is seen also in the work of Purvis and Ehrlich (*197*), Ehrlich (*130*), and Ehrlich and Henry (*136*) who noted that while a minimal effective dose was 3.5 ppm nitrogen dioxide for 2 hours when bacterial infection is the indicator, it required 3 months of continuous or 6 months intermittent exposure to 0.5 ppm to achieve the same results. In view of these striking CT differences, it is tempting to speculate that different mechanisms of action are operant at the acute and chronic levels.

Other data suggest that varying the dose schedules may be more effec-

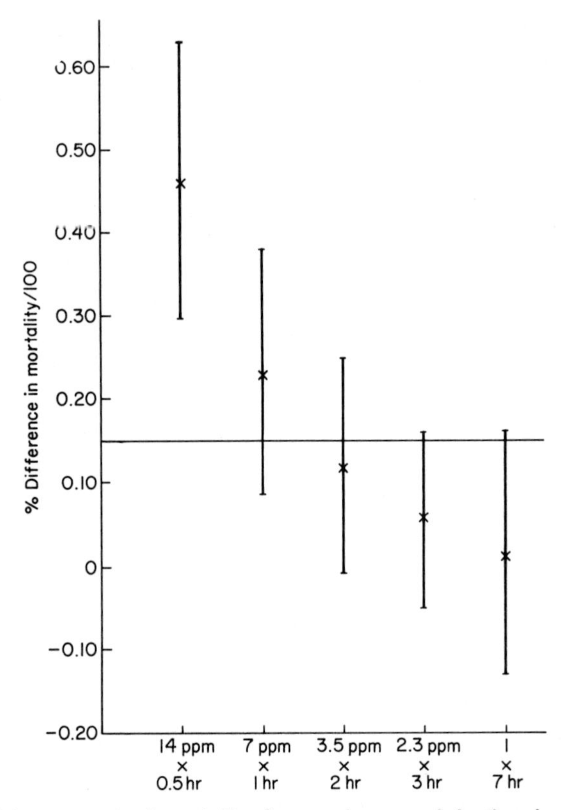

Figure 14. Enhancement of mortality from pulmonary infection in mice exposed to the same concentration × time (7 nitrogen dioxide ppm hours) for various exposure times. Note that the actual concentration has more influence than the length of exposure. The horizontal line indicates the predicted effect from the observed data if there was a direct relationship between time and concentration (*196*).

tive in stimulating certain biological responses than a continuous one or a regularly intermittent one at a constant concentration. In this connection, Ehrlich (*130*) exposed mice to varying schedules of nitrogen dioxide and examined them by various parameters reflecting resistance to infection. Mice exposed to a continuous base dose of 0.5 ppm nitrogen dioxide with intermittent daily 1-hour exposure to 2 ppm showed greater response than those exposed to 2 ppm daily or continuously to only 0.5 ppm. This suggests that short exposures at peak concentrations are possibly more significant than longer exposures yielding enormously greater CT values. The implications for air monitoring are that it is most important to record accurately the short peaks, and that daily, weekly, or monthly averages may have much less value from the toxicological standpoint.

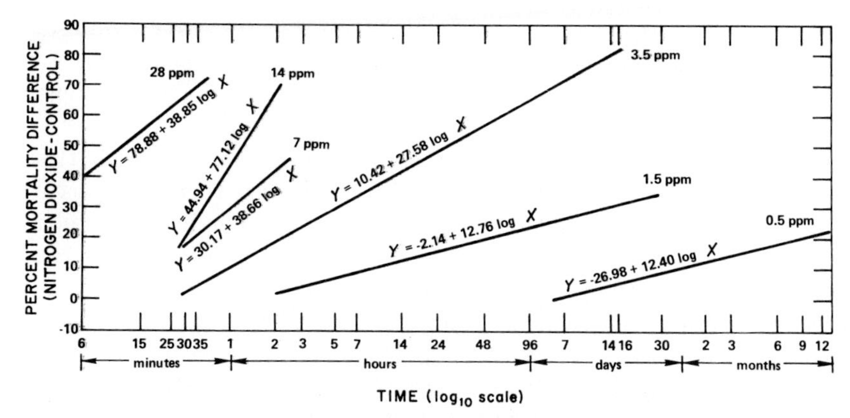

Figure 15. Percent mortality of mice versus length of continuous exposure to various nitrogen dioxide concentrations prior to challenge with streptococci. Note that there is a straight-line relationship between effect and duration of exposure at any given concentration of nitrogen dioxide (*196*).

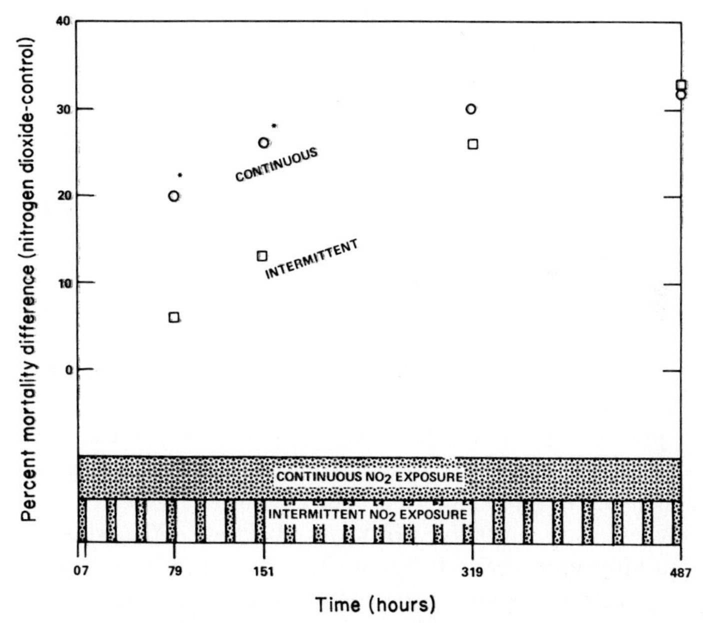

Figure 16. Percent mortality of mice versus length of either continuous or intermittent exposure to 1.5 ppm nitrogen dioxide prior to challenge with streptococci. A comparison of the effect of intermittent and continuous exposures indicates that effects were similar on the basis of concentration × time. In the figure the effect of the intermittent exposure lags until sufficient concentration × time is achieved to equal a minimal effective dose. The results then appear to equate with the continuous on the basis of concentration × time. Continuous and intermittent treatment means are significantly different at $p < 0.05$ (*196*).

Furthermore, the toxicological implications are that the toxic manifestations of nitrogen dioxide are best shown by exposures simulating the natural occurrence of the gas in the atmosphere, i.e., fluctuating doses may be an important determinant in the toxicological system.

The subject of tolerance has been discussed in detail in Section II,G. It should suffice here to point out that this phenomenon may play a role in toxicological manifestations of nitrogen dioxide. For instance, tolerance induced by repetitive exposures or even during the course of continuous exposure may explain the striking flattening of the dose–response curve with time. Too little is known of this phenomenon's induction mechanism to assess accurately its possible role in resistance to toxic effects of nitrogen dioxide at or near ambient concentrations. However, if inference may be made from recent studies of ozone, tolerance is not an important factor at low concentrations of exposure.

Healing of anatomic and chemical lesions seems to be an important feature of exposure of animals to nitrogen dioxide. (This may be a nonspecific basic feature of the lung.) When the squamous bronchiolar alveolar epithelial cell (type I) is injured by nitrogen dioxide, it is quickly replaced by rapidly proliferating granular pneumonocytes (type II), which, in turn, give rise to normal squamous epithelial cells if exposure is discontinued. If, on the other hand, exposure continues, the newly proliferated granular pneumonocytes persist, thus providing provisional repair of the denudated basement membranes of the alveoli and, theoretically speaking, conferring temporary protection against the entrance of infection, carcinogens, and the like. As has been suggested by DeLucia *et al.* (*63*) for ozone, these newly formed epithelial cells, which may become hyperplastic, might provide a real source of defensive enzyme, thus being a possible mechanism for "tolerance." While long continued exposure leads to an apparently irreversible emphysema-like state, the lesion is preceded by a chemical and physical alteration of the connective tissue of the lung, which results in a loss of contractility or lung recoil and leads to stretching and enlargement of the whole organ. This phase, too, appears reversible. However, Bartlett (*90*), while observing this lesion after exposure to 0.2 ppm ozone, failed to see a similar phenomenon after exposure to the same concentration of nitrogen dioxide. This would be expected, since the known nitrogen dioxide equivalent for 0.2 ppm ozone would be at least 2.0 ppm, which is more or less the lower minimal dose producing emphysematous lung changes. However, according to Freeman *et al.* (*54*), continued exposure eventually leads to fibrosis, scarring, and septal breaks from which presumably the lung could not be wholly restored. Thus, the lung appears subject to an interplay of destructive lesions, hyperplasia, regeneration, physical stretching, and fibrotic change in a highly complex manner during exposure to nitrogen dioxide. If the exposure is continued long enough and at sufficient concentration, full blown emphysematous change results.

There has been considerable variation of results concerning the production of emphysema by exposure to nitrogen dioxide. Thus, while Freeman and co-workers report the lesion from continuous exposure of rats to nitrogen dioxide, Hine *et al.* (*198*) and Boren (*199*) failed to see the lesion from intermittent exposure of several species exposed for 1–5 months at similar levels (which was probably insufficient time for the development of the lesion), whereas, this lesion has been reported to occur after intermittent exposure to nitrogen dioxide by Kleinerman and Wright (*200*) in guinea pigs exposed to 15–20 ppm, Ehrlich (*130*) in mice exposed to 0.5 ppm, and Port and Coffin (*176*) in mice exposed to base dose of 0.5 ppm with a superimposed hourly pulse of 1 ppm. Since it is difficult to compare such data by different investigators done under varying conditions and methods of evaluation, it appears reasonable to give most weight as the basis for air quality criteria to data from schedules that mimic the potential exposures in ambient air.

IV. Sulfur Oxides

Sulfur dioxide is the emission product of the combustion of fuels containing sulfur and of many industrial processes. It is therefore ubiquitous in populated communities, its concentration roughly equating total fuel consumption, fuel sulfur content, and the degree of sulfur-emitting technological development. Decrements to human health are ascribed to pollution of the air with sulfur dioxides on the basis of clinical observation and the epidemiological association of certain impaired health states with the sulfur dioxide concentrations in the atmosphere (*201–205*).

Sulfur dioxide is an irritating gas for conjunctiva and upper respiratory tract and throat. Despite this fact, experimental animals have been singularly resistant to persistent major injury from exposure to any reasonable concentration of this gas. The physiological and pathological effects of high concentrations of sulfur dioxide are well documented (Table VIII).

Concentrations of pure gaseous sulfur dioxide 100 or more times ambient values are required to kill small animals; concentrations 50 or more times ambient produce little distress. Mortality is associated with congestion and hemorrhage of the lungs, pulmonary edema, thickening of the interalveolar septa, and other relatively nonspecific alterations of the lungs.

Table VIII Selected List of Responses to Sulfur Oxides with and without Aerosols (Up to 10 PPM Sulfur Dioxide)

Concentration (ppm or mg/m³)	*Length of exposure*	*Observed effect(s)*	*Species*	*Reference*
0.0035–0.1 ppm (SO_2)	3 years and longer	Excessive acute respiratory disease in communities heavily polluted with sulfur dioxide and sulfates	Man	*(206)*
0.003–0.02 mg/m³ H_2SO_4 (aerosols)	3 years and longer	Excessive acute respiratory disease in communities heavily polluted with sulfur dioxide and sulfates	Man	*(206)*
>0.02 ppm (SO_2)	260 days	Community mortality 2% greater than expected	Man	*(207)*
0.03–0.5 ppm (SO_2)	Daily	Condition of bronchitic patients worsened by high levels of sulfur dioxide and smoke	Man	*(205)*
0.1, 1.0, 5.0 ppm (SO_2)	52 weeks, continuous	No adverse effects in animals exposed for 52 and 78 weeks, respectively	Guinea pigs	*(208)*
0.5 mg/m³ fly ash	78 weeks, continuous	No adverse effects in animals exposed for 52 and 78 weeks, respectively	Monkeys	*(208)*
0.13, 1.0, 5.7 ppm (SO_2)	12 months, continuous	Moderate swelling and vacuolation of hepatocytes, 5.7 ppm level only, with lower incidence and severity of spontaneous disease	Guinea pigs	*(209)*
0.16–10.0 ppm (SO_2) ± NaCl aerosol	1 hour	Increased pulmonary flow resistance changes in tidal volume, respiratory frequency, and minute volume	Guinea pigs	*(210)*
0.2 ppm (SO_2) + 0.6 mg/m³ H_2SO_4	3, 6, 9, 12, 15 minutes	Increased optical chronaxie	Man	*(211)*
0.3–1.0 ppm (SO_2)	20 seconds	Attenuation of alpha wave in encephalographic measurement	Man	*(211)*
0.3–6.4 ppm (SO_2)	15 minutes	Increased light sensitivity	Man	*(212)*
0.3, 1.0, 3.0, 4.2, 6.0 ppm SO_2	Up to 120 hours continuous	Positive correlation of S-sulfonate plasma levels with atmospheric sulfur dioxide	Man	*(213)*
0.52 ppm (SO_2)	15 minutes	Increased eye sensitivity to light during dark adaptation	Man	*(212)*
0.35–5 mg/m³ (H_2SO_4)	5–15 minutes	Reflex bronchoconstriction, all levels, shallow rapid breathing; decreased minute volume, 5 mg/m³ level	Man	*(275)*

0.6–2.4 mg/m^3 (H_2SO_4)	60, 90, 120 minutes	Transient increased sensitivity to light	Man	(*214, 215*)
1.0, 5.0 ppm (SO_2)	10 minutes, 15–20 minutes rest; 30 minutes, 1 month apart	Increased pulmonary airway resistance	Man	(*216*)
1.0–10 ppm (SO_2)	20–40 minute	Decreased compliance and increased resistance	Dogs	(*217*)
1.0, 5.0, 25 ppm (SO_2)	6 hours	Nasobronchial reflex bronchoconstriction, all levels; decrease in nasal mucus flow rate, 5 and 25 ppm; no change in closing volumes	Man	(*218*)
1.6 ppm (SO_2)	—	Lowest concentration resulting in bronchoconstriction	Man	(*219*)
1.6–10 ppm (SO_2)	5 minutes ± NaCl aerosol	Increased pulmonary airway resistance	Man	(*220*)
2.0 ppm (SO_2) ± carbon	102 days, 192 days	Enhancement of antibody production; suppression of antibody response; carbon enhanced the effect over sulfur dioxide alone	Mice	(*221*)
2.0, 5.0 ppm (SO_2)	3–10 minutes	Increased airway resistance	Man	(*222*)
2.4 and 4.8 mg/m^3 (H_2SO_4) (3.6 μm; 0.73 μm particle size)	78 weeks continuous	Alteration of lung structure and deterioration of pulmonary functions	Monkeys	(*223*)
3.0–30 mg/m^3 (H_2SO_4)	10 minutes, 60 minutes, ± water vapor	Humidity increases irritancy of sulfuric acid mists	Man	(*224*)
4.7 ppm (SO_2) ± 0.9 mg/m^3 (H_2SO_4)	620 days, 21 hours daily	Increased nitrogen washout times, sulfur dioxide alone; reduction of carbon monoxide diffusion capacity, residual volume, sulfuric acid + sulfur dioxide	Dogs	(*225*)
7.0 ppm (SO_2)	15 minutes, 20 minutes rest, repeated	Increased nasal and laryngeal resistance	Dogs	(*226*)
10 ppm (SO_2)	Several days, continuous	Half-life of plasma clearance rate of S-sulfonates several days	Rabbits	(*228*)
10 ppm (SO_2)	Up to 72 hours	Edema, necrosis, desquamation of respiratory and olfactory epithelium in "defined flora" strain; more severe responses with upper respiratory tract infection	Mice	(*229*)

Particulate interaction and elevations of relative humidity enhance the reactivity of animals to sulfur dioxide. Sulfuric acid mist and certain particulate sulfates elicit more response in animals than sulfur dioxide itself. This suggests that alteration of sulfur dioxide to a higher oxidation state increases its irritability in animal models. Since the rate of conversion of sulfur dioxide to its oxidation products is influenced by photochemical activity and the presence of metallic catalysts and other factors, these interactions have importance for air pollution control because it is possible for the rate of conversion of sulfur dioxide to acid sulfates to have greater health significance than the concentration of sulfur dioxide in the air. The discussion that follows, therefore, includes sulfuric acid and acid sulfates.

A. Pathological Effects of Sulfur Oxides

Few acute pathological changes have been reported in the respiratory tract at reasonable toxicological concentrations. O'Donoghue and Graesser (*230*) noted pulmonary hemorrhage and hyperinflation of the lungs in swine acutely exposed at 5 ppm of sulfur oxides. These changes were associated with salivation, lacrimation, and rapid shallow ventilation. The latter signs suggest that the pulmonary lesion may have been the result of tachypnea rather than the irritant effect of sulfur dioxide per se. Changes in the upper airway were reported by Giddens and Fairchild (*229*), who induced inflammatory change in the nasal mucosa of mice exposed to 10 ppm sulfur dioxide for 24 hours. The lesions were noted in both the ciliated olfactory epithelium and the pseudostratified columnar epithelium in the anterior portion of the nasal chambers. After an exposure of 72 hours with the same concentration, there was necrosis and sloughing of the nasal epithelium. The lesions were more severe in animals with evidence of preexisting infection. Inflammatory lesions of the lung were also noted in swine exposed to 33 ppm by Martin and Willoughby (*231*).

Numerous long-term studies have been carried out with sulfur dioxide and sulfuric acid mist exposure in a variety of species of experimental animals (see Table VIII). These data are difficult to interpret because of differences in exposure modes, particle size (in the case of the mist), and the presence of intercurrent disease. For instance, no lesions were observed in guinea pigs exposed to 0.1, 1.0, or 5 ppm of sulfur dioxide alone or in combination with fly ash (2.7–5 μm in size) after 52 weeks of exposure, while rats exposed to 32 ppm sulfur dioxide had shortened lifespans. Effects at lower concentrations were not convincing (*232*). However, another investigator claimed the production of tracheitis, bronchitis, and focal interstitial pneumonia in rats after exposure of 65 days to as little as 1.74 ppm (*233*). It is possible that the latter changes may have resulted from presence of intercurrent infectious disease, possibly with *Mycoplasma pulmonis,* a common rat infection.

Sulfuric acid mist has been the subject of considerable investigation. Extremely high levels of sulfuric acid aerosol are required to induce histopathological alterations. Treon *et al.* (*234*) exposed a variety of animals to sulfuric acid mist aerosol of mass media diameter <2 μm at concentrations of 87×10^3 to 1600×10^3 μg/m^3. He noted a descending order of sensitivity as follows: guinea pigs > mice > rats > rabbits. Mathur and Olmstead (*235*) observed that mice were less sensitive than guinea pigs. They also observed that young animals were more sensitive than those above 18 months of age.

Experiments in which sulfuric acid mist has been administered chronically have been conducted. The toxicity of sulfuric acid mist at concentrations varying from 1 to 4 μg/m^3 at three particulate sizes (0.6, 0.9, and 4 μm) has been evaluated by Thomas *et al.* (*236*) in guinea pigs exposed for 18–40 days. Changes consisted of slight edema of the larynx and trachea and a slight increase in desquamated cells in the small bronchi. The 0.9 μm sized aerosol appears the most effective; older animals are more resistant to the exposure.

Bushtueva (*214, 215, 237*) has noted that guinea pigs exposed to 2 mg/m^3 of sulfuric acid mist of unspecified particle size develop pulmonary edema with thickening of the alveolar walls after 5 days exposure and that those exposed at 1 mg/m^3 for 2 weeks or longer show minimal cartarrhal inflammation of the trachea and bronchi, pulmonary interstitial proliferations, and perivascular and peribronchiolar lymphocytic infiltration. Exposure at 0.5 mg/m^3 produces only minimal changes. Lewis *et al.* (*225*) observed no histopathological lesions in dogs exposed for 21 months to 0.9 mg/m^3 of sulfuric acid mist, although functional decrements were observed. Alarie *et al.* (*223*) exposed cynomalogus monkeys to sulfuric acid mist of two particle sizes (2.15 μm and 0.54 μm m.m.d.) at concentrations of 0.38 and 0.48 mg/m^3. They observed bronchiolar epithelial hyperplasia and thickening of the walls of the respiratory bronchials and alveolar septa. The larger sized aerosol was more effective in eliciting these lesions. Guinea pigs exposed under similar conditions did not develop any lesions of the lung.

B. Carcinogenic Influence of Sulfur Oxides

Laskin *et al.* (*238*) have shown that sulfur dioxide influences the development of pulmonary cancer in rats when combined with benzo[*a*]-

pyrene. Their studies indicate that exposure to either 10 ppm sulfur dioxide for six hours a day with a superimposed treatment for 1 hour with an atmosphere containing 3.5 ppm sulfur dioxide + 10 mg/m^3 aerosolized benzo[*a*]pyrene or to the later treatment alone will induce bronchogenic squamous cell carcinomas in rats. A similar role for nitrogen dioxide exposure has also been demonstrated by Laskin *et al.* (*239*). In the latter experiments, a limited number of rats dying after chronic exposure to 25 ppm nitrogen dioxide combined with benzo[*a*]pyrene have shown squamous cell bronchogenic carcinoma. The role of pulmonary irritants as cofactors in carcinogenesis has been reviewed by Laskin and Sellikumar (*240*).

C. Pathophysiological Aspects

Sulfur dioxide is rapidly removed from the respiratory air stream and distributed systemically. According to Balchum *et al.* (*241, 242*), 19% of the inhaled $^{35}SO_2$ was removed in dogs from the airstream in the nasopharynx. When animals inhaled through a tracheal cannula a greater amount of the labeled sulfur was found in the trachea and lungs. Frank *et al.* (*243*), also working with dogs, noted a 95% absorption in the nasopharynx. These authors also observed that the absorption in the upper respiratory tract was inversely related to the flow rate, and, at a rate of 3.5 liters per minute, 99% of the gas was absorbed in the nasopharynx. Similar results have been noted in rabbits (*244, 245*) and in man (*246*). Absorption of sulfur dioxide in the upper airway is also inversely proportional to its concentration. Thus, relatively more is absorbed when high concentrations are inhaled than when they are low. From Strandberg's data (*247*) (Table IX), it can be seen that at lower concentration exposures, proportionately greater amounts reach the deeper structures. Sulfur dioxide produced from sulfur-35 indicate that the sulfur is readily

Table IX Absorption of Sulfur Dioxide by the Upper Respiratory Tract of Rabbits *(247)*

Sulfur dioxide ppm	*Absorption* (%)	*Range* (%)
100	95	90–98
20	95	75–95
5	80	50–90
0.5	50	10–60
0.1	5	1–10

absorbed into the blood and is distributed systemically. Sulfur-35 complexed with protein can be detected 11 days after exposure, with traces persisting in the lung for 3 weeks (*248*).

Perhaps the first effect of sulfur dioxide exposure is an increase in flow resistance in the nasal cavity, which, according to Frank and Speizer (*226*) accounts for approximately 50% of the increased respiratory flow resistance. Total respiratory flow resistance decrements have been the most widely employed parameters for the detection of effects of sulfur dioxide (*249*). Flow resistance in guinea pigs is increased 10% by exposure for 1 hour to 0.16 ppm and 265% by exposure for the same period to 912 ppm (*250*). Flow resistance returns to normal within 1 hour after termination of the exposure to the gas. Other irritants, such as formaldehyde and various oxides of sulfur, likewise elicit this reaction, and it is usually presumed to be a nonspecific reflex response of the bronchial tree to an irritant. Similar responses to sulfur dioxide have been reported in dogs (*217, 226, 241, 251*) and cats (*252*).

While it can be shown that exposure to pure sulfur dioxide produces small changes in respiratory mechanics at ambient concentrations, the alterations are in a sense trivial and perhaps not greater than those produced by meteorological variables or minor nonspecific irritants. However, this method has been a very useful tool in elucidating the various interactants that are synergistic when used in connection with exposure to sulfur dioxide, since no other functional parameter is as immediately responsive to the effects of this gas. The effects of interacting particles, sulfuric acid mist, specific sulfates, and meteorological variables have been demonstrated by this means. Amdur (*210, 253, 254*), Amdur and Corn (*255*) and Amdur and Underhill (*256*) showed that the inclusion of sodium chloride particles in the airstream with sulfur dioxide augmented the effect of the gas to produce increased pulmonary flow resistance. The concentration of sodium chloride used by Amdur in the above experiments was 4–10 mg/m^3, and the concentration of sulfur dioxide was between 2.0 and 265 ppm. Similar potentiation of the effect of sulfur dioxide on pulmonary resistance was not seen with solid particles such as carbon, ferric iron oxide, fly ash, and manganese dioxide at comparable levels. On the other hand, potassium chloride, ammonium thiocyanite, ferrous iron oxide, and magnesium and vanadium salts augmented the action of sulfur dioxide. Protection was conferred by a number of particulate substances when administered prior to the sulfur dioxide (*256*).

Augmented effects are produced in animals exposed to a combination of sulfur oxide and sulfuric acid mist when the sulfur dioxide is 82 ppm and the sulfuric acid 20 mg/m^3. Smaller sized aerosols of sulfuric acid produced the greatest effect in these mixtures, the smallest size used

(0.29 μg m.m.d., mass median diameter) exerting the maximum effect (*257*). The degree of potentiation of the effect of sulfur dioxide by particles appeared related to the solubility of sulfur dioxide in the salts employed (*256*). Studies have been reported by Corn *et al.* (*252*) in which cats were exposed to sulfur dioxide alone and in combination with sodium chloride aerosol. Since there were only two reactors out of twenty cats exposed, the authors suggested that guinea pigs rather than cats might be the model for the sensitive segment of the human population (*252*).

The effects of sulfur dioxide, sulfuric acid mist, and of several specific sulfates have been studied by Amdur (*255, 258, 259*) and Amdur and Corn (*255*). When sulfuric acid mist was employed at particle sizes of 0.8, 2.5, or 7.0 μm m.m.d., the largest particle size produced both increased flow resistance and compliance changes; whereas, the 0.8 μm particles produced increased flow resistance without decrements in compliance. Flow resistance appears to be related to both particle size and concentration. At high concentrations of sulfuric acid mist (40 mg/m^3), greater response was elicited by the 2.5 μm particles, whereas, the 0.8 μm particles produced a greater response at a mist concentration of 2 mg/m^3 sulfuric acid. The mode of reactivity varied, with much quicker response being noted for the smaller particles.

Another feature of the response to sulfuric acid aerosol or a combination of inert aerosols and sulfur dioxide has been persistence of the increased resistance after cessation of the exposure, suggesting that the particles maintain their irritant property for some time after their deposition in the respiratory tract (*258*).

Amdur and Corn (*255*) have studied the influence of aerosols of zinc ammonium sulfate, zinc sulfate, and ammonium sulfate in eliciting resistance changes. They note that the irritant potency as measured by increased flow resistance is inversely proportional to particle size. Thus, a small increase in mass concentration produces a larger increment in biological response as particle size decreases. Therefore, mass concentration alone is not a true measure of the toxic potential of these substances. The influence of particle size in mass concentration for zinc ammonium sulfate is shown in Figure 17. It is thus difficult to rank the various sulfates as to toxicity unless both particle size and mass concentration are strictly controlled. However, in spite of this confounding by particle size, Amdur and Underhill (*256*) have ranked certain sulfates according to their toxicity at 1 mg/m^3, at which concentration ferric sulfate increased flow resistance 77%, but ferrous sulfate and manganous sulfate produced no change from normal.

In a comparative study of the irritant properties of various compounds based on their sulfur content as derivatives of sulfur dioxide, Amdur

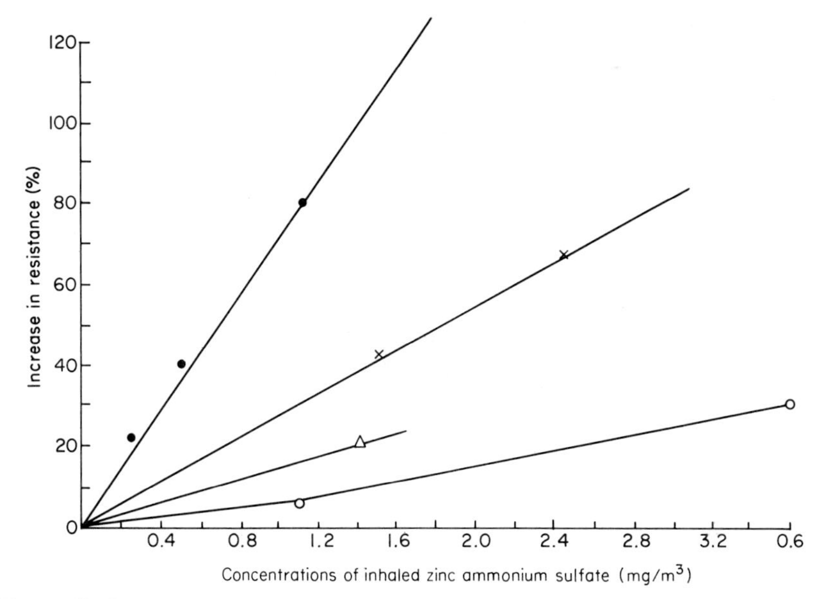

Figure 17. Dose–response curves for zinc ammonium sulfate of different particle sizes. ● = 0.3 μm, × = 0.54 μm, △ = 0.7 μm, ○ = 1.4 μm (255).

(*259*) has proposed the following theoretical differences based on atmospheric sulfur dioxide at 1 ppm (equivalent sulfur 1.3 mg S/m³), which produced an increase of 15% in flow resistance. When converted by this sulfur equivalent to sulfuric acid at 0.7 μm particle size, the expected resistance change would be 60%, and when converted to zinc ammonium sulfate of 0.3 μm particle size, the expected increase would be 300% (Fig. 18). In recent experiments by McJilton *et al.* (*260*) in which guinea pigs were exposed to a sulfur dioxide–sodium chloride aerosol mixture as employed by Amdur, but in which the relative humidity was controlled either low (below 40%) or high (above 85%), pulmonary flow resistance was increased only at the high level of humidity. It has been postulated that the enhanced reactivity to sulfur dioxide when particles are included relates to the ability of the particle to provide a nucleus and a means for conversion of sulfur dioxide to acid products by virtue of its water of condensation.

The effect of sulfur dioxide on ciliary action and mucous flow have been studied beginning in 1942 with Cralley (*261*), who noted that cessation of ciliary beat was dose related to concentration of sulfur dioxide in ranges from 18 to 20 ppm and 8 to 12 ppm. Dalhamn (*262*) and Dalhamn and Rohdin (*263*) measured ciliary beat and mucous flow in rats by noting the time necessary to block ciliary motility and by the time

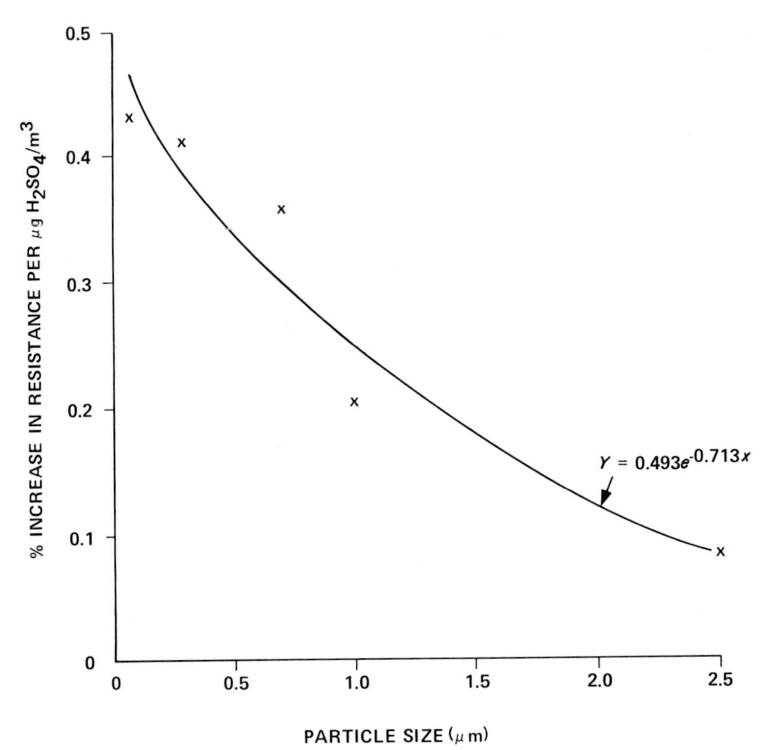

Figure 18. Irritant potency, as measured by the percent increase in pulmonary flow resistance per microgram of inhaled sulfate per cubic meter of air, can be seen to decrease as the size of the inhaled particle increases. The regression analysis was based upon the data of Amdur (from *259*).

required for a marker to progress a given distance in the isolated trachea. They noted that reduction in ciliary activity was accomplished by short exposures for 4–6 minutes to 35 mg/m³, but that longer exposures (18–67 days) at lower concentrations (3 mg/m³), while producing only transient reduced ciliary beat, resulted in a slowing or complete cessation of mucus flow. The slowing of the mucus transport was associated with as much as a fivefold increase in thickness in the mucus blanket, apparently due to increased mucus secretion. Problems with mucus flow persisted as long as 1 month after cessation of exposure to sulfur dioxide. Studies were also made in rabbits by an *in vitro* method in isolated tracheas. Rabbits exposed during spontaneous breathing required 30 ppm to produce cessation of ciliary beat, whereas in isolated tracheas, only 7–10 ppm were required. Fairchild *et al.* (*264*), utilizing labeled bacteria, studied the effect of sulfuric acid aerosol on particle deposition and clearance from the respiratory tract of mice. Exposure to 15 mg/m^3 (3.2 μm count median diameter) after exposure to sulfuric acid aerosol for 4 hours resulted in the reduction of clearance of nonviable streptococci from the lungs and noses of mice. Exposure to 15 mg/m^3 (3.2 μm count median diameter) for 4 days prior to the aerosolized bacteria resulted in a reduced rate of clearance from the nose but not the lungs. Exposure to 1.5 mg/m^3 (0.6 μm count median diameter) caused no alterations in rate of clearance from the nose or lungs. No alteration in clearance of viable streptococci was noted in this experiment. Similar effects have been noted for cigarette smokers, bronchitics, and asthmatics, suggesting a similar pattern of deposition for such exposures (*214, 215*).

D. Biochemical Effects of Sulfur Oxides

Experiments investigating the biochemical effects of sulfur oxides have been infrequently reported and have mainly concerned its influence on enzyme activity, incorporation into serum or tissue proteins, and stimulation of histamine release. P'an and Jegier (*265*) found that concentrations of 4 ppm and above employed in *in vitro* studies on bovine erythrocytes resulted in reduction of acetylcholinesterase. Gunnison and Benton (*266*) indicate that during sulfur dioxide exposure, sulfite enters the blood and forms S-sulfonite groups with plasma constituents. Gunnison and Palmes (*213*) postulate that such a mechanism would be protective against exposure to sulfur dioxide. Douglas *et al.* (*267*) studied the release of histamine in guinea pigs exposed to sulfur dioxide, nitrogen dioxide, cigarette smoke, and histamine aerosols. Histamine release caused by a 1 hour exposure to 34 ppm sulfur dioxide was insignificant. Charles and Menzel (*268*) have studied histamine release from slices of guinea pig lungs exposed to sulfates and nitrates of various cations. Their data indicate that of the compounds tested those with ammonium cations were most effective in promoting histamine release (Fig. 19). Histamine release from exposure of guinea pigs to combined sulfuric acid mist and sulfur dioxide, but not to sulfuric acid mist alone, was also reported by Bushtueva (*214, 215, 237*).

E. Exposure of Human Volunteers

A number of investigators have exposed human volunteers to sulfur dioxide, sulfuric acid mist, and aerosols of sulfate. Lack of consistency concerning method of exposure, concentration, duration of exposure, and parameters examined makes interpretation of results difficult. The parameters that have been employed range from odor perception threshold,

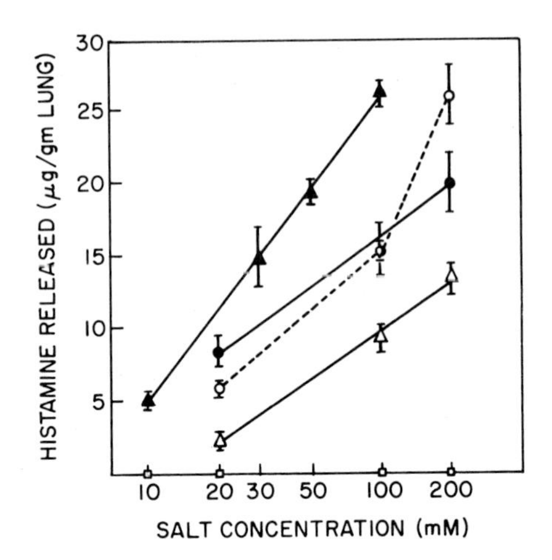

Figure 19. Release of histamine from lung fragments by salts. Each point represents the mean ± standard error; ▲ = ammonium sulfate, ○ = ammonium nitrate, ● = ammonium acetate, and △ = ammonium chloride (*268*).

visual performance (light sensitivity, optical chronaxie), effect on alpha rhythm, cardiovascular effects, mucociliary clearance, and influence on pulmonary function.

Reports, mainly from the Soviet Union, concern neurobehavioral effects of sulfur oxides that are of uncertain significance, since they constitute rather specific minor reversible physiological responses that may merely indicate perception of the substance (*269*). Dubrovskya (*212*) determined the odor threshold for sulfur dioxide as 0.5–0.7 ppm for the most sensitive individuals. Bushtueva (*270*) determined the odor threshold for combined sulfur dioxide and sulfuric acid mist exposure as above 0.38 ppm sulfur dioxide and 400 $\mu g/m^3$ sulfuric acid mist. However, Popov *et al.* (*271*) were unable to show odor detection by human volunteers below 1.4 ppm sulfur dioxide. The work reported by Bushtueva has generally been confirmed with similar studies on odor detection by United States workers (*272*).

The influence of sulfur dioxide on dark adaptation has been reported by Dubrovskaya (*212*) who claims alterations of adaptation from 0.23 to 1.53 ppm of sulfur dioxide. Other Soviet work on sulfur dioxide largely is confirmatory of the above. Bushtueva (*270, 273*) studied the influence of sulfuric acid mist combined with sulfur dioxide on sensitivity to light, noting increase in sensitivity after exposure of volunteers to concentrations as low as 1.4 ppm sulfur dioxide and 700 $\mu g/m^3$ sulfuric acid mist. Bushtueva *et al.* (*211*) and Bushtueva (*274*) reported desynchronization of alpha rhythm by means of electroencephalography for sulfur dioxide at 0.23 ppm, for sulfuric acid mist at 400 $\mu g/m^3$, and for the two agents combined at 0.34 ppm and 600 $\mu g/m^3$, respectively.

F. Pulmonary Physiological Alteration in Human Volunteers

As in experimental animals, sulfur dioxide is rapidly absorbed in the nasopharynx of man. As determined by Frank and Speizer (*216*) and Speizer and Frank (*246*), following exposure to 16 ppm sulfur dioxide, concentration of the gas in air in the airway decreased 14% of the initial concentration 2 cm behind the external nasal orifice and had declined to zero at the level of the posterior pharynx. They also noted that on exhalation a nasal concentration of approximately 2 ppm was detectable, suggesting desorption from the nasopharynx during exhalation.

The effect of sulfur dioxide on the clearance of mucus from the airway was early studied by Cralley (*261*) who noted a dose related slowing of clearance after exposure to 10 ppm for 1 hour.

Sulfur dioxide exposure alters the mode of respiration in man similarly to animals. Thus, increased frequency, diminished tidal volume, and decreased respiratory and expiration flow rates have been noted for exposure to sulfur dioxide and sulfuric acid mist (*275, 276*). Frank *et al.* (*277*) and Frank (*278*) found slightly detectable increased flow resistance at 1 ppm in only one subject. It was definite and uniform at 5 ppm and appeared dose related. Essentially similar results were obtained by Nadel *et al.* (*222, 279*) and by Melville (*280*). Nadel *et al.* (*279*) was able to block the response with atropine.

Snell and Luchsinger (*281*) compared the effects of sulfur dioxide alone and in conjunction with a sodium chloride aerosol. They noted that maximum flow rate at one-half vital capacity was reduced significantly by concentrations of both 1 ppm and 5 ppm. No differences were noted after inclusion of the aerosol.

As might be expected from the high absorbance of sulfur dioxide in the nose, greater effect was noted during mouth breathing by Snell and Luchsinger (*281*), Speizer and Frank (*282*), and Melville (*280*). Bates and Hazucha (*94*) in interaction studies between ozone and sulfur dioxide have demonstrated that a concentration of 0.75 ppm sulfur dioxide produced a decrement in maximal expiratory flow rate after 30 minutes in subjects undergoing moderate exercise.

Amdur *et al.* (*275*) studied the effect of exposure to sulfur dioxide on pulmonary function in human subjects. Concentrations as low as 1 ppm were associated with increased respiratory frequency and decreased tidal volume in some subjects, while a concentration of 5 ppm produced more pronounced changes from normal in all subjects. These changes occurred within 2 minutes after the beginning of the exposure and peaked in about 8 minutes. Experiments with sulfuric acid mist at 500 $\mu g/m^3$ by Amdur *et al.* (*276*) demonstrated increased respiratory rate, decreased respiratory and expiratory flow rates, and decreased tidal volumes within 15 minutes after the beginning of the exposure. These findings could not be confirmed by Lawther (*227*).

Other workers have had varying results from exposures of human volunteers. Burton *et al.* (*283*) found no changes in airway resistance or lung compliance after exposure of subjects to 1–3 ppm sulfur dioxide with or without the simultaneous presence of sodium chloride aerosol, while Toyama (*284*), Toyama and Nakamura (*285*) and Nakamura (*286*) have demonstrated synergism and hightened airway resistance with sulfur dioxide and aerosols of water, saline, and hydrogen peroxide beginning at exposures of 1.9 ppm sulfur dioxide and becoming more pronounced at 9 ppm and above.

G. Discussion of the Effects of Sulfur Oxides

Alterations of the state of human health most likely to be attributable to exposure to sulfur dioxide, sulfuric acid mist, or acid sulfates are acute irritation of the upper respiratory tract and conjunctiva, acute aggravation of acute cardiopulmonary disease, and possibly exacerbations of asthmatic attacks and chronic obstructive pulmonary disease. Acute irritating effects are not prominent in animal models. Most laboratory animals do not possess highly developed lacrimal glands and thus do not tear. Furthermore, since animals cannot report nasal or pharyngeal irritation subjectively, they are unsuitable for detecting slight irritant effects not productive of anatomic, functional, or biochemical lesions. A possible exception to the above appears to be the effect on swine from exposure to sulfur dioxide (*230*) in which irritation of the eyes and excessive salivation was noted after exposure for 8 hours at 11 mg/m^3 (5 ppm) and above. It is possible that these animals may react more like human beings than other laboratory animals with regard to the irritant effect of the gas, or they respond more patently through exaggerated signs.

As previously mentioned, exposure to sulfur dioxide, sulfuric acid mist, and acid sulfates is productive of increased pulmonary flow resistance. However, these changes are minimal near ambient concentrations and would not be expected to contribute greatly per se to the severity of an asthmatic attack. Two modes of action of sulfur oxides on restriction of pulmonary ventilation may be postulated; one, through reflex action by stimulation of receptors high in the respiratory tract; the other, by release of bronchoconstrictive mediators through chemical reaction in the lower airways. It has been shown by *in vivo* experiments that exposure to sulfuric acid mist combined with sulfur dioxide results in histamine release. *In vitro* experiments demonstrate that acid sulfates are also potentially capable of such action (*268*).

Neither chronic obstructive pulmonary disease nor chronic bronchitis has been produced in animals by exposure to sulfur dioxide, although the apparent epidemiological association has led to some attempts at the construction of such a model (*205*). The human disease appears to be characterized by recurring episodes of excessive mucus production and productive cough. Lesions consist of hyperplasia of the bronchial mucus glands and thickening of the mucosa of the bronchial tree. During acerbations, a variety of bacterial and viral agents may be isolated. While infection appears to be an important component, the increased risk of the disease with urban residence and polluted air has led to the speculation that pollutants may contribute to the etiology or exacerbations of established disease. Historically, sulfur dioxide has been a prime suspect.

Experiments by Reid (*287, 288*) attempting to replicate the human disease in animals showed that exposure of rats to 300–400 ppm sulfur dioxide by intermittent schedule for 3 months did in fact stimulate hyperplasia of mucus-secreting cells and promote their occurrence at deeper levels in the bronchial tree than they normally occur. These changes were accompanied by excessive amounts and stagnation of flow of mucus.

While Reid reported the infrequent occurrence of pathogens in her exposed rats, she noticed a reduction in response in a strain of animals less prone to bronchiectasis. This lesion is one of the characteristics of chronic murine pneumonia, now known to be the result of infection with *Mycoplasma pulmonis*, a ubiquitous pathogen in most rat colonies. While the concentration required to achieve the biological results reported above were at highly unrealistic concentrations, they at least established the fact that exposure to sulfur dioxide in rats is capable of inducing an experimental model simulating certain lesions of chronic obstructive bronchitis in man. Subsequent investigators working at lower levels have had variable results in the production of lesions of the bronchial tree and lung of the rat, possibly pointing to the interaction with some endemic factor in certain strains. One investigator, for instance, reported the presence of tracheitis, bronchitis, and focal pneumonia after exposure of rats for 65 days to as little as 1.7 ppm sulfur dioxide (*233*).

The so-called air pollution episodes in which acute mortality has occurred in man (*289*) and domestic animals (*290*) are well known. However, our knowledge of the conditions which brought about the mortality is based mostly on *post hoc* speculation. It is clear however that sulfur dioxide, sulfuric acid mist, and various other sulfur compounds together with nitric oxide, heavy particulate loading, water vapor (fog), and relatively low ambient temperature were associated in these atmospheres. In view of information concerning the heightened effects of certain acid sulfates on pulmonary flow resistance and other parameters, it seems tenable to speculate that the meteorological conditions and the contribution of primary pollutants might favor the development of sulfuric acid mists and specific acid sulfates of small particle size peculiarly capable of eliciting sufficient pulmonary resistant changes to cause death in susceptible people and in cattle whose respiratory reserve was possibly reduced because of obesity (*290*). Some analytical data are available for the Donora, Pennsylvania, 1948 episode from studies by Hemeon (*291*) of particulate collected on an air conditioner filter in operation during the smog period, and of the total airborne particulate by Schrenk *et al.* (*292*). These data have been discussed by Amdur and Corn (*255*). Analysis of the air conditioner particulate showed that 22% of the particulate matter was water soluble and that, of this, 58% was zinc ammonium sulfate and 21% was zinc sulfate. Total particulate during the smog was approximately 2.5 mg/m^3, leading to Amdur and Corn's estimate of 0.3 mg/m^3 zinc ammonium sulfate and 0.1 mg/m^3 zinc sulfate atmospheric loading. These compounds were subsequently studied experimentally and showed a striking potential for eliciting increased pulmonary flow resistance (Fig. 17).

As can be seen, problems associated with interpreting data from animal experimentation are compounded by a number of variables. These include (a) differing susceptibility of various species; (b) within-species difference due to interacting disease; (c) failure to control particle size of various compounds; (d) failure to control relative humidity; and (e) the sparsity of knowledge concerning the characterization of the sulfur compounds in the atmosphere. While biological variables may have troubled experimenters in the past, enough is known at this time regarding the differences in susceptibility to altered pulmonary functional parameters to use these differences by experimental design rather than by accident. For instance, the ability of the guinea pig to react to low concentrations of sulfur dioxide suggests that in this regard these animals may mimic hypersensitive people.

The problems of particle size constantly plague those who employ sulfur acid mist and acid sulfates experimentally. Due to the hygroscopicity of these compounds, growth of particles via absorption of water occurs with time and by passage into an atmosphere of greater relative humidity (i.e., the bronchial tree). For instance, if two concentrations of sulfuric acid mist are generated at exactly the same particle size, the one of greater concentration will grow larger under similar conditions, so that the end result will be the appearance of greater reaction reactivity of the lower concentration due to the fact that smaller particles elicit greater response. Reaction appears more related to the number of particles and their mass median diameter than to their total mass per cubic meter. Thus, unless strict attention is paid to such factors as particle size, humidity, temperature, and residence time of particle prior to exposure of animals, little comparability of results from one experiment to another can be expected. Thus, a high state of sophistication of aerosol generation is required for meaningful experimentation in this field.

V. Carbon Monoxide

From the point of view of the total amount in the air, carbon monoxide is the most important gaseous pollutant. It is an odorless and colorless gas and evokes no potent warning signs before actual poisoning occurs. Thus, severe poisoning by carbon monoxide through accidental exposure to high levels is of fairly common occurrence. With increasing concentrations, the usual sequence of signs and symptoms are headache, dizziness, lassitude, flickering before the eyes, ringing in the ears, nausea, vomiting, difficulty of breathing, muscular weakness, collapse, unconsciousness and death (Tables X and XI). No overt signs are to be expected from exposure to carbon monoxide at even the highest concentrations found in the ambient air (50–100 ppm). However, great concern has been expressed regarding the possibility that covert effects may decrease physical or mental acuity or interfere with the function of organs already suffering oxygen deficiency, such as the heart in coronary disease.

The mechanism of action for carbon monoxide is now regarded as principally an interference with oxygen transport brought about by the combination of the gas with hemoglobin. Since carbon monoxide has a greater affinity for combination with hemoglobin and less affinity for dissociation, carboxyhemoglobin gradually accumulates and, with time, eventually reaches a plateau of equilibrium at any given concentration unless death occurs. Thus, the level of carboxyhemoglobin is a function of both the concentration of the gas and the duration of exposure. Another factor in ambient exposure is that persons at risk are not usually in the atmosphere continuously, and the carboxyhemoglobin diminishes to background when the exposed persons leave the area of contamination,

Table X Summary of Effects of Carbon Monoxide in Animals[a]

Carbon monoxide level	*Length of exposure*	*Carboxyhemoglobin (COHb) (%)*	*Effect*[b]	*Species*	*Reference*
55 mg/m³ (48 ppm)	—	3.7–4.7	No decrement in performance, even at simulated altitude of 27,000 feet	Monkeys	*(302)*
58 mg/m³ (50 ppm)	6 hours/day, 5 days/week, for 6 weeks	2.6–5.5	Brain: Mobilization of glia and dilatation of lateral ventricles; necrosis and demyelination absent	Dogs	*(292a)*
	24 hours/day, 7 days/week, for 6 weeks		Heart: 10/15 developed EKG changes in third week		
58 mg/m³ (50 ppm)	3 months–2 years	—	No changes noted in fertility, fetal survival, body growth, food intake, weight and water content of various organs, EKG, or blood chemistries	Mice	*(301)*
58 mg/m³ (50 ppm)	24 hours/day, 7 days/week, for 3 months	Rats 1.8, rabbits 3.2, dogs 7.3	Dogs: No changes in EKGs and pulse rates; no histologic difference between exposed and control animals; significant increases in hemoglobin levels, hematocrits, and RBC counts	Rats, rabbits, dogs	*(293)*
58 mg/m³ (50 ppm)	24 hours/day, 7 days/week, for 3 months	—	EKG changes in the first 2 weeks returning to normal in third week; slight increase in hemoglobin levels, hematocrits, and RBC counts	Rats	*(294)*
58 mg/m³ (50 ppm)	1 hour	—	Changes in EEG increasing in severity with length of exposure	Rats	*(295)*
	1–5 hours/day for 4 days	—	Progressive deterioration in EEG returning to normal 48 hours after end of exposure		
115 mg/m³ (100 ppm)	6 hours/day, 5 days/week, for 6 weeks	7–12	Brain: Mobilization of glia and dilatation of lateral ventricles; necrosis and demyelination absent	Dogs	*(292a)*
	24 hours/day, 7 days/week, for 6 weeks		Heart: 8/8 developed abnormal EKGs after about 2 weeks. Autopsy: dilatation of right heart and occasionally of left heart; some degeneration of heart muscle		

115 mg/m^3 (100 ppm)	$5\frac{3}{4}$ hours/day, 6 days/week, for 11 weeks	Up to 21	Brain: No changes in EEG or in peripheral nerves; consistent disturbance of postural reflexes and gait. Autopsy: 6/6 showed some indication of cortical damage. Heart: 1/4 had inverted T-wave after second week; 2/4 had inverted T-wave by tenth week. Autopsy: 4/4 showed degenerative changes in muscle fibers.	Dogs	(*296*)
115 mg/m^3 (100 ppm)	6 weeks, continuous and intermittent	7–12	EKG changes, more marked with continuous than intermittent exposure; dilatation of right cardiac chamber; dilatation of lateral ventricles of the brain	Dogs	(*300*)
58 mg/m^3 (50 ppm)	6 weeks, continuous and intermittent	2.6–5.5	Same changes as noted with exposure to 115 mg/m^3, except EKG changes were not as prominent	Dogs	(*300*)
115–11,500 mg/m^3 (100–10,000 ppm)	Up to 48 minutes	—	Impairment in time discrimination	Rats	(*297*)
195 mg/m^3 (170 ppm)	8 weeks	19.7 at end of first week	Decrease in percent COHb from 19.7 to 15.1, increase in hemoglobin level in first 5 weeks followed by stabilization, exposure to 400 mg/m^3 (350 ppm) caused further increase in Hb; rats were cholesterol fed, increased total cholesterol in aortic tissue in rabbits	Rabbits	(*298*)
920–1150 mg/m^3 (800–1000 ppm)	6–8 hours/day, 7 days/week, for 36 weeks	—	Increased RBC counts, hemoglobin, and blood volume; challenge with 575 mg/m^3 (500 ppm) in these dogs produced the same % COHb as in nonacclimatized controls	Dogs	(*299*)
$>$1150 mg/m^3 (1000 ppm)	—	3–40	Prominent EKG changes; high mortality rate; alterations in cerebral and myocardial histology	Dogs	(*300*)

[a] From "Air Quality Criteria for Carbon Monoxide," AP62 Public Health Service, U.S. Department of Health, Education, and Welfare, Washington, D.C., 1970.

[b] EKG = electrocardiogram, EEG = electroencephalogram, RBC = red blood cell, Hb = hemoglobin, 10/15 = 10 out of 15.

Table XI Summary of Effects of Carbon Monoxide in Humans

Carbon monoxide level	*Length of exposure*	*Carboxy-hemoglobin (COHb) (%)*	*Effect*[a]	*Number of subjects*	*Reference*
Up to 115 mg/m^3 (100 ppm)	8 hours	11–13	No impairment in psychomotor test performance	18	*(310)*
29–1150 mg/m^3 (25–1000 ppm)	Up to 16 hours	Up to 31.8	No impairment of time estimation; no change in blood chemistry	18	*(310)*
35 mg/m^3 (30 ppm)	24 hours	5	Approximate equilibrium value of COHb reached by 12 hours; 60% of equilibrium value reached in 2 hours, 80% in 4 hours, and the remainder over 8 hours	10	*(304)*
58 mg/m^3 (50 ppm)	5 hours	Not measured	Impairment of auditory vigilance between the eightieth and one hundred twenty-fifth minute of exposure; no changes in psychomotor tests	12	*(313)*
58, 115, and 174 mg/m^3 (50, 100, and 150 ppm)	110 minutes	Not measured	Impairment of auditory vigilance: percentage of missed signals is correlated to carbon monoxide concentration in inspired air, impairment of EEG	20	*(312)*
58–230 mg/m^3 (50–250 ppm)	Up to 2 hours	2.5	Significant impairment in time-interval discrimination after exposure to 58 mg/m^3 (50 ppm) for 90 minutes	18	*(297)*
58–230 mg/m^3 (50–250 ppm)	Up to 2 hours	3	Consistent impairment in 3 or 4 parameters of visual function after exposure to 58 mg/m^3 for 50 minutes	4	*(308, 309)*
115 mg/m^3 (100 ppm)	$3\frac{1}{2}$ hours	7.2	Impairment of visual perception, of manual dexterity, of learning, and impairment of performance of certain intellectual tasks	42	*(311)*
115–23,000 mg/m^3 (100–20,000 ppm)	Up to 5 hours	Up to 35.0	Uptake of carbon monoxide increases with concentration of carbon monoxide, length of exposure, and ventilation rate; rate of uptake of carbon monoxide as measured by increase in blood % COHb is constant up to one-third of equilibrium COHb; rate of uptake decreases	7	*(303)*

			with increased P_{O_2} and apparently increases with decreased P_{O_2} due to hyperventilation; when the latter is corrected for, rate of uptake is unaltered		
100% (100–300 ml)	10–15 minutes	Up to 20	Impairment of visual function detectable at approximately 4.5% COHb and increases with increase in COHb; recovery lags behind excretion of carbon monoxide; latter depends on composition of gas during post-exposure period; experiments at simulated altitudes gave a similar pattern of results; data given for one subject only	4	*(305)*
1 Cigarette	—	2	Impairment of visual function	1	*(306)*
3 Cigarettes	—	4	Impairment of visual function similar to that produced by an altitude of 8000 feet (2400 m)	1	
—	Up to 2 minutes	Above 5	Decreased P_{a0_2} and P_{v0_2}	5	*(314)*
—	Up to 2 minutes	Above 5 (average 8.96)	Increased oxygen extraction in the presence of COHb	26	*(315)*
—	Up to 2 minutes	Above 5 (average 8.5–9)	Increased lactate production in patients with coronary heart disease; increased coronary blood flow, and fall in mixed venous and coronary sinus P_{0_2}	111	
—	—	Up to 20.4	Impairment in response to certain psychomotor tests, detectable at 5% COHb; no effect on pulse, respiratory rate, blood pressure, neurological reflexes, muscle persistence, or static steadiness test	49	*(307)*

[a] COHb = carboxyhemoglobin, EEG = electroencephalogram, P_{0_2} = partial pressure of oxygen, P_{a0_2} = partial alveolar oxygen pressure, P_{v0_2} = partial venous oxygen pressure.

as in traffic or when the ambient levels subside at night. Since the effects to be expected from the exposure of normal individuals to carbon monoxide in the ambient air are minimal and are expected to be mediated through subtle neurological reaction, man appears to be the best subject for experimentation. Furthermore, there is general agreement that exposures to low levels in normal individuals are entirely reversible, evoking no lasting dysfunction of any sort. Thus, experimental exposure of human subjects under carefully controlled conditions has been often practiced. Carbon monoxide has been the subject of extensive review (*316*, *317*) and is discussed in the following chapter; it will not be covered in detail here.

As might be expected, when subtle neurological parameters are being utilized, considerable variability has occurred among results of various investigators. For instance, in a study in which subjects were isolated in booths depriving them of sight and sound and other normal attention-getting stimuli, decrements occurred in the subjects' ability to detect differences in the duration of a tone after exposure to as little as 50 ppm carbon monoxide over a period of 2 hours (*297*). It is estimated that the exposure should have yielded carboxyhemoglobin levels only 2% above background. In contrast, another study, in which the subjects were in a less isolated situation, indicated that decrements in performance occurred at levels of carbon monoxide yielding approximately 20% carboxyhemoglobin (*310*). A number of other studies have been reviewed concerning the influence of carbon monoxide exposure of human subjects on vigilance, reaction time, discrimination and estimation of time, coordination and tracking, sensory processes, and complex intellectual behavior. Experimental behavioral alterations with carbon monoxide appear to be small and variable, the most reliable effects being noted in decrements in vigilance (*317*).

A. Influence of Carbon Monoxide on Impaired Human Volunteers

Anderson *et al.* (*318*) studied the influence of carbon monoxide on human subjects with stabilized angina pectoris, who were exposed to 50 or 100 ppm carbon monoxide intermittently on 4 or 5 consecutive days. The criterion used was the time to onset of pain when the persons were given a standardized exercise regimen on a treadmill. The time to onset of pain was reduced by both 100 and 50 ppm carbon monoxide in exposures that were equivalent to 4.5% and 2.9% carboxyhemoglobin. Experiments by Aronow and Isbell (*319*) were similar. Aronow *et al.* (*320*) in studies with peripheral vascular disease indicate that patients with intermittent claudication showed a reduction in the time of onset of leg pain when exercised on a bicycle ergometer while breathing 50 ppm carbon monoxide as compared to the same individuals breathing bottled air.

B. Animal Studies

Experimental carbon monoxide exposures of various laboratory animals have been performed. These are probably of greater value for determinations of various mechanistic processes, since concentrations can be employed greater than can be used for experimental exposure of humans, than for evaluation of the more subtle effects of ambient concentrations (Table X). Thus, Musselman *et al.* (*293*) exposed rats, rabbits and dogs to 50 ppm carbon monoxide on a continuous basis for months in which the average carboxyhemoglobin concentrations achieved were 1.8% for the rats, 3.2% for the rabbits, and 7.3% for the dogs. The dogs had significant increases (10–12%) in hemoglobin, hematocrit, and erythrocyte numbers. This same increase in erythrocyte parameters has been seen by numerous other workers in animals undergoing both intermittent and continuous exposure to carbon monoxide (*317*).

Vascular diseases have been described in animals undergoing exposure to carbon dioxide. Astrup *et al.* (*298*) showed that rabbits fed a cholesterol–enriched diet developed greater aortic atheroma than control rabbits on similar diets. Continuously exposed rabbits with carboxyhemoglobin levels of 15% developed a 30% increase in cholesterol deposits. Similar findings were noted in rabbits exposed 8 hours a day. The authors reported similar but less severe effects from simple hypoxia.

C. Discussion of the Effects of Carbon Monoxide

Experimental studies on human volunteers have shown variable results for differing conditions and investigators. Effect on vigilance appears evident at low carboxyhemoglobin concentrations (*317*). Changes of other behavioral parameters at low levels of carbon monoxide appear unclear at this time. The deleterious influence of carbon monoxide on subjects suffering cardiac or peripheral vascular impairment appears real as measured by reduced time of onset of pain during exercise of panels of human beings with stabilized angina or intermittent claudication.

Animal studies indicate that exposure to moderately high levels of carbon monoxide results in increased red cell parameters, suggesting the possibility of adaptation similar to that occurring from hypoxia from other causes. Exposure to carbon monoxide also favors the development of athersclerosis in the rabbit hypercholesterol diet model.

VI. Particulate Material

By virtue of their external and internal surface area, particles may adsorb chemicals such as carcinogens and by thus increasing their penetration into the lungs or prolonging their residence time there, enhance the effect of the adsorbed agents. Particles may also serve as condensation nuclei for water and other vapors to produce droplets in which hygroscopic gases such as sulfur dioxide may be absorbed as acids, thereby augmenting the biological effect of these gases. Heavy loading of the respired air with particles may overtax the mucociliary apparatus, thus reducing the rate of removal of irritant chemicals, infectious agents, and carcinogens from the lung.

Fly ash particles are composed of stable elements or compounds that are usually not considered directly toxic in concentrations found in ambient air. However, it is well known that silica in exposures of sufficient concentration elicits pulmonary fibrosis by means of a complex interaction of lung macrophages with fibroblasts. It has also been shown that particles composed of carbon or iron oxide enhance pulmonary tumor production when used experimentally in conjunction with polycyclic hydrocarbon carcinogens. It is possible that the chemical nature of the particle is unimportant in this reaction so long as there is sufficient surface area for adsorption of the carcinogen (*321*).

From the standpoint of specific toxicity, the chief importance of fly ash may be the proclivity of trace metallic element vapors to condense on it as an effluent gas stream cools (*322*). These trace elements include many of known biological activity, e.g., cadmium, vanadium, nickel, mercury, and manganese. Solid particles may initiate toxicity when inhaled or by the alimentary route when toxic compounds accumulate on forage, food, or in nasal or bronchial mucus that is ingested. Natusch (*323*) has reviewed the subject of fly ash and pointed out the importance of selective adsorption of metals on the ash surface in accordance to their boiling points. Substances whose boiling point is lower than the surface temperature of the ash tend to concentration on the smaller particles (Table XII). This effect probably occurs with vapor phase organic compounds as well. This fact has important toxicological implications, since smaller particles have greater penetration and deposition in the respiratory tract. Aerosols derived directly from manufacturing processes may be liquids (e.g., sulfuric acid mist) or solids (e.g., carbon and silica).

Comparative studies made in man and various animal species indicate that there is general agreement regarding aerosol deposition (Fig. 20). Observations on man, monkey, and guinea pig indicate that total retention is higher for the guinea pig and alveolar deposition is much lower

Table XII Concentration of Certain Metallic Elements in Airborne Fly Ash according to Particle Size *(323)*

Particle diameter (μm)	*Element concentration* ($\mu g/gm$)									*Sulfur concentration* (*wt %*)
	Pb	*Tl*	*Sb*	*Cd*	*Se*	*As*	*Ni*	*Cr*	*Zn*	
11.3	1100	29	17	13	13	680	460	740	8100	8.3
7.3–11.3	1200	40	27	15	11	800	400	290	9000	—
4.7–7.3	1500	62	34	18	16	1000	440	460	6600	7.9
3.3–4.7	1550	67	34	22	16	900	540	470	3800	—
2.1–3.3	1500	65	37	26	19	1200	900	1500	15000	25.0
1.1–2.1	1600	76	53	35	59	1700	1600	3300	13000	—
0.65–1.1	—	—	—	—	—	—	—	—	—	48.8

for 1.5 μm particles, principally because of greater removal of these size particles in the tracheobronchial region *(324)* (Fig. 20). Fairchild *(325)* has studied this problem in mice. It appears that there is a cut-off at a smaller particle size in small rodents as compared to man, but that these differences are not great enough to preclude effective employment of biological models for biomedical purposes using these animals. Both chemical, physical, and physiological factors may effect deposition of particles. Particle shape, in the case of long narrow fibers or crystals, favors penetration of relatively large masses deep into the lung, e.g., asbestos fibers. Hygroscopic particles such as concentrated sulfuric acid droplets or

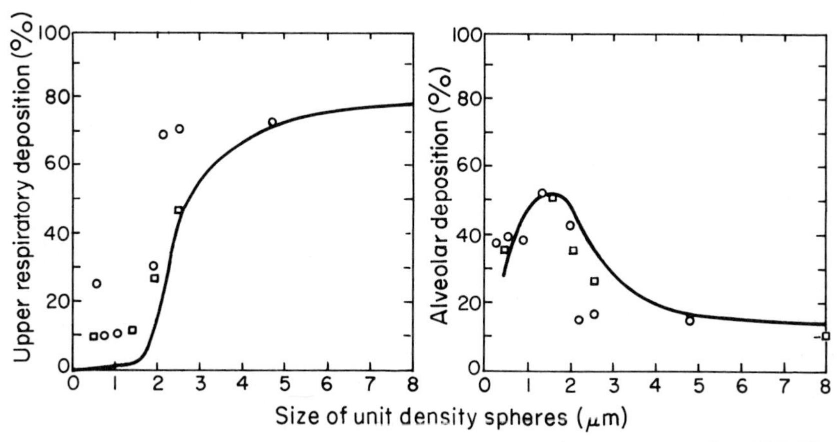

Figure 20. Deposition versus particle size of inhaled particles in the upper respiratory tract and in the lungs of the guinea pig (○) and monkey (□) compared with man (curve) (324).

calcium sulfate, or even sodium chloride, tend to grow after inhalation by the addition of water by virtue of the high relative humidity in the respiratory tract and thus are deposited higher in the tract than their diameter prior to inhalation would suggest. Dautrebande and Walkenhorst (*326*) compared the deposition of coal dust and dry sodium chloride particles of the same size and found it necessary to correct by a factor of seven to account for the growth of salt particles in the respiratory tract (*327*).

Physiological differences affect deposition of particles (*328–330*). Slower, deeper breathing results in greater deposition of the larger particles in the lungs as compared to rapid, shallow respiration. In experiments comparing 1.6 μm and 0.14 μm particles, the larger were most affected by alterations in breathing, whereas the smaller, in which deposition would be presumably the result of Brownian movement, were much less affected (*330*). Constriction of the upper airway as the result of exposure to irritant chemicals tends to result in greater deposition in the nasopharyngeal region and consequently less penetration into the lungs. In this connection, Fairchild noted that after exposure of mice to ozone there was a greater deposition of virus particles in the nose (*331*). It therefore seems logical that considerable variation of particle deposition would occur because of spontaneous diseases and other factors. However, as can be seen above, there appears to be little effect from such factors on the deposition of particles below 1.6 μm.

A. Respiratory Tract Clearance

Particles insoluble in the respiratory tract deposited in the ciliated portion of the respiratory tract, are removed rather rapidly via the mucociliary apparatus (so-called fast clearance) and eventually reach the gastrointestinal tract. Similar particles deposited in the deeper air spaces are much more slowly cleared via the phagocytic apparatus (so-called slow clearance) but have the same gastrointestinal fate. Therefore, inhaled chemicals, even if insoluble in the lung, eventually reach the gastrointestinal tract, where again there is a chance for solubilization in gastric fluids and subsequent chemical reaction. Spontaneous disease and exposure to many inhaled chemicals slows clearance by inhibiting the movement of cilia or through clogging the respiratory tract by excessive production of mucus. However, a portion of the insoluble material is slowly cleared from the lung by penetration of the respiratory epithelium, thereby gaining entrance to the interstitial area and lymphatics. This reaction is readily seen in lungs of individuals exposed to carbon or iron oxide. Chemicals soluble in the respiratory tract are quickly absorbed at all levels of the tract from the nose to the pulmonary alveoli, as evidenced by absorption of sulfuric acid and other chemicals, many of which are absorbed more readily from the respiratory tract than from the alimentary tract. Therefore, exposure to chemicals by inhalation results in possible toxic interaction at many levels and compartments of the body.

B. Chemical Effect of Particles

Many trace elements have important biological activity and thus are potential health hazards. A vast literature has accumulated on the effects of such elements as vanadium, cadmium, nickel, maganese, and mercury. The absorption, deposition, and fate of compounds of these elements has been reviewed in a number of publications which generally indicate that toxic problems would be associated with greater exposures than would be expected from ambient air (*332–336*).

C. Carbon

A relatively foreign pigmentation of the lung is associated with urban residence in man and the lower animals. This pneumoconiosis is a function of the carbon portion of smoke. Its degree roughly equates with the degree of pollution of the air by combustion of carbonacous fuels and the length of exposure to such atmospheres. Studies in animals have indicated no measurable pathological effect of carbon alone (*337*). It is generally considered relatively innocuous.

D. Iron Oxide

Iron oxide constitutes 2–26% of fly ash and occurs in polluted atmospheres as hematite (Fe_2O_3) or magnetite (Fe_3O_4) (*338*). Little biological effect is ascribed to community air pollution exposure to iron oxide. Although iron pigmentation (siderosis) is frequently associated with pulmonary fibrosis, no causal relationship is suspected and the pigmentation is generally supposed to be endogenously derived from hemoglobin degradation. A finding that the effects of benzo[*a*]pyrene in the induction of lung cancer in experimental animals can be intensified by introducing benzo[*a*]pyrene with iron oxide (Fe_2O_3) has been reported by Saffiotti *et al.* (*339*) and with carbon by Shabad *et al.* (*340*). It is unclear whether the hematite and carbon act as carriers or play more specific roles in the phenomenon.

E. Asbestos

That so-called "asbestos bodies" are prevalent in the lungs of the general population of city dwellers is well known, but the association of these bodies with asbestos fibers has been a moot point, since other fibers can develop into these structures after residence in the tissue. However, critical examination of twenty-eight human lungs from autopsy material from New York, New York, by Selikoff *et al.* (*341*) have shown chrysotile fibers in all of them (Table XIII) and a strong correlation between these fibers and asbestos bodies (Table XIV) (*342*). Similar studies in London have demonstrated chrysotile fibers in 80% of the lungs examined.

The discovery of asbestos fibers in lungs of human beings not occupationally exposed indicates that a potential exists for risk to the general population. Until more surveys are performed, it is difficult to assess the extent of this risk. Asbestos has come into widespread use only in the past 100 years. There has been a thousandfold increase in its use in the twentieth century. There has been an increase in the amount of fibers deposited in the ice cap in Greenland since the beginning of the twentieth century (*343*).

1. Pulmonary Lesions

There are several types of pulmonary lesions associated with asbestos inhalation—fibrosis, carcinoma, and mesothelioma. It is possible that the mechanism for these different lesions are themselves different.

Table XIII Chrysotile Fibers in Human Lungs from New York, New York *(341)*[a]

Group	*Number of chrysotile fibers + fibrils*	*Cases (out of 28)*	*Male*	*Female*
1	≤9	4	3	1
2	10–50	11	5	6
3	91–99	6	1	5
4	100–200	4	4	0
5	≥201	3	3	0
"Blank grids"	≤9	—	—	—

[a] Chrysotile fibers in 28 cases studied by electron microscopy.

Table XIV Correlation of Inorganic Fibers with Asbestos Bodies in Human Lungs *(352)*[a]

A. Percentage of cases showing asbestos bodies in a lung in relation to occurrence of thin inorganic fibers in that lung

Number and percent of lungs with the number of thin inorganic fibers shown at head of column

Number of asbestos bodies in a lung	*0*		*1–4*		*5–14*		*15+*	
	Number	*%*	*Number*	*%*	*Number*	*%*	*Number*	*%*
0	1168	59.5	332	38.3	35	28.0	16	33.3
1–4	705	35.9	424	49.0	52	41.6	13	27.1
5–14	76	3.8	89	10.3	21	16.8	6	12.5
15+	13	0.8	20	2.4	17	13.6	13	27.1
Total	1962	100.0	865	100.0	125	100.0	48	100.0

B. Percentage of cases showing thin inorganic fibers in a lung in relation to occurrence of asbestos bodies in that lung

Number and percent of lungs with number of asbestos bodies shown at head of column

Number of thin inorganic fibers in a lung	*0*		*1–4*		*5–14*		*15+*	
	Number	*%*	*Number*	*%*	*Number*	*%*	*Number*	*%*
0	1168	75.3	705	59.0	76	39.6	13	20.6
1–4	332	21.4	424	35.5	89	46.3	20	31.8
5–14	35	2.3	52	4.3	21	10.9	17	27.0
15+	16	1.0	13	1.2	6	3.2	13	20.6
Total	1551	100.0	1194	100.0	192	100.0	63	100.0

[a] Lungs from 3000 consecutive autopsies in New York, New York, 1966–1968. Correlation was done by optical microscope study.

a. Pulmonary Fibrosis (Asbestosis). Fibrosis is usually related to massive dose and has the shortest period of latency. Thus, asbestosis is associated with industrial exposure to all forms of asbestos fibers (*344*). There is some suggestion that smoking increases the risk of asbestosis (*341*). Another variety of fibrosis (parietal pleural calcification) appears strongly associated with industrial exposure to asbestos and with residence in regions where asbestos is produced. It has been reported in Finland from exposure to anthophyllite (*344*) and in Czechoslovakia from exposure to chrysotile (*345*), both from naturally occurring deposits.

The exact initiating mechanisms for fibrosis are still obscure, but there is evidence that lysosomal injury is involved (*346*). Long fibers stimulate more and quicker fibrosis than short ones (*347*), possibly because they are less well phagocytized, since transport begins with fibers of 20 μm and increases as the fibers become shorter. Short fibers are engulfed by macrophage, long ones are not (*348*). Chrysotile asbestos was shown to elicit pulmonary fibrosis as early as 1925 (*344*). Subsequently, amosite and crocidolite were also shown to have this property (*349*). It now appears that most types of asbestos are fibrogenic. Asbestosis due to urban air pollution can be disregarded, because in humans it associates only with high occupational exposures and residence near asbestos mines.

b. CARCINOMA. Exposure of animals by both inhalation and intratracheal instillation of asbestos has been productive of pulmonary carcinoma in animals. In one study employing rats given chrysotile, one-third of the animals surviving 16 months or longer had malignancies of the lung, divided approximately evenly between adenocarcinoma and other tumors consisting of squamous cell carcinoma and fibrosarcoma. One mesothelioma was seen (*350, 351*). It is interesting to note that the authors ascribe their tumor yield to the concomitant presence of trace metals from their grinding methods, since they contaminated their asbestos with nickel, cobalt, and chromium. However, other investigators who have produced tumors of the lung have ascribed them to the influence of asbestos alone.

S.P.F. rats were exposed to three types of amphibole asbestos (amosite, anthophyllite, and crocidolite) and two types of chrysotiles (Canadian and Rhodesian) by inhalation (*352*). Retention was cumulative with length of exposure. After animals were removed from exposure, the amount of dust in their lungs declined. There appeared to be a better retention of amphibole asbestos (amosite, crocidolite, anthophyllite) than with the chrysotiles. (Fig. 21). Tumor induction was found with all forms of asbestos. No tumors occurred before 300 days after start of exposure, and more tumors occurred when exposure was over 6 months, but there was no difference between 12 months and 2 years of exposure (Table XV).

The significant findings in the above study, beyond tumor induction by the inhalation route were (a) difference in retention rate for amphibole over chrysotile, and (b) the fact that all the forms of asbestos tested were approximately equal in their ability to induce pulmonary tumors.

c. MESOTHELIOMA. Mesothelial tumors are induced by all forms of asbestos via the inhalation route (Table XV). A 1962 report showed that crocidolite and chrysotile produced mesothelial tumors in mammals upon intrapleural innoculation of the fibers (*353*). Experiments in mice,

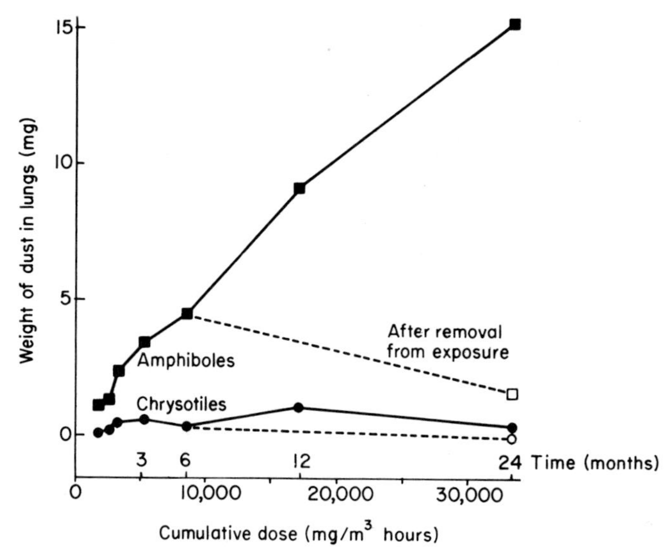

Figure 21. Deposition and clearance of two types of asbestos from the lungs of rats after exposure by inhalation to same concentration. Mean weight of dust in rat's lungs in relation to dose and time (*352*).

rats, fowl, and hamsters suggest that the fibrous nature of the material may have a bearing on the tumorogenicity by this route and that a hypothesis could be advanced that the carcinogenicity of the material is related to the ability of the fiber to react with the cell membrane without destroying the cell. The presence of trace metals or organics on the fiber was not a determinant. A relationship between the smallness of the fiber and the number of mesotheliomas could be seen (*354, 355*). It is interesting to note that mesothelioma in these experimental animals occurred with but one day's exposure to amosite and crocidolite.

Tumors (sarcomas) of the mesothelial tissues have been elicited with nonasbestos fibers such as fiberglass and aluminum oxide (Al_2O_3) when placed on a fiberglass mesh and implanted on the pleural surfaces of the rats. On this basis, it has been theorized that a particular configuration of the crystal or fiber is the determinant for tumorogenesis, i.e., diameters of 5 μm or less and lengths of 20 μm or more. However, studies with Canadian chrysotile of extremely small fiber length have produced mesotheliomas in rats (*354*). The significance of these reports is in question, because the tumors induced were sarcomas rather than classic mesotheliomas, which are tumors with both mesenochymal and epithelial characteristics. The histogenesis of pleural and peritoneal tumors is complex. The earliest neoplastic response often consists of small pedunculated nodules with a central core of reticulin surrounded by a surface of non-

Table XV Number of Animals with Lung Tumors or Mesotheliomata *(352)*

Exposure	*Number of rats at risk*[a]	*Number with lung tumor*	*Type of lung tumor*				*Number with mesothelioma*
			Adenoma	*Adenomatosis*	*Adenocarcinoma*[b]	*Squamous carcinoma*[b]	
Amosite							
1 day	45	3	3	0	0	0	1
3 months	37	10	7	3	0	0	0
6 months	18	2	1	0	1	0	0
12 months	25	10	5	4	1	0	0
24 months	21	13	3	1	3	6	0
Total	146	38	19	8	5	6	1
Anthophyllite							
1 day	44	2	2	0	0	0	0
3 months	37	6	6	0	0	0	0
6 months	18	6	3	1	1	1	0
12 months	28	20	9	6	4 (1)	1	1
24 months	18	16	2	5	3	6	1
Total	145	50	22	12	8 (1)	8	2
Crocidolite							
1 day	43	6	5	0	1	0	1
3 months	36	14	10	2	1	1 (1)	1
6 months	18	4	2	2	0	0	0
12 months	26	18	5	4	3	6	2
24 months	18	13	4	5	2 (1)	2 (1)	0
Total	141	55	26	13	7 (1)	9 (2)	4
Chrysotile (Canadian)							
1 day	42	1	0	0	1 (1)	0	0
3 months	34	18	15	0	3	0	0
6 months	17	5	2	2	0	1	0
12 months	23	11	1	3	6 (1)	1 (1)	3
24 months	21	10	2	3	1 (1)	4 (2)	1
Total	137	45	20	8	11 (3)	6 (3)	4
Chrysotile (Rhodesian)							
1 day	45	5	4	0	1	0	0
3 months	36	16	11	2	3 (1)	0	0
6 months	19	8	2	3	3 (1)	0	0
12 months	27	19	2	4	7 (2)	6 (4)	0
24 months	17	11	0	1	5 (2)	5	0
Total	144	59	19	10	19 (6)	11 (4)	0
Control							
1 day	44	4	4	0	0	0	0
3 months	40	3	3	0	0	0	0
6–24 months	42	0	0	0	0	0	0
Total	126	7	7	0	0	0	0

[a] Rats which survived at least 300 days after start of exposure.
[b] Numbers in parentheses are those with metastases.

differentiated connective tissue. These may spread diffusely over the mesothelial surface as a sheet of connective tissue cells with epithelial cells on the surface. It is supposed that the tumors arise from undifferentiated mesenchymal cells in the submesothelial tissue and are differentiated giving rise to epithelial cells on the surface and spindle-shaped connective tissue cells in the depths (*356*).

Evidence has been accumulating to suggest that brief or even casual exposure to asbestos may result in the development of diffuse mesotheliomas of the pleura or peritoneum after a period of latency of 30–40 years or more (*357*). On the basis of retrospective diagnosis of material derived from necropsies, or in some instances, biopsies, an epidemiological relationship has been noted by a number of investigators. Mesotheliomas in humans were first reported in South Africa by Wagner (*358*) and Wagner *et al.* (*359*). Subsequently, they were reported from the same country by Webster (*360*), from London, England, by Newhouse and Thompson (*361*), from Belfast, Northern Ireland, by Elmes and Wade (*362*), and from Liverpool, England, by Owen (*342*). It has been possible to demonstrate asbestos bodies in the lungs or sputum of these cases, but their demonstration within the tumors has not been so common. Hourihane (*363*) reporting observations made on biopsy specimens of seventy-four mesotheliomas from London, England, identified asbestos crystal structures in about one-fourth of the cases after heating tissue sections on microscope slides. In view of the slight or brief known occupational exposure in many of the reported cases of mesothelioma, it has been suspected that exposures of sufficient magnitude to induce these cancers might be possible from industrial contamination of the ambient air. In support of this thesis, Thompson (*364*) has published data demonstrating that in two consecutive series of 500 autopsies, one in Capetown, South Africa, and one in Miami, Florida, approximately 30% of the men and 20% of the women were positive for the presence of asbestos bodies in the lung. For this study, smears prepared from the cut surface of the base of each lung were airdried and fitted with a cover glass and subjected to microscopic examination. A negative report was indicated when examination of 300 low-power fields showed no asbestos bodies. The findings suggested to Thompson that in the 85% of the positive cases where the asbestos bodies were scanty, the exposure was the result of community air pollution; and that in the 6% of the males, where the bodies were numerous, exposures were probably of occupational origin. In view of the latency associated with the prime low-level lesion (mesothelioma) and the apparent increase of potential exposure via air pollution, the risk of mesothelioma development cannot be disregarded.

2. Intratracheal Studies

Intratracheal studies have been advocated as a convenient method to test the potential of a chemical to induce pulmonary neoplasms; particularly to show response to a cofactor in conjunction with a chemical carcinogen (*339, 340*). Comparative studies using chrysotile have been made with intratracheal injection, inhalation and intrapleural injection (*365*).

Intratracheal injection shows a dose of response according to the frequency of the injections. The induction of benzo(*a*)pyrene was required to elicit tumors in 29% of the rats.

Inhalation for 13.5 months on a schedule of 2 hours a day, 5 days a week, at 230 mg/m³ resulted in precancerous lesions in 20% of the rats on chrysotile alone, 59% with added benzo[*a*]pyrene and 38% with added tobacco smoke. A total of 29% of the rats had tumors, and carcinoma of the lungs occurred only in the chrysotile–benzo[*a*]pyrene group. Previous experiments demonstrated slowing of excretion of benzo[*a*]-pyrene with asbestos (*366*).

Intrapleural injection was performed by injecting chrysotile three times per month into the right pleural cavity (20 mg fibers in 0.5 ml saline). Almost all the rats had local proliferation of the mesothelium that was presumably preneoplastic. Mesothelioma appeared after 8 months and increased with time, 13 of 47 rats at 8–17 months, and 15 of 21 between 8 and 24 months. Forty-six percent of all rats showed mesotheliomas (*365*).

As mentioned previously, qualitative differences in fiber size and shape appear to be the determinant in induction of mesotheliomas. Carcinogenesis appears unrelated to the trace metal content or the presence of organic matter on the fibers. The possible induction of tumor by material other than commercial asbestos has also been considered, and the experimental data suggest that a number of inorganic crystals below 5 μm in diameter may be equally effective in inducing these tumors. The implication of these findings is that small fine particles are the effective component for induction of these tumors.

3. Synergistic Effects

Reference has been made earlier to the remarkable association of asbestos and smoking in induction of bronchiogenic carcinoma in exposed workers in the asbestos industry. Animal exposure by inhalation of chrysotile for 16 months has produced up to 33% pulmonary tumor adenocarcinomas, fibrosarcomas, squamous cell carcinoma, and one mesothelioma (*350*). The carcinogens responsible for these were thought to be contaminated by metallic substances from hammer milling. The same author reported similar findings in a second inhalation experiment (*351*).

The role of benzo[*a*]pyrene in promoting lung tumors from exposure of animals to various forms of asbestos is known, although the evidence for such effect is at present mainly for chrysotile (*365*). The other forms of asbestos have been insufficiently studied in this regard (*351*). Contamination by organics including polycyclic hydrocarbons due to geological and industrial factors has also been suggested as a causative factor in pulmonary carcinoma, but the use of extensively extracted purified fibers has shown no difference in this regard (*367*).

Evidence presently at hand suggest that either (a) some direct action of the asbestos fiber or interaction with naturally occurring cofactors is responsible for tumorigenesis of the lung in experimental exposure of animals, (b) the addition of benzo[*a*]pyrene or tobacco smoke enhances tumorigenesis with chrysotile, or (c) all forms of commercial asbestos are more or less equally carcinogenic.

4. *In Vitro* Effects

A great deal of the basic biological effect of asbestos can be learned from studies *in vitro* (*348*). Many of the actions reported for asbestos are similar to silica. These are (a) lysis of erythrocytes, (b) cytotoxicity of pulmonary macrophages, and (c) stimulation of collagen synthesis.

F. Discussion of the Effects of Particulate Material

While it is convenient to consider "particulates" as a single class for control purposes, it is impossible to discuss the toxicological potential of source material as though it were a single entity, as for instance nitrogen dioxide. The particles in the air are derived from innumerable natural and man-made sources, each contributing differing physical and chemical properties to the mix. While portions of the particulate milieux consist of relatively nonreactive substances such as carbon or calcium, they differ in their biological activity by virtue of the presence of adsorbed chemicals that may be added through condensation of volatile chemicals in the stack as the flue gases cool, or through interactions with other pollutants in the atmosphere. Therefore, the potential for biological activity of a given particle is governed not only by physical and chemical nature of the fuel, the combustion process, but also by its opportunistic ability to interact with other environmental pollutant substances capable of chemical or physical interaction.

The biological activity of the average of the particles in various locations then must of necessity vary because of differing pollutant source profiles for any given community. These variations are expressions of both quantitive and qualitative differences, as for instance the relative amount of sulfuric acid mist, sulfates, or other reactive substances in the particulate mix or the relative amounts of specific carcinogenic compounds in the organic fraction of airborne particulate. An example of the former is the association of health effects in human beings with the quantity of particulate sulfate collected from air and of the latter quantitive and qualitative differences in tumor induction when animals are treated experimentally with identical amounts of organic extracts from various cities. Such reactivity is probably the result of a complex interplay between carcinogens, anticarcinogens, and various cofactors whose relative amounts in a given atmosphere will be governed by the type of fuel consumed and temperature of combustion. For example, this factor is evidenced by the difference of the relative quantity of benzo[*a*]-pyrene in particulates derived from Los Angeles, California, and Gary, Indiana, either expressed as a ratio to total amounts of collected particulate or to the amount of crude organic extract. Other particles, such as asbestos, have such a high potential for reaction but occur in such minute amounts in the air that their presence or absence contribute nothing to the overall loading of the atmosphere, and even their detection and identification among particles collected on filters is fraught with the utmost difficulty.

It therefore can be seen that the evaluation of the biological activity ascribable to "particulates" is complex and depends not only on the total quantity, size range and intrinsic physical or chemical properties, but also on their chance for interaction in the polluted air. The opportunity for variation in biological activity is enormous, and at this time it would seem unwise to attempt to relate health effects to the total amount of respirable particulate material in the atmosphere without regard for their qualitative differences.

VII. Interaction of Air Pollutants and Biological Agents

It is well known that many if not most disease states of man, animals, and even plants are the result of several deleterious influences acting in concert. For the establishment of infection, for instance, there are required not only the specific microorganism, but accessory factors that make the particular individual vulnerable to its invasion. Such supplementory stresses are frequently called predisposing influences. They may consist of many things that weaken resistance, such as the preexistence of other infections, deficiency of certain dietary elements, the presence of toxicants, meteorologic variables, and many other conditions peculiar to a particular host or to a particular environment. The operation of such factors in conjunction with the prime etiological agents in the infectious diseases has been termed the principle of multiple causality and has been discussed in detail by Top (*368*).

Air pollutants have been shown to serve as supplemental enhancing factors for infectious disease, e.g., the sulfur oxide–particulate atmosphere already discussed (*369–371*). The role of air pollutant gases in this regard is well known from outbreaks of spontaneous infectious pneumonias in rodents while under exposure to such gases (*105, 125, 372*).

Experimental models to demonstrate the propensity of toxicants in general and air pollutants in particular to alter the susceptibility to infectious disease via the portal of entry of the lung have approached the subject from two aspects; on the one hand, indirectly by looking at increased residence time in the lung for introduced bacteria of little or no pathogenicity to the species, and on the other hand, by directly examining the subject by the introduction of pathogenic bacteria into the lung and observing alterations in pathogenic mode, morbidity, mortality, and the like as a result of such introduction.

Employing the first method, Stillman (*373*) noted that intoxication by alcohol delayed the disappearance of *Diplococcus pneumoniae* from the lungs of mice. Later studies by Laurenzi *et al.* (*374, 375*) and Green and Kass (*376, 377*) in studies designed primarily to elucidate pulmonary defense mechanisms, showed that a variety of deleterious influences delayed the rate of so-called bacterial clearance from the lung. This phenomenon was shown to be related to loss of viability of the deposited organisms to a greater degree than their physical removal from the lung. The mechanism was attributed to alteration by toxicant action on the pulmonary alveolar macrophage. Decrements of clearance were noted after exposure to agents such as ethanol, reduced oxygen tension, and tobacco smoke. Slight reduction of the rate of clearance has also been noted after exposure to carbon black and a combination of coal dust and sulfur dioxide (*378*). Green and Kass (*377*) showed that the clearance phenomenon consists of two parts, a decline in bacterial viability and physical removal of the bacterial cells; the greater portion, that resulting from the loss of bacterial viability amounting to approximately 85% of the clearance at 4 hours, the remainder being explained by slowing of physical removal of the bacterial cells from the lung. They interpreted the reduction of viability as resulting from macrophage action and the physical removal as that resulting from mucociliary clearance. These estimates were determined by a method that measured bacterial viability loss by cultural methods and physical removal by monitoring the decline in radioactivity following the introduction of labeled bacteria.

Direct observation on the role of toxicants in altering infection was noted by McDonnell (*379*) in 1930 in a study of the influence of ozone exposure on enhancement of infection with *Mycobacterium tuberculosis*. Subsequent studies by a number of investigators have indicated that exposure to ozone (*5, 21, 44*), nitrogen dioxide (*380*), synthetic auto smog (*381*), lead, and cadmium chloride (*382*) remarkably enhances the mortality from subsequent or prior administration of pathogenic bacteria by aerosol. These studies involved a variety of animal species, including the mouse, rat, hamster, and squirrel monkey, and utilized several kinds of microorganisms including streptococci, *Klebsiella pneumoniae, Diplococcus pneumoniae,* and influenza virus.

Exposure of mice to ozone by this system has been reported to elicit enhancement of infection after an exposure of approximately 0.08 ppm and above for 3 hours (*5*). The minimal effective dose for nitrogen dioxide has been reported as 3.5 ppm for 2 hours (*136*). Positive results were yielded by intermittent exposure to nitrogen dioxide at 0.5 ppm for 3 months (*136*) and to continuous exposure for 15 days (*380*). An effective enhancement of mortality was noted after exposure of 4 hours to synthetic auto smog in which total oxidizing activity equivalent to 0.15 ppm ozone was administered (*383*). Few studies have been reported for other gases. No significant enhancement of streptococcal mortality was noted after exposure of mice to sulfuric acid mist of a mean diameter of 1 μm (*384*) and to sulfur dioxide, both at several times ambient concentrations.

Four observations have been reported on the influence of metals on infections. Exposure of mice to lead nitrate by intraperitoneal injection enhanced mortality from intraperitoneal introduction of *Salmonella* (*381*); and aerosol exposure of mice to manganese (*385*), nickel (*386*), and cadmium chloride (*382*) enhanced their mortality from aerolized infectious agents. In the later study, the minimal effective aerosolized dose was less than 100 $\mu g/m^3$ for 2 hours, resulting in a retention of approximately 0.1 μg cadmium chloride per lung (*382*).

A. Mechanism of Action

The infective system has been most thoroughly studied for ozone, though more limited experiments suggest that the mechanisms are essentially the same for nitrogen dioxide. Coffin and Gardner noted that striking alterations in bacterial retention, decline, replication, and invasion of the blood occur after exposure to ozone (*44*). In these studies, exposure for 3 hours at approximately 0.35 ppm and above results in reduced bacterial retention as compared to controls, whereas below this concentration there appeared to be increased retention (Fig. 22). Possible explana-

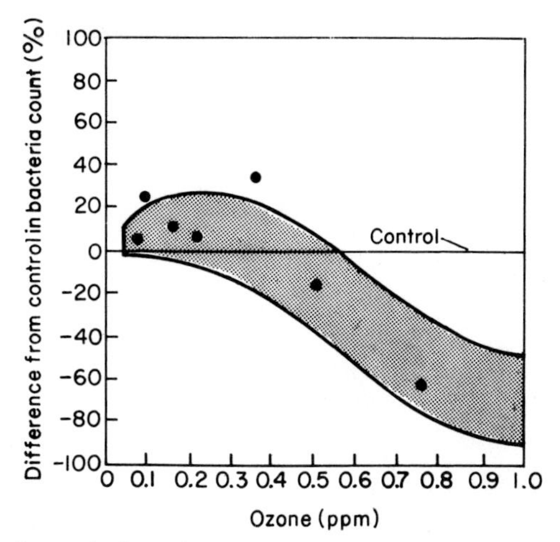

Figure 22. Experimental data fitted to a mathematical model calculated from previous data indicates that there is a reduction in immediately retained viable bacteria at ozone concentrations greater than 0.4 parts/10^6. Conversely, at lower concentrations there is an increase in retention of viable organisms (*44*).

tions for the diminished retention might be (a) that alteration in breathing patterns and breath holding and the like might reduce ventilation in the deep lung, or (b) that constriction of the upper airway might favor deposition there rather than in penetration to the deep lung. Experimental evidence indicating increased deposition of particles in the nose following ozone exposure gives credence to the latter view. (*331*). The apparent increased deposition below 0.35 ppm suggests either increased initial penetration or possibly that normal deposition plus slowed bacterial clearance during the aerosol exposure resulted in greater accumulation of viable bacterial cells at the termination of this period.

The rate of removal and subsequent growth of pathogenic bacteria are markedly deterred by exposure to ozone. Figure 23 demonstrates the influence of exposure as a function of dose where pathogenic streptococci were employed. At lower concentrations of ozone, there is an initial decline with a subsequent increase in numbers of viable bacterial cells, whereas at higher concentrations no initial decline in numbers is apparent. There is an influence of lower doses of ozone on *in situ* bacterial multiplication even though clearance rates are identical to controls (Table XVI). Experiments conducted comparing the phenomenon of clearance and growth of a pathogenic streptococcus with that of the nonpathogenic *Serratia marscesens* indicate that while there is striking gain in the number of streptococcal cells, little or no difference in rate of clearance was noted for the nonpathogenic *S. marscensens* (Fig. 21). The above

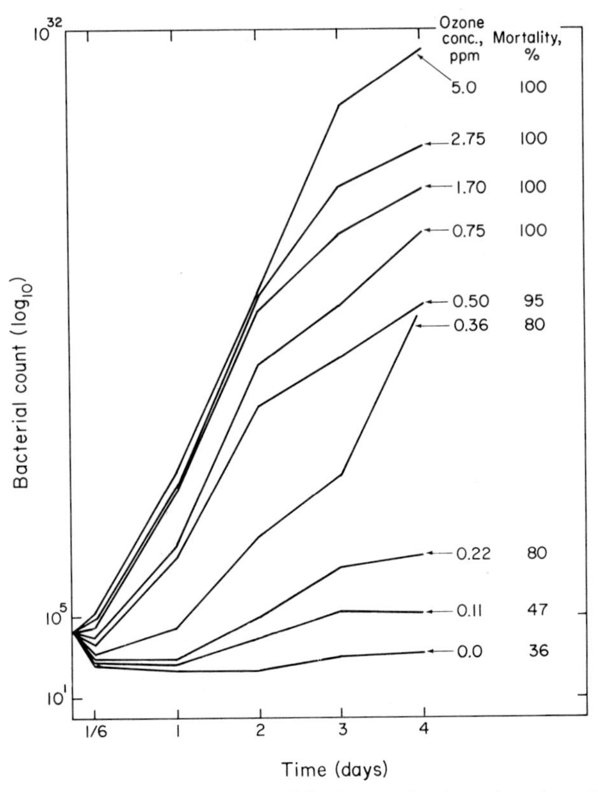

Figure 23. Number of viable streptococci in lungs of mice at various times following 3 hours of ozone treatment and bacterial aerosol (*290*).

facts show the increased sensitivity of an actual infective system employing pathogenic bacteria over a system that employs only bacterial clearance of a nonpathogen as an indicator of toxicity. In the streptococcus model, focal pneumonia develops from centrolobular foci, which tend to coalesce forming large necrotizing areas. Mortality appears associated with penetration of the bacteria into the blood stream. Septicemia mimics the mode of mortality approximately a day sooner. In the experiment summarized in Figure 25 in which blood cultures were made at daily intervals, mortality in both the controls and the ozone-treated groups exactly conformed to positive blood cultures. It is thus tenable to assume that augmented pneumonia led to earlier invasion of the blood by the bacteria, although the possibility exists that direct invasion of the blood prior to development of significant pneumonic lesions may have occurred in the ozone-treated groups. The well-known proclivity of ozone and nitrogen dioxide to induce desquamative lesions in the central lobular

Table XVI Number of Bacteria (Log Percent of Initial Deposition) at 4 Hours and 4 Days as a Function of Ozone Concentration[a]

Ozone (parts/10⁶)	*4 hours*	*4 days*
5.0	2.426	30.07
2.75	2.347	24.74
1.7	2.156	23.10
0.75	1.886	20.65
0.36	0.908*	18.74
0.22	0.377*	12.26
0.095	0.450*	13.39
0.08	0.555*	14.39
0	$\bar{x}$ 0.554*	$\bar{x}$ 4.84
Controls	±0.07	±1.18

[a] *Note:* The comparative importance of clearance and growth enhancement as a function of ozone exposure is demonstrated in the above table in which the logarithm of the number of bacteria (percent of initial deposition) is compared at 4 hours and 4 days. It will be noted that while little difference exists from controls at the 4-hour period for the three lowest ozone concentrations, a striking difference is noted at 4 days. A decrease from zero time is represented by the data marked* whereas the rest of the data represent an increase in bacterial growth at the period. Values stated for controls are mean ± standard error (*44*).

pulmonary areas might be conducive to such direct invasion into the blood stream.

It should be noted here that the rapid expression of mortality before the establishment of appreciable pneumonic lesions at high dosage in the cadmium model argues for the toxicant-inducing direct penetration of the blood, perhaps independently of the development of pneumonia (*382*). In support of this premise it is known that exposure to cadmium salts increases the permeability of blood vessels.

B. Alteration of the Macrophage System

It is generally believed that pulmonary macrophages are the principal deterrent to the development of infection of the deep lung from intro-

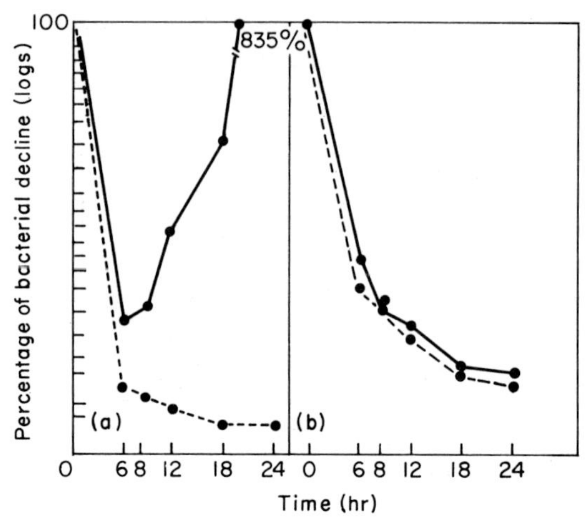

Figure 24. Bacterial clearance and growth curves determined from plate counts indicate that the enhancement in streptococcal pneumonia due to ozone exposure appears to be related to the ability of the inhaled agent to multiply rather than to affect general pulmonary clearance. In the figure, note that little difference is seen in the clearance curves for the nonpathogenic *S. marcescens,* whereas an eightfold increase is evident with pathogenic streptococci at the same concentration of ozone. (a). solid curve = ozone + streptococci; dashed curve = ambient + streptococci. (b). solid curve = ozone + *S. marcescens;* dashed curve = ambient + *S. marcescens* (*44*).

duced pathogenic bacteria. Thus considerable attention has been paid to this cell in relationship to pulmonary defense impairment by various environmental agents. Green and Kass (*377*) have indicated that macro-

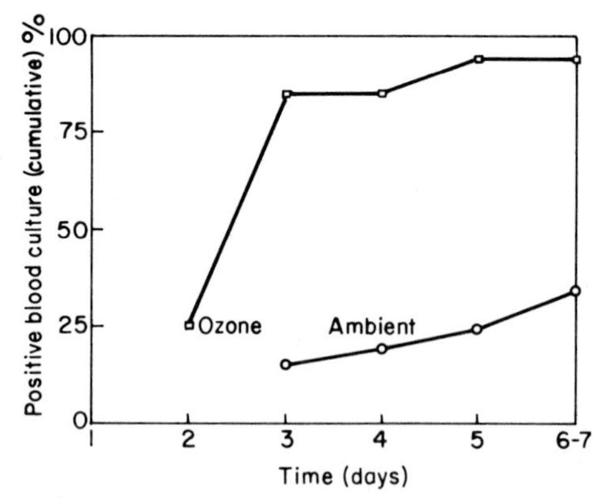

Figure 25. A marked acceleration in the rate of bacterial invasion of the blood is produced by exposure to ozone. Indicated above is the difference between exposure to 1 part/10^6 and controls as determined by daily intravenous cultures; $p = 0.01$ (*44*).

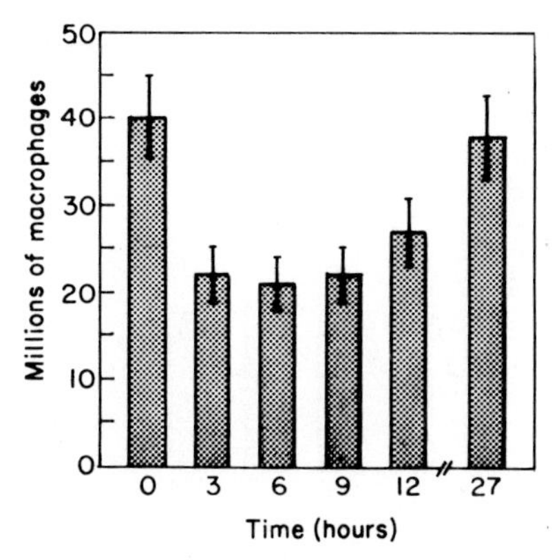

Figure 26. Ozone exposure elicits a reduction in the number of macrophages obtainable by pulmonary lavage. This reversible effect peaks 6 hours after exposure to 5 parts/10^6 ozone for 3 hours (*44*).

phage activity is the principal factor in bacterial clearance in the noninfectious model by the observation that the introduced bacterial cells at first lie free in the respiratory bronchioles and alveolar spaces and later are incorporated within the cytoplasm of macrophages. In conjunction with air pollutants, Coffin *et al.* (*24*) and Gardner *et al.* (*387*) have indicated that exposure to ozone, and to a lesser extent nitrogen dioxide, produces decrements in numbers, phagocytic activity, and certain other parameters in macrophages lavaged from the lung of the rabbit. An influx of polymorphonuclear leukocytes was concomitant with the above change. Hurst *et al.* (*388*) demonstrated a reduction in activity of three lysosomal enzymes in these same cells, and Dowell *et al.* (*17*) noted increased osmotic fragility resulting after *in vivo* exposure of rabbits. Figures 26–28 summarize these macrophage effects. Bingham *et al.* (*389*) have also noted reduction of lavagable alveolar macrophages in rats after exposure to lead sesquioxide.

C. In Vitro Studies

Studies performed on surviving alveolar macrophages *in vitro* have replicated some of the above effects and have extended observations to the relative toxicity of trace metals for this cell. Green and Carolin (*390*) noted that exposure of mice to cigarette smoke depressed the *in vitro* antibacterial activity of alveolar macrophages. Coffin *et al.* (*24*) noted reduction in phagocytic activity and loss of bacterial viability in rabbit alveolar macrophages from tests made after exposure of the animals to ozone. Hurst and Coffin (*65*) noted that *in vitro* exposure of rabbit

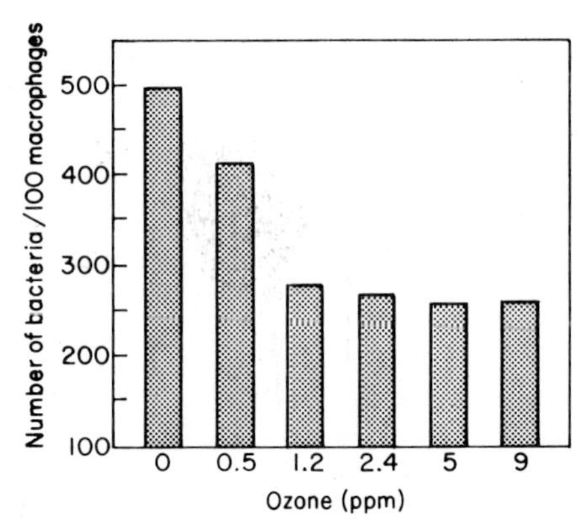

Figure 27. Ozone exposure elicits a reduction of the phagocytic capabilities of alveolar macrophages as determined *in vivo*; $p < 0.001$ (*44*).

alveolar macrophages replicated *in vivo* results on lysosomal enzymes. Waters *et al.* (*391–394*), in a series of studies, have examined the relative toxicity and mode of action of a number of trace metals for rabbit alveolar macrophage. Briefly, their results indicate that for any given metal, cellular toxicity varies with the solubility of the compound employed. In the case of vanadium oxides (*391*), the particulate pentoxide was approximately twice as toxic as the trioxide and three times as toxic as the dioxide, using cell death as an endpoint. However, when these compounds were dissolved in tissue culture media prior to exposure of

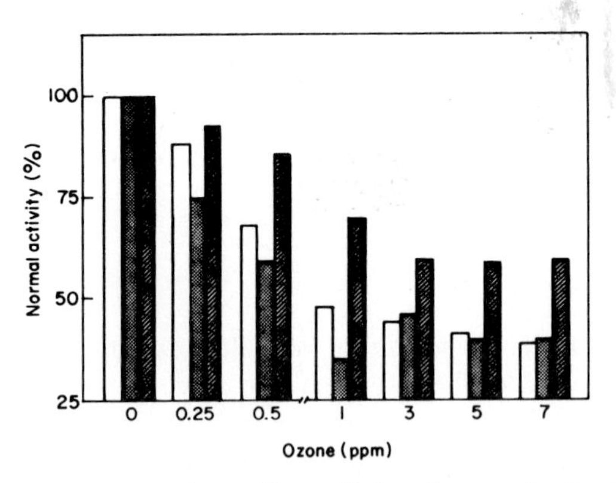

Figure 28. Ozone exposure reduces the activity of several alveolar macrophage lysosomal enzymes. Normal activity is represented by 100%, and loss of activity after a 3-hour exposure demonstrated a dose–response relationship. Plain bars indicate acid phosphatase, shaded bars indicate lysozyme, and hatched bars indicate β-glucuronidase; $p < 0.001$ (*44*).

cells, cytotoxicity was determined to be a function of solubility. Other indices of cellular toxicity, including lysis, phagocytic activity, hydrolase activities, and electron microscopic examination of surface morphology (*392*), have been employed to rank the relative toxicity of trace metals for alveolar macrophages. Similar parameters, including those relating specifically to DNA, RNA, and protein biosynthesis have been used to estimate toxicity of trace metals for human lung fibroblasts in culture (*394*). Findings that have emerged from these investigations include the following. Exposure to certain divalent cations, notably those of cadmium, mercury, chromium, and zinc, confers unusual stability to the plasma membrane such that death of alveolar macrophages is not followed by lysis as usual. Graham *et al.* (*395*) and Waters *et al.* (*393*) also found that certain other metals, including Ni^{2+}, Cd^{2+}, Hg^{2+}, Cu^{2+}, Zn^{2+}, and Pt^{4+} preferentially inhibit the phagocytic activity of alveolar macrophages—an effect possibly mediated by alteration in energy metabolism as evidenced by depressed total cellular ATP levels (Fig. 29). Selective interference in DNA synthesis is a result of Pt^{4+} exposure of human lung fibroblasts; and secondary alterations of RNA and protein biosynthesis are also observed. Concomitant decreases in all three biosynthetic activities result from exposure to Cd^{2+} and V^{+5}. Unlike the case of Pt^{4+}, the reversibility of the biosynthetic inhibitory effects of the latter two metals is highly dependent upon the concentration of the initial exposure (*394*).

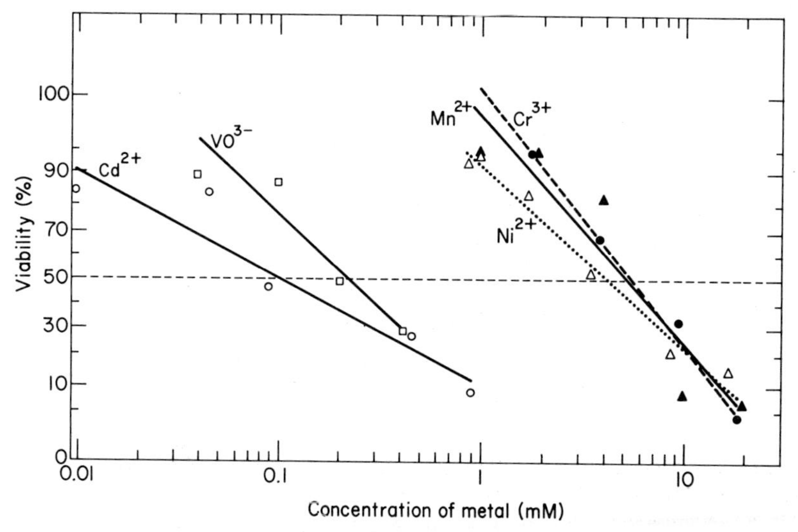

Figure 29. Influence of metallic chlorides and ammonium vanadate on viability of rabbit alveolar macrophages in *in vitro* exposure for 20 hours (*393*).

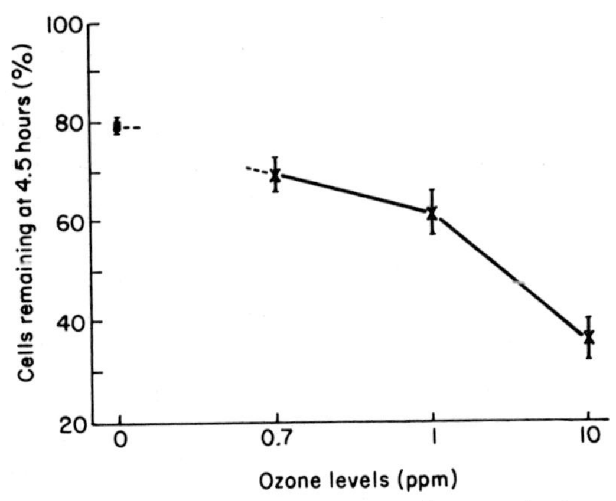

Figure 30. Dose–response change in percent of normal alveolar macrophages when maintained for 4.5 hours in normal lung lavage fluid (LWS^{-1}) following *in vitro* exposure to ozone for 30 min. ■ = normal cells in normal LWS^{-1}; × = normal cells in ozone LWS^{-1} (*44*).

D. Effect of Ozone on the Noncellular Milieu of the Lung

According to Gil and Weibel (*396*), the alveolar lining of surface active material also encloses the macrophages. Gardner (*43*) and Gardner *et al.* (*93*) have examined the possibility that this or similar substances may mediate the action of ozone on the pulmonary macrophage. In these experiments, they were able to isolate a material in the lung lavage of normal rabbits that prolonged the *in vitro* life of alveolar macrophages beyond that of artificial media. However, when lavage fluid from rabbits previously exposed to ozone was employed, no prolongation was apparent. This result could be duplicated *in vitro* by exposure of the lavage fluid or its extract to ozone (Fig. 30). This "protective factor" was located in the same centrifuge fraction as the surface active material, though no reduction of the surface activity by ozone was noted in these experiments. Lentz and DiLuzio have reported similar effects on the protective factor by the action of cigarette smoke (*397*).

E. Discussion of Interaction of Air Pollutants and Biological Agents

Frank manifestations of disease are uncommon in environmental toxicology at levels of toxicants commonly found in nature. Work with infective systems indicates that profound alteration of the lungs' defenses against introduced bacteria can be effected by levels of oxidant pollutants well within the range of those occurring in polluted ambient air. While the bacterial system has been most sensitive in mice, there is reason to believe that the difference in levels of pollutants required to infect other animals—rats, hamsters, squirrel monkeys, and even man—may lie more in lack of virulence of the bacterial invaders selected than in any basic difference in the action of pollutants on the bacterial defense mechanisms. For instance, while mechanistic studies have yielded decrements in many bacterial defense mechanisms in rabbits, no organism capable of regularly infecting the lung acutely with the production of subsequent mortality has been studied with this species; and despite a high virulence for mice by other routes, pneumococci have shown a low potential for the production of mortality when the exposure was by means of an aerosol. The difference in threshold for nitrogen dioxide after a 2-hour exposure in the mouse and hamster infected by *K. pneumoniae* at 3.5 ppm and 35 ppm, respectively, may be best explained by the very low susceptibility of the hamster to this organism (*130*).

It would indeed appear that differences in response (i.e., infection and subsequent death) to the combined treatment with the various oxidant pollutants and infectious agents among the various species of animals hinge more on the virulence and invasiveness of the infectious organism chosen than on the ability of the gas to attack the defense systems of the lung. The *sine qua non* for the successful operation of this system as presently used in animals is the availability of a microorganism that is capable of invading the lung with the production of subsequent mortality. The pollutant then enhances the mortality by depressing the host defense mechanism.

Man is commonly susceptible to several bacteria that cause acute pneumonia, notably *D. pneumoniae* and *K. pneumoniae,* and to influenza virus. These agents have thus far been effective in producing infection in animals in the model under discussion. It appears reasonable to speculate that exposure to sufficient ozone, oxidant from photochemical smog, or nitrogen dioxide derived from combustion from various sources might predispose human beings to such acute infections of the lung. A report of an association between high ambient nitrogen dioxide values and human influenza is available (*398*).

The fact that increased susceptibility to acute bacterial infection is associated with subtle anatomical and chemical lesions in the terminal airway raises the point that exposure to oxides of nitrogen may be suspected in the causation of chronic obstructive lung disease in human beings.

VIII. Real and Synthetic Atmospheres

A. Effects of Temperature and Humidity

Elevations in environmental temperature are known to enhance the toxicity of certain gases. Stokinger (*2*) has reported that raising the environmental temperature from 75° to 90°F will double the toxicity of ozone for rats and mice. Middleton *et al.* (*399*) found that LC_{50} values for peroxyacylnitrate (PAN) were reduced from 128 to 92 ppm when the temperature was increased from 70° to 90°F.

Trujillo and Spalding (*400*) reported reduced survival from irradiation injury in mice exposed to cold, and Miraglia and Berry (*401*) have described the activation of the bacterial carrier state from resident streptococcus. Coffin and Blommer (*402*) have shown an enhancement of mortality in mice by cold exposures at 50°F in association with ozone and streptococcus treatment. The data suggest that this additional mortality was a result of interaction rather than simply an additive effect. The same investigators have also noted enhancement of mortality and shortening of the course to fatal termination when mice were exposed to a temperature of 93°F as compared to those exposed at 75°F in a system employing interaction of ozone and streptococcus aerosol (*403*).

Swann *et al.* (*404, 405*), in a study of pulmonary function of guinea pigs, have noted increased resistance on cold, damp days. It thus appears that while the data in this field of investigation are extremely limited and tentative at this time, there is certainly enough evidence of the adverse effect of temperature variations in both the epidemiological and experimental data to warrant further animal experimentation.

B. Real Atmospheres

Study of the effect of naturally polluted atmospheres on experimental animals appears on first inspection to be a feasible and promising type of experimentation in the air pollution field. However, the toxicological problems imposed by exposure of animals to ambient air limit the fruitfulness of such study. Few of the conditions of man likely to be attributable to air pollutants are of the type for which obvious signs will be induced by simple exposure of small animals. These conditions are minimal in their clinical signs or complicated in their pathogenesis. Evaluation in animals is difficult, since detectable effect can only be possible at the highest possible dose, i.e., that present in the atmosphere at the time of exposure, and it is impossible to go to higher levels to plumb the possibility of establishing clear-cut dose effects. For situations that reflect the complex disease state of man in which air pollutants are possibly interacting with other environmental influences, such as smoking, one is hard pressed to find suitable animal models.

Studies have consisted chiefly of comparing colonies of various normal laboratory animals exposed to ambient air, to like animals exposed at the same site to filtered air. Ambient air studies have been carried out on two natural atmospheres (a) the photochemical smog of Los Angeles, California, and (b) the mixed urban atmosphere of Detroit, Michigan. These atmospheres differ in many ways, particularly in total oxidant content, which was high in Los Angeles and relatively low in Detroit.

Changes in total respiratory flow resistance in guinea pigs exposed to ambient Los Angeles air during their entire lifetime have been reported by Swann *et al.* (*405*). This was not a continuous phenomenon, but was noted at certain periods associated with peak values for total oxidant of 0.30 ppm or higher. They employed a modification of Admur and Mead's technique (*406*) using a face mask pneumotachograph and a body chamber plethysomograph and imposing sine wave pressure superimposed upon the breathing flow pattern. Also noted by Swann *et al.* (*405*) were statistically significant increases in expiratory flow resistance in animals in the filtered air room. This occurred when values for oxidant in outside air were 0.51 and 0.56 ppm, respectively, and carbon monoxide was 38 ppm. Since oxidant and the larger hydrocarbons were removed by the filters, the authors suggest that some other factor may have been responsible for the difference. In their study, certain animals had no response, some responded maximally but showed quick recovery, while others responded maximally but had slow recoveries or no recovery and death. The exaggerated response in certain animals could be explained by demonstrable abnormality of the lung, while in others, equally responding, no lesions were detectable, and the response seemed unrelated to impairment of the lungs. Resistance changes appeared to be greater in older animals in accord with previous reports by Swann *et al.* (*405*) noting gradual increase for pulmonary resistance in guinea pigs with age, regardless of exposure history. Meteorological variables appeared to influence the resistance measurements of guinea pigs, since values were higher on cold, damp days and appeared to decrease as the temperature increased, even during smoggy days. However, when oxidant levels reached 0.30 ppm, resistance increased despite elevated temperature.

Another finding from ambient air exposure is that of Bils (*407*) by electron microscopy, who has shown an interesting sequence of alterations in the wall cells (type II pulmonary alveolar epithelia) lining the pulmonary alveoli of mice. This has consisted of fragmentations in the so-called lamellar bodies in the wall cells after the animals have been exposed to ambient air. These changes appear more severe as the length of exposure is increased and become appreciable in the 2-year-old mice. Trial exposures of animals to synthetic mixtures simulating auto exhaust atmosphere but containing higher levels of oxidant have yielded similar changes, resulting in frank wall cell destruction. The authors' hypothesis is that the basis for the change is acceleration in the transformation of wall cell mitochondria into the so-called lamellar bodies or inclusions brought about by increased demand for surface-active lining material of the alveoli to replace that damaged by the air pollutant. This increased demand for surfactant results in frank pathological alteration or disaese of the alveoli in the older animals or at higher levels of pollutants.

C. Synthetic Atmospheres

Methods for producing simulated smog atmospheres for the exposure of laboratory animals, which today seem rather crude, are summarized in "Motor Vehicles, Air Pollution, and Health" (*408*). Ozonized gasoline vapors were used by Kotin and Falk (*409*) and Kotin and Thomas (*410*) for exposure of mice. This mixture which, according to the authors, simulated naturally formed smog in properties such as oxidant and eye irritation, was shown not only to be carcinogenic but also to have other biological effects on mice, notably reduction of growth rate, lowered fertility, and reduced neonatal survival. Boche and Quilligan (*411*), using a similar atmosphere, showed a reduction of spontaneous activity in mice housed in freely turning wheels. The temporary nature of this change was indicated by a resumption of normal activity in the experimental subject when the atmosphere was changed from smog to clean air. The same authors showed that when the smog atmosphere was imposed on animals previously treated with ANTU (a pulmonary edema-producing agent) mortality was enhanced over mice treated with ANTU only.

In the search for an atmosphere more realistically resembling photochemical smog, Stephens (*412*) found that a substance resembling natural smog in its major components—presence of oxidant, phytotoxicity, and aerosol haze—could be produced in the laboratory by irradiating diluted automobile exhaust with artificial light. This principle has been used to produce smog in chambers to establish a constant concentration for exposure purposes by Rose and Brandt (*413*) and a cycled or variable concentration in a dynamic flow system by Hinners (*414*). The acute effects of these atmospheres on pulmonary function in guinea pigs have been studied by Murphy *et al.* (*415*) and Murphy (*416*) employing the multianimal test system of Amdur and Mead (*406*), as modified by Murphy and Ulrich (*132*). In these studies, the concentration of exhaust gases in the experimental atmospheres was varied between levels that approximated ambient community levels and concentrations several times greater (air/exhaust ratios ranging from 1350:1 to 150:1 and total oxidant from 0.33 to 0.82 ppm). Exposure to irradiated diluted exhaust atmospheres consistently produced statistically significant decreases in respiratory rates proportional to the exhaust concentration. Total expiratory flow resistance was increased to twice the baseline level and persisted for the entire 4-hour exposure period at the highest level of exhaust, while smaller, but statistically significant, increases occurred during the first hour of exposure at levels simulating ambient levels. All pulmonary physiological alterations promptly returned to baseline levels after the cessation of exposure to the irradiated air–exhaust mixture. The same authors also reported on the influence of this mixture on spontaneous activity of mice, its effects on ANTU-dosed animals, and biochemical measurements. Individual mice had relatively consistent patterns of activity to which changed activity after exposure could be related. Exposure to the irradiated diluted auto exhaust caused reduced activity at the higher dilutions, which were stated by the authors to be about those of community air pollution levels. Activity returned to normal when the animals were removed to clean air. No significant differences were found in biochemical measurements (alkaline phosphatase, glutamic–oxaloacetic transaminase, cholinesterase, and oxygen consumption of lung slices) when animals were exposed for brief periods to exhaust levels approximating three to eight times those of community levels; however, increased carboxyhemoglobin levels were found. No great difference in mortality was noted when the exhaust was superimposed on animals previously treated with ANTU.

In a system employing bacterial interaction, Coffin and Blommer (*403*) showed enhancement of mortality in mice by a 4-hour exposure to irradiated diluted automobile exhaust above a threshold value of the dilution of the exhaust yielding 25 ppm carbon monoxide and 0.15 ppm oxidant.

Biological effects of a 2-year exposure to both irradiated and nonirradiated automobile exhaust atmospheres have been reported by Hueter *et al.* (*125*). For this study, four concentrations of auto exhaust–air mixtures were subjected to cyclic irradiation in four separate chambers so that diurnal peaks for oxidant and other constituents would reflect those seen in naturally produced auto smog. Peak concentrations ranged from 100 to 20 ppm carbon monoxide and from 1.0 to 0.2 ppm oxidant at the entrance to the animal exposure chambers. Comparable values for nitric oxide were 2.0 to 0.4 and nitrogen dioxide, 1.9, to 0.3 ppm. Trough values were approximately one-fourth of the peak values. At the site of animal exposure within the chambers, concentrations were reduced approximately

80% for oxidant, 10% for nitric oxide and 50% for nitrogen dioxide due to losses in the system.

Despite these rather low effective levels for oxidant, nitric oxide and nitrogen dioxide, guinea pigs showed a greater incidence, severity, and mortality to spontaneous bacterial pneumonias in irradiated exhaust as compared to nonirradiated exhaust or pure air. Also, reduction in spontaneous activity in mice as measured in freely turning wheels occurred, which persisted for approximately 3 weeks before returning to the baseline figure. This pattern suggested stress and adaptation. However, when the experiment was repeated and exposure to irradiated diluted exhaust was continued beyond 11 weeks, a significant decrease again occurred in LAF mice. According to Hueter *et al.* (*125*), this secondary depression of activity indicated a time-dependent detrimental effect after the initial adaptation period.

Reduction in fertility and infant survival were noted in the LAF mice, the only strain that reproduced under the conditions of the study. Mice exposed to the two highest concentrations of irradiated diluted exhaust showed two-fold diminution in pups born. The frequency of litters per mother and the number of infants surviving to weaning were also much reduced. This effect appeared to be independent of litter size, since no significant change occurred in the number of young per gestation.

Biochemical measurements for lipid showed reductions (expressed as micromoles per gram of dry lung weight) for all diluted irradiated auto exhaust treatments. The authors indicate that this parameter reflects a gain in dry lung weight, suggesting an increase in nonfunctional or abnormal lung tissue. Other biochemical parameters measured were unproductive, with the exception of a statistically significant elevation in bone lead at the highest concentration of nonirradiated diluted exhaust.

A study specifically designed to elucidate the alterations in fertility and infant survival mentioned above has been reported by Lewis *et al.* (*126*). In this study, preconditioning of males and females took place according to an experimental plan designed to show the relative importance of the male or female in the sterility factor. Incidence of nonfertility was 6% in females bred to nonexposed males, while an incidence of 13% was noted in females bred to males conditioned by exposure to irradiated diluted auto exhaust. An effect similar to that of the previous study was noted in infant survival.

D. Discussion of Real and Synthetic Atmospheres

Biological effects in exposed animals have been observed when exhaust concentrations and irridation have been sufficient to produce levels of oxidant and other constituents equal to peak ambient levels in both ambient and synthesized atmospheres. While certain of the manifestations appear transient, real toxic effects leading to serious alteration or death are observed in interaction with certain other stresses, or in young or aged animals. Inhibition of defense mechanisms by irradiated diluted exhaust resulted when animals were subsequently exposed to streptococcal aerosols. Electron microscopically visualized alterations in the alveolar wall cells suggest a stress on these cells beyond the aged individual's adaptation ability. Similar impairment probably accounts for the exaggerated pulmonary physiological response in aged animals. Small but probably significant nonspecific stress in normal individuals may be indicated by the pattern of spontaneous activity alterations, which simulates a stress adaptation–deadaptation sequence. The possibility of a simple stressing effect in terms of Selye's adaptation syndrome may also be inferred from the apparent additive effect, in terms of augmented ketosteroid excretion, when polluted ambient air was coupled with a frank psychomotor stress. Effect of synthetic atmospheres on survival of newborn animals again indicates stress on an animal delicately balanced in its ecological milieu.

From the evidence to date, it would appear that while exposures to both ambient and synthesized irradiated atmospheres have yielded evidences of toxicity to small animals, little appreciable toxic manifestations of nonirradiated diluted exhaust have been observed.

IX. Air Pollution and Domestic Animals

The influence of air pollutants on free-living domestic animals has been of interest because of the economic value of animals per se as well as their possible value when employed as monitors in the interest of human health. Because of their location in rural areas, few economic livestock are at maximum risk from urban air pollution. A considerable number of dairy cattle and chickens, however, are kept close enough to metropolitan complexes to suffer some degree of exposure to pollution. On the other hand, pet animals are at the same risk as their owners in cities and elsewhere. Animals have been reported to have suffered deleterious effects in the exposures in Donora, Pennsylvania, and Poza Rica, Mexico, (*417*, *418*), and also in the 1953 London, England, episode (*417*). In the London episode, which occurred at the same period as the Smithfield Cattle Show, exhibited cattle suffered considerable illness and mortality. Sickness and mortality in the above episodes was attributed to respiratory embarrassment.

A systematic survey conducted on city dogs in Columbus, Ohio, showed an association of anthracosis and pulmonary fibrosis with dust fall for the various localities (*419*). Several studies have been conducted on animals in the San Francisco district of California (*420*).

Air pollutants that have caused serious or widespread effects on livestock are confined chiefly to two substances—fluorine (fluoride) and arsenic. Of the two, fluorine has caused more widespread damage and has proved more troublesome to control.

A. Fluorine (Fluoride)

1. Sources of Fluoride

Sources of airborne fluoride are several kinds of heavy chemical industry, chief among which are the manufacture or production of phosphate fertilizer, aluminum, fluorinated hydrocarbons (refrigerants, aerosol propellants), fluorinated plastics (e.g., polytetrafluoroethylene), uranium, and certain other heavy metals. Phosphate rock, as the source of phosphate fertilizer, contains from 3.5 to 4% fluoride, one-third to one-half of which is evolved in the manufacture of the fertilizer, primarily as silicon tetrafluoride, which subsequently hydrolyzes to hydrogen fluoride as a gas or mist. In aluminum production from bauxite (Al_2O_3), cryolite (Na_3AlF_6) is fused as the electrolyte in amounts of 1 ton for every 270 tons of Al_2O_3. Fluorine is liberated chiefly as hydrogen fluoride and silicon tetrafluoride. Fluorinated hydrocarbon manufacture starts with hydrogen fluoride, fluorine, or some chlorinated hydrocarbon. In uranium processing, free elemental fluorine is used to prepare the intermediate uranium hexafluoride, with liberation into the air of fluorine and hydrogen fluoride. Scrubbing of fluorine with sodium hydroxide solution results, in part, in liberation of gaseous OF_2.

2. Fluoride Toxicity

The toxic effects of fluoride in livestock arise from ingesting forage on which the various fluoride forms have either settled or with which they have reacted with consequent accumulation to levels far greater than occur in the ambient air. Livestock, being foraging animals, may ingest relatively large quantities of fluoride. Heavily contaminated forage, however, acts as a brake on the intake of acutely toxic amounts by inducing voluntary refusal of such feed (*421*). Hence, acute fluoride poisoning in livestock under grazing conditions is rarely, if ever, seen. Rather, a lag period of a duration depending on the degree of fluoride contamination, fluoride type, and animal species is a dominant feature

Table XVII Results of Long-Term Intake by Cattle of Sodium Fluoride or Drinking Water Containing 5 ppm Fluorine

	Years of intake			
	2	*5.5*	*6*	*10*
Result	*mg F/kg body weight*	*ppm F in fodder*	*mg F/kg body weight*	*mg F/kg body weight*
"No effect" concentration	<1.0	—	>0.3	—
Tolerated concentration	1.5	<20	1	1
Toxic concentration	>2.5	40	1.5	—

of fluoride intoxication of livestock. Lag periods are commonly a matter of years. The understanding of chronic fluoride intoxication is greatly aided if one recognizes three characteristics of its metabolism:

a. Absorption. Poorly absorbed fluorine compounds (Na_3AlF_6, CaF_2) are tolerated better, dose for dose, than more soluble compounds (NaF), which are readily absorbed from the gastrointestinal tract into the blood (Tables XVII and XVIII).

b. Excretion. Elimination of absorbed fluoride proceeds through the kidney, being characterized by high rate of clearance almost independent of the fluoride concentration in the blood. An upper renal excretion limit is reached, however, when levels of fluoride become toxic to the kidneys.

c. Fixation in skeletal structures. Fluoride is fixed in the apatite crystals of bone and accumulates according to a quadratic exponential function (*422*).

From studies made in man, sodium fluoride is nearly completely absorbed, whereas calcium fluoride from bone meal, cryolite, and rock phosphate are absorbed respectively 62, 37, 77, and 87% (*423*). About

Table XVII Results of Long-Term Intake by Cattle of Sodium Fluoride or Drinking Water Containing 5 ppm Fluorine

	Years of intake			
	2.8	*3.5*	*4*	*5*
Result	*ppm F in fodder*	*mg F in fodder*	*mg F/kg body weight*	*ppm F in fodder*
"No effect" concentration	<50	25	<0.1	—
Tolerated concentration	<200	100	2.5–3.6[a]	—
Toxic concentration	300	—	5.5	<260

[a] The higher value refers to mineral phosphate from the Christmas Islands.

50% of the absorbed fluoride is excreted by the kidney, almost all of the remainder going to the skeleton; trace amounts only are retained by all soft tissues, except the kidneys, which tend to accumulate fluoride.

Irrespective of the nature of the fluoride, the effects are the same—an abnormal calcification of bone (hyperostoses) and tooth structures (fluorosis), owing to a large increase in fluoride in these structures. Milk production is also decreased. Animals lose weight, acquire a stiff posture, become lame, and the hair coat becomes rough. The pathophysiology of fluorosis and its relation to bone and tooth fluoride content are discussed elsewhere (*422, 424*).

3. Fluorine Tolerances

Safe levels of dietary fluorine have been experimentally determined for livestock for both soluble and less soluble fluoride forms (Table XIX) (*425*). The tolerances are given as ranges, because other dietary factors, such as minerals, magnesium, aluminum, and iron, influence the deposition of fluoride in bone. Species differences are to be noted. Normal concentrations of fluoride in bones of adult animals is of the order of 500 ppm; pathological changes in bone are just detectable at around 3000 ppm fluorine and changes detectable by X rays at and above 4000 ppm. The mean fluoride content of bone at which changes are uniformly observed is 5000 ppm (*422*).

B. Arsenic

Although more common in the past, gross contamination of forage by airborne arsenicals is still occurring (*421*). One of the earliest reports on

Table XIX Safe Levels of Fluorine in Total Ration of Livestock *(425)*

	Fluorine source	
Animal	*Sodium fluoride or other soluble fluoride (ppm)*	*Phosphatic limestone or rock phosphate (ppm)*
Dairy cow	30–50	60–100
Beef cow	40–50	65–100
Sheep	70–100	100–200
Swine	70–100	100–200
Chicken	150–300	300–400
Turkey	300–400	—

this subject, by Harkins and Swain (*426*), deals with widespread poisoning of cattle, horses, and sheep, which first became noticeable in 1902 at Anaconda, Montana. The epizoobiological evidence was the following: 625 sheep, of a flock of 3500, died after feeding on vegetation 15 miles (24 km) from a copper smelter after having been brought into the area from 28 miles (45 km) distant. Analysis of the grass and moss on which the animals fed showed 52 and 405 ppm arsenic trioxide (As_2O_3), respectively. Furthermore, horses kept in an area remote from the smelter but that fed on hay grown in a location on which smelter fumes had fallen, died. Arsenic analysis of the hay showed 285 ppm arsenic trioxide. The livers of the animals showed 1.3 mg arsenic trioxide per 100 gm of tissue. Finally, arsenic was found in large quantities (10–150 ppm) on all types of vegetation near the smelter; copper analysis of the same vegetation showed equally high values (128–1800 ppm). Analysis of eighty-two specimens comprising eight tissues from calves, cows, sheep, and horses feeding in the area showed values for arsenic ranging from traces (liver) to several hundred parts per million (hair). By the method of analysis used, control tissues failed to show the presence of arsenic. Although there was a real question of the involvement of copper in the poisoning cases, the signs and symptoms of poisoning that were most carefully evaluated agreed in all details with those of acute and chronic arsenic poisoning, and the diagnosis was confirmed as well by both gross and microscopic pathology. Experimental verification was also made by feeding animals graded doses of arsenic and then analyzing the tissues. A complete organ analysis of a horse chronically administered As is given. From 0.2 to 5.7 ppm arsenic trioxide was found in nine samples of milk from cows that had been feeding in that area. The arsenic content of ulcers of the nose of horses from the smelter regions showed values as high as 1015 ppm. The tail-hair of a horse feeding in the Anaconda region showed 58 ppm arsenic trioxide; that of a colt, 605 ppm. The content of the liver and bone was from one-tenth to one-hundredth that of the hair.

REFERENCES

1. "Air Quality Criteria for Photochemical Oxidants", U.S. Department of Health, Education, and Welfare, National Air Pollution Control Administration Publ. No. AP-63. Washington, D.C., 1970.
2. H. E. Stokinger, *Arch. Environ. Health* **10,** 719 (1965).
3. L. S. Jaffe, *Amer. J. Pub. Health* **57,** 1269 (1967).
4. H. E. Stokinger and D. L. Coffin, *in* "Air Pollution" (A. C. Stern, ed.), Vol. 1, pp. 445–546. Academic Press, New York, New York, 1968.
5. D. L. Coffin, E. J. Blommer, D. E. Gardner, and R. Holzman, *Proc. 3rd Ann. Conf. Atmos. Contam. Confined Spaces,* Dayton, Ohio pp. 71–80 (1968).

6. R. A. Jones, L. J. Jenkins, R. A. Coon, and J. Siegel, *Toxicol. Appl. Pharmacol.* **17,** 189 (1970).
7. R. Brinkman, H. B. Lamberts, and T. S. Venings, *Lancet* **1,** 133 (1964).
8. S. Watanabe, R. Frank, and E. Yokoyama, *Amer. Rev. Resp. Dis.* **108,** 1141 (1973).
9. R. N. Matzen, *Amer. J. Physiol.* **190,** 84 (1957).
10. S. D. Murphy, C. E. Ulrich, S. H. Frankowitz, and C. Xintaras, *Amer. Ind. Hyg. Ass. J.* **25,** 246 (1964).
11. A. Y. S. P'an, J. Beland, and Z. Jegier, *Arch. Environ. Health* **24,** 229 (1972).
12. R. J. Stephens, M. F. Sloan, M. J. Evans, and G. Freeman, *Exp. Mol. Pathol.* **20,** 11 (1974).
13. R. J. Stephens, M. F. Sloan, M. J. Evans, and G. Freeman, *Amer. J. Pathol.* **74,** 31 (1974).
14. D. E. Gardner, T. Lewis, S. Alpert, D. Hurst, and D. L. Coffin, *Arch. Environ. Health* **25,** 432 (1972).
15. M. J. Evans, W. Mayr, T. Bils, and C. Loosli, *Arch. Environ. Health* **22,** 450 (1971).
16. S. M. Alpert, D. E. Gardner, D. J. Hurst, T. R. Lewis, and D. L. Coffin, *J. Appl. Physiol.* **31,** 247 (1971).
17. A. R. Dowell, L. Lohrbaver, D. Hurst, and S. Lee, *Arch. Environ. Health* **21,** 121 (1970).
18. R. F. Bils, *Arch. Environ. Health* **20,** 468 (1970).
19. C. J. Dillard, N. Urribarri, K. Reddy, B. Fletcher, S. Taylor, B. de Lumen, S. Langberg, and A. L. Tappel, *Arch. Environ. Health* **25,** 426 (1972).
20. M. S. Palmer, R. Exley, and D. L. Coffin, *Arch. Environ. Health* **25,** 439 (1972).
21. S. Miller and R. Ehrlich, *J. Infec. Dis.* **103,** 145 (1958).
22. G. C. Buell, Y. Tokiwa, and P. K. Mueller, *Arch. Environ. Health* **10,** 213 (1969).
23. L. D. Scheel, O. J. Dobrogorski, J. T. Mountain, J. L. Svirbely, and H. E. Stokinger, *J. Appl. Physiol.* **14,** 67 (1959).
24. D. L. Coffin, D. E. Gardner, and R. S. Holzman, *Arch. Environ. Health* **16,** 633 (1968).
25. H. E. Stokinger, W. D. Wagner, and O. J. Dobrogorski, *AMA Arch. Ind. Health* **16,** 514 (1957).
26. J. N. Roehm, J. Hadley, and D. Menzel, *Arch. Environ. Health* **24,** 237 (1972).
27. R. J. Stephens, G. Freeman, J. F. Stara, and D. L. Coffin, *Amer. J. Pathol.* **73,** 711 (1973).
28. G. A. Bobb and E. J. Fairchild, II, *Toxicol. Appl. Pharmacol.* **11,** 558 (1967).
29. S. D. Murphy, H. V. Davis, and V. L. Zaratzian, *Toxicol. Appl. Pharmacol.* **6,** 520 (1964).
30. S. Werthamer, S. H. Schwarz, J. J. Carr, and L. Soskind, *Arch. Environ. Health* **20,** 16 (1970).
31. R. E. Easton and S. D. Murphy, *Arch. Environ. Health* **15,** 160 (1967).
32. R. S. Holzman, D. E. Gardner, and D. L. Coffin, *J. Bacteriol.* **96,** 1562 (1968).
33. B. D. Goldstein, *Arch. Environ. Health* **26,** 279 (1973).
34. R. G. Skillen, C. H. Thienes, J. Cangelosi, and L. Strain, *Proc. Soc. Exp. Biol. Med.* **107,** 178 (1961).
35. R. G. Skillen, C. H. Thienes, J. Cangelosi and L. Strain, *Proc.. Soc. Exp. Biol. Med.* **108,** 121 (1961).
36. B. D. Goldstein, B. Pearson, C. Lodi and R. D. Buckley, *Arch. Environ. Health* **16,** 648 (1968).

37. B. D. Goldstein and O. J. Balchum, *Toxicol. Appl. Pharmacol.* **27,** 330 (1974).
38. B. D. Goldstein, R. D. Buckley, R. Cordinos, and O. J. Balchum, *Science* **169,** 605 (1970).
39. E. G. Trams, C. Lauter, E. Brown, and O. Young, *Arch. Environ. Health* **24,** 153 (1972).
40. B. D. Goldstein, M. R. Levine, R. Cuzzi-Spada, R. Cardenas, R. Buckley, and O. Balchum, *Arch. Environ. Health* **24,** 243 (1972).
41. S. Mittler, D. Herrick, M. King, and A. Gaynor, *Ind. Med. Surg.* **25,** 301 (1956).
42. J. L. Svirbeley and B. E. Saltzman, *AMA Arch. Ind. Health* **15,** 111–118 (1957).
43. D. E. Gardner, Ph.D. Thesis, University of Cincinnati, Cincinnati, Ohio (1971).
44. D. L. Coffin and D. E. Gardner, *Ann. Occup. Hyg.* **15,** 219 (1972).
45. S. Watanabe, F. Yokoyama, E. S. Boatman, and R. Frank, *Amer. Rev. Resp. Dis.* **108,** 1141–1151 (1973).
46. E. S. Boatman and R. Frank, *Chest* **65,** Suppl., 9S-11S (1974).
47. M. J. Evans, L. J. Cabral, R. T. Stephens, and G. Freeman, *Amer. J. Pathol.* **70,** 199 (1973).
48. G. Freeman, R. J. Stephens, S. C. Crane, and N. J. Furiosi, *Arch. Environ. Health* **17,** 181 (1968).
49. R. J. Stephens, G. Freeman, and M. J. Evans, *Arch. Environ. Health* **24,** 160 (1972).
50. R. J. Stephens, G. Freeman, S. C. Crane, and N. J. Furiosi, *Exp. Mol. Pathol.* **14,** 1 (1971).
51. C. G. Plopper, D. L. Dungworth, and W. S. Tyler, *Amer. J. Pathol.* **71,** 375 (1973).
52. C. G. Plopper, D. L. Dungworth, and W. S. Tyler, *Amer. J. Pathol.* **71,** 395 (1973).
53. G. Freeman, R. J. Stephens, D. L. Coffin, and J. F. Stara, *Arch. Environ. Health* **26,** 209 (1973).
54. G. Freeman, L. T. Juhos, N. J. Furiosi, R. Mussenden, R. Stevens, and M. J. Evans, *Arch. Environ. Health* **29,** 203–210 (1974).
55. J. R. Dixon and J. T. Mountain, *Toxicol. Appl. Pharmacol.* **7,** 756 (1965).
56. R. M. Mendenhall and J. R. Dixon, unpublished results (1966).
57. E. J. Fairchild, II, G. A. Bobb, and G. E. Thompson, *Fed. Proc., Fed. Amer. Soc. Exp. Biol.* **25,** 692 (abstr.) (1966).
58. E. J. Fairchild, II, S. D. Murphy, and H. E. Stokinger, *Science* **130,** 861 (1959).
59. J. B. Mudd, R. Leavitt, R. Ogren, and T. T. McManus, *Atmos. Environ.* **3,** 669 (1969).
60. J. T. Mountain, *Arch. Environ. Health* **6,** 357 (1963).
61. M. E. King, Doctoral Dissertation, Delaware Institute of Technology, University Microfilms, Ann Arbor, Michigan (1961).
62. D. B. Menzel, *Arch. Environ. Health* **23,** 149 (1971).
63. A. J. De Lucia, P. M. Hoque, M. G. Mustafa, and C. E. Cross, *J. Lab. Clin. Med.* **80,** 559 (1972).
64. G. M. Green, *Science* **162,** 810 (1968).
65. D. J. Hurst and D. L. Coffin, *Arch. Intern. Med.* **127,** 1059 (1971).
66. H. V. Thomas, P. K. Mueller, and R. L. Lyman, *Science* **159,** 532 (1968).
67. B. D. Goldstein, C. Lodi, C. Collinson, and O. Balchum, *Arch. Environ. Health* **18,** 631 (1969).
68. J. N. Roehm, J. G. Hadley, and D. B. Menzel, *Arch. Environ. Health* **23,** 142 (1971).

69. D. B. Menzel, R. Slaughter, D. Donovan, and A. Bryant, *Pharmacologist* **15**, (abstr.) (1973).
70. R. Criegee, *Rec. Chem. Progr.* **18**, 111 (1957).
71. D. B. Menzel, R. J. Slaughter, A. M. Bryant, and H. O. Jauregui, *Arch. Environ. Health* **30**, 234 (1975).
72. R. H. Fetner, *Res. Eng.*, pp. 11–13 (1960).
73. R. E. Zelac, H. L. Cromroy, W. E. Bolch, Jr., B. G. Dunavant, and H. A. Bevis, *Environ. Res.* **4**, 262 (1971).
74. R. Brinkman and H. B. Lamberts, *Nature (London)* **181**, 504 (1958).
75. D. C. Peterson and H. L. Andrews, *Radiat. Res.* **19**, 33 (1963).
76. K. Hattori, N. Kato, M. Konoshita, S. Kimoshita, and T. Sunada, *Nature (London)* **198**, 1220 (1963).
77. G. C. Buell, E. Jeung, and W. Ferminger, "Chemical Changes in Respiratory Tissue Following Ozone Exposure," Proc. West Coast Sect., p. 140. Air Pollution Control Association, Monterey, California, 1963.
78. E. J. Fairchild, II, *Arch. Environ. Health* **6**, 79 (1963).
79. E. J. Fairchild, II, S. L. Graham, R. Killins, and L. D. Scheel, *Toxicol. Appl. Pharmacol.* **6**, 607 (1964).
80. E. J. Fairchild, II and G. A. Bobb, *Toxicol. Appl. Pharmacol.* **7**, 483 (abstr.) (1965).
81. T. R. Vaughan, L. F. Jennett, and T. R. Lewis, *Arch. Environ. Health* **19**, 45 (1969).
82. E. Yokoyama and R. Frank, *Arch. Environ. Health* **25**, 132 (1972).
83. C. E. McJilton, J. Thielk, and R. Frank, *Amer. Ind. Hyg. Ass., J.* **33**, 20 (1972).
84. A. S. Konigsberg and C. H. Bachman, *Int. J. Biometeorol.* **14**, 261 (1970).
85. T. R. Vaughan, W. J. Moorman, and T. R. Lewis, *Toxicol. Appl. Pharmacol.* **20**, 404 (1971).
86. R. Frank, *Nat. Res. Counc. Can.*, Publ. NRCC **14096**, 185–204 (1975).
87. S. M. Alpert and T. R. Lewis, *J. Appl. Physiol.* **31**, 243 (1971).
88. R. Frank, J. D. Brain and D. E. Sherry, *Amer. Ind. Hyg. Ass., J.* **31**, 31 (1970).
89. E. Yokoyama, *Amer. Rev. Resp. Dis.* **105**, 594 (1972).
90. D. Bartlett, Jr., *J. Appl. Physiol.* **37**, 92 (1974).
91. R. M. Mendenhall and H. E. Stokinger, *J. Appl. Physiol.* **17**, 28 (1962).
92. E. Yokoyama, *Proc. Jap. Soc. Air Pollut.* p. 123 (1972).
93. D. E. Gardner, E. A. Pfitzer, R. T. Christian, and D. L. Coffin, *Arch. Intern. Med.* **127**, 1078 (1971).
94. D. V. Bates and M. Hazucha, "The Short Term Effects of Ozone on the Human Lung". *in* Proceedings of Conference on Health Effects of Air Pollutants. National Academy of Sciences, p. 507, Committee on Public Works, U.S. Senate, Committee Print 93-15, 93rd Congress, 1st Session, U.S. Government Printing Office, Washington, D.C., Nov. 1973.
95. S. S. Griswold, L. A. Chambers, and H. L. Motly, *AMA Arch. Ind. Health* **15**, 108 (1957).
96. S. S. Clamann and R. W. Bancroft, *Adv. Chem. Ser.* **21**, 352 (1959).
97. G. Bennett, *Aerosp. Med.* **33**, 969 (1962).
98. J. M. Lagerwerff, *Aerosp. Med.* **34**, 479 (1963).
99. W. A. Young, D. B. Shaw, and D. V. Bates, *J. Appl. Physiol.* **19**, 765 (1964).
100. W. Y. Hallett, *Arch. Environ. Health* **10**, 295 (1965).
101. J. R. Goldsmith and J. A. Nadel, *J. Air Pollut. Cont. Ass.* **19**, 329 (1969).

101a. D. V. Bates, G. M. Bell, C. D. Burnam, M. Hazucha, J. Mantha, L. D. Pengelly, and F. Silverman, *J. Appl. Physiol.* **32,** 176 (1972).
102. M. Hazucha, F. Silverman, C. Parent, S. Field, and D. V. Bates, *Arch. Environ. Health* **27,** 183 (1973).
102a. F. J. Miller, Ph.D. Thesis, North Carolina State University, Raleigh, North Carolina, unpublished (1977).
103. R. E. Zelac, H. L. Cromroy, W. E. Bolch, Jr. *et al., J. Appl. Physiol.* **14,** 67 (1959).
104. R. Brinkman and H. B. Lamberts, *Nature* (*London*) **181,** 1202 (1958).
105. H. E. Stokinger, *AMA Arch. Ind. Health* **15,** 181 (1957).
106. O. S. Atwal and T. Wilson, *Arch. Environ. Health* **28,** 91 (1974).
107. D. E. Gardner, J. W. Illing, F. J. Miller, and D. L. Coffin. *Res. Commun. Chem. Pathol. Pharmacol.* **9,** 689–700 (1974).
108. A. Abudonia, D. E. Gardner, D. L. Coffin, and D. B. Menzel, United States Environmental Protection Agency, Research Triangle Park, North Carolina (unpublished data).
109. D. Harmon, *J. Gerontol.* **23,** 476 (1968).
110. H. R. Stokinger and L. D. Scheel, *Arch. Environ. Health* **4,** 327 (1962).
111. E. J. Fairchild, II, *Arch. Environ. Health* **6,** 79–85 (1963).
112. D. Henschler, E. Hahn, and W. Assman, *Naunyn-Schmiedebergs Arch. Exp. Pathol. Pharmakol.* **249,** 325 (1964).
113. E. J. Fairchild, II, *Arch. Environ. Health* **14,** 111 (1967).
114. D. Henschler, E. Hahn, H. Heymann, and H. Wunder, *Naunyn-Schmiedebergs Arch. Exp. Pathol. Pharmakol.* **249,** 343 (1964).
115. S. M. Alpert, B. B. Schwartz, S. D. Lee, and T. R. Lewis, *Arch. Intern. Med.* **128,** 69 (1971).
116. D. L. Coffin and D. E. Gardner, *Gig. Sanit.* **1,** 80–92 (1975).
117. M. J. Evans, R. J. Stephens, L. J. Cabral, and G. Freeman, *Arch. Environ. Health* **24,** 180 (1972).
118. D. M. Pace, P. A. Landoldt and B. T. Aftonomos, *Arch. Environ. Health* **18,** 165 (1969).
119. E. Christensen and A. C. Giese, *Arch. Biochem. Biophys.* **51,** 208 (1954).
120. R. Prat, C. Nofre, and H. Cier, *Ann. Inst. Pasteur, Paris* **114,** 595 (1968).
121. I. Davis, "Microbiological Studies with Ozone. Mutagenesis of Ozone for *Escherichia coli,*" U.S.A.F. School of Aerospace Medicine Report No. SAM-TR-61-66. Brooks Air Force Base, San Antonio, Texas, 1961.
122. N. Sachsenmaier, W. Siebs, and Tjong-an-Tan, *Z. Krebsforsch.* **67,** 112 (1967).
123. R. H. Fetner, *Nature* (*London*) **194,** 793 (1962).
124. R. H. Fetner, *Nature* (*London*) **181,** 504 (1958).
125. F. G. Hueter, G. L. Contner, K. A. Busch, and R. G. Hinners, *Arch. Environ. Health* **12,** 553 (1966).
126. T. R. Lewis, F. G. Hueter, and K. A. Busch, *Arch. Environ. Health* **15,** 26 (1967).
127. W. S. Wayne, P. F. Wehrle, and R. E. Carroll, *J. Amer. Med. Ass.* **199,** 901 (1967).
127a. "Proceedings of the First Japan/US Conference on Photochemical Air Pollution, Tokyo, Japan, 1973."
128. D. E. Gardner, J. W. Illing, F. J. Miller, and D. L. Coffin, *Toxicol. Appl. Pharmacol.* **29,** 129 (1974).
129. H. V. Thomas, P. K. Mueller, and R. L. Lyman, *Science* **150,** 532 (1968).

130. R. Ehrlich, *Bacteriol. Rev.* **30,** 604 (1966).
131. G. von Nieding and H. Krekeler, *Int. Arch. Arbeitsmed.* **29,** 55 (1972).
132. S. D. Murphy and C. E. Ulrich, *Amer. Ind. Hyg. Ass., J.* **25,** 28 (1964).
133. A. R. Dowell, K. H. Kilburn, and P. C. Platt, *Arch. Intern. Med.* **128,** 74 (1971).
134. P. K. Mueller and M. Hitchcock, *J. Air Pollut. Contr. Ass.* **19,** 670 (1969).
135. W. H. Blair, M. C. Henry, and R. Ehrlich, *Arch. Environ. Health* **18,** 186 (1969).
136. R. Ehrlich and M. C. Henry, *Arch. Environ. Health* **17,** 860 (1968).
137. W. D. Wagner, B. R. Duncan, P. G. Wright, and H. E. Stokinger, *Arch. Environ. Health* **10,** 455 (1965).
138. R. J. Stephens, G. Freeman, S. C. Crane, and N. J. Furiosi, *Exp. Mol. Pathol.* **14,** 1 (1971).
139. O. J. Balchum, R. D. Buckley, R. Sherwin, and M. Gardner, *Arch. Environ. Health* **10,** 274 (1965).
140. J. Kleinerman and G. W. Wright, *Amer. Rev. Resp. Dis.* **83,** 423 (1961).
141. C. P. McCord, P. C. Harrold, and S. F. Meek, *J. Ind. Hyg. Toxicol.* **23,** 200 (1941).
142. J. Mersch, B. J. Dyce, B. J. Haverback, and R. P. Sherwin, *Arch. Environ. Health* **27,** 94 (1973).
143. R. P. Sherwin and D. A. Carlson, *Arch. Environ. Health* **27,** 90 (1973).
144. C. Chen, S. Kusumoto, and T. Nakajima, *Proc. Osaka Pref. Inst. Pub. Health, Ed. Ind. Health,* **10,** 1–9 (1972).
145. T. Nakajima, C. Chen, S. Kosumoto, and K. Okamoto, *Proc. Osaka Pref. Inst. Pub. Health, Ed. Ind. Health* **7,** 35 (1969).
146. S. Hattori, R. Tateishi, T. Horai, T. Nakajima, and T. Miura, *Jap. J. Thorac. Dis.* **10,** 16 (1972).
147. T. Nakajima and S. Kusumoto, *Proc. Osaka Pref. Inst. Pub. Health, Ed. Ind. Health* **6,** 17 (1968).
148. G. Freeman, N. J. Furiosi, and G. B. Haydon, *Arch. Environ. Health* **13,** 454 (1966).
149. S. Kosmider and A. Misiewicz, *Int. Arch. Arbeitsmed.* **31,** 249 (1973).
150. J. Fenters, R. Ehrlich, J. Findlay, J. Spangler, and V. Tolkatz, *Amer. Rev. Resp. Dis.* **104,** 448 (1971).
151. L. S. Mitina, *Gig. Sanit.* **27,** 3 (1962).
152. N. J. Furiosi, S. C. Crane, and G. Freeman, *Arch. Environ. Health* **27,** 405 (1973).
153. R. P. Sherwin, J. Dibble, and J. Weiner, *Arch. Environ. Health* **24,** 43 (1972).
154. R. P. Sherwin, J. B. Morgolick, and S. P. Azen, *Arch. Environ. Health* **26,** 297 (1973).
155. E. C. Arner and R. A. Rhoades, *Arch. Environ. Health* **26,** 156 (1973).
156. M. C. Henry, J. Findlay, J. Spangler, and R. Ehrlich, *Arch. Environ. Health* **20,** 566 (1970).
157. A. M. Giordano and P. E. Morrow, *Arch. Environ. Health* **25,** 433 (1972).
158. R. P. Sherwin, V. Richters, M. Brooks, and R. D. Buckley, *Lab. Invest.* **18,** 269 (1968).
159. J. T. Davidson, G. A. Lillington, G. B. Haydon, and K. Wasserman, *Amer. Rev. Resp. Dis.* **95,** 790 (1967).
160. G. Freeman and G. B. Haydon, *Arch. Environ. Health* **8,** 125 (1964).
161. G. Freeman, S. C. Crane, R. J. Stephens, and N. J. Furiosi, *Arch. Environ. Health* **18,** 609 (1969).

162. R. D. Buckley and O. J. Balchum, *Arch. Environ. Health* **10,** 220 (1965).
163. R. D. Buckley and O. J. Balchum, *Arch. Environ. Health* **14,** 424 (1967).
164. R. D. Buckley and O. J. Balchum, *Arch. Environ. Health* **14,** 687 (1967).
165. G. Freeman, S. C. Crane, N. J. Furiosi, R. J. Stephens, M. J. Evans, and W. D. Moore, *Amer. Rev. Resp. Dis.* **106,** 564 (1972).
166. J. H. Riddick, K. I. Campbell, and D. L. Coffin, *Am. Soc. Clin. Pathol.* **11,** 87 (abstr.) (1967).
167. J. Kleinerman and C. R. Cowdrey, *Yale J. Biol. Med.* **40,** 579 (1968).
168. G. Freeman, S. C. Crane, R. J. Stephens, and N. J. Furiosi, *Yale J. Biol. Med.* **40,** 566 (1968).
169. G. Freeman, S. C. Crane, and R. J. Stephens, *Amer. Rev. Resp. Dis.* **98,** 429 (1968).
170. I. Y. R. Adamson and D. H. Bowden, *Lab. Invest.* **30,** 35 (1974).
171. E. R. Weibel, *Arch. Intern. Med.* **128,** 54 (1971).
172. R. J. Stephens, G. Freeman, and M. J. Evans, *Arch. Intern. Med.* **127,** 873 (1971).
173. C. Aranyi, C. D. Port, and D. L. Coffin, "Scanning Electron Microscopy Examination of the *in vivo* Effects of Air Pollutants on Pulmonary Tissue," Contract Report 68-02-0761, United States Environmental Protection Agency, Research Triangle Park, North Carolina, 1974.
174. G. Freeman, S. C. Crane, and N. J. Furiosi, *Amer. Rev. Resp. Dis.* **100,** 662 (1969).
175. G. Freeman, L. T. Juhos, N. J. Furiosi, R. Mussenden, and T. Weiss, *Amer. Rev. Resp. Dis.* **110,** 754 (1974).
176. C. D. Port, D. L. Coffin, P. Kane, and K. V. Ketels, *J. Toxicol. Environ. Health* **2,** 589 (1977).
177. S. Svorcova and V. Kant, *Cesk Hyg.* **16,** 71 (1971).
178. M. C. Henry, R. Ehrlich, and W. H. Blair, *Arch. Environ. Health* **18,** 580 (1969).
179. R. A. Williams, R. A. Rhoades, and W. S. Adams, *Arch. Intern. Med.* **128,** 101 (1971).
180. C. D. Cook, J. Mead, G. L. Schreiner, N. R. Frank, and J. M. Craig, *J. Appl. Physiol.* **14,** 177 (1959).
181. G. D. von Nieding, H. Krekeler, R. Fuchs, H. M. Wagner, and K. Koppenhagen, *Int. Arch. Arbeitsmed.* **31,** 61 (1973).
182. G. D. von Neiding, M. Wagner, H. Krekeler, U. Smidt, and K. Muysers, *Int. Arch. Arbeitsmed.* **27,** 234 (1970).
183. G. von Nieding, M. Wagner, H. Krekeler, U. Smidt, and K. Muysers, *Int. Arch. Arbeitsmed.* **27,** 338 (1971).
184. M. Tusl, V. Stolin, H. M. Wagner, and D. Ast, *in* "Adverse Effects of Environmental Chemicals and Psydiotropic Drugs: Quantitative Interpretation of Functional Tests," Proc. Int. Assoc. Occuptl. Health Study Group in Functional Toxicity (M. Horvath, ed.), Vol. I, p. 155–160. Elsevier, New York, New York, 1973.
185. H. M. Wagner, "Investigation of the Biological Effects of Nitrogen Oxides," Institute Feier, Wasser, Boden und Lufthygiene des Bundesgesundheitsamtes, Berlin, Federal Republic of Germany, 1973.
186. E. L. MacQuiddy, L. W. LaTowsky, J. P. Tollman, and A. I. Finlayson, *J. Ind. Hyg. Toxicol.* **23,** 134 (1941).
187. W. H. Pryor, H. L. Bitter, and R. J. Bertler, "The Effect of Nitrogen Dioxide—Nitrogen Tetroxide on Oxyhemoglobin Disassociation," SAM-TR-67-33. USAF

School Aerospace Medicine, Brooks Air Force Base, San Antonio, Texas, 1967.
188. H. Oda, S. Kusumoto, and T. Nakajima, *Arch. Environ. Health.* **30,** 453–55 (1975).
189. R. M. Estefan, E. M. Gause, and J. R. Rowlands, *Environ. Res.* **3,** 62 (1970).
190. D. B. Menzel, *in* "Free Radicals in Biology" (W. A. Prior, ed.), Vol. II, p. 181–202. Academic Press, New York, New York, 1976.
191. G. C. Buell, Y. Tokiwa, and P. K. Mueller, "Lung Collagen and Elastin Denaturation *In vivo* Following Inhalation of Nitrogen Dioxide," APCA Paper No. 66–7. Air Pollution Control Association, Pittsburgh, Pennsylvania, 1966.
192. R. P. Sherwin, S. Winnick, and R. D. Buckley, *Amer. Rev. Resp. Dis.* **96,** 319 (1967).
193. R. D. Buckley and C. G. Loosli, *Arch. Environ. Health* **18,** 588 (1969).
194. E. L. Gray, F. M. Patton, S. B. Goldberg, and E. Kaplan, *AMA Arch. Ind. Hyg. Occup. Med.* **10,** 423 (1954).
195. L. W. LaTowsky, E. L. MacQuiddy, and J. P. Tollman, *J. Ind. Hyg. Toxicol.* **23,** 129 (1941).
196. D. L. Coffin, D. E. Gardner, and E. J. Blommer, *Toxicol. Appl. Pharmacol.* **29,** 130 (abstr.) (1974).
197. M. R. Purvis and R. Ehrlich, *J. Infec. Dis.* **113,** 72 (1963).
198. C. H. Hine, R. D. Cavalli, and R. R. Wright, unpublished results (1968).
199. H. G. Boren, *Environ. Res.* **1,** 178 (1967).
200. J. Kleinerman and G. W. Wright, "The Effects of Prolonged and Repeated Nitrogen Dioxide Inhalation on the Lungs of Rabbits and Guinea Pigs," 5th Ann. Air Pollut. Med. Res. Conf. Berkeley, California, California State Dept. of Public Health, 1961.
201. W. S. R. Spicer, *U.S. Pub. Health Serv., Publ.* **1022,** 126–136 (1963).
202. C. M. Fletcher, *Amer. Rev. Resp. Dis.* **80,** 483 (1959).
203. P. J. Lawther, *Proc. Roy. Soc. Med.* **51,** 262 (1958).
204. P. M. Lambert and D. D. Reid, *Lancet* **1,** 853 (1970).
205. P. J. Lawther, R. E. Walker, and M. Henderson, *Thorax* **25,** 525 (1970).
206. J. G. French, G. Lowrimore, W. C. Nelson, J. F. Finklea, T. English, and M. Hertz, *Arch. Environ. Health* **27,** 129 (1973).
207. R. W. Buechley, W. B. Riggan, V. Hasselblad and J. B. Van Bruggen, *Arch. Environ. Health* **27,** 134 (1973).
208. Y. Alarie, R. J. Kantz, II, C. E. Ulrich, A. A. Krumm, and W. M. Busey, *Arch. Environ. Health* **27,** 251 (1973).
209. Y. Alarie, C. E. Ulrich, W. M. Busey, H. E. Swann, Jr., and N. H. MacFarland, *Arch. Environ. Health* **21,** 769 (1970).
210. M. O. Amdur, *Int. J. Air Water Pollut.* **1,** 170 (1959).
211. K. A. Bushtueva, E. F. Polezhaev, and A. P. Semenenko, *Gig. Sanit.* **25,** 57 (1960); *in* U.S.S.R. Literature on Air Pollution and Related Occupational Diseases" (trans. by B. S. Levine), Vol. 7, p. 137. NTIS 62-11103 U.S. Dept. of Commerce, Office of Technical Services, Washington, D.C., 1962.
212. F. I. Dubrovskaya, "Limits of Allowable Concentration of Atmospheric Pollutants," Book 3. Medgiz, Moscow, 1957; *in* "U.S.S.R. Literature on Air Pollution and Related Occupational Diseases—A Survey" (trans. by B. S. Levine), Vol. 1, p. 37. NTIS 59-21175 U.S. Dept. of Commerce, Office of Technical Services, Washington D.C., 1960.

213. A. F. Gunnison and E. D. Palmes, *Amer. Ind. Hyg. Ass., J.* **35,** 288 (1974).
214. K. A. Bushtueva, *Gig. Sanit.* **22,** 17 (1957).
215. K. A. Bushtueva, *in* "U.S.S.R. Literature on Air Pollution and Related Occupational Diseases—A Survey" (trans by B. S. Levine), Vol. 1, p. 63. NTIS 60 21049 U.S. Dept. of Commerce, Office of Technical Services, Washington, D.C., 1960.
216. N. R. Frank and F. E. Speizer, *Physiologist* **7,** 132 (1964).
217. O. J. Balchum, J. Dybicki, and G. R. Meneely, *Fed. Proc., Fed. Am. Soc. Exp. Biol.* **18,** 6 (1959).
218. I. Anderson, G. R. Lundquist, P. L. Jensen, and D. F. Proctor, *Arch. Environ. Health* **28,** 31 (1974).
219. Y. Tomono, *Jap. J. Ind. Health* **3,** 77 (1961).
220. T. Toyama, *Arch. Environ. Health* **8,** 153 (1964).
221. A. Zarkower, *Arch. Environ. Health* **25,** 45 (1972).
222. J. A. Nadel, D. F. Tierney, and J. H. Comroe. *Proc. Air Pollut. Med. Res. Conf., 3rd, 1959,* California State Dept. Health, Berkeley, California, pp. 66–74 (1960).
223. Y. Alarie, W. M. Busey, A. A. Krumm, and C. E. Ulrich, *Arch. Environ. Health* **27,** 17 (1973).
224. R. E. Pattle and F. Burgess, *J. Pathol. Bacteriol.* **73,** 411 (1957).
225. T. R. Lewis, W. J. Moorman, W. F. Ludmann, and K. I. Campbell, *Arch. Environ. Health* **26,** 16 (1973).
226. N. R. Frank and F. E. Speizer, *Arch. Environ. Health* **11,** 624 (1965).
227. P. J. Lawther, *J. Inst. Fuel* **36,** 241 (1963).
228. A. F. Gunnison and E. D. Palmes, *Toxicol. Appl. Pharmacol.* **24,** 266 (1973).
229. W. E. Giddens, Jr. and G. A. Fairchild, *Arch. Environ. Health* **25,** 166 (1972).
230. J. C. O'Donoghue and F. E. Graesser, *Can. J. Comp. Med. Vet. Sci.* **26,** 255 (1962).
231. S. W. Martin and R. A. Willoughby, *Arch. Environ. Health* **25,** 158 (1972).
232. C. O. T. Ball, R. M. Heyssel, O. Balchum, G. O. Elliott, and G. R. Meneely, *Physiologist* **3,** 15 (1960).
233. K. A. Bushtueva, *in* "Biological Effects and Hygienic Significance of Atmospheric Pollutants" (V. A. Ryazanov and D. O. Goldberg, eds.), Book 1/9; *in* "U.S.S.R. Literature on Air Pollution and Related Occupational Diseases" (trans. by B. S. Levine), Vol. 16, p. 168. U.S. Dept. of Commerce, Office of Technical Services, Washington, D.C., 1968.
234. J. F. Treon, F. P. Dutra, J. Cappel, H. Sigmon, and W. Younker, *Arch. Ind. Hyg. Occup. Med.* **2,** 716 (1950).
235. K. Mathur, Ph.D. Dissertation, Harvard School of Public Health, Boston, Massachusetts (1948).
236. M. D. Thomas, R. H. Hendricks, F. D. Gunn, and J. Critchlow, *AMA Arch. Ind. Health* **17,** 70 (1958).
237. K. A. Bushtueva, *in* "Limits of Allowable Concentrations of Atmospheric Pollutants" (Transl. by B. S. Levine), Book 5. U.S. Dept. of Commerce, Clearing house for Scientific and Technical Information, Springfield, Virginia, 1962.
238. S. Laskin, M. Kuschner, and R. T. Drew in "Inhalation Carcinogenesis Symposium," Ser. No. 18. U.S. Atomic Energy Commission, Washington, D.C., 1970.
239. S. Laskin, G. Katz, and M. Kuschner, "The Role of Nitrogen Dioxide in Multi-

factorial Chemical Carcinogenesis," Abstract Report of the Carcinogenesis Program, National Cancer Institute, U.S. Department of Health, Education, and Welfare, Washington, D.C., 1971.
240. S. Laskin and A. Sellikumar, *in* "Experimental Lung Cancer, Carcinogenesis, and Bioassays" (E. Karbe and J. F. Park, eds.), pp. 7–19. Springer-Verlag, Berlin, Federal Republic of Germany and New York, New York, 1974.
241. O. J. Balchum, J. Dybicki, and G. R. Meneely, *AMA Arch. Ind. Health* **21,** 564 (1960).
242. O. J. Balchum, J. Dybicki, and G. R. Meneely, *Amer. J. Physiol.* **197,** 1317 (1959).
243. N. R. Frank, R. E. Yoder, J. D. Brain, and E. Yokoyama, *Arch. Environ. Health* **18,** 315 (1969).
244. T. Dalhamn and L. Strandberg, *Int. J. Air Water Pollut.* **4,** 154 (1961).
245. T. Dalhamn and J. Sjoholm, *Acta Physiol. Scand.* **58,** 287 (1963).
246. F. Speizer and N. R. Frank, *Arch. Environ. Health* **12,** 725 (1966).
247. L. G. Strandberg, *Arch. Environ. Health* **9,** 160 (1964).
248. T. A. Bystrova, *Gig. Sanit.* **22,** 30 (1957); *in* "U.S.S.R. Literature on Air Pollution and Related Occupational Diseases—A Survey" (trans. by B. S. Levine), Vol. 1, p. 89. NTIS 60-21049 U.S. Dept. of Commerce, Office of Technical Services, Washington, D.C., 1960.
249. M. O. Amdur and J. Mead, *Nat. Air Pollut. Symp. 3rd, 1955* U.S. Dept. of Health, Education, and Welfare, Washington, D.C., p. 150 (1955).
250. M. O. Amdur, *Arch. Environ. Health* **12,** 729 (1966).
251. E. Yokoyamo and K. Ishikawa, *Jap. J. Ind. Health* **4,** 22 (1962).
252. M. Corn, N. Kotsko, D. Stanton, W. Bell, and A. P. Thomas, *Arch. Environ. Health* **24,** 248 (1972).
253. M. O. Amdur, "Inhaled Particles and Vapors" (C. N. Davies, ed.), p. 281. Pergamon, Oxford, England, 1961.
254. M. O. Amdur, *Amer. Ind. Hyg. Ass., Quart.* **18,** 149 (1957).
255. M. O. Amdur and M. Corn, *Amer. Ind. Hyg. Ass. J.* **24,** 326 (1963).
256. M. O. Amdur and D. Underhill, *Arch. Environ. Health* **16,** 460 (1968).
257. M. O. Amdur, *Pub. Health Rep.* **69,** 503 (1954).
258. M. O. Amdur, *Int. J. Air Pollut.* **3,** 201 (1960).
259. M. O. Amdur, *Proc. Amer. Phil. Soc.* **114,** 3 (1970).
260. C. McJilton, N. R. Frank, and R. Charlson, *Science* **182,** 503 (1973).
261. L. V. Cralley, *J. Ind. Hyg. Toxicol.* **24,** 193 (1942).
262. T. Dalhamn, *Acta Physiol. Scand.* **36,** Suppl. 123, 1 (1956).
263. T. Dalhamn and J. Rohdin, *Brit. J. Ind. Med.* **13,** 110 (1956).
264. G. A. Fairchild, P. Kane, B. Adams, and D. L. Coffin, *Arch. Environ. Health,* **30,** 538–545 (1975).
265. A. Y. S. P'an and X. Jegier, *Arch. Environ. Health* **21,** 498 (1970).
266. A. F. Gunnison and A. W. Benton, *Arch. Environ. Health* **22,** 381 (1971).
267. J. S. Douglas, P. Ridgway, and M. W. Dennis, *Arch. Environ. Health* **18,** 627 (1969).
268. J. M. Charles and D. B. Menzel, *Arch. Environ. Health* **30,** 314 (1975).
269. V. A. Ryazanov, *Arch. Environ. Health* **5,** 479 (1962).
270. K. A. Bushtueva, *in* "Limits of Allowable Concentration of Atmospheric Pollutants" (trans. by S. Levine), Book 3. U.S. Dept. of Commerce, Clearing house for Scientific and Technical Information, Springfield, Virginia, 1959.
271. I. V. Popov, Y. E. Cherkoson, and O. L. Trakhtman, *Gig. Sanit.* **5,** 16 (1952);

in "U.S.S.R. Literature on Air Pollution and Related Occupational Diseases—A Survey" (trans. by B. S. Levine), Vol. 3. p. 102. NTIS 60-21475 U.S. Dept. of Commerce, Office of Technical Services, Washington, D.C., 1960.

272. "Determination of Odor Threshold for 52 Commercially Important Organic Compounds," a report by Arthur D. Little to Manufacturing Chemists Ass., Washington, D.C., 1968.
273. K. A. Bushtueva, *in* "Limits of Allowable Atmospheric Pollutants—A Survey" (trans. by B. S. Levine), Vol. 3. U.S. Dept. of Commerce, Clearinghouse for Scientific and Technical Information, Springfield. Virginia, 1961.
274. K. A. Bushtueva, *in* "Limits of Allowable Concentration of Atmospheric Pollutants" (trans. by B. S. Levine), Book 5. U.S. Dept. of Commerce, Clearinghouse for Scientific and Technical Information, Springfield, Virginia, 1962.
275. M. O. Amdur, W. W. Melvin, and P. Drinker, *Lancet* **2,** 758 (1953).
276. M. O. Amdur, L. Silverman, and P. Drinker, *AMA Arch. Ind. Hyg. Occup. Med.* **6,** 305 (1952).
277. N. R. Frank, M. O. Amdur, J. Worcester, and J. L. Whittenberger, *J. Appl. Physiol.* **17,** 252 (1962).
278. N. R. Frank, *Proc. Roy. Soc. Med.* **59,** 1029 (1964).
279. J. A. Nadel, H. Salem, B. Tompkin, and Y. Takiwa, *J. Appl. Physiol.* **20,** 164 (1965).
280. G. N. Melville, *West Indian Med. J.* **19,** 231 (1970).
281. R. E. Snell and D. C. Luchsinger, *Arch. Environ. Health* **18,** 693 (1969).
282. F. E. Speizer and N. R. Frank, *Brit. J. Ind. Med.* **23,** 75 (1966).
283. G. G. Burton, M. Corn, J. B. L. Gee, D. Vossalo, and A. Thomas, *Arch. Environ. Health* **18,** 681–692 (1969).
284. T. Toyama, *Jap. J. Ind. Med.* **4,** 86 (1962).
285. T. Toyama and K. Nakamura, *Ind. Health* **2,** 34 (1964).
286. K. Nakamura, *Jap. J. Hyg.* **19,** 322 (1964).
287. L. Reid, *Brit. J. Exp. Pathol.* **44,** 437 (1963).
288. L. Reid, *Arch. Intern. Med.* **126,** 428 (1970).
289. "Air Quality Criteria for Sulfur Oxides," Nat. Air Pollu. Contr. Admin. Publ. No. AP-50, pp. 1–161. U.S. Government Printing Office, Washington, D.C., 1970.
290. D. L. Coffn, *in* "Environmental Science and Technology" (J. N. Pitts and R. L. Metcalf, eds.), Vol. 2, p. 1–38. Wiley, New York, 1971.
291. W. C. L. Hemeon, *AMA Arch. Ind. Health* **11,** 397 (1955).
292. H. H. Schrenk, H. Heimann, G. D. Clayton, W. M. Gafafer, and H. Wexler, *Pub. Health Bull.* **306,** (1949).

292a. R. Lindenberg, "An Experimental Investigation in Animals of the Functional and Morphological Changes from Single and Repeated Exposure to Carbon Monoxide," Annual Meeting Paper, American Industrial Hygiene Association, Akron, Ohio (1962).

293. N. P. Musselman, W. A. Groff, P. P. Yevich, F. T. Wilinski, M. H. Weeks, and F. W. Oberst, *Aerosp. Med.* **30,** 524 (1959).
294. A. Roussel, M. Stupfel, and G. Bouley, *Arch. Environ. Health* **18,** 613 (1969).
295. C. Xintaras, B. Johnson, C. Ulrich, R. Terrill, and M. Sobecki, *Toxicol. Appl. Pharmacol.* **8,** 77 (1966).
296. F. H. Lewey and D. L. Drabkin, *Amer. J. Med. Sci.* **208,** 502 (1944).
297. R. R. Beard and G. A. Wertheim, *Amer. J. Pub. Health* **57,** 2012 (1967).
298. P. Astrup, K. Kjeldsen, and J. Wanstrup, *J. Atheroscler. Res.* **7,** 343 (1067).

299. S. S. Wilks, J. F. Tomashefski, and R. T. Clark, *J. Appl. Physiol.* **14,** 305 (1959).
300. T. J. Preziosi, R. Lindenberg, D. Levy, and M. Christenson, *Ann. N.Y. Acad. Sci.* **174,** 369–384 (1970).
301. M. Stupfel, *Ann. N.Y. Acad. Sci.* **174,** 342–368 (1970).
302. K. C. Back and A. M. Dominguez, "Psychopharmacology of Carbon Monoxide Under Ambient and Altitude Conditions," AMRL-TR-68-175. Aerospace Medical Research Laboratories, Wright-Patterson Air Force Base, Dayton, Ohio, 1968.
303. W. H. Forbes, F. Sargent, and F. J. W. Roughton, *Amer. J. Physiol.* **147,** 352 (1946).
304. R. G. Smith, *J. Occup. Med.* **10,** 456 (1968).
305. M. H. Halperin, R. A. McFarland, J. I. Niven, and F. J. W. Roughton, *J. Physiol.* (*London*) **146,** 583 (1959).
306. R. A. McFarland, D. J. W. Roughton, M. H. Halperin, and J. I. Niven, *J. Aviat. Med.* **15,** 381 (1944).
307. J. H. Schulte, *Arch. Environ. Health* **7,** 524 (1963).
308. R. R. Beard and N. Grandstaff, *Proc. Conf. Biol. Effects Carbon Monoxide, 1970* New York Academy of Sciences, New York, New York (1970).
309. R. R. Beard and N. Grandstaff, *Ann. N.Y. Acad. Sci.* **174,** 396 (1970).
310. R. D. Stewart, J. E. Peterson, E. D. Baretta, R. T. Bachand, M. J. Hosko, and A. A. Herrmann, "Experimental Human Exposure to Carbon Monoxide," Contract CRC-APRAC, Proj. No. CAPM-3-68. Coordinating Research Council, Inc., New York, New York, 1969; see also *Arch. Environ. Health* **21,** 154 (1970).
311. W. Bender, M. Gothert, and G. Malony, *Staub-Reinhalt. Luft* **32,** 54 (1972).
312. E. Groll-Knapp, H. Wagner, H. Hauck, and M. Haider, *Staub-Reinhalt. Luft* **32,** 64 (1972).
313. C. C. Fodor and G. Winneke, *Staub-Reinhalt. Luft* **32,** 46 (1972).
314. S. M. Ayres, S. Giannelli, Jr., and R. G. Armstrong, *Science* **149,** 193 (1965).
315. S. M. Ayres, H. Mueller, J. Gregory, S. Giannelli, Jr., and J. Penny, *Arch. Environ. Health* **18,** 699 (1969).
316. "Air Pollution, Air Quality Criteria for Carbon Monoxide," Report of the Expert Panel on Air Quality Criteria, North Atlantic Treaty Organization. Obtainable from Air Pollut. Tech. Info. Center, U.S. Environmental Protection Agency, Research Triangle Park, North Carolina, 1974.
317. "Air Quality and Automobile Emission Control," Report of the Coordinating Committee on Air Quality Studies, National Academy of Science, Serial No. 93-24, U.S. Senate Committee on Public Works Vol. 2, pp. 1–182. U.S. Govt. Printing Office, Washington, D.C., 1974.
318. E. W. Anderson, R. J. Andelman, J. M. Strauch, N. J. Fortrum, and J. H. Knelson, *Ann. Intern. Med.* **79,** 46 (1973).
319. W. S. Aronow and M. W. Isbell, *Ann. Intern. Med.* **79,** 392 (1973).
320. W. S. Aronow, E. A. Stemmer, and M. W. Isbell, *Circulation* **49,** 415 (1974).
321. U. Saffiotti, F. Cefis, L. H. Kolb, and P. Shubik, *J. Air. Pollut. Contr. Ass.* **15,** 23 (1965).
322. D. F. S. Natusch, J. R. Wallace, and C. N. Evans, *Science* **183,** 202–204 (1973).
323. D. F. S. Natusch and J. R. Wallace, *Science* **186,** 699 (1974).
324. P. E. Palm, J. M. McNerney, and T. Hatch, *AMA Arch. Ind. Health* **13,** 355 (1956).

325. G. A. Fairchild, *Appl. Microbiol.* **24,** 812 (1972).
326. L. Dautrebande and W. Walkenhorst, *in* "Inhaled Particles and Vapours" (C. N. Davies, ed.), Vol. 1, pp. 110–121. Pergamon, Oxford, England, 1961.
327. "Task group on Lung Dynamics Deposition and Selection Models for Internal Dosimetry of the Human Respiratory Tract," *Health Phys.* **12,** 173 (1966).
328. C. E. Brown, *J. Ind. Hyg. Toxicol.* **13,** 285 (1931).
329. C. E. Brown, *J. Ind. Hyg. Toxicol.* **13,** 293 (1931).
330. B. Altschuler, L. Yarmus, E. D. Palmes, and N. Nelson, *AMA Arch. Ind. Health* **15,** 293 (1957).
331. G. A. Fairchild, *Amer. Rev. Resp. Dis.* **109,** 446 (1974).
332. "Medical and Biological Effects of Environmental Pollutants," Committee on Medical and Biological Effects of Environmental Pollutants; Nickel. National Academy of Sciences, Washington, D.C., 1975.
333. "Medical and Biological Effects of Environmental Pollutants," Committee on Medical and Biological Effects of Environmental Pollutants: Chromium. National Academy of Sciences, Washington, D.C., 1974.
334. "Medical and Biological Effects of Environmental Pollutants," Committee on Medical and Biological Effects of Environmental Pollutants: Manganese. National Academy of Sciences, Washington, D.C., 1973.
335. "Medical and Biological Effects of Environmental Pollutants," Committee on Medical and Biological Effects of Environmental Pollutants: Vanadium. National Academy of Sciences, Washington, D.C., 1974.
336. "Medical and Biological Effects of Environmental Pollutants," Committee on Medical and Biological Effects of Environmental Pollutants: Particulate Polynuclear Organic Matter. National Academy of Sciences, Washington, D.C. 1972.
337. C. A. Nau, J. Neal, V. A. Stembridge, and R. N. Cooley, *Arch. Environ. Health* **4,** 415 (1962).
338. W. S. Smith and C. W. Gruber, "Atmospheric Emissions from Coal Combustion and Inventory Guide," Publ. 999-AP-24 Pub. Health Serv. U.S. Dept. of Health, Education, and Welfare, Washington, D.C. (1966).
339. U. Saffiotti, F. Cefis, L. H. Kolb, and M. J. Grote, *Proc. Amer. Ass. Cancer Res.* **4,** 59 (1963).
340. L. M. Shabad, L. W. Pylev, and T. S. Kolesnichenko, *J. Nat. Cancer Inst.* **33,** 135 (1964).
341. I. J. Selikoff, W. J. Nicholson, and A. M. Langer, *Arch. Environ. Health* **25,** 1 (1972).
342. W. G. Owen, *Ann. N.Y. Acad. Sci.* **132,** 674 (1965).
343. A. M. Langer, A. N. Rohl, and M. Wolf, "Secular Changes in the Crystaline Asbestos Content of the Greenland Ice Sheet, 1750–1920," 1st annual report to the Nat. Inst. Environ. Health Sci. for the Study of the Biological Effects of Environmental Agents. pp. 236–240. N.I.E.H.S., Research Triangle Park, North Carolina, 1974.
344. W. J. Smither and H. C. Lewinsohn, "Working Group Meeting on Biological Effects of Asbestos," Preprint 26. Int. Agency Cancer, Lyon, France, 1972.
345. M. Navratil and F. Trippe, *Environ. Res.* **5,** 210 (1972).
346. K. Miller and J. S. Harrington, *Brit. J. Exp. Pathol.* **53,** 397 (1972).
347. J. M. G. Davis, *Brit. J. Exp. Pathol.* **53,** 190 (1972).
348. J. S. Harrington, A. C. Allison, and D. V. Badami, *in* "Advances in Pharmacology and Chemotherapy," Vol. 12 (S. Grattini, A. Goldin, F. Hawkins, and I. J. Kopin, eds.), p. 292. Academic Press, New York, New York, 1975.

349. A. J. Vorworld, T. M. Durkan, and P. C. Pratt, *Arch. Ind. Hyg. Occup. Med.* **3**, 1 (1951).
350. P. Gross, R. T. P. de Treville, E. B. Talker, M. Kaschak, and M. A. Babya, *Arch. Environ. Health* **15**, 343 (1967).
351. P. Gross, R. T. P. de Treville, and L. J. Cralley, *in* "Pneumoconiosis" (H. A. Shapiro, ed.), p. 220. Oxford Univ. Press, London, England, and New York, New York, 1970.
352. J. C. Wagner, G. Berry, J. W. Skidmore, and V. Timbrell, *Brit. J. Cancer Res.* **29**, 252 (1974).
353. J. C. Wagner, *Nature (London)* **196**, 180 (1962).
354. J. C. Wagner and G. Berry, *Brit. J. Cancer* **23**, 567 (1969).
355. J. S. Harrington, "Working Group on Biological Effects of Asbestos," Preprint 45. Int. Agency Res. Cancer, Lyon, France, 1972.
356. J. M. G. Davis, *J. Nat. Cancer Inst.* **52**, 1823 (1974).
357. H. E. Whipple, ed., "Biological Effects of Asbestos," *Ann. N.Y. Acad. Sci.* **132**, Part 1. N.Y. Acad. Sci., New York, New York, 1965.
358. J. C. Wagner, *Proc. Cont. Pneumoconiosis, Johannesburg, South Africa* (A. J. Grenstien, ed.), pp. 373–382. Churchill, London, England, 1960.
359. J. C. Wagner, C. A. Sleggs, and P. Marchand, *Brit. J. Ind. Med.* **17**, 260, (1960).
360. I. Webster, *Ann. N.Y. Acad. Sci.* **132**, 623 (1965).
361. M. L. Newhouse and H. Thompson, *Ann. N.Y. Acad. Sci.* **132**, 579 (1965).
362. P. C. Elmes and O. L. Wade, *Ann. N.Y. Acad. Sci.* **132**, 549 (1965).
363. D. O'B. Hourihane, *Ann. N.Y. Acad. Sci.* **132**, 196 (1965).
364. J. G. Thompson, *Ann. N.Y. Acad. Sci.* **132**, 196 (1965).
365. L. M. Shabad, L. N. Pylev, L. V. Krivosheeva, T. F. Kulagina, and B. A. Nemenko, *J. Nat. Cancer Inst.* **52**, 1175 (1974).
366. L. M. Pylev, F. J. Roe, and G. P. Warwick, *Brit. J. Cancer* **23**, 103 (1969).
367. J. C. Wagner, *Cancer Res.* **39**, 37 (1972).
368. F. H. Top, *Arch. Environ. Health* **9**, 699 (1964).
369. W. W. Holland and D. D. Reid, *Lancet* **1**, 445 (1965).
370. J. W. Douglas and R. E. Waller, *Brit. J. Prev. Soc. Med.* **20**, 1 (1966).
371. J. E. Lunn, J. Knowelden, and A. J. Handyside, *Brit. J. Prev. Soc. Med.* **21**, 7 (1967).
372. E. Ronzoni, *Arch. Hyg.* **67**, 287 (1908).
373. E. G. Stillman, *J. Exp. Med.* **28**, 117 (1923).
374. G. A. Laurenzi, J. Guarneri, R. Endriga, and J. Carey, *Science* **142**, 1572 (1963).
375. G. A. Laurenzi, L. Berman, M. Finst, and E. H. Kass, *J. Clin. Invest.* **43**, 759 (1964).
376. G. A. Green and E. H. Kass, *J. Clin. Invest.* **43**, 769 (1964).
377. G. A. Green and E. H. Kass, *Brit. J. Exp. Pathol.* **46**, 360 (1965).
378. R. Rylander, *Arch. Environ. Health* **18**, 551 (1969).
379. H. B. McDonnell, *J. Ass. Offic. Agr. Chem.* **13**, 19 (1930).
380. D. L. Coffin, D. E. Gardner, and E. J. Blommer, *Environ. Health Perspect.* **13**, 11 (1976).
381. F. F. Hemphill, M. L. Kaeberle, and W. B. Buck, *Science* **172**, 1031–1032 (1971).
382. D. E. Gardner, J. Illing, and J. M. Kurtz, *Proc. Fed. Amer. Soc. Exp. Biol.* **36**(3), 211 (1976).

383. D. L. Coffin and E. J. Blommer, *Arch. Environ. Health* **15,** 36 (1967).
384. D. L. Coffin, "The Relationship of Infectious Agents and Air Pollutants," Work. Doc. Inter-Reg. Symp. Air Quality Guides. World Health Organization, Geneva, Switzerland, 1970.
385. R. Z. Maigetter, J. D. Fenters, D. E. Gardner, and R. Ehrlich, *Environ. Res.* **11,** 386 (1976).
386. D. P. Port, J. D. Fenters, R. Ehrlich, D. L. Coffin and D. E. Gardner, *Environ. Health Perspect.,* **10,** 268 (abstr.) (1975).
387. D. E. Gardner, R. S. Holzman, and D. L. Coffin, *J. Bacteriol.* **98,** 1041 (1969).
388. D. J. Hurst, D. E. Gardner, and D. L. Coffin, *J. Reticuloendothel. Soc.* **8,** 288 (1970).
389. E. Bingham, E. A. Pfitzer, W. Barkley, and E. P. Radford, *Science* **162,** 1297 (1968).
390. G. M. Green and D. Carolin, *N. Engl. J. Med.* **276,** 421 (1967).
391. M. D. Waters, D. E. Gardner, and D. L. Coffin, *Toxicol. Appl. Pharmacol.* **28,** 253 (1974).
392. M. D. Waters, D. E. Gardner, C. Aranyi, and D. L. Coffin, *Environ. Res.* **9,** 32 (1975).
393. M. D. Waters, D. E. Gardner, T. O. Vaughan, J. A. Campbell, F. J. Miller, and D. L. Coffin, *In Vitro* **10,** 342, (abstr.) (1974).
394. M. D. Waters, D. R. Abernathy, H. R. Garland, and D. L. Coffin, *In Vitro* **10,** 343 (abstr.) (1975).
395. J. Graham, D. E. Gardner, M. D. Waters, and D. L. Coffin, *Infec. Immunity* **11,** 1278 (1975).
396. J. Gil and E. R. Weibel, *Resp. Physiol.* **8,** 13 (1970).
397. D. E. Lentz and N. R. DiLuzio, *J. Appl. Phys.* (in press), 1977.
398. C. M. Shy, J. R. Creason, M. Pearlman, K. E. McClain, F.B. Benson, and M. M. Young, *J. Air Pollut. Contr. Ass.* **20,** 582 (1970).
399. J. T. Middleton, L. O. Emik, and O. C. Taylor, *J. Air Pollut. Contr. Ass.* **15,** 476 (1965).
400. T. T. Trujillo and J. F. Spalding, Lams 2780-102, No. 9. Los Alamos Sci. Lab. Los Alamos, New Mexico, 1962.
401. G. J. Miraglia and L. J. Berry, *J. Bacteriol.* **85,** 345 (1963).
402. D. L. Coffin and E. J. Blommer, *J. Air Pollut. Contr. Ass.* **15,** 523 (1965).
403. D. L. Coffin and E. J. Blommer, unpublished data (1967).
404. H. E. Swann, Jr., D. Brunol, and O. J. Balchum, *Arch. Environ. Health* **10,** 25 (1965).
405. H. E. Swann, D. Brunol, L. G. Wayne, and O. J. Balchum, *Arch. Environ. Health* **11,** 765 (1965).
406. M. O. Amdur and J. Mead, *Amer. J. Physiol.* **192,** 264 (1958).
407. R. F. Bils, *Arch. Environ. Health* **12,** 689 (1966).
408. "Motor Vehicles, Air Pollution, and Health," Doc. No. 489, pp. 89–124. U.S. Dept. of Health, Education, and Welfare, Washington, D.C., 1962.
409. P. Kotin and H. L. Falk, *Proc. Nat. Air Pollut. Symp., 3rd, 1955* p. 141. (1955).
410. P. Kotin and M. D. Thomas, *Arch. Environ. Health* **16,** 411 (1957).
411. R. D. Boche and J. J. Quilligan, *Science* **131,** 1733 (1960).
412. E. R. Stephens, *Proc. Conf. Air Pollution Research on Atmospheric Reactions of Constituents of Motor Vehicle Exhaust, 5th, 1961*. California State Dept. of Health, Berkeley, California, 1966.
413. A. H. Rose and C. S. Brandt, *J. Air Pollut. Contr. Ass.* **12,** 527 (1962).

414. R. G. Hinners, *J. Air Pollut. Control Ass.* **12,** 527 (1962).
415. S. D. Murphy, J. K. Leng, C. E. Ulrich, and H. V. Davis, *Arch. Environ. Health* **7,** 60 (1963).
416. S. D. Murphy, *J. Air Pollut. Contr. Ass.* **14,** 303 (1964).
417. E. J. Catcott, *in* "Air Pollution," p. 221–231, Monograph Series, No. 46, World Health Organization, Geneva, Switzerland, 1961.
418. C. A. Mills, "Air Pollution and Community Health." Christopher Publishing House, Boston, Massachusetts, 1954.
419. C. R. Cole, R. L. Farrell, and R. A. Griesemer, "The Relationship of Animal Disease to Air Pollution," Contract SAPH 69436. National Air Pollution Control Administration, U.S. Dept. of Health, Education, and Welfare, Ohio State Research Foundation, Columbus, Ohio, 1964.
420. C. R. Dorn, *J. Nat. Cancer Inst.* **40,** 295 (1968).
421. P. H. Phillips and J. W. Suttie, *Arch. Environ. Health* **21,** 343 (1960).
422. J. Bronch and N. Grieser, *Berlin. Muenchen. Tieraerztl. Wochenschr.* **77,** 401 (1964).
423. E. J. Largent, *AMA Arch. Ind. Health* **21,** 318 (1960).
424. J. L. Shupe, M. L. Miner, and W. Binns, *J. Amer. Vet. Med. Ass.* **92,** 195–201 (1955).
425. P. H. Phillips, D. A. Greenwood, C. S. Hobbs, and C. F. Huffman, *Nat. Acad. Sci.—Nat. Res. Counc., Publ.* **381,** pp. 1–19 (1955).
426. W. D. Harkins and R. E. Swain, *J. Amer. Chem. Soc.* **30,** 928 (1908).

6

Organic Particulate Pollutants—Chemical Analysis and Bioassays for Carcinogenicity

Dietrich Hoffmann and Ernst L. Wynder

I. Introduction

Incomplete combustion involving fossil fuels or, more generally, carbon hydrogen-compounds is the major source of organic particulates in the respiratory environment. The variety of compounds is increased by the release of aerosols from industrial activities, from natural sources, and by secondary reactions in the air. This chapter will be primarily concerned with those agents that one can expect in urban pollutants.

A. Particle Size

In polluted air, particles occur in sizes ranging in diameter from 0.001 to 10,000 μm. In general, controlled combustion produces particles of the 0.1 to 10 μm range. Within this group, the "lung-damaging" components are considered to be particles of 0.25–10 μm in diameter (*1–4*). Particles with diameters greater than 10 μm lodge primarily in the upper respiratory tract and generally do not reach the lung.

The analysis and bioassay of organic particles require the use of high-volume samplers with fiberglass filters (without binders) and with high collection efficiency (>99.9%) (*5*). Only a few studies have explored the distribution of specific organic agents in airborne particles of different sizes. According to one such study, approximately 60% of the pentacyclic aromatic hydrocarbons are associated with particulates smaller than 1.1 μm, 25% with those between 1.1 and 7.0 μm, and the remainder with nonrespirable particulates (>7.0 μm). The mass median diameter ranged from 0.69 to 0.71 μm (*6*). This observation reflects a relatively uniform distribution of pentacyclic hydrocarbons in particuates in urban air (*7*).

B. Sampling Sites

Samples should be collected at nose level for at least 24 hours, so that representative samples for analysis, as well as for assay of the exposure of man to pollutants, are obtained. Furthermore, pollution levels fluctuate widely in urban and suburban environments, depending on the nature and distribution of sources, meteorological conditions, wind directions, and seasonal and annual variations (*8*). Such factors are known to be decisive in determining distribution profiles of polynuclear aromatic hydrocarbons (PAH) and pesticides in urban air (*9–15*). A guideline for comparing pollution in various cities has been presented by Waller and Commins (*16*) for PAH in the air based on their study conducted in European cities. The investigators divided these regions, each of which had populations numbering approximately one half million, into five areas nearly equal in population, and collected particulates for PAH analysis over a period of 12 months. Waller and Commins subsequently assessed, from their analytical data, the exposure of man to carcinogens by inhalation of urban air.

C. Collection Systems

The choice of a collection system is primarily determined by the physicochemical nature of the compounds and by their expected con-concentrations in polluted air. In the United States and Canada, a high. volume sampler with rectangular fiberglass filter (20×25 cm) of high collection efficiency and low resistance has been widely used (*17*). Sampling devices, including most of the European equipment (*16*), collect particles up to 10 μm in size, the coarser particles being removed before passage through the system. For the collection of large samples, mobile collection systems (140 m^3/minute) have been developed (*18*). All samplers should be equipped with sun roofs or sun shields to avoid photooxidation (*19–21*). The relative instability of some organic pollutants precludes the use of electrostatic precipitators. For some compounds, e.g., naphthylamines, special collection systems are required.

At ambient temperature, high-volume samplers quantitatively collect only high molecular weight organic compounds (*15, 22*). Pretreatment of glass fiber filters significantly improves the efficiency of the high-volume sampler (*23*). The efficiency of a collection device should be established before it is utilized for the quantitative analysis of organic pollutants.

Because deposition patterns of air pollution particles in the lung differ according to particle size, information is needed on the differences in the chemical composition of air pollutants of various diameters. The Andersen cascade impactor has been used to classify respirable particles into four size groups and to separate these from the nonrespirable particles (>7 μm) (*7, 24*). This sampler has been employed for establishing the particle size distribution of some PAH (*6*). For highly concentrated aerosols, as may occur indoors, centrifugal aerosol collectors appear to be most suitable for separating particles of various sizes (*25*).

D. Extraction and Storage

Several studies have been conducted on the effects of the choice of solvent mixtures with regard to types and amounts of organic matter and bioactive agents extracted from urban particulate matter (*26–28*). It has been shown that benzene will extract benzo[*a*]pyrene quantitatively, but benz[*c*]acridine (63%) of 7*H*-benz[*de*]anthracene-7-one only partially (42%) (*28*). In studies of air pollution carcinogenesis, extraction with benzene and benzene–methanol mixtures have yielded the most bioactive extracts (*29–36*). The investigator is advised, therefore, to find the appropriate solvent or solvent mixture before he starts his assay.

The analyst should also explore the relative speed and efficiency of extraction by the ultrasonic and Soxhlet methods. For example, ultrasonic extraction of PAH from particulates is complete within half an hour, whereas Soxhlet extraction requires 8 hours (*37*). Since one cannot disregard the possibility that some losses of carcinogenic activity may occur during continuous Soxhlet extraction, a comparative study needs to be made between the activity of a Soxhlet extract and that of an ultrasonic extract.

The organic extracts should be refrigerated and stored in glass vessels in complete darkness. When organic agents cannot be identified immediately after collection, the loaded filters should be stored in aluminum foil, rather than in plastic bags or paper, in order to avoid losses and/or contamination.

II. Standard Laboratory Conditions

Analysis and bioassay of toxic environmental agents require special precautions. All known carcinogenic PAH and azaarenes absorb light at wavelengths of 350–400 μm and may be oxidized in the presence of oxygen (*20, 21, 38, 39*). There are similar instabilities with phenolic compounds (tumor promoters) and alkylating agents. It is therefore recommended that such analyses and bioassays be performed in laboratories illuminated with yellow light, i.e., no radiation below 450 μm (*39, 40*). During enrichment steps involving concentration from organic solvents, losses can be reduced or even eliminated by working below 45°C. When large volumes of solvent need to be concentrated, evaporation should take place in distilling columns operated with at least a 2:1 reflux ratio.

Despite precautions, one cannot avoid loss of certain organic particulates. In order to overcome uncertainties it is best to use internal standards, compounds of similar chemical nature and polarity as the agents to be assayed but known to be absent from the test material. ^{14}C-labeled agents are the preferred internal standards (*41*).

As in trace analysis generally, all chemicals and laboratory equipment used must be free of contamination and unable in any way to generate compounds of the type to be assayed. Solvents for extractions, fractionations, and analyses must be free of fluorescent materials as well as of residues and especially of plasticizers. These latter, which often originate from solvent containers, have polarity similar to that of PAH and can therefore, affect the analysis and influence the carcinogenic activity of organic pollutants (*42*).

Finally, one should be aware that most available reference compounds, especially PAH, azaarenes, and phenols, or starting materials for their synthesis, originate from coal tar or petroleum products and contain impurities with structures similar to the reference compounds. Commercial pyrene and chrysene, for example, contain methylated derivatives of the parent compounds as well as oxidation products, and azaarenes (*43, 44*), and may have contributed to misleading carcinogenicity data (*45*). It is recommended that only synthetic products be used for reference purposes.

III. Fractionation

For the identification of biologically active constituents, fractions and subfractions are assayed separately. The active subfractions are taken for detailed chemical analytical studies that will lead to the isolation of those agents that contribute to the observed *in vivo* or *in vitro* activity. In the final step, quantitative methods have been developed for the determination of the toxic organic particulate compounds in polluted air and for the assessment of the exposure of man to such carcinogens, mutagens, teratogens, and cytotoxic agents.

In the scheme utilized by Hueper and his associates (*32*) for identifying tumorigenic agents benzene was used to extract the organic matter from urban particulates and classical methods were employed to fractionate the crude benzene extract into weak and strong acidic, basic, and neutral portions (Fig. 1). The neutral portion was chromatographed and yielded aliphatic, aromatic, and oxygenated fractions. The experimental details of this procedure have been presented by Tabor *et al.* (*46*). The same fractionation system was applied to composites of air pollutants from eight United States cities. The crude samples (benzene extracts), as well as the three neutral end fractions, were assayed for tumorigenicity. Several of the crude benzene extracts and end fractions were also tested for their

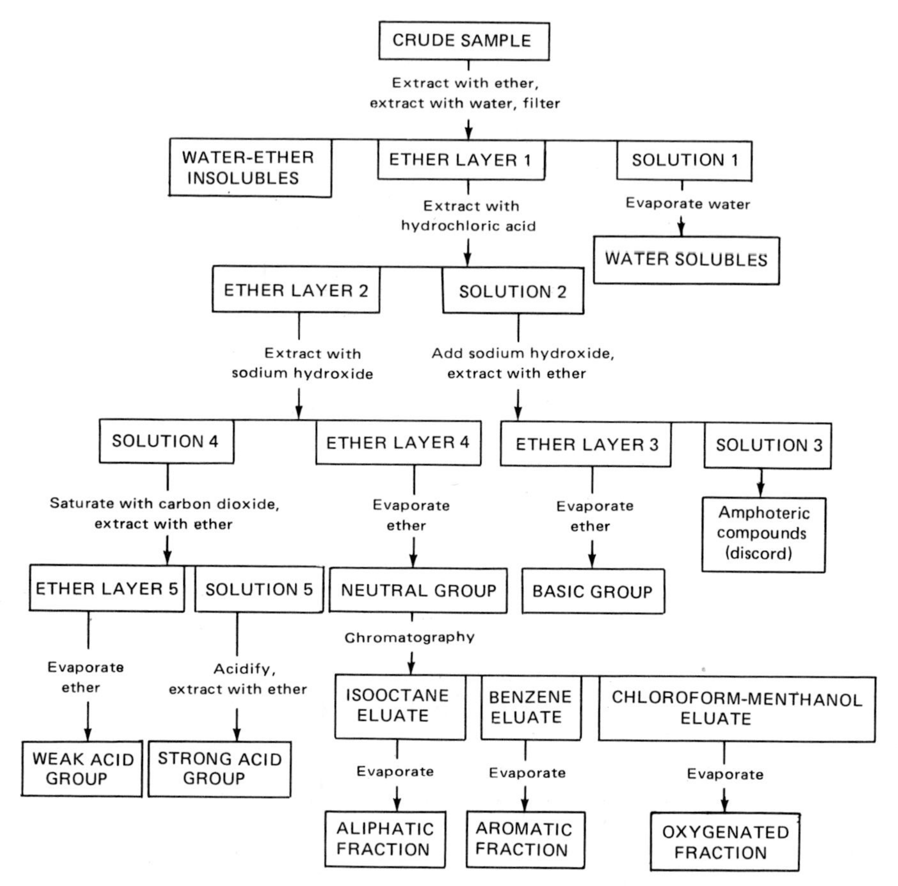

Figure 1. Separation scheme for organic particulate matter by Hueper *et al.* (*32*).

photodynamic activity (see Section VII,C) (*47*). Epstein and his co-workers have applied the same fractionation scheme, with only slight modification, to a large air pollution sample from New York, New York, and injected the fractions subcutaneously into infant mice (see Section VII,D,4,a) (*48*).

The benzene soluble portion of particulates collected at the campus of the University of Antwerp, Belgium, was also fractionated according to Tabor *et al.* (*46*) and resulted in 59.3% neutral compounds, 22.5% acidic compounds, 9.7% basic compounds, and 8.5% water solubles (*48a*). These fractions were directly analyzed for their major constituents.

A somewhat different fractionation scheme has been employed by our group at the Sloan-Kettering Institute (Fig. 2) (*33*). Our experience suggests that for the extraction of the basic portion, diluted sulfuric acid

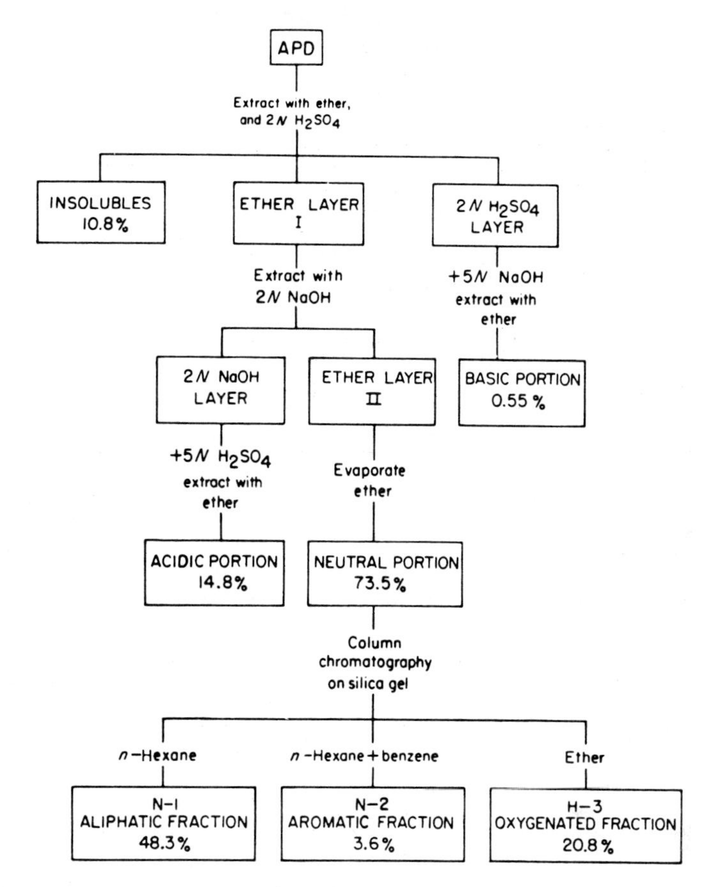

Figure 2. Separation scheme of the particulate matter from a Detroit, Michigan, sample (*33*).

should be used instead of diluted hydrochloric acid, since sample changes such as hydrochloric acid addition, are known to occur with certain terpenes (*49*). For large-scale chromatographic fractionation, neutral alumina, activity II, results in the lowest degree of chemical changes, although certain isomerizations may still occur. Adsorbents, such as silica gel and Florosil in particular may contribute to saponifications of esters.

The classification of organic pollutants into neutral, acidic, and basic components is chosen mainly for didactic reasons and is based primarily on our experience with urban pollutants. The analytical chemist has developed and will develop specific enrichment procedures for each group of compounds. Although he may bypass the classic separation scheme, it has proven most helpful for the bioassay of environmental and naturally occurring toxic materials.

IV. Neutral Fraction

For bioassays one subfractionates the neutral portion of the organic particulate matter by column chromatography as outlined in Figures 1 and 2. Employing alumina (neutral, activity II [Woelm]; "tar" to absorbent 1:100) and *n*-hexane with increasing amounts of benzene as solvents, one obtains under reproducible conditions three subfractions. The grouping of the subfractions is directed either by thin-layer chromatography (tlc) spot tests or by labeled markers ([^{14}C]phenanthrene and [^{14}C]cholesterol). The aliphatic fraction N-1 is eluted from the column until spot tests indicate phenanthrene to coronene (spot test or beginning of β-activity of [^{14}C]cholesterol) and is followed by the oxygenated fraction N-3. N-1 contains as major constituents nonvolatile saturated and unsaturated hydrocarbons, terpene hydrocarbons and chlorinated hydrocarbon insecticides.

A. Saturated Hydrocarbons

1. Precursors

A considerable portion of the organic particulates in urban pollution is made up of aliphatic and saturated hydrocarbons (20–65%) (*46, 50*). The major sources of these substances include combustion products of petroleum, gasoline, and diesel fuel. During the incomplete combustion of these materials, traces of paraffins are emitted unchanged, partially cracked, or partially pyrosynthesized to longer chain hydrocarbons. Hydrocarbons are emitted into the air from gasoline engines through (a) carburetor and fuel tank evaporation, (b) crankcase blow-by, and (c) tailpipe exhaust. While gasoline and diesel fuel do not contain substantial amounts of alkanes in excess of 16 and 20 carbon atoms, one does find normal alkanes in the C_{16}–C_{36} range in the exhaust fumes of these fuel engines (Table I) (*50*).

In addition, experiments indicate that irradiation of exhaust gases may further increase alkanes in polluted air beyond the C_{20} range (*50*). Indoors, tobacco smoke may serve as a major contributor of alkanes. Each combustion material gives rise to a specific spectrum, or "fingerprint," of normal and isoalkanes. Improvements in automatic analytical equipment interphased with calculators or computers should enable us to define the

Table I Summary Data for Aliphatic Fraction in Air Pollution Source Effluents and in Air Pollutants (*50*)

Sample	*50% Breakpoint*[a]	*n-Alkane*[b] (%)	*Ratio*[c]
Diesel fuel	$<C_{19}$	—	—
Gasoline	$<C_{19}$	—	—
Raw auto exhaust	C_{24}	8.9	0.10
Diesel soot	C_{24}	54.4	1.19
National Air Surveillance Network			
nonurban ambient air samples	C_{24}	—	—
Irradiated auto exhaust	$C_{25}+$	22.4	0.29
Amboy, California	$C_{25}+$	41.7	0.72
Coal soot	$C_{25}+$	—	—
Coal extract	C_{26}	—	—
Motor oil	$C_{26}+$	—	—
Birmingham, Alabama	$C_{28}+$	4.0	0.04
Indoors	$C_{28}+$	8.4	0.09
Los Angeles, California			
September	$C_{28}+$	6.4	0.07
January	$C_{28}+$	13.2	0.15

[a] Aliphatic hydrocarbons with carbon number at which 50% of sample is greater or less by weight.
[b] Normal alkanes in aliphatic fraction.
[c] Ratio of normal alkanes to isomeric alkanes in fraction.

major sources of a given air pollutant by measuring specific inert agents with known sources. The fairly inert, nonvolatile alkanes may well serve for this purpose.

2. Chemical Analysis

In general, high-volume samplers collect, along with particulate matter, hydrocarbons with boiling points above 150°C. Therefore the benzene extract of the particulates results in a semiliquid residue that contains the saturated and unsaturated hydrocarbons.

Methods for the analysis of normal and isoalkanes originate primarily from studies of petroleum fuels and tobacco (*51–54*). Only a limited number of studies have been concerned with these hydrocarbons in polluted air.

In earlier investigations the organic matter of a dust sample was partitioned between cyclohexane, methanol/water (4:1), (*55*) and subsequently between cyclohexane and nitromethane (*56*). The residue of the cyclohexane layer, free of hydrophilic materials (methanol/water layer) and PAH (nitromethane layer), contained components with retention times comparable to those of normal hydrocarbons from $C_{18}H_{38}$ to $C_{30}H_{62}$ (when directly separated by gas–liquid chromatography, glc). The first more detailed analysis was reported by Sawicki *et al.* (*57, 58*), who enriched the nonaromatic hydrocarbons from the benzene-soluble organic matter by chromatography, removed the unsaturated hydrocarbons by bromination, and separated the normal and isoalkanes by chromatography on molecular sieves. The normal alkanes were determined by temperature-programmed glc. The results for nonvolatile normal paraffins from composite air pollutant samples from 100 United States communities are presented in Table II.

Hauser and Pattison (*50*) refined these methods (*57, 58*), especially glc, with respect to separation efficiency and the time needed for the glc separation. Their reported values for normal alkanes in Los Angeles, California, air are presented in Table III. The major findings of their study are summarized in Table IV, illustrating that although gasoline and diesel fuel contain almost no normal alkanes with more than 19 carbon atoms, the engine exhaust gases contain normal alkane spectra that peak around C_{24}, indicating some pyrosynthesis of higher alkanes during the combustion of fuels. Artificial irradiation of alkanes from gasoline ex-

Table II Concentrations of Nonvolatile Normal Paraffins in the Average American Urban Atmosphere[a] (*50*)

		μg compound per gm of		
n-Paraffin	*Formula*	*Benzene-soluble fraction*	*Airborne particulates*	*μg of compound in 1000 m³ of air*
Heptadecane	$C_{17}H_{36}$	240	20	2.5
Octadecane	$C_{18}H_{38}$	1400	110	14
Nonadecane	$C_{19}H_{40}$	1900	160	20
Eicosane	$C_{20}H_{42}$	2200	180	23
Heneicosane	$C_{21}H_{44}$	3900	320	40
Docosane	$C_{22}H_{46}$	5800	480	60
Tricosane	$C_{23}H_{48}$	7400	620	77
Tetracosane	$C_{24}H_{50}$	5800	480	60
Pentacosane	$C_{25}H_{52}$	5800	400	60
Hexacosane	$C_{26}H_{54}$	1000	85	11
Heptacosane	$C_{27}H_{56}$	3100	260	32
Octacosane	$C_{28}H_{58}$	4100	340	43
Total ($C_{17}H_{36}$–$C_{28}H_{58}$)		42640	3525	442.5

[a] Composite air pollutant sample from downtown of approximately one hundred communities (1963).

Table III Concentration of Normal Alkanes in Los Angeles, California, Air (*50*)

	Concentration ($\mu g/m^3$)	
Normal alkane	*September, 1969*	*January, 1970*
C_{17}	—	0.1
C_{18}	—	0.8
C_{19}	—	4
C_{20}	—	8
C_{21}	—	11
C_{22}	3	18
C_{23}	5	28
C_{24}	10	34
C_{25}	11	43
C_{26}	16	39
C_{27}	23	41
C_{28}	24	38
C_{29}	22	35
C_{30}	12	34
C_{31}	9	26
C_{32}	8	19
C_{33}	4	10
C_{34} and higher	1	8
Total	148	396.9

haust changes their relative composition, a fact that may have some implication where a high rate of solar irradiation affects urban air. As expected, the alkane fraction of indoor pollutants is greatly influenced by tobacco smoke.

Hauser and Pattison (*50*) also report that the aliphatic fraction is devoid of unsaturated materials—an observation that needs reconfirmation, since Sawicki *et al.* (*59*) earlier found appreciable amounts of olefinic compounds in their aliphatic fractions. Tobacco smoke is known to contain small amounts of C_{10}–C_{32} alkanes (*60*).

Despite the remarkable progress in the analysis of alkanes in organic particulates, further improvements are still needed. First, such analyses should not be undertaken without internal standards, preferably with ^{14}C- and ^{3}H-labeled normal and isoalkanes (*61*). Second, whenever possible, the glc analysis should be supplemented with mass spectral (ms) analysis. Future studies must also be directed toward the determination of benzene derivatives with long alkane side chains and toward the determination of hydrogenated naphthalenes and other cyclic saturated

Table IV A Carbon Atom Range (%) of Aliphatic Fraction *(50)*

Carbon number range	*Birmingham, Alabama*	*Los Angeles, California*		*NASN*[a]	*Amboy, California*[b]	*Indoors*
		September	*January*			
<19	1.1	0.06	0.2	1.7	25.3	0.3
19–22	3.4	2.1	2.1	16.7	9.4	4.5
22–24	5.1	4.6	5.5	21.7	7.3	9.5
24–26	10.9	12.0	11.5	19.3	7.8	15.4
26–28	16.8	19.7	17.2	14.1	11.3	14.2
28–30	18.4	19.5	19.5	11.3	11.4	14.6
30–32	17.1	18.2	18.0	7.4	10.1	20.5
32–36	17.4	19.4	19.8	6.7	12.9	16.8
>36	9.7	5.0	6.3	1.1	4.5	4.2

[a] National Air Surveillance Network nonurban ambient air samples.
[b] Nonurban Sample.

hydrocarbons, since such agents are known to occur in certain engine oils or fuels, and may affect the carcinogenicity of PAH (*45*).

Cantreels and Van Cauwenberghe (*48a*) solved some of the analytical problems by separating the aliphatic fraction of the organic particulates (Fig. 1) by glc and subsequently monitoring the mass 85 ($C_6H_{13}^+$ ion) with a total ion current chromatogram. The authors present some evidence that by comparison with other alkyl ions, the one with mass 85 shows high sensitivity and low probability of interference by other compounds in aerosol samples. A 24-hour suburban sample from Antwerp, Belgium yielded data for unbranched hydrocarbons from *n*-tetradecane ($C_{14}H_{30}$) to *n*-pentatriacontane ($C_{35}H_{72}$) with a maximum concentration of 351 ppm for *n*-pentacosane ($C_{25}H_{52}$) in dry particulates (100 mg/600 m^3 polluted air) (*48a*).

While nonvolatile hydrocarbons may generally be regarded as biologically inactive, there are indications that paraffins from decane to hexadecane may influence the tumorigenicity of carcinogenic hydrocarbons on epithelial tissues. The C_{10}–C_{16} alkanes applied in high concentrations increase the activity of carcinogenic PAH and may act as weak tumor promoters (*62–64*). On the other hand, normal alkanes with approximately 30 carbon atoms may inhibit the carcinogenicity of PAH (*65*).

B. Unsaturated Hydrocarbons

Although considerable knowledge has been gathered on the presence of volatile unsaturated hydrocarbons in urban air and in other polluted

Table IV B Carbon Atom Range (%) of Aliphatic Fraction *(50)*

Carbon number range	*Auto exhaust*		*Auto oil*	*Diesel*		*Coal*	
	Raw	*Irradiated*		*Soot*	*Fuel*	*Extract*	*Soot*
<19	1.2	1.4	0.5	4.0	85.7	11.6	2.3
19–22	22.0	13.3	5.3	20.0	11.4	11.0	15.2
22–24	16.7	11.7	12.5	9.1	2.0	8.7	13.5
24–26	20.2	14.4	22.3	2.4	0.3	12.5	17.0
26–28	16.8	20.4	22.5	3.2	0.1	12.8	15.8
28–30	10.7	13.9	15.3	1.6	0.2	13.4	16.4
30–32	6.6	9.4	11.1	8.3	0.1	10.2	8.3
32–36	4.4	14.1	8.3	9.3	0.2	13.4	9.1
>36	1.3	1.5	1.8	2.1	—	6.4	2.4

respiratory environments, little is known about nonvolatile unsaturated hydrocarbons as air pollutants.

In particulate matter, olefins and unsaturated terpenes without functional groups would be expected in the aliphatic and aromatic fractions (N-1 and N-2) of Figures 1 and 2. In fact, for the analysis of alkanes, it is necessary to separate the unsaturated hydrocarbons by bromination. Sawicki *et al.* (*59*) developed an analytical method for determination of higher molecular weight olefinic substances in organic particulates in which the unsaturated compounds are converted by oxidation to aldehydes which are then condensed with 3-methyl-2-benzothiazoline to colored formazans. The formazans are assayed spectrophotometrically. Based on this study, it appears that the aliphatic fractions in urban pollutants can contain considerable amounts of olefinic compounds. So far, no other method has been developed for the assay of nonvolatile unsaturated hydrocarbons in polluted air.

The lack of information about nonvolatile olefins is rather surprising, since the aliphatic fraction (N-1) shows some tumorigenic activity upon subcutaneous injection into adult as well as infant mice (*32, 48*). Furthermore, several authors consider it possible that photodecomposition will occur not only with volatile unsaturated hydrocarbons, but also with nonvolatile olefins in the particulate matter of urban air (*66*).

Some of these decomposition products are expected to be epoxides and peroxides. Several representatives of these two groups of compounds are known to be tumorigenic to the experimental animal (*45, 66, 67*). In fact, the oxygenated fractions of air pollutants that may contain epoxides and peroxides were shown to be tumorigenic to the skin and subcutaneous tissue of mice (*32*). Several of these oxygenated components are ciliatoxic and thus inhibit lung clearance in experimental animals. In view of this, information is needed on the exact chemical nature of nonvolatile olefinic agents. Accordingly, data on terpenes and terpene esters in indoor and outdoor air also appear important.

In the countryside, away from highways and close to forests, the predominant sources of air pollution are plant products. Rasmussen and Went (*68*) employed glc for the analysis of fresh air and found concentrations of pollutants to range from 2 to 10 ppb, depending on meteorological conditions and the density and activity of plant cover. Among the major pollutants identified were the terpenes, α- and β-pinene, and myrcene. One can assume that the organic particulate matter analyzed was sampled in or near forests. Terpenes found indoors derive mostly from the combustion products of wood and tobacco (*65, 69*).

C. Benzene, Naphthalene, and Their Derivatives

Polluted air contains benzene and naphthalene and a wide spectrum of their derivatives. Gasoline, diesel fuel, internal combustion engine exhaust, and organic combustion can be regarded as the major sources for these pollutants. Furthermore, burning of plant products with peak temperatures below 600°C results in a large number of alkylated naphthalenes (*70*). Since most of the benzenes and naphthalenes are volatile at ambient temperatures, only traces of their alkyl derivatives are expected as particulate matter in filter samples. For example, we found traces of *n*-decylbenzene and *n*-dodecylbenzene in the N-1 fraction of large samples from collection sites with high traffic density (*33*).

Some of the long-chain alkylated benzene derivatives are weak tumor promoters (*62*). Sawicki in his review article "Airborne Carcinogens and Allied Compounds" (*71*) lists naphthalene and 2-methylnaphthalene as compounds identified in polluted air. In addition, alkylated naphthalenes are expected as indoor pollutants (*65*).

D. Polynuclear Aromatic Hydrocarbons

Polynuclear aromatic hydrocarbons are the most studied trace compounds in polluted air as well as in personal pollution. The great interest in these compounds lies in the fact that several are known animal carcinogens and some are suspected carcinogens for man (*41, 45, 72*). In addition, some studies have correlated PAH, especially benzo[*a*]pyrene (BaP), in polluted air with the urban factor for lung cancer (*41, 73*). The major potentially carcinogenic PAH so far identified with certainty in air pollutants are aromatic hydrocarbons with four, five, and six rings (Fig. 3). In addition to their parent hydrocarbons, polluted air may also contain traces of some carcinogenic alkylpolycyclics (see Section IV,D).

In Figure 3 and Table V, we have identified the relative carcinogenic activities of the PAH as strong carcinogens (++++), moderately strong carcinogens (+++), weak carcinogens (++), very weak carcinogens (+), and noncarcinogens (—). This designation is based on our own laboratory assays on the skin of Swiss mice (*43, 74–80*). Calculating the relative tumor potency numbers (TPN) according to Blanding *et al.* (*81*), we have designated with (++++) those PAH which with a 0.1% solution gave a TPN >100, (+++) for whose that gave a TPN of 50–100, (++) for those that gave a TPN from 20 to 50, (+) for those

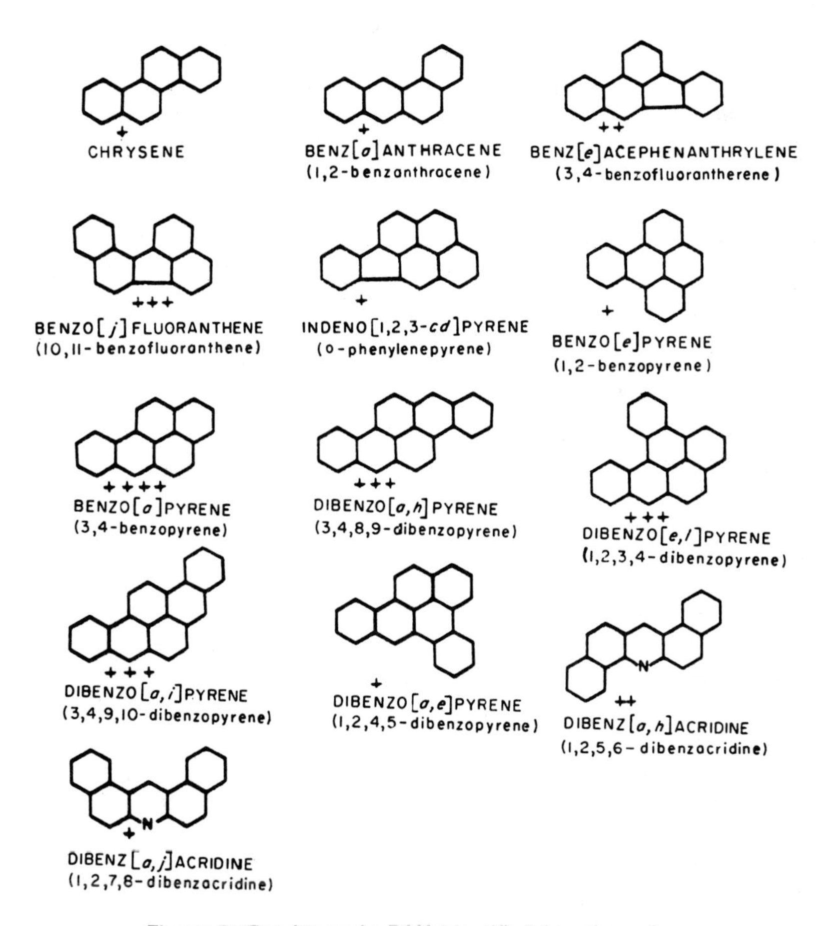

Figure 3. Carcinogenic PAH identified in urban air.

Table V Polynuclear Aromatic Hydrocarbons in Urban Atmosphere[a,b]

No.	*Ring system and name*[c]	*Carcino-genicity*[d]	*Concentration* (μg *per 1000* m^3)	*Location*	*Reference*
1.	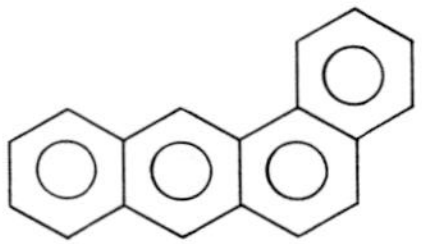 Benz[*a*]anthracene (1,2-Benzanthracene)	+	0.4–21.6	3 Sites in Detroit, Michigan (20 samples)	(*83*)
			0.36–6.25	1 Site in Sidney, New South Wales, Australia (12 samples)	(*84*)
			0.1–16.0	4 Sites in New York, New York (16 samples)	(*10*)
			0.2–7.2	4 Sites in Los Angeles, California (52 samples)	(*11*)
			0.04–3.1	4 Sites in Los Angeles, California (4 samples)	(*12*)
			1.1–3.1	5 Samples from Rome, Italy (winter 1965–1966)	(*85*)
			0.1–13.1	14 Samples from Rome, Italy	(*86*)
			~4	Average United States urban atmosphere[e]	(*87*)
2.	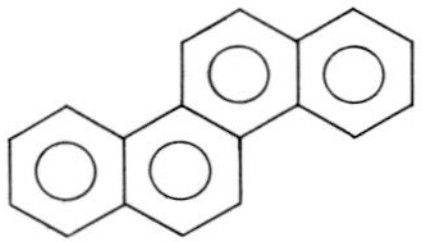 Chrysene (1,2-Benzophenanthrene)	+	1.3–11.6	3 Sites in Detroit, Michigan (6 samples)	(*39*)
			0.2–6.65	1 Site in Sidney, New South Wales, Australia (12 samples)	(*84*)
			0.4–39.0	14 Samples from Rome, Italy	(*86*)
			1.3–4.1	5 Samples from Rome, Italy (winter 1965–1966)	(*85*)

	Compound		Concentration	Location	Ref.
3.	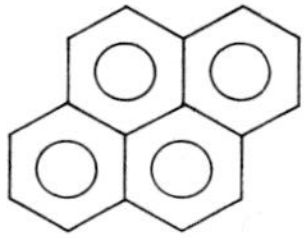 Pyrene	—	Trace–35	12 United States cities	(*88*)
			0.9–15.0	3 Sites in Detroit, Michigan (6 samples)	(*83*)
			0.08–4.0	1 Site in Sidney, New South Wales, Australia (12 samples)	(*84*)
			0.4–12.0	20 Sites in Ontario, Canada (1961–1963; 5 samples at each location)	(*13*)
			1.0–54	20 Sites in Ontario, Canada during inversion period (1962)	(*13*)
			0.4–17.0	14 Samples from Rome, Italy	(*86*)
			2.4–5.1	5 Samples from Rome, Italy (winter 1965–1966)	(*85*)
			0.2–3.8	4 Sites in Los Angeles, California (4 samples)	(*12*)
4.	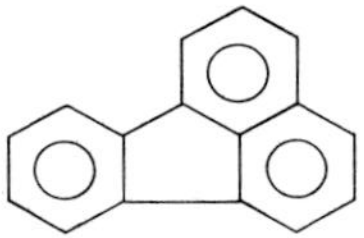 Fluoranthene	—	0.9–15.0	3 Sites in Detroit, Michigan (6 samples)	(*83*)
			0.3–10.6	20 Sites in Ontario, Canada (1961–1963; 5 samples on each location)	(*13*)
			0.6–41	20 Sites in Ontario, Canada, during inversion period (1962)	(*13*)
			0.06–2.6	1 Site in Sidney, New South Wales, Australia (12 samples)	(*84*)
			1.0–18.0	14 Samples from Rome, Italy	(*86*)
			2.1–4.5	5 Samples from Rome, Italy (winter 1965–1966)	(*85*)
			0.1–3.4	4 Sites in Los Angeles, California (4 samples)	(*12*)
			~4	Average United States urban atmosphere[e]	(*87*)

(*Continued*)

Table V *(Continued)*

No.	*Ring system and name*[c]	*Carcino-genicity*[d]	*Concentration (μg per 1000 m³)*	*Location*	*Reference*
5.	Benzo[*a*]pyrene (3,4-Benzopyrene)	++++	0.2–17.0	3 Sites in Detroit, Michigan (20 samples)	(*83*)
			0.6–8.2	1 Site in Sidney, New South Wales, Australia (16 samples)	(*84*)
			20–39	London, England, street	(*89*)
			0.1–9.4	4 Sites in New York, New York (16 samples)	(*10*)
			5.7	Average United States urban atmosphere[e]	(*87*)
				For other benzo[*a*]pyrene values see Table VII	
6.	Benzo[*e*]pyrene (1,2-Benzopyrene)	+	1–25	12 United States cities (est. from data)	(*88*)
			0.3–11	20 sites in Ontario, Canada (1961–1963; 5 samples each location)	(*13*)
			4.3–42	20 Sites in Ontario, Canada (during inversion period)	(*13*)
			8	Copenhagen, Denmark mean annual value	(*90*)
			1–37	2 Sites in London, England (1954–1964)	(*91*)
			740	1 Site in London, England (inversion period, December, 1957)	(*91*)
			16–25	5 Sites in Belfast, Northern Ireland (1961–1962)	(*16*)
			6–14	5 Sites in Dublin, Ireland (1961–1962)	(*16*)

			5–8	5 Sites in Oslo, Norway (1962–1963)	(*16*)
			3–5	5 Sites in Helsinki, Finland (1962–1963)	(*16*)
			0.8–7.1	1 Site in Sidney, New South Wales, Australia (12 samples)	(*84*)
			0.1–10.5	14 Samples from Rome, Italy	(*86*)
			0.1–3.2	4 Sites in Los Angeles, California (4 samples)	(*12*)
			5.0	Average United States urban atmosphere[e]	(*87*)
7.	Perylene	–	Trace–5.0	12 United States cities (est. from data)	(*88*)
			0.3–0.6	4 Samples from Rome, Italy (winter 1965–1966)	(*85*)
			0.01–1.2	4 Sites in Los Angeles, California (4 samples)	(*12*)
			0.7	Average United States urban atmosphere[e]	(*87*)

(Continued)

Table V *(Continued)*

No.	*Ring system and name*[c]	*Carcino-genicity*[d]	*Concentration (μg per 1000 m^3)*	*Location*	*Reference*
8.	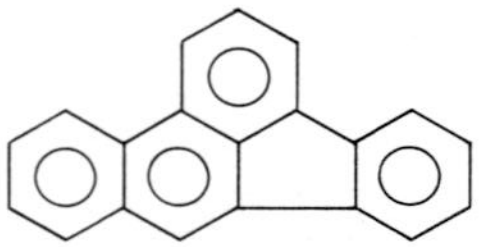 Benz[*e*]acephenanthrylene (Benzo[*b*]fluoranthene, 3,4-benzofluoranthene)	++	2.3–7.4	3 Sites in Detroit, Michigan (6 samples)	(*83*)
			0.1–1.6	4 Sites in Los Angeles, California (4 samples)	(*12*)
9.	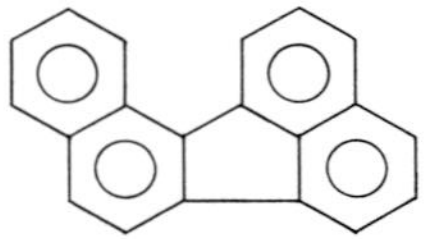 Benzo[*j*]fluoranthene (10,11-Benzofluoranthene)	+++	0.8–4.4	3 Sites in Detroit, Michigan (6 samples)	(*83*)
			0.01–0.8	4 Sites in Los Angeles, California (4 samples)	(*12*)

	Compound		Concentration	Location	Ref.
10.	Benzo[k]fluoranthene (11,12-Benzofluoranthene)	—	0.5–20	12 United States cities (est. from data)	(88)
			1.1–15.7	2 Sites in Detroit, Michigan (6 samples)	(83)
			1.3–14.0	20 Sites in Ontario, Canada (during inversion period, 1962)	(13)
			0.03–1.3	4 Sites in Los Angeles, California (4 samples)	(12)
11.	Benzo[ghi]perylene (1,12-Benzoperylene)	—	1.0–31	2 Sites in London, England (1954–1964)	(91)
			1000	1 Site in London, England (during inversion period, December, 1957)	(91)
			6.3–46	20 Sites in Ontario, Canada (during inversion period 1962)	(13)
			45	Liverpool, England mean value	(90)
			18–37	5 Sites in Belfast, Northern Ireland (1961–1962)	(16)
			5–15	5 Sites in Dublin, Ireland (1961–1962)	(16)
			0.4–0.8	5 Sites in Oslo, Norway (1962–1963)	(16)
			3–7	5 Sites in Helsinki, Finland (1962–1963)	(16)
			12–46	London, England, street	(87)
			2.3–11.5	3 Sites in Detroit, Michigan (6 samples)	(83)
			2–35	12 United States cities (est. from data)	(88)

(Continued)

Table V *(Continued)*

No.	*Ring system and name*[c]	*Carcino-genicity*[d]	*Concentration* (μg *per 1000* m^3)	*Location*	*Reference*
			0.9–9.7	1 Site in Sidney, New South Wales, Australia (12 samples)	(*84*)
			1.4–3.6	4 Samples from Rome, Italy (winter 1965–1966)	(*85*)
			0.2–9.2	4 Sites in Los Angeles, California (4 samples)	(*12*)
12.	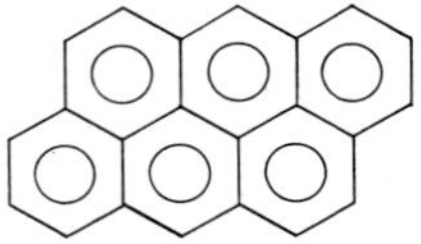 Dibenzo[*cd,jk*] pyrene (Anthanthrene)	−	1.1	Detroit, Michigan	(*83*)
			2–6	London, England, street	(*89*)
			Trace–3	12 United States cities (est. from data)	(*88*)
			0.01–1.2	4 Sites in Los Angeles, California (4 samples)	(*12*)
			0.4–0.8	4 Samples from Rome, Italy (winter 1965–1966)	(*85*)
			0.26	Average United States urban atmosphere[e]	(*87*
13.	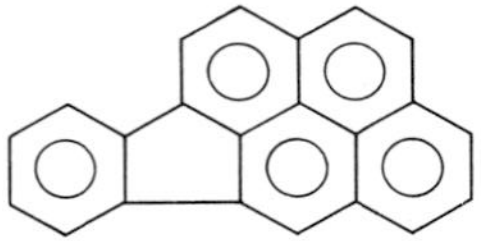 Indeno(1,2,3-*cd*)pyrene (*o*-phenylenepyrene)	+	1.5–8.2	3 Sites in Detroit, Michigan	(*83*)
			0.03–1.2	4 Sites in Los Angeles, California	(*12*)

14.	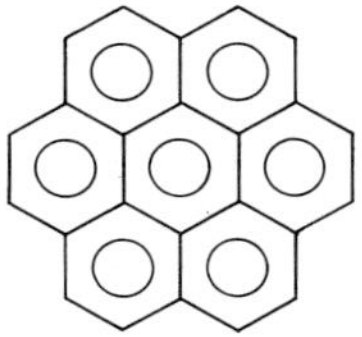 Coronene	—	2.3–12.2	2 Sites in Detroit, Michigan	(*83*)
			4–20	London, England, street	(*89*)
			0.7–7	5 Sites in Belfast, Northern Ireland (1961–1962)	(*16*)
			3–6	5 Sites in Dublin, Ireland (1961–1962)	(*16*)
			4–8	5 Sites in Oslo, Norway (1962–1963)	(*16*)
			1–4	5 Sites in Helsinki, Finland (1962–1963)	(*16*)
			3–48	20 Sites in Ontario, Canada	(*13*)
			0.2–6.4	4 Sites in Los Angeles, California	(*12*)
			0.7–1.0	4 Samples from Rome, Italy (winter 1965–1966)	(*85*)
			2	Average United States urban atmosphere[e]	(*87*)

[a] Only four or more ring aromatic hydrocarbons.

[b] Also identified in urban air were traces of 11*H*-benzo[*b*]fluorene (—), 5*H*-benzo[*a*]fluorene (—), 7*H*-benzo[*c*]fluorene (—), benzo-(*ghi*)fluoranthene (—), dibenzo(*ah*)pyrene (+++), dibenzo(*ai*)pyrene (+++), dibenzo(*ae*)pyrene (+), and indeno(1,2,3-*cd*)-fluoranthene (?).

[c] Ring orientation and names in first line given by IUPAC system. In the case of benzo[*e*]acephenanthrylene, the old IUPAC system's name, benzo[*b*]fluoranthene, is given.

[d] The relative carcinogenicities: ++++ = strong, +++ = moderately active, ++ = weakly active, + = very weakly active, — = inactive, and ? = unknown, derive from tests on the skin of ICR female mice (*74–80*).

[e] Values from composite of samples collected in 1963 from downtown areas of approximately one hundred communities (*87*).

that gave a TPN <20, and (—) for those PAH that were inactive on mouse skin. Since the literature points to additional carcinogens in polluted air that were not tested by us, we have given their relative carcinogenicity in parenthesis. Some investigators may disagree with these designations because some hydrocarbons are more active in the connective tissue than on the skin. We have chosen the squamous epithelium of the skin because it is somewhat comparable to the type of cells affected in man; and this bioassay system is less influenced by artifacts than is the bioassay using connective tissue (*82*).

Table V (*10–13, 16, 39, 83–91*) lists parent PAH with four or more rings that have been identified in urban pollutants. Several other PAH have been identified with certainty. However, quantitative data have not been given. These include 11*H*-benzo[*b*]fluorene (—) (*71, 92*), 5*H*-benzo[*a*]fluorene (—) (*71, 92*), 7*H*-benzo[*c*]fluorene (—) (*71, 92, 93*), benzo[*ghi*]fluoranthene (—) (*71, 92, 93*), dibenzo[*ah*]pyrene (+++) (*71, 93*), dibenzo[*ai*]pyrene (+++) (*71, 93*), dibenzo[*ae*]pyrene (+) (*94*), and indeno[1,2,3-*c,d*]fluoranthene (?) (*71, 92*). Although urban pollutants contain traces of several alkylated PAH, none have been identified individually, with the possible exception of 2-methylpyrene (*86, 93, 95*), a noncarcinogenic hydrocarbon.

1. Pyrosynthesis of PAH

Oxygen-deficient reducing atmospheres exist within a burning flame or a combustion chamber. Such environments are favorable for the pyrosynthesis of PAH. Primarily two mechanisms have been proposed for the pyroformation of PAH, the C,H-radical mechanism being the predominant one. At elevated temperatures (>450°C) in a reducing atmosphere, organic components are partially cracked into smaller, unstable molecules (pyrolysis). These fragments, mostly radicals, recombine to larger, thermodynamically favored, and relatively stable aromatic ring hydrocarbons (PAH) and heterocyclic hydrocarbons (pyrosynthesis).

Badger and his associates at the University of Adelaide, Australia, have studied the C,H-pyrosynthesis of PAH, especially of the carcinogenic polycyclics (*96–98*). Their initial working hypothesis is reproduced in Figure 4. Based upon their data and those of other investigators, the following concept emerges: The temperature that exists during the burning of organic matter easily breaks single carbon and carbon–hydrogen bonds and yields free radicals. These radicals combine and are dehydrogenated to aromatic ring systems that are known to be relatively stable; only the carbon–hydrogen bonds of these components are broken down to any significant extent. Long-chain paraffins, that are widely found in

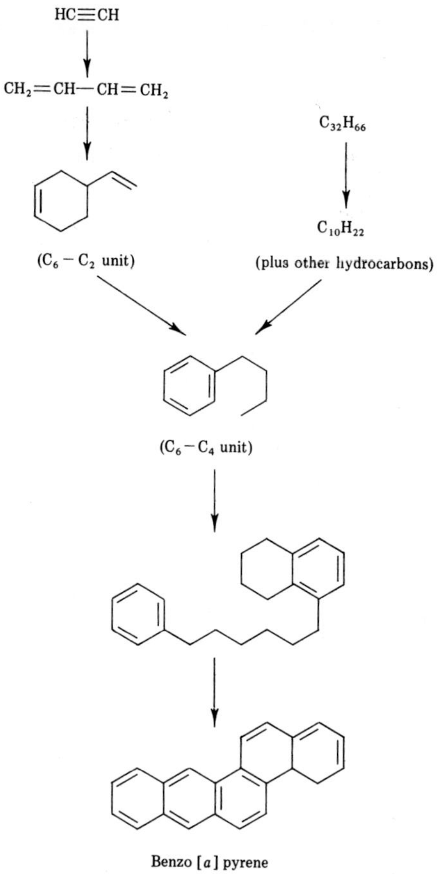

Figure 4. Benzo[*a*]pyrene synthesis as suggested by Badger (*96*).

fuels and plants, serve as special precursors for PAH. Badger's original concept of PAH formation was proved correct by extensive, carefully designed and executed experiments utilizing ^{14}C-labeled precursors. Figure 5 summarizes the pathways for BaP pyrosynthesis as they are suggested from the experimental findings. The above concept can be extended to hydrocarbons with uneven carbon numbers since methane also serves as a precursor for PAH (*99*).

Another aspect was also briefly considered by Badger and his group (*98*), namely the temperature dependence of the pyrosynthesis of PAH. It appears from the study with *n*-butylbenzene, one intermediate in the formation of BaP and benzo[*b*]fluoranthene, that the maximal yields of these, and most likely also other PAH, occur beyond 650°C, (Fig. 6)

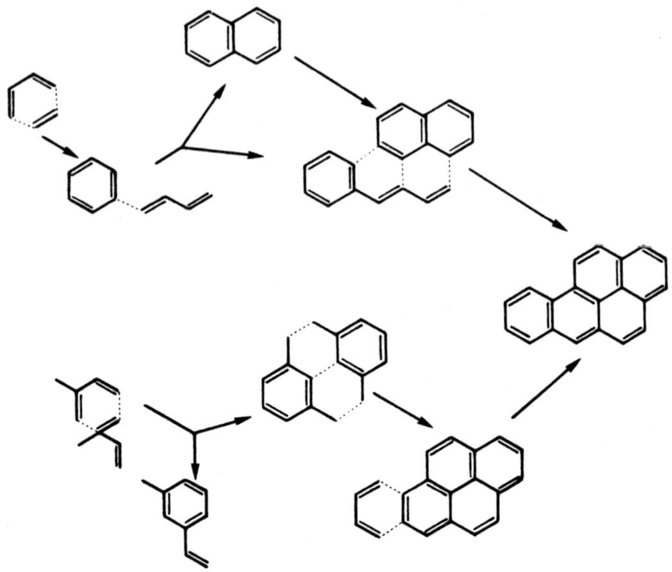

Figure 5. Pathways for the pyrosynthesis of benzo[*a*]pyrene.

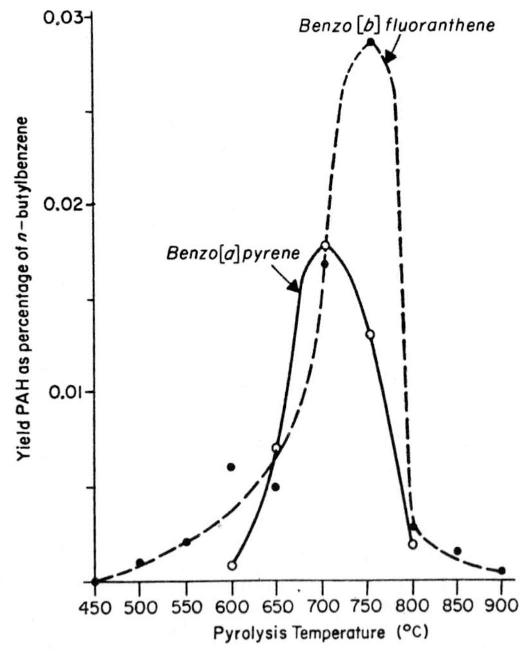

Figure 6. Temperature dependence of the pyrosynthesis in nitrogen of benzo[*a*]pyrene and benzo[*b*]fluoranthene (*96–98, 101*).

and that significant differences may exist in the temperature at which individual PAH are pyrosynthesized with maximal yields.

This aspect may be reflected in the differences in the relative ratio of individual PAH to each other in the combustion products of various emission sources. It is hoped that future model experiments will elucidate on the specific peak temperatures for the pyrosynthesis of individual PAH. With such information at hand, one may be able to reduce the release of carcinogenic hydrocarbons into the environment by changing the temperature profile of certain combustion processes. Pyrolysis studies have repeatedly shown that natural products that already have compounds with six ring structures can serve as specific precursors for PAH. Phytosterols, for example, result in high yields of chrysene and BaP (*100, 101*) and even of methylchrysenes (*80*).

A mechanism other than C,H-radical formation has been suggested for some carcinogenic hydrocarbons (*79, 102*). Combustion products of plants, especially of leaves, contain a relatively high concentration of alkylated PAH. This observation and the fact that leaves contain a considerable amount of terpenes, which upon incomplete combustion deliver isoprene and other volatile dienes, led to the suggestion that a Diels–Adler addition (diene synthesis) followed by dehydrogenation may lead to the formation of alkylated PAH. Pyrolysis experiments with volatile dienes and indenes or acenaphthylenes, respectively, supported this hypothesis (Fig. 7).

Most pyrolysis experiments have been conducted in a nitrogen atmosphere and, for this reason, have been criticized as being unrelated to conditions prevailing during the actual burning of organic matter. The fact remains, however, that these studies are qualitatively in agreement with actual burning conditions of organic matter. The cited studies have been aimed mainly at finding the major precursors and pathways of the

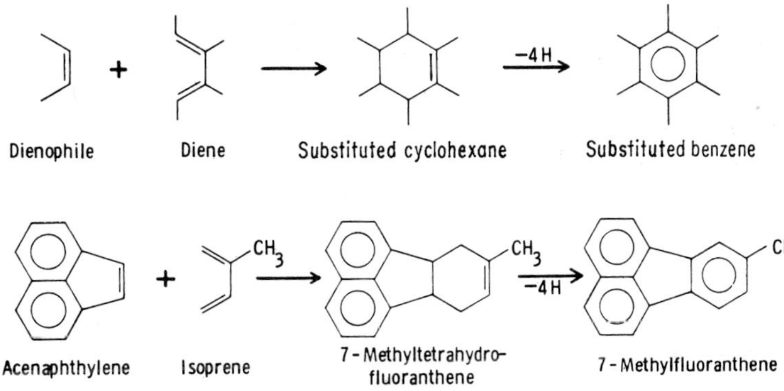

Dienophile + Diene → Substituted cyclohexane —$-4H$→ Substituted benzene

Acenaphthylene + Isoprene → 7-Methyltetrahydrofluoranthene —$-4H$→ 7-Methylfluoranthene

Figure 7. Diels–Alder addition to PAH (*79*).

pyrosynthesis of PAH rather than directed toward obtaining quantitative data.

2. Enrichment of PAH from Air Pollutants

For PAH analysis, air pollutant samples are usually extracted with benzene in Soxhlet (*42*), or by ultrasound (*23*). It has been suggested that PAH be sublimated directly from a representative portion from the loaded filter (*85, 93*). The direct sublimation of PAH from a filter is a rather questionable technique, but sublimation from the dry benzene extract ("organic matter") appears promising, provided the PAH are determined from the sublimate by combined glc/ms or high-speed liquid chromatography followed by ms.

PAH are enriched mostly by chromatography on alumina, silica gel, or Florisil. A major drawback of this clean-up step is that the PAH fractions are rich in "nonspecific background" (*103*), which can yield incorrect results. The background is due to the presence of lubricant and/or fuel oil components (*104*), which are primarily nonvolatile alkanes, alkylated benzenes, and ketones. These agents can be easily separated by distribution between cyclohexane and nitromethane (*56*). Other solvent systems for PAH enrichment are cyclohexane and dimethylsulfoxide, acetonitrile and *n*-hexane, and nitromethane and carbon disulfide (*105, 106*).

A valuable technique for the enrichment of the major and minor PAH in organic matter is countercurrent distribution. A 1000-cell Craig countercurrent distribution apparatus with a solvent system of *n*-hexane and aqueous monoethanol–ammonium deoxycholate, or of cyclohexane and methanol–water (9:1) containing 0.83% tetramethyluric acid (*107, 108*) is suggested. While this technique may lead to the unambiguous identification of hitherto unknown carcinogenic hydrocarbons, it is not practical for the routine analysis of PAH.

PAH concentrates are best chromatographed on alumina or Florisil of uniform particle size, either 80–100 or 100–120 mesh, and in columns with diameter–length ratios of at least 1:25. Silica gel is less practical, since the elution of the hydrocarbon fractions requires a relatively long time. The separation of PAH concentrates into PAH subfractions has been successfully achieved on the lipophilic gel LH-20 (*108a, 108b*). The individual fractions are determined either by spot tests, or in the case of ^{14}C-labeled internal standards, by the radiogram of the eluate (*109*). The purity of solvents and adsorbents must be ascertained for the analysis of PAH, since both materials may contain not only plasticizers, but also traces of PAH.

When only small samples are available for the PAH analysis and the alkylated PAH are not of interest, the concentrate may be applied directly to paper chromatography sheets. Two systems give relatively satisfactory separation—Whatman no. 1 paper saturated with dimethylformamide using *n*-hexane as solvent (*110*), and acetylated Whatman no. 1 paper (20–25% acetate content) with different solvent systems (*56, 111*).

Thin-layer chromatography (tlc) has increasingly replaced the more time consuming and laborious column and paper chromatography techniques. Sawicki and his associates (*112–114*) have skillfully and extensively applied tlc to air pollution research and have demonstrated repeatedly that BaP can be isolated almost pure by tlc within 4–6 hours from organic particulates, or, more appropriately, from column chromatography fractions. Thin-layer chromatographic separations do not serve equally well for other PAH but, in the hands of an experienced analyst, tlc is a quick and helpful technique.

When separating carcinogenic PAH, other than BaP, one must be cautious. Analyzing the extracts of the tlc spots or bands only by spectroanalytical techniques can be erroneous. Mass spectrometry, gas chromatography, and/or liquid chromatography have shown that the isolated materials contain often not only one but several PAH and alkylated PAH in addition to nonaromatic compounds. As in paper chromatography, the PAH on thin layers decompose rapidly. Therefore, internal standards should be used to control losses during analytic procedures.

3. Instrumental Analysis

a. Spectroscopic Methods. Ultraviolet absorption spectroscopy has been applied for more than two decades to PAH analysis of polluted air. Extensive compilation of UV-spectra has been very useful in air pollution research, especially in PAH analysis (*115, 116*). Available book collections of UV-spectra are relatively expensive and often lack data on PAH, especially alkylated PAH (*117, 118*). The American Society for Testing and Materials (ASTM) offers punched computer cards listing the information on published UV-spectra of PAH (*119*).

The ultraviolet absorption spectra of all carcinogenic, nonalkylated PAH and azaarenes are highly structured and indicate that the parent polycyclic hydrocarbons and azaarenes are planar in the ground state (*120*). However, it is misleading to assume that all carcinogenic PAH present in air pollutants are planar (*41*). Certain carcinogenic, alkylated PAH, such as 12-methylbenz[*a*]anthracene and 5-methylchrysene have less structured UV-spectra and are most likely not planar (*80, 121*). These and similar alkylated PAH are possibly present in our respiratory environment.

Clar has chosen an empirical classification of the bands in the UV-spectra of PAH (*116*). The α bands are weak and usually on the low energy (long wavelength) side of the spectrum and are sometimes covered by the neighboring p bands. The β bands occur at shorter wavelengths, are more intense, and show fewer details than the α bands. Although the total UV-spectrum is unique for a given parent PAH, individual bands or features are not. For example, in cyclohexane, BaP and benzo[*k*]fluoranthene have α bands at 401 and 402 nm, respectively, and BaP and benzo[*j*]fluoranthene both have p bands at 383 nm. Therefore, identification or quantitation of PAH on the basis of only a few peaks is not recommended, especially when compounds are not clearly separated in chromatographic steps.

Inert, nonpolar solvents such as *n*-hevane, or cyclohexane are recommended, since they refine spectral features of PAH. For qualitative analysis the UV-spectra of isolated polycyclics should be recorded down to at least 250 nm. Often, individual aromatic hydrocarbons are isolated together with impurities of other PAH and/or background material (see Section IV,D,2). Therefore it is recommended that qualitative data be derived from peak heights with corrected baselines.

Luminescence spectrophotometry is a very sensitive and widely used analytical method for PAH (*41, 121*), especially for the trace analysis of aromatic systems in organic particulates (*122*). In general, emission and excitation spectra are recorded in solution at room temperature. Cyclic or acyclic alkane spectra at lower temperatures are often very characteristic, showing many bands. Sawicki *et al.* have explored a new technique, quenchofluorometry, for multicomponent mixtures (*123*). Using this technique, a quenching agent (e.g., nitromethane) is added which effects a more pronounced quenching for one type of PAH in comparison to others.

Phosphorescence spectra are measured in solid solution at low temperatures. Phosphorometry appears to be the most sensitive optical technique for PAH and is more selective than fluorescence spectroscopy. This method has been explored in great detail by Dikun and his group and has been widely applied in the Soviet Union for the analysis of PAH, especially BaP (*124*).

By comparison luminescence spectrophotometry of PAH offers high sensitivity but only limited accuracy ($\pm 10\%$). The polycyclic hydrocarbon isolated from air pollutants must be highly purified prior to analysis, in order to lead to reliable quantitative data. New techniques that circumvent extensive purifications (*123*) appear encouraging and should be pursued.

The Intersociety Committee recommends a tentative method for monitoring exposure to BaP in airborne particulates (*124a*). Particulates are collected on a high-volume sampler, a portion of the filter extract is chromatographed on alumina tlc plates, and the BaP band is extracted with a mixed solvent. The residue is then dissolved in sulfuric acid and the BaP content is measured fluorometrically near 520 nm excitation and 540 nm emission wavelengths. Lannoye and Greinke improved this fluorometric method by a comprehensive study of each individual step of the analytical procedure (*124b*) and they established the best conditions for obtaining reproducible data for BaP in pollutants ($\pm 6.2\%$). This refined method has a sensitivity down to 15 ng and should be employed in control laboratories interested in rapid methods for reproducible BaP data as an indicator of the concentration of carcinogenic PAH in the respiratory environment.

Infrared spectrophotometry and nuclear magnetic resonance (nmr) spectrometry are valuable for the synthesis and structure elucidation of polycyclics. In general the sensitivity of proton nmr was poor compared to other spectrophotometric techniques, and its use in the identification of traces of methylated PAH in mixtures such as organic pollutants, severely restricted. However, with the advent of rapid spectrum accumulation through Fourier transform methods, proton nmr can be facilitated for the identification of small quantities of methylated PAH (*125*). The latter development becomes even more important since certain methyl derivatives of PAH (e.g., chrysene, benz[α]anthracene and BaP have now been found to be strong carcinogens, whereas other isomeric methylated derivatives of the same parent PAH are only weakly carcinogenic or even inactive (*125a, 125b*).

b. High-Speed Liquid Chromatography. Column chromatography had remained unattractive as a method for the separation of PAH because of its long elution time, poor column efficiency, and the need for large volumes of solvents. Since 1967 (*126*), however, the speed and efficiency of liquid chromatography (lc) has significantly increased, and because of the development of new instruments, lc has partially replaced paper chromatography, tlc, and gas–liquid chromatography. For PAH, lc is expected to gain a predominant role (*127*). In contrast to gas–liquid chromatography, high-speed liquid chromatography operates at low temperatures (generally below 50°C), thereby preventing thermal rearrangements and decompositions. An additional advantage is that the detector systems do not consume materials, the separation columns can be used for an unlimited time and unknowns as well as solvents can be recycled.

Liquid chromatographic separations are based on specific interactions of sample molecules with the stationary and mobile phases. A test solution is injected via the sample introduction system into the solvent or solvent mixture (carrier) and is pumped under high pressure through tightly packed columns filled with particles of small size (2–40 μm) and desired surface characteristics; the sample is finally passed through a small-volume detector. Ultraviolet detectors operating at wavelengths between 200 and 400 nm with sensitivities of 10^{-10} to 10^{-9} gm of PAH are preferred for this type of analysis. This basic system is constantly being improved by achievement of higher resolutions and higher sensitivities, as well as by the development of specific detector systems.

Liquid chromatograph pumps have pressure outputs of 50–400 atm to overcome the high resistance of long, densely packed columns. They are pulse free and deliver a constant solvent flow. Originally, column separations were achieved with one solvent system. However, higher resolutions can be achieved by gradient elution. For this purpose, the sample is eluted from the column by a solvent system that varies gradually. Maximal separations can be achieved by recycling the solutions with the unknowns (*128–130*).

High-speed liquid chromatography has been very helpful for the separation of the metabolites of carcinogenic PAH (*131*). We recommend, therefore, that an investigator interested in the analysis of PAH consult the latest literature (*131a-c*), including that of major instrument manufacturers, in order to avail himself of an optimal separation scheme.

High-speed liquid chromatography should not be employed as the sole method of PAH identification. Each recorded maximum in the chromatogram requires analysis by UV-absorption, fluorescence spectrophotometry, and/or mass spectrometry. It is expected, however, that within a few years lc will be interphased with UV-spectrophotometers thereby allowing a detailed analysis of each eluted band. This would make it possible to concurrently analyze all major PAH pollutants after a simple clean-up step.

c. Gas–Liquid Chromatography. For gas–liquid chromatographic separation, PAH concentrates are vaporized in an inert gas and carried through columns containing a liquid or solid phase (columns being 1–50 m long and isothermal or temperature programmed). The PAH are transferred along the stationary phases and separated according to their relative volatilities and their affinities to the phases. At the column exit, the individual compounds or mixtures of compounds pass through a sensor, which, for PAH, is usually a flame ionization detector or an electron capture detector with sensitivities down to about 10^{-9} gm.

The resolution capability and sensitivity of glc systems has very significantly increased. The use of the glc method revealed that combustion products, and especially organic pollutants, contain many more hydrocarbons of various ring structures than were previously assessed (*86, 92, 109, 132*). The many maxima in the gas chromatograms of PAH concentrates, however, should not be identified only by comparison with reference compounds, but also by additional criteria, preferably UV-spectra. Our experience has shown that a single glc peak will resolve into several isomeric PAH or even into different individual PAH with background materials. Satisfactory separation of individual PAH is a major prerequisite for estimating the carcinogenic potential of air pollutant samples. We can only partially separate the six isomeric dibenzopyrenes (*92*), hexacyclic aromatic hydrocarbons with different carcinogenic activities (see Section IV,D + VI,D + VII,D). Although one can separate chrysene from benz[*a*]anthracene by glc (*109*), one cannot separate methylchrysenes from methylbenz[*a*]anthracenes. These are alkylated PAH having different or no carcinogenic activities (*45, 80*). The greatest advances in separation have been achieved by the development of new column phases. Whereas at one time the chemist had to be satisfied with the limited separation of some PAH, e.g., BaP from benzo[*e*]pyrene, (BeP), or chrysene (Chr) from benz[*a*]anthracene (BaA) (*86, 133*), such inadequacies are no longer acceptable. For a time, the higher resolutions achieved by pretreated or untreated capillary columns made them superior to packed columns despite their limited lifetime. With the emergence of new liquid phases, packed columns now yield equally good resolutions, and are preferred to capillary columns because of their better reproducibility at higher temperatures (>200°C), and their greater capacity for large volumes of solutions. Although we do not expect major future improvements in the separation of PAH by glc, one would be well advised to consult the manufacturers for the latest developments in column packings.

d. Combination of Separation Methods. Compared with such environmental agents as tobacco and tobacco smoke, marijuana smoke, and petroleum, most PAH concentrated from polluted air can be easily separated and identified. In general, the analysis of organic particulates begins with a cleanup step, which is primarily column chromatography. The PAH containing fractions are subsequently separated by paper chromatography or gas chromatography. Although the fluorescence and R_f values of the separated bands or the retention times of the maxima in the chromatograms are helpful for the identification of individual PAH, these criteria must be supplemented by others for positive identification. The most reliable identification has been obtained from fluorescence or ultraviolet absorption spectra of the isolated polycyclics.

In the analysis of PAH chromatographic fractions, mass spectrometry possesses the advantage over ultraviolet spectroscopy of detecting all organic compounds present, not just the major PAH. However, the molecular stability of six-ring PAH upon ionization is so great, that the excess kinetic energy imparted by 70 eV electrons can be accommodated within the ionized molecule. The result is a minimum of molecular fragmentation in the mass spectra and, except for molecular ions with single and multiple charges, a minimum of distinguishing qualitative features among spectra of isomeric PAH. Thus, mass spectrometry is primarily useful for determining molecular weights and purities of isolated PAH rather than for their identification.

The choice of a method of determination for carcinogenic PAH depends primarily on the degree of accuracy required, on the availability of a professional and technical staff, and on funds for equipment. In our judgement, a combination of column chromatographic cleanup yielding three to four PAH fractions, followed by parallel separation with glc and lc, and final identification by UV-spectra offers the most promising and most practical route. We realize that this analytical scheme is time-consuming and requires special skills. It is expected that high-speed liquid chromatography equipment interphased with ultraviolet spectro-photometers will be marketed in the near future and will offer more reliable quantitative data as well as a speedy procedure for comprehensive PAH analysis.

4. Quantitative Analysis

Quantitative PAH analysis requires specific precautionary laboratory conditions, since PAH are easily photooxidized and are often present in trace amounts as impurities in solvents, absorbents, and commercial reference compounds (see Section II). Whenever possible, synthetic references should be used (*134*); if they are not available, spectral data from the literature should be employed for comparison.

The use of internal standards for quantitative trace analysis is highly recommended. Internal standards are essential for the quantitative analysis of tobacco smoke comprised of more than two thousand compounds, but PAH analysis of the less complex urban organic pollutants is to some degree reproducible without internal standards. A well-experienced staff and frequent dual analyses are prerequisites for a reliable determination of microgram amounts of PAH without the use of internal standards.

Based on our own experience, we prefer ^{14}C-labeled compounds as internal standards. They guarantee quantitative data for the major PAH ($\pm 8\%$) as well as reproducible PAH fractionation. We employ ^{14}C-

labeled BaA, BaP, and cholesterol for the BaA, BaP, and post-BaP fractions. Aliquots of these fractions are analyzed by glc, lc, and paper chromatography. The gas chromatograms and liquid chromatograms of each of the three PAH fractions represent the PAH profile, or PAH fingerprint, for a given air pollution sample. The losses during the analysis are below 15%.

In our earlier studies of air pollutants in more than a hundred PAH analyses, we quantitated only BaA and BaP with their labeled counterparts as internal standards. The organic particulates were chromatographed on alumina columns and the PAH fractions were used for the isolation of BaA and BaP by paper chromatography. The bands with the two hydrocarbons were extracted, and the extract was analyzed by UV-spectrophotometry and liquid scintillation counting. We found 15–34% losses during this type of analysis (with major losses during paper chromatography). BaA losses were at least 5% higher than BaP losses. It must be assumed that similar techniques with tlc as the final separation step would cause similar losses.

Sawicki and his co-workers introduced an analytical procedure that appears quite simple and permits the assay of an air pollution sample for at least four PAH within a reasonable time (*135*). The benzene-soluble pollutants are chromatographed on alumina, and the eluted, fluorescent column fractions are used for the determination of individual hydrocarbons. BeP is assayed by dissolving the fraction in sulfuric acid and recording its fluorescence spectrum at an activating wavelength of 362 nm. BaP is determined by dissolving the fraction in sulfuric acid and recording the spectrum at an activating wavelength of 500 nm. Perylene is quantitated by dissolving the sample in pentane and recording the fluorescence spectra at an activating wavelength of 430 nm. Benzo[*k*]-fluoranthene is determined by its maxima at 401 nm in the UV-spectrum of the PAH fractions. This method is especially recommended for obtaining a simple PAH profile of air pollutants, since it requires only a short time and no special laboratory equipment.

The majority of published quantitative methods employ column chromatography as a first step and various techniques for the final isolation of individual PAH. Tlc is the most popular technique for isolation. From the extract of the fluorescent bands, PAH are determined by UV- or fluorescence spectra. Gas chromatography has been increasingly utilized. Although modern liquid phases, such as OV-17, have significantly improved the resolution for several individual PAH, other maxima remain to be resolved. Peak heights are affected by background material; therefore, glc data have to be confirmed by separations with other means, namely tlc, paper, or liquid chromatography. One major goal of air pollution research and control is to gather, within a reasonable time, quantitative data on toxic and carcinogenic agents in the respiratory environment. PAH are one group of compounds associated with the urban factor for lung cancer. Although significant progress has been made, the methods for PAH determination are still too time consuming and need improvement in accuracy. It is expected that new instrumentation will alleviate these problems.

5. Sources in Polluted Air

In a previous section (IV,D,1), we noted that the majority of the polynuclear aromatic hydrocarbons in our environment derive from incomplete combustion of carbon- and hydrogen-containing compounds. The amount of PAH pyrosynthesized will vary widely with the efficiency of combustion. Four major sources are prime contributors to the PAH emission into our respiratory environment: (a) heat and power generation, (b) refuse burning, (c) coke production, and (d) motor vehicle emissions. We must add to these technological sources uncontrolled combustions, such as forest fires. Although uncontrolled burning may at times contribute significantly to the air pollution situation of a locality, it does not play an appreciable role in the total air pollution of industrialized nations.

The estimated emission data for BaP in the United States are presented in Table VI. The figures derive primarily from assessments by Hangebrauck and his associates (*135–139*), the United States Government (*140–144*), and industrial research (*145*) and were compiled by the National Academy of Sciences (*41*). We have excluded agricultural and coal refuse burning from Table VI. Based on our own data on BaP in air pollutants from refuse coal burning in Pennsylvania, we calculated an emission of <50 tons of BaP per year ($<120{,}000$ tons of refuse coal burned for the United States) (*146*).

In summary, we estimate that, during one year, from 900 to 1400 tons of BaP are emitted into the atmosphere of the continental United States. Calculations for other carcinogenic PAH are rather speculative since only limited data have been published. In the case of the weak carcinogens, BaA and chrysene, we expect at least the same quantities as for BaP and possibly as much as twice the BaP concentration. Figures for BeP a weak carcinogen suggest that the BeP to BaP ratio varies between 0.5 and 3.1. Values for the other carcinogenic PAH such as benzo[*j*]fluoranthene, benzo[*b*]fluoranthene, and indeno[1,2,3-*cd*]pyrene could not be calculated because of the scarcity of reported data (see Table V).

These data were discussed to give the reader an idea of the magnitude

Table VI Benzo(*a*)pyrene—Sources and Estimated Emission in the United States[a]

Major sources	*BaP emission (tons/year)*
Vehicles	40–50
Gasoline powered	20–25
Diesel fuel powered	20–25
Heat and power generation	450–500
Coal	400–450
Oil	2–3
Gas	44
Wood	40
Refuse burning	300–500
Enclosed incinerators	30–35
Open burning	200–250
Coal refuse burning[b]	<50
Industrial emission	200–220
Petroleum	6
Coke production	200
Total estimated emission	1000–1200

[a] From Hangebrauck *et al.* (*136–139*); U.S. Government (*140–144*), *Industry* (*145*).
[b] Compilation of National Academy of Sciences (*41*); 340 tons/year was based on unpublished information.

of emissions of carcinogenic PAH into the air of the United States and, similarly, into the air of other industrialized nations. Whereas vehicle exhaust appears to contribute less than 5% to the total PAH pollution in the United States, it represents a major PAH source in cities such as New York, New York, and Los Angeles, California (*10–12*). In London, England, domestic coal combustion was, at least until 1955, the major source for pollution by polycyclics (*91*). In Detroit, Michigan, nonautomotive industrial polluters are a major source (*9*).

In Table VII, we have summarized the ranges of particulates and BaP concentrations found in various sources and atmospheres. Again, BaP was chosen because the published data are more reliable. It is formed by the same mechanism as other PAH, and it represents the most active carcinogenic hydrocarbon in combustion products and polluted air.

Differences in collection methods and analytical techniques do not permit an absolute comparison of the values reported by various groups, and we have presented only the highest values reported in the literature up to and including *Chem. Abstr.* **79,** 1973.

Table VII (*28, 65, 71, 139, 140, 147–151*) compares some values for PAH obtained from the undiluted aerosols as arising from the combustion

Table VII Range in Particulate and Benzo(*a*)pyrene Concentrations in Air Pollution Source Effluents and Various Atmospheres[a]

Location or source	*Benzene-soluble material ($mg/1000\ m^3$)*	*Particulates ($mg/1000\ m^3$)*	*µg BaP per gm benzene-solublematerial*	*µg BaP per gm particulate*	*µg BaP per 1000 m^3 air*	*Reference*
Source effluent						
Coke oven, loading point[b]	—	—	—	—	33,000,000	(*147*)
Coal tar pitch kettle, 310°C 20 cm from surface	420,000	420,000	14,000	14,000	6,000,000	(*148*)
Petroleum catalytic cracking, catalyst regeneration[b]	—	—	22,700	—	2,800,000	(*139*)
Gasworks, retort fumes	—	—	—	—	2,300,000	(*149*)
Stack—home heating, coal[b]	150,000	470,000	9,400	3,100	1,400,000	(*71*)
Incinerator, garbage	98,000	820,000	14,000	1,700	1,400,000	(*139*)
Outdoor burning—floor mats, auto seats, etc.	40,000	460,000	4,300	380	170,000	(*71*)
Incinerator, auto parts[b]	300,000	1,300,000	580	130	170,000	(*71*)
Incinerator, vegetable matter	3,100	200,000	4,500	70	14,000	(*71*)
Outdoor burning-grass, leaves	93,000	140,000	29	45	4,200	(*71*)
Municipal incinerator, refuse	63,000	91,000	29	41	2,600	(*71*)
Power plant stack	1,500	620,000	220	0.5	320	(*71*)
Motel, space heater (gas)	1,400	—	51	—	70	(*71*)
Air polluted mainly by one type of pollutant						
Sidewalk tarring	17,300	17,300	4,500	4,500	78,000	(*139*)
Coal tar pitch working area	50,000	50,000	5,200	5,200	75,000	(*140*)
Roof tarring area	740	740	19,000	19,000	14,000	(*71*)
Blackwell tunnel—mainly auto traffic	—	930–2,400	—	91–151	120–290	(*71*)
Boston, Massachusetts, Sumner tunnel						
Outgoing air	240	610	300	110	69	(*71*)
Incoming air	5.4	100	500	26	2.6	
Bus garage	—	1,400	—	18	80	(*28*)
Indoor and personal pollution						
Hut I in mountain of Kenya	2,580	2,700	33	31.5	85	(*150*)
Hut II in mountain of Kenya	2,750	3,625	106	71	291	(*150*)
Room polluted by tobacco smoke	—	—	—	—	28–87	(*151*)
Cigarette smoke[c]	82,000,000	95,000,000	1.2	1.0	95,000	(*65*)
Cigarette sidestream smoke[c]	—	—	3.6	3.0	—	(*65*)

[a] Most values taken from Sawicki (*71*).
[b] Highest value found.
[c] An 85 mm United States cigarette without filter tip was tested and smoked once a minute with a 35 ml puff of 2 seconds duration. To reach a 23 mm butt, an average of 10.5 puffs had to be taken.

of various types of organic matter and from air pollution samples. The extreme values of 33 gm of BaP per 1000 m^3 of air at the loading platform of a coke oven during recharging originates from Czechoslovakia (*147*); absolute PAH values for coke oven effluents from the United States have not been reported, although the problem has been studied analytically (*132*). In this connection, it is of more than academic interest that, workers, employed for five or more years on the top side of coke ovens, have 6.9 times the lung cancer rate expected (*152, 153*). This study by Lloyd *et al.* supports the concept that certain PAH can be carcinogens in the lungs of man when inhaled in high doses.

Table VIII (*139, 144, 154, 155*) reveals that pollution of the respiratory environment is not limited to industrialized nations, but occurs also inside the dwellings of certain inhabitants of developing nations (*150*).

Very high PAH values have been reported during temperature inversions. During the 1956 air pollution episode in London, England, an extreme value of 2200 μg of BaP per 1000 m^3 was found (*91*). Due to air pollution control regulations in all industrial nations, we should not find such extreme conditions ever again. Nevertheless, relatively high PAH concentrations will be present during inversion periods, in spite of strictly enforced air pollution control measures as is exemplified by the air pollution episode of November 25 to December 5, 1962 in the Province of Ontario, Canada (*13*). The highest BaP values there were 40–50 μg per 1000 m^3 compared to 5–20 μg found under standard weather conditions during November and December of other years.

Several other studies have demonstrated significant seasonal variations in the concentration of organic particulate matter and PAH for given locations (*9–11, 13, 16, 84, 88, 91, 137*). For example, BaP and BaA values (μg/1000 m^3 air) for one location in a commercial area of Los Angeles, California, varied between 0.42 and 0.71 in July 1964 and between 11.0 and 11.5 in January 1965 (*11*). For Birmingham, Alabama, Sawicki *et al.* reported daily variations between summer and winter; for BaP, 6.4 and 25.0 μg/1000 m^3 air; BeP, 5.9 and 10.0; benzo[*g, h, i*]-perylene (BPer), 8.2 and 18.0; and coronene, 2.4 and 3.5 (*88*). For Sydney, New South Wales, Australia, even higher seasonal variations have been reported: BaP 0.8 and 7.7; BeP, 1.0 and 5.9; BPer, 1.5 and 8.6; and coronene, 0.9 and 4.8 (*84*). These seasonal variations in particulate matter and PAH may be attributed to more frequent inversions, lower wind velocity, and lower atmospheric destruction rates (photooxidation), as well as to more active emitters in winter.

Several calculations have indicated that automotive exhaust contributes less than 10% to the organic particulate pollution, especially in terms of carcinogenic PAH pollution, in the air of industrialized nations. The

Table VIII Chemical Analytical Data—Situations Analyzed During Cooking in Huts in Kenya (*150*)

Geographic location	*Ethnic group*	*Tribe and hut number*	*Sample size*[a]	*Total particulate matter extracted (mg)*	*Total organic matter extracted (mg)*	*Total organic matter (mg/m^3)*	*BaP ($\mu g/1000\ m^3$)*	*BaA ($\mu g/1000\ m^3$)*
Mountain	Bantu	Nyeri 2	A	731.2	632.0	6.763	166	515
Mountain	Bantu	Nyeri 3	A	252.4	240.6	2.575	85	79
Coast	Bantu	Wadigo 1	A	140.2	75.5	0.808	24	29
Coast	Bantu	Wadigo 1[b]	B	33.0	31.9	0.304	None found	16
Mountain	Nilohamitic	Nandi 1	B	431.0	289.3	2.754	291	268
Mountain	Nilohamitic	Nandi 2	B	592.0	409.5	3.898	140	225
Central plateau	Nilohamitic	Samburu	A	246.2	93.9	1.005	37	33

[a] Total sample size A = 3,300 ft^3 = 93.446 m^3; B = 3,710 ft^3 = 105.050 m^3.
[b] Sample was obtained in a bedroom adjacent to the kitchen.

figure for the United States is less than 5% (Table VI). A study by the automotive industry has attempted to determine the contribution of gasoline engine exhaust to the air pollution in Detroit, Michigan, New York, New York, and Los Angeles, California, (*9–11*). Carbon monoxide and other gaseous pollutants, lead, organic particulates, BaA and BaP were determined at four to five locations in each city during all four seasons, during day and night, during the week and at weekends, and during different traffic patterns. In the centers of these three cities, the investigators found that the "correlation coefficients of BaP and BaA with carbon monoxide and lead are significantly nonzero, indicating a possible causal relationship of polynuclear aromatic hydrocarbon emission and automotive traffic" (*11*). The higher annual BaP values for two locations in the center of Detroit, Michigan (5.8 and 6.9 μg/1000 m^3) as well as the somewhat elevated annual BaP levels for two New York, New York, locations (3.9 and 1.3), in comparison to these data for Los Angeles, California (1.6–2.1), suggest that a much higher proportion of BaP arises from nonautomotive sources in the first two cities (*11*). It was also calculated that automotive exhaust contributes significantly to the PAH pollution in residential areas of Detroit (1.0 μg BaP per 1000 m^3, estimated to be 42% of total BaP emission), as well as in New York (0.3 and 0.6 for two locations), and Los Angeles (0.3 and 0.5).

Table IX demonstrates that cars with emission control systems have significantly reduced PAH emission in contrast with older models and that further improvements can be expected. Proper engine maintenance is one important factor in controlling PAH emission. In one study it was observed that a newer car—5000 miles (8000 km) emits about one-fifth of the BaP of a car approaching the 50,000 mile (80,000 km) mark (*139*). The exhaust emission of PAH (pyrene, fluoranthene, BaA and BaP) increases 4–12 times with an increase in oil consumption from 1 quart (1 liter) per 1600 miles (2560 km) to 1 quart per 200 miles (320 km) (*156*).

Table IX Automotive BaP Emission Factors Influencing the Emission of BaP from Automobiles with Gasoline Engines

Source	*BaP (μg/gallon fuel consumed)*	*Reference*
Uncontrolled car (1956–1964)	170	(*139, 154*)
Uncontrolled car (1966)	45–70	(*144*)
Emission controlled vehicle (1968)	20–30	(*144, 154*)
Advanced systems	≦10	(*155*)

The composition of fuel is decisive in the emission of PAH (Table X). Predictably, increase in the aromatic portion (about 30% in standard fuel) elevates PAH emission significantly. This becomes an important consideration when, in order to achieve a high octane number, tetraethyllead in gasoline is reduced or eliminated and the aromatic portion of the fuel is increased (*157*).

Gasoline contains traces of PAH which are formed during petroleum refining (BaP up to 1 ppm). The PAH in gasoline may contribute to the concentrations in the exhaust gases. When gasoline was spiked with BaP-^{14}C it was calculated that between 15 and 30% of the emitted BaP originated from the fuel (*158*). These few data show that progress has been made in controlling the emission of carcinogenic PAH from internal combustion engines and that we now know several factors related to the emission rates of these hydrocarbons.

Several studies have attempted to pinpoint relations between the individual polycyclic hydrocarbons (e.g., BaP to coronene, BeP, BaA, or pyrene) as a function of their source (*41*). In one case, the concentration of coronene alone was correlated with traffic density (*12*). Based on our experience, this approach is oversimplified. The pyrosynthesis of individual PAH depends on a number of conditions, and therefore, the increase or decrease in the emission of one PAH is not precisely paralleled by a change in the emission of another. A more fruitful approach for determining the sources of organic pollutants appears to be a comparison of the profiles of nonvolatile alkanes of the major polluters with the alkane profile at the measured location.

In Table XI (*9–11, 13, 16, 84, 159–166*) we have summarized the BaP values found in the air of several communities. Although a few data appear unusually high and need confirmation, it is obvious from all reports that BaP is highest during winter. In general, European cities experience higher PAH pollution than cities in the United States. This may be due to the extensive use of coal for space heating in Europe.

Table X BaP and BaA Emission Rates from a Gasoline Engine Operated with Different Fuels[a] (*78*)

Fuel	*(μg) BaP*	*(μg) BaA*
Gasoline	9.6	17.3
2,3,4-Trimethylpentane	2.6	0.8
2,4,4-Trimethyl-1-pentene	0.7	0.4
50% *o*-Xylene plus 50% benzene	25.8	56.3
25% 2,4,4-Trimethylpentane plus 25% isopentane plus 50% xylene	9.0	32.3

[a] One-minute run.

Some of these data are 10–20 years old and in the meantime progress has been made in air pollution control in Europe. But the fact that BaP values of 10 $\mu g/1000\ m^3$ and more are still reported suggests that further reduction in organic particulates in urban air must be achieved.

E. Pesticides

In the United States alone, 300–400 pesticides are registered for use in food production, and among these are about 80 different chemicals utilized in the control of insects that affect crops, livestock, and households. DDT [1,1,1-trichloro-2,2-bis(*p*-chlorophenyl)ethane] has probably had a greater effect on the prevention of disease and on agricultural economics than any other man-made chemical (*167–169*). Nevertheless, we must also realize that some of the pesticides, especially chlorinated hydrocarbon insecticides, biodegrade only slowly and, therefore, can accumulate in the adipose tissue of man. The scientific literature contains conflicting reports on the carcinogenicity of some of these chemicals (*169, 170*). For example it appears that DDT, aldrin, and dieldrin induce some tumors in experimental animals when applied for extended periods in very high doses (*45*). In the widely used DDD [1,1-dichloro-2,2-bis(*p*-chlorophenyl)ethane], we find the carcinogenic *o,p′*-DDD as an impurity in industrial preparations (*45*). In developed countries, such observations have led to the ban or the voluntary withdrawal of some pesticides. In the United States, the Delaney Amendment to the Food, Drug, and Cosmetic Act (Public Law 85-929) forbids the use of any chemical additive to food that induces cancer in man or animal (*171*).

Analytically, DDT, DDE (DDT hydrochloride), DDD, and their isomeric technical by-products will be found in the neutral aromatic fraction N-2 of the particulate matter (Figs. 1 and 2). The most advanced technique for the quantitative analysis of chlorinated hydrocarbon insecticides and other pesticides appears to be microcoulometric glc in which the pesticide-containing "clean-up" is injected into a gas chromatography column with an electron capture detector, and with hydrogen as carrier gas. The portion of the effluent that presumably carries a pesticide is passed through a 950°C quartz tube in which the organic compounds are reduced to hydrocarbons, water, hydrogen chloride, hydrogen sulfide, and hydrogen phosphide. The last three compounds precipitate silver ions and, therefore, register in a microcoulometric titration cell. Insertion of a short silica gel column removes hydrogen chloride; an alumina column eliminates hydrogen chloride and hydrogen sulfide, and thereby measurements of hydrogen phosphide can be made. Application of appropriate combinations enables one to determine all

Table XI Benzo(*a*)pyrene Content of Urban Air

City	Benzo[a]pyrene ($\mu g/1000\ m^3$)				Year	Reference
	Spring	Summer	Autumn	Winter		
North America						
Atlanta, Georgia	2.1–3.6	1.6–4.0	12–15	2.1–9.9	1960	(159)
Birmingham, Alabama	6.3–18	6.1–10	20–74	23–34		
Cincinnati, Ohio	2.0–2.1	1.3–3.9	14–18	18–26		
Detroit, Michigan	3.4–12	4.1–6.0	18–20	16–31		
Los Angeles, California	0.4–0.8	0.4–1.2	1.2–13	1.1–6.6		
Nashville, Tennessee	2.1–9.0	1.4–6.6	30–55	25		
New Orleans, Louisiana	2.6–5.6	2.0–4.1	3.6–3.9	2.6–6.0		
Philadelphia, Pennsylvania	2.5–3.4	3.5–19	7.1–12	6.4–8.8		
San Francisco, California	0.8–0.9	0.2–1.1	3.0–7.5	1.3–2.4		
Detroit, Michigan (central sites)						
Commercial	7.2	—	—	5.0–17.0	1965	(9)
Freeway	—	4.0–6.0	3.4–7.3	9.2–13.7		
Residential	—	0.2	—	0.9–1.9		
Los Angeles, California						
Commercial	0.58	0.42	0.7	6.6	1970	(11)
Freeway	0.44	0.56	0.53	5.0		
Residential	0.46	0.13	0.15	0.51		
New York, New York						
Commercial	0.5–8.1	0.7–3.9	1.5–6.0	0.5–9.4	1971	(10)
Freeway	0.1–0.8	—	3.3–3.5	0.7–1.3		
Residential	0.1–0.6	—	0.6–0.8	0.5–0.7		
Toronto, Ontario (May–Oct.; Nov.–March)	—	9.5	—	5.4	1966	(13)
Ottawa, Ontario (May–Oct.; Nov.–March)	—	4.0	—	14		

Europe						
England						
London	25–48	12–21	44–122	95–147	1952	(*160*)
Sheffield	20–44	21–33	56–63	74–78		
Cannock	4–16	6–11	27–31	27–32		
London (in traffic background)	20	11	57	68	1965	(*89*)
	11	1	38	42		
Belfast, Northern Ireland	—	9	—	51	1967	(*16*)
Dublin, Ireland	—	3	—	23		
Oslo, Norway	—	0.5	—	14		
Helsinki, Finland	—	1.6	—	5		
Copenhagen, Denmark	6.0	5.0	14	15	1956	(*161*)
Hamburg, Federal Republic of Germany (1962)	14–72	10–26	66–296	94–388	1964	(*162*)
Hungary						
Budapest	—	17–32	—	72–141	1963	(*163*)
Debrecen	—	6.8	—	13.4	1969	(*164*)
Africa						
Pretoria, South Africa	—	10	—	22–28	1964	(*165*)
Johannesburg, South Africa	—	—	—	22–49		
Durban, South Africa	—	—	—	5–28		
Asia						
Osaka, Japan						
Commercial	5.7	2.7	9.4	14	1966	(*166*)
Residential	3.3	1.4	3.8	6.7		
Australia						
Sydney, New South Wales	0.6–2.4	0.6–1.8	2.5–7.4	3.8–8.2	1965	(*84*)

the pesticides that contain chlorine, sulfur, and/or phosphorus. The sensitivity of the method approaches 0.1 μg of pesticides.

In 1963 and again in 1964, Tabor collected particulate air pollutants on glass filters in eight agricultural communities and in four communities with active insect control programs (*15*). The filters were extracted with pentane and subsequently with benzene. The residues of these extracts were analyzed by glc. An electron capture detector, which is sensitive to chlorinated hydrocarbon insecticides and thiophosphates, was used, succeeded by a sodium thermoionic detector, which was regarded as responding solely to thiophosphates. Experimental details were not given. Nevertheless, the results are meaningful as minimum values. DDT, chlordane, aldrin, toxaphene, and malathion were determined. The data were obtained from pesticide-polluted air collected distant from the sprayed area and were reported as "minimum values." The highest value found for DDT was 11 μg/1000 m^3 (mean 5 μg). However, during fogging (spray application) of DDT, as much as 8000 μg/1000 m^3 was detected. Concentrations for malathion ranged up to 140 μg, chlordane up to 5.6 μg, aldrin up to 3 μg, and toxaphene up to 15 μg/1000 m^3 air. Assuming that no decomposition of insecticides occurs during fractionation, one would expect to find the chlorinated hydrocarbon insecticides (DDT, DDD, chlordane, aldrin, and toxaphene) in the N-2 fraction and the thiophosphor ester (malathion) in the N-3 fraction (Fig. 2).

Tabor concluded from his studies that except on days of insecticide fogging, man takes up less of the insecticides by respiration than by dietary ingestion (*14, 15*). However, this conclusion applies only to nonsmokers. Persons who smoke twenty cigarettes daily inhale with the smoke about 15 μg of DDT, 7 μg of DDE, and 35 μg of DDD (*172, 173*). It is estimated that in the United States daily human exposure by food intake amounted to 19 μg (1968), 16 μg (1969), and 15 μg (1970) of DDT, and 15, 11, and 10 μg of DDE, and 11, 9, and 4 μg of DDD, respectively (*174*). The concentrations of these chlorinated hydrocarbons in food and tobacco have significantly decreased during the last several years because of less frequent use of the related pesticides.

One group of investigators analyzed the concentration of DDT in the particulate matter of the air in Pittsburgh, Pennsylvania, and reported summer values up to 2.5 μg/1000 m^3 (*175*). Trace amounts of DDT, DDE, DDD, and dieldrin have been reported in the air of London (*176*). During two separate studies, DDT and DDE were found in airborne dust in the northeast Atlantic Ocean and in the trade winds at Barbados, West Indies (*177, 178*).

Jegier reported on the concentrations of thirteen pesticides in the air in orchards surrounding Montreal, Canada (*179*). During fogging, he found pesticide concentrations of 0.05–7.7 $\mu g/m^3$ of air. These figures will assist health authorities in evaluating the possible toxic and even carcinogenic effects of pesticides in the occupational environment of farm workers, which is additional to the amounts ingested in food. In a study of four cities in California, DDT concentrations during October 31 and December 5, 1963 were found to be 0.5–19 $\mu g/1000\ m^3$ of air (*180*). Pesticides in air derive not only from fogging of pesticide preparations but also from vaporization into the air of insecticides from field soils (*180*). Some insecticides in the air may become chemically converted or become metabolically activated into substances toxic or carcinogenic to the mammalian cell. The insecticides heptachlor and aldrin, for example, may be oxidated in air or *in vivo* to epoxides (*181, 182*).

F. Oxygenated Neutral Compounds

This section discusses ascertained or suspected oxygenated compounds in the neutral subfractions N-2 and N-3 (Fig. 2). These are O-heterocyclic hydrocarbons, epoxides, peroxides, aldehydes, ketones, sugars, quinones, esters, and lactones. Some of these compounds, particularly O-heterocyclic hydrocarbons, esters, and quinones, will appear in both fractions, N-2 and N-3. This is due to the specification of N-2 as the neutral subfraction that contains all PAH and is not caused by any lack of technique in concentrating the "oxygenated" neutral compounds in one specific fraction. In our experience, the separation of PAH and neutral O-, N-, and S-heterocyclic hydrocarbons from certain esters and ketones by column chromatography of neutral air pollutants is rather fruitless, whereas fractionated crystallization of N-2 from acetone below 20°C often leads to an enrichment of esters of fatty acids and long-chain ketones. Another method that should be helpful for the isolation and identification of these oxygenated compounds is countercurrent distribution with at least three hundred cells, e.g., between cyclohexane and nitromethane as the solvent pair.

Based on our knowledge of the combustion of plant products and the large-scale industrial use of plasticizers (phthalates, etc.), one would expect a considerable number of nonvolatile oxygenated compounds to exist in our respiratory environment. So far, however, very few chemical studies have dealt with these air pollutants. The paucity of chemical data in this area is perhaps a consequence of lack of information on the effect on biological systems of subfractions containing oxygenated substances.

1. Cyclic Ketones and Quinones

Compounds of this type are known to occur in urban pollutants. In addition, one would suspect that O-heterocyclic hydrocarbons occur in the N-2 and N-3 fractions (Fig. 2). Our knowledge of the composition of coal tar (*183, 184*) and tobacco smoke (*65, 185, 186*) leads us to anticipate that atmospheric organic particulates should contain benzo-, dibenzo-, and/or naphthofurans and -xanthenes. The absence of any known carcinogenic activity for O-heterocyclic hydrocarbons may account for lack of interest in these agents.

In the group of S-ring compounds, only benzothiazole has been identified in urban pollutants from New York City (see Table XIIIa, *186a*). Otherwise, S-ring compounds have been identified thus far only in petroleum products, where they may contribute to the carcinogenic activity of some petroleum oils (*184, 187*). These compounds could become airborne during the incomplete combustion of S-ring-containing fuels. Some of them are carcinogenic to the experimental animal (*45*). Special glc detectors for S-heterocyclic hydrocarbons are available. Investigation into their presence in air pollution is desirable.

In Figure 8, we show the ring structure of the quinones and ketones that have been reported to be present in urban pollutants. These six compounds were isolated by thin-layer chromatography and determined qualitatively and quantitatively by spectrofluorimetry (*71, 188–190*). Since the techniques of glc, lc, and ms are available in most laboratories,

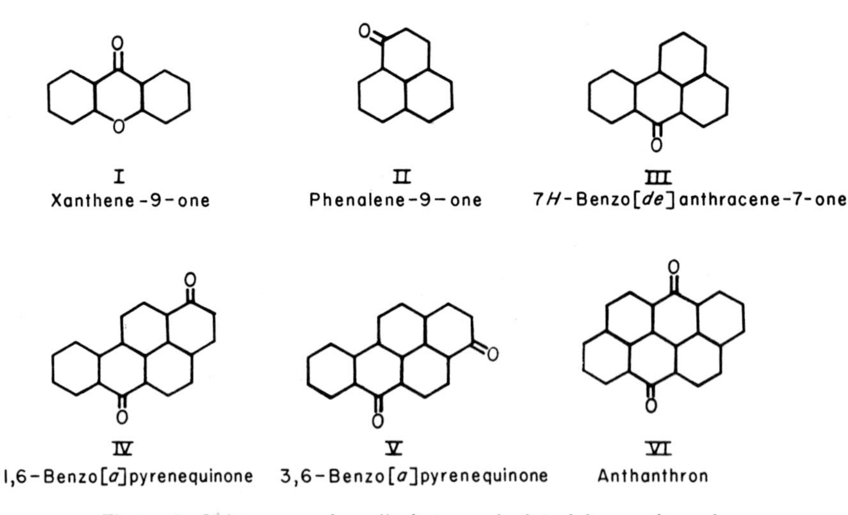

Figure 8. Quinones and cyclic ketones isolated from urban air.

analytical results should no longer be based on spectrofluorimetric measurements alone.

The concentration of 7*H*-benz[*de*]anthracene-7-one (benzanthrone) (III, Fig. 8) in an average United States urban atmosphere was found to be 8 μg/1000 m^3 (*71*). In air samples from an urban site in the United States during January 1968, Stanley *et al.* found values for benzanthrone of 2–48 μg/1000 m^3 air and for phenalene-9-one (II, Fig. 8), of 0.3–17 μg/1000 m^3 air (*189*). Xanthene-9-one (I, Fig. 8), the two benzo[*a*]pyrenequinones (IV and V), and anthanthrone (VI) have been only qualitatively determined (*71, 190*). Phenalene-9-one, 7*H*-benz[*de*]-anthracene-7-one and anthanthrone, as well as an anthraquinone were also identified in an urban pollution sample from Antwerp, Belgium (*48a*). None of these oxygenated ring compounds have been found to be carcinogenic. In one review article, benzanthrone has been considered as a carcinogen (*71*); however, when one applies the standards for carcinogens of the United States Academy of Sciences and the World Health Organization (*191, 192*), this cyclic ketone must be considered inactive. This was confirmed in our laboratories, with a 0.1% acetone solution of benzanthrone being inactive on mouse skin. The quinones of benzo[*a*]pyrene (IV and V) have not yet been tested for carcinogenic activity. Based on studies with other quinones, one many expect the benzo[*a*]pyrenequinones also to be inactive when tested alone. However, they may have a moderate effect when applied together with carcinogenic PAH. This area of "cocarcinogenesis" needs detailed investigation.

Several of the oxygenated neutral compounds are toxic to the experimental animal. These include some mutagenic and tumorigenic lactones, especially γ-lactones with conjugated endo- or exocyclic double bonds, steroids, epoxides, and peroxides (*45, 193, 194*), and perhaps certain esters with tumor-promoting properties (*195*). Several phthalates in high concentrations are known to be skin irritants to animals and possibly to man (*196, 197*).

2. Epoxides and Peroxides

Unsaturated hydrocarbons can be decomposed under various environmental conditions to a variety of compounds, including peroxides and epoxides. The latter belong to the group of experimentally proved mutagens and carcinogens (*45, 193, 198*). Kotin and Falk suggested that photochemical reactions may occur during smog formation, involving unsaturated hydrocarbons, nitrogen oxides, oxygen, and solar radiation, and may lead to the formation of volatile and nonvolatile peroxides and epoxides. These may be significant in air pollution carcinogenesis (Fig. 9) (*199*).

In order to test their hypothesis, Kotin *et al.* (*194, 200–202*) exposed mice to an atmosphere of ozonized gasoline ("synthetic smog"). As a second experimental variable, they used mice that had been pretreated with influenza virus. The finding of an increase in lung adenomas and adenocarcinomas for nonpretreated mice exposed to ozonized gasoline was confirmed by Nettesheim *et al.* in a large-scale inhalation study (*203*). However, mice that were pretreated with influenza virus and exposed to ozonized gasoline showed a reduction in lung adenomas compared to the nonpretreated mice in both studies (*201, 203*).

Kotin *et al.* (*194, 200–202*) observed morphological changes in the bronchi with epithelial hyperplasia and squamous metaplastic changes in the group of mice exposed to smog alone. In combination with influenza virus, synthetic smog produced not only increased metaplastic changes but also lung tumors, which the investigators considered to be squamous carcinomas of a type similar to that seen in man (*201*). The latter observation was not confirmed by Nettesheim *et al.* In fact, these authors

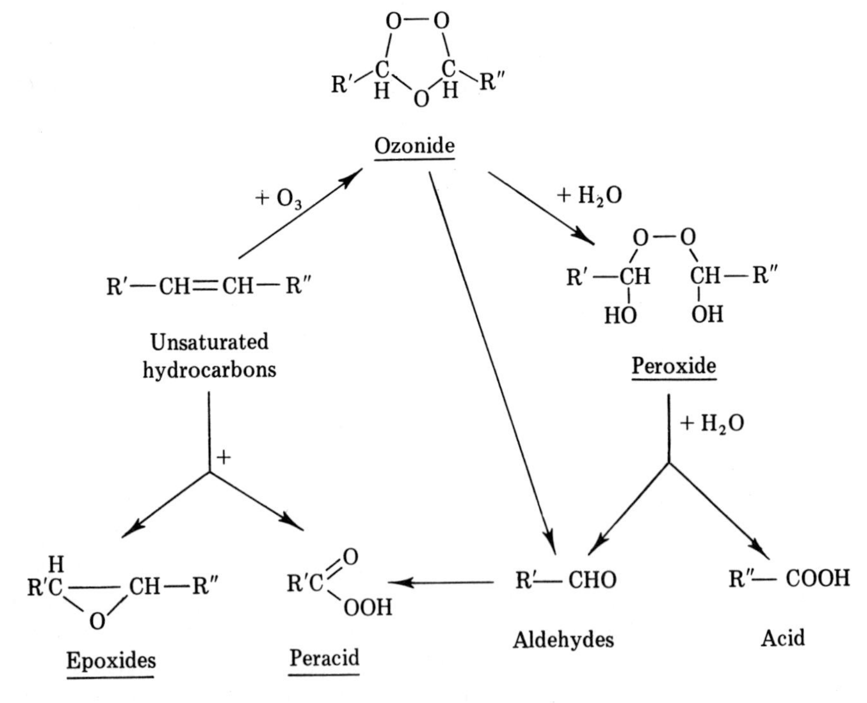

Figure 9. Photochemical reactions of carcinogenic significance in polluted air as postulated by Kotin and Falk (*199*).

observed neither squamous metaplastic changes nor squamous cell tumors in the bronchi (*203*). However, the Nettesheim group treated their mice with virus only once, whereas Kotin *et al.* exposed their mice to virus several times.

The testing of the most polaric neutral fraction of the organic matter of air pollutants from Los Angeles, California, on mouse skin resulted in tumors (*204*). This "oxygenated" fraction was free of carcinogenic hydrocarbons. Chemical studies with this portion of the organic pollutants indicated the presence of epoxides and peroxides. Similar experiments with "oxygenated" fractions from pollutants of United States cities gave positive results only for samples from Birmingham, Alabama, and San Francisco, California (*33*). Epstein and co-workers bioassayed an air pollution sample from New York, New York, and upon injection of the oxygenated neutral portion into infant mice, observed an increased number of lymphomas and lung adenomas (*48*). Our studies with samples from Detroit, Michigan, and New York, New York, did not result in tumor formation on mouse skin when the "oxygenated" subfractions were tested alone, but did demonstrate tumor-promoting activity on initiated skin (see Table XIX) (*33*). These data suggest some tumorigenic activity of the "oxygenated" neutral subfraction of air pollutants and indicate the need for a combined chemical–biological study as to the chemical nature of the active agents. One may speculate that the active agents will include peroxides and epoxides.

3. Steroids, Esters, and Lactones

Di-*n*-butylphthalate was identified in a Detroit, Michigan sample (*33*). Mass spectrometry afforded the identification in the air of Antwerp, Belgium of di-*i*-butyl-, di-*n*-butyl-, benzylbutyl- and di-2-ethyl-hexyl-phthalate (*48a*). It must be remembered though that these and other phthalates are widely used as plasticizers. Their presence in many industrial products, such as solvents, glass fiber filters, plastic laboratory ware, etc. may lead to contamination of the pollution samples.

No study has been concerned with the identification of steroids, esters, and lactones in the benzene soluble portion of air pollutants. Some representatives of these chemical groups are active as mutagens and some as tumorigenic agents (*45, 193*). One may find these organic compounds in the air near the burning of plant products such as hay, bark, wood, and leaves, and indoors in tobacco smoke-filled rooms. All of these materials are known to contain phytosterols, fatty acids, esters, and in some instances lactones, such as coumarin and angelica lactones. Some esters, such as phthalates and sebacates, will be found in the vicinities of industries and shops utilizing plasticizers. The increasing use of plasticizers (United States >300,000 tons annually) supports the possibility of finding certain of these esters in urban air. Some synthetic and naturally occurring lactones, for example, β-propiolactone and certain unsaturated γ-butyrolactones, are known animal carcinogens (*205*) and may occur in pollutants at specific sites.

4. Water-Soluble Organic Compounds

In the study of organic compounds in urban pollutants, major emphasis has been placed on the benzene-soluble portion, while the water-soluble compounds have been widely neglected. The benzene-insoluble material may consist of such highly polaric compounds as mono- and polysaccharides, amino acids, proteins, glucoproteins, nucleotides, nucleic acids, glycosides, glucuronides, and sugars. Biologically, some of these compounds, especially reducing sugars, are of interest, since they may participate in the manifestation of allergy and asthma. Sawicki and Engel studied analytical methods for these water-soluble organic particulates (*206*). They developed a method for the determination of furfural and of those sugars that yield furfural when treated with sulfuric acid. The furan-2-aldehyde is condensed with azulene and results in an intense color. The method is claimed to have a sensitivity of 1.7 μg of furfural. For the analysis of pollutants, the particulate matter is extracted with boiling water and, after the addition of concentrated sulfuric acid and azulene, the color is measured at 492 nm using the baseline correction method. In 1 gm of the effluent from an industrial coffee-roasting operation in New Orleans, Louisiana, the authors determined 11 mg of furfural precursors assayed as ribose. In 1000 m^3 of air of eight United States cities, the investigators found between 1 and 2.5 mg of furfural precursors (*207*).

Another method for identifying furfural precursors was developed by Sawicki *et al.* for the determination of all water-soluble compounds that have a 1,2-glycol structure, including polyhydroxyalcohols, aldoses, ketoses, and α-hydroxyketones. These compounds are oxidized with periodic acid to formaldehyde and are then condensed with 3-methyl-2-benzothiazoline hydrazone. The developed color is measured spectrophotometrically. The method was applied to particulates from eleven United States cities and gave data between 2 and 7 mg of water-soluble α-glycolic compounds in 1000 m^3 of urban air (*207*).

The above methods and those used for other water-soluble agents (*208*) are quite helpful in indicating the concentration of these types of compounds in pollutants. However, like many colorimetric methods, they are unspecific and may give misleading results because of interference by other agents. It is hoped, therefore, that more specific methods for identification and quantitation of individual water-soluble organic particulates will be developed. Gas chromatography and high-speed liquid chromatography appear to be the methods of choice. It will be necessary, however, to silylate or otherwise derivatize these water-soluble agents for such instrumental analysis.

G. N–Nitrosamines

In 1954, Barnes and Magee reported that dimethylnitrosamine (DMN) induces cirrhosis of the liver in man (*209*). This observation led to toxicity tests in rats and other small mammals, resulting in the induction of malignant tumors in the kidney and liver by DMN (*210*). In subsequent studies throughout the world, it was shown that more than eighty nitrosamines are carcinogenic to experimental animals (*210–212, 212a*). The smallest amounts of nitrosamines (DMN—12.5 μg per rat) induce malignant tumors and also exhibit organ specificity for each animal species tested. Similarities in the metabolism of DMN in human and animal tissues *in vitro* suggest the possibility that some nitrosamines and nitrosamides are carcinogenic to man (*213*).

N-Nitrosamines are formed not only from secondary amines but also from tertiary amines and quaternary ammonium salts *in vitro* as well as *in vivo* (Fig. 10) (*214–218*). Methods have been developed that enable the detection of parts per billion levels of these carcinogens in solids, aerosols, and gases (*214, 219–221*). Whereas most nitrosamines in polluted air are expected to occur in the gas phase (*222*), we know from studies on tobacco smoke that this type of compound can also occur in the particulate phase.

The nonvolatile *N*-nitrosamines and nitrosamides without basic or acidic functional groups are expected in the neutral N-3 subfraction (Fig. 2). It is suggested, however, that the search for nitrosamines and nitrosamides should begin with the material collected on the filter and that the extraction be completed in the presence of ascorbic acid to inhibit the artifactual formation of these carcinogens during the extraction with organic solvents. The extract, preferably obtained with chloroform, is concentrated and chromatographed on deactivated alumina. The search should be directed toward those fractions that can contain the nitro-

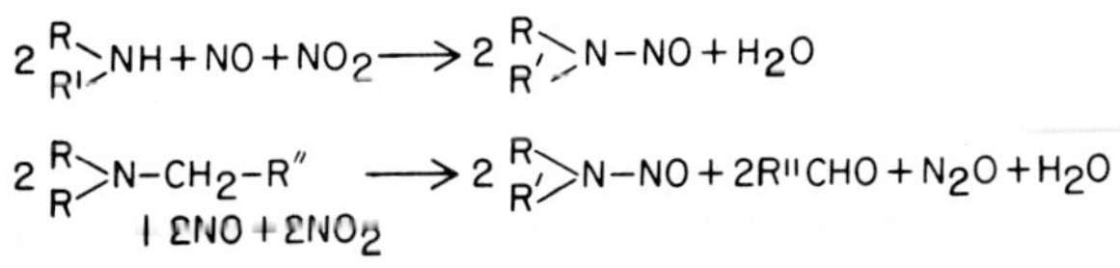

$$2\,{}^{R}_{R'}\!\!>\!NH + NO + NO_2 \longrightarrow 2\,{}^{R}_{R'}\!\!>\!N{-}NO + H_2O$$

$$2\,{}^{R}_{R}\!\!>\!N{-}CH_2{-}R'' \underset{+\,2NO\,+\,2NO_2}{\longrightarrow} 2\,{}^{R}_{R'}\!\!>\!N{-}NO + 2R''CHO + N_2O + H_2O$$

Figure 10. Formation of *N*-nitrosamines.

samines of the major secondary and tertiary amines. (The amines themselves are found in the basic portion of the particulate matter.) Subsequent clean-up should be carried out by tlc, and suspicious bands should be analyzed by glc/ms, or by high-speed liquid chromatography and mass spectra. Mass spectral analysis is essential, since chromatographic methods have been proven to be ambiguous.

Highly sensitive methods have been developed for volatile and nonvolatile *N*-nitrosamines (*214, 219–222*). The "Thermal Energy Analyzer" which has a detector system that is highly sensitive to nitrosamines enabled Fine *et al.* to detect trace amounts of dimethylnitrosamine in the polluted air of some U.S. cities (*222a*): up to 0.08 ppb DMN was found in urban air, and levels up to 0.84 ppb were recorded in the vicinity of a rocket fuel manufacturing plant. The air in bar cars of commuter trains contained between 0.1 and 0.13 ppb DMN, and 0.24 ppb were found in the air in a suburban bar. These data indicate that an individual inhales 50-200 ng DMN per hour in such environments (Brunnemann, K. D. and Hoffmann, D., unpublished data).

V. Acidic Fraction

The acidic portion of urban pollutants (Tables I and II) may account for 10–26% of the benzene-soluble organic matter (reported 2–10 $\mu g/m^3$) (*32, 33, 48, 48a, 223*). In addition, acidic compounds are present in the water-soluble portion of organic particulates. Nonvolatile fatty acids and phenols have been identified as constituents of the benzene-soluble portion of the acidic fractions; other agents, such as aromatic carboxylic acids and acidic enols, are suspected to be present. Since only limited information is available regarding the biological activity of the acidic fractions, little interest has been shown for the qualitative and quantitative composition of these fractions.

A. Nonvolatile Fatty Acids

Incomplete combustion of unrefined materials from plant and animal sources will yield nonvolatile fatty acids in the generated aerosol. It is therefore not surprising that fatty acids are found in urban particulates. Quantitative data for the major saturated and unsaturated fatty acids for one location each in Detroit, Michigan, and New York, New York, are summarized in Table XII. The major acids measured had unbranched chain lengths from C_{12} to C_{22}, with the saturated palmitic (C_{16}) and stearic (C_{18}) acids found in the highest concentrations. As expected, we found much lower concentrations of the unsaturated oleic and linolenic acids (both C_{18}). This might be due, in part, to a lower quantity of free

Table XII Free Nonvolatile Fatty Acids as Determined in the Particulate Matter of Air Pollutant Samples[a]

	Sampled in Detroit, Michigan, on freeway interchange Oct.–Nov., 1963		*Sampled in New York, New York, at "high-traffic" city location, Feb., 1964*	
Fatty acid	*% in acidic portion*[b]	*μg per 1000 m³ air*	*% in acidic portion*[b]	*μg per 1000 m³ air*
Saturated				
Lauric	0.29	13.7	0.43	39.4
Myristic	0.47	17.5	0.53	48.6
Palmitic	3.10	146.8	2.40	220.0
Stearic	2.42	113.7	1.50	137.5
Behenic	1.31	61.6	1.30	119.2
Unsaturated				
Oleic	0.68	32.2	1.34	122.8
Linolenic	0.56	26.5	Trace	Trace

[a] Values represent averages from three analyses.
[b] For definition see Figure 2.

and esterified unsaturated acids in the material burnt and, in part, to a lower transfer rate of these compounds from combustible matter into air pollutants and also due to various degrees of chemical reactivity of compounds in the air. Secondary reactions of unsaturated organic compounds can lead to epoxide and peroxide formation (see Section IV,F,2), and since these agents can be mutagenic and/or carcinogenic, we need more information on the chemical nature and the fate of nonvolatile fatty acids in urban air.

Cantreels and Van Cauwenberghe esterified the fatty acids of the acidic portion of organic pollutants (Fig. 1) by glc/ms, and by recording the parent ion as well as m/e 74, corresponding to the MacLafferty rearrangement fragment of the methyl esters, they identified 19 fatty acids (*48a*). These ranged from n-$C_{11}H_{23}COOH$ to n-$C_{29}H_{59}COOH$ (Table XIIa). The even numbered fatty acids were always present in much higher concentrations than the odd numbered ones. Oleic acid ($C_{17}H_{33}COOH$) was the only unsaturated fatty acid identified. Based on retention times and, primarily, ms analysis, the authors identified also a number of aromatic acids (Table XIIa). The ms identification of tetrachlorophenol and pentachlorophenol in this fraction was supported by the relative abundance of the 5 and 7 parent ions for the two Cl-components as given by the isotope distribution of ^{35}Cl and ^{37}Cl.

Modern analytical methods permit the determination of trace amounts of fatty acids in our respiratory environment. One analyzes the acidic

Table XIIa Acidic Compounds in the Particulate Matter of Antwerp, Belgium *(48a)*

Acids		*Dry particulate matter (ppm)*[a]
Name	*Structure*	
Aliphatic acids		
Lauric	$C_{11}H_{23}COOH$	91
Myristic	$C_{13}H_{27}COOH$	103
Pentadecanoic	$C_{14}H_{29}COOH$	61
Palmitic	$C_{15}H_{31}COOH$	452
Margaric	$C_{16}H_{33}COOH$	78
Oleic	$C_{17}H_{33}COOH$	44
Stearic	$C_{17}H_{35}COOH$	510
Nonadecanoic	$C_{18}H_{37}COOH$	42
Arachidic	$C_{19}H_{39}COOH$	196
Heneicosanoic	$C_{20}H_{41}COOH$	74
Behenic	$C_{21}H_{43}COOH$	250
Tricosanoic	$C_{22}H_{45}COOH$	92
Lignoceric	$C_{23}H_{47}COOH$	180
Pentacosanoic	$C_{24}H_{49}COOH$	35
Cerotic	$C_{25}H_{51}COOH$	110
Heptacosanoic	$C_{26}H_{53}COOH$	18
Montanic	$C_{27}H_{55}COOH$	71
Nonacosanoic	$C_{28}H_{57}COOH$	9
Melissic	$C_{29}H_{59}COOH$	26
Aromatic acids		
Benzoic	C_6H_5COOH	9
Phthalic	$C_6H_4(COOH)_2$	25
Isophthalic	$C_6H_4(COOH)_2$	25
Terephthalic	$C_6H_4(COOH)_2$	150
Methylbenzoic	C_7H_7COOH	—
Methylphthalic	C_7H_6COOH	3
2-Hydroxybenzoic	$C_6H_4(OH)COOH$	6
4-Hydroxybenzoic	$C_6H_4(OH)COOH$	66
3,4-Dihydroxybenzoic	$C_6H_3(OH)_2COOH$	8
Naphthalene carboxylic[b]	$C_{10}H_7COOH$	18
Phenanthrene carboxylic[b]	$C_{14}H_9COOH$	13
Anthracene carboxylic[b]	$C_{14}H_9COOH$	42
Pyrene carboxylic[b]	$C_{16}H_9COOH$	3
Hydroxyphenanthrene or hydroxyanthracene[b]	$C_{14}H_8(OH)COOH$	2
Hydroxypyrene or hydroxyfluoranthene[b]	$C_{16}H_9O_4$	—
Tetrachlorophenol[b]	C_6HCl_4OH	20
Pentachlorophenol	C_6Cl_5OH	75

[a] Average 100 mg/600 m^3 polluted air per 24 hours.
[b] Exact structure not known.

portion of urban pollutants as free acids, or esterified, by gas chromatography or liquid chromatography. Internal standards, [9,10-^{14}C]stearic acid for the saturated fatty acids and [9,10-^{3}H]oleic acid for the unsaturated fatty acids, secure exact data. In our analytical procedures, somewhat higher losses (5–10%) were encountered for the unsaturated fatty acids. Dual channel counting makes it easy to distinguish the β activity of ^{14}C- and ^{3}H-labeled internal standards.

B. Nonvolatile Phenols

Many materials, especially polysaccharides and aromatic hydrocarbons, can give rise to phenols during incomplete combustion. For example, the aromatic portion of gasoline gives rise to phenol and alkylated phenols in the exhaust (*78*). However, these simple phenols will rarely be found in the particulate phase of pollutants. Sawicki *et al.* examined the acidic portion of coal tar pitch and of an air pollution particulate sample from New York, New York, for polynuclear aromatic and heteroaromatic phenols (*224*) by means of electrophoretic techniques and thin-layer chromatography of the nonvolatile phenols. The extracts of the spots from the separations were examined by fluorimetry in neutral or alkaline solution. This study showed that urban pollutants contain trace amounts of nonvolatile phenols such as naphthols, fluorenols, and hydroxycarbazoles. In another study, the organic soluble phase was coupled with diazotized *p*-nitroaniline. The dye extract was separated by several chromatographic steps, and finally, the individual colored spots on the tlc plates were determined with a spectrophotometer. The samples were found to contain 0.19–1.21 μg of 2-naphthol per 1000 m^3 of polluted air (*225*). Tentatively identified were also hydroxyphenanthrene and/or hydroxyanthracene and hydroxypyrene and/or hydroxyfluoranthene (Table XIIa). Polychlorinated phenols in urban air originate most likely from herbicides and/or decachlorobiphenyl (Table XIIa; *48a, 92*).

As a next step, methods have to be developed for the quantitative assay of these weakly acidic compounds. At present, high-speed liquid chromatography interphased with ultraviolet spectrophotometry appears the most promising choice. Since some of the agents may be biologically active, the identification of unknowns has to be supported by mass spectral data.

C. Polyphenols

Essential constituents of plants include polyphenols (tobacco leaves, 0.5–7%) with scopoletin and its 0-glycoside, scopolin, as the major

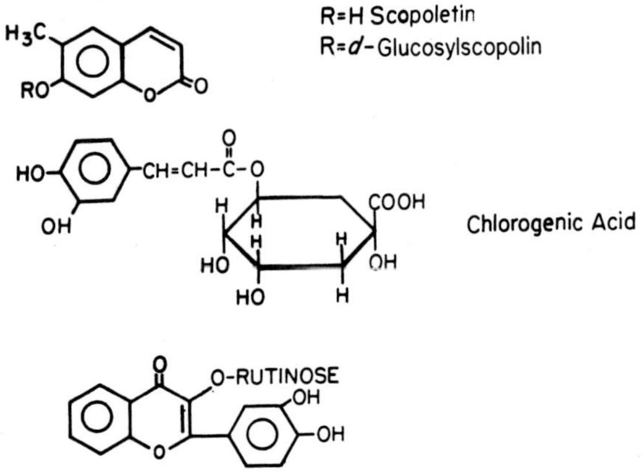

R=H Scopoletin
R=*d*-Glucosylscopolin

Chlorogenic Acid

O-RUTINOSE

Figure 11. Major plant polyphenols as present or suspected to be present in air pollutants.

coumarins, chlorogenic acid as the major caffetannin, and rutin as the major flavonoid (Fig. 11). It appears likely that during the decomposition and/or burning of plants, especially of the leaves, polyphenols are released into our respiratory environment. Whereas studies with tobacco smoke have shown that personal pollution contains many polyphenols (*65*), little is known as to their presence in outdoor and indoor pollution.

Sawicki and Golden developed a fluorimetric method for scopoletin in organic particulates in which the organic particulate phase is chromatographed on silica gel tlc plates, and the filtered extract of the polyphenol spot is analyzed by fluorimetry (*226*). Using R_f values in different tlc systems, mobility values of electrophoretic separations, and fluorescence spectra, the authors ascertained the presence of scopoletin in urban pollutants in an indoor sample and in the effluent of a coffee-roasting plant. The reported values for urban air of four cities (0.2–1.3 μg/1000 m^3) and for an indoor sample (0.7 μg/gm dust) can serve as guidelines for future studies. Based on experience with polyphenols in tobacco and tobacco smoke (*65*), it is suggested that more specific methods (glc with ms and/or lc with ms) be used to assure quantitative values for individual polyphenols, rather than values for a mixture of polyphenols.

VI. Basic Fraction

The basic fraction of urban pollutants constitutes only a small percentage of the organic matter (Fig. 2). Its chemical composition is only partially known, azaarenes (N-heterocyclic aromatic hydrocarbons) and nonvolatile amines being the only groups studied. There has been very little bioassay of this fraction. This is due, in part, to the small proportion of the fraction in particulate matter. Cutaneous and/or subcutaneous applications to adult animals require amounts of test material that can be supplied only through sampling of vast volumes of air. Because it requires relatively small amounts of material, subcutaneous testing on infant small mammals has been undertaken and has suggested that the basic portion contains carcinogenic agents (*48*). These results should be further substantiated and should be complemented by the detailed analysis of the basic portion of organic pollutants.

A. Azaarenes

The concentrations of the *N*-heterocyclic hydrocarbons identified in urban air are presented in Table XIII (*87, 227, 228*). In addition, an alkylacridine (*227*), and alkylated benzo[*b*]quinolines, benzo[*h*]quinolines, benz[*a*]acridines, and benz[*c*]acridines, as well as an alkylated dibenz[*aj*]acridine and 1-azafluoranthene have been identified (*71*). Of the identified azaarenes, dibenz[*ah*]acridine and dibenz[*aj*]acridine are known animal carcinogens (*45*). In addition, one may find that some of the alkylated benzacridines and alkylated dibenz[*aj*]acridine, if tested individually, might also be carcinogenic.

Table XIII Concentration of Azaarenes in the Urban Air of United States Cities[a]

	Concentration		
Azaarene	*In benzene-soluble fraction* ($\mu g/gm$)	*In airborne particles* ($\mu g/gm$)	*In air* ($\mu g/1000\ m^3$)
Acridine[b]	49	—	—
Benzo[*f*]quinoline	20	2	0.2
Benzo[*h*]quinoline	30	3	0.3
Benz[*a*]acridine	20	2	0.2
Benz[*c*]acridine	50	4	0.6
Benz[*c*]acridine (51 United States cities)[c]	0–200	—	0–1.5
11*H*-Indeno[1,2-*b*]quinoline	10	1	0.1
Dibenz[*ah*]acridine	7	0.6	0.08
Dibenz[*aj*]acridine	4	0.3	0.04

[a] All data from Sawicki *et al.* (*87*) unless otherwise noted.
[b] From Sawicki and Engel (*227*).
[c] From Stanley *et al.* (*228*).

Table XIIIa Azaarenes in New York City Suspended Particulate Matter ***(186a)***

	Concentration (ng/1000 m³)	
Azaarene	*Sample I*	*Sample II*
Quinoline	69	22
Methylquinolines[a]	35	33
Dimethylquinolines[a]	48	44
Ethylquinolines[a]	14	22
3*C*-Quinolines[a]	10	
Isoquinoline	180	140
5- or 8-Methylisoquinoline	310	170
Other methylisoquinolines[a]	76	70
Dimethylisoquinolines[a]	62	+
Ethylisoquinolines[a]	160	68
3*C*-Isoquinolines[a]	28	+
Acridines	41	40
Methylacridines[a]	7	
Benzo[*h*]quinoline	10	13
Benzo[*f*]quinoline	11	10
Phenanthridine	22	18
Benzo[*f*]isoquinoline	110	34
5*H*-Indeno[1,2-*b*]pridine	5	5
11*H*-Indeno[1,2-*b*]quinoline	+[d]	+[d]
Benzo[*lmn*]phenanthridine	21[b]	22[b]
Indeno[1,2,3-*ij*]isoquinoline	5[b]	5[b]
Benzothiazole	14	20
Caffeine[c]	3400	7000

[a] Exact structure not known

[b] Includes other isomers.

[c] Other six values from New York City varied between 700–6000 ng/1000 m³ (*233b*).

[d] +, present but not quantitated.

In Table XIV, data as reported by Sawicki *et al.* (*229*) are summarized on the concentrations of azaarenes in the effluent gases of several polluters. One will most likely find additional azaarenes in coal burning effluents, in view of the identification of several N-heterocyclics in coal tar (*183*). Indoor air polluted by tobacco smoke may also contain dibenz[*aj*]acridine and neutral dibenzo[*cg*]carbazole as well as *N*-methylcarbazoles (*65, 230, 231*).

Sawicki and his group, as well as others, have identified azaarenes after chromatographic enrichment by spectrofluorimetric means (*87, 227, 228*). Five of these procedures are compared and discussed in a communication (*232*). These procedures have given composite quantitative results for several azaarenes, rather than quantitative results for individual compounds. We should, therefore, take advantage of new analytical instrumentation and recheck earlier methods. Ray and Frei have made an effort in this direction by exploring high-speed liquid chromatography for the separation of N-heterocyclic hydrocarbons (*233*). The authors found satisfactory separation on a polaric column. Progressive development of polaric column materials and ion exchange columns will lead to even better separations. In any case, the detection of individual azaarenes in the basic portion of organic particulates (extracted from their solution by concentrated acids) should be achieved by lc interphased with UV-spectrophotometry. Cantreels and Van Cauwenberghe reported a detailed study of the basic fraction by mass fragmentography (*48a*). After glc separation, they identified 15 bands which contained azaarenes, including benzacridines and dibenzacridines. Dong *et al.* prefractionated the basic fraction by lc into 8 subfractions and characterized each subfraction independently by glc/ms and lc reversed phase followed by UV and fluorescence spectroscopy of the isolated peak (*186a*). The results are summarized in Table XIIIa. The reported quinoline and isoquinoline values in organic particulate samples from New York City represent isolated quantities from material not purposely sampled for azanaphthalenes which are present mainly in the gas phase. The induction of hepatomas in mice with quinoline (*233a*) should encourage methods for the quantitative analysis of quinolines and isoquinolines in polluted air. Table XIIIa shows that azaarene concentrations are much lower in the

Table XIV Aza Heterocyclic Compounds in Air Pollution Source Effluents *(229)*

	From effluent gases (mg/1000 m³)			
Compound	*Coal heating I*[a]	*Coal heating II*[a]	*Industrial coal furnace*	*Coal tar pitch*
Acridine	111	3.30	65.0	0.87
Benzo[*f*]quinoline	57	96.00	140.0	0.42
Benzo[*h*]quinoline	38	200.00	410.0	0.26
Phenanthridine	32	200.00	240.0	0.02
Benz[*a*]acridine	26	7.70	18.0	0.20
Benz[*c*]acridine	15	18.00	60.0	0.12
Benzo [*lmn*]phenanthridine	+[b]	+	≃48.0	+
Indeno[1,2,3-*ij*]isoquinoline	≃17	+	+	0.03
11*H*-Indeno[1,2-*b*]quinoline	24	<0.17	34.0	0.19
Dibenz[*ah*]acridine	17	<0.12	0.7	0.01
Dibenz[*aj*]acridine	2	<0.15	1.8	0.001

[a] Domestic coal combustion stack.
[b] +, present but not quantitated.

urban air sample, than in the concentrations of the corresponding PAH (see Table V). The unexpected finding of relatively high concentrations (3.4 and 7.0 μg/1000 m^3) of caffeine in New York City air was finally shown to be a likely consequence of the ubiquity of coffee roasting plants in the city as well as in adjacent New Jersey (*233b*). Tabor *et al.* had earlier isolated caffeine in crystalline form from the organic matter of pollutants from New Orleans, Louisiana (*46*). It is therefore recommended that quantitative data should be considered reliable only when derived from studies that use internal standards.

It is likely that organic pollutants will contain traces of neutral azaarenes such as iminophenanthrenes, benzocarbazoles, and dibenzocarbazoles. In the fractionation scheme of organic pollutants (Fig. 2) one would expect them to be enriched with the neutral N-2 subfraction and to be eluted from a chromatographic column after the PAH. The likely occurrence of neutral azaarenes in organic pollutants is supported by the fact that these agents have been found in coal tar (*183*) and tobacco smoke (*65, 185, 231*).

B. Nonvolatile Amines

Nonvolatile aromatic amines (e.g., 2-naphthylamine, 4-aminobiphenyl, and benzidine) are well established urinary bladder carcinogens in man (*234*). Since trace amounts of 2-naphthylamine are already considered to be carcinogenic in man (*235*), this and other aromatic amines are proposed to be banned from use in the United States (*236*). This consideration, and the facts that 2-naphthylamine, and probably other carcinogenic amines, are pyrolysis products of amino acids and protein (*237*) and have been found in tobacco smoke (*238, 239*), place importance on the need for reliable analytical data for aromatic amines and other nonvolatile amines in urban air. This request appears even more urgent because secondary and tertiary amines and amides can serve as *in vivo* precursors for carcinogenic *N*-nitrosamines or nitrosamides, as discussed earlier (see Section IV,G).

Analytical methods have been published for the trace analysis of aromatic amines (*238–240*). In the only study specific to the analysis of aromatic amines in air pollution, it was suggested that nonvolatile aromatic amines be enriched by column and/or thin-layer chromatography and that the spots be subsequently identified by excitation and emission spectra (*241*). Aromatic amines, especially aminobenzenes and aminonaphthalenes, are very unstable, easily oxidized, and subject to condensation with carbonyl compounds. Therefore, they are probably subject to oxidation during the collection of the particulate matter and decompose during the analytical procedure. These problems are best avoided by forming stable derivatives of the amines. One should select those that are easily formed and that can be detected in minute amounts. The most promising lead appears to be the utilization of trifluoroacetamides, pentafluoropropionamides, or similar derivatives. These amides are easily formed in high yields from amines and fluorinated carboxylic acid anhydrides. Furthermore, after glc separation, they induce high response from electron capture detectors (detection limit $\leq 10^{-9}$ gm) (*238–240*). Since aromatic amines are easily photooxidized it is advisable to collect them near their suspected source.

VII. Bioassay

Biological evaluations are essential to give realistic meaning to chemical data on pollutants suspected of affecting man.

A. Host Entry

The principal passages for environmental agents into the human body are the respiratory system, the digestive tract, and the skin. The majority of studies pertaining to particulate matter have dealt with its passage, deposition, and retention in the respiratory tract.

1. Particle Size

The particle size of pollutants in urban air can range from 0.001 to 10,000 μm. In general, combustion processes produce particles of 0.1–10 μm in diameter (*7*). The biological importance of particle size is based primarily on three factors. First, particles above 10 μm in diameter are largely retained in the mucous membranes of the nose, accessory nasal sinuses, pharynx, and oral cavity (*242*). Second, particles from 0.25 to 10 μm in diameter are largely retained in the alveolar region, bronchi and/or trachea (*243*). Third, only a small percentage of particles less than 0.25 μm in diameter are retained in the lung. During normal breathing, maximal lung retention is about 80% for 1 μm particles, and is less than 5% for particles smaller than 0.1 μm and particles larger than 15 μm (*41*).

2. Retention and Elution of Particles

Ordinarily, the time interval during which particles remain in the lung is too short to permit elution of their toxic components to a significant degree. The tracheobronchial cells are protected by a mucus blanket that is propelled by the underlying cilia. Up to 90% of coal, carbon black, and iron oxide particles that enter alveolar regions are cleared out with the mucus. If, however, mucus functioning is impaired, the increased time of retention of these particles may permit some degree of extraction of any toxic agents they may carry. Impairment of mucus flow may appear as a change in its viscosity or in the rate of ciliary movement. In air pollution and tobacco smoke carcinogenesis, Hilding proposed that without inhibition of ciliary movement and concomitant mucus stagnation, histopathological changes leading to epithelial cancer are not likely to occur (*244*). This concept is widely accepted.

A large number of chemical, physical, and microbiological factors, including air pollutants, can induce alterations and disturbances of ciliary mucus flow (*245, 246*). One air pollutant incriminated as enhancing the absorption of toxic pollutants is sulfur dioxide (*247*). Studies with benzo[a]pyrene mist, with and without concurrent sulfur dioxide exposure, have shown that bronchiogenic carcinoma in animals is induced with concurrent exposure to sulfur dioxide (*248*). The elution of toxic agents from particles deposited in the lung has been studied elaborately with respect to polynuclear aromatic hydrocarbons. In principle, similar mechanisms apply to other constituents of the particulate matter.

Falk and Kotin *et al.* have studied such elution phenomena with soot particles as absorbents and plasma proteins as vehicles (*249, 250*). They demonstrated experimentally that BaP and other PAH are readily eluted from soot particles recovered from the lung.

Certain asbestos fibers, especially the γ-chrysotile form, can absorb an oily film containing PAH, which can thereby be transported into the lungs, retained there with the asbestos fibers, readily eluted, and then retained in the tissue (*251*). When the respiratory air contains ciliatoxic agents, or other irritants, the retention time of the particles in the lung is increased.

B. Measurement of Mucus Viscosity and Respiratory Cilia

The ciliated epithelium of the respiratory tract and its mucus layer are continuously in contact with respiratory air. Short-term exposure to high doses or long-term exposure to low doses of contaminants in air may alter the properties of the mucus and/or affect the underlying ciliated cells (*252*). *In vivo* studies on cats have shown mucolytic activity for a high concentration of pollutants such as cigarette smoke, acrolein, formaldehyde, and volatile phenols (*253, 254*). Certain bases were found to possess higher mucolytic activity as salts than in free form. This observation could be of importance with respect to the particulate matter of urban air.

Several studies have dealt with the ciliatoxic effects of gaseous and particulate constituents of personal and urban air pollutants (*65, 246, 252, 255, 256*). Table XV presents a list of some agents considered to inhibit cilia movement that are definitely or probably present in the respiratory environment (*65, 198*). This list is certainly incomplete, especially with respect to nonvolatile pollutants. It must be understood that *in vitro* techniques measure mainly volatile components. Of the many techniques that have been developed as assays for ciliatoxic agents in urban air, it appears that those that employ *in vivo* measurement of cilia toxicity in the trachea of cats, rats, and rabbits are more realistic (*252*).

C. Microbiological Tests

Most microbiological tests for the bioassay of respiratory environments are designed to evaluate the toxicity of polluted inhalants as a whole and are not applicable to the assessment of particulate matter and its constituents. These methods include the tests developed by Estes for the inhibition of growth of *Escherichia coli* following exposure to air pollutants (*257, 258*).

Ames *et al.* (*259, 260*) have described a rapid and inexpensive bacterial test system for detecting carcinogens as frameshift mutagens. The test

Table XV Compounds in or Related to Constituents of Polluted Urban Air Showing Cilia Movement Inhibition[a]

Paraffins	Carboxylic acids
2-Methylpentane	Formic acid
Olefins	Acetic acid
2-Methylbutene-2	Phenols
2-Methylpentene-2	Phenol
Aromatic hydrocarbons	Cresols
Benzene	Peroxides
Toluene	Acetyl peroxide
Xylenes	Peracetic acid
Aldehydes	Epoxides
Formaldehyde	Propylene oxide
Acetaldehyde	Cyclohexene oxide
Propionaldehyde	Inorganic gases
Acrolein	Sulfur oxides
	Nitrogen dioxide

[a] From Kotin and Falk (*198*) with some additions.

combines a particularly sensitive histidine-requiring mutant of *Salmonella typhimurium* with a liver homogenate fraction for activating the carcinogen to its proximate form. With this system, as little as a few micrograms of benzo[*a*]pyrene may be detected. This bioassay system, requiring only trace amounts of material, should be helpful for estimating the carcinogenic potential of organic pollutants and their fractions, although it should be realized, that all known carcinogens give positive results, and that quantitatively, mutagenic activity and carcinogenic activity for a given compound are not always in good agreement. Azaarenes, for example, are relatively strong mutagens and relatively weak carcinogens in contrast to PAH. Talcott and Wei (*260a*), applying the Ames test, found that organic particulates sampled in Buffalo, New York, contained mutagens requiring liver enzyme activation as well as direct acting mutagens, whereas a sample from Berkeley, California contained only the latter.

A method applicable to particulate matter utilizes the photodynamic response of *Paramecium caudatum* (*261–264*). The assay measures concentrations of photosensitizing PAH in organic extracts of air pollutants and reflects the ability of these compounds to sensitize cells to the otherwise nontoxic effects of long-wave ultraviolet light.

Cloned cultures of *P. caudatum* are maintained at about 28°C in a semisynthetic medium. About thirty cells of this unicellular species are mixed with aqueous suspensions of the test material (organic pollutants and their fractions) and placed on a microscope slide. After brief incubation, the admixture is irradiated with long-wave ultraviolet light. The duration of irradiation required to induce 90% immobilization is chosen as the factor of photodynamic activity of the tested material. For the known carcinogen BaP, it takes a concentration of 1 μg/ml to achieve 90% immobilization after 1 minute. A 10 ng/ml solution of BaP effects 90% immobilization after about 18 minutes. The photodynamic activities of 240 PAH measured with this assay system have been compared with their *in vivo* induction of oxazolamine hydroxylase activity in rats, and a highly significant association was demonstrated (*264*). Epstein *et al.* (*265*) also found a statistically significant association between photodynamic activity and carcinogenicity. However, the photodynamic assay cannot identify a particular polycyclic compound as being carcinogenic or noncarcinogenic.

The application of the photodynamic assay to six samples of crude benzene extracts of air pollutants from cities in the United States showed photodynamic activity in every case with an evident correlation between apparent relative potencies as obtained with the test and the BaP concentration. "Aliphatic fractions" (Fig. 1) were inactive as they were earlier shown to be by skin test (*33*). The one "aromatic fraction" assayed was highly active, and the three "oxygenated fractions" showed photodynamic activity, although they were free of PAH. In recovery experiments, neither the benzene extracts nor the oxygenated fractions, interfered with assays for added BaP. The presence of PAH other than BaP in the air pollutant samples did not contribute substantially to the potencies of these samples. In a more recent study, the neutral fraction of organic extracts of pollutants was chromatographically separated into 217 fractions, and the distribution of photosensitizing PAH was determined with the photodynamic bioassay using *P. caudatum* (*266*).

Epstein contends that the photodynamic assay is a "rapid, simple, and economical biological index of potential carcinogenic hazard attributable to polycyclic hydrocarbons" (*262*). An important aspect of this technique lies in the small amount of test material required for the assay. Not even a quantitative chemical analysis for PAH could be carried out with 1 mg or less of particulate matter. Furthermore, the experiment itself can be completed in a relatively short time in any chemical or biological laboratory without the maintenance of a large animal colony.

However, we do not, as yet, have a satisfactory correlation between this photodynamic assay and other bioassays for carcinogenicity. Such positive correlation could at best be established merely on an empirical basis, since on the cellular level carcinogenicity and photodynamic activity are vastly different biochemical processes (*267*). Nevertheless, photodynamic assay may prove to be a useful technique for the evaluation of the carcinogenic hazard attributable to PAH in air pollutants.

D. Assays for Carcinogenicity of Air Pollutants

1. Certain Aspects of Bioassays

The majority of bioassays on the carcinogenicity of organic air pollutants have utilized the subcutaneous tissue and skin of adult mice as testing sites. Recently, newborn mice have been used to estimate the carcinogenic potential of air pollutants. In addition, a number of other tests have aided in the evaluation of the carcinogenicity of chemicals and environmental agents. In 1964, Hueper, and more recently, J. H. Weisburger reviewed most of the bioassays in detail and discussed their merits (*268, 269*). In this section, we limit discussion to those assays considered to be helpful in studies with air pollutants.

a. Assays Using Non-Neoplastic Criteria. The photosensitizing effects of PAH and organic pollutants on *P. caudatum* just discussed represent one method of assaying with non-neoplastic criteria. Another promising test, a combination of liver homogenates for activation and bacteria for detection of carcinogens was also briefly mentioned (*259, 260*).

In this context, the short-term test on mouse skin should also be cited. As early as 1940, Pullinger observed that only a few topical applications of carcinogens produced changes on the surface of the skin of mice, including the suppression of sebaceous glands (*270*). Since then, these changes have been explored in detail (*271*) and have been found to be of value in determining the tumorigenic potential of several fractions of catalytically cracked petroleum oils, coal tar (*272*), and tobacco smoke condensate (*273–275*).

Bock and Mund have explored the possible correlation between sebaceous gland destruction and carcinogenicity of PAH and found high levels of suppression activity to be associated with the benz[*a*]anthracene ring structure (*276*). The bioassay consists of applying benzene or acetone solution of the material being assayed onto the shaved backs of mice in the second resting phase of the hair cycle (7–12 weeks of age). Applications are given on days 1, 3, 5, and 7, and the animals are killed on day 9. The treated area of the skin is removed and prepared for histological evaluation. The suppression or destruction of the sebaceous glands, as well as the formation and degree of hyperplasia, indicates the tumorigenic potential of the test agent (*277*). We found that the application of an acetone–benzene solution of 12.5% of organic air pollutants from Detroit, Michigan, (three samples) and New York, New York, (one sample), and their neutral subfractions N-2 (Fig. 2), produced complete suppression of the sebaceous glands and hyperplasia. These observations are in good agreement with long-term tests for these samples, which showed a relatively high degree of tumorigenicity for the four whole samples and their N-2 fractions (*34*). In spite of this agreement, a greater number of comparisons of short- and long-term assays for air pollutant samples is required before this assay can be termed a valid technique in air pollution carcinogenesis.

Several other bioassays with non-neoplastic criteria are in use or are suggested. These include the "newt test" (*278, 279*) and the tetrazolium-reduction test in mouse skin (*280*). These two assays and others appear presently of only limited value for estimating the carcinogenic potential of air pollutants. Several groups have developed and applied chemical methods for rapid assays of carcinogenicity. Epstein *et al.* (*281*) found "no particular association between complex formation and carcinogenicity" in applying four PAH tests that were claimed to be based on charge-transfer complex formation with iodine, chloranil, trinitrobenzene, and acridine.

b. CELL AND ORGAN CULTURES. The transformation of normal cells in culture into cancer cells has for many years been known to be induced by

DNA- and RNA-oncogenic viruses. In 1963, Berwald and Sachs reported characteristics morphological alterations by carcinogenic PAH in the growth patterns of hamster fibroblasts in culture (*282*). Later, the same authors reported that mass cultures of morphologically transformed cells gave rise to tumors in hamster cheek pouches, whereas the control hamster cells did not (*283*). Subsequently, quantitative cloning techniques were developed and, using BaP as the carcinogen, quantitative dose–response curves were observed, indicating that *in vitro* transformation of embryonic hamster fibroblasts was a one-hit process and was separate from the toxicity exerted by the carcinogens (*284*). Individual clones of transformed cells give rise to tumors in hamsters (*285*).

Lasnitzki found that carcinogenic PAH produce histological alterations in organ cultures of mouse ventral prostate (*286*), a result confirmed by others (*287*). The implantation of the altered pieces of prostate into isologous mice, however, did not produce tumors. Although the prostate fragments of C3H mice treated with carcinogenic hydrocarbons in organ culture were not malignant, Heidelberger and Iype succeeded in obtaining from them lines of cells that grew in a disoriented fashion and produced tumors on inoculation into isologous mice (*288*).

Methods were subsequently developed for growing cell lines from C3H mouse ventral prostate that did not elicit tumors upon injection into irradiated C3H mice. The prostate cells hardly transform spontaneously to malignancy as mouse embryonic cell lines do, and when treated with the carcinogenic hydrocarbon 3-methylcholanthrene, the cell culture forms piled up multilayers and produces fibrosarcomas upon injection in C3H mice (*289*). These results led to the development of a quantitative system with the prostate cells in which transformation and toxicity can both be measured in sparsely plated cells. It was shown that there was a good correlation between the number of piled up, malignant, transformed colonies and the carcinogenic activities of PAH on mice. Furthermore, there was no dose–response relation between the process of transformation and toxicity. Therefore, it was suggested that systems like this might be useful for screening the carcinogenic potential of air pollutants and their fractions (*41*).

Since animal tests are time consuming and require relatively large amounts of fractionated material, organ culture tests could be helpful in assaying the carcinogenic potential of air pollutants. This technique maintains organs or pieces of organs *in vitro* so as to retain normal histological associations among cell types and preserve cell differentiation and growth at rates resembling those *in vivo*. Adult pieces or organs can be maintained up to 4 weeks and are regarded as responsive to test materials similar to the behavior of such cells *in vivo*. Several advantages are seen with such organ culture tests, namely, need for only small amounts of test material, good control of dose and duration of exposure to suspected carcinogen, and the possibility for testing human tissue. Results in organ culture have shown a parallelism between organ responses *in vivo* and *in vitro*. For example, respirating (*290*) and prostatic tissues (*287, 291*) undergo epithelial metaplasia, pleomorphism, or devitalization in approximate proportion to the toxic and carcinogenic activity of PAH as established in animals.

Environmental agents assayed in organ culture include cigarette smoke condensate and its fractions (*65, 292*), the benzene-soluble portion of pollutants, and a number of PAH (*293*). These tests established the comparability of *in vivo* and *in vitro* responses. Based on experiments with rodent respiratory tract tissues (*292, 294*), such tests could be of great value in studying chemical, physical, and physiological factors that play a role in lung carcinogenesis. The organ culture system also offers means of directly comparing the effects of air pollutants on human and animal lung tissues.

c. Two-Stage Carcinogenesis. In 1947, Berenblum and Shubik formulated the two-stage theory of initiation and promotion in chemical carcinogenesis (*295*). They suggested that the initiating agents irreversibly transform a certain number of normal cells into latent or dormant tumor cells. The promoting agent is unable to induce this transformation but can cause the latent tumor cells to develop into frank tumors. This concept has been confirmed by a large number of studies on mouse skin (*296, 297*), including tests with tobacco tar and its fractions (*65, 298–300*) and air pollutants and their fractions (see Section VII,D,4b). One of the most important characteristics of two-stage carcinogenesis is its potentially insidious character. An organ or tissue may have received a subthreshold dose of a carcinogenic hydrocarbon in one or several doses, leading to a latent tumor cell but not to a frank tumor. The persistence of the initiating effect has been demonstrated for mouse skin. With an interval between initiation and promotion of more than 1 year, tumors will still arise rapidly after promoter application (*301*). In addition to mouse skin, the principle of two-stage carcinogenesis has been demonstrated for the skin of rabbits, the vulva of mice, forestomach of mice, and the urinary tract of mice and rats (*297*).

d. Factors in Experimental Carcinogenesis. In most experimental settings, carcinogenic PAH have been used as tumor initiators being applied once, or repeatedly, to mouse skin. The dose given was below the effective level of PAH needed to elicit a significant tumor response by itself. The tumor initiator need not necessarily be a subthreshold dose of a complete carcinogenic PAH, but may also be a PAH that is by itself an incomplete carcinogen, e.g., dibenz[*ac*]anthracene and 6-methylanthanthrene (*301, 302*), and 1-, 4-, and 6-methylchrysene (*80*). The lowest dose of a tumor initiator to induce latent tumor cells in mouse skin has been, in our setting, 30 μg of 5-methylchrysene or of benzo[*a*]pyrene (*146*). Although there are other known tumor initiators besides PAH and azaarenes, e.g., ethylurethane, none of them has as yet been implicated as playing a role in environmental carcinogenesis.

The classic tumor promoter used in carcinogenesis was croton oil. After its isolation by Hecker and Kubinyi (*303*) as the active ingredient in croton oil, the powerful promoter B1 [tetradecanoyl phorbol acetate (TPA)] has been widely used in model studies of carcinogenesis. Several other chemicals, although less powerful, have been shown to be active as tumor promoters. These include anthralin, Tweens, and Spans (*301*). Several weakly active tumor promoters have been identified in respiratory environments, among them *n*-dodecane, some volatile phenols, and some nonvolatile fatty acids. The particulate phases of tobacco smoke and urban pollutants are known to induce tumors on mouse skin pretreated with tumor-initiating PAH but thus far the major tumor promoters have not been identified in these inhalants.

The concurrent application of low doses of carcinogenic hydrocarbons with agents that are inactive by themselves can lead to a significantly higher tumor yield than expected from the low carcinogen dose alone. The second type of agent is called a cocarcinogen, or sometimes, tumor accelerator. The first report relating to air pollution cocarcinogenesis originates from Horton *et al.* (*304*). The authors demonstrated that certain inactive hydrocarbons, especially *n*-dodecane, when used as solvents for carcinogenic PAH, significantly accelerate tumor yield, a finding confirmed by Bingham and Falk (*305*). The latter group also found cocarcinogenic activity for 1-dodecanol and 1-phenyldodecane in mouse skin tests with a low concentration of a carcinogenic hydrocarbon.

The topical application to mouse skin of mixtures of carcinogenic PAH with inactive PAH may lead to a different tumor yield than anticipated with the carcinogen alone. The cocarcinogenic effect depends not only on the structure of the tested PAH but also on its molar ratio to the applied carcinogenic hydrocarbon. Cocarcinogenic activities have been reported for pyrene, fluoranthene, benzo[*ghi*]perylene, and the weakly carcinogenic benzo[*e*]pyrene (*78, 306*). A neutral agent of similar polarity to PAH and active as a cocarcinogen is *trans*-4,4′-dichlorostilbene, a major pyrolysis product of the insecticides DDD and DDT (*298*).

Anticarcinogenic agents are mostly, though not exclusively noncarcinogenic. They inhibit the activity of carcinogens in various manners. One environmental agent, a weakly active carcinogen that inhibits the carcinogenic activity of BaP on mouse skin and in connective tissue is benz[*a*]anthracene (*78, 307*). In a mixture with other PAH, as found in air pollutants, we have no evidence that BaA inhibits the activity of the carcinogenic PAH present.

2. Test Sites

Mouse skin is the most frequently chosen test site for the bioassays of organic environmental agents with suspected carcinogenic activity. The major advantage of mouse skin as a test organ is the relatively low cost of maintaining groups large enough for statistically significant data, while demand for test materials (pure chemicals or environmental agents) is relatively low. The high susceptibility of certain inbred strains of mice to environmental carcinogens is an added advantage. Several readily available review articles and books (*82, 269, 277, 308, 309*) have described the details of experimental conditions for mouse skin bioassays. In brief, the more important factors are: standardized test procedures, selection of mouse strain and sex, dosage, and mode of application.

Subcutaneous injection into mice and especially into rats is the most widely used form of parenteral administration in testing chemical carcinogens. However, it is not too frequently used in air pollution carcinogenesis (see Section VII,D,4a). The connective tissue test offers advantages in assaying highly reactive or unstable components in that it is necessary to inject pure chemical only once for the evaluation of its carcinogenic potential and that it is possible to dose the amount of test materials rather exactly. The disadvantages of the connective tissue test are its limit of 100 mg per application and the possibility that insoluble environmental agents may induce sarcomas by way of a physical effect rather than by biochemical reaction of the test material (*82, 269, 277, 309*).

The most logical method for testing respiratory carcinogens would be by inhalation. As discussed earlier, this technique has been explored with mice inhaling polluted air and artificial smog (see Section IV,F,2). The results of two such studies were rather conflicting and have underscored the observation that inhalation studies are not practical for the bioassay of the carcinogenic potential of polluted urban air. Daily inhalation of a very densely polluted inhalant, namely, dilute cigarette smoke, for 18 months induced in hamsters benign and malignant tumors of the larynx, but no lung tumors (*310*).

In view of their negative results in inducing bronchiogenic carcinomas in laboratory animals by inhalation of polluted air (*311*) or tobacco smoke (*65, 310*), investigators turned to the development of intratracheal instillation techniques for the assay of chemical and environmental carcinogens (*312–317*). BaP and other carcinogenic hydrocarbons adsorbed on hematite (Fe_2O_3) particles produced benign and malignant tumors in the respiratory tract of Syrian hamsters when injected via the trachea. The carcinogen–hematite–Syrian hamster model leading to carcinomas of the tracheobronchial tree and lung parenchyma mimics those carcinomas occurring in man that are thought to be at least partially due to respiratory carcinogens. The carrier for the carcinogen can also be carbon particles (*313, 314*) or asbestos (*318*). The instillation of carcinogenic hydrocarbons alone did not produce any tumors. Thus, it is postulated that the carcinogenic effect of the hydrocarbons adhering to the surface of the hematite particles was brought about by their continuous slow release into the tissue. The instillation of organic pollutants with hematite into the lung of hamsters may offer a model for the induction of bronchiogenic carcinomas and lead to further explorations in this area.

BaP on lead oxide instilled into Syrian hamster lungs induced tumors in the peripheral area of the lung, indicating bronchio-alveolar origin (*319*). BaP instilled together with furfural in the trachea of Syrian hamsters has induced bronchiogenic carcinoma (*320*).

3. Tumorigenicity of Urban Pollutants

The carcinogenicity of various potential air pollutants has been established for a number of concentrated environmental agents. Passey (*321*) and Campbell (*322*) proved carcinogenicity for extracts from chimney soots. Campbell also reported carcinogenicity for extracts from road dust (*323*). The tumorigenicity of "tars" from gasoline engine exhaust has been demonstrated in experiments by Kotin *et al.* (*324*), Gurinov *et al.* (*325*), and Wynder and Hoffmann (*156, 326*); that of diesel engine exhaust "tars" by Kotin *et al.* (*327*). Extracts from soots from aviation engines, both piston and turbine, have also been found to be highly carcinogenic to mouse skin (100% carcinoma yield) (*37*).

Leiter *et al.* (*328*) and Leiter and Shear (*29*) were among the first to induce subcutaneous sarcomas following injection of the organic matter of atmospheric dust and soot from various cities in the United States. Skin cancers, or subcutaneous tumors (sarcomas), in mice have also been produced with organic pollutants by Clemo and Miller (*329*), Clemo *et al.* (*330*), Kotin *et al.* (*31*), and Kotin and Falk (*204*). Ridgon and Neal injected the benzene-soluble portion of air pollutants that originated primarily from a collection station in the vicinity of several petroleum plants and induced fibrosarcomas (*35*). Bogacz and Koprowska applied organic air pollutants to the uterine cervix of mice and observed cellular abnormalities and histologic lesions that were morphologically indistinguishable from those accompanying the development of BaP-induced carcinomas (*331*).

4. Bioassays with Air Pollutants

a. COMPLETE CARCINOGENICITY. Several attempts have been made to correlate tumorigenicity of urban pollutants with the activities of the known animal carcinogens that constitute a portion of these pollutants. Hueper and co-workers (*32*) separated the benzene-soluble organic matter from air pollutants sampled in eight United States cities according to the fractionation scheme of Tabor *et al.* (*46*) (Fig. 1) and tested the whole organic matter as well as the aliphatic, aromatic, and oxygenated neutral end fractions individually. The test materials were dissolved in tricaprylin or ethyl laurate. Four milligrams of the crude benzene extracts were dissolved in 0.1 ml of tricaprylin and applied once a month by subcutaneous injection to each mouse. The dose of the material from the aromatic fraction was proportionately derived from 4 mg of the whole benzene-soluble sample; it was applied in this concentration for 11 months. Thereafter, the dose of crude benzene extract and aromatic fraction was doubled. The monthly dose for the aliphatic fraction was 0.5 mg, and that for the oxygenated fraction, 0.1 mg in 0.1 ml ethyl laurate.

Table XVI presents test results together with median BaP values for the crude benzene extract. The organic matter of the air pollutants induced a significant number of sarcomas for six of the eight tested samples (*32*). The eight aliphatic fractions were inactive, while two aromatic and two oxygenated neutral subfractions induced tumors. The observations of tumorigenic activity for some of the oxygenated fractions is of considerable importance, since it demonstrates that PAH-free fractions can contain tumorigenic agents. It appears that, in this experimental setting, a minimum of 40 μg of BaP in the sample is necessary to induce a significant number of tumors. The method failed, however, to show a correlation between the amount of BaP in the applied crude extract and the observed tumor yield.

Considering this result and some criticism on the use of repeated subcutaneous injections (*191*)—largely because of the well-confirmed Oppenheimer–Nothdurft effect (*65*)—and the absence of significantly higher numbers of tumors at sites distant from the point of injection (*32*), repeated topical application to mouse skin for the testing of organic air pollutants would appear preferable (*33*).

Compared to subcutaneous testing the mouse skin assay requires much more test material, which necessitates sampling of large volumes of air.

Table XVI Length of Latent Periods and Cancer Rates in Mice Injected with the Crude Benzene-Soluble Fraction and Aromatic Hydrocarbon Fraction of Air Pollutants[a] *(32)*

City	*Fraction*	*Latent period (months)*	*Number of survivors*	*Number of survivors with cancers*	*Percentage of survivors with cancers*	*Percentage BaP (median value)*
Atlanta, Georgia	Crude benzene	15	41	6	15	0.07
	Aromatic	12	55	3	6	—
	Aliphatic	13	—	0	—	—
	Oxygenated	—	—	0	—	—
New Orleans, Louisiana	Crude benzene	12	55	1	2	0.035
	Aromatic	12	57	2	4	—
	Aliphatic	—	—	0	—	—
	Oxygenated	18	22	2	7	—
Birmingham, Alabama	Crude benzene	9	66	4	6	0.12
	Aromatic	12	55	5	10	—
	Aliphatic	15	50	1	2	—
	Oxygenated	14	51	3	6	—
Los Angeles, California	Crude benzene	—	—	0	—	—
	Aromatic	—	—	0	—	—
	Aliphatic	—	—	0	—	—
	Oxygenated	—	—	0	—	—
San Francisco, California	Crude benzene	12	55	3	5	0.023
	Aromatic	9	68	1	2	—
	Aliphatic	—	—	0	0	—
	Oxygenated	20	22	4	18	—
Cincinnati, Ohio	Crude benzene	12	55	3	5	0.12
	Aromatic	12	58	1	2	—
	Aliphatic	19	51	1	2	—
	Oxygenated	—	—	0	—	—
Detroit, Michigan	Crude benzene	15	35	5	14	0.15
	Aromatic	—	—	0	—	—
	Aliphatic	—	—	0	—	—
	Oxygenated	16	49	1	2	—

[a] Groups without tumors do not show numbers in columns 3, 4, and 6; control mice (solvents alone), no tumors.

Our mouse-skin testing is carried out as follows. The residual organic matter of urban pollutants obtained by extraction of the filters with benzene–methanol (4:1) is dissolved in acetone–benzene (9:1) to constitute a 12.5% solution. This solution is applied to the shaved backs of female mice three times a week (each application 12.5 mg of organic matter) for at least 12 months. The tested fractions and subfractions of the organic matter are obtained according to the schematic outline in Figure 2. Chemical analytical, as well as biological findings of our studies are presented in Tables XVII through XIX.

Table XVII Chemical Analysis of Organic Matter and Fractions of Air Pollutants Tested on Mouse Skin[a] (33)

City and location	Year and season	Organic matter ($\mu g/1000\ m^3$)	Material[b]	(% of whole sample)	BaP[c] (% whole sample or fraction)	BaA[c] (% whole sample or fraction)	Palmitic acid[c] (% whole sample or fraction)	Oleic acid[c] (% whole sample or fraction)
Detroit, Michigan								
Major business area	1962, Winter	23	W.S.	100.0	0.035	0.06	Not analyzed	
	1963	32	W.S.	100.0	0.0185	0.034	0.46	0.10
			N-1	48.2	—	—	—	—
			N-2	3.6	0.50	0.92	—	—
			N-3	20.8	—	—	—	—
			A	14.8	—	—	3.10	0.68
			I	10.8	—	—	—	—
			B	0.55	—	—	—	—
Residential area	1962–1963, Winter; 1963. Summer	7	W.S.	100.0	0.014	0.27	(Not analyzed)	
New York, New York								
Major business area (Herald Square)		70.5	W.S.	100.0	0.0112	0.0169	0.31	0.17
			N-1	42.0	—	—	—	—
			N-2	1.9	0.58	0.87	—	—
			N-3	22.5	—	—	—	—
			A	13.0	—	—	2.40	1.34
			I	8.0	—	—	—	—
			B	3.4	—	—	—	—

[a] For biological data see Table X.
[b] *Abbreviations:* W.S. = whole sample; N-1 = aliphatic subfraction; N-2 = aromatic subfraction; N-3 = oxygenated subfraction; A = acidic fraction; B = basic fraction; I = insoluble portion.
[c] Quantitative values obtained with the isotope dilution technique.

Table XVIII Polynuclear Aromatic Hydrocarbons in Organic Matter of Three Detroit, Michigan, Samples

Polynuclear aromatic hydrocarbon	*Relative carcinogenic activity*[a]	*Concentration (ppm)*		
		Sample I	*Sample II*	*Sample III*
Benzo[*a*]pyrene	++++	354[b]	185[b]	137[b]
Benzo[*j*]fluoranthene	+++	200	138	118
Benzo[*b*]fluoranthene	++	492	564	313
Benz[*a*]anthracene	++	602[b]	339[b]	273[b]
Indeno[1,2,3-*cd*]pyrene	+	457	295	188
Chrysene	+	728	347	184
Benzo[*k*]fluoranthene	—	256	156	124
Benzo[*ghi*]perylene	—	664	510	317
Coronene	—	61	—	245
Pyrene	—	1045	157	180
Fluoranthene	—	820	139	118

[a] Relative carcinogenic activities on mouse skin (ICR Swiss female mice): ++++ = strong; +++ = moderately active; ++ = weakly active; + = very weakly active, — = inactive.
[b] Quantitative values ±5%, determined by the isotope dilution method, all other PAH values are semiquantitative.

When we applied material to the skin of mice, only the whole organic matter and the aromatic N-2 subfraction, which contains all known PAH, induced tumors. Subfractions N-1, N-3, and the acidic and "insoluble" portions were inactive. The basic portions could not be tested, since they were not available in sufficient quantities. The relative tumor potency number, according to Blanding *et al.* (*81*), which is given in Table XIX, was calculated for the tests in an attempt to evaluate the possible importance of BaP in the overall activity of the test materials. Figure 12 expresses the correlation between the concentration of BaP applied to the skin of mice and the TPN. The TPN values for BaP were obtained under conditions that resembled those for the pollutants and fractions as closely as possible. By placing the TPN of the four organic pollutant samples and the two N-2 subfractions into the BaP–TPN curve, one can estimate the approximate concentration of BaP required to induce comparable results. This comparison of activities of BaP and the BaP-content of a test material give a rough estimate of the importance of BaP in the test sample. One can deduce that about 55, 45, 45, and 50% of the tumorigenic activity of the pollutant samples 1 through 4, respectively, (Tables XVII and XIX) may be explained by the effects of BaP. For "aromatic subfractions" 2-B and 4-B, the corresponding percentages are 50 and 55%.

Table XIX Tumorigenicity of Air Pollutant Organic Matters on Mouse Skin[a]

City and location	*Fraction tested*[b]	*Concentration (%)*	*Number of mice started*	*First tumor appearance (month)*	*Tumor yield (%)*	*Average latency (days)*	*First carcinoma appearance (months)*	*Carcinoma yield (%)*	*Blanding tumor potency number*
Detroit, Michigan									
Major business area	W.S.	12.5	30	5	87	198	7	73	46
Major traffic area	W.S.	12.5	20	4	75	264	10	55	35
	N-1	12.0	10	—	—	—	—	—	—
	N-2	0.9	20	2	95	170	5	95	58
	N-3	5.2	10	—	—	—	—	—	—
	A	3.7	10	—	—	—	—	—	—
	I	2.15	10	12	10	350	—	—	—
Residential area	W.S.	12.5	30	6	70	370	—	—	—
New York, New York									
Major business area (Herald Square)	W.S.	12.5	20	9	30	365	13	15	18
	N-1	5.25	10	—	—	—	—	—	—
	N-2	0.475	10	6	70	290	9	70	34
	N-3	2.8	10	—	—	—	—	—	—
	A	1.6	10	—	—	—	—	—	—
	I	1.0	10	—	—	—	—	—	—

[a] ICR female mice. The fractions were tested at twice their concentration in the whole sample.

[b] For chemical data on samples and meaning of abbreviations, see Table XVI.

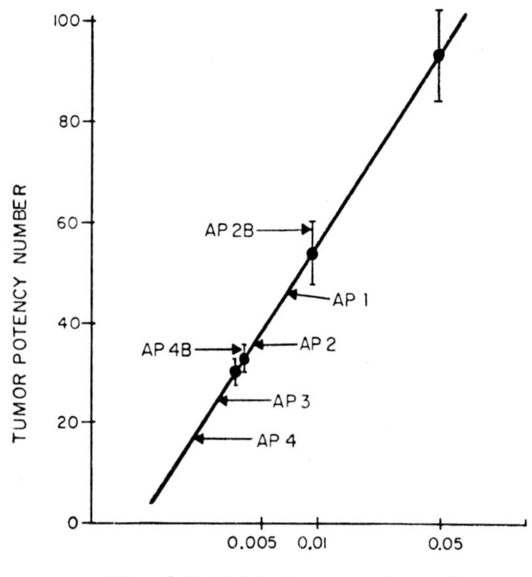

Figure 12. Method used for estimating the contribution of BaP to the tumorigenicity of organic air pollutants (*33*). The air pollution samples or fractions (AP) are defined in Table XVIII.

An assay such as outlined is of course limited to a rough estimate of the contribution of the major known carcinogenic air pollutants, the sum of PAH expressed as BaP, to the overall tumorigenicity of a test sample. The relative contributions, and, in particular, the obvious interrelationship of tumor initiators, tumor promoters, and tumor inhibitors cannot be fully understood with this type of assay. Studies by Ridgon and Neal (*35*) using subcutaneous injection of organic pollutants support our observation that BaP can explain only part of the observed carcinogenic activity of air pollutants.

Epstein and his group (*48*) collected the particulate matter on the air conditioner filters at the fifth floor level of the Pan Am office building in the center of New York, New York (August 1965 to February 1966). The crude sample from the filter (about 10 kg) was extracted with benzene, resulting in the organic phase (1079 gm). The latter was fractionated according to Tabor *et al.* (*46*) (Fig. 1). In addition, the oxygenated neutral subfraction (19.2% of the total organic matter) was separated by column chromatography into three subfractions. The investigators assayed for carcinogenicity by subcutaneous injections of 0.1, 0.1, and 0.2 ml. of suspensions of the organic matter and its subfractions in tricaprylin on days 1, 7, and 14 of life of Swiss mice.

Table XX summarizes the results from the bioassay for carcinogenicity

Table XX Tumors Induced in Swiss Mice Following Neonatal Subcutaneous Injections of Suspensions of a Benzene-Soluble Organic Extract of Particulate Atmospheric Pollutants and Fractions and Subfractions Thereof (48)

			Number of mice													% Weaned mice with any tumors		% Weaned mice with multiple tumors	
						Tumors													
								Pulmonary adenomas											
	Total organic material	Total dosage		Weaned		Hepatomas		Solitary		Multiple		Lymphomas							
Treatment groups	%	(mg)	*Treated*[a]	M	F	M	F	M	F	M	F	M	F			M	F	M	F
Uninjected			81(8)	26	23	0	0	0	0	0	0	0	1			0	4.3	0	0
Tricaprylin			86(8)	31	35	3	0	2	1	0	0	0	2			16.1	8.6	0	0
Benzene-soluble organic extract	100.0	10	71(6)	19	28	3	0	0	3	0	0	1	1			—	—	—	—
		20	53(5)	13	17	4	0	3	1	1	0	0	1			28.2	12.5	7.7	0
		40	70(6)	7	3	2	0	0	0	0	0	0	0			—	—	—	—
Acid fraction	17.5	10	50(5)	23	18	1	0	0	0	1	0	0	0			—	—	—	—
		20	57(5)	7	3	0	0	0	0	0	0	0	0			6.5	0	0	0
		40	67(6)	1	2	0	0	0	0	0	0	0	0			—	—	—	—
Basic fraction	1.0	10	51(5)	12	15	5	0	1	3	1	0	1	5			—	—	—	—
		20	61(5)	16	8	7	1	2	0	1	1	1	2			46.4[b]	43.5[c]	21.4	4.3
		40	54(6)	0	0	0	0	0	0	0	0	0	0			—	—	—	—
Neutral fraction	45.5	10	63(6)	24	29	0	0	1	1	1	0	0	1			—	—	—	—
		20	54(5)	16	11	2	0	1	3	2	0	0	0			19.1	20.8	2.9	7.5
		40	67(6)	28	13	4	0	2	3	1	1	1	4			—	—	—	—

Aliphatic fraction	19.0	10	66(6)	27	28	0	0	0	1	1	0	4	9	—	—	—	—
		20	46(6)	20	17	1	0	0	0	0	1	0	1	11.9	31.8[c]	3.0	1.5
		40	44(6)	20	21	1	0	1	1	0	0	1	5	—	—	—	—
Aromatic fraction	7.3	10	60(6)	23	30	0	0	1	1	1	2	0	5	—	—	—	—
		20	65(5)	30	29	2	0	2	3	2	1	2	5	22.4	40.7[c]	5.2	8.6
		40	66(5)	23	22	1	0	2	4	4	8[d]	3	5	—	—	—	—
Oxyneutral																	
Total	19.2																
Pentane–9% ether subfraction		10	73(6)	24	18	0	0	0	1	0	0	2	4	—	—	—	—
		20	58(5)	10	20	1	0	4	2	1	0	0	2	32.7	25.4	5.4	10.2
		40	46(4)	21	21	2	0	6	3	0	3	5	6	—	—	—	—
Pentane–12% ether subfraction		10	63(6)	32	27	2	0	0	2	0	3	1	4	—	—	—	—
		20	64(5)	18	17	2	0	4	2	3	0	1	4	23.6	35.4[c]	2.6	6.2
		40	68(6)	22	21	2	1	2	2	2	4	0	3	—	—	—	—
Pentane–36% ether subfraction		10	60(6)	17	12	1	0	3	1	0	0	0	0	—	—	—	—
		20	62(6)	16	14	2	0	2	3	0	0	0	3	22.0	24.4	4.0	7.2
		40	53(6)	17	15	2	0	0	1	2	3[e]	1	2	—	—	—	—
Insoluble fraction	11.2	10	65(6)	21	20	1	0	1	1	0	0	2	2	—	—	—	—
		20	61(6)	25	20	2	0	2	0	0	0	0	2	15.5	17.2	1.7	1.7
		40	60(6)	12	18	0	0	0	0	0	1	2	3	—	—	—	—

[a] Numbers in parentheses indicate number of litters.
[b] $p < 0.05$. Significantly greater than tricaprylin controls.
[c] $p < 0.01$.
[d] Two with a coexistent pulmonary carcinoma.
[e] One with a coexistent pulmonary carcinoma.

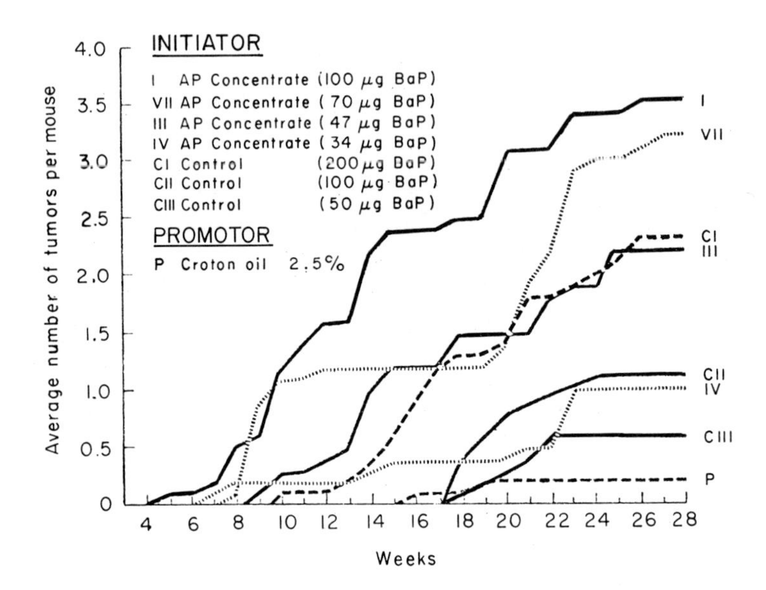

Figure 13. Tumor-initiating activity on mouse skin by PAH concentrates from urban pollutants (BaP present in sample in parenthesis). Three doses of BaP (200,100, and 50 μg) served as positive controls. Croton oil (2.5%) served as tumor promoter. From Wynder and Hoffmann (*33*).

on infant mice of organic particulates and their subfractions from a collection site in New York, New York. Before forming conclusions on the data, we have to recall that the sample was collected at the fifth floor level and that high toxicities were observed with the 20 and 40 mg doses, especially with the acidic and basic fractions, and that all fractions were applied in 10, 20, and 40 mg total doses, independent of their proportionate yields from the total organic particulate matter. Despite these factors, we can conclude that the study has confirmed the activity of the aromatic neutral subfraction, and that it has shown activity for the oxygenated neutral subfractions and for the basic fraction when applied in high doses. It is hoped that these results will encourage studies on the nature of the carcinogenic agents in the oxygenated neutral fraction and in the basic fraction.

b. Activity as Tumor Initiators and as Tumor Promoters. As discussed earlier, carcinogenic PAH can serve as tumor initiators in doses below their threshold level as complete carcinogens; furthermore, some noncarcinogenic PAH are active as tumor initiators (see Section VII,-D,1,d). In order to demonstrate that the PAH as present in air pollutants are tumor initiators, the following experiments were carried out (*33*). The neutral N-2 subfractions of ten air pollution samples from 80,000 m^3 of Detroit, Michigan, air were applied in ten subdoses as tumor initiators to mouse skin. Whereas no tumors were observed with the initiators alone (and solvent as promoter), secondary treatment with 2.5% croton oil as tumor promoter for 28 weeks lead to the induction of tumors. As seen in Figures 13 and 14, however, the amount of BaP in the N-2 fractions could not by itself account for the tumor initiator potential of the samples.

Air pollutants were also found to serve as initiators in experiments on mouse skin when cigarette smoke condensate was applied as tumor promoter (*332–334*). Although the results of these experiments prove only that on mouse skin the carcinogenic effects of urban pollutants and cigarette smoke condensate are additive, or even synergistic, they are in

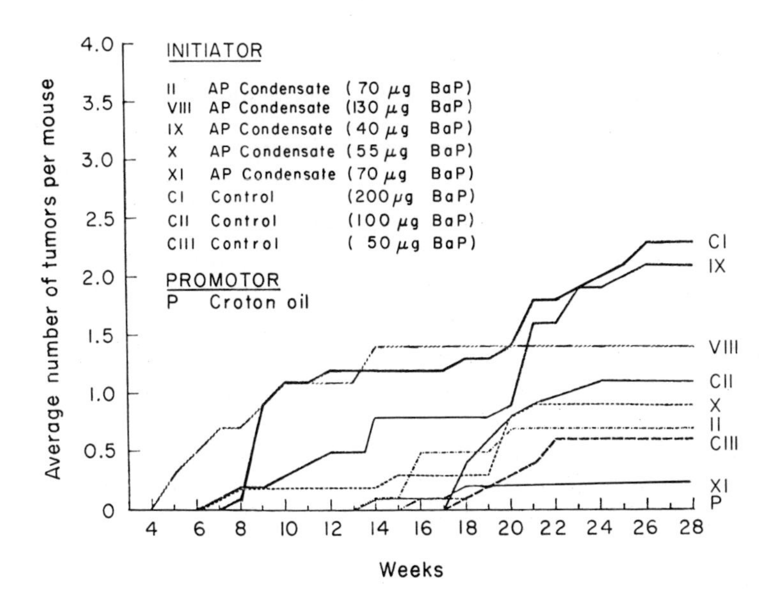

Figure 14. Tumor-initiating activity on mouse skin by PAH concentrates from urban pollutants (BaP present in sample in parenthesis). Three doses of BaP (200,100, and 50 μg) served as positive controls. Croton oil (2.5%) served as tumor promoter. From Wynder and Hoffmann (*33*).

line with the observation that cigarette smokers in cities have a somewhat higher lung cancer risk than cigarette smokers in rural areas (*45, 335*). This "urban factor" for lung cancer diminishes, although it does not disappear, after adjustments are made for differences in smoking habits betwen urban and rural smokers, for the occurrence of certain occupations with higher lung cancer risks in cities, and for differences in migration and immigration patterns between the two populations.

Whether organic air pollutants can act as tumor promoters on mouse skin was investigated by testing the neutral N-1 and N-3 subfractions, the acidic fractions and the "insoluble" fractions of two samples. Tumor-promoting activities were indicated for the N-1 and N-3 fractions and, in only one case, also for the acidic fraction (Table XXI). Certain nonvolatile fatty acids may, at least in part, be responsible for the promoter activity of the acidic fractions. Other tumor promoters in the acidic fractions as well as those in the neutral N-1 and N-3 subfractions need to be identified.

In summary, organic matter of urban air pollutants has been shown to be carcinogenic to mouse skin and to newborn mice and can produce sarcomas in the skin of mice upon repeated applications. Organic pollutants applied to the cervix of mice can also induce pathological changes

Table XXI Tumor-Promoting Activities of Fractions from Organic Matter of Air Pollutants[a]

City and location	Fraction tested	Concentration tested (%)	Number of mice per group	Tumor yield (%) Singularly applied	Tumor yield (%) Applied as a promoter
Detroit, Michigan					
Major traffic area	N-1	12.0	10	0	30
	N-3	5.2	10	0	80
	A	3.7	10	0	60
	I	2.15	10	0	0
New York, New York					
Major business area					
(Herald Square)	N-1	5.25	10	0	50
	N-3	2.8	10	0	30
	A	1.6	10	0	10
	I	1.0	10	0	10
Control (300 μg) 7,12-dimethylbenz[*a*]-anthracene (DMBA)	—	—	30	7	—

[a] For chemical data, see Table XVI; 300 μg DMBA were applied in a single dose to each group as a tumor initiator.

at this site. The tumorigenic and carcinogenic activities of air pollutants appear to a considerable degree to be caused by certain PAH. Based on tests with newborn mice, the basic portion and the "oxygenated" neutral subfraction seem to contain tumorigenic agents also. The N-1 and N-3 subfractions and the acidic fraction may act as tumor promoters.

Subcutaneous or cutaneous application of air pollutants to mouse skin may either be misleading or too costly in terms of materials. Testing with newborn mice appears useful for the bioassay of small samples of pollutants and their fractions. The limitation of all these tests particularly as they relate to effects on the health of man needs to be clearly understood.

VIII. Outlook

Our present knowledge of the chemical nature and the quantities of nonvolatile organic agents in polluted air must still be regarded as limited. This applies especially to our knowledge of known or suspected mutagens and/or carcinogens, such as epoxides and peroxides, alkylating agents, azaarenes, aromatic amines, and *N*-nitrosamines.

We continue to stress the importance of conducting interrelated chemical and biological studies. Isolated chemical data are of relatively little significance unless they can be evaluated in terms of their biological effects. Biological experimentation is of necessity dependent upon and limited to test systems other than the human setting. Experimental data that may relate to man must ultimately be evaluated in terms of epidemiological experience (*41, 336, 337*).

ACKNOWLEDGMENTS

The authors' research positions were supported by a grant from the National Cancer Institute CA-12376. We thank Mrs. Erika Paseka, Mrs. Marion Foulds, and Mrs. Ilse Hoffmann for their expert assistance in preparing and editing the manuscript.

REFERENCES

1. M. Eisenbud, *AMA Arch. Ind. Hyg. Occup. Med.* **6,** 214 (1952).
2. P. Kotin, *Cancer Res.* **16,** 375 (1956).
3. L. Dautrebande, "Microaerosols," pp. 58–85. Academic Press, New York, New York, 1962.
4. M. Lippman and R. Albert, *Amer. Ind. Hyg. Ass., J.* **30,** 257 (1969).

5. American Public Health Association, *Health Lab. Sci.* **7**, 279 (1970).
6. A. Albagli, L. Eagan, H. Oja, and L. Dubois, *Atmos. Environ.* **8**, 201 (1974).
7. R. E. Lee, Jr. and S. Goranson, *Environ. Sci. Technol.* **6**, 1019 (1972).
8. M. Katz, "Measurement of Air Pollutants. Guide to the Selection of Methods," p. 123. World Health Organ., Geneva, Switzerland, 1969.
9. J. M. Colucci and C. R. Begeman, *J. Air Pollut. Contr. Ass.* **15**, 113 (1965).
10. J. M. Colucci, and C. R. Begeman, *Environ. Sci. Technol.* **5**, 145 (1971).
11. J. M. Colucci and C. R. Begeman, *in* "Proceedings of the Second International Clean Air Congress" (H. M. Englund and W. T. Berry, eds.), pp. 28–35. Academic Press, New York, New York, 1971.
12. R. J. Gordon and R. J. Bryan, *Environ. Sci. Technol.* **7**, 1050 (1973).
13. G. E. Moore, M. Katz, and W. B. Drowley, *J. Air Pollut. Contr. Ass.* **16**, 492 (1966).
14. E. C. Tabor, *Trans. N.Y. Acad. Sci.* [2] **28**, 569 (1966).
15. E. C. Tabor, *J. Air Pollut. Contr. Ass.* **15**, 415 (1965).
16. R. E. Waller and B. T. Commins, *Environ. Res.* **1**, 295 (1967).
17. S. Hochheiser, F. J. Burmann, and G. B. Morgan, *Environ. Sci. Technol.* **5**, 679 (1971).
18. C. R. Begeman and J. M. Colucci, *Natl. Cancer Inst., Monogr.* **9**, 17 (1962).
19. H. L. Falk, I. Markul, and P. Kotin, *AMA Arch. Ind. Health* **13**, 13 (1956).
20. M. Kuratsune and T. Hirohata, *Natl. Cancer Inst., Monogr.* **9**, 117 (1962).
21. B. D. Tebbens, J. F. Thomas, and M. Mukai, *Amer. Ind. Hyg. Ass., J.* **27**, 415 (1966).
22. D. Ramdia, *Int. J. Air Water Pollut.* **9**, 113 (1965).
23. A. Brockhaus, *Atmos. Environ.* **8**, 521 (1974).
24. H. O'Donell, T. L. Montgomery, and M. Corn, *Atmos. Environ.* **4**, 1 (1970).
25. C. H. Keith and J. C. Derrick, *J. Colloid Sci.* **15**, 340 (1960).
26. L. Zechmeister and B. K. Koe, *Arch. Biochem. Biophys.* **35**, 1 (1952).
27. D. Hoffmann and E. L. Wynder, *Natl. Cancer Inst., Monogr.* **9**, 91 (1962).
28. T. W. Stanley, J. E. Meeker, and M. J. Morgan, *Environ. Sci. Technol.* **1**, 937 (1967).
29. J. Leiter and M. J. Shear, *J. Natl. Cancer Inst.* **3**, 155 (1943).
30. G. R. Clemo, *Chem. Ind.* (*London*) p. 957 (1953).
31. P. Kotin, H. L. Falk, P. Mader, and M. Thomas, *AMA Arch. Ind. Hyg. Occup. Med.* **9**, 153 (1954).
32. W. C. Hueper, P. Kotin, E. C. Tabor, W. W. Payne, H. L. Falk, and E. Sawicki, *Arch. Pathol.* **74**, 89 (1962).
33. E. L. Wynder and D. Hoffmann, *J. Air Pollut. Contr. Ass.* **15**, 155 (1965).
34. S. S. Epstein, S. Joshi, J. Andrea, N. Mantel, E. Sawicki, T. Stanley, and E. C. Tabor, *Nature* (*London*) **212**, 1305 (1966).
35. R. H. Rigdon and J. Neal, *Tex. Rep. Biol. Med.* **29**, 109 (1971).
36. L. M. Shabad and G. A. Smirnov, *Atmos. Environ.* **6**, 153 (1972).
37. G. Chatot, M. Castegnaro, J. L. Roche, and R. Fontanges, *Anal. Chim. Acta* **53**, 259 (1971).
38. W. W. Payne, *Natl. Cancer Inst., Monogr.* **9**, 75 (1962).
39. D. Hoffmann and E. L. Wynder, *Cancer* **15**, 93 (1962).
40. Y. Masuda and D. Hoffmann, *Anal. Chem.* **41**, 650 (1969).
41. National Research Council, "Particulate Polycyclic Organic Matter," p. 361. Natl. Acad. Sci., Washington, D.C., 1972.
42. American Public Health Association, *Health Lab. Sci.* **7**, Part 2, 31 (1970).

43. D. Hoffmann, E. L. Wynder, *Z. Krebsforsch.* **68,** 137 (1966).
44. D. Hoffmann, W. E. Bondinell, and E. L. Wynder, *Science* **183,** 215 (1974).
45. U.S. Public Health Service, "Survey of Compounds which have been Tested for Carcinogenic Activity," *U.S., Pub. Health Serv., Publ.* No. 149 (1951); No. **149,** Suppl. 1 (1957); Suppl. 2 (1969); DHEW (NIH) 72-35 (1972); DHEW (NIH) 73 Sects. I and II (1973); DHEW (NIH) 74-453 (1974).
46. E. L. Tabor, T. E. Hauser, J. P. Lodge, and R. H. Burttschell, *AMA Arch. Ind. Health* **17,** 58 (1958).
47. S. S. Epstein, M. Small, E. Sawicki, and H. L. Falk, *J. Air Pollut. Contr. Ass.* **15,** 174 (1965).
48. S. Asahina, J. Andrea, A. Carmel, E. Arnold, Y. Bishop, S. Joshi, D. Coffin, and S. S. Epstein, *Cancer Res.* **32,** 2263 (1972).
48a. W. Cantreels and K. Van Cauwenberghe, *Atmos. Environ.* **10,** 447 (1976).
49. B. L. Van Duuren and F. L. Schmidt, *Chem. Ind. (London)* p. 1006 (1958).
50. T. R. Hauser and J. N. Pattison, *Environ. Sci. Technol.* **6,** 549 (1972).
51. D. R. Long, *Anal. Chem.* **35,** 111R (1963).
52. C. M. Gambrill, *Anal. Chem.* **37,** 143R (1965).
53. D. L. Camin and A. J. Raymond, *J. Chromatogr. Sci.* **11,** 625 (1973).
54. J. D. Mold, R. K. Stevens, R. E. Means, and J. M. Ruth, *Biochemistry* **2,** 605 (1973).
55. A. Liberti, G. P. Cartoni, and V. Cantuti, *J. Chromatogr.* **15,** 141 (1964).
56. D. Hoffmann and E. L. Wynder, *Anal. Chem.* **32,** 295 (1960).
57. E. Sawicki, T. W. Stanley, T. R. Hauser, H. Johnson, and W. C. Elbert, *Int. J. Air Water Pollut.* **7,** 57 (1963).
58. S. P. McPherson, E. Sawicki, and F. T. Fox, *J. Gas Chromatogr.* **4,** 156 (1966).
59. E. Sawicki, C. R. Engel, and M. Guyer, *Anal. Chim. Acta* **39,** 505 (1967).
60. I. D. Entwistle and R. A. W. Johnstone, *Chem. Ind. (London)* p. 269 (1965).
61. A. W. Spears, C. W. Lassiter, and J. H. Bell, *J. Gas Chromatogr.* **1,** 34 (1963).
62. A. W. Horton, D. T. Denman, and R. P. Trosset, *Cancer Res.* **17,** 758 (1957).
63. E. Bingham and H. L. Falk, *Arch. Environ. Health* **19,** 779 (1969).
64. M. Saffiotti and P. Shubik, *Natl. Cancer Inst., Mongr.* **10,** 489 (1963).
65. E. L. Wynder and D. Hoffmann, "Tobacco and Tobacco Smoke." Academic Press, New York, New York, 1967.
66. P. Kotin and H. L. Falk, *Radiat. Res., Suppl.* **3,** 193 (1963).
67. B. L. Van Duuren, L. Langseth, L. Orris, and B. M. Goldschmidt, *J. Natl. Cancer Inst.* **39,** 1217 (1967).
68. R. A. Rasmussen and F. W. Went, *Proc. Natl. Acad. Sci. U.S.* **53,** 215 (1965).
69. T. C. Tso, *Prev. Med.* **3,** 294 (1974).
70. I. Schmeltz, D. Hoffmann, and E. L. Wynder, *Proc. Annu. Conf. Trace Substances Environ. Health* Vol. 8, p. 281 (1974).
71. E. Sawicki, *Arch. Environ. Health* **14,** 46 (1967).
72. W. C. Hueper, "Occupational and Environmental Cancers of the Respiratory System," p. 214. Springer-Verlag, Berlin, Germany, and New York, New York, 1966.
73. J. R. Goldsmith and L. Friberg, *in* "Air Pollution" (A. C. Stern, ed.), 3rd ed., Vol. II, p. 457. Academic Press, New York, New York, 1977.
74. E. L. Wynder, L. Fritz, and N. Furth, *J. Natl. Cancer Inst.* **19,** 361 (1957).
75. E. L. Wynder and D. Hoffmann, *Cancer* **12,** 1079 (1959).
76. E. L. Wynder and D. Hoffmann, *Cancer* **12,** 1194 (1959).
77. E. L. Wynder and D. Hoffmann, *Nature (London)* **192,** 1092 (1961).

78. D. Hoffmann and E. L. Wynder, *J. Air Pollut. Contr. Ass.* **13,** 322 (1963).
79. D. Hoffmann, G. Rathkamp, S. Nesnow, and E. L. Wynder, *J. Natl. Cancer Inst.* **45,** 1165 (1972).
80. S. S. Hecht, W. E. Bondinell, and D. Hoffmann, *J. Natl. Cancer Inst.* **53,** 1121 (1974).
81. F. H. Blanding, W. H. King, Jr., W. Priestley, Jr., and J. Rehner, Jr., *AMA Arch. Ind. Hyg. Occup. Med.* **4,** 335 (1951).
82. E. L. Wynder and D. Hoffmann, *Brit. J. Cancer* **24,** 574 (1970).
83. J. M. Colucci and C. R. Begeman, *J. Air Polut. Contr. Ass.* **19,** 255 (1969).
84. G. J. Cleary and J. L. Sullivan, *Med. J. Aust.* **56,** 758 (1965).
85. P. Valori, A. Grella, C. Melchiorri, and N. Vescia, *Nuovi Ann. Ig. Microbiol.* **17,** 383 (1966).
86. L. Zoccolillo, A. Liberti, and D. Brocco, *Atmos. Environ.* **6,** 75 (1972).
87. E. Sawicki, S. P. Pherson, T. W. Stanley, J. Meeker, and W. C. Elbert, *Int. J. Air Water Pollut.* **9,** 515 (1965).
88. E. Sawicki, T. R. Hauser, W. C. Elbert, F. T. Fox, and J. E. Meeker, *Amer. Ind. Hyg. Ass., J.* **23,** 137 (1962).
89. R. E. Waller, B. T. Commins, and P. J. Lawther, *Brit. J. Ind. Med.* **22,** 128 (1965).
90. P. Stocks, *Brit. J. Cancer* **20,** 595 (1966).
91. B. T. Commins and R. E. Waller, *Atmos. Environ.* **1,** 49 (1967).
92. R. C. Lao, R. S. Thomas, H. Oja, and L. Dubois, *Anal. Chem.* **45,** 908 (1973).
93. H. Matsushita, Y. Esumi, and K. Yamada, *Bunseki Kagaku* **19,** 951 (1970).
94. D. F. Bender, *Environ. Sci. Technol.* **2,** 204 (1968).
95. R. S. Thomas and G. E. Moore, *Atmos. Environ.* **2,** 145 (1968).
96. G. M. Badger, *Natl. Cancer Inst., Monogr.* **9,** 1 (1961).
97. G. M. Badger, "The Chemical Basis of Carcinogenic Activity," p. 16. Thomas, Springfield, Illinois, 1962.
98. G. M. Badger, R. W. L. Kimber, and J. Novotny, *Aust. J. Chem.* **17,** 778 (1964).
99. I. E. Burrows and A. J. Lindsay, *Chem. Ind.* (*London*) p. 1395 (1961).
100. E. L. Wynder, G. P. Wright, and J. Lam, *Cancer* **11,** 1140 (1958).
101. G. M. Badger, J. K. Donelly, and T. M. Spotswood, *Aust. J. Chem.* **18,** 1249 (1965).
102. D. Hoffmann and G. Rathkamp, *Anal. Chem.* **44,** 899 (1972).
103. J. L. Monkman, L. Dubois, and C. J. Baker, *Pure Appl. Chem.* **24,** 73 (1970).
104. L. Dubois, A. Zdrojewski, P. Jénnawar, and J. L. Monkman, *Atmos. Environ.* **4,** 199 (1970).
105. E. O. Haenni, J. W. Howard, F. L. Joe, Jr., *J. Ass. Offic. Agr. Chem.* **45,** 67 (1962).
106. J. H. Bell, S. Ireland, and A. W. Spears, *Anal. Chem.* **41,** 310 (1969).
107. R. R. Demisch and G. F. Wright, *Can. J. Biochem. Physiol.* **41,** 1655 (1963).
108. J. D. Mold, T. B. Walker, and L. G. Veasey, *Anal. Chem.* **35,** 2071 (1963).
108a. H. J. Klimisch, *Z. Anal. Chem.* **264,** 275 (1973).
108b. M. Novotny, M. L. Lee, and K. D. Bartle, *J. Chromatogr. Sci.* **12,** 606 (1974).
109. D. Hoffmann, G. Rathkamp, K. D. Brunnemann, and E. L. Wynder, *Sci. Total Environ.* **2,** 157 (1973).
110. D. S. Tarbell, E. G. Brooker, A. Vanterpool, W. Conwoy, C. J. Claus, and T. J. Hall, *J. Amer. Chem. Soc.* **77,** 767 (1955).
111. T. M. Spotswood, *J. Chromatogr.* **2,** 90 (1959).
112. C. R. Sawicki and E. Sawicki, *Progr. Thin-Layer Chromatogr. Relat. Methods* **3,** 233 (1972).

113. E. Sawicki, *Chem. Anal.* **54,** 24, 56, and 88 (1964).
114. E. Sawicki, E. Elbert, T. W. Stanley, R. T. Hauser, and F. T. Fox, *Anal. Chem.* **32,** 810 (1960).
115. R. A. Friedel and M. Orchin, "Ultraviolet Spectra of Aromatic Compounds." Wiley, New York, New York, 1951.
116. E. Clar, "Polycyclic Hydrocarbons," Vols. 1 and 2. Academic Press, New York, New York, 1964.
117. H. E. Ungnade, "Organic Electronic Spectral Data." Wiley (Interscience), New York, New York, 1960.
118. L. Lang, ed., "Absorption Spectra in the Ultraviolet and Visible Region." Vols. 1 and 2. Academic Press, New York, New York, 1961.
119. American Society for Testing and Materials, "Numerical List of Abstracted Ultraviolet Spectra Indexed on Wyandotte—ASTM Punched Cards." Amer. Soc. Test. Mater., Philadelphia, Pennsylvania, 1954.
120. I. B. Berlman, *J. Phys. Chem.* **74,** 3085 (1970).
121. I. B. Berlman, "Handbook of Fluorescence Spectra of Aromatic Molecules." Academic Press, New York, New York, 1965.
122. E. Sawicki, *Talanta* **16,** 1231 (1969).
123. E. Sawicki, T. W. Stanley, and W. C. Elbert, *Talanta* **11,** 1433 (1964).
124. P. P. Dikun and S. G. Cushkin, *Vopr. Oncol.* **5,** 34 (1959).
124a. Intersociety Committee on Methods of Air Sampling and Analysis, "Methods of Air Sampling and Analysis," p. 139. Amer. Publ. Health Ass., Washington, D.C., 1972.
124b. R. A. Lannoye and R. A. Greinke, *Amer. Ind. Hyg. Ass., J.* **35,** 755 (1974).
125. M. L. Lee, M. Novotny, and K. D. Bartle, *Anal. Chem.* **48,** 405 (1976).
125a. S. S. Hecht, W. E. Bondinell, and D. Hoffmann, *J. Natl. Cancer Inst.* **53,** 1121 (1974).
125b. A. Dipple, *in* "Chemical Carcinogens" (C. E. Searle, ed.), p. 245. Amer. Chem. Soc., Washington, D.C., 1976.
126. J. F. K. Huber and J. A. R. Hulsman, *Anal. Chim. Acta* **38,** 305 (1967).
127. L. R. Snyder and J. J. Kirkland, "Introduction to Modern Liquid Chromatography." Wiley, New York, New York, 1974.
128. N. F. Ives and L. Giuffrida, *J. Ass. Offic. Anal. Chem.* **55,** 757 (1972).
129. G. C. Vaughan, B. B. Wheals, and M. J. Whitehouse, *J. Chromatogr.* **78,** 203 (1973).
130. B. L. Karger, M. Martin, J. Loheac, and G. Guiochon, *Anal. Chem.* **45,** 496 (1973).
131. J. K. Selkirk, R. G. Croy, and H. V. Gelboin, *Science* **184,** 169 (1974).
131a. M. Dong, D. C. Locke, and E. Ferrand, *Anal. Chem.* **48,** 368 (1976).
131b. M. A. Fox and S. W. Staley, *Anal. Chem.* **48,** 1383 (1976).
131c. A. M. Krstulovic, D. M. Rosie, and P. R. Brown, *Anal. Chem.* **48,** 1383 (1976).
132. T. D. Searl, F. J. Cassidy, W. H. King, and R. A. Brown, *Anal. Chem.* **42,** 954 (1970).
133. E. Sawicki, *Atmos. Environ.* **7,** 233 (1973).
134. American Petroleum Institute, "Standard Reference Materials 1972–1973," Suppl. 1–3, 4, 5, and 6. Mellon Inst., Pittsburgh, Pennsylvania, 1973–1974.
135. E. Sawicki, W. Elbert, T. W. Stanley, T. R. Hauser, and F. T. Fox, *Int. J. Air Pollut.* **2,** 273 (1960).
136. R. P. Hangebrauck, D. J. von Lehmden, and J. E. Meeker, *J. Air Pollut. Contr. Ass.* **7,** 267 (1964).

137. D. J. von Lehmden, R. P. Hangebrauck, and J. E. Meeker, *J. Air Pollut. Contr. Ass.* **15**, 306 (1965).
138. R. P. Hangebrauck, R. P. Lauch, and J. E. Meeker, *Amer. Ind. Hyg. Ass., J.* **27**, 47 (1966).
139. R. P. Hangebrauck, D. J. von Lehmden, and J. E. Meeker, *U.S. Pub. Health Serv., Publ.* **999-AP-33**, 48 (1967).
140. U.S. Department of Health, Education, and Welfare, Washington, D.C., *U.S. Pub. Health Serv., Publ.* **1886**, 4 (1969).
141. U.S. Department of Agriculture, "1968 Wildfire Statistics," p. 48. U.S. Dept. of Agriculture, Washington, D.C., 1969.
142. U.S. Department of Health, Education, and Welfare, "Nationwide Inventory of Air Pollutant Emissions, 1968," Natl. Air Pollut. Contr. Admin. Publ. AP-73, p. 36. US Govt. Printing Office, Washington, D.C., 1970.
143. U.S. Department of the Interior, Bureau of Mines, "Automobile Disposal, A National Problem," p. 569. US Govt. Printing Office, Washington, D.C., 1967.
144. G. P. Gross, "First Annual Report on Gasoline Composition and Vehicle Exhaust Gas Polynuclear Aromatic Content." U.S. Dept. of Health, Education and Welfare, Durham, North Carolina, 1970.
145. C. R. Begeman and J. M. Colucci, *SAE* Paper No. 700469, (Soc. Automot. Eng., Inc.), Detroit, Michigan, May 1970.
146. D. Hoffmann and E. L. Wynder, unpublished data.
147. V. Masek, *Gig. Sanit.* **33**, 72 (1968).
148. J. Bonnet, *Natl. Cancer Inst., Monogr.* **9**, 221 (1962).
149. P. J. Lawther, B. T. Commins, and R. E. Waller, *Brit. J. Ind. Med.* **22**, 13 (1965).
150. D. Hoffmann and E. L. Wynder, *Proc. Int. Symp. Identif. Meas. Environ. Pollut.* (I. Hoffmann, ed.), p. 9 (1971).
151. V. Galuskinova, *Neoplasma* **11**, 465 (1964).
152. J. W. Lloyd, *J. Occup. Med.* **13**, 53 (1971).
153. C. K. Redmond, A. Ciocco, J. W. Lloyd, and H. W. Rush, *J. Occup. Med.* **14**, 621 (1972).
154. C. R. Begeman and J. M. Colucci, *SAE* Paper No. 700469, (Soc. Automot. Eng., Inc.), Detriot, Michigan, May 1970.
155. S. C. Hoffman, Jr., R. L. Willis, G. H. Patterson, and E. S. Jacobs, *Meet. Sect. Petr.* No. 52, Amer. Chem. Soc., Washington, D.C. (1971).
156. D. Hoffmann, E. Theisz, and E. L. Wynder, *J. Air Pollut. Contr. Ass.* **15**, 162 (1965).
157. F. G. Padrta, P. C. Samson, J. J. Donahue, and H. Skala, *Meet. Sect. Petr.*, No. 50, Amer. Chem. Soc., Washington, D.C. (1971).
158. C. R. Begeman and J. M. Colucci, *Science* **161**, 271 (1968).
159. E. Sawicki, W. C. Elbert, T. R. Hauser, F. T. Fox, and T. W. Stanley, *Amer. Ind. Hyg. Ass., J.* **21**, 433 (1960).
160. R. E. Waller, *Brit. J. Cancer* **6**, 8 (1952).
161. J. M. Campbell and J. Clemmensen, *Dan. Med. Bull.* **3**, 205 (1956).
162. H. D. Hettche, *Int. J. Air Water Pollut.* **8**, 185 (1964).
163. M. Saringer, *Egeszsegtudomany* **7**, 25 (1963).
164. F. Peter, E. Toth, and E. Jeney, *Egeszsegtudomany* **13**, 304 (1969).
165. G. Louw, *Amer. Ind. Hyg. Ass., J.* **26**, 520 (1965).
166. H. Wanatabe and K. Tomita, *in* "Proceedings of the First International Clean Air Congress," pp. 226–228. London, England, 1966.
167. S. W. Simmons, *in* "DDT. The Insecticide Dichlorodiphenylchloroethane and its

Significance" (P. Muller, ed.), Vol. II. p. 251. Birkhaeuser, Basel, Switzerland, 1959.
168. T. J. Jukes, *Amer. Sci.* **51,** 355 (1963).
169. T. J. Jukes, *Prev. Med.* **2,** 133 (1973).
170. U. Saffiotti, *Prev. Med.* **2,** 125 (1973).
171. U.S. Department of Health, Education, and Welfare, "Federal Food, Drug, and Cosmetic Act, as Amended," p. 26. U.S. Govt. Printing Office, Washington, D.C., 1969.
172. D. Hoffmann, G. Rathkamp, and E. L. Wynder, *Beitr. Tabakforsch.* **5,** 140 (1969).
173. J. J. Domanski, P. L. Haire, and T. J. Sheets, *Tobacco Sci.* **18,** 111 (1974).
174. R. E. Duggan, *Proc. Int. Symp. Identif. Meas. Environ. Pollut. 1971* (I. Hoffmann, ed.), p. 239. Campbell Printing, Ottawa, Ontario 1972.
175. P Antommaria, M. Corn, and L. Demaio, *Science* **150,** 1476 (1965).
176. D. C. Abbot, R. B. Harrison, J. O. G. Tatton, and J. Thomson, *Nature (London)* **211,** 259 (1966).
177. R. W. Risebrough, R. J. Huggett, J. J. Griffin, and E. D. Goldberg, *Science* **159,** 1233 (1968).
178. D. B. Seba and J. M. Prospero, *Atmos. Environ.* **5,** 1043 (1971).
179. Z. Jegier, *Ann. N.Y. Acad. Sci.* **160,** 143 (1969).
180. J. H. Caro, A. W. Taylor, and E. R. Lemon, *Proc. Int. Symp. Identif. Meas. Environ. Pollut. 1971* p. 72. Campbell Printing, Ottawa, Ontario 1972.
181. B. Davidow and J. L. Radomski, *J. Pharmacol. Exp. Ther.* **107,** 226 (1953).
182. H. L. Falk, S. J. Thompson, and P. Kotin, *Arch. Environ. Health* **10,** 847 (1965).
183. K. F. Lang and I. Eigen, *Fortschr. Chem. Forsch.* **8,** 91 (1967).
184. E. Proksch, *Z. Anal. Chem.* **178,** 23 (1966).
185. R. L. Stedman, *Chem. Rev.* **68,** 153 (1968).
186. D. Hoffmann and V. Mazzola, *Beitr. Tabakforsch.* **5,** 183 (1970).
186a. M. W. Dong, D. Locke, and D. Hoffmann, *J. Environ. Sci.,* in press (1977).
187. Medical Research Council, London, "The Carcinogenic Action of Mineral Oils: A Chemical and Biological Study," *Med. Res. Comm. (Gt. Brit.), Spec. Rep. Ser.* **306,** 251 (1968).
188. E. Sawicki and H. Johnson, *Microchim. Acta* **2,** 435 (1964).
189. T. W. Stanley, M. J. Morgan, and J. E. Meeker, *Environ. Sci. Technol.* **3,** 1188 (1969).
190. J. Jaeger, *Z. Anal. Chem.* **255,** 281 (1971).
191. Food Protection Committee, Food Nutrition Board, *Natl. Acad. Sci.—Natl. Res. Counc., Publ.* **749,** (1959).
192. World Health Organization, *Tech. Rep. Ser.* **220** (1961). World Health Organ., Geneva, Switzerland.
193. L. Fishbein, H. L. Falk, and W. G. Flamm, "Chemical Mutagens." Academic Press, New York, New York, 1970.
194. P. Kotin and H. L. Falk, *Cancer* **9,** 910 (1956).
195. B. L. Van Duuren, N. Nelson, L. Orris, E. D. Palmes, and F. L. Schmitt, *J. Natl. Cancer Inst.* **31,** 41 (1963).
196. P. Holsti, *Naturwissenschaften* **48,** 459 (1961).
197. F. S. Mallette and E. van Haam, *AMA Arch. Ind. Hyg. Occup. Med.* **6,** 231 (1952).
198. P. Kotin and H. L. Falk, *Advan. Cancer Res.* **7,** 475 (1963).
199. P. Kotin and H. L. Falk, *Annu. Rev. Med.* **15,** 233 (1964).
200. P. Kotin, H. L. Falk, and C. J. McCammon, *Cancer* **11,** 473 (1963).

201. P. Kotin and D. V. Wisely, *Progr. Exp. Tumor Res.* **3,** 186 (1963).
202. P. Kotin, *Can. Cancer Conf.* **6,** 475 (1966).
203. P. Nettesheim, M. G. Hanna, Jr., D. G. Doherty, R. F. Newell, and A. Hellman, *AEC Symp. Ser.* **18,** 305 (1970).
204. P. Kotin, H. L. Falk, and M. Thomson, *Cancer* **9,** 905 (1956).
205. B. L. Van Duuren, S. Melchionne, R. Blair, B. M. Goldschmidt, and C. Katz, *J. Natl. Cancer Inst.* **46,** 143 (1971).
206. E. Sawicki and C. R. Engel, *Anal. Chim. Acta* **38,** 315 (1967).
207. E. Sawicki, R. Schumacher, and C. R. Engel, *Microchem. J.* **12,** 377 (1967).
208. E. Sawicki and R. A. Carnes, *Anal. Chim. Acta* **41,** 178 (1968).
209. J. M. Barnes and P. N. Magee, *Brit. J. Ind. Med.* **11,** 167 (1954).
210. P. N. Magee and J. M. Magee, *Advan. Cancer Res.* **10,** 163 (1967).
211. D. Druckrey, R. Preussmann, S. Ivankovic, and D. Schmähl, *Z. Krebsforsch.* **69,** 103 (1967).
212. International Agency for Research on Cancer (Lyon, France), *Int. Ag. Res. Cancer, Monogr.* **1,** 95 (1972).
212a. P. N. Magee, R. Montesano, and R. Preussmann, *in* "Chemical Carcinogens" (C. E. Searle, ed.), p. 491. Amer. Chem. Soc., Washington, D.C., 1976.
213. R. Montesano and P. N. Magee, *Nature* (*London*) **228,** 173 (1970).
214. N. P. Sen, *in* "Toxic Constituents of Animal Foodstuffs" (I. E. Liener, ed.), 2nd ed., p. 131. Academic Press, New York, New York, 1973.
215. W. Lijinsky, L. Keefer, E. Conrad, and R. Van de Bogart, *J. Natl. Cancer Inst.* **49,** 1239 (1972).
216. S. S. Mirvish, *J. Natl. Cancer Inst.* **46,** 1183 (1971).
217. J. Sander, "Environment and Cancer," p. 109. Williams & Wilkins, Baltimore, Maryland, 1972.
218. W. Lijinsky, H. W. Taylor, C. Snyder, and P. Nettesheim, *Nature* (*London*) **244,** 176 (1973).
219. G. Neurath, *Experientia* **23,** 400 (1967).
220. D. Hoffmann, G. Rathkamp, and Y. Y. Liu, *Int. Ag. Res. Cancer, Sci. Publ.* **9,** 159 (1975).
221. D. Hoffmann, S. S. Hecht, R. M. Ornaf, and E.L. Wynder, *Science* **186,** 265 (1974).
222. K. Bretschneider and J. Matz, *Arch. Geschwulstforsch.* **42,** 36 (1973).
222a. D. H. Fine, D. P. Rounbehler, N. M. Belcher, and S. S. Epstein, *Science* **192,** 1328 (1976).
223. J. S. Rhim, R. J. Gordon, R. J. Bryan, and R. J. Huebner, *Int. J. Cancer* **12,** 485 (1973).
224. E. Sawicki, M. Guyer, R. Schumacher, W. C. Elbert, and C. R. Engel, *Mikrochim. Acta* **5,** 1025 (1968).
225. H. Weisz and A. Brockhaus, *Zentralbl. Bakteriol., Parasitenk., Infektionskr. Hyg., Abt. 1:Orig., Reihe B* **157,** 28 (1973).
226. E. Sawicki and C. Golden, *Microchem. J.* **14,** 437 (1969).
227. E. Sawicki and C. R. Engel, *Mikrochim. Acta* **6,** 91 (1969).
228. T. W. Stanley, M. J. Morgan, and E. M. Grisby, *Environ. Sci. Technol.* **2,** 699 (1968).
229. E. Sawicki, J. E. Meeker, and M. J. Morgan, *Int. J. Air Water Pollut.* **9,** 291 (1965).
230. B. L. Van Duuren, J. A. Bilbao, and C. A. Joseph, *J. Natl. Cancer Inst.* **25,** 53 (1960).

231. D. Hoffmann, G. Rathkamp, and S. Nesnow, *Anal. Chem.* **41,** 1256 (1969).
232. E. Sawicki, T. W. Stanley, and W. C. Elbert, *J. Chromatogr.* **26,** 72 (1967).
233. S. Ray and R. W. Frei, *J. Chromatogr.* **71,** 451 (1972).
233a. K. Hirao, Y. Shinohara, H. Tsuda, S. Fukushima, M. Takahashi, and N. Ito, *Cancer Res.* **36,** 329 (1976).
233b. M. Dong, D. Hoffmann, D. C. Locke, and E. Ferrand, *Environ. Lett.,* in press (1977).
234. W. C. Hueper, "Occupational and Environmental Cancers of the Urinary System," p. 465. Yale Univ. Press, New Haven, Connecticut, 1969.
235. R. Doll, *in* "Second World Conference on Smoking and Health," (R. G. Richardson, ed.), pp. 10–23, Pitman Medical, London, England, 1972.
236. Occupational Safety and Health Administration, *Chem. Ind.* (*London*) **52,** (6), 12 (1974).
237. Y. Masuda, K. Mori, and M. Kuratsune, *Int. J. Cancer* **2,** 489 (1967).
238. M. Pailer, W. J. Huebsch, and H. Kuhn, *Fachliche Mitt. Oesterr. Tabakregie* **7,** 109 (1967).
239. D. Hoffmann, Y. Masuda, and E. L. Wynder, *Nature* (*London*) **221,** 254 (1969).
240. Y. Masuda and D. Hoffmann, *J. Chromatogr. Sci.* **7,** 694 (1969).
241. E. Sawicki, H. Johnson, and K. Kosinski, *Microchem. J.* **10,** 72 (1966).
242. P Kotin and H. L. Falk, *Cancer* **12,** 147 (1959).
243. W. J. Bair, *AEC Symp. Ser.* **18,** 77 (1970).
244. A. C. Hilding, *N. Engl. J. Med.* **256,** 634 (1957).
245. S. Jakowska, *Ann. N.Y. Acad. Sci.* **130,** Art. 3, 869 (1966).
246. K. H. Kilburn and J. V. Salzano (Symp. eds.), *Amer. Rev. Resp. Dis.* **93,** (No. 3, part 2) 1–184 (1966).
247. T. Dalhamn and L. Strandberg, *Int. J. Air Water Pollut.* **7,** 517 (1963).
248. S. Laskin, M. Kuschner, and R. T. Drew, *AEC Symp. Ser.* **8,** 321 (1970).
249. H. L. Falk, A. Miller, and P. Kotin, *Science* **127,** 474 (1958).
250. H. L. Falk, P. Kotin, and I. Markul, *Cancer* **11,** 482 (1958).
251. National Research Council, "Asbestos," p. 40. Natl. Acad. Sci., Washington, D.C., 1971.
252. L. J. Casarett, *Annu. Rev. Pharmacol.* **11,** 425 (1971).
253. S. Carson, R. Goldhamer, and M. S. Weinberg, *Ann. N. Y. Acad. Sci.* **130,** Art. 3, 935 (1966).
254. S. Carson, R. Goldhamer, and R. Carpenter, *Amer. Rev. Resp. Dis.* **93,** 86 (1966).
255. S. E. Dahlgren, H. Dalen, and T. Dalhamn, *Virchows Arch. Abt. B Zellpathol.* **11** (3), 211 (1972).
256. A. M. Giordano and P. E. Morrow, *Arch. Environ. Health* **25,** 443 (1972).
257. F. L. Estes, *Anal. Chem.* **34,** 998 (1962).
258. F. L. Estes and C. H. Pan, *Arch. Environ. Health* **10,** 207 (1965).
259. B. N. Ames, F. D. Lee, and W. E. Durston, *Proc. Natl. Acad. Sci. U.S.* **70,** 782 (1973).
260. B. N. Ames, W. E. Durston, E. Yamasaki, and F. D. Lee, *Proc. Natl. Acad. Sci. U.S.* **70,** 228 (1973).
260a. R. Talcott and E. Wei, *J. Natl. Cancer Inst.* **58,** 449 (1977).
261. S. S. Epstein and M. Burroughs, *Nature* (*London*) **193,** 337 (1962).
262. S. S. Epstein, M. Small, J. Koplan, N. Mantel, H. L. Falk, and E. Sawicki, *Arch. Environ. Health* **7,** 531 (1963).

263. S. S. Epstein, *J. Air Pollut. Contr. Ass.* **16**, 545 (1966).
264. S. S. Epstein, N. P. Buu-Hoi, and D. P. Hien, *Cancer Res.* **31**, 1087 (1971).
265. S. S. Epstein, M. Small, H. L. Falk, and N. Mantel, *Cancer Res.* **24**, 855 (1964).
266. S. S. Epstein, N. Mantel, and T. W. Stanley, *Environ. Sci. Technol.* **2**, 132 (1964).
267. M. Wilk and W. Girke, *in* "The Jerusalem Symposia on Quantum Chemistry and Biochemistry (E. D. Bergmann and P. Pullman, eds.), Vol. 1, p. 91. Israel Acad. Sci., Jerusalem, Israel, 1969.
268. W. C. Hueper and W. D. Conway, "Chemical Carcinogenesis and Cancers," p. 744. Thomas, Springfield, Illinois, 1964.
269. J. H. Weisburger, *in* "Chemical Carcinogenesis" (C. E. Searle, ed.). Amer. Chem. Soc., Washington, D.C. (in press).
270. B. D. Pullinger, *J. Pathol. Bacteriol.* **52**, 463 (1940).
271. V. Suntzeff, E. V. Cowdry, and A. B. Croninger, *Cancer Res.* **15**, 637 (1955).
272. W. E. Smith, *Bull. N.Y. Acad. Med.* [2] **32**, 71 (1956).
273. V. Suntzeff, A. B. Croninger, E. L. Wynder, E. V. Cowdry, and E. A. Graham, *Cancer* **10**, 250 (1957).
274. P. Healey, L. E. Mawdesley-Thomas, and D. H. Barry, *Nature (London)* **228**, 1006 (1970).
275. I. Chouroulinkov, P. Lazar, C. Izard, C. Libermann, and M. Guérin, *J. Natl. Cancer Inst.* **42**, 981 (1969).
276. F. G. Bock and R. Mund, *J. Invest. Dermatol.* **26**, 475 (1956).
277. E. L. Wynder and D. Hoffmann, *Methods Cancer Res.* **4**, 3 (1968).
278. S. Neukomm, *Acta Unio Int. Contra Cancrum* **15**, 3 (1959).
279. E. Arffmann, *Acta Pathol. Microbiol. Scand.* **60**, 13 (1964).
280. O. I. Iversen and B. C. Christensen, *Acta Pathol. Microbiol. Scand., Suppl.* p. **156** (1962).
281. S. S. Epstein, I. Bulon, J. Koplan, M. Small, and N. Mantel, *Nature (London)* **204**, 750 (1964).
282. Y. Berwald and L. Sachs, *Nature (London)* **200**, 1182 (1963).
283. Y. Berwald and L. Sachs, *J. Natl. Cancer Inst.* **35**, 641 (1965).
284. E. Huberman and L. Sachs, *Proc. Natl. Acad. Sci. U.S.* **56**, 1123 (1966).
285. J. A. DiPaolo, R. L. Nelson, and P. J. Donovan, *Science* **165**, 917 (1969).
286. I. Lasnitzki, *Natl. Cancer Inst., Monogr.* **12**, 381 (1963).
287. M. R. Röller and C. Heidelberger, *Int. J. Cancer* **2**, 509 (1967).
288. C. Heidelberger and P. T. Iype, *Science* **155**, 214 (1967).
289. C. Heidelberger, *Fed. Proc., Fed. Amer. Soc. Exp. Biol.* **32**, 2154 (1973).
290. T. T. Crocker and B. I. Nielsen, *Arch. Environ. Health* **10**, 240 (1965).
291. I. Lasnitzki, *Brit. J. Cancer* **9**, 434 (1955).
292. I. Lasnitzki, *Brit. J. Cancer* **22**, 105 (1968).
293. E. R. Dirksen and T. T. Crocker, *Cancer Res.* **28**, 906 (1968).
294. T. T. Crocker, B. I. Nielsen, and I. Lasnitzki, *Arch. Environ. Health* **10**, 240 (1965).
295. I. Berenblum and P. Shubik, *Brit. J. Cancer* **1**, 383 (1947).
296. A. Graffi and H. Bielka, "Probleme der experimentellen Krebsforschung," Akad. Verlagsges. Leipzig, Germany, 1959.
297. D. B. Clayson, "Chemical Carcinogenesis," p. 467. Little, Brown, Boston, Massachusetts, 1962.
298. D. Hoffmann and E. L. Wynder, *Cancer* **27**, 848 (1971).
299. B. L. Van Duuren, A. Sivak, C. Katz, and S. Melchionne, *J. Natl. Cancer Inst.* **47**, 235 (1971).
300. F. G. Bock, A. P. Swain, and R. L. Stedman, *J. Natl. Cancer Inst.* **47**, 429 (1971).

301. B. L. Van Duuren, *Progr. Exp. Tumor Res.* **11**, 31 (1969).
302. B. L. Van Duuren, A. Sivak, B. M. Goldschmidt, C. Katz, and S. Melchionne, *J. Natl. Cancer Inst.* **44**, 1167 (1970).
303. E. Hecker and H. Kubinyi, *Z. Krebsforsch.* **67**, 176 (1965).
304. A. W. Horton, D. T. Denman, and R. P. Trosset, *Cancer Res.* **17**, 758 (1957).
305. E. Bingham and H. L. Falk, *Arch. Environ. Health* **19**, 779 (1969).
306. B. L. Van Duuren, C. Katz, and B. M. Goldschmidt, *J. Natl. Cancer Inst.* **51**, 703 (1973).
307. H. L. Falk, P. Kotin, and S. Thompson, *Arch. Environ. Health* **9**, 169 (1964).
308. R. K. Boutwell, *Progr. Exp. Tumor Res.* **4**, 207 (1964).
309. J. C. Arcos, M. F. Argus, and G. Wolf, "Chemical Induction of Cancer," Vol. 1, p. 491. Academic Press, New York, New York, 1968.
310. W. Dontenwill, H.-J. Chevalier, H.-P. Harke, U. Lafrenz, G. Reckzeh, and B. Schneider, *J. Natl. Cancer Inst.* **51**, 1781 (1973).
311. M. Kuschner, S. Laskin, E. Christofano, and N. Nelson, *Proc. 3rd Natl. Cancer Conf.*, p. 485. J. P. Lippincott Co., Philadelphia, Pennsylvania, 1957.
312. G. Della Porta, L. Kolb, and P. Shubik, *Cancer Res.* **18**, 592 (1958).
313. L. N. Pylev, *Acta Unio Int. Contra Cancrum* **19**, 688 (1962).
314. L. M. Shabad, *J. Natl. Cancer Inst.* **28**, 1305 (1962).
315. U. Saffiotti, F. Cefis, and L. H. Kolb, *Cancer Res.* **28**, 104 (1968).
316. U. Saffiotti, *Progr. Exp. Tumor Res.* **11**, 302 (1969).
317. A. R. Sellakumar, R. Montesano, U. Saffiotti, and D. G. Kaufman, *J. Natl. Cancer Inst.* **50**, 507 (1973).
318. International Agency for Research on Cancer "Evaluation of Carcinogenic Risk of Chemicals to Man," Vol. 2, pp. 17–47. Int. Ag. Res. Cancer, Lyon, France, 1973.
319. N. Kobayashi and T. Okamoto, *J. Natl. Cancer Inst.* **51**, 1605 (1974).
320. V. J. Feron, *Cancer Res.* **32**, 28 (1972).
321. R. D. Passey, *Brit. J. Med.* **2**, 1112 (1922).
322. J. A. Campbell, *Brit. J. Exp. Pathol.* **15**, 287 (1934).
323. J. A. Campbell, *Brit. J. Exp. Pathol.* **20**, 122 (1939).
324. P. Kotin, H. L. Falk, and M. Thomas, *AMA Arch. Ind. Hyg. Occup. Med.* **9**, 164 (1954).
325. B. P. Gurinov, V. N. Tugarinova, O. I. Vasilieva, M. V. Nifontova, and L. M. Shabad *Gig. Sanit.* **28**, 19 (1962).
326. E. L. Wynder and D. Hoffmann, *Cancer* **15**, 103 (1962).
327. P. Kotin, H. L. Falk, and M. Thomas, *AMA Arch. Ind. Hyg. Occup. Med.* **11**, 113 (1955).
328. J. Leiter, M. B. Shimkin, and M. J. Shear, *J. Natl. Cancer Inst.* **3**, 155 (1942).
329. G. R. Clemo and E. W. Miller, *Chem. Ind.* (*London*) p. 38 (1955).
330. G. R. Clemo, E. W. Miller, and R. C. Pybus, *Brit. J. Cancer* **9**, 137 (1955).
331. J. Bogacz and I. Koprowska, *Acta Cytol.* **5**, 311 (1961).
332. G. R. Clemo and E. W. Miller, *Brit. J. Cancer* **14**, 651 (1966).
333. F. J. C. Roe, F. Kearns, and B. T. Commins, *Brit. Emp. Cancer Campaign, Annu. Rep.* **42**, 26. London, England, 1964.
334. F. J. C. Roe, *in* "Alkylierend wirkende Verbindungen," (K. H. Weber, ed.), p. 110. Wissenschaftliche Forschungsstelle im Verband der Cigaretten-Industrie, Hamburg, Germany, 1968.
335. E. C. Hammond and D. Horn, *J. Amer. Med. Ass.* **166**, 1159 and 1294 (1961).
336. E. L. Wynder and E. C. Hammond, *Cancer* **15**, 79 (1962).
337. Royal College of Physicians, "Air Pollution and Health." Pitman, London, England, 1970.

7

Effects of Air Pollution on Human Health

John R. Goldsmith and Lars T. Friberg

I. Introduction

A. Importance of Air to Health

The average adult male requires about 30 pounds (13.64 kg) of air each day compared with less than 3 pounds (1.37 kg) of food and about $4\frac{1}{2}$ pounds (2.05 kg) of water. Compared with the other necessities of life, obligatory continuous consumption is a unique property of air. The insensible, intimate interpenetration of air, which courses in and out of the lungs, gives to air pollution its essential importance. It has been estimated that a man can live for 5 weeks without food, for 5 days without water, but for only 5 minutes without air. Air is essential to the senses of sight, smell, and hearing, and its pollution assaults the first two of these.

While initial studies of health effects associated with air pollution have properly concentrated on looking for excess mortality or increased numbers of cases with well-defined diseases, the absence of these certainly does not mean that no health effects have occurred. Health effects also occur when sensory irritation is present. For a scientific approach to defining health effects, three requirements must be met: (a) pollution, or an index of it, must be measured in a valid fashion; (b) one or more effects must be measured in a valid fashion; and (c) a relationship of air pollution to the defined effect must be conditioned by the validity of the measurement of pollution and of the putative effect(s) and the strength of the association between the two. The systematic study of naturally occurring associations between health effects and pollution has been called the "epidemiology of air pollution" (*1*). Whether or not the association between pollutant exposure and possible effect is thought to be causal will depend both on the variety, specificity, and strength of associations shown by epidemiological studies, as well as their concordance with experimental studies in animals or man.

Natural variability in responsiveness to air pollution is observed in all populations. Generally speaking, susceptibility is great among premature infants, the newborn, the elderly, and the infirm. Those with chronic diseases of the lungs or heart are thought to be at particular risk. Because of the wide variation in sensitivity to air pollution of different groups in the population, data concerning health effects on healthy persons may not be as important as the responses of the individuals most likely to be sensitive. Preschool and school children appear to be both sensitive and specifically reactive to air pollution health effects (see Section III,C). The control of air pollution, to the extent that it is based upon health effects, should be based on the most sensitive groups of persons. This principle requires that these sensitive groups be definable in terms of age and/or medical status.

The effects of air pollution on personal or community health are (*2*): (a) acute sickness or death; (b) insidious or chronic disease, shortening of life, or impairment of growth or development; (c) alteration of important physiological functions, such as ventilation of the lung, transport of oxygen by hemoglobin, sensory acuity, time interval estimation, or other functions of the nervous system; (d) impairment of performance, such as in athletic activities, motor vehicle operation, or complex tasks such as learning; (e) untoward symptoms, such as sensory irritation, which in the absence of an obvious cause, such as air pollution, might lead a person to seek medical attention and relief; (f) storage of potentially harmful materials in the body; and (g) discomfort, odor, impairment of visibility, or other effects of air pollution sufficient to cause annoyance or to lead individuals to change residence or place of employment.

Objectionable odor, visibility interference, or vegetation damage are useful guides to the likelihood or severity of health effects. A gray pall over a city or industrial area can have a depressing effect and impair the enjoyment of life. Survey methods, carefully applied, are capable of measuring such reactions and effects. Adaptation to many pollution effects is likely, even expected and relied upon; nevertheless, adaptive capabilities, while innate in human populations, are not to be used as an excuse to permit exposures to unnecessary pollution, since adaptive reactions are associated with a cost that is hard to estimate. A public exposed to sensible pollution will not wait for the demonstration that an unpleasant atmosphere causes disease. They will insist that the air be free of substances that interfere with visibility, have a bad odor, or cause irritation. There is no doubt, however, that the urgency with which steps are taken to improve air quality will depend very much on how severe or serious the risk of ill health from air pollution is thought to be.

In Sections III–V, the effects of specific pollutants on health are discussed. In Section VI, the relationship between pollution and disease states is discussed. To some extent, these interests and this treatment will overlap. We have, however, chosen for emphasis in Sections III–V pollutants for which studies have yielded information concerning dose–response relationships. In Section VI, we have cited those studies in which the quantitative relationships between exposure and effects are not well established, but a qualitative relationship was sought and evaluated.

First priority has properly been given to whether air pollution kills, next to whether it causes disease, and only somewhat later to whether it is a contributory factor along with other exposures to aggravating disease. Although laymen have long recognized the effects of pollution on the impairment of well-being and scientists have recognized the interference with normal functions of the body, only recently has a systematic study been undertaken of the effect of air pollution on annoyance reactions and on the more subtle biological or physiological changes of an unfavorable nature.

B. Relation of Air Pollution Effects to Other Exposures

1. Occupation

Many common air pollutants are also substances to which persons are exposed in their occupations. This is true, for example, of sulfur dioxide, carbon monoxide, and lead, as well as of smoke and many dusts. The downtown traffic policeman, the automobile mechanic, and the truck driver in a large metropolitan area may all have substantial exposures to carbon monoxide and lead in association with their occupations. Since these substances may also be found in community pollution, cessation of work does not necessarily terminate their exposure. Therefore, such individuals have an unusually high risk from exposure to community air pollution. Workers in the fields of occupational medicine and industrial hygiene have given us a wealth of information about the effects on human health of specific contaminants. This information has been organized and evaluated in the so-called threshold limit values (TLV) or maximum allowable concentrations (MAC), which really are neither thresholds, nor, for certain groups, allowable.

While industrial health experience is often relevant to community air pollution exposures, the quantitative relationships between these two types of exposure are neither constant nor dependable. The American

Conference of Governmental Industrial Hygienists (*3*), which has for years sponsored the publication of threshold limit values (Table I) (*4*), states:

> Threshold limit values refer to airborne concentrations of substances and represent conditions under which it is believed that nearly all workers may be repeatedly exposed day after day without adverse effect. Because of wide variation in individual susceptibility, however, a small percentage of workers may experience discomfort from some substances at concentrations at or below the threshold limit; a smaller percentage may be affected more seriously by aggravation of a preexisting condition or by development of an occupational illness. . . . These limits are intended for use in the practice of industrial hygiene and should be interpreted and applied only by a person trained in this discipline. They are not intended for use, or for modification for use, (1) as a relative index of hazard or toxicity, (2) in the evaluation or control of community air pollution nuisances, (3) in estimating the toxic potential of continuous uninterrupted exposures. . . .

2. Cigarette Smoking

The dose of pollutants from cigarette smoking is high and intermittent, relative to that from community air pollution. For example, 400 ppm (440 mg/m^3) carbon monoxide for 5 minutes with frequent repetition may be a representative cigarette smoking dose, while from 10 to 30 ppm (11–33 mg/m^3) carbon monoxide for 4–8 hours may represent a frequent community air pollution exposure. Each of these alone may produce inactivation of up to 5% of the circulating hemoglobin. Together, of course, the effect is greater, but because of the dynamics of carbon monoxide uptake and excretion, it is less than additive.

The exposure to oxides of nitrogen present in cigarette smoke may be assumed to be about 100 ppm each of nitric oxide and nitrogen dioxide (*5*). The average exposure to severe air pollution over periods of about an hour will not exceed about 1.5 ppm (2.8 mg/m^3). However, cigarette smoking produces intermittent exposures, while community air pollution is more constant. Such continuous exposure to nitrogen dioxide may produce effects different from acute exposure (*6*). Exposure to other substances from cigarettes is more variable, but at least include carcinogenic tars, aldehydes, and hydrogen cyanide, as well as lead and other metals.

As far as causing lung cancer or chronic disabling pulmonary disease in the whole population, effects of cigarette smoking are more important than effects of air pollution (*7*), but when both factors are present, they apparently have a more than additive effect. This means that cigarette smokers are at unusual risk if they live in areas with substantial air pollution and that the effects of air pollution on chronic disabling pul-

Table I Comparison of Industrial Threshold Limit Values with Selected Maximal Air Pollution Values

Substance	*Industrial threshold limit values (3, 4)*		*Maximal community air pollution levels*	*Place and date where observed*
	ppm	*mg/m³*		
Gases				
Acrolein	0.1	0.25	0.011 ppm	Los Angeles, California, 1960
Benzene[a]	10	30	0.057 ppm	Los Angeles, California, 1967
Carbon monoxide	50	55	360 ppm	London, England, 1956
Formaldehyde[a]	2	3	1.87 ppm[b]	Pasadena, California, 1957
n-Heptane	400	1600	4.66 ppm[c]	Los Angeles, California, 1957
n-Hexane	100	360	0.04 ppm	Los Angeles, California, 1966
Hydrogen sulfide	10	15	0.9 ppm	Santa Clara, California, 1949–1954
Methane	—	—	3.69 ppm	Los, Angeles, California, 1963
Nitric oxide	25	30	3.7 ppm[e]	Los Angeles, California, 1966
Nitrogen dioxide[a]	5	9	1.3 ppm	Los Angeles, California, 1962
Ozone	0.1	0.2	0.90 ppm	Los Angeles, California, 1955
Sulfur dioxide	5	13	3.16 ppm	Chicago, Illinois, 1937
Particulates				
Arsenic	—	0.25	0.069 mg/m³	Zemianske Kostol', Czechoslovakia, 1964
Asbestos[f]	(5 fibers/ml >5 μm in length)		1 fiber/ml	San Lucas, California, 1972
Beryllium[d]	—	0.002	0.0011 mg/m³	Pennsylvania, 1958
Fluoride (as F)	—	2.5	0.56 mg/m³	Ural Alum, USSR
Lead (inorganic)	—	0.15	0.042 mg/m³	Los Angeles, California, 1949–1954
Sulfuric acid	—	1	0.7 mg/m³	London, England, 1956
Vanadium (V_2O_5)	—	—	0.0007 mg/m³	Mihama, Japan, 1964
Dust	—	0.5	—	—
Fume[d]	—	0.05	—	—

[a] Ceiling value, rather than a time-weighted 8-hour average.
[b] Aldehydes as formaldehyde.
[c] Total hydrocarbons.
[d] Neighborhood, i.e., ambient levels (threshold limit values) also have been set uniquely for this substance at 0.00001 mg/m³ as a 30-day average.
[e] Combined oxides of nitrogen.
[f] Current discussion is concerned with lowering this or including submicroscopic fibers. Air pollution level based on electron microscopic counting of submicroscopic fibers.

monary diseases are more likely to occur in cigarette smokers. Such interpretations are based in part on studies carried out in the United Kingdom (*8, 9*) and on evidence from the United States (*10*). Nevertheless, nonsmokers may be the group in which air pollution effects can be most readily detected. This is because response indicators (impairment of lung function, symptoms of cough or sputum production) are less variable in nonsmokers than in smokers.

3. Domestic Pollution*

Home heating and cooking are capable of generating a group of air pollutants (carbon monoxide, sulfur oxides, oxides of nitrogen, sooty and oily aerosols) whose health effects have commonly been overlooked. During periods of low winds and still weather, the dissipation of such domestic pollution will be impaired to a similar extent as the dissipation of outdoor air pollution. In general, oxidants are about half as concentrated within buildings in polluted areas as they are outside, but carbon monoxide and nitric oxide are likely to occur at similar concentrations indoors and out, since they are influenced neither by the walls of rooms nor by air conditioners and filters (*11*). If smoking or cooking occurs indoors, exposures to carbon monoxide and nitrogen oxides may be quite high.

C. Routes of Absorption of Particulates and of Gases

Absorption may be defined as entry into the body by the passage of a substance across a membrane. The substance may be retained at the local site of absorption for long periods of time or may be transported to other parts of the body. This definition is in accord with a statement for metals by the Task Group on Metal Accumulation (*12*). Pulmonary absorption is the most important route for the entry of air pollutants into the human body. Varying amounts of substances deposited in the pulmonary tract may be transported to the gastrointestinal tract via the mucociliary escalator. This comprises the mucus lining of the conducting part of the respiratory system. The mucus is moved by tiny hairlike cilia toward the back of the throat from which it is usually swallowed.

Air pollution may contaminate grain, vegetables, soil and water and secondarily give rise to an exposure via ingestion. Absorption through the skin is not an important route for an air pollutant to enter the body. Eye irritation from pollutants implies impact on the conjunctiva, but little is known about absorption via the conjunctiva.

* See also Chapter 3, this volume.

The chemicophysical properties of an aerosol or a gas determine whether or not the substance will be inert or give rise to local injuries or systemic effects. Ninety percent or more of sulfur dioxide is normally removed in the upper airways (*13, 14*), while if the gas is adsorbed, e.g., on soot particles, the deposition may take place to a great extent in the deeper airways, depending on the size of the particles. This may well change the absorption rate and thus the toxicity of the gas as well as of the particle itself; synergistic effects have been proven in animal experiments for mixtures of sodium chloride aerosols and sulfur dioxide (*15*). However, synergism for such mixtures has not been proven in human experiments (*16*).

1. Absorption by Inhalation

a. Absorption of Particles. Absorption of particulates is influenced by three different mechanisms—deposition, mucociliary clearance, and alveolar clearance. A model of the three processes (Fig. 1) (*12, 17*) shows the various routes of deposition and subsequent fate that an inhaled aerosol can undergo. D_1 and D_2, respectively, represent the aerosol in the inhaled air. D_3, D_4, and D_5 depict the deposition in the nasopharyngeal compartment (nose and throat), the tracheobronchial compartment (the windpipe and large airways) and the pulmonary compartment (the gas-exchanging portions of the lung itself), respectively. The letters a–j stand for the different translocations that deposited particles can undergo within the lung or systemically. These processes are of fundamental importance in evaluating pulmonary absorption of particles.

i. *Deposition of particles in the respiratory tract.* Three mechanisms govern the deposition of an aerosol in the airways—impaction, sedimentation, and diffusion. The relative importance of these parameters varies with differences in the anatomy of the airways and with ventilatory parameters, but is, above all, dependent upon the aerodynamic properties of the aerosols.

Impaction means that the particle, due to its inertia, is deposited on the mucous membranes, perferentially at a site where the airstream changes its direction. It depends on the particle mass (size and density) and shape, but it is also dependent on the velocity of the airstream within the respiratory tract. Impaction will be of particular importance in the nose, mouth, and upper respiratory tract for unit density (unit density means that the specific gravity is 1.0) particles larger than a few microns in diameter.

Sedimentation is dominant for large, high density particles particularly

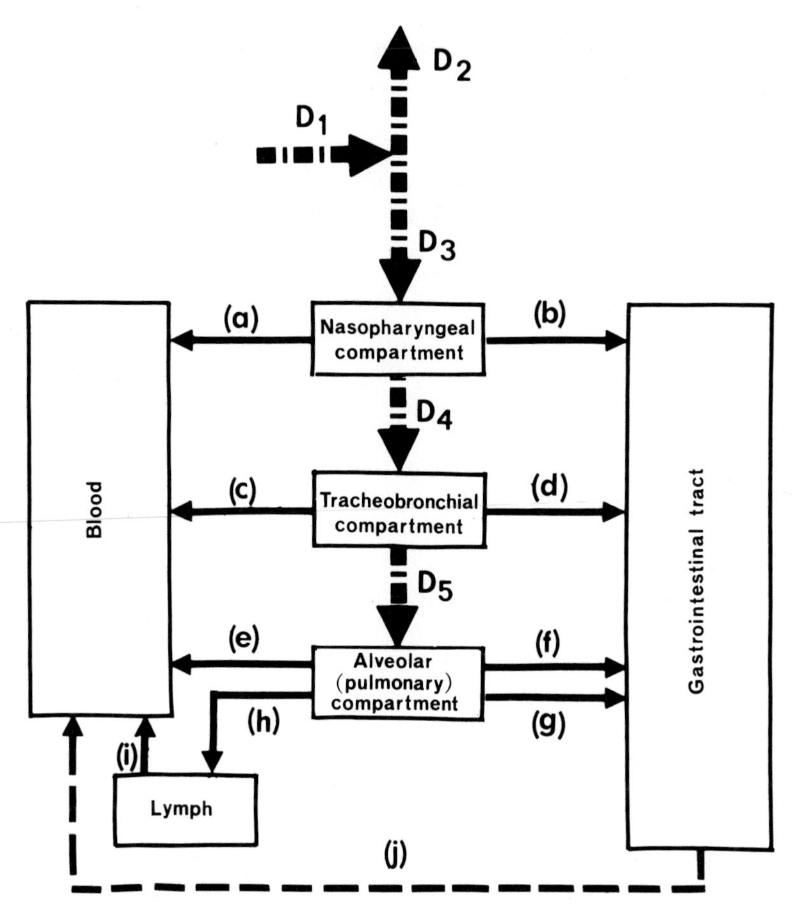

Figure 1. Respiratory tract clearance model for inhaled dust (*12, 17*): D_1 = total dust inhaled, D_2 = dust in exhaled air, D_3 = amount of dust deposited in nasopharyngeal compartment, D_4 = amount of dust deposited in tracheobronchial compartment, D_5 = amount of dust deposited in alveolar compartment. Mechanism: (a) rapid uptake of material deposited in nasopharyngeal compartment directly into the blood; (b) mucociliary clearance from nasopharyngeal compartment to gastrointestinal tract; (c) rapid uptake of material deposited in tracheobronchial compartment directly into the blood; (d) mucociliary clearance; (e) direct absorption of dust from the alveolar compartment to blood; (f) inclusion of alveolarly deposited particles by macrophages and their transport by the mucociliary escalator to the gastrointestinal tract; (g) same as f, but slower process; (h) slow removal of particles by the lymph system; (i) transport of dust cleared by h into the blood; (j) gastrointestinal absorption of particles cleared to the gastrointestinal tract by processes b, d, f, and g.

in regions where the airstream is moving slowly. This means that gravitational sedimentation is of importance mainly for large particles (above a few microns).

Diffusion, which is dependent on Brownian motion of the air molecules, is of importance only for small particles; it is negligible for unit density particles above about 0.5 μm in diameter. Diffusion is especially important for deposition in the alveoli.

Much of our knowledge of areosol deposition in the lungs is based on experimental studies by Wilson and LaMer (*18*) and Brown *et al.* (*19*) and on basic theoretical studies on deposition by Findeisen (*20*) and Landahl (*21*). Deposition in the nasopharyngeal, tracheobronchial, and pulmonary compartments of the body of the International Commission for Radiation Protection (ICRP) "standard man" are shown in Figure 2 for a tidal volume (volume per breath) of 1450 ml and 15 respirations per minute, which represents moderate activity. With larger tidal volumes at the same rate, or with slower respiration at the same volume per breath, there tends to be more deposition deeper in the lung as well as a higher fraction retained. Ambient air aerosols usually have a log normal particle size distribution (Chapter 3, Vol. I). It has been shown (*17*) that the deposition in the different compartments can be predicted using values for mass median aerodynamic diameter. The estimated deposition is not influenced substantially by differences in geometric standard deviation of particle size distribution. Junge (*22*) and Friedlander and Wang (*23*) have shown that particle agglomeration in the air is such that most

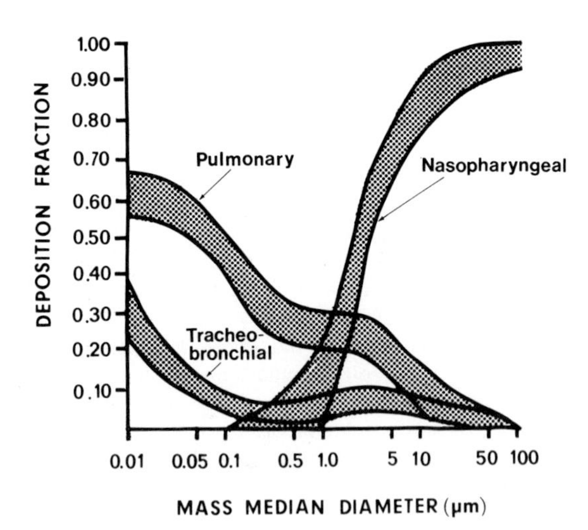

Figure 2. Deposition in various sections of the lungs for various particle populations (*17*). Each of the shaded areas (envelopes) indicates the variability of deposition for a given median (aerodynamic) diameter in each compartment when the distribution parameter σ varies from 1.2 to 4.5 and the tidal volume is 1450 ml.

natural aerosols "age" in the air to a predictable particle size distribution (Chapter 3, Vol. I). This adds to the importance of such deposition predictions. Studies with monodisperse aerosols in man have shown that in the 2–5 μm particle size range, deposition patterns fit the above model, but that there may be large individual variations (*24*). Individual differences in aerosol deposition have also been shown in animals (*25,26*).

ii. *Mucociliary clearance of insoluble particles.* Mucociliary clearance involves the transport of particles upward in the respiratory tract by means of the mucus flow and ciliary activity in the tracheobronchial and nasopharyngeal compartments. The ultimate fate of transported particles is either that they reach the gastrointestinal tract or are expectorated (*25*). Studies of inhaled insoluble particles tagged with radioactive isotopes have shown that the clearance of particles deposited in the tracheobronchial compartment usually is completed within about 8 hours (*26*). The mucus flow in the upper respiratory tract is faster than in the lower part. This and the fact that larger particles are deposited higher in the tracheobronchial tree result in faster clearance for larger particles than for smaller ones.

There are considerable differences among individuals with respect to mucociliary clearance. Some subjects have very slow clearance, while others have a mucociliary clearance that is complete within about 1 hour. Such differences to some extent seem to have a constitutional origin. Monozygotic twins have very similar clearance rates, while dizygotic twins differ almost as much as individuals within the general population (*27*). The medical significance of differences in clearance rates is not known in any detail, but it may well be that subjects with a slower clearance rate are more at risk from harmful effects of particles, as this may indicate a slow mucociliary transport and a higher deposition in the alveolar region in such persons.

Clearance of inhaled aerosols of bacteria and viruses are of particular importance. Their deposition will follow ordinary laws governing deposition. Clearance, however, will be influenced also by bactericidal properties of the tissue cells and the secretions of the lung. By studying the clearance of viable inhaled bacteria and killed bacteria tagged with radionuclides, Rylander (*28*) in animal experiments has estimated how much of the total clearance is due to mechanical clearance and how much to bactericidal action. Similar studies have been done by Laurenzi *et al.* (*29*) and Goldstein *et al.* (*30*).

iii. *Alveolar clearance.* Alveolar clearance can take place by transport of particles from the alveoli to the mucociliary escalator. The different mechanisms for this transport are not known, but uptake by alveolar scavenger cells is considered the most important mechanism (*12*). After passing the alveolar membrane, the particles may either remain within the pulmonary tissue or pass into blood and lymph for further translocation to different organs within the body. If the particle is weakly soluble or capable of surface reactions (as is true, for example, for fractured silica) these reactions tend to occur in the lung tissue. Mercer (*31*) has proposed a solubility model that indicates that the more soluble the particle, the larger the part that is translocated directly from the alveoli to the systemic circulation. Few hard data exist for most insoluble particles either in humans or animals. Concerning quantitative aspects of alveolar clearance, it is not known how much is absorbed systemically directly from the alveoli and how much is translocated to the mucociliary escalator. Asbestos particulates, with their long fibers, behave differently with respect to both deposition and clearance than spherical or near-spherical particulates. This tends to increase the likelihood of alveolar reactions and migration of ultramicroscopic fibrils of asbestos to the outer surfaces (pleura) of the lung (see Section V,A). These processes are just beginning to be studied by electron microscopy (*32*).

b. Absorption of Gases. As a general rule the solubility of a pollutant gas determines what proportion is deposited in the upper airway and what proportion reaches the terminal air sacs of the lung, the alveoli. For example, sulfur dioxide, which has high water solubility, is absorbed high in the airways and has as its primary reaction pattern irritation and increased airway resistance (*33*). Frank and Speizer (*14*) have shown that at relatively high concentrations, the largest proportion of sulfur dioxide is removed in the upper airways, i.e., the nasopharynx. However, Strandberg (*13*) has shown that at low concentrations, a high proportion is carried deep into the lung. Nitrogen dioxide and ozone are examples of relatively water-insoluble gases. They have their major biological reactions at the alveolar level. Depending on their ability to penetrate epithelial membranes, gases are absorbed to a varying degree. Since the alveolar cells are thought to be covered by a phospholipid layer, lipid solubility is probably of importance for penetration of the alveolar membrane.

2. Absorption by Ingestion

Substances entering the gastrointestinal tract may do so directly as contaminants of food and water. A substantial part, however, may be a result of the translocation from the lungs by means of the mucociliary clearance. The rate of absorption of substances from the gastrointestinal tract is highly dependent on the substance in question. For example, methylmercury is absorbed to about 95%, irrespective of whether it is administered as a salt in water solution or in protein-bound form (*34*), whereas mercuric mercury, orally administered, in tracer dose experiments is absorbed only to about 15% (see Section V,D). With respect to their rates of absorption, salts of metals are important only to the extent that ionization increases or decreases the absorption of the compound in the gastrointestinal tract. After metals have reached the small intestine, they are mainly bound to organic molecules, so that the nature of the anion is not highly important.

Great interindividual differences in absorption and variations related to sex and age may exist. Hursh and Suomela (*35*) measured lead absorption in three human volunteers using lead-212 and found that the absorption ranged from about 1 to 25%. Wetherill *et al.* (*36*), in a metabolic study using stable isotopes lead-204 and/or lead-207 to label food, studied two adult males. They found 6–9% absorption from the gut and a residence time in the metabolic pool of 35–50 days. Several factors may interact with the absorption of ingested air pollutants; e.g., reduction of dietary calcium increases the absorption of lead and cadmium (*37–39*). Six and Goyer (*40*) have further found that a lowering of dietary iron increases lead absorption. Increased lead uptake has also been found related to the protein content in food (see Section V,B).

II. Acute Air Pollution Effects and Their Detection

A. Air Pollution Episodes

1. Characteristics Common to Episodes

Several disastrous episodes have focused attention upon air pollution as a health problem. These episodes have made it obvious that the air quality of a community may deteriorate enough to damage the health of its citizens. The observed relationship of air pollution disasters to the presumptive exposures permits several general conclusions. The toll of excess mortality and morbidity in disasters has never been appreciated at the time of the episode; therefore, protective measures were usually not taken. The episodes have always occurred under extraordinary meterological conditions that reduced the effective volume of air in which the pollutants were diluted. Most have occurred under circumstances in which small water droplets were present and, therefore, it is likely that a combination of aerosols and gaseous pollutants was involved. Only since 1952 have valid contemporary measurements been made during episodes.

2. Meuse Valley, Belgium, 1930

On Monday, December 1, 1930, the narrow valley of the Meuse River was afflicted by an unusual and widespread weather condition that persisted during the remainder of the week. In this valley, 15 miles (24 km) in length and with hills about 300 feet (90 m) high on either side, a thermal inversion confined emitted pollutants to the air volume contained in the valley. In the valley were located a large number of industrial plants, including coke ovens, blast furnaces, steel mills, glass factories, zinc smelters, and sulfuric acid plants. On the third day of this unusual weather, a large number of people became ill with respiratory tract complaints and before the week was over, 60 had died. In addition, there were deaths in cattle. Older persons with previously known diseases of the heart and lungs had the greatest mortality; however, illness affected persons of all ages and was best described as an irritation of all exposed membranes of the body, especially those of the respiratory tract. Chest pain, cough, shortness of breath, and eye and nasal irritation were the most common symptoms. Treatment with antispasmodic drugs was of some help. Frequency of symptoms decreased strikingly on December 5, but fatalities occurred both on December 4 and 5. Autopsy examinations showed only congestion and irritation of the tracheal mucosa and large bronchi. However, there was some black particulate matter in the lungs, mostly within the phagocytes.

The chemical substances responsible for the illness and fatalities have been disputed. In the original report on the episode (*41*), it was estimated (since no measurements had been made during the event) that the sulfur dioxide content of the atmosphere was from 25 to 100 mg/m^3 (9.6–38.4 ppm). Assuming oxidation of the sulfur dioxide, high sulfuric acid mist concentrations might have resulted. Some have raised the question as to whether fluorides were possibly the cause of the episode (*42*). It is generally felt now that a combination of several pollutants may have been associated with this, as well as with other community disasters. Certainly strong suspicion attaches to sulfur oxides, but it is more likely that sulfur dioxide, when dissolved or combined with water droplets and in the presence of a multiplicity of other pollutants, was oxidized to sulfuric acid mist with a particle size sufficiently small to penetrate deeply into the lung. Firket (*41*) remarked prophetically that "the public services of London might be faced with the responsibility of 3200 sudden deaths if such a phenomenon occurred there."

3. Donora, Pennsylvania, 1948

The impact of the Donora disaster has been eloquently described by Breton Roueché (*43*):

> The fog closed over Donora on the morning of Tuesday, October 26. The weather was raw, cloudy and dead calm, and it stayed that way as the fog piled up all that day and the next. By Thursday, it had stiffened adhesively into a motionless clot of smoke. That afternoon it was just possible to see across the street, and except for the stacks, the mills had vanished. The air began to have a sickening smell, almost a taste. It was the bittersweet reek of sulfur dioxide. Everyone who was out that day remarked on it, but no one was much concerned. The smell of sulfur dioxide, a scratchy gas given off by burning coal and melting ore, is a normal concomitant of any durable fog in Donora. This time it merely seemed more penetrating than usual.

During this period, again, temperature inversion and foggy weather affected a wide area. Donora is located on the inside of the horseshoe-shaped valley of the Monongahela River. The city contained a large steel mill, a sulfuric acid plant, and a large zinc production plant, among other industries. The hills on either side of the valley are steep, rising to several hundred feet. At the time there were about 14,000 people living in the valley. About half of the population was made ill during the episode. Curiously, many of the persons who were not ill were unaware of the extent of ill health. Cough was the most prominent symptom, but all portions of the respiratory tract and the eyes, nose, and throat were irritated in at least some persons. Many complained of chest constriction, headache, vomiting, and nausea. The frequency and severity of illness increased with the age of the population. Most of those who became ill did so on the second day of the episode; of the 20 deaths, most occurred on the third day. Among those who died, pre-existing cardiac or respiratory system disease was common. A meticulous health survey of the population was made within a few months of the episode (*44*). The investigation was directed at the health effects that occurred among people and animals, the nature of the contaminants, and the meterological conditions. Interviews were obtained from persons who were ill and from physicians in the community. X-rays and blood tests were taken, and teeth, bone, and urine samples were studied to determine whether fluorides might have been involved. From examinations made for fluorides, it was felt that fluorine was probably not involved. Retrospective examination of mortality records indicated that a similar event might have occurred in April of 1945. Autopsy examinations of the 1948 fatalities gave nonspecific findings, but there was abundant evidence of respiratory tract irritation.

Environmental measurements had not been made during the episode, but it was inferred that sulfur dioxide had ranged between 1.4 and 5.5 mg/m^3 (0.5 and 2.0 ppm). Particulate matter was undoubtedly present. The calls for medical assistance in Donora ceased rather abruptly on Saturday evening despite the fact that the fog remained quite dense. This suggests that some change in the physical nature of the fog droplets may have occurred; for example, the particles may have increased sufficiently in size so that they were deposited in the upper airway instead of penetrating deeply into the lung. The population affected was restudied in 1952 and 1957 and was found to have a less favorable mortality and morbidity experience than persons not affected in 1948 (*45*). This could either be because those susceptible in 1948 would have had a less favorable experience in any event or because the exposure in 1948 had long-term effects.

4. Poza Rica, Mexico, 1950

Another type of community disaster resulting from the discharge of a toxic gas from a single source befell the small town of Poza Rica, Mexico (*46*). Here, a new plant for the recovery of sulfur from natural gas put a portion of its equipment into operation on the night of November 21, 1950. One of the steps in the process was the removal of hydrogen sulfide from natural gas. In order to do this, the hydrogen sulfide was concentrated in a system in which it was intended to be burned. During the early morning of November 24, the flow of gas into and through the plant was increased; the weather was foggy with weak winds and a low inversion layer. Between 4:45 a.m. and 5:10 a.m., hydrogen sulfide was released inadvertently and spread into the adjacent portion of the town. Most of the nearby residents were either in bed or had just arisen. Many were afflicted promptly with respiratory and central nervous system symptoms. Three hundred and twenty were hospitalized and 22 died. The characteristic manner in which the hydrogen sulfide affected these individuals was to produce loss of sense of smell and severe respiratory tract irritation (see also Section IV,E). Most of the deaths occurred in persons who had such central nervous system symptoms as unconsciousness and vertigo. A number of affected individuals also had pulmonary edema. Persons of all ages were affected and pre-existing disease did not seem to have much influence on which persons were afflicted.

5. London, England, 1952

From December 5 through 9, 1952, most of the British Isles was covered by a fog and temperature inversion. One of the areas most severely affected was London, which is located in the broad valley of the Thames. During this period, an unusually large number of deaths occurred and many more persons were ill. The illnesses were usually sudden in onset and tended to occur on the third or fourth day of the episode (*47*). Shortness of breath, cyanosis, some fever, and evidence of excess fluid in the lungs were observed. Most of those seriously ill were in the older age groups. Admissions to hospitals for the treatment of respiratory diseases were increased markedly, but so were admissions for heart disease. An increase in mortality among all ages was observed. However, the very old, those in the seventh and eighth decades, had the highest increment. The most frequent causes to which deaths were ascribed were chronic bronchitis, broncho-pneumonia, and heart disease. Of particular interest was the fact that mortality remained elevated for several weeks after the weather had improved. This observation has led to speculation that acutely increased pollution may start in motion a process that could continue to operate even after pollution has returned to normal. One such process could be lowered resistance to such infectious processes as influenza. The total excess was between 3500 and 4000 deaths, distributed as shown in Figure 3. Autopsy examination did not reveal any characteristic mode of death other than evidence of respiratory tract irritation. Measurements were available for the amount of suspended smoke and sulfur dioxide. The highest values reported were 4.46 mg/m^3 of smoke and 1.34 ppm (3.75 mg/m^3) of sulfur dioxide.

Search of the past records of meteorology and mortality indicated that periods of excessive mortality had occurred previously. Three hundred excess deaths occurred in the winter of 1948; detectable increases in mortality associated with fog were found in December 1873, January 1880, February 1882, December 1891, and December 1892. Subsequent episodes have occurred in 1959 and 1962 (*48, 49*). None of the other episodes, however, was quite as severe as the one in 1952. For further discussion of the 1962 episode, see Section 9, below.

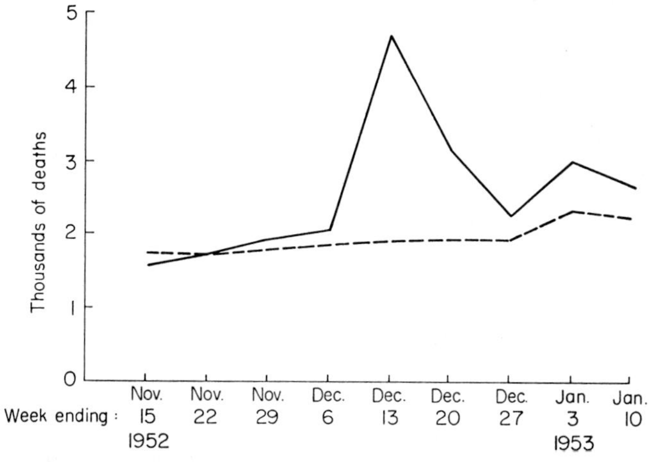

Figure 3. Deaths registered in greater London, England, associated with air pollution episode of December 5–8, 1952. Broken line indicates average deaths for 1947–1951.

6. New York, New York, 1953

From November 18 to 22, 1953, there was recorded a substantial increase in sulfur oxide and pollutant levels associated with a widespread atmospheric stagnation affecting much of the eastern United States (*50, 51*). Compared with the deaths by day for New York City during the month of November in 1950–1952 and in 1954–1956, there appeared to be an excess for 1953 during the period November 15 to 24.

7. New Orleans, Louisiana

In 1958 a series of unusual episodes of increased frequency of asthma was reported in certain districts of New Orleans. Frequency of visits to Charity Hospital increased to a maximum of 200 per day compared with the expected 25. Gentle winds, usually from the same direction, have been associated with these episodes (*52, 53*). It is thought that a single or closely grouped source of pollutants is likely to be involved because of the geographic distribution of cases.

When these episodes were first investigated, it was found that they had been occurring at least since 1955, that they tended to occur in October and November, with occasionally lesser episodes in June and July. The characteristic pattern of the outbreaks is shown in Figure 4. Use of skin tests on subjects who had acute asthmatic reactions, using as antigens

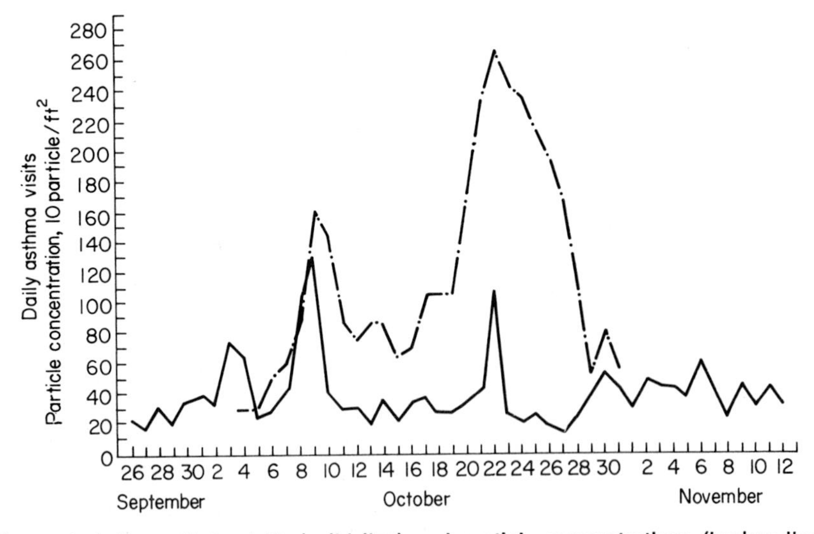

Figure 4. Asthma clinic visits (solid line) and particle concentrations (broken line) in New Orleans, Louisiana, autumn 1963.

extracts from air pollution samples, gave some valuable suggestions. It is thought possible that dust from a flour mill may have been involved (*54*). At any event, particulate matter in certain size ranges correlated well with the outbreaks (*55*). Efforts to find similar outbreaks in other cities have been unsuccessful (*56*). The study of asthma outbreaks in New York City has suggested only that they tend to occur at the time when the weather suddenly becomes cold (*57*).

8. Minneapolis, Minnesota, 1956

The Minneapolis campuses of the University of Minnesota are surrounded by storage and processing plants of the grain industry. Outbreaks of asthma have occurred there from time to time, and it has been thought that perhaps this is due to some pollutants from the neighboring plants. Two hundred and eighty eight records of asthmatic attacks were reviewed to see if there was any systematic pattern. This small number made it difficult to detect a pattern. Skin tests with grain dust were helpful (*54*) and showed cross reactions with the particulates from New Orleans. Further, subjects affected in New Orleans reacted to extracts from Minneapolis grain dust.

9. Worldwide Episode, November–December, 1962

A remarkable air pollution episode with serious health effects swept from west to east across certain portions of the earth between November 27 and December 10, 1962. It illustrates, among other things, that health effects are likely to be found where trained personnel collect the right kind of information and undertake a careful analysis of it.

In the eastern United States, November 27–December 5, pollution levels were noted to be high in Washington, D.C.; Philadelphia, Pennsylvania; New York, New York; and Cincinnati, Ohio (*58*). In two different studies that were made in New York City, an increase in respiratory symptoms was detected. Among a group of elderly persons, an increase in upper respiratory complaints was noted (*59*). In a continuing study of respiratory and other conditions in a study area in Manhattan, New York City, an increase in persistent cough was observed (*60*).

In London, December 5–7, the sulfur dioxide levels exceeded those of the episode in 1952, but the particulate pollution level was somewhat lower. Over this period of time and subsequently 700 excess deaths were reported in London, and there was also an increase in morbidity (*49*).

In Rotterdam, The Netherlands, December 2–7, the sulfur oxide level increased to nearly five times the normal level, and there was a small increase in mortality and hospital admissions, especially in people over 50 years of age with cardiorespiratory conditions. There was also a report of an increase in sickness absenteeism (*61, 62*).

In Hamburg, Federal Republic of Germany, December 3–7, sulfur dioxide levels rose to more than five times the usual level, and dust pollution was more than double. Small increases, though not statistically significant, were thought to have occurred in mortality due to heart disease, though there was no change in respiratory disease frequency (*63*).

Increased pollution levels were also recorded in Paris, France; the Ruhr and Frankfurt, Federal Republic of Germany; and Prague, Czechoslovakia, but concomitant health studies have not been reported.

In Osaka, Japan, December 7–10, pollution levels were high. Mortality studies were under way in Osaka and demonstrated 60 excess deaths (*64*).

This episode did not occur in all parts of the world. The data available for Sydney, Australia, and the west coast of the United States indicate that there were no increases in pollution levels. Such an episode may be related to large-scale anomalous meterological conditions thought by Namias (*65*) to be related to an episode of warming of waters of the North Pacific, which occurred earlier that year.

10. Combined Oxidant and Sulfate Episode, Tokyo, Japan, 1970

Since March of 1967, elevated measurements of oxidants had been reported by the monitoring station of the Tokyo Metropolitan Public Health Research Laboratory. The level had exceeded 0.15 ppm during the daytime in summer on 12 days in 1967, 14 days in 1968, and 9 days in 1969 (*66*). On July 18, 1970, a thick fog was present in the early morning. As the sun rose, the fog dispersed, but visibility was less than 2 km (1.2 miles) even in the center of the city. Between 11:00 a.m. and 1:00 p.m., complaints of eye irritation mounted, and in one school 45 youngsters complained of smarting eyes, sore throat, and difficulty with breathing. Several had to be taken to hospitals where they were treated for "smog poisoning." The pollutant concentrations being measured by the Metropolitan Public Health Research Laboratory rose quite abruptly between 8:00 a.m. and noon, reaching a peak of photochemical oxidant of 0.34 ppm and an hourly average of 0.295 ppm. Characteristic vegetation damage was also observed, and at 5:30 p.m. that day, the Metropolitan Government issued a statement that the symptoms and impairment of visibility were caused by photochemical pollution. During several subsequent days, additional persons were affected. The number of persons reporting smart-

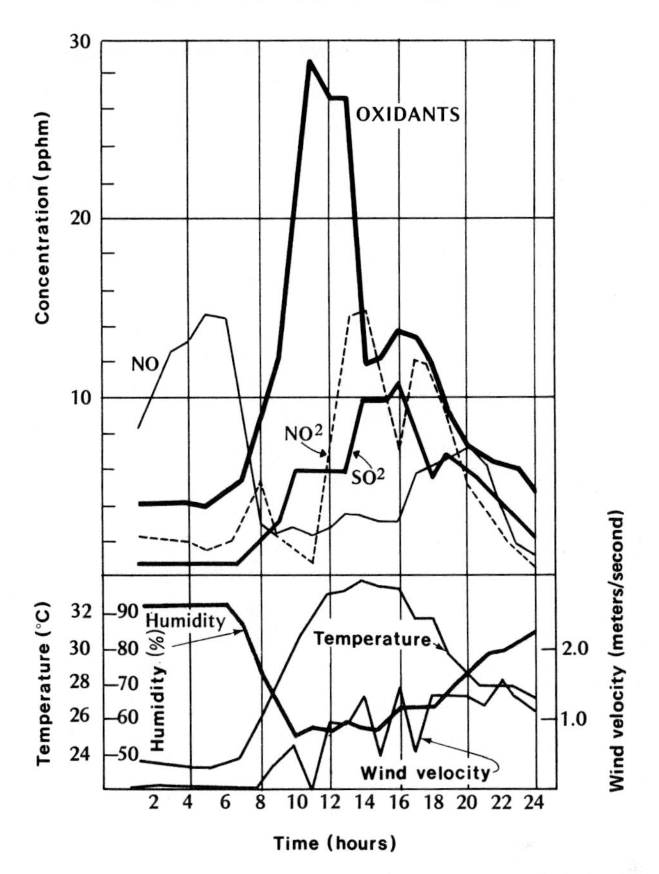

Figure 5. Hourly variation of pollutants and meteorological variables during July 18, 1970, in Tokyo, Japan. The complementary change in nitric oxide and oxidant level is characteristic of photochemical oxidant pollution. The presence of substantial amounts of sulfur dioxide makes it likely that sulfur trioxide, sulfuric acid, and/or acid sulfate aerosols were formed.

ing eyes and sore throat exceeded 6000. Sulfur dioxide reached a maximum of 0.39 ppm. It is inferred from this that the photochemical oxidant reacted with sulfur dioxide to produce a sulfuric acid mist. Figure 5 shows the time course of the pollution during this episode.

11. Methodological Problems of Detecting Acute Air Pollution Health Effects

Methods for detecting acute pollution effects have been developed and presented by Martin and Bradley (*48*), Martin (*67*), Greenburg *et al.* (*59*), McCarroll and Bradley (*68*), Cassell *et al.* (*69*), McCarroll *et al.*

(*70*), Hechter and Goldsmith (*71*), Ipsen *et al.* (*72*), Hexter and Goldsmith (*73*), Schimmel and Greenburg (*74*), and Buechley *et al.* (*75*). All make use of daily mortality or morbidity data. Martin has presented correlations of pollution levels and deviations from moving averages of daily mortality. On the basis of these and similar data, Lawther (*76*) has concluded that when both sulfur dioxide level exceeds 0.71 mg/m^3 and suspended smoke exceeds 0.75 mg/m^3, under conditions prevailing in London, increased mortality can be predicted to occur. Greenburg *et al.* (*50*) have compared the weeks before and after an episode with the same dates and years before and after. They have shown that periods with unusually high pollution are associated with small increases in mortality. Hechter and Goldsmith (*71*) have shown the applicability of auto- and cross-correlation to the time series problems of acute air pollution reactions. More extensive use of these methods has been made by McCarroll *et al.* analyzing data for New York, New York (*70*).

Ipsen *et al.* (*72*) have studied regression coefficients of environmental variables on sickness absence data from the Philadelphia, Pennsylvania, area and used canonical functions for combining sickness absence data for a series of days. This analysis, based on a relatively small population, did not support the hypothesis of an effect of air pollution on respiratory disease morbidity independent of climatic effects. The World Health Organization Symposium on Air Pollution Epidemiology (*77*) has pointed out the need for having a very large population (at least 500,000 and preferably several million) in order to detect acute effects on mortality.

Tucker (*78*), examining data on morbidity and mortality in Los Angeles, California, nursing homes, assumed that the number of deaths occurring in any day is distributed as a Poisson distribution (an often asymmetrical distribution applying to rare events occurring in a large population, with the mean and the variance or squared standard deviation of identical magnitude).

If unusual weather conditions caused an increase in mortality, the expected Poisson distribution would not be the same for the affected population as it would in other periods, and a test for this lack of homogeneity of the distribution was applied to the daily nursing home mortality data. No effect of air pollution was found. The sensitivity of this method partly depends on the choice of days for study as well as on other factors.

Other factors that complicate the interpretation of morbidity data are seasonal fluctuations, day-of-week biases in hospital admissions, concurrent infectious diseases occurring within the community, and a general scarcity of morbidity data. Another problem is the determination of how well recent reports of pollutant concentrations reflect ambient exposures of human populations. Evaluation of Continuous Air Monitoring Program (CAMP) data (Chapters 2 and 3, Volume I) will help in determining the relationship among such measurements as peak or mean values of pollutants for varying periods of time. Such systematic evaluations in conjunction with morbidity data will point the way toward more medically meaningful reporting of the ambient concentrations of pollutants.

Lebowitz *et al.* (*79*) have presented over fifteen reports on the association of various morbidity reactions to environmental and air pollution factors based on a 3-year diary study of daily symptoms, pollution, and weather in midtown Manhattan (New York, New York) for 1962 to 1965. In general, the reports indicate that symptom frequencies increase with a set of changes in pollution concentrations and meteorological circumstances that makes it difficult to identify any disease state or specific symptom with a specific pollutant. However, their data show convincing evidence that there are joint variations in the prevalence of a variety of respiratory symptoms and the associated increases in pollutants. In general, a 36-hour lag time has been observed between the peak of pollution and the peak of symptoms.

A series of studies have been reported of the possible relationships of daily mortality to air pollution, including one by Hexter and Goldsmith (*73*), Los Angeles, California, 1963–1965, and one by Schimmel and Greenburg (*74*), New York, New York, 1963–1968. A number of technical problems have been identified by the investigators, but when those problems are suitably treated, the capacity to find an association between pollution concentrations and mortality in very large urban areas appears to be established. However, the urban area must, in general, have a population base of several million, and a number of adjustments in the computations are necessary. For example, the roles of temperature, time of year, trend in the base population, wind, relative humidity, and sky cover should be accommodated for or represented in some way or other. The dominant form of pollution in New York City, that due to sulfur oxides and particulate matter, has been shown by Schimmel and Greenburg to have a significant association with fluctuations in daily mortality. Schimmel and Greenburg have used ten different mortality classes, and all except infant diseases were significantly correlated with the pollution variables, whether they were used in a crude, temperature-corrected, or a "fully adjusted" form taking into account variations due to time of year.

Buechley *et al.* studied daily mortality during the years 1962–1966 in the New York–New Jersey–Connecticut air quality control region. The population of the area in 1960 was nearly fourteen million persons. The mean number of deaths per day was 413.04. For each day, a ratio of observed daily deaths to this number was computed. Other variables included (a) influenza prevalence, (b) data from a single monitoring station for sulfur dioxide and (c) for particulate matter by COH (coefficient of haze) units, (d) the day's temperature difference, (e) the week's temperature difference, (f) the three-day extreme temperature (exponential) index, (g) holidays, (h) day of week, and (i) time of year reflected by a smoothed temperature cycle. After computing the regression weights for the full set of nine variables, the authors removed the contribution of sulfur dioxide variation, resulting in what they designated as a partial predicted mortality. Subtracting these from the observed mortality by classes of day within sulfur dioxide levels, a relationship was observed in which, with increasing sulfur dioxide there were first negative then positive deviations of residual mortality. For example, during the 260 days (out of 1826 total) on which sulfur dioxide exceeded 500 $\mu g/m^3$, the mortality deviations were from 1.75% to slightly more than 2% higher than the mean. This was equivalent to from about 7 to about 8 deaths per day in the population of fourteen million.

In reviewing data for subsequent years during which sulfur dioxide concentrations were much lower, a similar relationship is found (*78a*). The most likely explanation is that sulfur dioxide is behaving as an index of other pollutants, and not as a specifically causal agent.

The studies of Hexter and Goldsmith (*73*) dealing with Los Angeles, California, were adjusted for time of year by using Fourier terms and the trend in the size of the base population. After these adjustments, it was found that temperatures with lags of several days and polynomial forms with interactions accounted for a very large part of a remaining variance. The logarithm of carbon monoxide accounted for a small fraction of the variance.

Oechsli and Buechley (*80*) have reported on excess mortality associated with three Los Angeles, California, September heat spells in 1939, 1955, and 1963. The one in 1955 was also associated with excessive levels of photochemical smog. They show by fitting exponential equations to maximal temperature that age specific mortality excess in 1955 can be explained by temperature alone. The excess mortality was more clearly seen in deaths due to vascular diseases of the central nervous system than in other causes.

B. Determining Cause–Effect Relationships

In the years between 1965 and 1972, a major effort was undertaken to review data on effects of specific pollutants. The results are compiled in various reports on air quality criteria (*81–88, 88a*) for sulfur oxides, particulate matter, oxidants, carbon monoxide, hydrocarbons, and oxides of nitrogen. In developing statements of the relationship between air pollution levels and the effects caused by these levels, many critical decisions have to be taken as to the existence or extent of cause–effect relationships. These relationships may be quite indistinct when interpreting mortality and morbidity data on people and animals exposed for their lifetime to a multiplicity of stresses, only one of which is air pollution. For many of these stresses, including air pollution, there are inadequate measurements. Extrapolation to man of results from experiments involving exposures of experimental animals is a procedure that raises doubts. Doubt also exists on how to extrapolate short-term square-wave experiments (experiments where pollutant exposures are abruptly increased or decreased), using relatively pure substances, to the exposure of humans and animals to the fluctuating aged and irradiated mixture of precursors and reaction products that constitutes our real atmosphere.

In the process of identifying cause-and-effect relationships between health impairment and air pollution, certain questions continually recur. One is, what is meant by health impairment, injury, or damage? As experimental techniques improve, we are increasingly able to detect subtle physiological and psychological deviations from the norm, some of which can be attributed to pollution. The norm in this case is the health status of an individual or a population in the absence of exposure to polluted air. The deviation observed may be reversible when exposure to the pollutant stops. Some will argue that only irreversible deviations should be considered and that any reversible deviation should be considered benign until proved deleterious. However, the fact that a deviation is reversible upon cessation of exposure to the pollutant is not, of itself, assurance that allowing many such deviations to occur is without lasting harm. Prudence would argue for considering measurable, repeated deviations from the norm as deleterious until proved benign. If a temporary reduction in sensory perception or reaction time occurs, for example, during operation of a motor vehicle or other machines with moving parts, fatal accidents could occur. Thus, temporary, presumably reversible, effects of air pollution should not be dismissed as of no consequence to health, even though, with present data, we can demonstrate neither disease aggravation nor fatal potential.

The question is sometimes raised as to what percentage of the population (i.e., 99.9%, 99.99%, etc.) we should seek to protect by ambient air quality control; and what percentage (i.e., 0.1%, 0.01%) we should take care of by other means, such as their relocation to areas or structures having a cleaner air supply or by supportive medical treatment. Epidemiology is the special discipline with competence to obtain and interpret information relevant to this question. The question must ultimately be referred for answer to value-oriented, that is political, authorities, and the role of medical and other scientists is to inform, but not to decide. The answer to the question is a political one, because it should be based on alternative uses of resources and the values attached to health, to technology, to convenience, etc.

While it is well known that pollution exposures are to a heterogenous, poorly characterized mixture of materials, it remains a critical assumption that specific, measurable, physical or chemical agents are responsible for most of the detectable effects on health. If this assumption be valid, it becomes possible to apply toxicology to air pollution control in a manner that prevents or abates effects on health. If the assumed specificity were not valid, while we could attach blame to sources such as automobiles, a factory, or to coal burning, we would not know other than by trial and error, in what way, and in what magnitude, we should make modifications in the automobile, factory, or combustion chamber to abate or prevent specified health effects.

The application of this assumption, which occurs throughout Sections III–V, should not obscure the artificial nature of specificity. Pollutants do not usually act alone, nor are all of the effects of a given pollutant specific to that pollutant. For example, geographic factors, such as latitude and time of day or year affect the energy spectrum of photochemically active sunlight at the earth's surface. Time of year and climatological and meteorological variables affect human reactions and defences against a given pollution exposure.

C. Relationship of Epidemiological and Toxicological Evidence

Epidemiological and toxicological research are complementary approaches in studying and interpreting the effects of pollution exposures on health. While the most important evidence on health effects concerns the effect on natural populations of humans under realistic conditions, for reasons given above, we have difficulty in making precise, quantitative, universal statements based on such epidemiological evidence. However, the precise, quantitative, specific data from toxicological experiments may not reflect the complexity of real life. In general, interpretations of causation from epidemiological studies should not be based on a single study, but on the convergence of evidence derived from several independent epidemiological studies and related experiments. The conclusion that cigarette smoking was a major cause of cancer of the lung was based on such convergence (7).

Experimental data generally do not include long-term exposures of human subjects and therefore cannot allow us to estimate accurately the magnitude of long-term pollution effects. Epidemiological studies generally are affected by a large number of coexisting and fluctuating variables and often do not yield the specificity that control strategies seek. Epidemiological studies tend to produce nonspecific but realistic data. Inferences relating control to prevention of effects must utilize both types of data.

III. Combined Effects of Sulfur Oxides and Black Suspended Matter*

Sulfur oxides and suspended particulates are considered together because most of the epidemiological studies reported represent a combined exposure; furthermore, in ambient air they are found to fluctuate together. Thus, epidemiological data will not necessarily reveal whether or not an observed effect associated with emissions from a combustion source can be referred to one pollutant alone.

For particulate matter, most of the epidemiological studies reported, including those referred to here, have been concerned with particles produced by burning fossil fuels. Particulate matter may arise from several other sources, industrial as well as natural. Its effects may differ widely, depending not only on its chemical composition, but also on particle size and on its physical form. Epidemiological studies concerned with particles produced by burning fossil fuels should be extrapolated with caution to situations where sulfur oxides occur in combination with other forms of particulate matter.

A. Acute Effects

With very high concentrations, such as occurred during the well-known episodes in London, Donora, and New York (Section II,A), immediate effects were detectable in terms of increased mortality and morbidity. Effects were primarily seen in those already ill, old, or enfeebled. Excess mortality and increased hospital admissions have been identified when levels of suspended smoke particles and sulfur dioxide for 24 hours each has exceeded 500 $\mu g/m^3$ (*88*).

* The term "black suspended matter" is used because most of the epidemiological studies that have given results relevant to a discussion of dose–response relationships have used the British Standard Procedure, which measures black suspended material. The term "total suspended matter" describes what is measured when high-volume samplers are used; this is the commonest measure used in the United States.

Lawther *et al.* (*89*) report on the exacerbations of bronchitis in relation to air pollution. During several periods between 1954 and 1968, diaries were sent to patients with chronic bronchitis in London and Manchester. The patients had to report daily over several months, whether or not their symptoms were worse than usual. Daily measurements of sulfur dioxide [hydrogen peroxide method—titrimetric DSIR (Department of Scientific and Industrial Research, England)] and smoke (smoke stain index) were made. Data were given as 24-hour average values. It could be shown that an association occurred between high concentrations of sulfur dioxide and particulate matter on the one hand and an exacerbation of symptoms on the other hand (Fig. 6). There was no clearcut association between visibility, temperature, or humidity and these exacerbations. The lowest values where effects were clearly seen were 250–500 $\mu g/m^3$ of sulfur dioxide together with about 250 $\mu g/m^3$ of smoke.

During the later years of the studies, high peaks of sulfur dioxide and particulates were not found. On the other hand, even during the winter period of 1967–1968, when the daily data were treated statistically, a significant correlation (5% level) was found with both smoke and sulfur dioxide, with mean smoke concentrations of only 68 $\mu g/m^3$ (standard deviation = 48), and 204 $\mu g/m^3$ (standard deviation = 100) of sulfur dioxide.

The method used by Lawther *et al.* has not been validated, in the sense that objective evidence of aggravation has supported the subjects' own

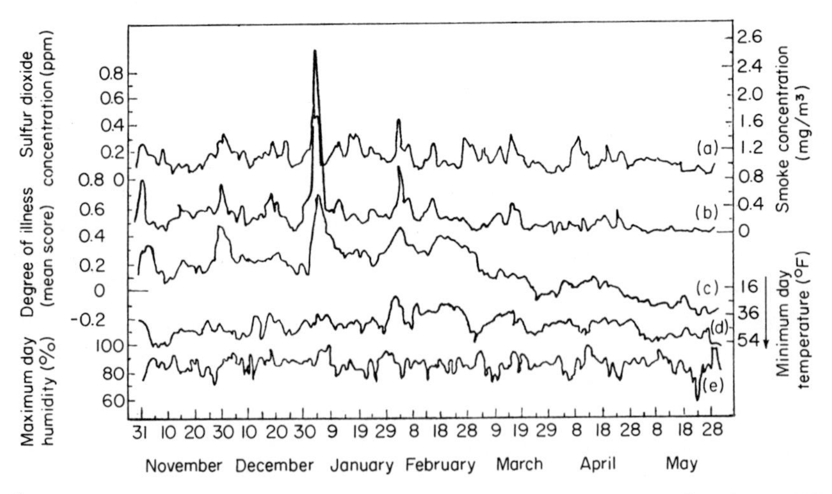

Figure 6. Degree of illness (c) of a group of 180 patients in greater London (1955–1956) with chronic bronchitis plotted with smoke (b) and sulfur dioxide (a) concentrations, temperature (d), and humidity (e) (89).

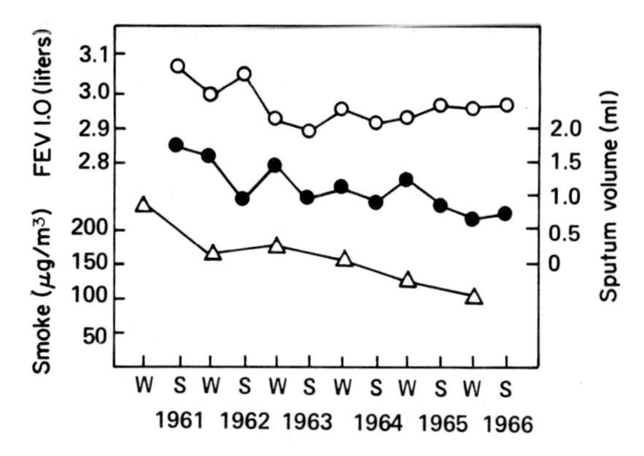

Figure 7. Mean values of $FEV_{1.0}$ and sputum volume in working men aged 30–59 in London with winter smoke concentrations (△) (*91*) (1951–1956). ○ = mean $FEV_{1.0}$ of all men who attended at least one of the last four surveys; ● = mean sputum volume of all men who attended at least one of the last four surveys; × = Mean concentration of black smoke during winter at seven sampling sites in London; W = winter; S = Summer.

report of symptoms. In summary, though, it is generally accepted that the association shown between high levels of smoke and sulfur dioxide on the one hand and exacerbation of chronic bronchitis in bronchitis patients on the other hand is real (*90*). When two winters (1959–1960 and 1964–1965) were compared in terms of the associations of exacerbation with sulfur dioxide, the general impression was of a slightly reduced and less consistent effect during the later period. The mean smoke concentrations had then decreased from a mean of 342 μg/m^3 in 1959–1960 to 129 in 1964–1965, while the sulfur dioxide concentrations had decreased only from 296 to 264 μg/m^3.

Fletcher (*91*) reported some data on a group of men aged 30–60 working in London from 1961 to 1966 that support the hypothesis of the beneficial effects of reducing particulates. Mean sputum volume decreased steadily even in men who had not changed their smoking habits. Over this period, the mean smoke concentration in seven sampling stations in London had dropped from approximately 420 μg/m^3 in 1959 to approximately 100 μg/m^3 in 1965. There was no corresponding drop of the concentration of sulfur dioxide (300 in 1959 and 260 μg/m^3 in 1965) (Fig. 7).

B. Chronic Effects on Adults

Holland *et al.* (*92–94*) studied post office and telephone workers in England and the United States. The prevalence of respiratory symptoms was studied using the British Medical Research Council's short questionnaire for respiratory symptoms. For measuring lung functions, the $FEV_{1.0}$* was used. Aerometric data were scant. Table II shows, however, that both mean and highest 24-hour values are higher in London, England, than in the rural towns, which, in turn, had higher values than United States cities for suspended particulates as well as for sulfur dioxide. Brasser *et al.* (*95*) have furnished some further sulfur dioxide values; London (St. Pancras) summer average 100 $\mu g/m^3$, winter average 500 $\mu g/m^3$; Gloucester, Petersborough and Norwich, England, 75 $\mu g/m^3$ and 200 $\mu g/m^3$, respectively, for summer and winter averages, representative of rural areas in the United Kingdom.

These studies show a close relationship between smoking habits and respiratory dysfunction and a gradient with regard to place of residence. Respiratory disorders were most prevalent in London, less in rural England and still less in the United States (Fig. 8). When smoking habits were taken into consideration, an urban effect was seen in smokers, but not in nonsmokers and exsmokers. The deterioration of lung function, however, was evident also in nonsmokers and exsmokers (Fig. 9). Considering the difference in effects between English rural towns and the United States cities, one may interpret the data as if the lowest values at which effects have been observed are about 75 $\mu g/m^3$ for sulfur dioxide together with about 200 $\mu g/m^3$ for suspended particulate matter (see also Section VI,A).

Reid *et al.* (*9*) report on an Anglo-American comparison of the prevalence of bronchitis. The British study was carried out on a random sample of men and women aged 40–64 drawn from the practice lists of 92 physicians working in urban and country areas of Britain. The United States data are a probability sample of residents in Berlin, New Hampshire, for whom it was possible to get a similar sex and age distribution. The British Medical Research Council's questionnaire was used for diagnosing chronic bronchitis and the Wright peak flow meter for measuring lung function. The results show that there is a difference in prevalence of bronchitic symptoms, tending to show that after standardization for age and lifetime cigarette consumption, the United States data for men would be about equal to the British rural data for men. Considerably higher prevalence was found in United Kingdom towns and cities. It is not known whether differences in socioeconomic status could have been factors of importance. Also, the standardization for lifetime cigarette smoking is

* $FEV_{1.0}$ stands for forced expiratory volume in 1 second, which is defined as the amount of air one can expel in 1 second with maximal effort after maximal inspiration.

Table II Comparison of Respiratory Findings among Outside Workmen Age 40–59 in the United Kingdom and the United States and Representative Pollutant Levels for Each Area (*93*)

	United Kingdom				*United States*			
	London		*Rural areas*		*Baltimore, Maryland; Washington, D.C.; and Westchester County, New York*		*San Francisco and Los Angeles, California*	
	Age 40–49	*Age 50–59*	*Age 40–49*	*Age 50–59*	*Age 40–49*	*Age 50–59*	*Age 40–49*	*Age 50–59*
Number of men examined	113	137	267	159	396	229	361	119
Persistent cough and phlegm (%)	25.7	38.7	24.0	18.9	22.2	25.8	21.6	24.4
Persistent cough and phlegm and chest illness episode (%)	10.6	10.9	7.5	5.0	6.8	7.0	4.0	7.6
$FEV_{1.0}$ (liters) mean values[a]								
Nonsmokers	2.8	2.6	3.0	2.8	3.5	3.1	3.7	3.3
Cigarette smokers								
1–14/day	2.6	2.3	2.8	2.6	3.4	3.0	3.4 (all smokers)	2.8 (all smokers)
15–24/day	2.6	2.2	2.8	2.5	3.2	2.8		
25 or more/day	2.5	2.1	2.7	2.5	3.2	2.9		
Sputum volume, 2 ml or more (%)	28.9	42.9	22.1	23.5	7.1	10.0	8.6	14.3
Suspended particulates ($\mu g/m^3$) 24 hour								
Mean	220		200		120		120	
Maximum	4000		3000		500		340	
Sulfur dioxide (ppm) 24 hour								
Mean	0.1		0.02		0.04[b]		0.01[c]	
Maximum	1.3		0.26		0.25		0.06	

[a] $FEV_{1.0}$ = forced expiratory volume in 1 second.
[b] Washington, D.C., 1962–1963.
[c] San Francisco, California, 1962–1963.

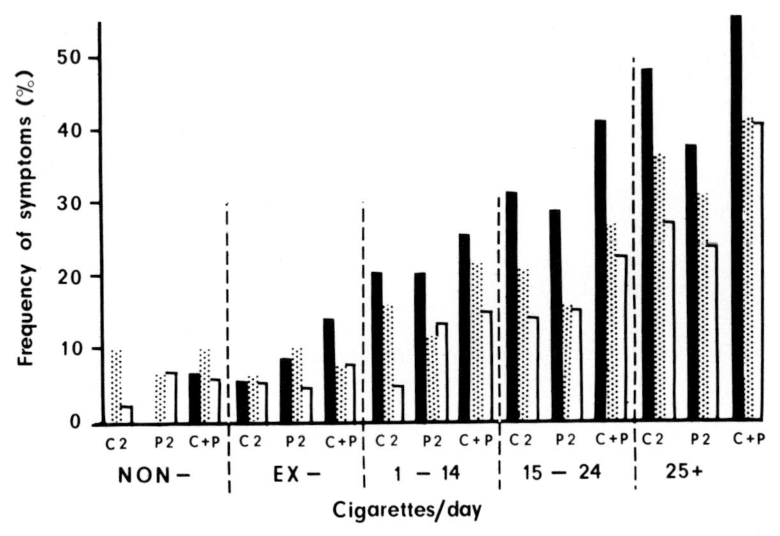

Figure 8. Frequency of cough grade 2 (C2), phlegm grade 2 (P2), and persistent cough and phlegm (C + P) in men from London (solid bars), three rural (shaded bars) towns, and the United States (white bars) by smoking habits (1961–1962) (*93*). Grade 2 means that the symptom continues during day or night for as much as 3 months each year.

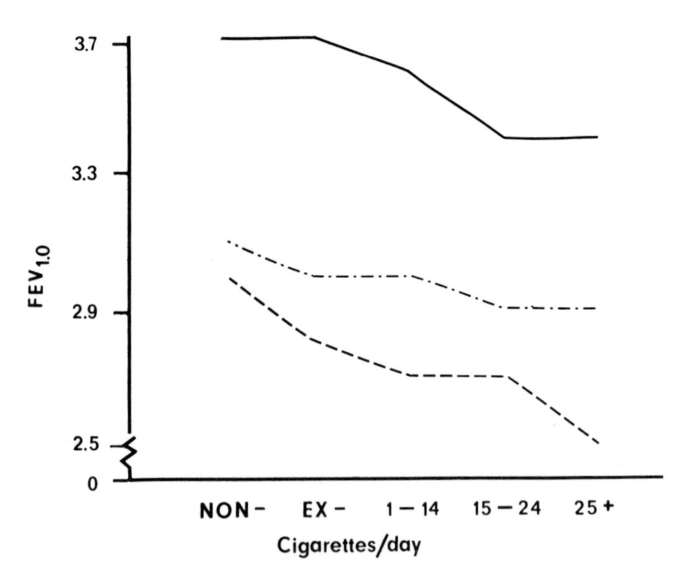

Figure 9. The $FEV_{1.0}$ (liters) standardized to age 40 in men from London (–––), rural towns (–•–•–), and the United States (——) by smoking habits (*93*). 1961–1962.

questionable, in view of the fact that the bronchitic symptoms seem to be closely related to fairly recent exposure.

In the first Berlin, New Hampshire, study (in 1961), nonsmokers in the more polluted area of the town had a higher prevalence of chronic respiratory disease than did nonsmokers in less polluted areas (Table III) (*96*). Indeed, the prevalences among male nonsmokers in the polluted area resemble those for pack-a-day smokers. As Ferris (*97*) points out the two- to threefold difference in pollution may have had effects that are understated by Table III. He believes that internal migration may have occurred among families that initially lived in the more polluted area, who, after experiencing pollution effects, moved to other areas. The pollutant exposure in Berlin is largely that due to a kraft paper mill and may not be representative of the more common types of pollution effects from sulfur oxide and black suspended matter.

Ferris and Anderson (*98*) in 1963 studied a random sample from Chilliwack, British Columbia, Canada, and compared results with the more polluted city, Berlin, New Hampshire. Respiratory symptoms, when standardized for age and cigarette smoking, tended to be higher in Berlin than in Chilliwack. Tests of pulmonary function ($FEV_{1.0}$, peak expiratory

Table III Prevalence of Nonspecific Respiratory Disease in Nonsmokers by Relative Pollution at the Place of Current Residence, Berlin, New Hampshire, 1961 (*96*)

	Least polluted		*Intermediate*		*Most polluted*	
Subjects	*Number*	*Percent*	*Number*	*Percent*	*Number*	*Percent*
Males						
Population	36	100.0	65	100.0	23	100.0
Chronic nonspecific respiratory disease	6	16.7	10	15.4	9	39.1[a]
Chronic bronchitis	3	8.3	9	13.9	7	30.4
Irreversible obstructive disease	4	11.1	2	3.1	5	21.7[a]
Females						
Population	106	100.0	178	100.0	86	100.0
Chronic nonspecific respiratory disease	21	19.8	27	15.2	18	20.9
Chronic bronchitis	11	10.4	16	9.0	8	9.3
Irreversible obstructive disease	14	13.2	11	6.2	12	13.9

[a] Differences significant comparing most polluted and both other locations for males. All other differences are not significant.

flow rate) after standardization for age, height, sex, and smoking category also showed some differences. There are several difficulties in interpreting these data. It seems reasonable that slight changes in respiratory symptoms and pulmonary function were related to pollution levels, sulfur oxides, and particulates. Particulate matter was estimated by high volume (Hi-Vol) samplers (see footnote on p. 483). The studies were repeated in 1966–1967 in Berlin, New Hampshire, and in this period, there was a lower prevalence of respiratory symptoms than in 1961 (*99*). Sulfation levels had also dropped, but total suspended particulate matter had not changed significantly.

In a number of papers (*100–102*), data from extensive studies in the Ruhr industrialized area of Germany covering about 8000 people are reported. Due to several biasing factors, e.g., high nonresponse rate and different sampling procedures used for different areas with large differences in occupation, the risk that selective factors have been operating seems to be high. There is some experimental evidence that ozone may have a comparable role to black suspended matter in promoting health impairment due to SO_2 (see Section IV,B,1).

C. Chronic Effects on Children

Douglas and Waller (*103*) report on the prevalence of upper and lower respiratory symptoms in children during an 11-year period beginning in 1952. The children were followed medically by health visitors and by medical examination at school. Levels of air pollution were estimated by the amount of coal consumed in a given area. In 1962, a validation of the estimation procedure was made using available aerometric data from a number of the study areas. Douglas and Waller noticed an increased occurrence of lower respiratory tract infections in areas with increased air pollution. The data can be interpreted as if effects were seen already at concentrations of about 140 $\mu g/m^3$ (± 9) of smoke and about 130 $\mu g/m^3$ (± 9) of sulfur dioxide, in both cases expressed as mean annual concentration. Aerometric data were based on the hydrogen peroxide method (sulfur dioxide) and smoke stain index (particulates).

Lunn *et al.* (*104*) report on the prevalence of respiratory symptoms in children 5 years of age and living in four different areas of Sheffield, England, with different air pollution exposure. Air pollution was measured (hydrogen peroxide method and smoke stain index) for the 3 years of the study. They found increased chronic upper respiratory infection in relation to the level of air pollution. Certain differences in social structure between the study areas were evident. However, for several symptoms, the gradient between the areas would hold true even when the groups were divided into social class, number of children in house, and so on. Data, divided into social class, for persistent or frequent cough for the different areas are given in Table IV. Comparing aerometric data with the medical findings yields the conclusion that symptoms may occur at concentrations (for both smoke and sulfur dioxide) of about 200 $\mu g/m^3$, but are not expected to occur when these concentrations are about 100 $\mu g/m^3$. A follow-up study by Lunn *et al.* (*105*) showed that with the reduction of smoke levels in Sheffield, England, the respiratory condition differences between the groups of children had decreased.

Holland *et al.* (*106*) and Colley and Reid (*107*) each have carried out studies on about 10,000 children in different parts of England. The object was to measure respiratory disease and associate the disease with different environmental factors including air pollution and social class. Holland *et al.* were studying primarily peak expiratory flow rate, and Colley and Reid were studying upper and lower respiratory tract symptoms and also peak expiratory flow rate. Levels of smoke were measured in Holland's study during 5 months in the winter of 1966–1967. The mean for that period in the most polluted area, Rochester, England, was 70 and for the least, 35 $\mu g/m^3$. The averages for the worst month were respectively 96 and 70 $\mu g/m^3$, and the highest daily concentration was 282 $\mu g/m^3$. In

Table IV Relationship of Persistent or Frequent Cough in British Children of Sheffield by Area, Pollution Levels, and Social Class (*104*)

	Greenhill	*Longley*	*Park*	*Attercliff*	*All areas*
Mean daily pollution levels ($\mu g/m^3$)					
Smoke	97	230	262	301	
Sulfur dioxide	123	181	219	275	
Number of days with readings over 500 $\mu g/m^3$					
Smoke	4	30	40	45	
Sulfur dioxide	1	11	16	32	
Percent of children with cough (numbers in parenthesis)					
Total	22.6 (350)	36.4 (162)	37.1 (105)	50.7 (71)	31.0 (688)
Social Class					
I and II (highest)	0.0 (23)	60.0 (5)	50.0 (2)	50.0 (2)	15.6 (32)
III	24.2 (260)	33.9 (109)	30.9 (55)	45.7 (46)	29.4 (470)
IV and V (lowest)	23.9 (67)	39.6 (48)	43.8 (48)	60.9 (23)	37.5 (186)

Colley and Reid's study, the levels of sulfur dioxide were measured by means of a lead peroxide sulfation method with results reported as sulfur dioxide using a conversion factor. Mean values of from 33 μg/m³ to 150 μg/m³ of sulfur dioxide were found in the different areas studied. Holland *et al.* found an association between impaired peak expiratory flow rate and living area and between previous respiratory diseases and social class. In the most exposed area (Rochester), the lowest levels of peak expiratory flow were found to be independent of parent's social class, family size, and past history of respiratory illness. Mean values of smoke during the winter months of about 70 μg/m³ appear to be associated with occurrence of these effects, but values of 30–50 μg/m³ did not appear to be. No data on SO_2 were reported.

In Colley and Reid's study, the most pronounced gradient was between respiratory symptoms and social class by area, at least for social classes IV and V (*107*) (Fig. 10). The data can be interpreted to mean that sulfur dioxide levels of about 100 μg/m³ give rise to effects that are not observed when sulfur dioxide levels are about 30 μg/m³. Colley and Reid found an association with air pollution for lower respiratory tract infection but not for upper respiratory tract infections. This study included no data on smoke. Their data thus agree more with those of Douglas and Waller (*103*) than with those of Lunn *et al.* (*104*).

Zapletal *et al.* (*108*) studied pulmonary function among 111 children, 10–11 years old, living in the vicinity of Most, Czechoslovakia. High

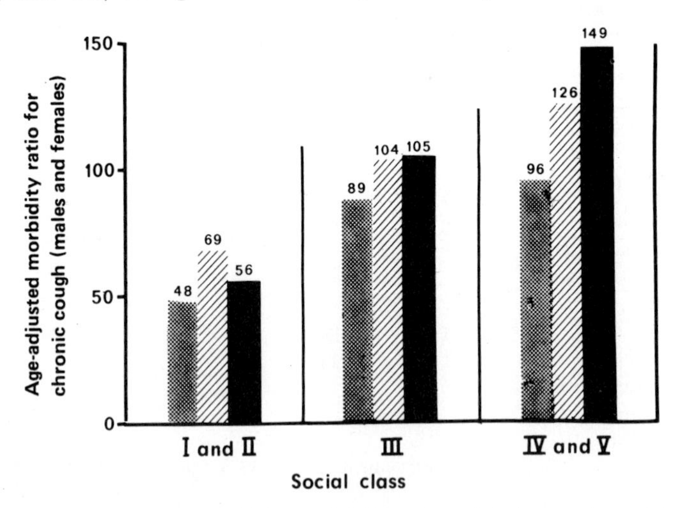

Figure 10. Bronchitis among children of different social classes in rural (shaded) and urban [Bristol and Reading (hatched); Newcastle and Bolton (solid)] England (*107*). Classes I and II are high income, and IV and V low. Numbers of subjects at head of each bar.

pollution levels occur here with daily averages of sulfur dioxide and particulate matter exceeding 240 $\mu g/m^3$ on more than 7 days a month during the winter. Screening spirometric tests showed no extraordinary findings. Nineteen children with the lowest values of forced expiratory volume were studied further, using more sensitive methods. In 6 of the 19, significant reduction in maximal flow rates at low lung volumes were found. This indicates unfavorable changes in small-caliber airways.

Kerrebijn *et al.* (*108a*) studied two groups of communities in an area with intense use of greenhouses for floriculture and horticulture. Most were heated by oil which produce high levels of SO_2 and smoke pollution; in the low-pollution area, gas was used. Fifth and fourth grade children were studied. Depending on criteria used, chronic respiratory disease prevalence in the more polluted area ranged between 5.3 to 12.7%, while in the clean area the range was 3.3 to 9.8%. The differences between the areas were statistically significant. Ventilatory function tests showed no differences.

IV. Effects of Gaseous Pollutants

A. Sulfur Dioxide

1. Absorption of Sulfur Dioxide and Experimental Exposures in Man

Sulfur dioxide, being a very soluble gas, is nearly completely absorbed during quiet breathing in the nose and upper airway. It was at first assumed that, under these conditions and with the low to moderate levels of exposure that occur in community air pollution (up to 2 ppm), no effects on the function of the lung could occur. This assumption must be qualified or discarded on the bases of the experimental work in cats by Widdicombe (*109*) and Nadel *et al.* (*110*) showing that many of the airway caliber changes produced by pollutants are mediated by reflexes. They have shown, for example, that sulfur dioxide or other stimuli given to the upper airway, which is anatomically isolated from the lower respiratory tract, will nevertheless produce increased airway resistance in the lung via the vagal nerve and that this can be abolished by cutting or otherwise interrupting the transmission of impulses in that nerve.

In addition, variability of response of different persons to sulfur dioxide exposure under experimental conditions has been documented (*82, 110–112*) (Table V) by a number of investigators. This is manifested by unusually severe impairment of lung function with exposures to between 2

Table V Effects of Breathing Sulfur Dioxide on the Airway Resistance in Normal Subjects (*112*)

		Mean airway resistance ($cm\ H_2O/liter/second$)				
Subject	*Sex*	*Control*	*After sulfur dioxide*	*After subsequent bronchodilator*	*Sulfur dioxide concentration (ppm)*	*Exposure time (minutes)*
B.B.	F	1.37	2.33	1.32	2	3
J.B.	F	1.13	1.67		2	6
J.B.	F	1.54	1.87	1.14	5	4
P.S.	F	1.80	1.82	1.61	5	4
D.H.	M	0.69	0.76		2	3
R.M.	M	1.15	8.54	1.22[a]	5	10
J.N.	M	1.10	2.74	0.98	5	10

[a] After previous bronchodilator, value was 0.92.

and 5 ppm or, in some cases, the development of a moderately severe attack of asthma in an otherwise healthy subject with no recent history of asthma. Whether these reactions are due to variability in sensitivity or to an exaggerated response to a dose to which any subject might react is not clear. Among the healthy subjects studied, about one in ten may manifest such an exaggerated reaction. We do not know how these data apply to persons with a history of chronic diseases. Nor do we know whether the same or similar phenomena apply to other gaseous pollutants. It is a reasonable hypothesis that soluble, odorous pollutants generally behave as does sulfur dioxide, producing some of their effects by reflex action initiated in the nose and with a proportion of the population manifesting exaggerated reactions. Similarly, mouth breathing whether due to exertion or disease is likely to lead to greater effects for such pollutants.

The effects at low concentrations include the distasteful odor, detectable at about 0.5 ppm (*113*). At slightly higher concentrations, the irritating effects lead to increased airway resistance. Chronic cough and mucus secretion may result from repeated exposures, although most of the exposures to which these effects appear to be related are probably not due to sulfur dioxide alone.

2. Epidemiological Evidence

Smelters are large sources of sulfur dioxide, and in studies done in the vicinity of Port Kembla, Australia, Bell and Sullivan (*114*) documented the excess occurrence of chronic respiratory symptoms in those populations with the greatest exposure. This study appears to represent predominant effects of sulfur dioxide. Other epidemiological studies that appear to reflect effects of sulfur dioxide are most likely due to its interactive effects with other pollutants, whether measured or not.

B. Ozone and Other Oxidants

The reaction of nitric oxide with hydrocarbon vapors in the presence of sunlight leads to production of ozone and nitrogen dioxide in the atmosphere, along with many other compounds. When such photochemical pollution is of more than mild intensity, ozone is the dominant species, but other active agents, such as nitrogen dioxide, peroxyacyl nitrate, and peroxybenzoyl nitrate account for some effects. Collectively, they have been estimated by their oxidizing properties and are designated "oxidants." Experimentally, the reproduction of this mixture and its transient products is difficult. Hence, most experimental data are reported as an effect of ozone (or peroxyacyl nitrate, peroxybenzoyl nitrate, etc.), while most of the epidemiological data are related to measurements of oxidant.

The experience of human populations with oxidant air pollution is likely to be variable, because among other reasons, the photochemical mixture is composed of different constituents at different times and at different locations. The human "symptoms" of photochemical smog, respiratory and eye irritation, were first observed in the Los Angeles, California, basin. The "diagnosis" can be confirmed because of the presence of the three other manifestations—production of ozone, interference with visibility, and characteristic forms of vegetation damage. Experience with occupational exposure to ozone and with human experimental exposures to ozone are helpful in separating the effects due to ozone and those due to other portions of the photochemical mixture. For example, eye irritation at oxidant concentrations observed in the Los Angeles, California, area cannot be reproduced experimentally by ozone; therefore, ozone itself is thought not to be the eye irritating ingredient in photochemical pollution. The exact substances that cause eye irritation are still not defined. They are now thought to include formaldehyde, acrolein, peroxybenzoyl nitrate, and peroxyacyl nitrate. The respiratory effects of photochemical pollution, some of which can be produced by the amount of ozone present, are thought to be aggravated by these other substances.

In a series of studies, Huess and Glasson (*115*) and Glasson and Tuesday (*116*) have sought to obtain and define a relationship between eye irritation potency of various photochemically reacted materials and their chemical reactivity. The eye irritation potency is estimated from the

Table VI Average Results of Irradiation of 2 PPM of Various Hydrocarbons and 1 PPM of Nitric Oxide (115, 116)

Class and compound	Nitrogen dioxide production (ppb/minute)	Percent hydrocarbon reacted[a]	Product yields (ppm)			Eye irritation average threshold time (seconds)	Eye irritation index (0 to 3 scale)
			Ozone maximum	Formaldehyde	PAN[b]		
Alkanes							
n-Butane	4.6	5	0.16	0.15	0	240	0.0
n-Hexane	6.5	18	0.17	0.19	0	240	0.0
Isooctane	4.7	10	0.19	0.25	0	219	0.2
Benzene	1.6	13	0.05	0.08	0.01	216	0.2
Alkylbenzenes							
Toluene	5.7	39	0.30	0.14	0.10	114	2.0
Ethylbenzene	7.2	29	0.21	0.11	0	138	1.6
n-Propylbenzene	5.7	30	0.21	0.10	0.01	111	2.5
Isopropylbenzene	6.6	33	0.19	0.28	0.01	202	0.6
n-Butylbenzene	6.0	33	0.24	0.11	0	86	2.3
Isobutylbenzene	5.0	28	0.17	0.10	0	104	2.5
sec-Butylbenzene	7.6	21	0.26	0.16	0	197	0.6
tert-Butylbenzene	3.8	28	0.13	0.15	0.01	219	0.3
Multialkyl benzenes							
o-Xylene	13.6	52	0.32	0.45	0.40	186	0.9
m-Xylene	15.5	54	0.39	0.39	0.50	171	1.3
p-Xylene	7.7	42	0.26	0.22	0.40	180	1.1
1,3,5-Trimethylbenzene	26.0	66	0.46	0.70	0.67	166	1.0

Olefins							
Ethylene	5.8	45	0.28	0.83	0.01	216	0.3
Propylene	12.1	73	0.54	1.17	0.35	148	1.2
1-Butene	13.1	85	0.47	0.90	0.05[c]	210	0.7
1-Hexene	9.6	80	0.41	0.78	0.02	157	1.2
Internal olefins							
trans-2-Butene	38.0	85	0.44	0.75	0.63	186	0.5
cis-2-Butene	28.0	90	0.44	0.69	0.36	201	0.6
2-Methyl-2-butene	50.0	95	0.49	0.68	0.85	195	0.5
Tetramethylethylene	170.0	95	0.60	0.63	0.65	207	0.7
Polyolefins							
1,3-Butadiene	25.0	92	0.48	0.80[d]	0.02	73	3.0
Estimated reproducibility[e] (%)	14	10	19	10	10[f]	8	22

[a] Corrected for dilution.
[b] PAN is peroxy-acetyl nitrate.
[c] PPN (peroxy-propionyl nitrate) also formed.
[d] 0.73 ppm of acrolein also formed.
[e] Calculated from the average of the average deviations from the mean for all hydrocarbons listed.
[f] Only one determination for each hydrocarbon. Reproducibility estimated from calibration.

time interval from exposure of experimental human subjects to their awareness of eye irritation. The chemical reactivity is estimated by either rate of hydrocarbon disappearance or rate of conversion of nitric oxide to nitrogen dioxide (Table VI). Within the several classes, there seems to be a relationship of chemical reactivity to eye irritation potency.

1. Effects of Ozone, Based on Toxicological Evidence

At a concentration of 0.02 ppm (40 $\mu g/m^3$), ozone can be detected by odor (*117*). When first detected, the odor is not an objectionable one. This has made it possible for ozone to be used for such purposes indoors as deodorization, which unfortunately often exposed residents to concentrations of ozone we now know to be excessive. This use was based on an apparent decrease in sensitivity to other odorants when a subject was exposed to low levels of ozone. It is not clear whether the ozone reacts primarily with the odorant or with the odor-detecting membranes in the nose.

At concentrations of about 0.3 ppm (590 $\mu g/m^3$) with exposures of a few minutes, the irritating effects of ozone are detected (*118*). These are readily detected at half a part per million by most individuals and may be detected at lower concentrations by some (*119–121*). The irritation includes a sense of dryness of the throat. Such irritation may occasionally be observed in places where ultraviolet lamps are used, and these can, of course, produce ozone. Under normal conditions of breathing through the nose, the irritation occurs high in the respiratory tract and does not affect the gas-exchanging part of the lung to the same extent. If the ventilation rate is sufficiently high, and mouth breathing occurs, the mechanisms of scrubbing the incoming air to remove the relatively insoluble gas are inefficient and the ozone may reach the deeper parts of the lung.* A more severe effect of ozone exposure is alteration in the airway resistance. This was shown by studies of Goldsmith and Nadel (*121*) to be detectable at concentrations between 0.1 and 1.0 ppm, but there was not a consistent dose–response relationship in the four individuals who were studied (Fig. 11). Subsequently, Mohler *et al.* (*122*) exposed three nonsmoking subjects to 0.6 to 0.8 ppm ozone for 60 minutes with spirometry and plethysmography at 30 minute intervals. No physiologically significant change in any subject occurred. The subjects had headache, nausea, and anorexia. With exercise, the subjects had coughing and chest pain. The experiments of Goldsmith and Nadel were carried out in an exposure chamber; four subjects were exposed for an

* For discussion of respiratory transport and absorption of ozone, see Chapter 7 of NAS–NRC report (*121a*), particularly Tables 7-1 and 7-4, which compare ozone and other gaseous pollutants. Figure 7-5 provides a comparison of dosage of sulfur dioxide and ozone per breath by level in the respiratory tract.

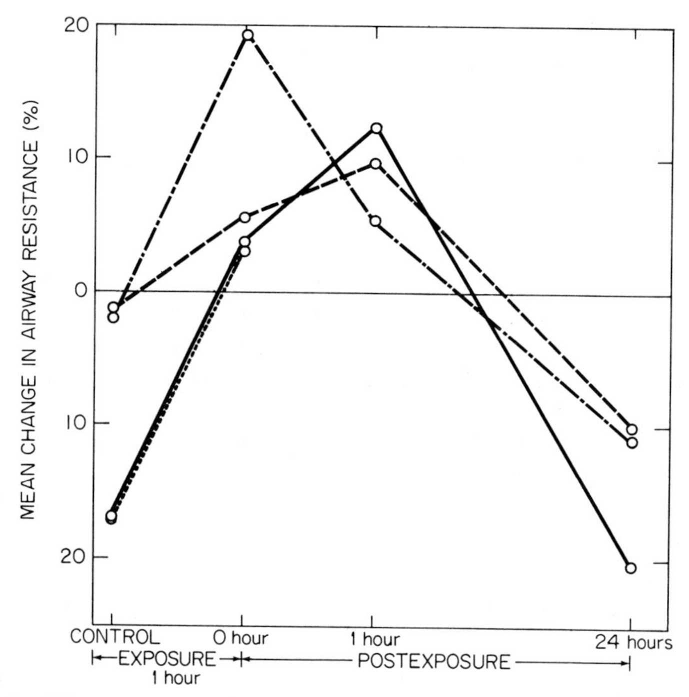

Figure 11. Effect of ozone on airway resistance, in four, presumably normal, male subjects (*121*). With 0.1 (····) and 0.4 (——) ppm, only one subject had significant increases, but with 1.0 ppm (–•–•–) all four did have significant increases; (–––) indicates 0.6 ppm.

hour simultaneously and then they were required to walk several hundred feet to a body plethysmographic laboratory on another floor. These circumstances of the experiment could have permitted a transient effect to be missed; perhaps there was a reaction to removal to ozone-free air. Nevertheless, the differences in methods should have led the results of Goldsmith and Nadel's study to be less positive. More definitive studies of lung function following 0.75 ppm ozone inhalation for 2 hours in subjects who were exercised have been published by Bates *et al.* (*123*). Most of the 10 subjects had chest soreness and a sensation of shortness of breath. Maximal transpulmonary pressure dropped, as did maximal flow at half of the vital capacity. Exercise accentuated the effects. Pulmonary resistance, most of which is in the larger airways, increased. Results varied among the subjects, however. Ozone at 0.37 ppm for 2 hours produced little change in lung function, but when combined with 0.37 ppm sulfur dioxide, marked and persisting changes were found by Bates and Hazucha (*124*).

Representative of the four lung function tests reported (*124*) are results for the $FEV_{1.0}$ resulting from 2-hour exposures with intermittent exercise in healthy adults. At 0.37 ppm of SO_2 there was on average no change in $FEV_{1.0}$ (although changes did occur with more sensitive tests). Exposure to 0.37 ppm ozone produced 8% decrement. Exposure to 0.75 ppm ozone produced an average decrement of 22%, 0.75 SO_2 an average decrement of 10%, and the combination of 0.37 ppm ozone and 0.37 ppm SO_2 a decrement of 19%. It thus appears that above a minimal ozone exposure level, the joint presence of the two pollutants affects lung function as though all the dose was of ozone. This impression is of special concern in areas with both photochemical pollution and sulfur oxide pollution, since it suggests that oxidant can have a comparable effect to black suspended matter in promoting the harmful effects of sulfur dioxide.

The most characteristic toxic effect of relatively high-level ozone exposure is pulmonary edema (*125*), by which is implied a leakage of fluid into the gas-exchanging parts of the lung. This is that part of the lung in which gas diffuses through what is normally a short distance between the terminal alveoli and the capillary. After exposure to sufficient ozone, the amount of water present in the tissues is increased, presumably through some change in cell permeability, and the path for gas diffusion is increased. Estimation of this effect has been done by Young *et al.* (*126*), who have calculated the amount of fluid that might be required to produce the interference with diffusion of gases which they have observed. This effect occurs at concentrations only slightly in excess of what is likely to be produced by community air pollution in Los Angeles, California. Pulmonary edema, if severe, can be crippling and fatal.

The ability of animals exposed to a few tenths of a part per million to tolerate subsequent exposures within a few hours of 6–10 ppm, which would otherwise be fatal (*127*), had not been shown for man.* With continued (i.e., several hours a day, several days a week for as long as a year) exposure to ozone at moderately elevated levels, 1960 $\mu g/m^3$ (1 ppm) in experimental animals, some species, but not all, react by developing fibrotic lung changes (*128*). The question of whether man has this type of reaction to similar exposures is unresolved. While ozone exposures have been reported to cause the appearance of tumors in tumor-prone animals at an earlier period in their life cycle and with a higher frequency than they would otherwise occur, a role of ozone in tumor production of man has not been shown.

A great deal of discussion promoted by these observations concerns the possibility that ozone is a radiomimetic agent. By this is meant that

* Several lines of evidence suggest that the same process occurs in man. The evidence is reviewed in the report by the National Academy of Science, National Research Council (*121a*).

it produces effects that mimic the effects of exposure to X-ray and other forms of radiation. To some extent, this is a valid and provocative possibility, but in other respects the exposures to ozone are not radiomimetic. For example, no increased level of leukemia occurs from ozone exposure as has been observed from exposure to radiation; no effects on the production of white blood cells occur, nor are there effects on other rapidly dividing tissues with ozone exposure, but these have been reported with radiation. By contrast, pulmonary edema is not a common low-level radiation reaction, nor is respiratory irritation. Some of these differences depend on how the agent is distributed and interacts with the cells that are first affected. Ozone's relative insolubility leads to a rather restricted site of impact, namely, the surface layers of the respiratory system, in contrast to radiation, at least of some sorts, which is penetrating. In some ways, ozone exposures can mimic effects of radiation, but the overlap is not sufficient so that they can be treated as equivalent (*83, 129*). Ozone has been shown (*129a*) to produce chromosomal abnormalities in human lymphocytes in subjects exposed to 0.5 ppm. The implications of this are yet to be determined.

2. Epidemiological Studies of Photochemical Oxidants and Ozone

a. DAILY MORTALITY IN RELATION TO PHOTOCHEMICAL OXIDANT VARIABILITY. The occurrence of episodes of photochemical smog in the fall of 1954 in Los Angeles, California, following, as it did, the widespread discussion of a London, England, smog disaster, led to understandable anxiety as to whether there might be a fatal effect of photochemical smog. The association of photochemical oxidant with high temperatures has made it difficult to study this possibility. In the fall of 1955, for example, a very substantial heat wave occurred that did lead to excess mortality (*80, 130, 131*). Despite early reports that excess mortality was due to photochemical oxidants, more careful analysis of the data suggests that the elevation of mortality coincides with, or very slightly follows, the occurrence of excessive temperatures. In particular, the occurrence of high ozone concentrations, that is ozone alerts with pollution levels in excess of 0.5 ppm, for several days following the peak of the heat wave could not be shown to lead to an increase in mortality. It could be argued that the most susceptible groups in the population had been reduced in number by the excessive temperatures, and second, if there had been a small increase in mortality due to ozone it could not have been detected. Neither of these arguments can be refuted; however, neither can one demonstrate that there is convincing positive evidence that excess mortality during this episode was associated with elevated levels of photochemical oxidants. Accordingly, the analysis of heat waves has been an important feature of the effort to identify excess mortality associated with photochemical oxidants. The effort of Hexter and Goldsmith (*132*), for example, failed to identify a contribution of photochemical oxidants to variation in daily mortality in 1962–1965 in Los Angeles, California, although a detectable effect of carbon monoxide was shown. Other methods that have been attempted to look for this include the so-called "two community" method of Landau *et al.* (*133*), which consists of attempting to divide the area of Los Angeles County, California, into regions that had expected similar autumn temperatures but differing oxidant levels; these two regions were then examined to see whether there was any difference in daily mortality attributable to differences in oxidant concentration. None was found. This is an example of the so-called temporaspatial strategy that has also been applied by Toyama, (*134*), and by Cohen *et al.* (*135*). While photochemical oxidant could have had a role in mortality, no role has so far been proven.

b. Epidemiological Studies Relating Photochemical Pollution and Disease Causation. Pulmonary emphysema is the major illness for which an epidemiological association of oxidant pollution has been sought. This is based on laboratory experiments with oxides of nitrogen that show this substance to be capable of producing a condition in animals that simulates emphysema in man (*136*). So far, no adequate test of this hypothesis has been undertaken. Monitoring of the number of emphysema deaths by year in Los Angeles County and other parts of California has been undertaken, and no excess has been observed in Los Angeles. Data collected by Buell *et al.* (*137*), however, suggests that there could

Table VII Total Chronic Respiratory Disease Mortality in an American Legion Study Population—California, 1958–1962[a] *(137)*

	Los Angeles County		*San Francisco Bay Area and San Diego County*		*All other counties*	
Residency (years)	*Mortality rate[b]*	*Total deaths*	*Mortality rate[b]*	*Total deaths*	*Mortality rate[b]*	*Total deaths*
10+	38.4	31	28.3	15	45.6	40
<10	41.2	14	45.6	8	41.3	17
Unknown	139.1	12	59.8	3	39.7	4
	46.7	57	34.0	26	44.4	61

[a] Age and smoking adjusted by the direct method to the total study population.
[b] Per 100,000 man-years.

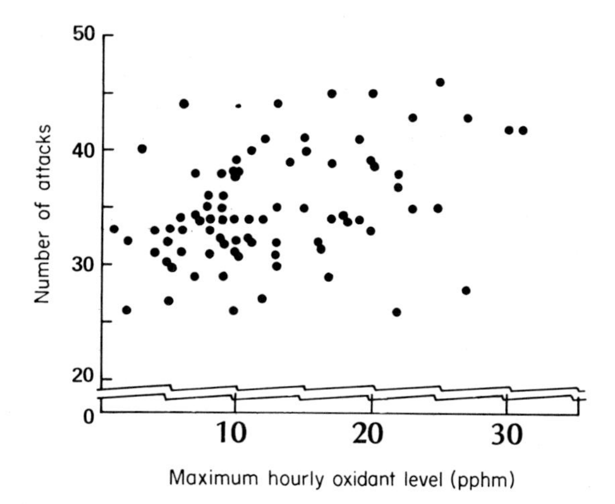

Figure 12. Association of the number of attacks of asthma of a panel in Pasadena, California, with the maximum hourly oxidant concentration (potassium iodide method) for the preceding day (*138*) (1956).

be a contribution of living in Los Angeles to excess mortality from chronic respiratory disease (Table VII). When respiratory disease mortality among this population of military veterans and their spouses living in Los Angeles and in the San Francisco Bay Area are compared with that of residents of other parts of California, the data, while suggestive, are not conclusive.

c. Epidemiological Studies of Disease Aggravation. Schoettlin and Landau (*138*) examined the occurrence of asthmatic attacks in a population in Pasadena, California, during the fall of 1956. Individuals in the study reported each week the occurrence of asthma attacks and the factors they felt might be related to them. A calculation of the association between smog levels estimated by oxidant and asthma attacks showed a statistically significant association and suggested that approximately 8 individuals out of 137 might have been responding to variation in severity of photochemical smog. There was a statistically significant excess of asthma attacks on days where the peak oxidant, measured by the potassium iodide method, was greater than 0.25 ppm (500 $\mu g/m^3$). This corresponds during the days actually measured to an hourly average of 0.20 ppm (400 $\mu g/m^3$) (Fig. 12).*

* In 1974 a discrepancy in oxidant calibration procedures was detected. According to the California Air Resources Board all oxidant monitoring data prior to June 1, 1975 (excluding Los Angeles County data) should be adjusted downward by about 20%. (See *140a* for more complete discussion.)

Remmers and Balchum (*139*) and, later with the same data, Ury and Hexter (*140*) examined impairment of respiratory function in a population of patients with established chronic respiratory disease at Los Angeles, California, County Hospital. The patients stayed in air conditioned rooms, and during alternate weeks in one with an activated charcoal filter that kept photochemical oxidants to a very low level. The most sensitive physiological index of impairment appeared to be airway resistance, which is normally elevated in chronic respiratory conditions. Among the smokers and the nonsmokers taken separately, there is a statistically significant association between the photochemical oxidant and airway resistance (Fig. 13). This is interpreted as being disease aggravation, since airway resistance is a valid indicator of the severity of disease in this syndrome. Los Angeles residents with chronic respiratory disease report aggravation during periods of photochemical pollution. Hammer *et al.* (*141*) studied acute illness in nursing students in two California areas, Santa Barbara and Los Angeles County, and compared the daily maximum oxidant level with the recorded acute symptoms and illnesses. Eye discomfort was the only symptom that occurred in close association with levels of pollution Table VIII, (Fig. 14).

Studies were undertaken in Los Angeles, California, from September through December 1954 to see whether hospital admissions themselves are related to photochemical pollution. While the data show some seasonal trend, particularly for certain diseases of the lung, no association between oxidant air pollution and hospital admission rates due to diseases of the the cardiovascular respiratory system could be found (*142*). A study of the relationship between cardiovascular admissions and pollution was

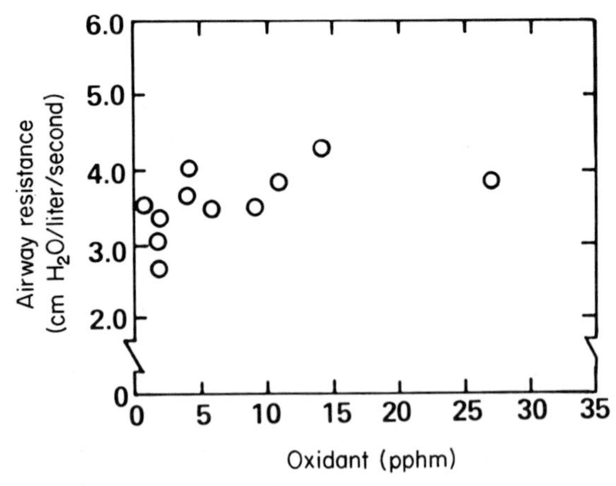

Figure 13. Airway resistance regression on afternoon (2 p.m.) oxidant levels for a typical patient with chronic pulmonary disease, a nonsmoker. The subject shows increases in resistance with oxidant. Correlation coefficient; $\gamma = 0.64$.

Table VIII Relationship of Average Daily Percentage of Symptoms to Photochemical Oxidant Levels (668 days, November 1961–May 1964) (*149*)

Daily maximum hourly oxidant level (ppm)	*Number of days*	*Average daily percentage of symptoms reported*[a]			
		Headache	*Eye discomfort*	*Cough*	*Chest discomfort*
≤0.08	413	10.6	5.2	9.5	1.8
0.09	35	10.6	5.6	10.2	1.9
0.10–0.14	176	11.0	5.9	9.4	1.8
0.15–0.19	144	11.4	6.9	9.7	1.7
0.20–0.24	63	11.6	9.2	9.1	1.6
0.25–0.29	25	11.5	11.2	9.6	2.0
0.30–0.39	9	13.4	17.8	11.7	2.3
0.40–0.50	3	15.0	31.8	16.9	5.8
Overall average		10.9	6.3	9.5	1.8

[a] All days on which the symptom was reported along with "feverish," "chilly," or "temperature" are excluded.

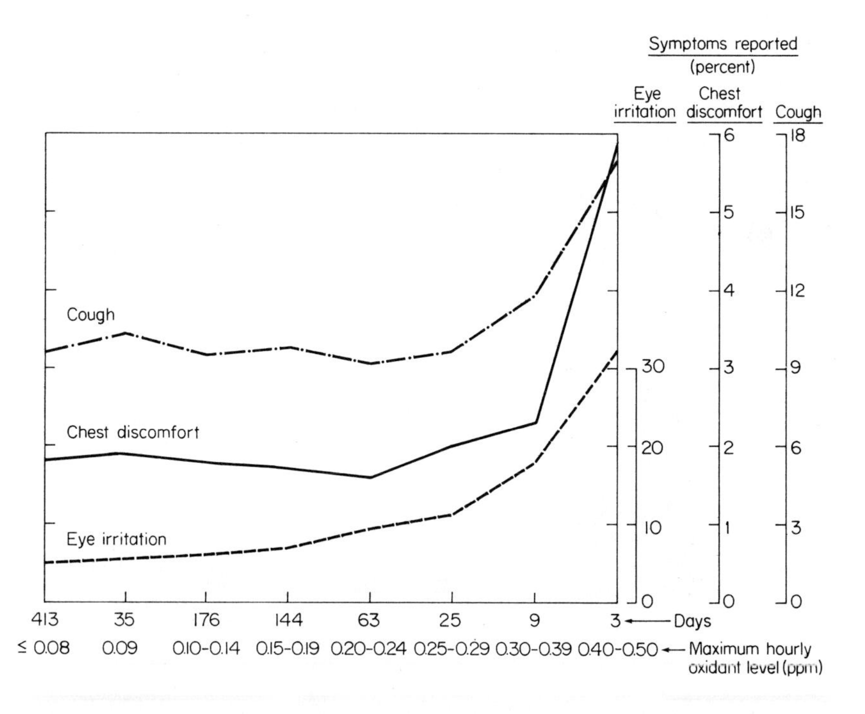

Figure 14. Proportion of a student nurse panel reporting eye irritation and respiratory symptoms at various levels of oxidant (reported as ozone) in Los Angeles, California, 1961–1964. (*149*). (For data see Table VIII.)

made in collaboration with the Los Angeles, California, County Medical Society in thirty-five hospitals for the year 1958. No significant findings relevant to oxidant were obtained. However, for associations of this population with carbon monoxide, see Section IV,C.

Schoettlin (*143*) studied a population of 166 men in a veteran's domiciliary unit in West Los Angeles, California, half of whom appeared to have signs or symptoms of chronic respiratory disease. The other half had similar ages and smoking histories but no evidence of chronic respiratory disease. These men were studied throughout the fall of the year. Maximal oxidant and oxidant precursor consistently accounted for more of the variations in signs and symptoms in the diseased group than in the nondiseased group. For example, maximal oxidant precursor accounted for 30% of the variations of symptoms in subjects with chronic respiratory conditions, compared to an insignificant 4% in the control population.

Rokaw and Massey (*144*) carried out a study of persons in a Los Angeles, California, chronic disease hospital to see whether there was any systematic variation in pulmonary function in association with fluctuations in photochemical oxidant. No positive findings were reported. Among outside plant telephone workers, Deane *et al.* (*145*) determined the prevalence of respiratory symptoms and lung function using the British Medical Research Council questionnaire. The same methods, equipment, and interviewers were used in both Los Angeles and San Francisco California, and the east coast United States cities, as well as in several cities in the United Kingdom. In the Los Angeles area, among the age group 50–59, a significant excess of persons with cough was observed, compared with San Francisco and other United States cities (Fig. 15 and

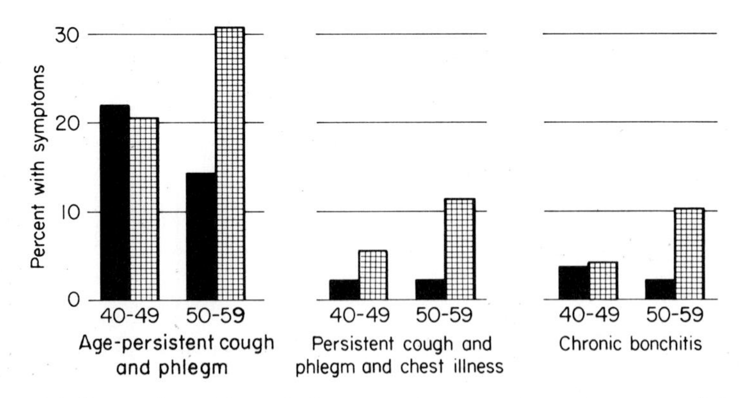

Figure 15. Comparison of respiratory symptom prevalence by age and location in a study of outside telephone workers in San Francisco (solid), Los Angeles (hatched), California in 1963. The proportions of smokers in the groups were similar (*145*).

Table II). These differences could not be explained by social class, occupation, or smoking habits. There were no differences detected in pulmonary function, comparing Los Angeles and other United States populations.

Kagawa and Toyama (*145a*) have reported effects of photochemical pollution on lung function among Japanese school children. Twenty normal 11-year-old Tokyo school children were studied weekly from June to December 1972. Ozone was significantly associated with increased airway resistance. Two of the subjects appeared to be unusually reactive to environmental factors. Temperature strongly affected lung function tests.

d. ACUTE RESPIRATORY DISEASE AND POLLUTION IN CALIFORNIA. Durham (*146*) has reported a comparative study of illness reported to student health services at seven California universities during the 1970–1971 academic year. Using temporospatial strategy, factor analysis, and time series analysis, he found associations of pollutant elevations with increased pharyngitis, bronchitis, colds, and sore throats. From the analysis, the effects of oxidants cannot be isolated from the contributions of sulfur oxides and nitrogen oxides. Comparing selected high and low pollution days, a 16.7% difference in these complaints can be shown for the Los Angeles schools situated in places with lowest and highest contaminant concentrations.

e. EYE AND RESPIRATORY IRRITATION FROM PHOTOCHEMICAL OXIDANTS. Eye irritation remains the most common relationship between photochemical oxidants and health status. The data first obtained during the Air Pollution Foundation Study by Renzetti and Gobran (*147*) and from panel studies indicate that above about 0.14 ppm, there is an increasingly impressive association between eye irritation and photochemical oxidant. A similar association was observed by Richardson and Middleton (*148*) and Hammer *et al.* (*149*) (Table VIII, Fig. 14) and has been reported by a number of other authors in the California area. Similar associations are not yet clearly established for other locations. The data from Tokyo, Japan, certainly indicate that eye and respiratory irritation occurs with photochemical oxidants. Based on data obtained in the California Health Survey (*150*), while eye irritation may be the most frequent cause of difficulty, there also was a high proportion of respondents (about 5% in some areas) who had a feeling of difficulty in breathing during smog episodes. Table VIII and Figure 14 show characteristic results obtained by Hammer *et al.* in a study of student nurses at Los Angeles County Hospital in California.

f. ATHLETIC PERFORMANCE. Wayne *et al.* (*151*) studied high school track athletes to determine whether there are factors in the environment that

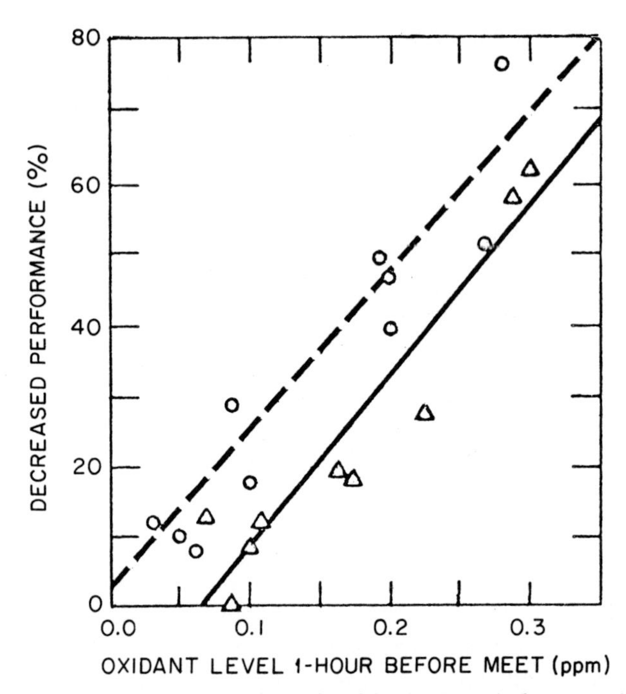

Figure 16. Relationship between oxidant level in the hour before an athletic event and percent of team members with decreased performance (*151*). Solid line is for 1959–1961, broken line for 1962–1964; $r = 0.945$.

affect their track performance. The data on performance were obtained in advance of any effort to see whether there was an association with photochemical pollution. It was assumed that these athletes would improve their performance steadily. The analysis shows that the proportion who failed to improve showed an association with photochemical oxidants (Fig. 16). The range of points runs from an oxidant level of about 0.5 to 0.3 ppm. However, only when points above 0.2 ppm oxidant are added to values at lower concentrations can a regression be said to be present.*

g. Oxidants and Motor Vehicle Accidents. Ury *et al.* (*152*) studied the association of motor vehicle accident frequency by hour and day in Los Angeles, California, with both oxidant levels and carbon monoxide during summer and winter periods from 1963 to 1966. They identified a

* Twenty-eight subjects were exercised during ozone exposures of 0.37, 0.5, and 0.75 ppm for 2 hours. Altered respiratory patterns were observed, with increase in rate and decrease in depth of breathing, presumably due to irritation, which was highly correlated to ozone dose (*151a*).

number of confounding variables, of which the number of vehicles at risk was the most obvious. By assuming that for a given hour of day and day of the week the number of vehicles at risk was only subject to random variation, they applied a simple nonparametric statistical test for adjacent week pairs. They found that more often than not ($p < 0.0001$) increasing oxidant levels are accompanied by increasing frequency of motor vehicle accidents, but such a significant effect was not found for carbon monoxide.

C. Carbon Monoxide (Goldsmith)

Carbon monoxide and lead are examples of pollutants whose effects depend on how much is in the body, and this, in turn, depends on a balance between uptake and excretion. Both are also pollutants for which substantial exposures to the public occur from sources other than air pollution.

1. Uptake, Excretion, and Body Burden

Carbon monoxide is not irritating. Experimental carbon monoxide studies usually expose man or animals to square-wave doses, that is, with an abrupt increase above or decrease to background, which permits a smooth prediction of resulting carboxyhemoglobin. However, most real carbon monoxide exposures fluctuate quite widely (Fig. 17). Its effects are related to both the amount in any given breath and to the amount held in the body most of which is combined with hemoglobin.

Carbon monoxide is produced under natural conditions in the human body. Sjöstrand showed that the body breaks down hemoglobin (a process that permits the recycling of the iron it contains) with the release of four molecules of carbon monoxide for each molecule of hemoglobin broken down (*153, 153a*). When red blood cell destruction is greater, as for instance, with a transfusion reaction, the amount of carbon monoxide produced is increased (*154, 155*). As a matter of fact, the rate of breakdown of red blood cells and other related molecules (*156, 157*) can be estimated by the production and excretion of carbon monoxide. Hemoglobin binds carbon monoxide and carries it through the body. The amount of carbon monoxide carried by the hemoglobin is used as an estimate for how much carbon monoxide is in the body. Nearly all the carbon monoxide in the hemoglobin is excreted unchanged, via the lungs, with the expired air. It is relatively easy to determine the amount of carbon monoxide in the body by measuring the carbon monoxide concentration of the expired air after breath holding (*156, 158–160*).

Since the body's hemoglobin provides the critical mechanism for carrying oxygen from the lungs to tissues and carbon dioxide from tissues to

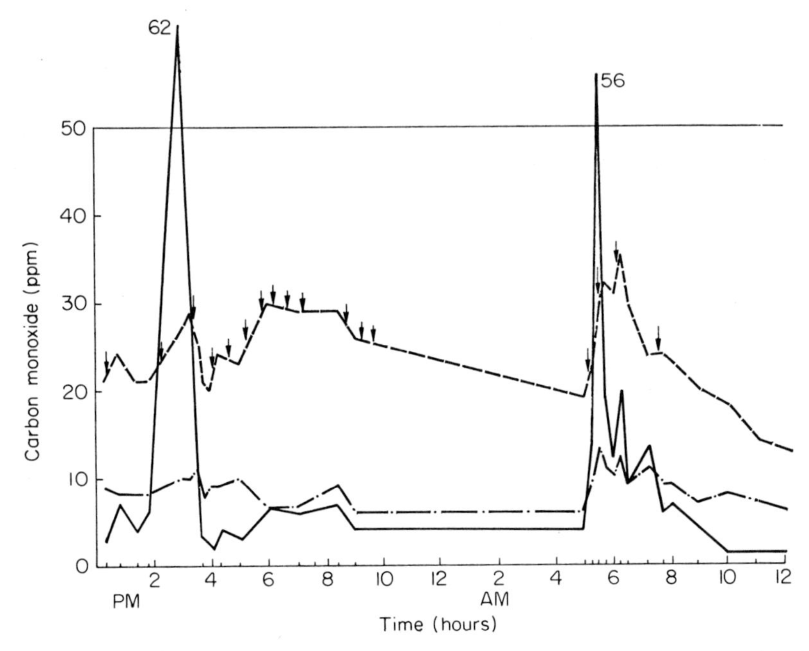

Figure 17. Carbon monoxide levels of the ambient air (—) and expired air after 20-second breath-holding of a smoker (---) and a nonsmoker (–•–), Los Angeles and Pasadena, California. Arrow represents the time of cigarette smoking. The exposure between 2 and 4 p.m. is in a parking garage, and between 5 and 8 a.m. during morning traffic congestion on streets and freeways. The subjects spent the night at a Pasadena motel.

lung, the resulting interference with the functioning of hemoglobin is important. Carbon monoxide is bound to hemoglobin 245 times more firmly than is oxygen. This means that if a gas mixture of oxygen and carbon monoxide is bubbled through blood in such proportions that half the hemoglobin will be combined with oxygen and half with carbon monoxide, the gas mixture must contain 245 molecules of oxygen for every molecule of carbon monoxide. If we know the ratio of oxygen to carbon monoxide in the air in the lung, we can find the ratio of the carbon monoxide to the oxygen combined with the hemoglobin molecules, assuming sufficient time for equilibrium to occur. This relationship was first expressed by the British physiologist J. B. S. Haldane and is known by his name (*161–163*) The Haldane equation for carbon monoxide at equilibrium is:

$$\frac{\%\mathrm{COHb}}{\%\mathrm{O_2Hb}} = M\,\frac{p\mathrm{CO}}{p\mathrm{O_2}} \tag{1}$$

where M is usually cited as 245, %COHb and $\%O_2Hb$ represent the amount of hemoglobin combined with carbon monoxide and oxygen, respectively, and pCO and pO_2 represent the proportions of gas molecules in air, which are, respectively, carbon monoxide and oxygen. This equation not only permits us to estimate the equilibrium percent of carboxyhemoglobin, but also (a) the effect of high altitude (low pO_2) on carbon monoxide uptake—it will increase it; (b) how to help the body get rid of excess carboxyhemoglobin—increase the oxygen content in the inspired air; and (c) the effects of carbon monoxide—it blocks hemoglobin's oxygen transport function. Since in the arterial blood leaving the lung the gas binding sites on hemoglobin are normally almost full of oxygen or carbon monoxide, the sum of $\%O_2Hb$ and %COHb is expected to be 100. Thus, if %COHb in the arterial blood is increased, $\%O_2Hb$ must decrease. That is the first effect of breathing carbon monoxide.

The Haldane equation describes only what happens when the situation is in equilibrium. Since all the body's blood is not in the lung at the same time, and since carbon monoxide transport across the lung and red cell membranes takes time, and since the amount of carbon monoxide leaving the gas phase is not replaced until the next breath, it takes time for the body's hemoglobin to reach the equilibrium shown by Equation (1). In fact, it takes about 4 hours for a resting adult male to reach half saturation value; it takes about 3 hours for females, because of their smaller body size and lesser blood volume. In a person who is exercising, circulation and breathing occur faster, and the time taken to reach equilibrium is shorter. Similarly, an infant or child because of smaller body size takes less time to reach equilibrium. Since there is both uptake and excretion at the same site—in the capillary vessels in the wall of the alveoli (the ultimate air sacs of the lung)—the transfer rate of carbon monoxide is a net result of the amount in the gas in the alveoli and the amount in the blood. During uptake, since the greater the amount in the blood, the less leaves the alveoli each time the blood passes through the lung, the uptake is exponential. The same is true for excretion (Fig. 18) (*164*).

2. Quantitative Relationships

For continuous exposures to concentrations of less than 100 ppm, the California State Department of Public Health (*165*) has derived, from published data, the equilibrium condition equation

$$[CO] \times 0.16 = \%COHb \tag{2}$$

where [CO] is the concentration of carbon monoxide in parts per million by volume, and %COHb is the percentage of carboxyhemoglobin *at*

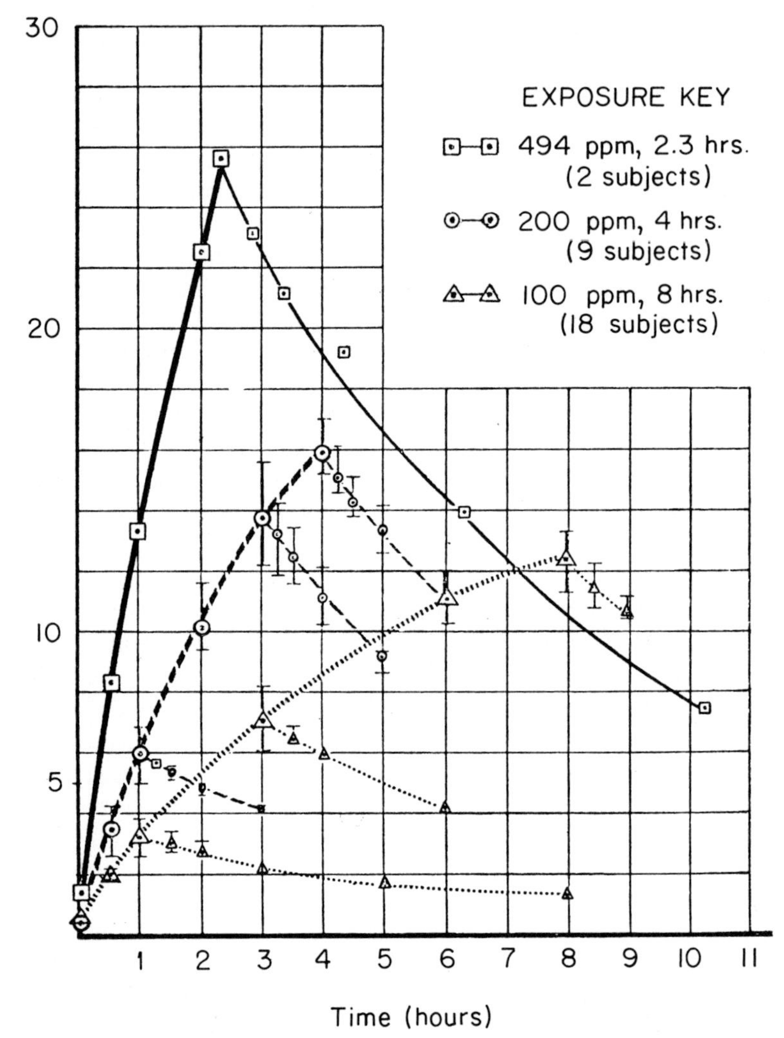

Figure 18. Carbon monoxide absorption and excretion in healthy, sedentary, non-smoking, Caucasian males (*164*).

equilibrium at sea level for healthy subjects. This equilibrium relationship is, as noted, not reached for several hours. Forbes *et al.* (*166*) studied the rate of uptake of carbon monoxide for exposures of from 100 to 2000 ppm.

A more complex relationship has been expressed by Coburn *et al.* (*166a*) called the CFK equation. Peterson and Stewart (*166b*) have shown its validity, and applications to community pollution are appropriate.

Since the effect of carbon monoxide is an impairment of transport of oxygen to the tissues, at high altitudes and in other situations where oxygen tensions are low, the effects of a given concentration of carbon monoxide will be correspondingly more severe. California and Nevada have adopted an 8-hour air quality standard of 6 ppm carbon monoxide for high altitude areas, i.e., about two-thirds the sea level standard. In California it applies to the Lake Tahoe Air Basin; in Nevada to all areas above 5,000-feet elevation.

3. Uptake of Carbon Monoxide from Community Air

The relationship between the exposure to carbon monoxide pollution and the resulting carboxyhemoglobin are not immediately obvious from the measured data because of the time delay between exposure and the build-up of carboxyhemoglobin levels in the blood and its effect on the tissue. Chovin studied (*167*) the levels of carbon monoxide before and after work in a group of traffic policemen in Paris, France. During their work, the policemen were not permitted to smoke cigarettes. The data can be divided, therefore, into two population groups—cigarette smokers and nonsmokers. The averaged occupational exposure was between 10 and 12 ppm. The nonsmokers, as a population, showed an increase during the 5 hours between coming on duty and leaving their place of work, but some of the smokers showed a decrease; that is they had a net excretion of carbon monoxide. The extent to which a smoker and a nonsmoker absorb carbon monoxide has been demonstrated by two medical students who took a car from a Los Angeles, California, garage, drove on the freeway, spent the night in a motel in Pasadena, California, during which both decreased their carboxyhemoglobin, and then after the cigarette smoker had had his morning smoke, returned to the freeways. Both subjects showed detectable effects of freeway exposure and of exposure in the parking garage, but the influence of cigarette smoking was greater than that of exposure in the garage or on the freeway (Fig. 17).

Stewart *et al.* (*167a*) obtained blood samples from persons who appeared at various blood banks in a number of urban areas in the United States. Table IX shows a sample of such data for smokers and nonsmokers in Los Angeles, California; Chicago, Illinois; Denver, Colorado; and New York, New York. The locations chosen for presentation in Table IX are large cities, their airports, and a less urban town in the vicinity. The nonsmokers showed lower COHb in the suburbs, and at some airports; urban donors had higher COHb than in the suburban areas. Smokers showed similar patterns. The fact that urban populations have carboxyhemoglobin levels so much above the expected background level of about 0.5 to 0.9% COHb suggests that there is a systematic contribution from pollution. In addition, the difference between the three subpopulations for

Table IX Median Percent Carboxyhemoglobin and 90% Range for Smoking and Nonsmoking Blood Donors in Selected Cities and Adjacent Airports and Suburbs (*164*)

	Nonsmokers			*Smokers*		
Place and date	*Number*	*Median*	*90% Range (COHb%)*	*Number*	*Median*	*90% Range (COHb%)*
Chicago, Illinois, November 1970						
Downtown	30	2.7	2.2–3.7	34	6.9	3.2–9.3
O'Hare Airport	32	2.5	1.8–3.0	16	6.6	5.2–11.1
Suburban—Palatine	41	1.4	0.8–4.4	16	4.8	1.5–7.7
Los Angeles, California, May–June 1972						
Downtown	166	2.7	1.0–3.2	108	6.0	2.0–9.4
Airport	213	1.4	1.0–2.1	75	5.6	1.2–9.6
Suburban—Huntington Beach	72	1.6	1.2–2.3	37	6.1	2.0–9.0
New York, New York, December 1970 and October–December 1971						
Downtown	841	1.4	0.8–2.3	813	5.2	1.4–9.2
JFK Airport	38	2.1	1.5–2.8	46	6.9	2.3–10.9
Suburban—Croton	197	1.0	0.4–3.7	113	4.1	1.2–8.1
Denver, Colorado, April–May 1971						
Blood Center	676	2.0	1.0–3.7	884	5.5	2.0–9.9
Stapleton Airport	42	1.5	0.8–2.5	16	5.8	2.7–9.1
Suburban—Boulder	27	1.2	0.6–2.1	13	4.5	1.2–2.7

each urban area should be noted. The samples from Los Angeles were taken at a time of year when atmospheric carbon monoxide levels are usually low. Among nonsmokers, 76% in Los Angeles, California, 74% in Chicago, Illinois, 35% in New York, New York, and 26% in Milwaukee, Wisconsin, had carboxyhemoglobin in excess of 1.5% (*164*).

In a Los Angeles, California, study by Deane *et al.* (*168*), automobile drivers manifested a detectable increase of carboxyhemoglobin during the morning commuting period.

4. Carbon Monoxide Uptake from Smoking, Occupation, and Other Sources

In Chovin's reports (*167*) for policemen in Paris, France, the cigarette smokers can be divided into two populations—one, which arrived on duty with a relatively high carboxyhemoglobin, excreted some carbon monoxide through their lungs and ended up after work with a lower carboxyhemoglobin level; the other which arrived on duty with a low carboxyhemoglobin and had some uptake during the day. There is a range of carboxyhemoglobin at which no apparent uptake or excretion occurred (from about 4 to 7% COHb).

The magnitude of exposure to carbon monoxide from smoking has been estimated in a population of longshoremen (*169*) examined prior to their work shift and during a time when there was little community air pollution. The results, therefore, reflect primarily the effects of smoking. Exposure estimates were based on measurements of carbon monoxide in expired air after the individual had held his breath for 20 seconds. The results are shown in Table X. The median value for moderate cigarette smokers who inhale is 5.9% COHb. The relatively low levels in pipe smokers and cigar smokers (see below for exception) are due to the fact that less smoke is usually inhaled when tobacco is used in these forms. Inhaling clearly increases the uptake of carbon monoxide. Landaw (*170*) has accurately measured uptake and excretion of carbon monoxide during and after smoking. He calculated that 8.6 ml of carbon monoxide was taken up with each cigarette based on blood measurements (7.10 ml based on expired air measurements). Some cigar smokers who were formerly cigarette smokers (*171*) continue to inhale with resulting high uptake of carbon monoxide. Apparently, this is more likely with so-called little

Table X Proportion of Smokers and Median Expired Carbon Monoxide among 3311 Longshoremen (*169*)

	Smoking pattern (%)	*Median CO in expired air (ppm)*		*Median COHb estimated from regression (%)*	
Never smoked	23.1	3.2		1.2	
Ex-smoker	12.1	3.9		1.4	
Pipe and/or cigar smoker only	13.4	5.4		1.7	
		Inhaler	*Non-inhaler*	*Inhaler*	*Non-inhaler*
Cigarette smokers					
Light smoker (half pack or less)	13.0	17.1	9.0	3.8	2.3
Moderate smoker (more than half pack, less than 2 packs)	31.3	27.5	14.4	5.9	3.6
Heavy smoker (2 packs or more)	7.0	32.4	25.2	6.8	5.6

cigars. Russell *et al.* (*172*) studied uptake of carbon monoxide by nonsmokers and smokers in poorly ventilated rooms. With a duration of exposure of 78 minutes to an average carbon monoxide concentration of 38 ppm, there was an average increase in carboxyhemoglobin of nonsmokers of from 1.6 to 2.6%. The cigarette smokers increased from 5.9 to 9.6% COHb. Rylander (*172a*) has reviewed a series of studies on the effect of environmental tobacco smoke on nonsmokers. Carbon monoxide is often used as the indicator substance in such studies (*172b*).

Breysse and Bovee (*173*) used expired air samples to estimate occupational exposures of stevedores and gasoline-powered forklift truck drivers and winch operators. Forty (5.7%) out of 700 determinations of carboxyhemoglobin were in excess of 10%. Carboxyhemoglobin levels in excess of 5% were found in 30% of smokers but in only 2% of nonsmokers.

The present maximum allowable atmospheric concentration, or threshold limit value (TLV), for occupational exposure in industry in the United States is 50 ppm for 8 hours. The limit was reduced from 100 ppm in 1964 because of new evidence of possible adverse effects, mostly on the central nervous system, from exposures in the 50–100 ppm range. The United States National Institute of Occupational Safety and Health criteria report proposes a 35 ppm threshold limit value and notes the importance of exposure hazard in men with heart disease.

Open fires and charcoal braziers produce a substantial amount of carbon monoxide (*174, 175*). Carbon monoxide is a particular risk for residents in houses heated by improperly vented kerosene or gas stoves. A survey by the United States Department of Health, Education, and Welfare (*176*) concludes that "there is a definite health hazard from carbon monoxide poisoning in dwelling units and in vehicles during cold weather." In selected communities, 3–30% of housing units were found to have evidence of a carbon monoxide problem. This form of indoor air pollution has been neglected. Cigarette smoking may also produce a form of indoor air pollution with carbon monoxide.

5. Effects of Carbon Monoxide on Cardiovascular Disease

When air quality standards were first set by the state of California in 1959, the effects of carbon monoxide on the limitation of oxygen transport function were predictable from physiological experiments (*177*). It was thought that among persons who were at unusual risk were those with cardiovascular disease, particularly those who had had acute episodes and required a maximal capacity to transport oxygen to tissues for their health or survival. Among groups thought to be vulnerable were persons with recent myocardial infarctions. This hypothesis was further elaborated by Permutt and Fahri (*178*), who emphasized that the heart muscle (myocardium) might be particularly vulnerable because it had a high oxygen extraction ratio, that is, because it ordinarily removed a high proportion of the oxygen brought to it by the blood. Most of the organs of the body that need to increase their oxygen supply can do so by increasing the amount they remove from the blood brought to them by the arteries. This is not the case for the heart muscle, because under normal circumstances, a high proportion of the oxygen brought to it is removed. The only way the oxygen supply can be increased is by the increased flow of blood. In persons with coronary heart disease, because of the rigid blood vessels supplying the heart muscle, it is hard for the circulation to deliver more oxygen when needed.

Long-term exposure experimental studies in animals were undertaken by Lewey and Drabkin (*179*), Stupfel and Bouley (*180*), and Preziosi *et al.* (*181*). These investigators did not always produce results that were entirely consistent, but their data suggest that for exposures over long periods to 50 or 100 ppm, there were some circumstances under which one observed transient changes in the electrocardiogram and loss of tissue in the heart muscle. Some experiments also demonstrated loss of central nervous system tissue.

In another series of experiments, Astrup and colleagues (*182–184*) showed that rabbits that were exposed to elevated levels of carbon monoxide (producing about 11% COHb) over a sufficient period of time and were fed diets high in cholesterol manifested increased fatlike deposition in the walls of the great blood vessels. This fatty material resembled the early process of what is called arteriosclerosis or hardening of the arteries. The same result could be reproduced if the animals were exposed to simulated high altitude, that is they breathed air containing a decreased amount of oxygen.

In a series of experiments, Ayres (*185, 186*) and his colleagues have shown that the experimental administration of carbon monoxide sufficient to increase the carboxyhemoglobin to around 9% in human subjects diminishes the amount of oxygen removed from the blood by the heart. The effect in general is greater than the relative binding of carbon monoxide represented by the percent of carboxyhemoglobin. In addition, they were able to show that the subjects who had already had disease of the arteries affecting the heart muscle were unable to increase the rate of blood flow to the heart muscle sufficiently to keep the metabolism in good working order. The effect they observed was a shift to a type of energy production that builds up a deficit requiring increased amounts of oxygen at a later period. This deficit is called "oxygen debt." In the second set of studies with humans (*186*), effects were seen to be more moderate when a given level of carboxyhemoglobin (9–10%) was approached slowly with exposure to 0.1% carbon monoxide than when it was approached rapidly by a high dose (5% carbon monoxide for 60–90 seconds). This discrepancy suggests that some of the effects on the heart muscle may be produced by carbon monoxide combining with substances other than hemoglobin, substances that are within the muscle tissue itself. Patients with coronary vascular disease showed significant metabolic changes in the heart muscle with elevation of carboxyhemoglobin above 6%. According to Ayres, dogs were less sensitive than man, and significant changes were not observed in dogs until the carboxyhemoglobin saturation exceeded 25%. Chevalier *et al.* (*187*) exposed nonsmokers to high concentrations of carbon monoxide (5750 mg/m^3) for a few minutes in order to produce a carboxyhemoglobin level similar to that in smokers. This resulted in carboxyhemoglobin levels of 3.95%. Increase in the amount of oxygen consumed after exercise (oxygen debt) was observed after exposure.

EPIDEMIOLOGICAL STUDIES OF CARBON MONOXIDE EFFECTS ON THE CARDIOVASCULAR SYSTEM. A large body of data has already been assembled concerning the relation of cigarette smoking to diseases of the heart. In general, this shows that there is a substantially higher rate of heart disease in moderate to heavy cigarette smokers than is true for lighter cigarette smokers, and, in turn, these people have a higher mortality than those who are nonsmokers. The gradients are especially steep in younger smokers (Table XI). In addition, smoking tends to aggravate manifestations of a disease called angina pectoris. This is a disease in which mild exercise, because of insufficient oxygen supplied to the heart muscle, produces symptoms of pressure or pain in the chest. The person then either must stop walking or stair-climbing or take medication to cause the blood vessels to dilate. In persons who are exposed to cigarette smoking, it has been shown systematically by Aronow *et al.* (*188*) that there is a decrease in the amount of exercise that can be tolerated before the chest pain of angina occurs. At first Aronow's studies implicated nicotine in this reaction, but even in subjects who are smoking cigarettes that do not contain nicotine this effect will occur (*189*) when the carbon monoxide becomes sufficiently elevated.

To test the effects of urban carbon monoxide pollution, Aronow and his colleagues (*190*) drove subjects with angina pectoris around streets and freeways in Los Angeles, California, for 90 minutes. The samples of air taken contained approximately 50 ppm carbon monoxide. After this exposure, there was an increase in the carboxyhemoglobin of the subject. Two hours later, carboxyhemoglobin had come part way back to normal.

Table XI Relationship of Median Percent Carboxyhemoglobin[a] (COHb) in Smokers (California Health Department) and Coronary Heart Disease Mortality Ratio (CHDR) by Age (Hammond) Relative to Nonsmokers (*169*)

		Smoking class for current cigarette smokers					
		Fewer than 10 cigarettes/day		*Cigarettes/day*			
Age	*Nonsmokers COHb*	*CHDR*	*COHb*	*10–19 CHDR*	*10–39 COHb*	*20+ CHDR*	*40+ COHb*
<45	*0.89*	—[b]	*3.8*	—[b]	*6.0*	—[b]	*6.7*
45–54	*0.89*	2.4	*2.6*	3.1	*5.3*	3.2	*6.7*
55–64	*0.89*	1.5	*3.5*	1.9	*5.0*	2.0	*5.3*
65–74	*0.89*	1.3	*2.6*	1.6	*3.3*	1.6	—[c]
75–84	*0.89*	1.2	—[c]	1.4	*2.9*	1.1	—[c]

[a] COHb percent estimated from expired air carbon monoxide (CO) using the regression COHb = 0.21 + 0.19 (CO).
[b] No subjects available in this age group in Hammond's study.
[c] COHb data cited only if 10 or more subjects were tested.

There was a statistically significant association between the increase in carboxyhemoglobin and the decrease in the exercise capacity, both immediately after exposure and 2 hours later (Fig. 19). Aronow *et al.* have further demonstrated (*191*), in a double blind study, that exposure to 50 ppm of carbon monoxide for 2 hours will reproduce, in general, the effects observed in the angina patients exposed on freeways and streets. They have also shown (*192*) that the same exposure will decrease exercise performance in persons with sclerotic changes in the arteries of the legs, presumably due to impairment of the delivery of oxygen to leg muscles.

A study of the survival of persons having myocardial infarctions admitted to thirty-five hospitals in Los Angeles, California, has been reported by Cohen *et al.*, (*135*). The study used a set of data obtained for another purpose, and therefore was dependent on the available air pollution monitoring data for the year in which the hospital admission data were obtained, the year 1958. The authors divided the hospitals according to whether they were in an area thought to have high air pollution or low air pollution, based on Los Angeles, California, Air Pollution Control District data. A comparison was then made between the case fatality rate in high and low pollution areas. The case fatality rate is the proportion of persons admitted to the hospital with acute myocardial infarction who died before being discharged. The case fatality rate was higher in the higher pollution area, and this elevation occurred only during the fourth

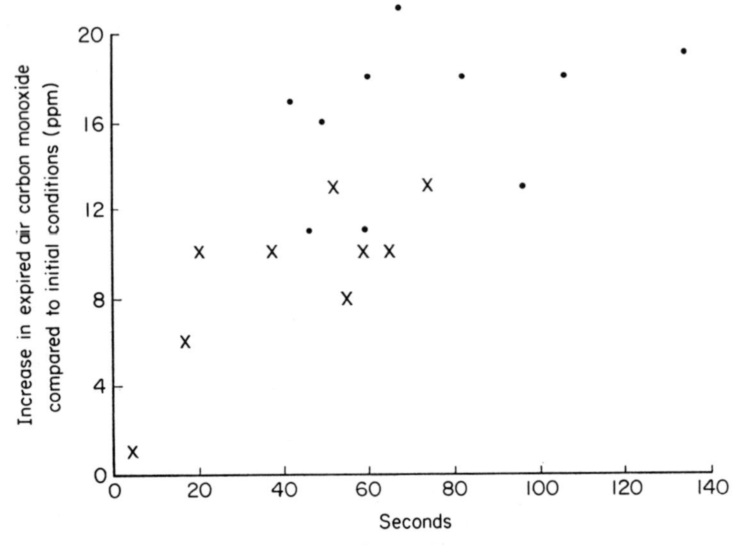

Figure 19. Relation of increases in carbon monoxide in expired air to impairment of exercise tolerance in men with angina pectoris after 90-minute freeway exposure (•) and 2 hours after such exposure (×) (*190*).

quartile of weeks of the year in which pollution by carbon monoxide was high (Table XII). In a later study, Hexter and Goldsmith (*73*) compared the environmental factors that contributed to variation in the daily mortality in Los Angeles County, California, for the years 1962–

Table XII Proportion of Weeks That Case Fatality Rates for Hospitalized Myocardial Infarction Cases Are Greater in More and Less Polluted Areas by Quartile of Weekly Average Carbon Monoxide[a], Los Angeles, California, 1955 (*135*)

	Quartile 1	*Quartile 2*	*Quartile 3*	*Quartile 4*
Average weekly carbon monoxide (ppm)	5.36–5.87	5.96–6.95	6.97–8.43	8.49–14.53
Number of weeks with higher case fatality rate in:				
More polluted area	8	6	9	12
Less polluted area	5	7	4	1
Significance	Not significant	Not significant	Not significant	$p < 0.01$

[a] Weekly average carbon monoxide at downtown station tested by sign test.

1965. After adjusting for trend and using trigonometric terms to represent the contribution of time of year, a major contribution to reducing the variance was made by temperature and various polynomial and lagged terms concerning temperature. In addition, the logarithm of the basin average of carbon monoxide concentration made a statistically significant contribution to explaining the variability in mortality. The authors concluded that this was suggestive but not conclusive evidence of a role of carbon monoxide in daily mortality. Most of the variation in daily mortality associated with carbon monoxide occurred in deaths attributed to arteriosclerotic heart disease. Thus, this study tends to support the possible role of carbon monoxide in heart disease mortality suggested by the observations of Cohen *et al.* (*135*).

Anderson *et al.* (*193*) have studied effects of 4-hour exposures to 50 and 100 ppm of carbon monoxide on 10 men with angina pectoris. At 50 ppm (58 mg/m^3) there was an average increase of 1.5% carboxyhemoglobin (1.4 to 2.9%). At 100 ppm (115 mg/m^3) there was an average increase of 2.9% (1.6 to 4.5%). Mean duration of exercise before pain was significantly shortened with both levels of exposure, and the duration of pain was significantly prolonged after 100 ppm. Gordon and Roger (*194*) studying fire fighters exposed to carbon monoxide while fighting fires in the Denver, Colorado, area observed that carbon monoxide exposure produced changes in the electrocardiogram of persons without known heart disease.

The possibility that carbon monoxide exposures repeated frequently may contribute to the occurrence, severity, or progress of the arteriosclerotic process is suggested by the work of Astrup (*195*), Kjeldsen (*196*), and Wald *et al.* (*197*). Kjeldsen observed a number of factory workers and found that compared with people of the same age and smoking history, patients with heart or peripheral vascular disease had substantially higher carboxyhemoglobin levels. Wald *et al.* (*197*), using Kjeldsen's data, found that among moderate cigarette smokers (smoking 10–20 cigarettes per day) aged 30–69, the proportion who had atherosclerotic disease increased with the amount of carboxyhemoglobin. For example, of 122 persons in this category with a carboxyhemoglobin of between zero and 3.9%, two, or 1.6% of the total had atherosclerotic disease, whereas of the 20 persons with 8% carboxyhemoglobin or more, four, or 20%, had atherosclerotic disease. In the intermediate group, between 4 and 7.9% carboxyhemoglobin, 5 of 126 persons had atherosclerotic disease, i.e., 4%. Wald *et al.* are cautious, however, in their interpretation and it is made clear in the discussion that carbon monoxide may not be causally associated with the effects; several other factors might have operated.

Interpretation. From these reports, we may hypothesize that there are four possible ways by which carbon monoxide can affect the heart: (a) decreasing the probability of survival of persons who have a myocardial infarction and possibly other cardiovascular diseases; (b) producing aggravation of existing circulatory conditions, particularly angina pectoris and peripheral arteriosclerosis, as shown by Aronow *et al.* (*188–192*) and Anderson *et al.* (*193*) [Angina affects about 2% (3,800,000) of the United States population (*199*)]; (c) causing unfavorable changes in a healthy person (*194*); (d) as a factor in the onset and progression of cardiovascular disease, a possibility for which evidence is still only suggestive. These data are therefore suggestive of an association of carbon monoxide exposure with cardiovascular disease. Materials other than carbon monoxide, inhaled with cigarette smoke, could also be partially responsible for the findings associated with smoking. The role of genetic factors in cardiovascular reactions of smokers has been shown to be of importance by twin studies by Friberg *et al.* (*200*) and Cederlöf *et al.* (*201*). Carbon monoxide is at most only one factor in the pathogenesis of cardiovascular disease, but the importance of this agent is increasingly recognized (*202*, *203*).

6. Effects of Low Concentrations on the Central Nervous System

In work done during World War II on the effects of carbon monoxide on visual threshold, McFarland (*198*) detected an effect at a carboxyhemoglobin concentration of about 5%. Schulte (*204*) studied the effects on fireman of exposure to low concentrations of carbon monoxide. He evaluated pulse rate, respiratory rate, changes in blood pressure, and neurological reflexes and conducted a battery of psychomotor tests. In some of these tests, significant changes in response were found after exposure to carbon monoxide. For some of the tests, variations in performance were found at carboxyhemoglobin levels well below 5%, possibly even at levels as low as 2%. Schulte predicted that similar studies in a larger group of subjects might show significant variations in performance at even lower carboxyhemoglobin levels. However, at the lowest carboxyhemoglobin levels, there was a somewhat erratic change in response as concentrations were increased, a fact that casts some doubt on the validity of this prediction. In any event, the tests used are rather complex and are not readily related to other types of behavior.

Beard and Wertheim (*205*) have reported distinct effects upon the ability to perceive differences in the duration of auditory stimuli among healthy subjects exposed for 90 minutes to carbon monoxide at concentrations as low as 50 ppm (Fig. 20). In supplementary tests to determine whether the effect was due to impairment of hearing or to impairment of temporal discrimination, the subjects were asked to estimate time intervals of 10 and 30 seconds; these tests showed that discrimination was impaired, not hearing. The results for the test subjects were significantly different from those for controls following exposure of the test subjects to carbon monoxide concentrations as low as 50 ppm for 75 minutes. Such an exposure could have produced an increase of carboxyhemoglobin concentrations of less than 2%.

There have been some experimental efforts to repeat the findings of Beard and Wertheim. Stewart *et al.* (*205a*) and Mikulka (*206*) have not been able to confirm Beard and Wertheim's findings despite considerably higher carbon monoxide exposure. It is not known to what extent differences in study design and technique may be responsible. The possibility of faulty technique in the work of Beard and Wertheim has been raised. The internal consistency of the studies from this laboratory is impressive; however, for the time being this effect is considered controversial.

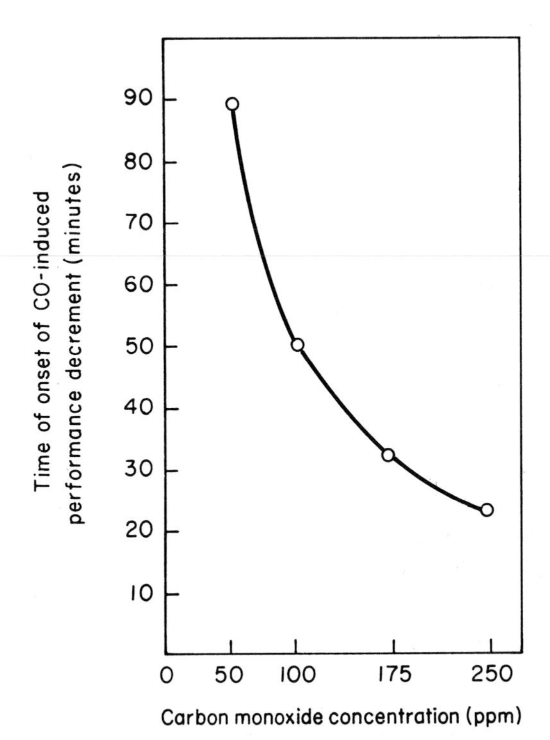

Figure 20. Dose–response relationship for impairment of estimation of time interval for individual subjects exposed to carbon monoxide in acoustical booths. The time of onset of significant decrement of performance below control conditions is shown. (*205*)

7. Carbon Monoxide Exposures and Motor Vehicle Accidents

We do not know whether possible effects of low concentrations of carbon monoxide on the nervous system could be of importance in connection with motor vehicle accidents.

Such effects, if present, should be detectable by epidemiological methods, and perhaps special attention should be paid to experiences of drivers in vehicles with which they were not familiar.

The possibility of an effect is suggested by the findings of Chovin (*167*) and Clayton *et al.* (*210*), although the work of neither of these investigators is conclusive. Chovin, for example, has shown in a 5-year population study by the Prefecture of Police in Paris, France, that drivers involved in and thought to be responsible for motor vehicle accidents had, on the whole, significantly higher carboxyhemoglobin levels in their blood than did two other populations studied in his laboratory. Without estimating the effect of alcohol ingestion, which is more common in smokers, we cannot attribute this effect to carbon monoxide alone. Wright *et al.* (*207*) concluded that an increase of 3.4% COHb is sufficient to prejudice safe driving, but McFarland (*208*) reports from a different set of experiments that 6% COHb did not appear to seriously affect the ability to drive motor vehicles.

Ury (*152, 211*) has presented a resourceful, nonparametric, statistical method for determining the possible association of motor vehicles accidents and urban air pollution, using data from the Los Angeles, California, area as a basis. The method depends on the comparison of the frequencies of accidents for a given hour of the day, and day of the week, across adjacent weeks, with comparable data for pollution concentrations. He has shown highly statistically significant evidence of an association of photochemical oxidants and accident frequencies. In the early papers in this series, the effects of carbon monoxide were not statistically significant. More recent analysis, using lagged terms, has given suggestive evidence that carbon monoxide may contribute to traffic accident frequencies.

Negative findings were reported by Waller and Thomas (*212*), who collected blood samples from individuals involved in accidents and from a control population in Vermont. Here, the role of carbon monoxide could not be detected against the background of the very substantial contributions of alcohol.

D. Nitrogen Dioxide and Other Nitrogen Oxides

Of the seven oxides of nitrogen known to exist in the ambient air, only two are thought to affect human health. These are nitric oxide (NO) and nitrogen dioxide (NO_2). At the present time there are no data either from animal or human studies showing that nitric oxide at the levels encountered in the ambient air is a health hazard. Nitric oxide is readily oxidized to nitrogen dioixde. This oxidation may occur in the atmosphere or in the membranes and tissues lining the airways.

1. Hemoglobin Reactions

Both nitric oxide and nitrogen dioxide, if transferred across the lung–blood barrier, can produce inactive forms of hemoglobin, the most important of which is called methemoglobin. Studies undertaken in Los Angeles, California, indicate that methemoglobin levels in school children and in commuting adults have ranged up to 5.2% and among groups of commuters average between 2.0 and 2.5% (*213*). This is higher than the 1% that had previously been thought normal. The methemoglobin levels appear to vary with time and place in a way that could be evidence of an effect of air pollution, but the role of cigarette smoking among adults is not detectable in these elevations. Since cigarette smoke contains high levels of oxides of nitrogen, this is negative evidence concerning inhaled oxides of nitrogen being responsible for this variation.

In vitro, nitric oxide is so closely bound to hemoglobin that it will readily replace carbon monoxide; however, the lack of observation of a human nitric oxide–hemoglobin complex *in vivo* implies that the combination occurs only under conditions in which oxygen is nearly absent. While some questions remain about hemoglobin reactions with oxides of nitrogen, there is no positive evidence that nitric oxide exposure is a health hazard associated with community air pollution.

2. Nitrogen Dioxide Reactions with Lung Tissue

Nitrogen dioxide is found as a pollutant in association with certain types of rocket fuel* and is known to cause occupational disease. Nitrogen dioxide effects tend to occur many hours after exposures have ceased. Persons occupationally exposed are often unaware of the severity of their exposure. The resulting pulmonary edema may progress for a number of hours after the exposure has ceased. It is very difficult both to estimate the magnitude of occupational exposure and to isolate the effect of nitrogen dioxide. The major basis for concern about oxides of nitro-

* Hatton, *et al.* (*213a*) report the biochemical evidence of breakdown of collagen, presumably from lung, in three U.S. astronauts resulting from an accidental exposure to NO_2 at an estimated concentration of 250 ppm for four minutes and forty seconds. Chest roentgenograms showed a chemical pneumonitis which cleared in five days.

gen as a low-level–long-term exposure hazard, therefore derives from animal studies (see Chapter 5, this volume).

3. Occupational Effects of Nitrogen Dioxide

Among occupations with nitrogen dioxide hazards are the manufacture of nitric acid (*214*), exposures of farmers to silage that has had high nitrate fertilization (*215–217*), electric arc welding (*218, 219*), and mining utilizing nitrogen compounds as explosives (*220*). In the manufacture of nitric acid, exposures estimated at 30–35 parts per million of nitrogen dioxide did not appear to show effects of injury (*221*); however, Russian studies (*222*) report chronic lung disease from lower levels of exposure, 3–5 years at concentrations below 3 ppm of nitrogen dioxide. Norwood *et al.* (*218*) and McCord *et al.* (*219*) reported that methemoglobin is elevated in workers using electric arc welding with coated rods. The maximum observed value was 3% methemoglobin. On the basis of the reports cited above in Section IV,D,1, there is doubt how much elevation this really represents. They estimated that the "nitric gas" was 13 ppm, expressed as nitrogen dioxide.

It is estimated that eye and nasal irritation will be observed after exposure to about 15 ppm of nitrogen dioxide and pulmonary discomfort after brief exposures to 25 ppm (*223*). It is likely that pathological changes can be detected on the basis of exposures of 25–75 ppm for short time periods (*216*). It is also thought that exposure to 150–200 ppm of nitrogen dioxide will lead to gradual production of fatal pulmonary fibrosis. Meyers and Hine (*223*) exposed individual volunteers experimentally to up to 50 ppm for 1 minute and produced significant respiratory tract irritation.

4. Human Experimental Studies at Low Levels (5 ppm and less)

Abe (*224*) studied sulfur nitrogen dioxide separately and together and had shown the lowest level of effect on normal subjects (see *397*a) Of the 5 subjects used, all were free from a history of respiratory disease and were from 21 to 40 years old; four of them were nonsmokers. Exposure was to 5 ppm of nitrogen dioxide for 10 minutes. Exposures led to increased inspiratory and expiratory resistance at 10 minutes after stopping the exposure; a further increase in resistance was observed 20 minutes after stopping, and a marked increase 30 minutes after cessation of exposure. Expiratory resistance increased on the average by 77% and inspiratory resistance by 92%. Effective lung compliance decreased 30 minutes after exposure to 40% of the control value. In these experiments, the effects of sulfur dioxide appeared to occur independently.

Shalamberidze (*225*) reports reactions to sulfur and nitrogen oxides on reflexes. According to Henschler, nitrogen dioxide is detectable by odor at 0.12 ppm (*226*). Rokaw *et al.* have experimentally exposed people to nitrogen dioxide (*227*). As can be seen in Figure 21 when the exposure of one subject was at 0.9 ppm for an hour, there was no change in airway resistance with exercise, but when the subject was exposed to 2.0 ppm with exercise, there was a substantial increase in specific airway resistance which exceeded that of the control. There appears to be individual variability in response to nitrogen oxide exposures that makes it quite difficult

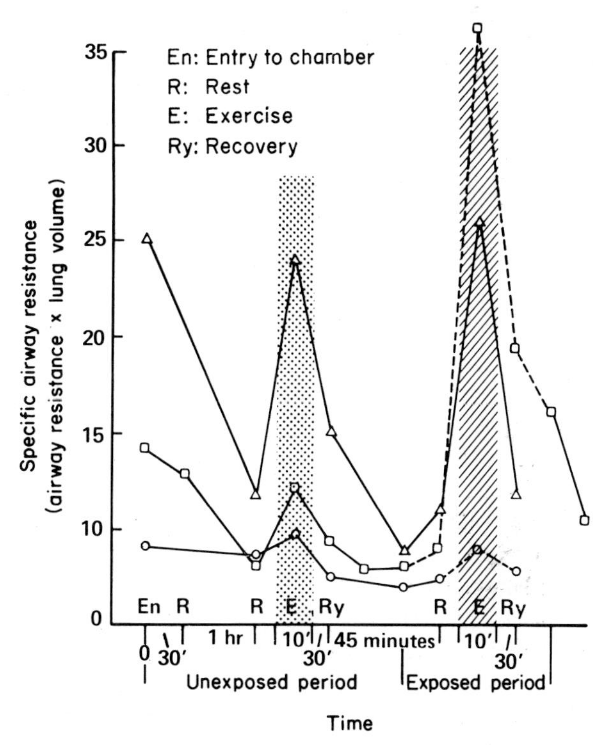

Figure 21. Airway resistance in response to nitrogen dioxide exposure and exercise in one subject on three occasions (*227*). Note that with exercise but no nitrogen dioxide, there is an increase in specific airway resistance (left-hand shaded area), and that recovery was incomplete after 30 minutes. There is substantial day-to-day variation in both resting and exercise measurements, but lesser within-day variation. With exposures to 0.9 and 2.0 ppm, no effect is noted at rest, but with exercise, a substantial increase in resistance occurs. Broken line indicates nitrogen dioxide exposure; △ = control day, ○ = 0.9 ppm exposure day, □ = 2.0 ppm exposure day.

to predict its specific effects. The delay in the occurrence of serious effects should be interpreted as a need for caution.

5. Effects of Nitrogen Dioxide on Pulmonary Antibacterial Defenses

A series of studies has examined the extent to which nitrogen oxides and, for that matter, other materials interfere with mechanisms of defense against infection. One basic method, that of determining survival in the lung of bacteria at different times after inhalation, was first demonstrated using bacterial aerosols by Laurenzi (*29*). Many other things can interfere with the pulmonary defense mechanisms, and if pulmonary edema occurs with impairment of oxygen uptake, this, of course, can produce some of these effects. There is no affirmative evidence concerning such a process in man, however. Further details of animal studies will be found in Chapter 5, this volume.

6. Epidemiological Studies of Nitrogen Dioxide as a Community Pollutant

The observations by Remmers and Balchum (*139*) of airway resistance alterations in persons with chronic respiratory disease in Los Angeles, California, has been reanalyzed to see whether the effects on airway resistance that were associated with oxidant exposure could have been a result of nitrogen dioxide. These analyses indicate that there was no significantly positive effect of nitrogen dioxide for either smoking or nonsmoking subjects. In nitrogen oxide and sulfur oxide polluted areas of Czechoslovakia, Petr and Schmidt (*228*) found in children elevated levels of methemoglobin (above an average of 2.5% methemoglobin in one community). In addition, these workers observed an apparent effect on lymphocytes and monocytes in peripheral blood, but it was not thought possible to separate the effects of nitrogen dioxide and sulfur dioxide on blood cells (see Section VI,F).

In a series of studies in the Chattanooga, Tennessee, area, Shy *et al.* (*229*) and Pearlman *et al.* (*230*) observed adverse effects on both adult and childhood respiratory illnesses and pulmonary function as well as acute lower respiratory illness in children living near TNT plants emitting nitrogen dioxide. The authors report what they believe to be an effect of oxides of nitrogen on respiratory disease prevalence both in children and in adults. The second grade children in all four areas were enrolled in the study and the forced expiratory volume in three-quarters of a second was measured weekly during the months of November 1968 to March 1969 and adjusted for height. This test value reported was significantly higher in control than in "high nitrogen dioxide" areas. There was no significant difference between the tests on children from the "particulate" area and the control areas. The incidence of acute respiratory illness was obtained for all household members, a total of 4445 individuals in 960 families. There was a similar respiratory illness pattern by time of year in all four areas, but "nitrogen dioxide" areas had more respiratory illness than the control area. In the "high nitrogen dioxide" area, 0.1 ppm of nitrogen dioxide was exceeded 40, 18, and 9% of the days in each of three stations of the area, but in one of the control areas, this level was exceeded on 17% of the days at one station. Heuss *et al.* (*231*) have pointed out that one of the schools, school 3, is included in the "high nitrogen dioxide" area but that the nitrogen dioxide exposure was apparently the same as that in control area 1. They have also noted that the differences in the pulmonary function tests are very small. One of the areas had a high particulate level, and this area had a slightly lower socioeconomic level than the other areas, although the authors felt this was not a significant difference. There have also been criticisms of the method used to measure nitrogen dioxide.

A World Health Organization Task Group on Environmental Health Criteria for Oxides of Nitrogen (*232*) concluded that a nitrogen dioxide concentration of 940 $\mu g/m^3$ (0.5 ppm) was established to be the lowest level at which adverse effects, due to short-term exposure, can be expected to occur. A maximum one-hour exposure of 190–320 $\mu g/m^3$ (0.10–0.17 ppm), no more frequently than once a month, was felt to be consistent with protection of the public health.

E. Hydrogen Sulfide and Mercaptans

Hydrogen sulfide is a colorless gas having a characteristic disagreeable odor often described as that of rotten eggs. The presence of low concentrations is evidenced by its odor, the discoloration of some paints, and the tarnishing of brass and silver. Exposures for short periods can result in fatigue of the sense of smell. This is particularly serious in industrial exposures, because the sense of smell is lost after 2–15 minutes of exposure to 100 ppm or more (*233*). Irritation of the mucosa is thought to be based on the formation of sodium sulfide (*234*). It is thought that the gas is absorbed as the sulfide ion.

One air pollution episode during which hydrogen sulfide caused deaths has already been discussed in Section II,A,4. A common human effect of hydrogen sulfide is sensory irritation.* Reports in the Russian literature indicate that the odor of hydrogen sulfide can be

* In Rotorua, New Zealand, a town situated in a geothermally active area, hydrogen sulfide emissions have collected in depressed areas, such as, cellars, and have caused fatalities.

detected at 0.072 ppm (*235*). The Soviet Union atmospheric standard for a maximum single value is 0.036 ppm, this being the odor threshold value (0.072 ppm) divided by two. Katz and Talbert (*236*) indicate that for a panel of observers, the odor threshold level is 0.13 ppm. The California State Department of Health conducted experiments on odor threshold concentrations for hydrogen sulfide with a panel of 16 persons. Each person was tested for the approximate concentration at which he detected odor in an air stream containing continuously increasing hydrogen sulfide. He was exposed to three inhalations per test of each of several air streams containing increasing concentrations of hydrogen sulfide in 0.01 ppm increments. The lowest concentration detected was recorded as the threshold odor concentration. The subject was exposed to the next higher concentration of hydrogen sulfide and invariably detected it also. The range of detection was 0.012–0.069 ppm, and the geometric mean and standard deviation was 0.029 ± 0.005 ppm.

Methyl mercaptan (CH_3SH) and ethyl mercaptan (C_2H_5SH) are among the most potent odorants. For practical purposes, they have no other effect on human health at the concentrations at which they are odor nuisances (see Section VI,E,3). The mercaptans are emitted in mixtures of pollutants from some pulp mills (*237, 238*), petroleum refineries, and chemical manufacturing plants. Mercaptans are often added to natural or manufactured gas supplies so that leakage of gas will be noticed.

F. Hydrocarbon Vapors

Some of the hydrocarbon vapors in the atmosphere have health implications. Among the aldehydes of importance are formaldehyde and acrolein, both potent irritants. Based on data from Los Angeles, California, formaldehyde is present in the atmosphere in a concentration of up to 0.1 ppm. Total aldehyde concentrations occur at concentrations up to 1.3 ppm. Among the aromatic hydrocarbons, benzene is present in concentrations up to 0.06 ppm; toluene and xylene to approximately 0.1 ppm. Formaldehyde is the pollutant present in concentrations nearest the level producing health effects under occupational conditions (Table I, Section I,B,1). Its effects are primarily irritating. It is thought to be a major contributor to the eye and respiratory irritation of photochemical smog. Formaldehyde is the leading candidate among pollutants that may have a health effect in photochemical air pollution and that are not being systematically monitored or studied.

Benzene was a popularly used solvent, but its industrial use is now restricted because sufficient exposure to it interferes with the formation of red blood cells in the bone marrow. Leukemia occurs in some individuals with long-term occupational exposures. The range in which any hazardous effects have been observed in occupationally exposed workmen has its lower bound between 5 and 25 ppm, which is two orders of magnitude greater than reported air pollution concentrations. However, the alterations in chromosomes from lymphocytes in persons who had recovered from occupational exposure to benzene, reported by Forni *et al.* (*239*), require that we treat any exposure as a serious matter.

V. Particulate Pollutants and Their Effects

A. Asbestos

The occupational hazards of asbestos were first recognized in 1924 (*240, 241*). For the mineralogy of asbestos, see Hendry (*242*) and Gaze (*243*). Observed reactions were primarily the deposition of asbestos in the lung with production of fibrotic reactions and stiffening of the lung, resulting in shortness of breath. The awareness that asbestos exposure is likely to cause increased cancer frequency dates from 1949 (*244–246*). A series of studies by Selikoff and Hammond (*247*) and Selikoff *et al.* (*248–250*) has presented the health experience of asbestos insulation workers who were followed prospectively. Among these workers, the frequency of cancer of the lung is exceptionally high, especially in cigarette smokers. The frequency of respiratory diseases, including cancer of the lung, is such as to cause the number of observed deaths of those who have worked with asbestos for 20 years or more to be nearly double the number of expected deaths. In addition, there is a rare type of tumor of the lining of the chest or abdominal cavity called malignant mesothelioma that is now recognized as being due, in most cases, to exposure to asbestos (*251–253*).

The risk of cancer in asbestos-exposed workers is apparently dose related. The occurrence of lung cancer is far more common among cigarette smokers exposed to asbestos than in cigarette smokers with no asbestos exposure. There is no detectable excess of lung cancer among asbestos-exposed nonsmokers. By contrast, the dose–response relationship for mesothelioma is poorly defined. It is felt that exposure to chrysotile (one of the mineral forms of asbestos) is less likely to lead to this type of tumor than is exposure to crocidolite. Cigarette smoking does not appear to be important in the development of mesothelioma.

Mesothelioma has occurred in persons who were not occupationally exposed but whose residence was near shipyards or other industries where asbestos is used abundantly; there is also an excess of asbestos-related disease among household members of workers heavily exposed to asbestos (*253a–253f*). Pulmonary reactions to inhaled asbestos have been widely observed in urban populations of South Africa, Great Britain, and the United States (*254*). For example, the frequency of observation of so-called asbestos bodies appears to have increased between 1936 and 1966 in London, England (*255*). A hundred consecutive autopsy examinations in 1936, 1946, 1956, and 1966 at hospitals in London have shown that the prevalence of observed asbestos bodies was 0, 3, 14, and 20%, respectively. The increase could not be explained by change in the age at death nor by other phenomena. However, in New York, New York, no increase in asbestos bodies was observed at autopsies performed in 1934 and 1967, respectively (*256*).

There is uncertainty as to whether light microscopy adequately detects the asbestos present in air samples. Electron microscopy reveals a larger number of particles than does light microscopy. Despite their length, the particles are so narrow that they have relatively small mass. A threshold limit value of 5 fibers/ml is equivalent to 12,500 ng/m^3 of 5 μm long fibers. The precision of the counting procedures for electron microscopy is relatively poor, so that even in a well-equipped and experienced laboratory, 10 ng/m^3 may mean anything between 3 and 30 ng/m^3.

Asbestos mines, mills, shipyards, fabricating plants, and spraying of steel framed skyscrapers with an asbestos-rich fire retardant mixture

Table XIII Occurrence of Mesothelioma in Occupational and Nonoccupational Asbestos Exposures in Selected Populations[a]

Total number of cases	*Clear occupational exposure*	*Home contact with occupationally exposed*	*Neighborhood nonoccupational exposure*	*No known exposure*	*Study location*	*Reference*
87	12	—	75	—	South Africa	(*253a*)
76	31	9	11	25	England	(*253b*)
21	17	—	—	4	Netherlands	(*253c*)
42	10	3	8	21	Pennsylvania	(*253d*)
17	15	—	2	—	New Jersey	(*252*)
80	50	—	6	24	Scotland	(*253e*)
165	41	14	1	109	Canada	(*253f*)
488	176	26	103	183		

[a] Populations are selected in various ways, and classification is inferred in some cases.

are likely to spread asbestos fibrils for a considerable distance. Brake lining commonly is made with asbestos, and wear of such brake lining adds to street dust. Most of the abraded material is assumed to be structurally altered so that it is no longer a health problem (*257*). That this may not be entirely true is shown by the threefold excess of asbestos fibrils found in air samples in New York City near the Hudson River Tunnel toll station compared to a control area (*258*).

Although direct, quantified, evidence of community asbestos exposure and specific asbestos-related disease is not available, the occurrence of mesothelioma in persons living in the vicinity of mines, mills, and shipyards suffices to indicate the need for careful handling of this hazardous substance. A comparable study of cancer of the lung among those with nonoccupational exposure would be difficult, owing to the concomitant role of cigarette smoking in this disease. Nevertheless, the occupational data indicate that the combined effect of asbestos exposure and smoking is more than additive.

From the evidence, it is accepted that serious reactions to asbestos may take up to 30 years between the first exposure and the appearance of a tumor. For this reason, it has been decided in the United States that there are "known serious effects of the uncontrolled inhalation of asbestos materials in industry and uncertainty as to the shape and character of a dose response curve in man. . . ." Accordingly, it is generally agreed that "it would be highly imprudent to permit additional contamination of the public environment with asbestos" and "the major sources of man-made asbestos emissions into the atmosphere should be defined and controlled" (*257*).

B. Lead

The major source of exposure to lead is from food and water, with estimated intake of 0.12–0.35 mg/day (*259*). From 5 to 10% of ingested lead in adults may be metabolically absorbed, whereas from 20 to 50% of the lead inhaled from community air pollution may be metabolically absorbed into the body (*36*). The percentage absorbed tends to be higher with smaller particle sizes. Acute forms of toxicity are manifested when blood levels in adult males are greater than 75 μg/100 gm. Kehoe provided guidelines (*260, 261*) that indicated that "absorption from all routes of 120 micrograms of lead a day might be harmful, while a total of 60 micrograms is presumably safe."

Three types of toxicity are documented—gastrointestinal cramps (lead colic), central and peripheral nervous system effects (lead encephalitis, wrist drop), and anemia. Kidney disease, excess frequency of hypertension, and vascular disease have been reported but are not universally accepted as long-term effects.

In children ingesting sufficient lead paint, anemia and effects on the brain are universally recognized. Mental retardation (*262, 263*) and hyperactivity (*264*) have been reported to be associated with high blood lead levels or history of lead intoxication, but proof of lead as the causative agent for these long-term effects is disputed, since mental retardation and hyperactivity may be a cause rather than the effect of ingestion of lead and other nonfood materials.

The so-called Three City Study (*265*) documented a substantial and persistent excess of lead in blood samples from urban dwellers as compared to rural populations. With increasing respiratory exposure, there tended to be increasing blood levels. In a 1966 study (*266, 267*) in Los Angeles, California, samples of blood were taken from two groups of 50 persons each, the homes of the first group being adjacent to a heavily traveled freeway; the homes of the second were near the sea coast, in an area with substantially less atmospheric lead. There was a consistently significant excess of blood lead among the population living near the freeway, compared to coastal residents, independent of age, sex, or ethnic status. However, the freeway values did not deviate markedly from the values found for other populations in the community as a whole (Los Angeles, California).

1. Relation of Body Burden to Ambient Air Level

Goldsmith and Hexter (*268*) found a logarithmic regression of mean blood lead of males on estimated atmospheric exposure for community and occupational groups exposed to vehicular pollution; this regression agrees closely with experimental data reported by Kehoe (*269*). The estimated nonoccupational mean exposures range from 0.12 $\mu g/m^3$ to 2.4 $\mu g/m^3$. Of seven urban population groups, only the long term exposures of those from Pasadena, California, and Philadelphia, Pennsylvania, exceeded an average of 2 $\mu g/m^3$. The groups occupationally exposed to motor vehicle exhaust had 24-hour average exposures estimated at 2.1–6.5 $\mu g/m^3$. According to the authors, these data imply that long-term increases in atmospheric lead will result in predictably higher blood lead levels in the exposed populations. A later (*259*) report by the United States National Academy of Sciences concluded that ". . . it is not possible, on the basis of epidemiological evidence, to attribute any increase in blood lead concentration to exposure to ambient air below a mean lead concentration of about 2 or 3 $\mu g/m^3$."

The so-called seven-city study (*270–272*) showed annual geometric mean values ranging from 0.14 $\mu g/m^3$ (Los Alamos, New Mexico) to 4.55

$\mu g/m^3$ (Los Angeles, California). One station in Cincinnati, Ohio, two in Houston, Texas, and two in Washington, D.C., had values averaging more than 2 $\mu g/m^3$. The data showed there were higher lead concentrations in most of the identical locations in 1968–1969 than in 1961–1962. For Los Angeles, which had the most consistent increase, the eight stations showed an increase of from 32 to 63%.

In the seven-city study, blood and some 24-hour fecal samples were obtained from women. Female smokers had higher blood lead values than nonsmokers. Urban women had higher blood lead levels than rural women and were exposed to higher levels of atmospheric lead. No significant area differences were found in fecal lead levels. Presumably, this reflects lack of differences in dietary lead ingestion to account for area differences. There was a tendency in locations with higher atmospheric lead for female residents to have higher levels of blood lead, with several communities being deviant. This result confirms for women, over a narrower range of exposure, what Goldsmith and Hexter showed for men (*268*). Tepper and Levin (*271*) concluded that "there was no clear association between air and blood lead levels. . . ," meaning that there was not a statistically significant tendency. However, Hasselblad and Nelson (*272*), carrying out an analysis for the United States Environmental Protection Agency, conclude. "The contributions of air lead gradient and urban–suburban contrast on blood levels are shown to be significant with each explaining approximately 18 percent of the total area variation in blood lead concentrations". Kehoe's experimental data have been amplified by the experimental exposure data of Knelson (*273*) at levels of 3.2 and 10.9 $\mu g/m^3$ for 18 weeks. Such levels are shown to regularly increase body lead burdens.

Regardless of epidemiological data, a theoretical approach shows that a considerable part of the body burden comes from inhaled lead. If we conservatively assume that 10% of ingested lead (about 150 μg/day) is absorbed and that 20% of inhaled lead is absorbed (2 $\mu g/m^3 \times 20$ m^3/day), this gives a contribution from ingested lead of 15 μg/day and from inhaled lead of 8 μg/day. Thus, inhaled lead would contribute about one-third of the total exposure. With lower exposure via inhalation, it would contribute less and vice versa.

2. Biochemical Effects

Hernberg *et al.* (*274*) have shown in various population groups that aminolevulinic acid dehydratase values are closely (inversely) correlated with blood lead, and Selander and Cramer (*275*) have shown that increased excretion of aminolevulinic acid in urine is evidence of increased lead exposure. The NAS–NRC report (*259*) suggests that several levels be defined (for adults). This implies that at blood lead levels below 40 μg/100 gm of blood, no demonstrable effects other than decreased aminolevulinic acid dehydratase are expected. No one has shown that this decreased aminolevulinic acid dehydratase has medical significance. From 40 to 60 μg/100 gm, a subclinical metabolic effect may be manifested by a slight increase in urinary aminolevulinic acid. Above these levels, shortened red blood cell life span can be observed. The CEC–EPA Symposium (*270*) did not produce any contrary evidence.

Children are more vulnerable to lead exposures than are adults. Their ingestion is distributed to a smaller tissue mass; their gastrointestinal absorption is estimated by Karhausen (*276*) to be 25%, as compared to 8% for adults. This estimate is consistent with the results reported by Alexander *et al.* (*277*). Some children are also exposed to ingestion of dirt and paint chips by their lesser discrimination as to what is edible; their more rapid growth rate and metabolism makes them more vulnerable. These considerations assume greater weight in light of the reported association of elevated levels of lead excretion in children with the behavior disorder called hyperactivity syndrome. These children may have blood lead levels below 40 μg/100 gm (*264*). The biological limits proposed by Zielhuis (*278*) to protect children and adults are shown in Table XIV.

Urban children have higher blood lead levels than rural ones. In one study of 230 rural children, 19 had elevated levels of blood lead, and 18 of the 19 lived in homes with at least one accessible paint surface containing more than 1% lead. A similar source of leaded paint could be identified in only 60% of the 68 out of 272 urban poverty-area children who had "elevated" levels. About one-fourth of the urban children had blood lead values above 40 μg/100 gm, compared to less than one-tenth of the rural children (*279*).*

Ninety percent of 1- to 5-year-old children living in a rural area near a lead smelter in El Paso, Texas, had blood lead values greater than 40 μg/100 gm. No paint exposure occurred in this area, and levels of lead in dirt (0.4–0.5%) are not markedly different from similar levels in city streets and parks. These elevations reflect either primary respiratory exposure or secondary exposure to re-entrained dust. Similar findings were reported by Loveless *et al.* (*280*). In two California communities with

* Efforts to detect and prevent excess lead exposure in children, based on screening for free erythrocyte porphyrins and blood lead, have been mounted in the U.S. Values in urban areas in excess of 35 μg/100 gm are so frequent that lead-based paint exposures are unlikely to account for them. Direct air pollution exposures, and orally ingested lead in dust as a secondary pollutant, are likely to be assigned a major role.

Table XIV Acceptable Limits[a] of Lead and Its Biochemical Indices Proposed as Guidelines for Public Health Hazard Appraisal (*278*)

	Individual upper limit	*Group average*	*Unit*
For adults			
Blood Lead (PbB)	≤40	≤25	μg Pb/100 ml
Urine ALA[b]	≤ 6	≤ 3	mg/liter urine
Blood ALA dehydratase	≥20	≥30	Percentage decrease from 100% at PbB = 10 μg Pb/100 ml
For children			
Blood Lead (PbB)	≤35	≤20	μg Pb/100 ml
Urine ALA	≤ 5	≤ 3	mg/liter urine
Blood ALA dehydratase	≥30	≥40	Percentage decrease from 100% at PbB = 10 μg Pb/100 ml

[a] These limits should not be regarded as fine lines discriminating innocuous from harmful environments; however, if these limits are exceeded, the possibility of undue acute effects and/or chronic sequelae increases. The data for population groups should be interpreted by experts who can judge the representativeness of the population sample. Single values moderately exceeding upper limits in individuals should cause awareness and caution; the measurement should be repeated.

[b] ALA–amino levulinic acid.

similar (high) socioeconomic status, children living in the community with higher lead values in air had higher blood levels (*281*) (Table XV).*

Thus, for acute lead poisoning in children, much of the increase in body burden of lead and most of the damaging effects to the nervous system are due to lead paint; air pollution, directly or through increased lead levels of dirt in streets, parks, and other areas where children play, can make a measurable contribution to elevated lead burdens in children.

Table XV Lead Concentration in Air and in Blood of Primary School Children in Two Comparable California Communities—1972 (*281*)

Community	*Lead concentrations in air* ($\mu g/m^3$)[a]	*Lead in blood of children* ($\mu g/100$ gm)[b] *Male*	*Female*
Burbank	3.27 ± 1.59 (10)	23.3 ± 4.70 (17)	20.4 ± 2.91 (19)
Manhattan Beach	1.87 ± 1.37 (10)	16.8 ± 4.01 (21)	17.1 ± 4.37 (19)

[a] Mean ± standard deviation; number in parentheses indicates number of samples.

[b] Mean ± standard deviation; number in parentheses indicates number of children. Differences between blood levels of both sexes are statistically significant at the 1% level.

* Roels *et al.* (*281a*) studied the biochemical impact of lead exposures on both children and adults near a smelter in Belgium. Their data supports the sensitivity of the test for free erythrocyte porphyrins (FEP) as an index of lead exposure. Based on this test they recommend a maximum lead in blood of children of 25 μg/100 gm.

Table XVI Weighted Average Weekly Exposure to Lead for Occupational and Nonoccupational Groups in Zagreb, Yugoslavia *(282)*

A. Lead-exposed occupational groups

Population group	Concentration (μg Pb/m³) × time (hours/week): Occupational	Nonoccupational: Outdoor	Nonoccupational: Indoor	Total concentration × time	Weighted average weekly exposure (μg/m³)
Lead smelter workers	650[a] × 42	40 × 35	24 × 91	30,884	184
Inhabitants of lead smelter areas	—	40 × 35	24 × 133	4,592	27
Lead article manufacturers	139[a] × 42	0.3 × 14	0.1 × 112	5,851	35
Customs officers	6.2 × 42	0.3 × 14	0.1 × 112	275.8	1.64
Traffic policemen	8.2 × 35	0.3 × 14	0.5 × 124	437.2	2.60
Streetcar drivers	3.9 × 42	0.3 × 14	0.5 × 112	308.0	1.83

B. Average citizen not occupationally exposed (on basis of location)

Location	Average Concentration (μg Pb/m³)	Time (hours/week)	Concentration × time	
Working place	1.2	42	50.4	
Outdoor activities	6.3	14	88.2	
Recreation	0.2	6	1.2	
At home				
Rest of day	0.7	22	15.4	
Night	0.3	48	14.4	
Weekends	0.5	36	18.0	
Nonoccupational exposure:				1.1

[a] Respirable fraction.

Fugas *et al.* (*282*) show a weighted average weekly exposure of 1.1 μg/m³ lead for citizens of Zagreb, Yugoslavia (Table XVI). This represents a useful model for evaluating community lead exposures.

C. Cadmium (Friberg)

1. Sources

The cadmium emitted from industrial and domestic sources either will be inhaled by humans or animals or will be deposited in soil, vegetation, or water. Irrigation with cadmium-contaminated water and the use of cadmium-containing fertilizers and sludge from sewage treatment plants can build up cadmium levels in soil so that uptake of cadmium in the growing plant will be increased. Wheat and rice have been shown to absorb significant amounts of cadmium (*283, 284*). Cadmium in paints and plastics may sooner or later end up in incinerators whose effluent may become airborne. Cadmium has been measured at up to 1% in children's plastic toys, but the risk for children who might chew on the toys is small, since the cadmium seems to be firmly bound to the plastic.

The normal concentration of cadmium in air is about 0.001 $\mu g/m^3$ and will not contribute significantly to the daily intake of cadmium. Cadmium concentrations in air around cadmium-emitting factories, however, may be several hundred times greater (*285*). This can increase the concentration of cadmium in moss, as has been shown near a factory producing copper–cadmium alloys in Sweden (*286*), and Kobayashi (*286a*) has shown the same process affecting mulberry leaves in the vicinity of a Japanese cadmium smelter.

Smoking was first mentioned as a source of cadmium exposure in 1969 (*287*). A mean content of 1.5 μg per cigarette was found in a comparison among eight types of West German cigarettes. Of this amount, 0.1–0.2 μg of cadmium could be inhaled by smoking one cigarette. In a pack-a-day smoker, this would add up to 2–4 μg of cadmium per day.

2. Uptake and Metabolism

A toxicological and epidemiological appraisal of cadmium in the environment has been published by Friberg *et al.* (*285*), and those who wish a comprehensive treatment of the subject are referred to that monograph. Absorption via inhalation will vary with different chemical forms of cadmium and different particle sizes. Some data from animal experiments tend to show that the respiratory uptake may vary between 10 and 40% (*288, 289*). The gastrointestinal absorption ratio is much lower (*290, 291*). Animal experiments show an absorption of about 1–3%. Rahola *et al.* (*292*) have studied the uptake of radioactive cadmium in humans after a single exposure. They found uptakes varying from 4.7 to 7.9%.

Once absorbed, cadmium will be found in blood, organs, and excreta. With chronic exposure, the cadmium in blood will be found primarily in the red cells (*293*). A major part of the cadmium will be transported to the liver and kidneys, where it accumulates. These two organs at chronic low-level exposure contain as much as 75% of the retained cadmium. High concentrations of cadmium are found also in the pancreas and the salivary glands. Only low concentrations of cadmium have been found in the bones and in the central nervous system (*285*).

Cadmium in the body is mainly bound to a low molecular weight (6000–7000) protein, metallothionein. This protein is involved both in the transportation and selective storage of cadmium. It can bind up to 11% metal as cadmium or zinc due to the fact that it has a large number of sulfhydryl groups. This protein is found in large amounts in liver from cadmium-exposed animals, but similar proteins also have been found in the gastrointestinal tract, and it is possible that the transport from the gastrointestinal tract takes place with cadmium bound to proteins similar to metallothionein. As metallothionein has a low molecular weight, it can pass the glomerular (filtering) membrane of the kidney and then be almost completely reabsorbed in the proximal tubules as long as kidney function is normal. This probably explains the rather selective accumulation of cadmium in the kidneys.

Normally, only small amounts of cadmium are excreted with the urine. When tubular dysfunction of the kidneys appears in experimental animals due to the accumulation of cadmium in the kidneys, the urinary excretion of cadmium increases drastically (*294, 295*). Normal human cadmium excretion in urine is usually less than 2 μg/day. Excretion via feces is probably of the same order of magnitude (*296–300*). The biological half-time for cadmium in the kidney is very long, probably of the order of decades. This means that for a continuous exposure, it will take a very long time to reach a steady state. Whole-body accumulation does not follow a one-compartment exponential model. There is reason to believe that, for example, the biological half-time of cadmium in the liver is shorter than in the kidneys (*12, 285*). The newborn are practically free of cadmium. Henke *et al.* (*301*) found a range of 4–20 ng/gm wet weight in kidneys and less than 2 ng/gm in the liver from the newborn, making a total body burden of less than 1 μg. This indicates that the placental barrier is effective at chronic low level exposure.

Normal human exposure results in an accumulation in the kidney such that at the age of 50 it will give rise to a kidney cortex concentration of about 25–50 μg/gm wet weight of cadmium in the United States males. As about one-third of the total body burden is in the kidney, this means a total body burden of about 15–30 mg. Corresponding values in Sweden seem to be about 20–30 μg/gm of kidney cortex; and in areas not particularly contaminated with cadmium in Japan, mean values of up to 100 μg/gm kidney cortex have been shown. After age 50, cadmium concentration in the kidneys decreases. The reason for this is not yet known (*285, 302, 303*).

From both animal and human occupational data, it appears that long-term exposure may be associated with very low cadmium values in urine as well as with high values. Similar evidence comes from a normal population in Japan (*304*). Available data seem to indicate that the urinary concentration reflects kidney levels rather than liver levels. Piscator (*305*) could not find a correlation between exposure time and urinary cadmium values in workers without proteinuria. Piscator (*298*) and Miettinen (*299*) report efforts to relate blood cadmium to exposure time and to accumulation in the kidney. Blood values may reflect the most recent exposure, but they decrease more rapidly than the body burden of cadmium.

3. Effects

Inhalation of high concentrations of cadmium dust or fumes in industry may lead to kidney damage and lung damage (*289*). The most commonly found lung damage is of an emphysematous nature. The kidney dysfunction in chronic cadmium poisoning starts as a tubular dysfunction with proteinuria of a tubular type (*305*). This proteinuria is the result of defective reabsorption of low molecular weight proteins in the glomerular filtrate, and albumin will constitute, therefore, only a minor part of the total urinary proteins. Common clinical methods for detecting proteinuria, such as dipstick, boiling test, or nitric acid test, will not give positive reactions for this type of proteinuria in its early stages. Trichloracetic acid or sulfosalicylic acid should be used when screening for tubular proteinuria. To confirm the diagnosis of tubular proteinuria, electrophoretic separation of urine protein should be performed.

Since tubular proteinuria is an early sign of intoxication, screening for this dysfunction may prevent more severe damage. On the other hand, when tubular proteinuria has appeared, it is usually permanent. In a limited number of workers with proteinuria, no progress was seen when the workers were transferred to "cadmium-free" jobs. Apart from proteinuria, there may be an increased excretion of amino acids, glucose, phosphorus, and calcium. The disturbance in calcium and phosphorus metabolism may lead to osteomalacia (thinning of the bones), in turn leading to a serious disease syndrome characterized by severe pains due to multiple fractures. Such effects have been seen in industry (*306–308*) and, due to an excessive exposure from the general environment, in nonindustrially exposed persons living near the Jintzu River, Toyama, Japan. The disease in Japan has been given the name itai-itai disease (ouch-ouch disease). The occurrence of itai-itai cases in the Toyama area is only one aspect of cadmium poisoning in Japan. A large number of cases with tubular proteinuria, but without bone changes, have been found. There may well be other contributory factors in Japan for the severe manifestations of cadmium poisoning. In the Toyama area, the calcium intake is low, the fat intake is low, and the number of hours of sunlight are few. In all probability there is thus a low intake of vitamin D. The relation between cadmium and the itai-itai disease has been discussed extensively in the monograph by Friberg *et al.* (*285*).

Both animal and human data favor a critical concentration for proteinuria of about 200 μg cadmium per gram kidney cortex (*285*). The concentration of cadmium in the renal cortex in normal persons, exposed workers, and itai-itai patients is seen in Figure 22. The figure clearly shows the increase with age in normal persons. The cadmium levels in exposed workers are either higher or of the same magnitude or even lower than those in normal persons. It has been possible to show that when tubular dysfunction appears, the cadmium excretion increases, which explains the fact that the more severe the intoxication, the lower the kidney levels. This is not true for the liver, where all values in cadmium intoxicated workers and persons with itai-itai disease are considerably higher than normal (Fig. 23).

Cooper *et al.* (*309*), in multiphasic examinations of former lead smelter workers who also have had a substantial cadmium exposure, find evidence of tubular dysfunction that may also have been contributed to

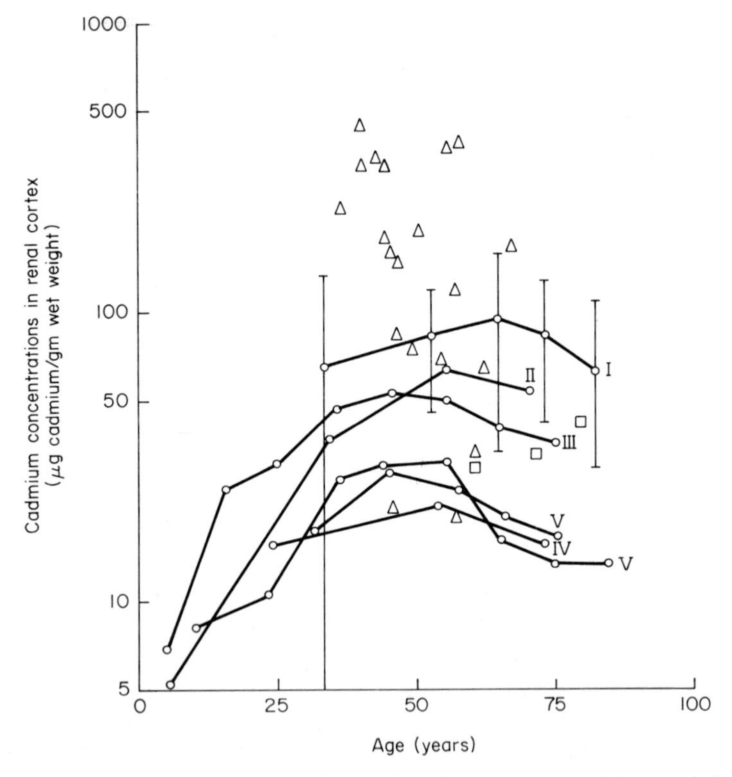

Figure 22. Cadmium concentrations in renal cortex from normal human beings in different age groups and different countries: I. Kanazawa, Japan; II. Kobe, Japan; III. United States; IV. United Kingdom; V. Sweden (two areas). Circles are mean values, vertical bars indicate range; △ = exposed workers (single values); □ = itai-itai patients (single values) (*285*).

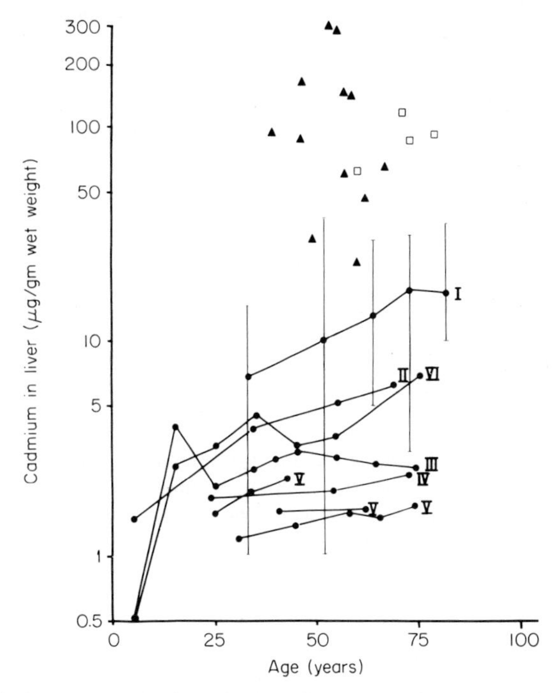

Figure 23. Cadmium concentrations in liver from normal human beings in different age groups and different countries: I. Kanazawa, Japan; II. Kobe, Japan; III. United States; IV. United Kingdom; V. Sweden (three areas); VI. Tokyo, Japan. Circles are mean values, vertical bars indicate range; ▲ = exposed workers (single values); □ = itai-itai patients (single values) (*285*).

by lead or other metals. Cadmium may also give rise to dysfunction of the liver and to anemia. In animal experiments, injection of single doses of cadmium can give rise to a complete necrosis of the testicles, but since such exposure does not occur in human beings, this type of damage probably does not occur in man. Industrial exposure to cadmium has been associated with an increase of lung and prostate cancer. The causal association with lung cancer is very weak and with prostate cancer is not conclusive. Animal data have shown that cadmium may give rise to a significant number of tumors at the site of injection.

Of paramount importance for the future is whether or not there is a risk that the cadmium concentrations in the kidney will increase from community exposure. If such is the case, steps must be taken very soon due to the very long biological half-time of cadmium. In order to make a prognosis, much more information is needed. There is a need to determine whether or not cadmium exposure will increase. More information on the biological half-time of cadmium is needed. The total cadmium exposure required for reaching a kidney cortex concentration of 200 μg/gm (the "critical concentration"), using different alternatives for biological half-time in the kidney cortex and for exposure times, is given in Table XVII, which is taken from Nordberg (*310*) and compiled from data by Friberg *et al.* (*285*).

D. Mercury (Friberg)

Mercury occurs in the environment in several physical and chemical forms. The most obvious distinction is between the inorganic forms and

Table XVII Cadmium Exposure Required for Reaching a Kidney Cortex Concentration of 200 μg Cd/gm Using Different Alternatives for Biological Half-Time in Kidney Cortex and Exposure Time (*285, 310*)

Basis of calculation[a]		*Exposure time (years)*	*Estimated biological half-time in kidney cortex (years)* 38	18
			Daily retention (μg)	
Constant daily retention during whole exposure time		10	36	39
		25	16	20
		50	10	13
			Industrial air concentration (μg/m³)	
25% pulmonary absorption, 10 m³ inhaled per work day, 225 work days/year		10	23	25
		25	11	13
			Daily cadmium intake (μg)	
Food exposure for 50-year-old person (2500 cal/day; 4.5% retention; changing caloric intake by age accounted for)		50	250	360
			Corresponding average concentration in foodstuffs (μg/gm)	
Total amount (w.w.) of food/day	300 gm	50	0.8	1.2
	600 gm		0.4	0.6
	1000 gm		0.25	0.35

[a] Assumption: One-third of whole body retention reached kidney and kidney cortex concentration 50% higher than average kidney concentration.

the organomercurials. The term "inorganic" here refers to elemental mercury vapor, mercurous and mercuric salts, and complexes in which mercuric ions can form reversible bonds to tissue ligands, e.g., thiol groups on proteins. "Organic mercury compounds" are those in which mercury is linked directly to a carbon atom by a covalent bond. The organic mercury compounds are further subdivided into the alkyl, aryl, and alkoxyalkyl groups. There is a great variation in toxicity, especially among organic forms. Two of the alkyls, methyl- and ethylmercury, for example, have been shown to be extremely toxic. Phenylmercury, an aryl form, is less toxic and could be compared to some of the inorganic forms in this respect. Phenylmercury, like inorganic forms, is converted to mercuric ion in the animal or human body, whereas the alkyls are not. A toxicological and epidemiological appraisal of mercury in the environment has been published by Friberg and Vostal (*317*). Those who wish a comprehensive treatise on the subject are referred to that monograph.

1. Sources

The largest source of atmospheric contamination is probably the burning of coals and fossil fuels (*311*). This is estimated to result in a world-wide total of 3000 tons of emitted mercury per year. Smelting and refining processes in connection with mercury mining can add to the environmental load, as can geothermal sources. The alkyl compounds have been used mainly in agriculture for seed dressing.

2. Uptake and Metabolism

A. Metallic Mercury. Metallic mercury enters the body primarily via inhalation. When elemental mercury is spilled in the work place or the home, vapor inhalation can occur. Elemental mercury is not absorbed to any significant degree by the gastrointestinal tract. Elemental mercury can easily diffuse from blood into tissues. Experiments in animals have shown high uptake of mercury by the brain after exposures to mercury vapor. Following absorption in the blood, metallic mercury is oxidized to mercurous ion, which is unstable and is converted into mercuric ion. Ionized mercury joins complexes with sulfhydryl groups and other ligands in the body tissues. Because this process takes some time, mercury is initially transported from the lungs to the tissues mostly as physically dissolved mercury vapor. After oxidation, the blood cells and plasma contain almost equal amounts of mercury. The placental transfer of elemental mercury has not been studied in detail. Since elemental mercury penetrates membranes easily, the placental barrier should not hinder it. Once ionized, it will act as other forms of inorganic mercury, which in animal studies have been shown to be stopped by the placental membrane.

There is uneven distribution of absorbed metallic mercury in the body. A considerable amount penetrates the blood–brain barrier. The brain is the critical organ after exposure to metallic mercury. The potential for a mercury accumulation in the brain after mercury exposure is apparent from autopsy data by Takahata *et al.* (*312*) and Watanabe (*313*), who studied brain specimens from two mercury mine workers who died of pulmonary tuberculosis. Though they died 6 and 10 years after the end of exposures, which had lasted a little over 5 years, the occipital cortex, parietal cortex, and substantia nigra still contained high concentrations (15 ppm or above in all cases) of mercury. Even within the brain the distribution is very uneven.

In a Swedish study (*314*), daily excretion in urine and feces as well as yearly retention were calculated for 30 workers exposed to mercury vapor in the chloralkali industry. Absorption was measured by 4-day running average sampling in the breathing zones of the workers. The calculated mean mercury exposure was 0.1–0.2 mg/m^3 in 10 workers' breathing zones. Their mean daily mercury outputs were 0.19 mg in urine and 0.14 mg in feces. Their yearly retention was calculated to be a mean of 51 mg.

b. Inorganic Mercury Compounds. Aerosols of inorganic mercury compounds are absorbed via the respiratory tract to a considerably lesser degree than is mercury vapor (*315*). On the other hand, there are reports of poisoning after exposure to inorganic mercury aerosols (*316*). The distribution of mercury following inorganic mercury compound exposure is also very uneven, but it differs from the distribution after exposure to metallic mercury in that the values in the brain are about ten times lower. The kidney, which generally contains the highest concentration of mercury, is the critical organ for exposures to inorganic mercury compounds. After exposure to inorganic mercury compounds, the risk of penetration of the placental barrier is considerably less than after exposure to metallic mercury.

c. Methylmercury. There are no quantitative data concerning the respiratory uptake of methylmercury compounds. Since several methylmercury salts vaporize relatively easily at room temperature, the risk may be considerable. Animal experiments as well as human data from industry (*317*) have shown without doubt that inhalation of methylmercury compounds can give rise to severe intoxication. Outside industry, gastrointestinal absorption is probably the most relevant route of methylmercury compound absorption. All recent poisonings have involved ingestion of contaminated food, mostly fish or grain. More than 90% of ingested methylmercury salts or proteinates can be absorbed, as seen in whole-body studies on human volunteers (*318, 319*). The biological halftime of methylmercury compounds seems to follow a single exponential model and averages about 70 days (*318, 319*).

3. Effects

a. METALLIC MERCURY. The most common effects after chronic exposure to metallic mercury are tremor and erethismus mercurialis (behavioral disturbances). Oral manifestations may include gingivitis and stomatitis (soreness and inflammation of mouth and gums). Occasionally, proteinuria including a nephrotic syndrome (*320, 321*) may occur. In the Soviet Union, much attention has been paid to the occurrence of micromercurialism (*322*). This is an asthenic–vegetative syndrome consisting of nonspecific signs such as abnormal tiredness, irritability, impaired memory, and loss of self-confidence. Included in the syndrome of micromercurialism are functional changes in organs of the cardiovascular, urogenital, and endocrine systems. Trachtenberg (*322*) believes that the symptomatology as a whole originates from damage to the cortical centers of the central nervous system.

For mercury in air and its effects, there are to date no large-scale environmental and epidemiological studies. In workmen occupationally exposed to mercury, some mercury may be found in exhaled breath samples, and skin and clothing may absorb mercury and disseminate it after the workman comes home to his family. Thus, an exposed workman may become a source of pollution in his own home. Industrial dose–response information is scanty. The most comprehensive study is by Smith *et al.* (*323*). They examined 567 male workers exposed to mercury vapor in chlorine production and 382 male controls. Signs and symptoms from the nervous system revealed a clear dose–response relationship. At even the lowest exposure (less than 0.01–0.05 mg/m^3), some symptoms were increased (Fig. 24).

In Russian studies (*322*), thyroid effects (increased uptake of iodine-131) have been stated to have occurred after some months of exposure to very low concentrations of mercury (0.01–0.02 mg Hg/m^3). Changes in conditioned reflexes have been reported at even lower concentrations in some Soviet studies. The Russian studies are discussed in detail by Trachtenberg (*322*) and by Friberg and Nordberg (*324*). As is noted in the latter, the medical significance of the reported changes is difficult to evaluate.

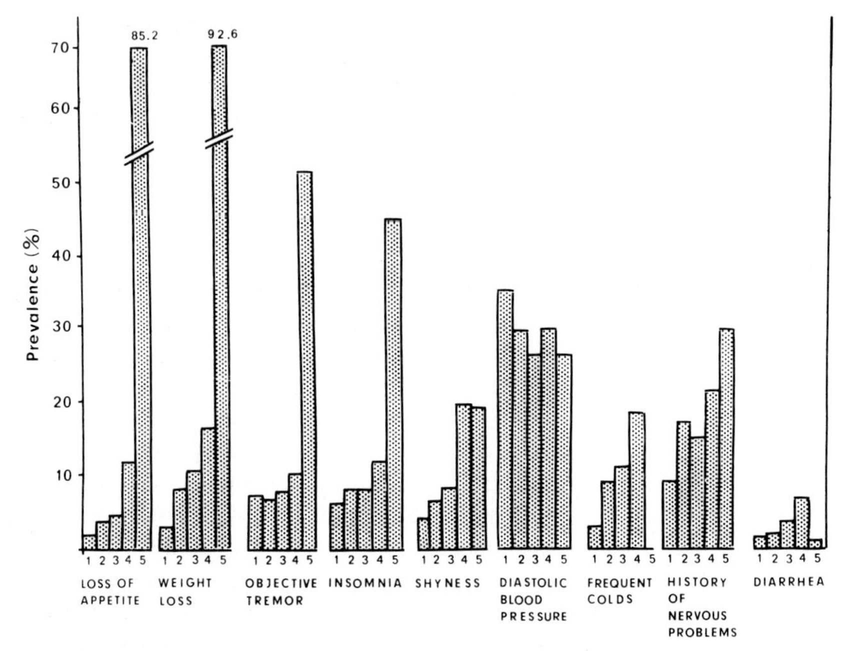

Figure 24. Percentage prevalence of certain signs and symptoms among workers exposed to mercury in relation to degree of exposure (*323*). Data on diastolic blood pressure probably mean percent below a certain level. This level is not given in the article. The numbers below each column refer to average exposures shown in milligrams per cubic meter: (1) control, (2) <0.01–0.05, (3) 0.06–0.10, (4) 0.11–0.14, (5) 0.24–0.27.

b. METHYLMERCURY. Methylmercury poisoning can occur prenatally, because methylmercury passes the placental barrier. Several cases from Minamata, Japan, are known. Symptoms are indistinguishable from those seen in cerebral palsy. In postnatal poisoning, cerebral and cerebellar symptoms dominate. Symptoms may include ataxia, constriction of visual fields, and sensory disturbances. Hearing may be severely impaired. The prognosis is often poor. Skerfving *et al.* (*326*) have reported an increased chromosome breakage in human lymphocyte cultures from persons without any symptoms, but with increased mercury blood levels due to excessively high consumption of fish with low-level mercury contamination. Perhaps the most well-known cases are the ones from Minamata and Niigata in Japan (*324a*), but the most extensive epidemic of methylmercury poisoning was in Iraq in 1971–1972, where 6,000 patients were admitted to hospitals and 500 deaths in hospitals were reported (*310, 324b*). A recent evaluation by a WHO task group on the environmental health criteria for mercury has shown that the first signs of methylmercury intoxication appear at concentrations of mercury (measured as total mercury) of 200–500 ng/ml in blood or 50–125 μg/gm in hair. These values correspond to a long-term daily intake of 3–7 μg/kg body weight of mercury as methylmercury (*325*).

4. Mercury in Urine and Blood as Indicators

There are no good biological indicators for evaluating long-term risks of intoxication after inhalation of mercury vapor. Neither mercury in blood nor in urine is satisfactory. On a group basis, there is an association between exposure and blood levels as well as urinary levels. Probably, the concentrations in blood and urine do not reflect the concentrations in critical organs but, instead, recent exposure. On an individual basis, the scatter is very large. For evaluating recent exposures, blood and urinary mercury probably are of importance. An exposure to about 0.1 mg Hg/m³ of air corresponds to an average of 0.2 mg Hg/liter of urine (*323*) (Fig. 25). For methylmercury, blood levels seem to reflect the concentration in the critical organ and thus are good indicators. There is also a good correlation between exposure and blood concentration. To reach a critical concentration in the brain and the blood, the corresponding daily intake of mercury as methylmercury will be about 300 μg for a 70 kg man, corresponding to about 4 μg/kg body weight.

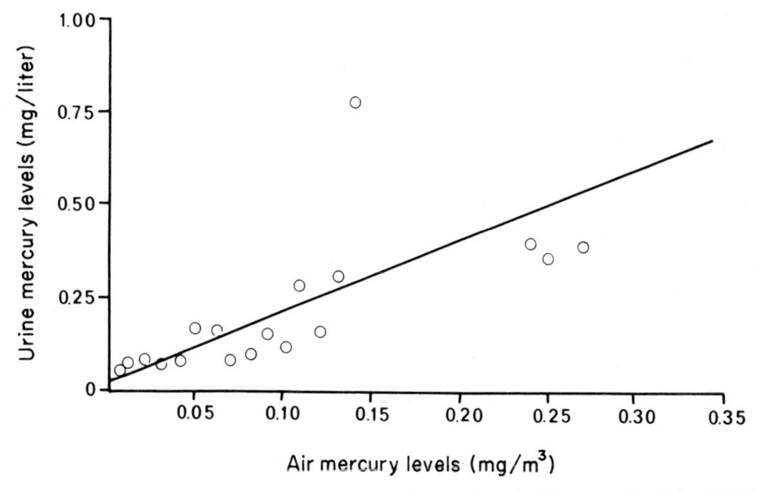

Figure 25. Concentrations of mercury in urine (uncorrected for specific gravity) in relation to time-weighted average exposure levels (*323*).

5. Normal Values of Mercury in Blood, Urine, and Hair as a Basis for Estimation of Exposure

It has been proposed in a World Health Organization report that 30 ng/ml of blood should be regarded as the highest acceptable "normal concentration" of mercury (*327*). This was based on a survey of 812 whole blood samples from fifteen countries. Not all the populations sampled were similar, nor were all individuals free from some excess ingestion of some pollutants. In no instance did the survey take fish consumption into account. All samples were analyzed in the same laboratory by atomic absorption. Of the subjects giving blood who were considered to have no occupational or unusual exposure, i.e., "normal," 85% had values below 10 ng/ml. In the same study, the general level of mercury in urine was 0.5 μg/liter based on the same method—atomic absorption spectrometry—1107 samples. Twenty micrograms per liter was regarded as the upper limit of normal levels. For blood and urine, no differences correlating with age, sex, or urban residence could be found.

No such international study on hair levels exists, but there are large samplings from several individual countries. In 776 samples in Canada, the range was 0–19 μg/gm scalp hair (*328*) by neutron activation analysis. The mean was $\sim$1.8 μg/gm. In a Japanese study (*329*) on 73 samples, using the dithizone method, the mean was 6 μg/gm while the range, 0.98–23 μg/gm, was similar to that of the Canadian study.

E. Beryllium

Occupational beryllium exposures produce a chronic granulomatous disease, primarily affecting the lung. An acute pneumonic disease is also known. Cases of chronic berylliosis have been reported with some frequency in the vicinity of plants using or manufacturing beryllium. Hardy, who first clearly established the long-term toxic nature of the material, maintained a beryllium registry. In one study, of 60 persons with nonoccupational berylliosis, 18 were exposed to polluted air in the vicinity of beryllium plants (*330*). Other nonoccupational exposures were to beryllium phosphors, at one time commonly used in fluorescent lamps; they are no longer used. Other studies (*331*) show a higher proportion of cases from neighborhood exposures. Extensive use of beryllium in nuclear reactors led to much experience by the United States Atomic Energy Commission, which established a limit for beryllium exposure in the ambient air of 0.01 $\mu g/m^3$ averaged over a 30-day period. As a result of this experience and the lack of any reported cases of disease when concentrations were held below this level, the United States National Academy of Science concluded that this has been proven to be a safe level (*332*). Beryllium has been classified by the United States Environmental Protection Agency as a "hazardous air pollutant" (*333, 334*). A possible source of community beryllium exposure has been associated with gasoline or gas lamp mantles, which are made of fabric containing beryllium oxides (*335*). Beryllium exposure by inhalation will induce lung cancer in rats and monkeys (*336*) but has not been thought to be a human carcinogen (*332, 337–339*). However, Mancuso has put forward some evidence to the contrary based on occupational exposure data (*340*).

F. Arsenic

Arsenic may occur in the pentavalent or the trivalent form, the former having low toxicity and the latter relatively high toxicity. Emissions to air will largely be in the trivalent form (As_2O_3). Elevated arsenic levels in particulate matter have been found in the vicinity of coal-burning power plants in Czechoslovakia, where average concentrations of 19 $\mu g/m^3$ were found (*341*), in India, near metal mines (*342*) or smelters, in the United States (*343–345*), and in Chile (*346*).

In Czechoslovakia, (*341*) destruction of bee colonies was the first evidence of emissions after a power plant burning high-arsenic coal went into operation. Later water pollution by arsenic, elevated levels of arsenic in the hair of nearby residents, and decrease in blood counts of exposed children were shown to occur. In the studies by Birmingham *et al.* in Nevada (*344*), pustular skin eruptions in children were found. Milham and Strong (*345*) found that urinary arsenic was elevated in persons living in the vicinity of a smelter near Tacoma, Washington. In Chile, significant absorption of potentially toxic material was noted. In addition, skin sensitization to sodium arsenate was shown by patch tests among those not occupationally exposed. Long term exposure to a high concentration of arsenic trioxide via inhalation or ingestion is known to cause cancer of the skin. There is also some evidence for arsenic causing cancer of the lung (*347*).

Arsenic is found in cigarettes, the amounts having been considerably higher some decades ago when arsenates were used as insecticides in tobacco fields (*348*). Tobacco grown in fields where arsenates had been used years previously still contained substantial amounts of lead and arsenic, and cigarette smokers who use cigarettes made from such tobacco manifest elevated levels of such metals.

Increased body burden of arsenic is the best early guide to the need for control. Of the various indices, measurement of the arsenic content of hair is fairly well standardized (*349*). Hammer *et al.* (*350*) showed that the geometric mean for arsenic content of hair from fourth grade school boys varied from 0.3 ppm in an unpolluted community, to a high of 9.1 in a community with a lead–zinc smelter. Elevated lung cancer rates have been reported to occur among both men and women living in U.S. counties with non-ferrous metal smelters, but it is not yet proven that exposure to arsenic was the cause (*350a*).

G. Fluoride

Fluoride concentrations in community air occasionally have reached levels where absorption by children leads to dental mottling (*351, 352*), but the major biological effects of fluoride emissions are on vegetation (see Chapter 4, this volume) and on livestock that consume fodder high in fluoride (see Chapter 5, this volume).

H. Chromium

Trivalent chromium has low toxicity and is the form in which chromium most often occurs naturally. Hexavalent chromium has high toxicity, and its presence is usually associated with industrial activities. Both forms can occur in air (*353*). Occupational exposure to hexavalent chromium compounds, which are both irritant and corrosive, has caused symptoms in both the respiratory tract and the skin. Nasal perforation has been a common finding in workers exposed to chromates. It is also well established that lung cancer may result from occupational exposure to chromates (*354*).

In one study of effects on the general population of emissions from a chromate-producing plant in Japan, "light" cases of pharyngitis were found at a concentration of chromic acid of $<1.5\ \mu g/m^3$ (*355*). Exposure to trivalent chromium compounds has not been found to cause any significant effects. Whereas in most human body organs chromium levels do not change or even decrease during a lifetime, there is an accumulation with age in the lungs, indicating that chromium has a long biological half-time in that organ. In the Soviet Union, the threshold concentration for chromium to cause nasal irritation and reflex activity has been found to be $2.5\ \mu g/m^3$.

I. Manganese

Manganese is considered as an essential element for man and other animals, which means that some manganese is critical to normal biochemistry and physiology. Both manganese deficiency disease and manganese toxicity are known, although the deficiency state has not been identified for man but has for several animal species.

It has been suggested (*356*) that similarities between a relatively common chronic nervous system disorder, Parkinson's disease, and chronic manganese poisoning are close enough to point to a possible role of manganese ingestion or metabolism in some cases of the disease. A form of lobar pneumonia was reported among inhabitants of a Norwegian town contaminated by emissions from a metallurgical plant producing ferromanganese. Although industrial exposures to manganese have been reported in the United States with neurological effects, no community exposures have been reported to have such effects in the United States. Persons industrially exposed manifest psychological disturbances and disorders of coordination, vaguely similar to patients with Parkinson's disease.

Manganese has been shown to be a useful additive to fuel oil for reducing particulate emissions. It also has some desirable antiknock properties as a gasoline additive as MMT (methylcyclopentadenyl manganese tricarbonyl). Although no community air pollution problems from manganese are now known, if its use as a fuel additive should increase, to as much as 0.5–2.0 gm of manganese per gallon of gasoline, the amounts in the air will surely increase. The possibility of such a problem should be evaluated before such additives are widely used.

VI. Air Pollution as a Causal Factor in Chronic Disease

A. Definition of Terms

For this discussion, "causation" is defined as a high probability that a specified air pollution exposure of a specified population will augment the incidence or prevalence of a specified condition. To meet this definition, two measurements are required—the exposure and the augmentation of the condition. Variation in population susceptibility must be assumed. Judgment of causation of long-term human reactions is made on data of association and not data of experimentation. Hence, absolute proof of causation of chronic human disease is not ethically possible. Judgment as to the magnitude of the probability and the extent to which extraneous variables have been controlled will always be an important element in interpretation of possible causation.

It will be convenient to express the probability of air pollution being a causal factor in terms of "likely," "possible," and "suggested." A "likely" causal factor would be one for which there is no substantial doubt as to causation. A "possible" causal factor would be one for which there is doubt on one or more points that additional information could clarify. A "suggested" causal factor would be one for which evidence is not yet adequate for a firm judgment.

Among the factors that make for a high probability of disease causation are:

a. The number of different populations for which a similar association is observed, including different types of people, locations, times of year, and climates.

b. A progressive augmentation in reaction with a progressive increase in exposure to air pollution.

c. A decreased frequency of the condition associated with decrease in exposure.

d. A mechanism that can be hypothesized for an observed association, and which correctly predicts other associated changes.

The lack of these supporting factors, when they have been looked for, correspondingly weighs against causation. We are not, in any event, dealing with simple causal systems, whereby some exposure is necessary and sufficient for the production of disease. Instead we are likely to be dealing with a causal network of many stages and factors. Some of these stages have been suggested in the case of environmental agents and cancer by Armitage and Doll (*357*) and by Neyman and Scott (*358*). Proposals for path analytical treatment of these problems have been put forward by Goldsmith and Berglund (*359*).

B. Bronchitis and Emphysema

The respiratory system has two classes of possible reactions to air pollution readily recognized by physicians: (a) acute reactions such as the irritative bronchitis that was found among the victims of the Donora, Pennsylvania, and London, England, episodes; the impairment of ventilatory function observed among bronchitic persons in Los Angeles, California, or episodic asthma such as found in New Orleans, Louisiana, and Minneapolis, Minnesota, and (b) chronic diseases such as chronic bronchitis and pulmonary emphysema.

Standardized questions and thoroughly tested instructions have come into use to determine by a uniform method whether a subject has a persistent cough and/or sputum, whether there is shortness of breath on hurrying on the level or walking up a slight hill, and whether there are acute respiratory conditions superimposed on this. In the United States, simple chronic bronchitis is defined as chronic or recurrent increase in the volume of sputum (mucoid bronchial secretion) sufficient to cause expectoration, while in the United Kingdom, it has become customary to consider that persistent cough and sputum with at least one episode of increased symptoms lasting 2 weeks or more within a period of 2 years is equivalent to a diagnosis of chronic bronchitis. Along with these symptoms, it is customary to estimate pulmonary function. It is conventional in the United States to consider that persisting cough and sputum with shortness of breath are evidence of the possible presence of emphysema or advanced bronchitis.

Emphysema is a condition with unevenly distended air sacs of the lung, usually with loss of lung tissue. Emphysema and chronic bronchitis are commonly associated. Pulmonary function tests are used for diagnostic purposes in both conditions. For emphysema, microscopic or macroscopic study of the lung is the basic criterion, while symptoms of cough and sputum production are the basic criteria for chronic bronchitis.

1. Epidemiological Patterns

The changes in respiratory function most sensitive to acute inhalation exposure are those in airway resistance and in the closing volume (*360–362*). Somewhat less sensitive are changes in the forced expiratory volume in 1 second (or $\frac{3}{4}$ second) and in the maximal rate of expiratory air flow (*363, 364*). Measurement of changes in diffusion capacity and carbon monoxide uptake are relatively elaborate and not proportionately more sensitive as indicators of chronic bronchitis and emphysema. Vital capacity is the least sensitive of the commonly used tests for detection of bronchitis and emphysema. Changes on X-ray are of some interest, but, by most techniques, are nonspecific.

Unfortunately, most epidemiological studies have not been accompanied by adequate estimates of the air pollution to which the populations were exposed. Rather, they have used location, usually of place of residence, an an index of severity of pollutant exposure. Often the only available measurements are of the town or district, and the sampling site may not have been representative even of the community air quality. Place of residence as a criterion for air pollution exposure should not be accepted uncritically, for it is clear that place of residence is also associated with (a) cultural and ethnic traits, (b) living standards, (c) climate, (d) occupation, and (e) exposure to infectious agents. The validity of using place of residence also is qualified by the duration of time during which a person lives in a given area. Such a problem becomes exceedingly troublesome when one wishes to study air pollution effects in a location such as California, to which a substantial proportion of the population moved after having lived a large part of their lives elsewhere. Mobility within the state is also high.

2. Representative Studies

The College of General Practitioners in Great Britain carried out a survey of symptoms and lung function among patients on the registry of their members (*365*). In addition to the diagnosis of chronic bronchitis given by the practitioners, the questionnaire also permitted the definition of chronic bronchitis based on morning sputum in the winter, attacks of cough and sputum lasting at least 3 weeks over the past 2 years, and breathlessness on walking at ordinary pace on the level (Table XVIII). It is clear from these data for both sexes that from three to four times as much chronic bronchitis is present in metropolitan as opposed to rural areas. Compared to social class or residential classification, smoking accounts for a factor of two to three times increase in the frequency of bronchitis. Buck and Brown (*366*) show that for bronchitis mortality in the United Kingdom there are significant correlations with smoke and sulfur dioxide (Table XIX). See Section VI,D for discussion of the data of this study relevant to lung cancer.

In Japan, Toyama (*367*) reported a significant correlation between bronchitis mortality and dustfall in twenty-one districts of Tokyo, but no correlation was found between dustfall and mortality from cardiovascular disease, pneumonia, or lung cancer. Age-standardized respiratory morbidity rates, based on interviews, were highest in the most polluted areas and showed a gradient with presumptive pollution exposures that ranged from 4.0 to 29.0 per 1000. Such gradients were not observed for other diseases. Similar patterns were observed in frequency of

Table XVIII Age-Adjusted Male Bronchitis Prevalence Rates Based on Diagnosis by General Practitioners by Smoking Habits and Area of Present Residence (*365*)

	Total		*Rural district*		*Urban district and municipal borough*		*County and metropolitan borough*	
Subjects	*Number*	*Percent*	*Number*	*Percent*	*Number*	*Percent*	*Number*	*Percent*
Nonsmoker	54	6	16	7	23	10	13	0
Exsmoker	141	13	27	11	64	7	30	27
Present smoker	592	18	115	13	268	18	209	24

Table XIX Correlations of Standard Mortality Ratios (SMR) for Lung Cancer and for Bronchitis with Pollution and Other Variables by Class of Area for Locations in the United Kingdom (*366*)

	London borough	*County borough*	*Metropolitan borough*	*Urban district*	*Rural district*
Number of locations	19	49	70	61	15
SMR Lung cancer	148	114	108	91	88
SMR Bronchitis	151	129	113	106	77
Persons/acre (1961)	48.9	15.6	13.3	6.4	0.8
Social index[a] (1951)	171	158	130	119	112
Average pollutant levels ($\mu g/m^3$) (March 1962)					
Smoke	168	281	203	185	85
SO_2	274	259	221	173	94
Significant correlations with male lung cancer[b]					
Persons/acre (1961)	—	0.67	0.65	0.52	—
Social index (1951)	0.69	0.49	—	—	—
Smoke (March 1962)	—	—	—	—	—
SO_2 (March 1962)	—	—	—	—	—
Significant correlations with male bronchitis[b]					
Persons/acre (1961)	—	0.46	—	—	—
Social index (1951)	0.85	0.51	0.56	—	—
Smoke (March 1962)	—	0.57	0.49	—	—
SO_2 (March 1962)	—	0.73	0.49	—	—

[a] Social index is the proportion of unskilled workers/1000 males of 15 years and over (1951 census).

[b] Significant values are said to be present when the probability of the true value being zero is less than one in a hundred. If no significant correlation is observed, no number is given.

respiratory diagnosis in National Health Insurance morbidity reports. Bronchitis is believed to be a large fraction of these cases. A high correlation of dustfall with morbidity from bronchitis and pneumonia in Ube, Japan, is reported by Nose (*368*).

In the winter of 1946, United States servicemen and their dependents stationed in the Tokyo–Yokohama area of Japan reported a high frequency of respiratory tract irritation with some asthmatic symptoms (*369*). The preceding year, some of the affected individuals reported a mild bronchitis. An allergic history was obtained infrequently, but almost all those affected were cigarette smokers. Subsequent studies (*370–375*) have confirmed the association of this condition with low winds and high pollutant levels, mostly by visual inspection (there are, regrettably, few measurements given in these reports). The condition is thought clinically and pathologically to resemble chronic bronchitis, notwithstanding the early asthmatic symptoms shown by some subjects. While acute symptoms were promptly relieved by leaving the area, some of those affected had persisting pulmonary function abnormalities consistent with emphysema. Evidence of allergic reactions in affected servicemen has been put forward by Meyer (*375*). The condition was originally thought not to occur among Japanese. However, Oshima *et al.* (*376*) reported an increase in chronic cough, sputum production, and diminished lung function among industrial employees in the Tokyo–Yokohama area compared with men having similar occupations in the less polluted area of Niigata. Symptoms were more common among cigarette smokers and those with allergic histories. They feel this pattern differed from the one affecting the servicemen.

Takahashi (*377*) has found in a population living in the city of Osaka three times as high a level of persistent cough and phlegm as among those living on a less polluted adjacent island. However, the frequencies of simple chronic bronchitis (persistent cough and phlegm) appear lower than in the United Kingdom or in the United States, but differences in methods could account for this.

Yoshida *et al.* (*378, 379*) have reported a high frequency of respiratory morbidity (including bronchitis) in Yokkaichi, Japan, seemingly related to air pollution from a power generation and petrochemical complex. Initial symptoms include asthmatic wheezing, but some of those affected had continuing symptoms and probably had the chronic bronchitis syndrome. Increased morbidity from respiratory conditions was reported by survey methods from the more polluted portions of the affected areas.

In a study of telephone workers in both the Eastern United States (*380*) and Japan the role of cigarette smoking was found in both to have similar effects, but no clearcut effects of place of residence were found.

Emphysema mortality in the United States is reported more frequently in urban than rural areas (*381*). It is also a condition that appeared to be increasing rapidly in frequency as a cause of death, but it is possible that greater diagnostic ability and interest may be responsible for some of the increase. Since 1967, the rate of increase has slowed in California. Hausknecht (*382*), in a study of a probability sample of California population, found a somewhat higher proportion with respiratory symptoms in Los Angeles than in other parts of the state. In a study using the Medical Research Council questionnaire and with methods as comparable as possible to those of Holland *et al.* (*383*), Deane *et al.* (*145*) found, among outside workmen, a greater frequency of symptoms of cough and sputum in Los

Angeles, California, than in San Francisco, California, especially among the older men. However, no differences in respiratory function were noted. The differences in symptom frequency between the two cities cannot be attributed to smoking.

Carnow *et al.* (*384*), using a number of monitoring sites in Chicago, Illinois, have followed a group of persons with chronic bronchitis, estimating for each person a weighted average of exposure at the residence and place of work. They report associations of sulfur dioxide with increased respiratory morbidity. No doubt other pollutants, including particulates, were also present. Spicer *et al.* (*385*) and Spicer (*386*) in Baltimore, Maryland, using the sensitive body plethysmographic method, observed that lung function changes among a group of chronic bronchitis patients tended to fluctuate together and to correlate with meteorological changes and with sulfur dioxide measured 38 hours previously. These studies also suggest that symptoms determined in a standardized manner may be about as sensitive an indicator of air pollution effects as are lung function tests. The data of McCarroll *et al.* (*387*) tend to support this. They have shown in longitudinal studies of a normal population in New York, New York, that cough has a lagged correlation with sulfur dioxide and particulate pollution, which is greater with a 24-hour and a 48-hour lag than it is with measurements on the same day. Winkelstein *et al.* (*388*) in Erie County, New York, have shown a correlation for total and chronic respiratory disease mortality with air pollution when economic status is controlled. No air pollution correlations were found to be significant for lung cancer.

Ishikawa *et al.* (*389*) have compared the prevalence and extent of emphysema in autopsied persons from St. Louis, Missouri, a city with a high level of air pollution, and Winnipeg, Manitoba, Canada, with much less pollution. Emphysema is more advanced and more frequent in St. Louis among both sexes and irrespective of smoking history. There is in each community more emphysematous change in smokers.

3. Interpretation

Data referrable to combined effects of sulfur oxides and particulate matter are discussed in Section III. Sufficient exposure to air pollution of various types is a likely causal factor in chronic bronchitis and a suspected one in emphysema; the likelihood clearly must be different in different locations. In no locality has its importance been better studied than in the United Kingdom, where the combined sulfur oxide and smoke pollution is thought to have synergistic (i.e., more than additive) effects with cigarette smoking. Thus, according to Reid (*390*), "It is, however, fair to say that . . . the accumulating epidemiological evidence satisfies at least some of the criteria of proof of causal relationship between air pollution and chronic bronchitis. We may thus reasonably conclude that although air pollution is certainly not the only cause, nor perhaps even the major initiating cause, it is almost certainly a promoting or aggravating factor of serious chronic lung disease." While this applies to the exposures in the United Kingdom it need not apply elsewhere, where pollutant dosage has been different.

On the question of air pollution causing respiratory disease in the Los Angeles, California, area, air pollution is a suspected causal factor of increased cough and sputum in older workmen and a likely cause of impairment of pulmonary function for persons with chronic respiratory conditions. There is suggestive evidence that populations of cigarette smokers are particularly susceptible to air pollution aggravation of their bronchitis, but that in nonsmokers the probability of the more serious consequences of bronchitis being caused by air pollution is low. There is as yet too little evidence for any firm conclusion concerning a causal relationship of community air pollution and emphysema. However, air pollution should be considered a suggested causal factor in emphysema.

C. Asthma

Acute outbreaks of asthma have been suspected as having a causal relationship to air pollution (Section II,A,7 and 8). It is commonly thought that inhaled pollens and certain dusts can produce classic asthma. Certain chemicals such as toluene 2,4-diisocyanate (TDI) are potent sensitizing agents on inhalation and can lead to asthmatic attacks in selected individuals (*391*). Accelerated impairment of lung function occurs with repeated exposure (*392*). Schoettlin and Landau showed that photochemical oxidants affected some adult asthmatics in Pasadena, California (*138*). Zeidberg *et al.* (*393*) studied populations in Nashville, Tennessee, drawn from the University Hospital Clinic. When the city was divided into areas classified according to high, medium, and low ranges of sulfation, there was a higher frequency of asthma attacks in adults, though not in children, in the more polluted areas. However, the population characteristics differed greatly by location in that 16 out of the 18 asthmatic adults living in the highly polluted area were nonwhite, whereas the total study population had about equal numbers of white and nonwhite persons.

Asthma emergency visits in Haifa, Israel, during 1969 and 1970 were analyzed by Peranio *et al.* (*394*), who found epidemic periods similar to those in New Orleans, Louisiana (see Section II,A,7), which tend to occur during times of low winds when pollutants tend to build up. No specific pollutant has been identified.

Possible sensitivity to sulfur dioxide is suggested by the report of Sim and Pattle (*395*): several human subjects of experimental exposure had a protracted period of cough, wheezing, and expectoration following exposure. Subsequent exposure to ambient pollution reactivated the symptoms. Workers in at least three other laboratories have reported development of acute asthmatic episodes when some subjects were exposed to sulfur dioxide at levels that produce only mild reactions in most subjects. In CHESS studies by EPA, sulfate aerosols are associated with increased likelihood of asthma attacks when the environmental temperature is low (*396*) or when oxidant pollution is also elevated.*

Interpretation

Pollens are almost certainly a causal factor in some cases of asthma. Other materials may, as community air pollutants, be causal factors in certain locations (New Orleans, Louisiana; Minneapolis, Minnesota). Generalized community air pollution is a suggested causal factor, but only for a small proportion of all adult asthma patients. Photochemical pollution is a likely causal factor in aggravation of asthma, in a portion of adult asthmatic patients. Sulfate aerosols are a suggested causal factor. Hyperreactivity of the airway caused by pollutants is likely to be an important mechanism in air-pollution aggravated asthma.

D. Air Pollution Exposures and Respiratory Cancer (Goldsmith)

1. Background of Smoking and Occupational Exposures

There are two types of evidence for judging the importance of population exposure to possible respiratory carcinogens—experimental animal data and human population data. Occupational experience is the most convincing evidence that inhalation can cause cancer (Table XX). In addition, there is an excess of lung cancer in some cities, which has been attributed to air pollution. Studies of migrants to and from polluted areas have helped in the analysis of the urban factor. Occupational exposures contribute to the urban excess and, along with the greater prevalence of long-term cigarette smoking in urban areas, may be sufficient to explain it.

Available evidence, however, is not convincing that exposures to common pollutants cause lung cancer. Part of the uncertainty

* Nadel (*397*) has analyzed the reflex nature of asthma and shown that ozone can greatly augment the reactivity of the airway as tested by histamine inhalation. Similar results for NO_2 have been found by Orehek *et al.* (*397a*) from exposures to 0.11 ppm NO_2 (210^2 $\mu g/m^3$) for an hour, in 13 of 20 subjects, using the bronchoconstrictor, carbachol, instead of histamine.

Table XX Inhalants Having Recognized or Suspected Carcinogenic Site(s) under Occupational Conditions

Material	*Recognized site(s) of carcinogenic effects*	*Suspected site(s) of carcinogenic effects*
Arsenic	Lung	Larynx
Asbestos	Lung	Digestive system
Chromium	Lung, nasal cavity	—
Nickel	Nasal cavity, sinuses, lung	—
Aromatic amines	Bladder	Lung
Isopropyl oil	Sinuses, larynx, lung	—
Mustard gas	Lung, larynx	—
Coal tar pitch	Larynx, lung	—
Creosote	Skin	Lung
Soots	Lung	—
Mineral oils	Skin	Lung
Petroleum, asphalt	Skin	Lung
Paraffin wax	Skin	Lung
Radon and ionizing radiations	Lung, bone, other sites	—
Bis-chloromethyl ether	Lung	—
Epoxides	—	Lung
Other chlorinated hydrocarbons	—	Lung
Beryllium	—	Lung
Iron (hematite)	—	Lung
Macromolecular polymers	—	Lung
Vinyl chloride	Liver	Kidney

can be attributed to the dominant role of cigarette smoking as a cause of lung cancer. Smoking is so important, that the lesser role of other inhalants is hard to detect. A second major cause of uncertainty is the long time spans involved. Based on what we know of smoking and occupational exposures, it usually requires several decades of human exposure before alterations in the occurrence of respiratory cancer are detectable. Residential mobility makes it difficult to find and study an adequately large population with several decades of residence and exposure in one location. All this is additional to the problems of estimating the exposure levels of populations.

It is difficult to apply occupational data to community exposure, since among other reasons, most urban exposures are usually to a mixture of agents. For several of the occupational carcinogenic agents (radon exposures in underground miners and asbestos exposures in insulation workers), the interaction with smoking exposure appears to be multiplicative rather than additive (*398, 399*). Since under occupational conditions such a wide variety of chemical inhalants can produce cancer of the respiratory tract in man (*400–407*), it is reasonable to suspect that populations exposed to similar agents as part of their general environment also share some increased risk of respiratory cancer (*408*).

2. Experimental Carcinogenicity of Atmospheric Pollutants

The carcinogenicity of atmospheric pollutants to mice has been demonstrated with organic extracts collected from various sources. Usually these have been applied to the mouse skin or injected. Administration of pollutant extracts to mouse skin, whether by painting or subcutaneous injection, has generally yielded local tumors (papillomas or carcinomas, sometimes accompanied by multiple pulmonary adenomas). A notable exception was when small concentrations of organic extracts of pollutants were injected subcutaneously in infant mice, which later showed a high incidence of distant tumors (hepatomas and lymphomas) in addition to multiple lung tumors, together with a virtual absence of local tumors (*409*). Marked variation has been noted in the carcinogenicity of organic extracts of pollutants from various urban sites, coincidentally with low activity of material from Los Angeles, California (*409, 410*).* In addition to the well known and much studied effects of benzo[*a*]pyrene, the role of other carcinogens in crude organic extracts is now generally accepted. Evidence for this includes tumor production by benzo[*a*]pyrene-free pollutants, such as aliphatic aerosols of synthetic smog (*411*) and by aliphatic and oxygenated fractions of organic extracts of particulate atmospheric pollutants. Lack of parallelism is noted between carcinogenicity of organic extracts of particulate pollution samples and their benzo[*a*]pyrene concentrations (*409, 410*). However, not all of the tumors reported are invasive. Noninvasive ones do not, in the opinion of some workers, deserve to be considered as cancers.

When bound to soot (*412*) or to hematite (*413*), pure chemical carcinogens, such as benzo[*a*]pyrene, known to be present in polluted air, have been shown to be carcinogenic for lungs of rodents by intratracheal instillation. There is evidence of more than additive effects between sulfur dioxide and benzo[*a*]pyrene following exposure by inhalation rather than by intratracheal instillation (*414*). Squamous carcinoma production by intratracheal instillation of benzo[*a*]pyrene absorbed on hematite (*413*) has suggested the importance of inert particles as carriers of carcinogens in the production of lung cancer in animals. The role of inert particles may be the same in man, though the particle may be a

* An authoritative review of migration and other epidemiological evidence from several countries on the possible association of air pollution and cancer has been published by the International Agency for Research on Cancer (*455a*).

liquid aerosol (tobacco smoke) rather than a solid one (hematite). Vitamin A appears to inhibit the carcinogenic effect of benzo[*a*]pyrene plus hematite and reduces the alteration of cell types in the hamster respiratory system. This potential protective mechanism against lung cancer deserves further study (*415*).

3. The Urban Industrialization Factor in Lung Cancer Epidemiology

Most of the relevant epidemiological studies show an association between urban residence and increased risk of lung cancer. This urban–rural difference is more evident among men than women (*137, 416–420*). These results lend support to the argument that air pollutants, found more often in an urban environment, are carcinogens affecting the general population. In Norway (*421*), the death rates for cancer of the respiratory system are more than three times greater for urban males than for rural males, with a smaller excess for females (Table XXI). In New York State (Table XXI), excluding New York City, the age-adjusted lung cancer rate is twice as high in urban portions of metropolitan standard statistical areas as in rural parts of nonmetropolitan areas. In Iowa, the urban–rural gradient is nearly threefold (Table XXI). However, in some rural parts of Finland, the age-adjusted lung cancer rate for males is higher than even in the most polluted cities (145.4/100,000) (*422*). (see also p. 569).

Mortality among different groups is often compared by means of the standardized mortality ratio. Usually, the standardization adjusts the observed data so that different age distributions of populations will not

Table XXI Rural–Urban Lung Cancer Death Rates[a] for Males and Females in Norway, New York State (excluding New York City), and Iowa

Subjects	*Norway (421)*		*New York State (1949–1951) (420) (age-adjusted)*		
	1959	*1969–71 (Age-adjusted)*	*Parts of metropolitan SMSA[b]*	*Parts of nonmetropolitan areas[b]*	*Iowa (418)*
Urban					
Male	40.7	50.7	29.2	20.8	29.0
Female	6.5	8.7	3.2	3.2	7.8
Rural					
Male	12.4	20.4	23.9	15.2	10.2
Female	4.3	5.1	3.5	2.4	5.3

[a] Rates are in units of deaths per 100,000 per year.
[b] SMSA stands for standard metropolitan statistical area. Each SMSA includes at least one city with a population of 50,000 or more.

Table XXII Lung Cancer Standard Mortality Ratios in a United States Sample Adjusted for Age and Smoking History *(435, 436)*

Population	*Male*	*Female*
Counties in SMSA[a]		
500,000 and over	123	132
50,000–500,000	111	92
10,000–50,000	164	137
2500–10,000	107	104
Counties not in SMSA[a]		
10,000–50,000	89	96
2500–10,000	84	96
Rural nonfarm	80	84
Farm	65	69

[a] Standard metropolitan statistical areas, which include at least one city with a population of 50,000 or more.

distort the comparison. Adjustment may also be made to allow for the influence of smoking history, occupation, race, education, etc. and usually leads to reports given in terms of what is called a standardized mortality ratio. A standardized mortality ratio of 100 means that the population (e.g., living in a certain district or city) is affected by mortality rates at the same intensity as the population used as a basis for standardization. A ratio of 164 means there is 64% excess mortality and a ratio of 65 implies a 35% deficiency in relation to the base population. Data for Great Britain in terms of age-standardized mortality ratios shows less than a twofold difference between rural districts and London (*366*) (Table XIX). Higgins (*423*) studying trends for male lung cancer mortality between 1956 and 1970 demonstrates that in London, rates at ages less than 65 have begun to decrease. The lung cancer rates for ages 45 to 64 remain higher in London than in other parts of Great Britain and the United States. The age-adjusted mortality from lung cancer in Great Britain is about twice as high as in the United States and this is true both for men and women. Table XXII, in which data are adjusted for smoking history in the United States, data show that cities with populations from 10,000 to 50,000 have the highest lung cancer mortality ratios. There is thus evidence for an urban factor that contributes to the excess risk of lung cancer. Environmental carcinogens that are inhaled from polluted urban atmospheres have been proposed as a possible explanation.

The effects of cigarette smoking and urban residence appear to be at least additive. In contrast, the combined effect of occupational asbestos inhalation and tobacco smoking is greatly in excess of the risk attributable to the sum of the two factors (*249, 399*). Similar relationships are found for the combined effect of uranium mining and cigarette smoking (*424*). Among asbestos workers who do not smoke, bronchogenic cancer is said to be rare, but among asbestos workers who smoke, the risk of bronchogenic cancer is about eight times that of smokers with no asbestos exposure (*249*). Attention has been focused by experimental work (*425*) upon the possibility that chemical carcinogens may interact with viruses. Thus, the risk of lung cancer might be increased in individuals who are exposed to polluted air containing carcinogens while being ill with influenza and other respiratory virus infections. The possibility has been raised that this is primarily due to the infectious process breaking down the defense mechanisms, permitting the carcinogens to react with the germinal cells of the respiratory tract.

There are a number of variables other than air pollution that condition the health effects of urbanization. These include (a) demographic factors, (b) meterological and climatic factors, (c) occupational exposures (including unemployment), (d) household crowding and household dilapidation, (e) land congestion, (f) use of household fuel for cooking and heating, (g) income and education variables, (h) spread and occurrence of infectious disease, (i) nutrition, and (j) smoking. Demographic factors include ethnic origin, age, family structure, and migration history. Household crowding and dilapidation are often interrelated with low income, educational status, recent urban migration, and the spread of infectious disease. Land congestion is likely to be related to fuel emissions both from transport and domestic heating and cooking. Use of household fuels for cooking and heating may be a major cause of air pollution and may further be a basis for high-level exposures in the household. Income, education, and ethnic practices affect nutrition.

Analyses for this complex of interacting variables have been published for mortality in forty-six of the United States standard metropolitan statistical areas by Schwing and McDonald (*426*), and by Lave and Seskin (*427, 428*). This has been done on a smaller scale for Los Angeles, California, by Chapman and Coulson (*429*), and by Menck *et al.* (*430*). Using over one hundred communities within Los Angeles county as a base, Goldsmith (*431*) does not find any association of air pollution complaints in 1956 with lung cancer mortality in 1970, but an association is found for population density. The Los Angeles data (by census tracts for 1966) analyzed by Chapman and Coulson show no significant correlations for lung cancer mortality with the ethnic and income variables, percent black, percent Spanish surnamed, or average family income.

Lave and Seskin have not used age-adjusted data; thus, the most obvious demographic factor has not been controlled for. Similar studies for classes of urbanized areas in Great Britain have been reported by Buck and Brown (*366*) (Table XIX). Buck and Brown find that "lung cancer mortality in administrative districts of England and Wales is not, in general, significantly associated with corresponding levels of smoke or sulfur dioxide in the residential areas of these districts. There is a positive association between lung cancer mortality and population density within the different classes of areas which is significant and accounts for differences between classes of areas in the average mortality rates."

Lave and Seskin (*427, 428*) find that for malignant disease of the respiratory system (I.C.D. 162, 163)* for 117 standard metropolitan statistical areas for 1960, the following variables have statistically significant coefficients ($p \leq 0.05$; $t \geq 1.7$): percent of the population over 65 years; percent of the population nonwhite; persons per square mile; percent of the population male; percent of workers employed in transportation, communication, and other public utilities; percent of families with incomes less than $3,000 (negative coefficient). Among variables that did not have significant correlations were biweekly measured levels of mean, minimum, and maximum of suspended particulates and sulfates and proportions employed in other occupational groups. When Lave and Seskin excluded occupational factors, pollutant variables showed significant coefficients. In a study of nonsmokers in England, however, Doll (*432*) found no effect of population density on lung cancer mortality. Possibly, therefore, population-dense areas merely have a higher proportion of heavy smokers.

The National Academy of Sciences—National Research Council's report (*433*) on particulate polycyclic organic matter notes that median winter–spring quarter of 1959 concentrations of benzo[*a*]pyrene in urban sites were 6.6 $\mu g/1000\ m^3$ (6.6 ng/m^3) and for nonurban sites 0.4 $\mu g/1000\ m^3$. Disregarding smoking, the report "roughly associates" the apparent doubling of lung cancer in urban as compared to rural areas to the 6.2 ng/m^3 excess of benzo[*a*]pyrene.

Carnow and Meier (*434*) have published a regression analysis for lung cancer (I.C.D. 160–164) for the forty-eight contiguous states using age-adjusted data for both sexes and for white and nonwhite populations; the "independent variables" used are tobacco sales (in dollars per person of age 16 or more) and benzo[*a*]pyrene in air (micrograms/1000 m^3). The benzo[*a*]pyrene measure, however, is apparently derived by multiplying the proportion of the state's urban population, by the average urban

* I.C.D. refers to the International Statistical Classification of Diseases, Injury, and Causes of Death. The classes include 160—cancer of the nose, middle ear, and nasal accessory sinuses; 161—cancer of the larynx; 162—cancer of the trachea, bronchus, and lung, specified as primary; 163—cancer of the lung, unspecified as primary or secondary; and 164—cancer of the mediastinum. (Seventh revision of I.C.D.)

benzo[*a*]pyrene level (presumably 6.6 ng/m^3) and the rural proportion by a much lower benzo[*a*]pyrene level (presumably 0.4 ng/m^3); thus, the independent variable represented as benzo[*a*]pyrene appears to be an indirect measure of urbanization and may not reflect differences in air pollution exposures by state. Their tabulation yields the following apparent anomalies. Illinois and Indiana have lower values for benzo[*a*]pyrene (2.38 and 2.37) than does Idaho (2.87). Oregon and Washington have higher values (2.03 and 1.72) than Massachusetts, Rhode Island, and Michigan (1.26, 1.49, and 1.04). In addition, the difference between states in dollars spent for tobacco inadequately reflects the differential effects of smoking on lung cancer rates. This analysis appears to have been relied upon by the authors of the NAS–NRC report (*433*). In the United States in 1958, the urban to rural standardized lung cancer mortality ratio for males (adjusted for age and smoking history) showed a 43% excess in urban as compared to rural areas, and for females a 22% excess (*435, 436*). The male urban-to-rural ratio for lung cancer mortality appears to be declining.

Among nonsmokers, urban residents have about 20% higher lung cancer rates than do rural residents for both sexes. The male and female rates are similar, and for rural residents the female rates are slightly higher than the rates for males. In the United States, 51% higher rates were observed in foreign born males than in native born, for females the rates are 52% higher. For those who migrate from farms to metropolitan areas, the ratio is higher than for those who lived all their lives in metropolitan counties, 71% excess for males, 110% excess for females.

Hitosugi (*437*) studied lung cancer death rates in an area near Osaka, Japan, based on a family interview for families of 259 persons dying of cancer and a random sample of 4500 adults. Generally, high lung cancer mortality was found for exsmokers, suggesting that onset of illness may have led some smokers to become exsmokers. Among nonsmokers, no pollution gradient was shown (although the number of deaths is small). Urban factors and smoking appear, however, to have interacted in this Japanese population as they seem to have done in United States and British and other industrialized population groups.

Following a recommendation by a World Health Organization study group on epidemiology of cancer of the lung (*438*), a comparison of lung cancer, atmospheric pollution, smoking, and other variables was carried out between Dublin, Ireland, and Belfast, Northern Ireland, and between Oslo, Norway, and Helsinki, Finland (*439*), and other European cities. This study appeared to show an association between solid fuel combustion and lung cancer death rates after differences in smoking habits were taken into account. Additional studies in Northern Ireland (*440*) led to an estimation (*441*) that if the death rate from lung cancer for a symptomless rural nonsmoker is taken as the irreducible minimum, the risk of death from the disease would be about doubled for a man living in an urban area, but increased twentyfold if he smoked more than 20 cigarettes a day. According to the Royal College of Physicians (*8*), such retrospective inquiries, "although they appear to point to an urban factor in the causation of lung cancer, they do not amount to an indictment of air pollution." The Royal College report concludes, "Pollution of British towns cannot therefore be the whole explanation of the adverse experience of this country with respect to lung cancer."

Pedersen *et al.* (*422*) carried out a personal interview study of the role of cigarette smoking pollution and other variables among a group of populations in rural and urban Norway and Finland. The study includes a rural Finnish region which has lung cancer mortality in excess even of that in London, and exceeding the levels in urban Norway and Finland. They were able to show that the type and magnitude of cigarette smoking and the age at starting to smoke were the most likely explanations for the wide variations observed. Neither Winkelstein *et al.* (*442*) nor Hagstrom *et al.* (*443*) found an association of pollution indices with lung cancer in community studies in Buffalo, New York, and Nashville, Tennessee, respectively. Neither of these studies estimated or adjusted for the effects of smoking. Menck *et al.* (*430*) report that in the south central portion of Los Angeles County, California, the age-adjusted lung cancer mortality for Caucasian males for 1968–1969 is greater than for other parts of the county. This area includes most of the industralized and low-income portions of the county. The differences persist when comparisons were made of occupational groups in the lower socioeconomic categories. Similar clustering for lung cancer is not found among women, though it should be if community air pollution were a potent factor. Since heavy cigarette smoking is more common among the lower economic classes, the authors looked to see if other smoking-associated cancers were high in this area, but they were not. This area includes the highest levels of benzo[*a*]pyrene in air and soil found among four sampling sites in the county. The authors, not finding any occupational exposures to account for these findings, feel that the "most likely explanation is a synergistic action between smoking and neighborhood air pollution." The area is not one that has had high photochemical pollution relative to other parts of the county. The authors suggest that the neighborhood pollution may be related to petroleum and chemical industries in the area. However, it is likely that occupational exposures are more intense although less persistent than neighborhood exposures.

Hammond (*444*), in a prospective study of lung cancer epidemiology, reports on the role of smoking, place of residence, and presence or absence of occupational exposure. The population was enrolled by volunteer workers of the American Cancer Society beginning in October 1959 and followed for 6 years. Data were obtained in 1121 counties in 25 states and included 16 of the 20 largest cities; 2063 of the subjects died of lung cancer. Compared to nonsmokers, the mortality ratio for smokers increased from 4.62 for men who smoked 1 to 9 cigarettes a day up to 18.77 for men who smoked 40 or more cigarettes a day. The analysis of urban residence and occupation was limited to men who, at the time of enrollment, said they had lived in their present neighborhood for at least 10 years. Adjustment for age and smoking habits was based on 5-year age groups and six classes of smoking—those who never smoked regularly and five categories of smokers by type and amount smoked. Within groups by place of residence, the experience was further subdivided according to whether the respondents said they "were or ever had been occupationally exposed to dust, fumes, vapors, gases, or x-rays." These statements cover a wide range of exposures and a variety of durations and intensities. Results are shown in Table XXIII.

Disregarding place of residence, the age–smoking adjusted mortality ratio for men with "occupational exposure" was 1.09 compared to 0.96 for men without occupational exposure. In large metropolitan areas, the occupational excess was greater than in smaller and nonmetropolitan areas. For men not occupationally exposed, the high mortality ratio (1.06) was for men living in cities in large metropolitan areas, and those living in towns and rural parts of smaller metropolitan areas (1.05). If air pollution is assumed to increase with size of city, then these data, according to Hammond, "give little or no support to the hypothesis that urban air pollution has an important effect upon lung cancer death rates." Data for Los Angeles, Riverside, and Orange Counties, California, with their high-level exposures to oxidant and carbon monoxide, show high lung cancer rates for men occupationally exposed. For those with no occupational exposure, the mortality ratio (0.96) for men in Los Angeles is equal to that for the whole unexposed population.

These data suggest that once the contribution of smoking is removed a large fraction of the urban excess in lung cancer is related to occupational exposure; among the men not occupationally exposed, there is nearly as much excess in towns (2500–50,000) as in great cities. This finding resembles those of Haenszel, who found when smoking was controlled for, that the highest lung cancer rates were among residents of such towns.

In the United States, coke oven workers in the steel industry, also pre-

Table XXIII Observed and Expected Number of Lung Cancer Deaths by Place of Residence and by Occupational Exposure to Dust, Fumes, Gases, or X-Rays[a] *(444)*

	Occupationally exposed			*Not occupationally exposed*		
Place of residence[b]	*Observed number*	*Expected number*	*Ratio*	*Observed number*	*Expected number*	*Ratio*
Total, all subjects	576	530.5	1.09	934	979.7	0.96
Metropolitan area (pop. 1,000,000+)	165	134.1	1.23	281	285.7	0.98
City	92	69.1	1.33	168	158.3	1.06
Rural or town	73	65.0	1.12	113	127.4	0.89
Metropolitan area (pop. <1,000,000)	166	145.4	1.14	271	280.5	0.97
City	92	83.3	1.10	170	184.0	0.92
Rural or town	74	62.1	1.19	101	96.5	1.05
Nonmetropolitan area	245	251.0	0.98	382	413.5	0.92
Town	102	104.9	0.97	200	199.1	1.00
Rural	143	146.1	0.98	182	214.4	0.85
Los Angeles, Riverside, and Orange Counties, California	30	21.9	1.37	38	39.6	0.96
Farmers	63	77.6	0.81	71	92.9	0.76
8 Cities—high particulates (130–180 $\mu g/m^3$)	45	32.9	1.37	66	73.9	0.89
11 Cities—moderate particulates (100–129 $\mu g/m^3$)	21	18.8	1.12	39	49.5	0.79
14 Cities:—low particulates (35–99 $\mu g/m^3$)	48	37.4	1.28	110	100.1	1.10
9 Cities—high benzene-soluble particulates (8.5–15.0 $\mu g/m^3$)	28	21.0	1.33	52	51.5	1.01
10 Cities—moderate benzene-soluble particulates (6.5–7.9 $\mu g/m^3$)	44	32.7	1.35	65	75.1	0.87
12 Cities—low benzene-soluble particulates (3.4–6.3 $\mu g/m^3$)	33	29.2	1.13	76	81.8	0.93

[a] Adjusted for age and smoking habits; confined to men who had lived in same neighborhood for last 10+ years.

[b] A "metropolitan area" is defined as a county or group of contiguous counties with at least one city or pair of cities with 50,000 or more inhabitants, according to the 1960 census. A "town" means a place with 2500–49,999 people, and "rural" refers to people living in a place of less than 2500 people or in the country.

sumably exposed to high levels of benzo[*a*]pyrene, have in certain groups a threefold excess of lung cancer (*45, 406*). With such massive exposures (hundreds to thousands of times the levels to which other persons are exposed), the excess mortality seems small if benzo[*a*]pyrene is a potent respiratory carcinogen in man. In Kenya, where indoor smoke contains about ten times the benzo[*a*]pyrene encountered in heavily polluted air, there is a reported excess of nasopharyngeal cancer, but not of lung cancer (*445*).

One of the best criteria of malignant changes in the lung has been the appearance of the cells lining the respiratory tract. Changes studied in autopsied persons have correlated so well with epidemiological studies of lung cancer prevalence in smokers of varying types, and with results of sequential observation of animals exposed to carcinogens, that these observations may be interpreted as evidence of a mechanism that leads to human lung cancer. A well-controlled examination for these cells has been carried out by Auerbach and his colleagues (*446–448*). From a group of 1007 men and 515 women from whom multiple histological slides had been prepared, pairs were matched by age, cause of death, occupation, residence and smoking habits, and comparisons of the frequency of histological changes in matched pairs were carried out. Table XXIV gives the comparison of matched women smokers and nonsmokers of both rural and urban residence. Compared with the striking differences between women smokers and nonsmokers, the differences between urban and rural nonsmoking women are small. Women were chosen for this comparison because of the difficulty of matching, by occupation, an adequate number of nonsmoking men.

The United States National Academy of Sciences report's conclusion concerning the role of benzo[*a*]pyrene relies heavily on two regression analyses. One was that of Carnow and Meier discussed above. The other regression analyzes male lung cancer deaths in nineteen countries, along with average cigarette consumption per person per year. The countries are not listed, but it is based on the original analyses by Stocks of these nineteen countries (*439*).

While it is reasonable to assume that variations may be small in current cigarette consumption of comparable urbanized populations, in recently industrialized and urbanized countries there is likely to be less long-term cigarette smoking in relation to current smoking. Hence, average current cigarette consumption in a country may not have the same relevance in newly urbanized as in older urban areas. Solid fuel consumption may reflect the degree of industrialization, and to some extent, urbanization. The regression analysis, according to the National Academy of Sciences report, leads to "approximately a doubling in the lung cancer rate corre-

Table XXIV Changes in Bronchial Epithelium in Matched Pairs of Female Cigarette Smokers and Nonsmokers and in Matched Pairs of Female Nonsmokers, Urban and Rural (*446, 448*)

			Female nonsmokers	
Parameter	*Smokers*	*Nonsmokers*	*Urban*	*Rural*
Number of subjects	72	72	26	26
Number of sections with epithelium	3326	3670	1310	1284
Sections with one or more epithelial lesions (%)	91.1	24.3	26.3	18.8
Sections with three or more cell rows with cilia present (%)	85.5	11.1	14.0	9.5
Sections with cilia absent (%)	17.5	14.4	14.0	10.4
Sections with atypical cells (%)	79.7	0.7	0.3	0.8
Sections with atypical cells present with cilia absent (%)	14.6	0.1	—[a]	0.1
Sections with entirely atypical cells with cilia absent[b] (%)	2.9	0	0	0
Sections with hyperplasia and goblet cells in glands (%)	67.9	13.7	14.4	10.4
Sections with ulceration (%)	9.7	19.5	16.3	15.4

[a] Less than 0.05.
[b] Carcinoma.

sponding to an increase in smoking of a pack (20 cigarettes) per day"; this compares with a five- to twelvefold excess of lung cancer in current cigarette smokers compared to nonsmokers in seven prospective studies (*441, 449*). The regression computation does not, therefore, adequately reflect the relative contribution of long-term smoking to lung cancer, but rather the relative national differences in current cigarette smoking, which are likely to be small for the countries studied.

After listing a number of factors which fail to support the hypothesis of a causal association of benzo[*a*]pyrene and lung cancer, the National Academy of Sciences report (*433*) concludes that increased urban pollution with this material has the effect of increasing lung cancer death rates. The report asserts that reducing urban levels of benzo[*a*] pyrene, e.g., from 6 ng/m^3 o 2 ng/m^3 will reduce lung cancer death rates by 20%. This is followed by, "These data, however, are not to be interpreted as indicating that benzo[*a*]pyrene is the causative agent for lung tumors." In fact, such a decrease in benzo[*a*]pyrene has occurred (from 6.6 ng/m^3 in 1959 to 2.5 ng/m^3 in 1967 for United States urban sites in the January–March quarter). If this decrease is to lead to reduction of lung cancer death rates, it will probably do so after a latent period. Based on the decrease in lung cancer in British physicians (*450*) whose smoking has decreased, a latent period of 2–5 years will be needed. We should also see the greatest decrease in the communities that previously had the highest pollution levels. Thus, the hypothesis put forward by the National Academy of Sciences report can be tested.

4. Studies of Migrants

Changed rates of lung cancer in migrants have been used to estimate a possible effect of air pollution on lung cancer rates. Migrants from the United Kingdom to New Zealand (*451*) and to South Africa (*452*) have higher lung cancer rates than the native born, and this is especially true for those who migrated after 30 years of age. Similar data for United Kingdom emigrants to the United States and to the Channel Islands have also been reported (*453, 454*). In dealing with the extensive migration data, the National Academy of Sciences report starts from the assumption, "if such migrants can be considered as random or representative samples of the populations of the home countries, then differences in death from those in the home countries can be ascribed to changes in environmental conditions." There is a good reason to question whether immigrants from an area are a random or representative group with respect to environmental conditions in the home countries; neither are they likely to have representative exposures in the place to which they migrate. Should they migrate to communities with less favorable housing or working conditions than the native born, immigrants may have multiple unfavorable environmental exposures in their new environments. For migrants to urban areas, this seems likely.

Mancuso and Coulter (*417*), studying immigrants from Italy and Great Britain to Ohio, show that while native-born Americans have lung cancer mortalities (for 25- to 64-year-old males) intermediate between mortality rates in Italy and the United Kingdom, immigrants from these countries into Ohio tend to approach the mortality rates of native born. For the British born, this means lower rates, for the Italian born, higher rates. Mancuso and Sterling (*454a*) have shown that black males who migrate into Ohio have much higher rates than comparable groups born in the state or lifetime residents in the states from which migrants come. The National Academy of Sciences report (*433*) says, "Lung cancer death rates of migrants are intermediate between those of native U.S. residents and persons in the home countries." However, we note that in the Reid *et al.* study (*455*) (Table XXV), Norwegian woman migrants to the United States have higher mortality than native-born United States women. This may reflect the likelihood that Norwegian immigrant women

Table XXV Age-Adjusted Death Rates for Lung Cancer[a] for Persons Born in Great Britain, Norway, and the United States, along with Rates for Migrants (*455*)

Residency	*Males*	*Females*
Great Britain residents	151.2	19.3
Great Britain-born United States residents	93.7	11.5
Native United States residents	72.2	9.8
Norway-born United States residents	47.5	10.7
Norway residents	30.5	5.6

[a] I.C.D. 162–163.

were likely to have an unusually unfavorable environmental exposure. It is difficult to believe that an excess, if real, could be associated with exceptional smoking or community air pollution exposures.

This exception to the notion that migrants have intermediate lung cancer experience between that of their origin and destination, taken with Haenszel's finding that rural migrants to urban areas have higher lung cancer rates than lifetime urban residents, indicates that migration may have more complex effects than is indicated by the National Academy of Science report.

Eastcott's studies (*451*) of migrants from the United Kingdom to New Zealand show that those who migrated before age 30 had 35% higher risk of lung cancer than native New Zealanders compared to 75% higher risk if migration had occurred after age 30.

Such an effect (not adjusted for smoking) could as well support the hypothesis of a contribution by a less favorable occupational, residential, and socioeconomic status of those migrating in later years of life as opposed to younger migrants, as it could a presumptive longer exposure to carcinogenic agents in the United Kingdom. In any event, a potent carcinogenic effect of environment in early life is not supported by these data. Possibly, therefore, the greater susceptibility of infant mice (*409*) is not applicable to man (see *409a*).

5. Principal Arguments for Air Pollution as a Causal Factor in Excess Urban Lung Cancer

a. There are potent carcinogenic agents in polluted atmospheres.

b. The urban excess of lung cancer can be associated with urban pollution.

c. Workmen occupationally exposed to benzo[*a*]pyrene, to asbestos, arsenic, and to other materials present as pollutants show excess lung cancer.

d. Cigarette smoking is accepted as a cause of both bronchitis–emphysema and lung cancer. Urban pollution contains similar ingredients to those found in cigarette smoke.

6. Principal Arguments Opposing Air Pollution as a Causal Factor in Excess Urban Lung Cancer

a. The urban gradient should be largest in those states and countries where there is the heaviest urban pollution. It is not.

b. If exposure to urban pollution causes an augmentation in lung cancer, then the rates should be higher in lifetime urban residents than in migrants to urban areas. They are not.

c. Correlations of lung cancer rates with measured pollution should be found by studies in the United Kingdom where both lung cancer rates are high and pollution has been great. A positive correlation is found with population density, but not with pollution.

d. If the urban factor were community air pollution, it should affect women at least as much as men. It does not appear to.

e. If urban pollution by benzo[*a*]pyrene makes an important contribution to the urban excess, lung cancer in the locations most polluted by this material should be highest, and when the agent decreases, lung cancer should do so as well. This has not been shown to occur.

7. Conclusion

There may be other explanations of the urban factor (greater smoking, domestic exposures, occupational exposures, population density, infections), but the evidence presently available does not support the conclusion that air pollution *per se* is the factor; neither is it possible to reject the possibility. Air pollution remains a "suggested" explanation for the urban excess of lung cancer.

In concluding that air pollution has not yet been shown to be the causal agent in the urban excess of lung cancer, there are two important practical considerations. The first is that control of soot, suspended particulate matter, and products of incomplete combustion are appropriately among the highest and most urgent goals of air pollution control, without regard for the interpretations of pollution–cancer data. This is because of the strength of evidence on chronic bronchitis and because of other undesirable features of such pollution. Reducing this type of pollution will *pari passu* decrease pollution by benzo[*a*]pyrene. Thus, there is reason independent of conclusive evidence of causation of cancer risk to reduce urban pollution by benzo[*a*]pyrene and to reduce whatever risks may be present.

However, to behave as if urban and migration factors in lung cancer mortality are due solely to air pollution could lead to neglecting the possibility that domestic agents and occupational exposures play an important role. This would not be advisable, since these might readily be controlled, once attention is focused on them and their role is better defined. These conclusions apply to general urban air pollution and not necessarily to point source exposures in the vicinity of industrial plants.

E. Air Pollution and Cardiovascular Disease

There are three possible mechanisms by which air pollution may affect the cardiovascular system on a long term basis. The first and most important, associated with the role of carbon monoxide, was discussed earlier in this chapter, as was the effect of cadmium on cardiovascular disease, in particular on cardiovascular renal disease. The third effect depends on the interaction of the respiratory effects of pollution and the cardiovascular system. If there is respiratory irritation with coughing, the resulting internal pulmonary pressure increases. Chronic respiratory disease, particularly where extensive fibrosis or loss of lung tissue occurs, tends to be associated with cor pulmonale, or heart disease secondary to lung disease. Cor pulmonale occurs with substantial frequency in Great Britain and has been reported by Sinnett and Whyte as a most common problem in the Highland populations of New Guinea (*456*). These individuals showed a decrease in lung function and frequent attacks of pneumonia. Of five persons in the survey with heart failure, in four illness was due to underlying lung disease. The authors say "the people spend up to 12 hours per day inside smoke filled houses, and 70% of the adult males and 20% of the females smoke home-grown tobacco."

Among populations with an increased frequency of chronic respiratory conditions it is likely that secondary heart disease will be found. In the populations affected by the acute air pollution episodes, a number of deaths occurred among people with pre-existing heart disease. Whether this is due to any direct effect is hard to determine.

F. Air Pollution and Nervous System Reactions

1. Physiological Principles

The direct effects of air pollution on central nervous system functions have been studied extensively by Russian physicians utilizing Pavlovian experimental procedures, pioneered by Ryazanov (*457*) and reviewed by Izmerov (*458*). Many of these studies use the "method of optical chronaxy determination," in which a weak electrical current is applied to the eyeball, leading to a sensation of a flash of light. For a given subject, there is a minimum intensity of electrical stimulation below which this sensation does not occur, no matter how long the stimulus continues. At intensity twice that of the minimum that just produces stimulation, the length of time to produce sensation is measured. This length of time is called the "optical chronaxy." After exposure to a pollutant, there is a change in the amount of time needed to produce the sensation, i.e., a change in chronaxy, which is taken as a measure of the effect of the pollutant on the central nervous system.

Other studies have used the minimal detectable light intensity and odor threshold techniques with exposure to butyl acetate (Fig. 26).

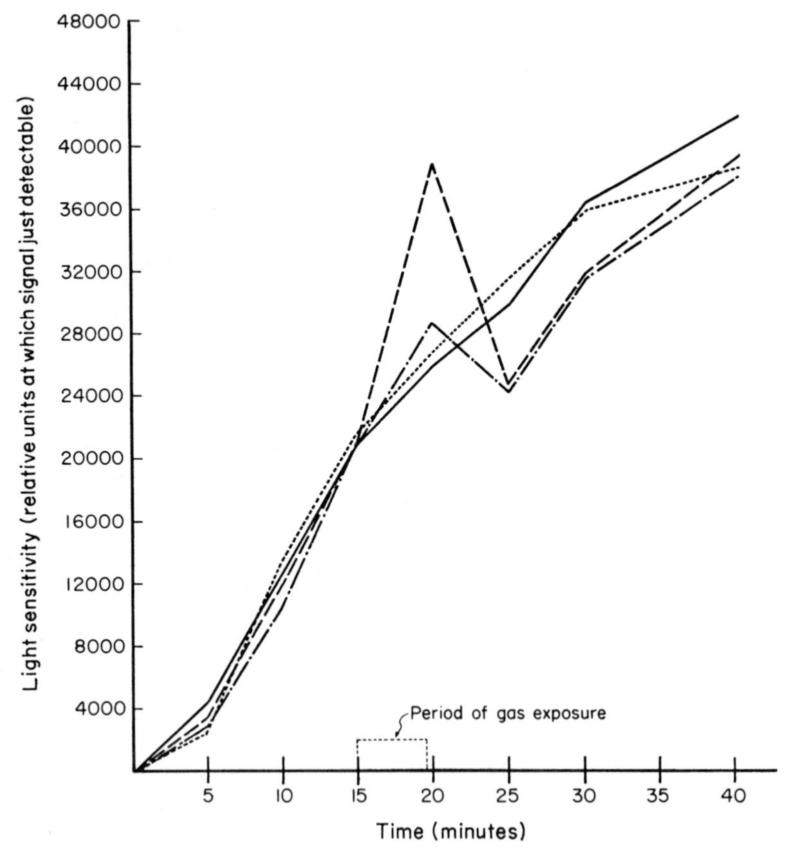

Figure 26. Changes in light sensitivity of the eye on inhalation of various concentrations of butyl acetate (*457*): (——) pure air; (---) 0.73 mg/m³; (-·-·-) 0.32 mg/m³; (····) 0.18 mg/m³. Note that the earliest effect is an increase in sensitivity (improvement in function) followed by impairment.

Ryazanov reported an increase followed by a decrease in minimum detectable light intensity. In this case, an exposure not detectable by odor produced first an increase in the ability to detect weak light stimuli; at a greater dose, a decrease in detection ability occurred. This biphasic effect can also occur with increasing duration of exposure. Other methods have used the alteration of electroencephalogram or brain waves. Such methods have been studied in the United States by Xinteras (*459*).

In addition to these reflexes, Nadel and Widdicombe (*109, 110, 112*), as mentioned earlier, have emphasized the importance of autonomic reflexes in many of the reactions of the respiratory system to irritants and particulate matter. Since the nervous system connections to the locations in the brain where odor is perceived are very rich, this leads to the assumption that odor must also cause reflex effects. The role of lead pollution in producing impairment of peripheral nerve conduction has been reported by Landrigan *et al.* (*460*). At this time, it is reasonable to state that central nervous system and autonomic nervous system reflexes are produced by air pollutants, but their significance has not been clearly defined. This area will benefit from further exploration.

2. Performance and Pollutant Exposures

The laboratory experiments of Beard and Wertheim (*205*) and Beard and Grandstaff (*209*) have raised the question as to whether carbon monoxide at commonly occuring exposure levels interferes with accurate time interval estimation. It has not been shown that such exposures interfere with performance of complex and unfamiliar tasks. Studies on the effect of carbon monoxide exposure on motor vehicle accidents has not produced decisive results; Wright (*207*) and McFarland (*208*), for example, present conflicting results.

Some studies have been done on school performance in relation to ozone exposure, but these, too, have failed to yield decisive results. In general, it is accepted that too much sensory irritation or too much discomfort will have an unfavorable effect on the learning process, but so far this has not been thoroughly studied. There is suggestive evidence of an effect of carbon monoxide on the performance of swimmers who are racing (*460*), and of oxidant on high school track meet performance (*151*), but beyond this, there are very few studies on these problems (*461*). It is reasonable to believe that if there were decisive evidence of community air pollution effects on performance in driving of motor vehicles or in learning in school, such data would be of great consequence for air pollution control.

3. Sensory Effects of Pollutants

a. EVALUATION OF ODORS. i. *Introduction* (see also Section IV,A and E). Exposure to odorous air pollutants constitutes an important problem in environmental health, particularly considering the number of people who may be exposed. A single point source such as a pulp mill may well give rise to serious odor problems miles away. In connection with a study around a Swedish pulp mill, it was found that 30–40% of the people interviewed reported "annoyance," even up to more than 10 miles (16 km) from the pulp mill (*462*).

If regulatory agencies are to take action in connection with odor exposure, information on relationships between the dose to which people are exposed and the annoyance reactions displayed must often be the point of departure. It is thus necessary to have methods that adequately describe the dosage as well as the response and to know the conditions under which the particular dose–response relationship is valid. Questions relating to odors in the ambient air have been discussed in detail at international symposia at Stockholm, Sweden, where methods for evaluating and measuring odorous air pollutants at the source and in the ambient air were discussed (*463*), and at Cambridge, Massachusetts, where evaluation of community air odor exposure was dealt with (*464*).

ii. *Definitions.* At the Stockholm meeting the following definitions were among those decided upon:

odor—a product of the activation of the sense of smell; an olfactory experience

odorant—any chemical compound that can stimulate the olfactory sense

iii. *Dose.* Since odor comes from a large number of substances with different chemical structures, it would be desirable if the estimation of odor could be based on certain physicochemical characteristics common to all odorous substances, for example, in analogy with measurement of sound level associations. Theoretically, it should be possible to translate chemical and physical data into psychophysical descriptions of odor intensity and quality, provided the odor intensity and quality of the individual substances as well as different mixtures of these substances are known. It is not possible to do this at present, although laboratory studies on the principles involved continue (*465–467a*). A deterrent to the use of physicochemical methods for this purpose is the very low concentrations of odorants. It is meaningless to estimate the average concentration of an odorous substance over a fairly long time period, since the nose responds not to an average odor, but to peaks that exceed the odor detection limit. Therefore, for instruments to be useful, they must have very short time period peaks, i.e., seconds. For most odorous substances, such instruments are not available. On the other hand, progress has been made on the analysis of certain sulfurous compounds (*468*) at very low concentrations.

The difficulties that may be encountered when comparing analytical data with sensory evaluation of odor can be illustrated by an example from the study at Morrum and Mönsteras pulp mills in Sweden (*469*) (Fig. 27). Odor threshold measurements (actually dilutions necessary to reach odor threshold) were carried out in parallel with chemical analysis for hydrogen sulfide, methylmercaptan, dimethyl monosulfide and dimethyl disulfide. The study aimed at seeing whether it was possible to find a single substance or a combination of substances that could be used as an index of odor. There was a correlation between the concentration of hydrogen sulfide and odor threshold. At both mills, a high concentration of hydrogen sulfide gave rise to a high odor threshold. But it is

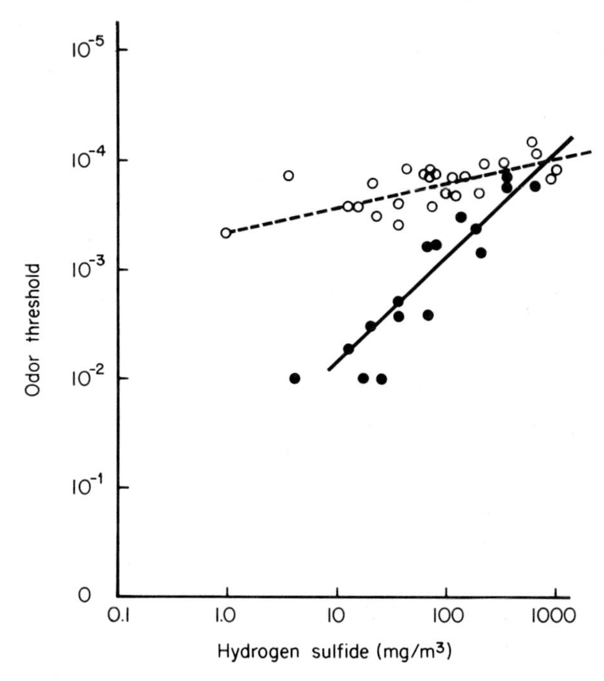

Figure 27. Correlation between hydrogen sulfide and odor threshold (*469*) near two pulp mills: ○ = Morrum (r = 0.74); ● = Mönsteras (r = 0.96). Odor threshold of 10^{-3} means that the exposure level was such that if the air sample were diluted one thousand times it would approximate the odor threshold.

s from the figure that the two mills differed with regard to low trations of hydrogen sulfide. In one, the odor threshold decreased sharply with a lowered hydrogen sulfide concentration, while in the other one, the odor threshold decreased rather insignificantly despite a decrease in hydrogen sulfide concentration from 1000 mg/m^3 to about 1 mg/m^3. Obviously, some substance other than hydrogen sulfide was responsible for the high odor thresholds at low concentrations of hydrogen sulfide.

iv. *Sensory methods.* Sensory methods, whether used for ambient air or stack gases, must include standardized experimental procedures guaranteeing an acceptable reproducibility and valid results. When measuring odors in the ambient air, an additional requirement is often a mobile odor laboratory. A large number of exposure and dosing devices, called olfactometers, using the principle of dilution of odorous gases with pure air or nonodorous gases have been adapted for both laboratory and field studies. One such is the odor chamber, in which the subject is entirely exposed to the odor to be studied. Many of the conventional olfactometers and odor chambers are not satisfactory for field work. Olfactometers should permit known and stable exposure concentration, rapid changes in concentration, and fairly natural respiratory conditions. Most investigators have chosen the odor hood as an alternative to the chamber. With the hood, only the nose, face, or head are exposed (*470–472*). By using odor hoods in air conditioned buses or trucks, it has been possible to construct high-quality mobile odor laboratories (*237; 471–476*).

A most important odor dimension is its acceptability. It is possible to study acceptability on the laboratory scale by applying direct scaling methods (*477*). Lindvall and Svensson (*478*) carried out studies in which the unpleasantness of hydrogen sulfide was compared with that of five different combustion gases. Acceptability has also been dealt with in United States field studies using a mobile laboratory for exposing test subjects to diesel exhaust gases (*479–481*).

There is no simple relationship between perception of threshold and suprathreshold intensity. Direct psychophysical scaling methods have been proposed (*482*). By these, sensation has been shown to grow as a function of stimulus intensity raised to a power. The evidence obtained in scaling intensity of many physical agents, including odors, supports the descriptive value of the power function

$$R = K(S - S_0)^n \tag{3}$$

when R is perceived intensity, S is physical intensity, S_0 is an estimate of threshold, K is a constant that depends on the units of measurement employed, and n is an exponent that depends on the sense modality. When studying intensity of odors, modern psychological scaling methods should be used (*472, 476, 483*).

Detection levels are of great importance; people may experience an odor as unpleasant at a low level of detection if it is frequent and prolonged. Particularly if the source of the odor is a point source, any change in odor intensity will influence the frequency at which the odor is perceived, i.e., the frequency with which the detection limit (absolute threshold) is exceeded in a particular area. Signal detection methodology (*484*) has been treated extensively by Lindvall (*472*), in the report from the Third Karolinska Institute Symposium on Environmental Health (*463*), and by Berglund *et al.* (*484a*). In contrast to classic threshold techniques, signal detection methodology allows for the estimation of not only the number of false positives but also the false negatives. Different modifications of the basic signal detection methodology have been used (*475, 484, 485, 486*).

b. Laboratory and Field Studies. The most extensive testing of single odorants has been done by Katz and Talbert (*487*) and more recently by Leonardos *et al.* (*113*). Psychophysical functions of perceived intensity for twenty-eight odorants have been presented by Berglund *et al.* (*488*). The exponent in the function is less than 1.0 (0.06–0.34), indicating that perceived magnitude of odor tends to be a negatively accelerated function of stimulus intensity. A model for summation of perceived intensities in odor mixtures has been suggested (*467*).

A mobile laboratory is usually required for field studies. Gas can be passed directly from the ambient air or other sources (e.g., stack gases) to the mobile laboratory. Another way is to collect samples in plastic bags, e.g., laminated Mylar (Dupont) or Hostaphen (Hoechst) (registered trademarks). Bag samples from some pulp mill effluents show only minor losses in odor intensity after a period of about 2 hours (*472*).

Sensory analysis can be used to study the relative importance of odor-generating processes. Taking such data into consideration together with the total volume of gases gives the factory and regulatory agencies valuable information as to where countermeasures should be focused (*469, 488*).

Sensory analysis can be used to study effects of countermeasures. In Table XXVI, the effects of black liquor oxidation and a chlorine scrubber on gases from a Swedish pulp mill are shown. It can be seen that black liquor oxidation brings about a reduction of odor strength (absolute odor threshold) by about ten times and chlorination a further decrease of two hundred times (since the data are based on logarithms, these are the antilogs of 1.1 and 2.3).

Table XXVI Effect of Control Measures in a Sulfate Cellulose Plant *(472)*

Source	*Odor threshold*[a]
Gas entering the oxidation tower	8.9
Gas emerging from the oxidation tower	7.8
Gas emerging from the chlorine scrubber	5.5

[a] The absolute odor thresholds are based upon group data and expressed as log dilution factors.

Several studies have been carried out in the United States on diesel exhaust gases. Emphasis has been upon the evaluation of odor intensity and quality when diluted to the suprathreshold levels typically encountered in urban areas (*474, 489–491*). In 1969 and 1970, over 5100 individuals in five cities (three or four sites per city) were exposed to diesel exhaust gases in a mobile laboratory and answered a questionnaire indicating to what extent the odor was pleasant, neutral, unpleasant, very unpleasant, or unbearable (*481*). Comparison was made with the United States Public Health Service odor quality–intensity kit. Studies have shown that the odor thresholds for limited samples of Swedish gasoline-powered cars did not differ considerably from those for diesels. Odor thresholds for the exhaust gases varied between dilutions of 10^3 and 10^4. Unfortunately, no studies on odor quality were performed. To provide an environment that will be free from odor from exhaust gases would be difficult to achieve in areas with heavy traffic (*474a*). In one Swedish study (*475*), odor levels in the ambient air were measured by signal detection methodology in a main street during rush hours. The index of detectability was found to vary in an expected fashion compatible with the traffic load.

c. Response to Odor Exposure. Possible responses involve disease and annoyance reactions. In addition, there is some evidence from laboratory studies in humans, as well as in animals, that exposure to odors may elicit transitory nervous system effects (*492, 493*). No evidence has yet shown that odors per se are related to disease states (*464*). However, some suggestive data keep this question open (*494–500*). What should be focused on particularly here is to what extent possible adverse effects are due to odors per se, to some specific odorous substance, or to some substances combined with the odorous substance.

i. *Annoyance reactions from odor* (see also Section VI E,4). The presence of an odorous substance in the air may or may not be associated with awareness or reactions. If the person is aware of the odor, he may or may not have a negative reaction. Negative reactions are generally considered annoyance reactions. In addition, it is possible that there are other reactions, such as irritation or reflex responses. Of great practical importance has been the determination of annoyance reactions by standardized questionnaires and survey procedures; these have been found to be the most valid and dependable ways of estimating the health importance of odor exposure.

Several interview surveys covering people exposed to odor from pulp mills, manure, oil refineries, and motor vehicle exhausts (*462, 474, 476, 481*) have shown that odor can give rise to pronounced annoyance reactions. For city dwellers who complain strongly about the air pollution situation, the exposure to odors is one major cause of such complaints, but exposure to other agents is certainly also important.

Some quantitative data can be given on prevalence of annoyance from odor. Studies near Swedish pulp mills have shown annoyance reactions in 12% to 62% of the population, depending on distance from the mill (*462, 469*). Similar frequencies have been observed in Eureka, California, and Clarkston, Washington (*498; 501*). Studies around a Swedish oil refinery show that up to about 40% of the population in the vicinity may report annoyance reactions.

During 1969, the Japanese government received 40,000 petitions concerning a variety of environmental problems. Of these, 8000 were on odor. For comparison, 18,000 petition complaints were received for the same time interval about noise and vibration (*502*). Spontaneous complaints are not always accurate indices of exposure. This is further discussed below.

ii. *Dose–response relationships.* A complete dose–response curve has not been shown for any odor exposure. A few points on the curve for certain odors can be derived from the previously noted interview surveys in Sweden, where about 12–62% of the population around a pulp mill [depending upon the distance from the plant, within a radius of about 10–15 miles (16–24 km)] reported annoyance reactions. The odor in the main stack was believed to correspond to a dilution factor of about 10^5 times above threshold. In a study of pulp mill odor problems in Eureka, California, it was possible to find a dose–response relationship for different areas in that annoyance reactions were more pronounced in areas with presumptively higher exposure (*498*). To some extent, the United States studies previously noted on diesel exhaust gases (*474, 481*) resulted in a type of dose–response curve that can be used for estimating what effects various levels of odor control would have on relative annoyance. The only caveat is that the ratings "pleasant" to "unbearable" were made in laboratory experiments, rather than in a real-life situation with continuous exposure.

No doubt the attitude of the public toward the need for protection against adverse effects from the environment will change considerably within the coming years. It may well be that what is acceptable today will be considered unacceptable within the not too distant future. Present dose–response relations will then no longer be valid. This is one reason that efforts should be focused on the dose and on simple psychophysical relations. The ability of the human nose to detect and to make perceptual evaluations in controlled laboratory experiments should not change. If information on the dose is available, it will always be possible to establish standards for odors based on known intensities and qualities.

4. Nonspecific Annoyance Reactions

Increased attention is paid nowadays not only to proven or suspected health hazards, but also to reactions only potentially related to disease states of uncertain health significance. Such effects include annoyance reactions resulting from sensory perception of pollutants (*503*). An international symposium in Stockholm (*503*) adopted the following definition of annoyance: "A feeling of displeasure associated with any agent or condition believed to affect adversely an individual or group." References in relation to annoyance from noise can be found in several reports (*503–509*). Annoyance from air pollution can arise after exposure to odors, particulates, and irritants, or because of impairment of visibility. Often it is not possible to distinguish which air pollutants have contributed most to the annoyance reactions found.

Odors are known to give rise to annoyance reactions, well known in the case of odors from pulp mills, fertilizer factories, oil refineries, and motor vehicles, as noted in the previous section. Irritants may cause severe annoyance. Sulfur dioxide constitutes an important source of irritation to the respiratory tract. In a Los Angeles, California, study, 75% of the population was bothered by air pollution, mainly because of eye and nasal irritation (*83*).

Although aerometric techniques and monitoring are constantly being improved and expanded, it is not necessarily true that a given objectively measured air pollution level will bring about a particular human annoyance reaction. This response depends upon sensory and sociopsychological variables. Some of these depend on age, sex, social class, length of residence, group ties, education, occupation, and economic relation to the possible source of annoyance. Likewise, these variables will also determine whether a feeling of annoyance will advance to an action. Annoyance reactions are a relatively new field of study. As brought out at the Stockholm symposium (*503*), the methods of measurement, especially for the response, are neither fully developed nor validated.

To date, socioepidemiological surveys have provided most quantitative data. This is because the only means by which we can get information on subjective symptoms and experience of annoyance is through information supplied by the individual himself. The methods used involve standardized questioning techniques that elicit responses expressed in terms relevant to groups. The criteria are the same for sampling the population to be examined as in other epidemiological studies. Since the objective of many surveys is to test differences among subgroups, defined by age, sex, and exposure, statistical design often involves stratification procedures. In investigating objective data, e.g., concentrations of substances in blood or urine, the method of measurement is almost invariably described. This is also necessary for measurements of subjective symptoms in epidemiological studies. Among possible sources of error in such studies, biases due to interviewer effect, respondent effect, and instrument effect are the most important.

The "interviewer effect" means that the interviewer does not register the response correctly. The reason may be anticipatory due to preconception of the subject's answer. If the answer is not entirely clear, the interviewer may probe or interpret it in accordance with his own preconception. The training of interviewers is important, and two handbooks on this subject are recommended—Cannel and Kahn (*510*) and "Manual for Interviewers" (*511*). A way of reducing or controlling the interviewer effect is to use several interviewers and to assign the subjects at random among them. This method was used in an epidemiological interview study of about 700 people living around a shale oil factory in Sweden. The prevalence of asthma and/or bronchitis as found by fourteen interviewers varied between 10% and 43% (*512, 513*). Such a result could be very misleading if the interviewers had been allotted to certain areas or certain strata of the population.

The respondent effect is due to differences in the respondents' frames of reference or to the conscious or unconscious desire of the respondent to bring about changes in the exposure. In the same survey around a shale oil plant at one time during the studies, part of the population interviewed were informed of the purpose of the investigation, while the rest were not (*514, 514a*). Table XXVII shows how the prevalence of certain symptoms seemed to differ between the two groups as a result. One way to avoid the respondent effect is to use an introduction as neutral and correct as possible. The subsequent questions should be as explicit as possible. Psycho-

Table XXVII Prevalence of Individuals in a Survey near a Swedish Shale Oil Plant with Certain Symptoms among Subjects "Not Informed" and "Informed" about the Purposes of the Survey *(513, 514)*

	Prevalence (%)	
Symptoms	*Not informed*	*Informed*
Smell noticed	52	69
Distressed by smell	13	20
Asthma and/or bronchitis	11	20

logical scaling methods have also been suggested for studies of the response criteria variations in different populations (*515*).

The instrument effect is related to the type of questionnaire used and to the wording of the actual questions. Of paramount importance is the use of proper question design (*516*). The questionnaire has to be simple to yield a satisfactory response rate, yet studies have shown that under certain circumstances, mail questionnaires may be a good alternative to personal interviews (*517, 518*).

The use of spontaneous complaints has been relied upon by governmental agencies in several countries. Most data show that the use of spontaneous complaints is a poor method, because so many factors influence whether or not a person will complain. The occurrence of a high frequency of spontaneous complaints may, of course, show that a problem exists, but it is not possible to draw from them any firm conclusions about the magnitude of the problem. It has been shown, for example, in American, British, and Swedish surveys that less than 10% of the population report any form of complaint by writing letters, telephoning, or making personal visits to officials (*505, 506, 519*). On the other hand, studies around a Swedish pulp mill have shown that of those who signed a petition against the pulp mill, only about 50% were annoyed by its odor.

G. Hematological Reactions

The effect of carbon monoxide in interfering with the oxygen transport function of the blood has been shown in cigarette smokers to affect the number of red blood cells produced. This is shown by a small increase in the average hematocrit (i.e., the fraction of red blood cells in the whole blood) and in the mass of circulating total red cell volume. However, there are no data indicating that such an effect occurs as a result of community air pollution. Similarly, there is no evidence of hematological effects from community exposure to levels of lead resulting from motor vehicle exhaust.

However, Kapalin (*520*) has summarized findings on blood changes in groups of children living in clean areas as opposed to those in areas polluted by smoke and sulfur dioxide (Fig. 28). Children from the polluted town of Litvinov in Czechoslovakia have statistically significant higher red cell counts and lower average cell volume than children from the clean town of Liebechov. There are no differences between the hematocrit and the total amount of hemoglobin levels among these communities. When the data are plotted on cumulative probability grids and studies are repeated over several years, these findings are striking in their persistence. When youngsters are taken to summer holiday camps, the blood picture generally returns to the clean air condition. Kapalin has also shown that there is a decrease in the rate of skeletal growth of children in the more polluted towns. It is difficult to estimate the amount and type of pollution that quantitatively produces such effects. This has led to some difficulty in interpreting these resuls, but there is no doubt as to their nature.

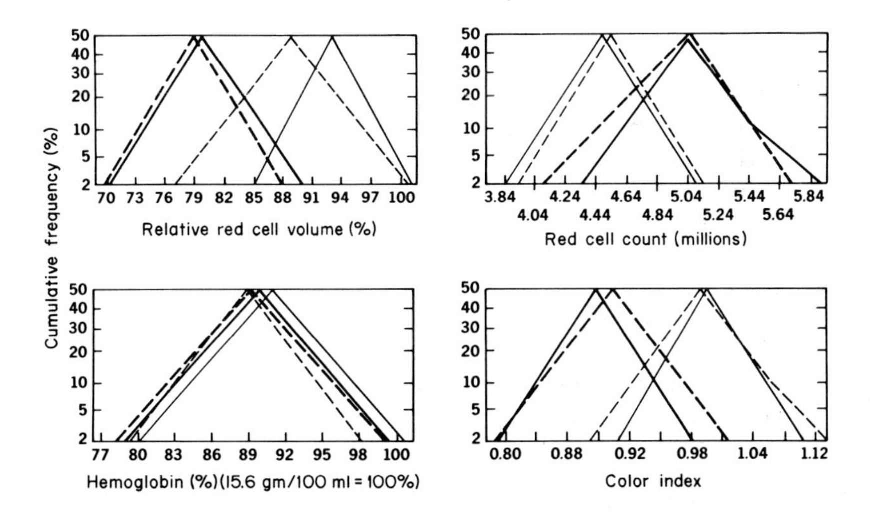

Figure 28. Distribution of values of red blood cell measurements for Czechoslovakian school children for 2 years—1962 (broken lines) and 1963 (solid lines)—in two towns Litvinov (heavy lines), a town with pollution by smoke, and Libechov (light lines), a comparable town with clean air. While no differences occur with hemoglobin measurements, in the polluted town the cell volumes are smaller and the cell counts higher.

H. Chemical Mutation

A suggestive link between photochemical pollution and possible lethal mutations was reported by Lewis *et al.* (*521*) in male mice exposed to irradiated auto exhaust during the full length of the spermatogenic cycle. Mertz *et al.* (*129a*) have shown that ozone exposure of humans led to detectable increased frequency of mutations in lymphocytes. Increased neonatal mortality was noted in mice born in irradiated automotive exhaust atmospheres containing commonly measured levels of photochemical pollutants. Male mice exposed during the spermatogenic cycle to irradiated exhaust containing ambient concentrations of pollutants sired fewer and smaller litters per mating (reduced fertility), and had more neonatal deaths than among litters living in clean air, suggesting lethal mutations (*521*). The same system used by Lewis *et al.* was employed by Coffin and Blommer (*522*) in demonstrating that irradiated automobile exhaust increased the lethality of bacterial infection after 4 hours of exposure to a pollutant mixture containing 25 ppm carbon monoxide and 0.15 ppm oxidant.

Potential Chromosome Damage and Birth Defects

Chromosome damage and human birth defects could be suspected on the basis that chemicals known to be suspected of having these effects in animals are present in the environment. Male infertility and increased early infant mortality have been described in mice exposed to irradiated auto exhaust, suggesting damage to chromosomes by simulated photochemical pollution. Possible lethal mutations and partial sterility in male mice and high mortality of infant mice suggest areas for more active study in human populations. Human males exposed to mutagenic pollutants might be expected to father children with an increase of genetically conditioned abnormalities. Such phenomena have not been documented but should be looked for in populations exposed to both occupational and community pollution.

Bacterial mutation tests have been greatly improved recently (*523*) and can be used for rapid screening of compounds and mixtures. Such screening is also thought to be relevant to carcinogenic hazards.

ACKNOWLEDGMENTS

Ms. Pamela Boston has greatly assisted the authors in verification and coordination of references and in editing.

REFERENCES

1. J. R. Goldsmith, *Arch. Environ. Health* **18**, 516–522 (1969).
2. "Health Hazards of the Human Environment," pp. 19–46. World Health Organ., Geneva, Switzerland, 1972.
3. "Threshold Limit Values for Chemical Substances and Physical Agents in the Workroom Environment with Intended Changes for 1976." Amer. Conf. Govtl. Ind. Hyg., Cincinnati, Ohio, 1976.
4. American Conference of Governmental Industrial Hygienists (ACGIH), *J. Occup. Med.* **16**, 39–49 (1974).
5. O. Brody, *Calif. Health* **23**, 85–86 (1965).
6. W. C. Cooper and J. R. Tabershaw, *Arch. Environ. Health* **12**, 522–530 (1966).
7. "Smoking and Health" (Report of the Advisory Committee to the Surgeon General of the Public Health Service), Pub. Health Serv. Publ. No. 1103, p. 194 and pp. 295–297. United States Department of Health, Education and Welfare, Washington, D.C., 1964.
8. Royal College of Physicians, "Air Pollution and Health." Pitman, London, England, 1970.
9. D. D. Reid, D. O. Anderson, B. G. Ferris, Jr., and C. M. Fletcher, *Brit. Med. J.* **2**, 1487–1491 (1964).
10. E. C. Hammond, *Nat. Cancer Inst., Monogr.* **19**, 127–204 (1966).
11. "Indoor-Outdoor Air Pollution Relationships Review," A Literature Publication AP-112. United States Environmental Protection Agency, Research Triangle Park, North Carolina, 1972.
12. Task Group on Metal Accumulation, *Environ. Physiol. Biochem.* **3**, 65–107 (1973).
13. L. G. Strandberg, *Arch. Environ. Health* **9**, 160–166 (1964).
14. N. R. Frank and F. E. Speizer, *Arch. Environ. Health* **11**, 624-634 (1965).
15. M. Amdur, *Int. J. Air Pollut.* **3**, 201–220 (1960).
16. N. R. Frank, M. O. Amdur, and J. L. Whittenberger, *Int. J. Air Water Pollut.* **8**, 125–133 (1964).
17. Task Group on Lung Dynamics, *Health Phys.* **12**, 173–208 (1966).
18. I. B. Wilson and V. K. LaMer, *J. Ind. Hyg. Toxicol.* **30**, 265 (1948).
19. J. H. Brown, K. M. Cook, F. G. Ney, and T. F. Hatch, *Amer. J. Pub. Health Nat. Health* **40**, 450 (1950).
20. W. Findeisen, *Pfluegers Arch. Gesemte Physiol. Menschen* Tiere **236**, 367–379 (1935).
21. H. P. Landahl, *Bull. Math. Biophys.* **12**, 43–56 (1950).
22. C. E. Junge, "Air Chemistry and Radioactivity." Academic Press, New York, 1963.
23. S. K. Friedlander and C. S. Wang, *J. Colloid Interface Sci.* **22**, 126–132 (1966).
24. M. Lippmann, R. E. Albert, and H. T. Peterson, *in* "Inhaled Particles" (W. H. Walton, ed.), 3rd ed., Vol. I, pp. 165–180. Unwin Brothers, London, England, 1971.
25. B. Holma, *Acta Med. Scand., Suppl.* **473** (1967).
26. R. E. Albert, M. Lippmann, and H. T. Peterson, *in* "Inhaled Particles" (W. H. Walton, ed.), 3rd ed., Vol. 1, pp. 165–180. Unwin Brothers, London, England, 1971.
27. P. Camner, K. Philipson, and L. Friberg, *Arch. Environ. Health* **24**, 82–87 (1972).
28. R. Rylander, *Acta Physiol. Scand., Suppl.* **306** (1968).

29. G. A. Laurenzi, L. Berman, M. First, and E. H. Kass, *J. Clin. Invest.* **43,** 759–768 (1964).
30. E. Goldstein, C. Eagle, and P. D. Hoeprich, *Arch. Environ. Health* **26,** 202–204 (1973).
31. T. T. Mercer, *Health Phys.* **13,** 1211–1221 (1967).
32. A. S. Teirstein, A. Miller, A. M. Langer, and I. W. Selikoff, *Amer. Rev. Resp. Dis.* **107,** 1113–1114 (1973).
33. N. R. Frank, *Proc. Roy. Soc. Med.* **57,** 1029–1033 (1964).
34. B. Åberg, L. Ekman, R. Falk, U. Greitz, G. Persson, and J.-O. Snihs, *Arch. Environ. Health* **19,** 478–484 (1969).
35. J. B. Hursh and J. Suomela, *Acta Radiol.* **7,** 108–120 (1967).
36. G. W. Wetherill, M. Rabinowitz, and J. D. Kopple, *Proc. Int. Symp. Recent Advan. Assessment Health Effects Environ. Pollut. 1974* Vol. II pp. 847–860. Published by the Commission of the European Communities Directorate General for the Dissemination of Knowledge Centre for Information and Documentation–CID, Luxemburg, Belgium, 1975.
37. K. M. Six and R. A. Goyer, *J. Lab. Clin. Med.* **76,** 933–943 (1970).
38. S.-E. Larsson and M. Piscator, *Isr. J. Med. Sci.* **7,** 495 (1971).
39. M. Piscator and S.-E. Larsson, *Int. Congr. Occup. Health, 17th.* Paper presented at Buenos Aires, Argentina, 1972.
40. K. M. Six and R. A. Goyer, *J. Lab. Clin. Med.* **79,** 128–138 (1972).
41. M. Firket, *Bull. Acad. Roy. Med. Belg.* **11,** 683–739 (1931).
42. K. Roholm, *J. Ind. Hyg. Toxicol.* **19,** 126–137 (1937).
43. B. Roueché, "Eleven Blue Men." Little, Brown, Boston, Massachusetts, 1953.
44. H. H. Schrenk, H. Heimann, G. D. Clayton, W. M. Gafafer, and H. Wexler, *Bull.* No. 306, Public Health Service, U.S. Department of Health, Education and Welfare, Washington, D.C. (1949).
45. D. J. Thompson and A. Ciocco, *Brit. J. Prev. Soc. Med.* **12,** 172–182 (1958).
46. L. C. McCabe and G. D. Clayton, *Arch. Ind. Hyg. Occup. Med.* **6,** 199–213 (1952).
47. Ministry of Health "Mortality and Morbidity During the London Fog of December 1952," Reports on Public Health and Related Subjects, No. 95, Table 1, p. 2. HM Stationary Office, London, England, 1954.
48. A. E. Martin and W. H. Bradley, *Mon. Bull. Min. Health Pub. Health Lab. Serv.* **19,** 56 (1960).
49. J. A. Scott, *Med. Off.* **109,** 250-252 (1963).
50. L. Greenburg, M. B. Jacobs, D. M. Drolette, F. Field, and M. M. Braverman, *Pub. Health Rep.* **77,** 7 (1962).
51. L. Greenburg and F. Field, *Arch. Environ. Health* **4,** 477–486 (1962).
52. R. Lewis, M. M. Gilkeson, Jr., and R. O. McCaldin, *Pub. Health Rep.* **77,** 947–954 (1962).
53. H. Weill, M. M. Ziskind, V. J. Derbes, R. J. M. Horton, R. O. McCaldin, and R. C. Dickerson, *Arch. Environ. Health* **10,** 148–151 (1965).
54. H. Weill, M. M. Ziskind, R. C. Dickerson, and V. J. Derbes, *J. Air Pollut. Contr. Ass.* **15,** 467–471 (1965).
55. P. A. Kenline, *Arch. Environ. Health* **12,** 295–304 (1966).
56. S. Booth, I. DeGroot, R. Markush, and R. J. M. Horton, *Arch. Environ. Health* **10,** 152–155 (1965).
57. L. Greenburg, C. L. Erhardt, F. Field, and J. I. Reed, *Arch. Environ. Health* **10,** 351–356 (1965).
58. D. A. Lynn, B. J. Steigerwald, and J. H. Ludwig, *Publ.* 999-AP-7, Public Health

Table I *(Cont.)*

Transactions of the British Ceramic Society
Transactions of the Institution of Chemical Engineers
Transactions of the Society of Heating, Air Conditioning, and Sanitary Engineers of Japan
Tribune du Cebedeau
Tsvetnye Metally (Non-Ferrous Metals)
Tue (Sicherheit und Zuverlässigkeit in Wirtschaft—Betrieb—Verkehr)

Uhli
Umschau in Wissenschaft und Technik
Umwelt (Düsseldorf)
Uspekhi Khimii
Uzbekskii Khimicheskii Zhurnal

V.D.I. Berichte (Veréin Deutscher Ingenieure)
V.D.I. Forschungsheft (Veréin Deutscher Ingenieure)
V.D.I. Nachirichten (Veréin Deutscher Ingenieure)
V.D.I. Zeitschrift (Veréin Deutscher Ingenieure)
VGB Kraftwerkstechnik
Vodosnabzhenie i Sanitarnaya Tekhnika
Voprosy Onkologii

W.H.O. Chronicle (World Health Organization)
Wärme Stoffubertraguny
Wasser Luft und Betrieb
Waste Age
Water, Air, and Soil Pollution
Water and Pollution Control
Water and Sewage Works
Weather
Werkstoffe und Korrosion
Wissenschaftliche Zeitschrif dter Technischen Universitäts Dresden
Work—Environment—Health (Helsinki)
World Petroleum

Yosui To Haisui (Water and Waste)

Zashchita Metallov
Zavooskaya Laboratory (Factory Laboratory)
Zeitschrift Für Analytische Chemie
Zeitschrift Für Erzbergbau und Metallhuttenwesen
Zeiteschrif Für die Gesamte Hygiene und Ihre Grenzgebiete
Zeitschrift Für Krebsforschung
Zeitschrift Für Lebensmittel-Untersuchung und Forschung
Zeitschrift Für Meteorologie
Zeitschrift Für Pflanzenkrankheiten Pflanzenpathologie und Pflanzenschutz
Zietschrift Für Physikalische Chemie (Leipzig)
Zeitschrift Für Physikalische Chemie, Neue Folge
Zeitschrift Für Präventivmedizin
Zeitschrift Für die Zucketindustrie
Zement—Kalk—Gips
Zentralblatt Für Arbeitsmedizin und Arbeitsschutz
Zentralblatt Für Bakteriologie, Parasitenkunde, Infektionskrankheiten und Hygiene, Abteilung I Originale
Zhurnal Analiticheskoi Khimii (Moscow)
Zhurnal Prikladnoi Khimii
Zucker

1. Journals Ranked by Frequency of Selection

Citations of the most frequently encountered journals within the APTIC (8) file are ranked in descending order in Table II. The significance of this frequency to the optimum utilization of resources for journal acquisitions should be apparent.

Table II Journals Most Frequently Productive in 1973 and 1974

Taiki Osen Kenkyu (Journal of the Japan Society of Air Pollution)	*Umwelt (Düsseldorf)*
Federal Register	*Gijutsu To Kogai (Technical Pollution)*
Atmospheric Environment	*Kankyo Gijutsu (Environment Conservation Engineering)*
Staub-Reinhaltung der Luft	*Netsu Kanri To Kogai (Heat Management and Public Nuisance)*
Environmental Science and Technology	*Kuki Seijo (Journal of the Japan Air Cleaning Association)*
Gigiena i Sanitariia	*Analytical Chemistry*
Journal of the Air Pollution Control Association	*VGB Kraftwerkstechnik*
Archives of Environmental Health	*Journal of Applied Meteorology*
Kogai To Taisakii (Journal of Pollution Control)	*Nature (London)*
Nenryo Oyobi Nensho (Fuels and Combustion)	*PPM (Japan)*
Kankyo Sozo (Environmental Creation)	*Sangyo Kogai (Industrial Public Nuisance)*
Nippon Koshu Eisei Zasshi (Japanese Journal of Public Health)	*Science*
American Industrial Hygiene Association Journal	*Kagaku Kojo (Chemical Factory)*
Pure and Applied Geophysics (Basel)	*VDI Berichte (Verein Deutscher Ingenieure)*
Wasser Luft und Betrieb	*Jidosha Gijutsu (Automobile Engineering)*

2. Journals Ranked by Frequency of Citation

All of the original works and reports published in the six issues of the *Journal of the Air Pollution Control Association* dated January through June 1974 were examined. The titles of the journals cited in their reference lists were recorded and counted. The top twenty listed in descending frequency order are shown in Table III.

Table III Journals Most Frequently Cited in the *Journal of the Air Pollution Control Association,* January–June, 1974

Journal title	*Citations*	*Journal title*	*Citations*
Journal of the Air Pollution Control Association	62	*Agronomy Journal*	2
Atmospheric Environment	10	*American Economic Review*	2
Environmental Science and Technology	8	*Analytical Chemistry*	2
Journal of Applied Meteorology	8	*Bryologist*	2
Archives of Environmental Health	5	*Bulletin of the American Meteorological Society*	2
Industrial and Engineering Chemistry	5	*Chemical Engineering*	2
Chemical Engineering Progress	3	*Journal of the American Medical Association*	2
Journal of the Institute of Fuel	3	*Journal of Applied Chemistry*	2
Nature (London)	3	*Proceedings of the American Power Conference*	2
Swedish Journal of Economics	3	*Science*	2

B. Books

Many books dealing with air pollution have been published. Proceedings of technical meetings, manuals, and multiauthored edited volumes are considered books (Tables IV, V, and VI). First, there are books on

Table IV Books on Environmental Pollution

A. General Information

"Cleaning Our Environment: The Chemical Basis of Action," American Chemical Society, Washington, D.C., 249pp (1969).
"Economics of Transboundary Pollution," OECD Publications Center, Washington, D.C., 218pp (1976).
"Electricity and the Environment—The Reform of Legal Institutions," West Publishing Company, St. Paul, Minnesota (1973).
"Handbook of Environmental Monitoring," Technomic Publishing Co., Westport, Connecticut (1974).
"Pollution Analyzing and Monitoring Instruments," Noyes Data Corp., Park Ridge, New Jersey, 354pp (1972).
"International Conference on Air Pollution and Water Conservation," British Non-Ferrous Metals Research Association, London, England, 197pp (1970).
"International Symposium on Environmental Health Aspects of Lead," The European Communities, Luxembourg, xic + 1168pp (1973).
"Proceedings of the International Symposium on Environmental Measurements," Beckman Instruments, Fullerton, California, 150pp (1974).
"Proceedings of Second Joint Conference on Sensing of Environmental Pollutants," Instrument Society of America, Pittsburgh, Pennsylvania, 377pp (1973).
"Lead in the Canadian Environment," National Research Council, Ottawa, Ontario, 116pp (1973).
"The Language of Pollution: A Glossary of Environmental Terms," New York Botanical Gardens, Bronx, New York, 31pp (1972).
"The Law and Practice Relating to Pollution Control in the Member States of the European Communities: A Comparative Survey," various editors, 9 vols., English language, Graham and Trotman, Ltd., London, England, 200–250 pages each (1976).
"The Monitoring of the Environment in the United Kingdom," H. M. Stationary Office, London, England, 66pp (1974).
"Restoring the Quality of Our Environment—Report of the Environmental Pollution Panel, President's Science Advisory Committee," U.S. Government Printing Office, Washington, D.C., xii + 317pp (1965).
Scientific and Technical Assessment Report (STAR series), U.S. Environmental Protection Agency, Washington, D.C., Particulate Polycyclic Organic Matter (PPOM)—EPA-600/6-75-001 (March 1975); Manganese—EPA-600/6-75-002 (April 1975); Cadium—EPA-600/6-75-003 (March 1975); Vinyl Chloride and Polyvinyl Chloride—EPA-600/6/75-004 (December 1975).
Study of Critical Environmental Problems (SCEP), "Man's Impact on the Global Environment," MIT Press, Cambridge, Massachusetts, 319pp (1970).
Arbuckle, J. G., Baum, R. L., Miller, M. L., Sullivan, T. F. P. *in* "Environmental Law

Table IV (*Cont.*)

Handbook," (R. A. Young, ed.), Government Institutes, Bethesda, Maryland, xii + 308pp (1975).
Barnes, D. S., "Environmental Chemistry in the Laboratory," Canfield, London, England, 166pp (1973).
Berezkin, V. G., and Tatarinskii, V. S., "Gas-Chromatographic Analyses of Trace Impurities," Plenum, New York, New York 177pp (1973).
Berthouex, P. M., and Rudd, D. F., "Strategy of Pollution Control," Wiley, New York, New York, ix + 579pp (1977).
Brebbia, C. A., (ed.), "Mathematical Models for Environmental Problems," Halsted Press, New York, New York, 537pp (1976).
Brittin, W. E., West, R., and Williams, R. (eds.), "Air and Water Pollution," Colorado Associated University Press, Boulder, Colorado, 625pp (1972).
Burchell, R. W., and Listokin, D., "The Environmental Impact Handbook," Rutgers University, New Brunswick, New Jersey, iii + 234pp (1975).
Coate, L. E., and Bonner, P. A. (eds.), "Regional Environmental Management," Wiley, New York, New York, xi + 348pp (1975).
Coppock, J. T., and Wilson, C. B. (eds.), "Environmental Quality—With Emphasis on Urban Problems," Wiley, New York, New York 207pp (1974).
Coulston, F., and Korte, F. (eds.), "Environmental Quality and Safety—Global Aspects of Chemistry, Toxicology and Technology Applied to the Environment," Vol. 1, 267pp (1972); Vol. 2, xviii + 333pp (1973), Academic Press, New York, New York.
Davies, J. C., III, "The Politics of Pollution," Pegasus, New York, New York, 231pp (1970).
Daetz D. and Pantell, R. H., "Environmental Modeling Analyses and Management," Dowden, Hutchinson and Ross, Stroudsburg, Pennsylvania, xv + 407pp (1974).
Deininger, R. A. (ed.), "Design for Environmental Information Systems," Ann Arbor Science, Ann Arbor, Michigan, xvi + 422pp (1974).
Deininger, R. A. (ed.), "Models for Environmental Pollution Control," Ann Arbor Science, Ann Arbor, Michigan (1973).
Estes, J. E., and Senger, L. W., "Remote Sensing Techniques for Environmental Analysis," Hamilton/Wiley, New York, New York, 340pp (1974).
Fishbein, L., "Chromatography of Environmental Hazards, Vol. 2: Metals, Gaseous and Industrial Pollutants," Elsevier, New York, New York, 652pp (1974).
Fuller, E. C., "Chemistry and Man's Environment," Houghton Mifflin, Boston, Massachusetts, 502pp (1974).
Giddings, J. C., and Monroe, M. B. (eds.), "Our Chemical Environment," Canfield, San Francisco, California, 367pp (1972).
Goldman, M. I. (ed.), "Controlling Pollution—The Economics of a Cleaner America," Prentice-Hall, Englewood Cliffs, New Jersey (1967).
Grob, R. L. (ed.), "Chromatographic Analysis of the Environment," Marcel Dekker, New York, New York, x + 752pp (1975).
Haefele, E. T., "Representative Government and Environmental Management," Resources for the Future, Baltimore, Maryland, xii + 188pp (1974).
Heaslip, G. B., "Environmental Data Handling," Wiley, New York, New York, xi + 203pp (1975).
Heffernan, P., and Corwin, R., "Environmental Impact Assessment," Freeman Cooper, San Francisco, California, 277pp (1975).

Table IV (*Cont.*)

Higgins, I. J. and Burns, R. G., "The Chemistry and Microbiology of Pollution," Academic Press, New York, New York, vii + 248pp (1976).

Inhaber, H., "Environmental Indices," Wiley, New York, New York, xiv + 178pp (1976).

Jimeson, R. M., and Spindt, R. S. (eds.), "Pollution Control and Energy Needs—Advances in Chemistry Series No. 127, Proceedings of Symposium," American Chemical Society, Washington, D.C., 249pp (1973).

Lapedes, D. N. (ed.-in-chief), "Encyclopedia of Environmental Science," McGraw-Hill, New York, New York, 754pp (1975).

Leh, F. K., and Lak, R. K., "Environment and Pollution—Sources, Health Effects, Monitoring and Control," Thomas, Springfield, Illinois, xx + 288pp (1974).

Mallet, L. C. M., Brison, J., *et al.*, "Environmental Pollution by 3,4-Benzopyrene-Type Polycyclic Hydrocarbons; Pollution des Milieux Vitaux par les Hydrocarbures Polybenzeniques du Type Benzo-3,4 Pyrene," Maloine, Paris, France, 195pp (1972).

Masters, G. M., "Introduction to Environmental Science and Technology," Wiley, New York, New York, xii + 404pp (1974).

McKnight, A. D., Marstrand, P. K., and Sinclair, T. G. (eds.), "Environmental Pollution Control—Technical, Economic and Legal Aspects," George Allan and Unwin, London, England, 342pp (1974).

Payne, C. (ed.), "Fuel and the Environment," Ann Arbor Science, Ann Arbor, Michigan, 174pp (1973).

Pearce, D. W., "Environmental Economics," Longman, New York, New York, ix + 202pp (1976).

Perloff, H. S. (ed.), "The Quality of the Urban Environment," Resources for the Future, Washington, D.C., 332pp (1969).

Pitts, J. N., Jr., Metcalf, R. L., and Lloyd, A. (eds.), "Advances in Environmental Science and Technology," Wiley/Interscience, New York, New York, Vol. I, 356pp (1969); Vol. II, vii + 354pp (1971); Vol. III, ix + 366pp (1974); Vol. IV, ix + 383pp (1974); Vol. V, ix + 371pp (1975).

Polunin, N. (ed.), "Proceedings of the International Conference on Environmental Future," Harper and Row, New York, New York, 660pp (1972).

Pratt, J. W. (ed.), "Statistical and Mathematical Aspects of Pollution Problems," Marcel Dekker, New York, New York, 392pp (1974).

Quellette, R. P., Greeley, R. S., and Orerbey, J. W., "Computer Techniques in Environmental Science," Mason/Charter, New York, New York, 248pp (1975).

Rabin, E. H., and Schwartz, M. D. (eds.), "The Pollution Crisis—Official Documents," Oceana, Dobbs Ferry, New York, 510pp (1971).

Rosen, S. J., "Manual for Environmental Impact Evaluation," Prentice-Hall, Englewood Cliffs, New Jersey, viii + 232pp (1976).

Ross, S. S., "Environmental Regulation Handbook," Environment Information Center, New York, New York, 1063pp (1973).

Sax, J. L., "Defending the Environment—A Strategy for Citizen Action," Knapp, New York, New York, 252pp (1971).

Selig, E. I., "Effluent Charges on Air and Water Pollution," Environmental Law Institute, Washington, D.C., vi + 102pp (1973).

Shabad, L. M., "Circulation of Carcinogens in the Surrounding Environment;" "O

Table IV (*Cont.*)

Tsirkulyatsii Kantserogenov v Okruzhayushchei Srede," Meditsina, Moscow, USSR, 367pp (1973).
Singer, S. F. (ed.), "Global Effects of Environmental Pollution," Springer-Verlag, New York, New York/Reidel, Dordrecht, Netherlands, 218pp (1970).
Sive, M. R. (ed.), "Environmental Legislation," Praeger, New York, New York, xxxiv + 561pp (1976).
Stoker, H. S., and Seager, S. L., "Environmental Chemistry: Air and Water Pollution," Scott, Foresman, Glenview, Illinois, 186pp (1972).
Vesilind, P. A., "Environmental Pollution and Control," Ann Arbor Science, Ann Arbor, Michigan, 232pp (1975).
Walker, C., "Environmental Pollution by Chemicals," Hutchinson Educ., London, England, 108pp (1971).
Wang, J. V., "Instruments for Physical Environmental Measurements," Millieu Information Service, San Jose, California, 2 Vols., 820pp (1976).
Yannacone, V. J., and Cohen, B. S., "Environmental Rights and Remedies," 2 Vols., Lawyers Co-operative Publishing Co., Rochester, New York (1971).

B. Control Technology

"Pollution Control Technology; Air Pollution Control, Water Pollution Control, Solid Waste Disposal," Res. Educ. Assn., New York, New York, 585pp (1973).
Cheremisinoff, P. N., and Young, R. A., "Pollution Engineering Practice Handbook," Ann Arbor Science, Ann Arbor, Michigan, xii + 1073pp (1975).
Dober, R. P., "Environmental Design," Krieger, Huntington, New York, 286pp (1969).
Lindner, G., and Nyberg, K. (eds.), "Environmental Engineering, A Chemical Engineering Discipline," Reidel, Dordrecht, Holland/Boston, Massachusetts (1974).
Lund, H. F. (ed.), "Industrial Pollution Control Handbook," McGraw-Hill, New York, New York (1971).
Young, R. A. (ed.), "Plant Engineers' Pollution Control Handbook," American Institute of Plant Engineers, Cincinnati, Ohio, 92pp (1973).

C. Health Effects

"Environmental Health Monitoring Manual," United States Steel Corp., Pittsburgh, Pennsylvania (1973).
"Health Hazards of the Human Environment," World Health Organization, Geneva, Switzerland, 387pp (1972).
Medical and Biological Effects of Environmental Pollutants, National Academy of Sciences/National Research Council, Washington, D.C. Asbestos (ISBN 0-309-01927-3), 40pp, 1971; Chromium (ISBN 0-309-02217-7), ix + 155pp, 1974; Fluorides (ISBN 0-309-01922-2), 295pp, 1971; Lead (ISBN 0-309-01941-9), xi + 330pp, 1972; Manganese (ISBN 0-309-02143-X), vii + 191pp, 1973; Nickel, vii + 277pp, 1975; Particulate Polycyclic Organic Matter (ISBN 0-309-62027-1), xiii + 361pp, 1972; Vanadium (ISBN 0-309-02218-5), vii + 177pp, 1974; Vapor Phase Organic Pollutants—Volatile Hydrocarbons and Oxidation Products (NTIS—PB 249 357), 1975; Arsenic (NTIS—PB 262 167); Chlorine and Hydrogen Chloride (NTIS—PB 253 196); Ozone and Other Photochemical Oxidants (PB—260 570–571); Selenium (NTIS—PB 251 318), 1976.

Table IV *(Cont.)*

Coulston, F., Korte, F., and Goto, M. (eds.), "New Methods in Environmental Chemistry and Toxicology," International Academic Printing Co., Tokyo, Japan, 319pp (1973).
D'Itri, F. M., "The Environmental Mercury Problem," CRC Press, Cleveland, Ohio, 124pp (1972).
Friberg, L., "Cadmium in the Environment," (2nd ed.), Chemical Rubber Co., Cleveland, Ohio, 150pp (1974).
Friberg, L., and Vostal, J. (eds.), "Mercury in the Environment: An Epidemiological and Toxicological Appraisal," CRC Press, Cleveland, Ohio, 215pp (1972).
Hamilton, A., and Hardy, H. L., "Industrial Toxicology," Public Science Group, Acton, Massachusetts, 575pp (1974).
Horita, H., "Pollution, and Toxic and Hazardous Substances," (Kogai to KoKu Kikenbutsu), General Theory (Souronhen), 240pp; Inorganic (Mukinhen), 308pp; Organic (Yukihen), 540pp, Sankyo Shuppan, Tokyo, Japan (1971).
Schroeder, H. A., "The Poisons Around Us: Toxic Metals in Food, Air and Water," Indiana University Press, Bloomington, Indiana, 144pp (1974).
Lazareu, N. V. (ed.), "Harmful Substances in Industry," Handbook for Chemists, Engineers, and Physicians, Pt. 2: Inorganic and Heteroorganic Compounds, (6th ed.); Vrednye Veshchestva v Promyshlennosti. Spravochnik dlya Khimikov, Inzhenerov i Vrachei, Ch. 2: Neorganicheskie i Elementoorganicheskie Soedineniya, Khimiya, Leningrad., USSR, 619pp (1971).
Waldbott, G. L., "Health Effects of Environmental Pollutants," C. V. Mosby, St. Louis, Missouri, x + 316pp (1973).

Table V Books on Air Pollution

A. General Information

"Air Conservation—Report of the Air Conservation Commission" (2nd ed.), AAAS Publication No. 80, American Association for the Advancement of Science, Washington, D.C. (1965).
"Air Pollution," Monograph No. 46, World Health Organization, Columbia University Press, New York, New York, 442pp (1961).
"Air Pollution Abatement Manual," (C. A. Gosline, ed.), Manufacturing Chemists' Association, Washington, D.C. (1951).
"Air Pollution Manual," Part I, Evaluation (2nd ed.), xiii + 259pp (1972); Part II, Control, ix + 150pp American Industrial Hygiene Association, Akron, Ohio, (1968).
"Guide to the Appraisal and Control of Air Pollution" (2nd ed.), American Public Health Association, New York, New York, 80pp (1969).
"Long Term Maintenance of Clean Air Standards," Proceedings of a Specialty Conference, Air Pollution Control Association, Pittsburgh, Pennsylvania, iv + 337pp (1975).
"Proceedings: International Clean Air Conference, London, 1959," National Society for Clean Air, London, England (1960).
"Proceedings: International Clean Air Conference, London, 1966," National Society for Clean Air, London, England (1966).

Table V (*Cont.*)

"Proceedings of the Third International Clean Air Congress, International Union of Air Pollution Prevention Associations, Dusseldorf," VDI-Verlag GmbH, Dusseldorf, Federal Republic of Germany (1973).

"Proceedings of the Fifth International Clean Air Conference of Australia/New Zealand, February 1975," two volumes, 728pp Ann Arbor Science, Ann Arbor, Michigan (1975).

"Proceedings: 1st National Air Pollution Symposium, Pasadena, California (1949)," 2nd (1952), 3rd (1955), Sponsored by Stanford Research Institute in cooperation with California Institute of Technology, University of California, and University of Southern California.

"Proceedings: U.S. Technical Conference on Air Pollution, 1950," McGraw-Hill, New York, New York (1952).

"Proceedings: National Conference Air Pollution, Washington, D.C., 1958," Publ. 654, U.S. Public Health Service, Washington, D.C. (1959).

"Proceedings: National Conference on Air Pollution, Washington, D.C., 1962," Publ. 1022, U.S. Public Health Service, Washington, D.C. (1963).

"Proceedings: National Conference on Air Pollution, Washington, D.C., 1966," Publ. 1669, U.S. Public Health Service, Washington, D.C. (1967).

"Proceedings of the National Conference on the Clean Air Act," Air Pollution Control Association, Pittsburgh, Pennsylvania, iii + 203pp (1974).

"Problems in the Control of Air Pollution, Proceedings: 1st International Congress on Air Pollution, 1955," Reinhold, New York, New York (1955).

Survey of U.S.S.R. Air Pollution Literature. (Translations) Nuttonson, M. Y. (ed.), 21 volumes, American Institute of Crop Ecology, Silver Spring, Maryland:

I. "Atmospheric and Meteorological Aspects of Air Pollution" (1969).

II. "Effects and Symptoms of Different Plant Species in Various Habitats, in Relation to Plant Utilization for Shelter Belts and as Biological Indicators" (1969).

III. "The Susceptibility or Resistance to Gas and Smoke of Various Arboreal Species Grown under Diverse Environmental Conditions in a Number of Industrial Regions of the Soviet Union" (1969).

IV. "Meteorological and Chemical Aspects of Air Pollution; Propagation and Dispersal of Air Pollutants in a Number of Areas in the Soviet Union" (1970).

V. "Effects of Meteorological Conditions and Relief on Air Pollution: Air Contaminants—Their Concentration, Transport, and Dispersal" (1970).

VI. "Air Pollution in Relation to Certain Atmospheric and Meteorological Conditions and Some of the Methods Employed in the Survey and Analysis of Air Pollutants (1971).

VII. "Measurements of Dispersal and Concentration, Identification, and Sanitary Evaluations of Various Air Pollutants, with Special Reference to the Environs of Electric Power Plants and Ferrous Metallurgical Plants" (1971).

VIII. "A Compilation of Technical Reports on the Biological Effects and the Public Health Aspects of Atmospheric Pollutants" (1971).

IX. "Gas Resistance of Plants with Special Reference to Plant Biochemistry and to the Effects of Mineral Nutrition" (1971).

Table V (*Cont.*)

X. "The Toxic Components of Automobile Exhaust Gases: Their Composition under Different Operating Conditions and Methods of Reducing Their Emissions" (1971).
XI. "A Second Compilation of Technical Reports on the Biological Effects and the Public Health Aspects of Atmospheric Pollutants" (1972).
XII. "Technical Papers from the Leningrad International Symposium on the Meteorological Aspects of Atmospheric Pollution." Part 1 (1972).
XIII. "Technical Papers from the Leningrad International Symposium on the Meteorological Aspects of Atmospheric Pollution." Part 2. (1972).
XIV. "Technical Papers from the Leningrad International Symposium on the Meteorological Aspects of Atmospheric Pollution." Part 3. (1972).
XV. "A Third Compilation of Technical Reports on the Biological Effects and the Public Health Aspects of Atmospheric Pollutants" (1972).
XVI. "Some Basic Properties of Ash and Industrial Dust in Relation to the Problem of Purification of Stack Gases" (1972).
XVII. "A Fourth Compilation of Technical Reports on the Biological Effects and Health Aspects of Atmospheric Pollutants" (1972).
XVIII. "Purification of Gases through High Temperature Removal of Sulfur Compounds (1972).
XIX. "Environmental Pollution with Special Reference to Air Pollutants and to Some of Their Biological Effects" (1973).
XX. "Catalytic Purification of Exhaust Gases" (1973).
XXI. "Atmospheric Pollutants in Relation to Meteorological Conditions: A Procedure for Calculating the Atmospheric Dispersal of Pollutants and the Feasibility of their Study by Means of Satellites" (1973).

"U.S.S.R. Literature on Air Pollution and Related Occupational Diseases" (translated by B. S. Levine), Vols. 1–17. National Technical Information Service, U.S. Dept. of Commerce, Springfield, Virginia. Vol. 1, TT-60-21019 (1960); Vol. 2, TT-60-21188 (1960); Vol. 3, TT-60-21475 (1960); Vol. 4, TT-60-21913 (1960); Vol. 5, TT-61-11149 (1961); Vol. 6, TT-61-21982 (1961); Vol. 7, TT-62-11103 (1962); Vol. 8, TT-63-11570 (1963); Vol. 9, TT-64-11574 (1964); Vol. 10, TT-64-11767 (1964); Vol. 11, TT-65-61965 (1965); Vol. 12, TT-66-61429 (1966); Vol. 13, TT-66-62191 (1966); Vol. 14, TT-67-60046 (1967); Vol. 15, PB 179-140 (1968); Vol. 16, PB 179-141 (1968); Vol. 17, PB 180-522T (1968).

Atkisson, A., and Gaines, K. S. (eds.), "Development of Air Quality Standards," Proceedings of a symposium, Merrill, Columbus, Ohio, ix + 220pp (1970).

Bach, W., "Atmospheric Pollution," McGraw-Hill, New York, New York, 144pp (1972).

Barton, K., "Protection Against Atmospheric Corrosion—Theory and Technology," "Schutz Gegen Atmosphaerische Korrosion. Theorie und Technik," Verlag Chemie, Weinhein, Federal Republic of Germany, 209pp (1972).

Benarie, M. M. (ed.), "Atmospheric Pollution—12th Colloquium Proceedings," Elsevier, Amsterdam, The Netherlands, x + 649pp (1976).

Bibbero, R. J., and Young, I. G., "Systems Approach to Air Pollution Control," Wiley, New York, New York, xi + 531pp (1974).

Bond, R. G., and Straub, C. P. (eds.), "CRC Handbook of Environmental Control—Vol. I, Air Pollution," CRC Press, Cleveland, Ohio (1973).

Table V (*Cont.*)

Bower, B. T., "Selecting Strategies for Air Quality Management," Dept. of Energy, Mines and Resources, Ottawa, Ontario, 47pp (1971).

Chovin, P., and Roussel, A., "La Pollution Atmospherique," Presses Universitaires de France, Paris, France, 128pp (1968).

Crenson, M. A., "The Un-politics of Air Pollution; A Study of Non-Decision Making in the Cities," Johns Hopkins Press, Baltimore, Maryland, 227pp (1971).

Cresswell, C. R., "Notes on Air Pollution Control," Lewis, London, England, 178pp (1974).

Dotreppe-Grisard, N., "La Pollution de l'Air," Editions Eyralles, Paris, France, 250pp (1972).

Downing, P. B. (ed.), "Air Pollution and the Social Sciences: Formulating and Implementing Control Programs," Praeger, New York, New York, xv + 270pp (1971).

Edelman, S. (comp.), "The Law of Air Pollution Control," Environmental Science Services, Stamford, Connecticut (1970).

Englund, H. M., and Berry, W. T. (eds.), "A Critical Review of Regulations for the Control of Sulfur Oxide Emissions and Particulate Emissions," Air Pollution Control Association, Pittsburgh, Pennsylvania, vi + 58pp (1974).

Englund, H. M., and Berry, W. T. (eds.), "Proceedings of the Second International Clean Air Congress, International Union of Air Pollution Prevention Associations, Washington, D.C., Dec. 1970," Academic Press, New York, New York, 1380pp (1971).

Faith, W. L., and Atkisson, A. A., "Air Pollution," (2nd ed.), Wiley, New York, New York (1972).

Farber, S. M., and Wilson, R. H. L. (eds.) (in collaboration with J. R. Goldsmith and N. Pace), "The Air We Breathe: A Study of Man and His Environment," Thomas, Springfield, Illinois (1961).

Feller, I., Engel, A. J., Friedman, R. S., Menzel, D. C., Jr., and Sacco, J. F., "Intergovernmental Relations in the Administration and Performance of Research on Air Pollution," Pennsylvania State University, Center for the Study of Science Policy, University Park, Pennsylvania, x + 205pp (1972).

Gilpin, A., "Control of Air Pollution," Butterworth, London and Washington, D.C., 514pp (1963).

Gilpin, A., "Air Pollution," University of Queensland Press, St. Lucia, Queensland, Australia, 67pp (1971).

Goldstein, E., *et al.*, "Air Pollution and Methods of Control," MSS Information Corp., New York, New York (1973).

Hagevik, G. H., Mandelkev, D. R., and Brail, R. K. "Air Quality Management and Land Use Planning: Legal, Administrative and Methodological Perspectives," Praeger, New York, New York, xiii + 336pp (1974).

Hagevik, G. H., "Decision Making in Air Pollution Control," Praeger, New York, New York, 217pp (1970).

Hagevik, G. H., "The Relationship of Land Use and Transportation Planning to Air Quality Management," Rutgers University, New Brunswick, New Jersey, 287pp (1973).

Harrison, D., Jr., "Who Pays for Clean Air: The Cost and Benefit Distribution of Federal Automobile Emission Standards," Ballinger, Cambridge, Massachusetts, xix + 169pp (1975).

Havinghurst, C. C., "Air Pollution Control," Oceana, Dobbs Ferry, New York, 230pp (1969).

Table V (*Cont.*)

Hertzendorf, M. S., "Air Pollution Control, Guidebook to U.S. Regulations," Technomic Pub. Co., Westport, Connecticut, vi + 266pp (1973).
Hesketh, H. E., "Understanding and Controlling Air Pollution," (2nd ed.), Ann Arbor Science, Ann Arbor, Michigan, xix + 416pp (1974).
Huschke, R. E., "Glossary of Terms Frequently Used in Air Pollution," American Meteorological Society, Boston, Massachusetts (1972).
Izmerov, N., "Control of Air Pollution in the USSR," World Health Organization, Geneva, Switzerland (1973).
Jarzebski, S., and Kapala, J., "Principles of Determining Indexes of the Emission of Atmospheric Air Pollutants from Industrial Processes, Vol. 5"; "Zasady Wynaczania Wskaznikew Emisji Zanieczyszczen Powietrza Atmosferycznego z Procesow Przemyslowych, T. 5," PAN, Warsaw, Poland, 95pp (1971).
Jones, C. O., "Clean Air: The Policies and Politics of Pollution Control," University of Pittsburgh Press, Pittsburgh, Pennsylvania, xii + 372pp (1975).
Kohn, R. E., "Air Pollution Control—A Welfare Economic Interpretation," Lexington, Lexington, Massachusetts, xxii + 155pp (1975).
Krier, J. E., "Environmental Law and Policy: Readings Materials and Notes on Air Pollution and Related Problems," Bobbs-Merrill, Indianapolis, Indiana, xxxii + 480pp (1971).
Ledbetter, J. O., "Air Pollution," 2 Vols.—Part A, Analyses, xii + 424pp (1972); Part B, Prevention and Control, xiii + 286pp, Marcel Dekker, New York, New York (1974).
Lynn, D. A., "Air Pollution: Threat and Response," Addison-Wesley, Reading, Massachusetts, x + 388pp (1976).
Mammarella, L., "Inquiramenti dell'Aria," Pensiero Sci., Rome, Italy, 501pp (1971).
McCormac, B. M. (ed.), "Introduction to the Scientific Study of Atmospheric Pollution," Reidel, Dordrecht, The Netherlands, 177pp (1971).
Magill, P. L., Holden, F. R., and Ackley, C. (eds.), "Air Pollution Handbook," McGraw-Hill, New York, New York (1956).
Meetham, A. R., "Atmospheric Pollution, Its Origin and Prevention," (3rd rev. ed.), Pergamon, Oxford, England, 302pp (1964).
Miernyk, W. H., and Sears, J. T., "Air Pollution Abatement and Regional Economic Development," Heath, Lexington, Massachusetts, xvi + 194pp (1974).
Painter, D. E., "Air Pollution Technology," Reston/Prentiss-Hall, Reston, Virginia, xii + 283pp (1974).
Perkins, H. C., "Air Pollution," McGraw-Hill, New York, New York, xiv + 407pp (1974).
Rickles, R. N., "Air Pollution Control Primer," Environmental Science Services Stamford, Connecticut (1969).
Ridker, R. G., "Economic Costs of Air Pollution—Studies in Management," Praeger, New York, New York, 228pp (1967).
Ross, R. D., "Air Pollution and Industry," Von Nostrand Reinhold, New York, New York (1972).
Rossano, A. T., Jr., "Air Pollution Control: Guidebook for Management," Environmental Science Services, Stamford, Connecticut (1969).
Ryazanov, V. A., and Goldberg, M. S., (eds.), "Limits of Allowable Concentrations of Atmospheric Pollutants," Books 1–7; "Maximum Permissible Concentrations of Atmospheric Pollutants"; Book 8: "Biological Effects and Hygienic Significance of Atmospheric Pollutants" (transl. by B. S. Levine)—National Technical Information Service, U.S. Dept. of Commerce, Springfield, Virginia. Book 1:

Table V (*Cont.*)

Doc. 59-21173 (1952); Book 2: Doc. 59-21174 (1955); Book 3: Doc. 59-21175 (1957); Book 4; Doc. 61-1148 (1960); Book 5; Doc. 62-11605 (1962); Books 6–10 appear in "USSR Literature, etc." (B. S. Levine, trans.) Vols. 9, 15–17; Books 10 and 11 appear in "Survey of USSR Literature, etc." (M. Y. Nuttonson, trans.) Vols. 11, 15 and 17.

Scorer, R. S., "Air Pollution," Pergamon, New York, New York, 151pp (1968).

Scorer, R. S., "Pollution in the Air: Problems, Policies and Priorities," Routledge and Kegan Paul, London, England, 148pp (1973).

Seinfeld, J. H., "Air Pollution Physical and Chemical Fundamentals," McGraw-Hill, New York, New York, xv + 523pp (1975).

Smith, V. K., "The Economic Consequences of Air Pollution," Ballinger, Cambridge, Massachusetts, xvi + 144pp (1976).

Sparrow, C. J., and Foster, L. T. (eds.), "An Annotated Bibliography of Canadian Air Pollution Literature," Ann Arbor Science, Ann Arbor, Michigan, xvi + 270pp (1976).

Spedding, D. J., "Air Pollution," Oxford University Press. Fairlawn, New Jersey, 90pp (1974).

Sproull, W. T., "Air Pollution and Its Control" (2nd ed.), Exposition Press, Jericho, New York (1972).

Stern, A. C., Wohlers, H. C., Boubel, R., and Lowry, W., "Fundamentals of Air Pollution," Academic Press, New York, New York, xiv + 492pp (1973).

Strauss, W. (ed.), "Air Pollution Control, Part I," Wiley, New York, New York (1970).

Strauss, W. (ed.), "Air Pollution Control, Part 2," Wiley, New York, New York, xi + 300pp (1972).

Suess, M. J., and Craxford, S. R. (eds.), "Manual on Urban Air Quality Management," WHO Regional Publications, European Series #1, World Health Organization, Copenhagen, Denmark, 200pp (1976).

Thring, M. W., "Air Pollution," Butterworth, London, England and Washington, D.C. (1957).

Tomany, J. P., "Air Pollution: The Emissions, the Regulations and the Controls," American Elsevier, New York, New York, xii + 475pp (1975).

Webb, J. C., and Leroux, J., "Detection and Control of Air Pollution," MSS Information Corp., New York, New York, 177pp (1974).

Williamson, S. J., "Fundamentals of Air Pollution," Addison-Wesley, Reading, Massachusetts (1973).

Wolozin, H. (ed.), "The Economics of Air Pollution," Norton, New York, New York, 318pp (1966).

Woodwell, G. M., and Severs, R. K., "Ecological and Biological Effects of Air Pollution," MSS Information Corp., New York, New York, 181pp (1974).

B. Aerosols

U.S. Office of Scientific Research and Development, National Defense Research Committee, "Handbook on Aerosols," U.S. Govt. Printing Office, Washington, D.C. (1950).

Allen, T., "Particle Size Measurement," (2nd ed.), Halsted, New York, New York, xviii + 454pp (1975).

Table V (*Cont.*)

Avy, P., "Les Aerosols," Dunod, Paris, France (1956).
Cadle, R. D., "The Measurement of Airborne Particles," Wiley, New York, New York, xi + 342pp (1975).
Cadle, R. D., "Particles in the Atmosphere and Space," Reinhold, New York, New York (1966).
Cadle, R. D., "Particle Size," Reinhold, New York, New York (1965).
Cadle, R. D., "Particle Size Determination," Interscience, New York, New York (1955).
DallaValle, J. M., "Micromeretics," Pitman, London, England (1948).
Davies, C. N. (ed.), "Aerosol Science," Academic Press, London, England, xviii + 468pp (1966).
Davies, C. N. (ed.), "Inhaled Particles and Vapours II," Proceedings of the Second British Occupational Hygiene Society Symposium, Pergamon, Long Island City, New York, 605pp (1967).
Davies, C. N., "Recent Advances in Aerosol Research, A Bibliographical Review," Pergamon Press, Oxford, England (1964).
Dennis, R. (ed.), "Handbook on Aerosols" (TID-26608), Energy Research and Development Agency, National Technical Information Service, U.S. Dept. of Commerce, Springfield, Virginia, v + 142pp (1976).
Dimmick, R. L., and Akers, A., "An Introduction to Experimental Aerobiology," Wiley (Interscience), New York, New York, 494pp (1969).
Drinker, R., and Hatch, T., "Industrial Dusts" (2nd ed.), McGraw-Hill, New York, New York (1954).
Fedoseev, V. A. (ed.), "Advances in Aerosol Physics" Nos. 4–7, Wiley, New York, New York, No. 4: vi + 148pp (1973); Nos. 5–7, vii + 186pp (1974).
Fett, W., "Der Atmospharischer Staub," VEP Deutcher Verlag der Wissenschaften, Federal Republic of Germany (1958).
Friedlander, S. K., "Smoke, Dust and Haze: Fundamentals of Aerosol Behavior," Wiley, New York, New York, xvii + 317pp (1977).
Fuchs, N. A., "The Mechanics of Aerosols," Pergamon, Oxford, England (1964).
Fuchs, N. A., and Sutugin, A. G., "Highly Dispersed Aerosols," (trans. by Israel Program for Scientific Translations), Ann Arbor Science, Ann Arbor, Michigan, 105pp (1970).
Green, H. L., and Lane, W. R., "Particulate Clouds, Dusts, Smokes, and Mists" (2nd ed.), Van Nostrand, Princeton, New Jersey (1964).
Herdan, G., "Small Particle Statistics," Academic Press, New York, New York (1960).
Hidy, G., and Brock, J. R., "The Dynamics of Aerocolloidal Systems," Pergamon, Oxford, England, 379pp (1970).
Hidy, G., and Brock, J. R. (eds.), "Topics in Current Aerosol Research," Pergamon, New York, New York, viii + 157pp (1971).
Irani, R. R., and Callis, C. F., "Particle Size, Measurement, Interpretation, and Application," Wiley, New York (1963).
Orr, C., Jr., and Dalla Valle, J. M., "Fine Particle Measurement, Size, Surface, and Pore Volume," Macmillan, New York, New York (1959).
Liu, B. Y. H. (ed.), "Fine Particles, Aerosol Generation, Measurement, Sampling, and Analysis," Academic Press, New York, New York, xiii + 837pp (1976).

Table V (*Cont.*)

Mercer, T. T., "Aerosol Technology in Hazard Evaluation," Academic Press, New York, New York, xi + 394pp (1973).

Mercer, T. T., Morrow, P. E., and Stober, W. (eds.), "Assessment of Airborne Particles, Fundamentals, Applications and Implications of Inhalation Toxicology," Thomas, Springfield, Illinois, 560pp (1972).

Middleton, W. E. K., "Vision through the Atmosphere," University of Toronto Press, Toronto, Ontario (1952).

Sanders, P. A., "Principles of Aerosol Technology," Van Nostrand-Reinhold, New York, New York, 418pp (1970).

Van de Hulst, H. C., "Light Scattering by Small Particles," Wiley, New York, New York (1957).

C. Ambient Air Measuring and Monitoring

"Air Sampling Instruments" (4th ed.), American Conference of Governmental Industrial Hygienists, Cincinnati, Ohio (1972).

"Air Quality: Standards and Measurement," SP-10, Proceedings of a Specialty Conference, 1974, Air Pollution Control Association, Pittsburgh, Pennsylvania, 154pp (1974).

"Ambient Air Quality Measurements," Proceedings of a Specialty Conference, Air Pollution Control Association, Pittsburgh, Pennsylvania, xviii + 376pp (1975).

"Instrumentation for Monitoring Air Quality," A Symposium sponsored by the Environmental Protection Agency, the National Center for Atmospheric Research, and Committee D-22. American Society for Testing and Materials, Philadelphia, Pennsylvania, 192pp (1974).

"Methods of Air Sampling and Analysis," Intersociety Committee, American Public Health Association, Washington, D.C., xvii + 480pp (1972), 2nd ed., 900pp (In Press, 1977).

"Selected Methods of Measuring Air Pollutants," Offset Publication No. 24, World Health Organization, Geneva, Switzerland, ix + 112pp (1976).

"Operations Manual for Sampling and Analysis Techniques for Chemical Constituents in Air and Precipitation," Publ. no. 299, World Meteorological Organization, Geneva, Switzerland (1971).

"Standards, Part 26, Gaseous Fuels; Coal and Coke; Atmospheric Analysis." American Society for Testing and Materials, Philadelphia, Pennsylvania, xvi + 828pp (1974).

"Survey of Instrumentation for Environmental Monitoring," Vol. I, Part 2—Air, Lawrence Laboratory, University of California, Berkeley, California, 700pp (1975).

Adams, D. F. (ed.), "Air Pollution Instrumentation," Instr. Soc. Am., Pittsburgh, Pennsylvania (1965).

Aidarov, T. K., and Razyapov, A. Z., "Spectral Methods for Determining Noxious Substances in Air and Biological Materials," "Spektral'nye Metody Opredeleniya Vrednykh Veshchestv v Vozdukhe i Biologicheskikh Materialakh," Izd. Kazan., Univ. Kazan, USSR, 179pp (1973).

Araki, S., Kanazawa, J., Yanagisawa, S., Yamagata, N., Yamashita, T., and Yamate, N. (eds.), "Pollution Analysis Manual" "Kogai Bunseki Shishin," Air Series (Taiki Hen), Kyoritsu Shuppan, Tokyo, Japan (1972). No. 1a, 76pp; No. 1b,

Table V (*Cont.*)

80pp; No. 1c, 95pp; No. 2a, 69pp; No. 2b, 102pp; No. 2c, 81pp; No. 3a, 71pp; No. 3b, 102pp; No. 3c, 89pp.

Bach, W., and Daniels, A., "Handbook of Air Quality in The United States," Oriental, Honolulu, Hawaii, ix + 235pp (1975).

Fosters, J. F., and Beatty, G. H., "Interlaboratory Cooperational Study of the Precision and Accuracy of the Measurement of," American Society for Testing and Materials, Philadelphia, Pennsylvania (1974):

Nitrogen Dioxide Content of the Atmosphere Using ASTM Method D1607, 78pp. Sulfur Dioxide Content of the Atmosphere Using ASTM Method D2914, iv + 78pp. Total Sulfation in the Atmosphere Using ASTM Method D2010, ii + 45pp. Particulate Matter in the Atmosphere Using ASTM Method D1704, iii + 63pp. Dustfall Using ASTM Method D1739, iii + 45pp.

(1975): Lead in the Atmosphere Using the Colorimetric Dithizone Procedure, 79pp Particulates and Collected Residues in Stacks, 152pp Oxides of Nitrogen in Gaseous Combustion Products (Phenol Disulfonic Acid Procedure Using ASTM Method D1608-60), 111pp.

Hinkley, E. D. (ed.), "Laser Monitoring of the Atmosphere," Springer-Verlag, New York, New York, xiii + 380pp (1976).

Iovenko, E. N., "Automatic Analyzers and Signaling Indicators of Toxic and Explosive Substances in Air," "Avtomaticheskie Analizatory i Signalizatory Toksichnykh i Vzryvoopasnykh Veshchectv v Vozdukhe," Khimiya, Moscow, USSR, 188pp (1972).

Israel, H., and Israel, G., "Trace Elements in the Atmosphere," Ann Arbor Science, Ann Arbor, Michigan, 160pp (1974).

Jacobs, M. B., "The Chemical Analysis of Air Pollutants," Wiley (Interscience), New York, New York (1960).

Katz, M., "Measurement of Air Pollutants—Guide to the Selection of Methods," World Health Organization, Columbia University Press, New York, New York (1969).

Leithe, W., "The Analysis of Air Pollutants," Ann Arbor Science, Ann Arbor, Michigan, 304pp (1970).

Mamantov, G., and Shults, W. D. (eds.), "Determination of Air Quality," Proceedings of an American Chemical Society Symposium, Plenum, New York, New York, xii + 197pp (1972).

McCrone, W. C., and Delly, J. G., "The Particle Atlas" (2nd ed.) 4 Volumes, Ann Arbor Science, Ann Arbor, Michigan, 1138pp (1973).

Nelson, G. O., "Controlled Test Atmospheres—Principles and Techniques," Ann Arbor Science, Ann Arbor, Michigan, 256pp (1970).

Pinta, M., "Detection and Determination of Trace Elements," Ann Arbor Science, Ann Arbor, Michigan, 588pp (1970).

Ruch, W., "Chemical Detection of Gaseous Pollutants," Ann Arbor Science, Ann Arbor, Michigan, 180pp (1966).

Ruch, W., "Quantitative Analysis of Gaseous Pollutants," Ann Arbor Science, Ann Arbor, Michigan, 242pp (1970).

Scales, J. W. (ed.), "Air Quality Instrumentation," Instrument Society of America, Pittsburgh, Pennsylvania (1972), Vol. 1 (1972), Vol. 2, 326pp (1974), Vol. 3 (1975).

Schneider, T. (ed.), "Automatic Air Quality Monitoring Systems," American Elsevier, New York, New York, xvi + 267pp (1974).

Table V (*Cont.*)

Stahl, W. H., "Compilation of Odor and Taste Threshold Values Data," American Society for Testing and Materials, Philadelphia, Pennsylvania, 250pp (1973).
Stevens, R. K., and Herget, W. F., "Analytic Methods Applied to Air Pollution Measurements," Ann Arbor Science, Ann Arbor, Michigan, 320pp (1974).
Thom, G. C., and Ott, W. R., "Air Pollution Indices," Ann Arbor Science, Ann Arbor, Michigan, xi + 163pp (1976).
Warner, P. O., "Analysis of Air Pollutants," Wiley, New York, New York, xi + 329pp (1976).
Yavorovskaya, S. F., "Gas Chromatography as a Method for Determining Harmful Substances in Air and in Biological Media," "Gazovaya Khromatografiya-Method Opredeleniya Vrednykh Veshestv v Vozdukhe i v Biologicheskikh Sredakh," Meditsina, Moscow, USSR, 207pp (1972).
Zielinski, E. (ed.), "Gas Chromatography with Special Reference to Air Analysis," "Chromatografia Gazowa ze Szczigolnym Uwzglednieniem Analizy Powietrza," Pol. Akad. Nauk, Warsaw, Poland, 264pp (1972).

D. Atmospheric Chemistry and Radioactivity

"Chemistry and Physics of the Stratosphere," American Geophysical Union, Washington, D.C., 171pp (1976).
"Photochemical Oxidant Air Pollution," Organization for Economic Cooperation and Development, Paris, France, 94pp (1975).
"Photochemical Smog and Ozone Reactions," Two American Chemical Society Symposia, American Chemical Society, Washington, D.C., 285pp (1972).
Bassow, H., "Air Pollution Chemistry: An Experimenter's Sourcebook," Hayden, Rochelle Park, New Jersey, 128pp (1976).
Bockris, J. O. M., "Electrochemistry of Cleaner Environments," Plenum, New York, New York, 296pp (1972).
Butcher, S. S., and Charleson, R. J., "An Introduction to Air Chemistry," Academic Press, New York, New York, xiii + 241pp (1972).
Cadle, R. D. (ed.), "Chemical Reactions in the Lower and Upper Atmosphere," Wiley (Interscience), New York, New York (1961).
Calvert, J. G., and Pitts, J. N., Jr., "Photochemistry," Wiley, New York, New York, xvii + 899pp (1966).
Eisenbud, M., "Environmental Radioactivity" (2nd ed.), Academic Press, New York, New York, xii + 542pp (1973).
Heicklen, J., "Atmospheric Chemistry," Academic Press, New York, New York, xiv + 406pp (1976).
Hidy, G. M. (ed.), "Aerosols and Atmospheric Chemistry," Proceedings of an American Chemical Society Symposium, Academic Press, New York, New York, xviii + 348pp (1972).
Junge, C. E., "Air Chemistry and Radioactivity," Academic Press, New York, New York (1963).
Leighton, A., "Photochemistry of Air Pollution," Academic Press, New York, New York (1961).
McEwan, M. J., and Phillips, L. F., "Chemistry of the Atmosphere," Halsted, New York, New York, ix + 301pp (1975).

Table V (*Cont.*)

Pryde, L. T., "Chemistry of the Air Environment," Cummings Pub. Co., Menlo Park, California, 59pp (1970).
Rasool, S. L. (ed.), "Chemistry of the Lower Atmosphere," Plenum, New York, New York, xi + 335pp (1973).
Tuesday, C. S. (ed.), "Chemical Reactions in Urban Atmosphere," Elsevier, New York, New York (1971).
Westberg, K., and Cheng, R. T., "The Chemistry of Air Pollution," MSS Information Corp., New York, New York, 200pp (1974).

E. Health Effects

"Have Health Effects of Air Pollution Been Oversold?" Proceedings of an APCA Specialty Session, sponsored by the TE-1 Bio-Medical Committee of the Technical Council, Air Pollution Association, Pittsburgh, Pennsylvania, viii + 40pp (1974).
Aharonson, E. F., Ben-David, A., and Klingberg, M. A. (eds.), "Air Pollution and the lung," Halsted, New York, New York, xxii + 314pp (1976).
Chovin, P., "Physicochimie et Physiopathologie des Pollutants Atmosphèriques," Masson, Paris, France (1973).
Duel, W. C., "The Physician's Guide to Air Pollution," American Medical Association, Chicago, Illinois, 44pp (1973).
Finkel, A. J., and Duel, W. C. (eds.), "Clinical Implications of Air Pollution Research," Publishing Sciences Group, Inc., Acton, Massachusetts, xxvii + 374pp (1975).
Gardner, M. B., and Bates, D. V., "Physiological Effects of Air Pollution," MSS Information Corp., New York, New York, 222pp. (1974).
Holland, W. W., "Air Pollution and Respiratory Disease," Technomic Pub. Co., Westport, Connecticut (1972).
Turk, A., Johnson, J. W., Jr., and Moulton, D. G. (eds.), "Human Response to Environmental Odors," Academic Press, New York, New York, 345pp (1974).
Venuti, G. C., Donelli, G., Maiani, L., Pocchiari, F., Sellerio, U., and Tabet, E., "Health Problems in the Siting of Heating and Electric Power Plants (Air Pollution Aspects)," "Riflessi Sanitari dell'Insediamento delle Centrali Termoelettriche," Ist. Super. Sanita, Rome, Italy, 39pp (1973).
Verveij, J. H. P., "Molybdeenovermaat bij het door luchtverontreiniging," Molybdenosis in Cattle by Air Pollution," Veip. Ter Hoeven, The Netherlands, 132pp (1970).
Zvonov, V. A., "Toxicity of (Exhaust from) Internal-Combustion Engines," "Toksichnost Dvigatelej Vnutrennego Sgoraniya," Mashinostroenie, Moscow, USSR, 200pp (1973).

F. Meteorology and Modeling

"Conference on Air Pollution Meteorology," American Meteorological Society, Boston, Massachusetts, v + 159pp (1971).

Table V (*Cont.*)

"Meteorological Aspects of Air Pollution," Specialty Meeting of Mid Atlantic States Section, Air Pollution Control Association, Pittsburgh, Pennsylvania, 107pp (1969).
Barry, R. G., and Chorley, R., "Atmosphere, Weather and Climate," Holt, New York, New York, 320pp (1970).
Berlyand, M. E. (ed.), "Air Pollution and Atmospheric Diffusion," Wiley, New York, New York (1973) Halsted Press, New York, New York, 1—221pp (1973), 2—242pp (1974).
Byers, H. R., "General Meteorology," McGraw-Hill, New York, New York, 540pp (1959).
Church, H. W., and Luna, R. E. (eds.), "Symposium on Air Pollution, Turbulence and Diffusion," New Mexico State University, Albuquerque, New Mexico, 398pp (1971).
Day, J., and Sternes, G., "Climate and Weather," Addison-Wesley, Reading, Massachusetts, 407pp (1970).
Derco, N. J., and Truhlar, E. J. (eds.), "Proceedings of the International Conference on Structure, Composition, and General Circulation of the Upper and Lower Atmosphere and Possible Anthropogenic Perturbations," Vol. 1, Atmos. Environ. Serv., Downsview, Ontario, 513pp (1974).
Donn, W. L., "Meteorology," (4th ed.), McGraw-Hill, New York, New York, x + 518pp (1975).
Flohn, H., "Climate and Weather," McGraw-Hill, New York, New York, 253pp (1969).
Forsdyke, A. G., "Meteorological Factors in Air Pollution," World Meteorological Organization, Geneva, Switzerland, 32pp (1970).
Frisken, W. R., "The Atmospheric Environment," Resources for the Future, Johns Hopkins University Press, Baltimore, Maryland, ix + 68pp (1973).
Gates, D. M., "Energy Exchange in the Biosphere," Harper, New York, New York, 151pp (1962).
Geiger, R., "The Climate Near the Ground" (4th ed.), Harvard University Press, Cambridge, Massachusetts, 611pp (1965).
Goody, R. M., "Atmospheric Radiation," Oxford University Press, London, England, 436pp (1964).
Griffiths, J. F., "Applied Climatology," Oxford University Press, London, England, 118pp (1966).
Hess, S. L., "Introduction to Theoretical Meteorology," Holt, New York, New York, 362pp (1959).
Kondratyev, K., "Radiation in the Atmosphere," Academic Press, New York, New York, 912pp (1969).
Kondratyev, K., "Radiative Heat Exchange in the Atmosphere," Pergamon, Oxford, England, 411pp (1965).
Lowry, W. P., "Weather and Life—An Introduction to Biometeorology," Academic Press, New York, New York, xiii + 305pp (1969).
Munn, R. E., "Biometeorological Methods," Academic Press, New York, New York, xi + 336pp (1970).
Munn, R. E., "Descriptive Micrometeorology," Academic Press, New York, New York, 245pp (1966).

Table V (*Cont.*)

Pasquill, F., "Atmospheric Diffusion" (2nd ed.), Ells Harwood, Chichester, England/ Halsted Press/Wiley, New York, New York, xi + 429pp (1974).

Sutton, O. G., "Micrometeorology," McGraw-Hill, New York, New York, 333pp (1953).

World Meteorological Organization, Geneva, Switzerland:

"Climatological Aspects of the Composition and Pollution of the Atmosphere," WMO 393, 44pp (1975).

"Dispersion and Forecasting of Air Pollution," WMO 319, 116pp (1972).

"Meteorological Aspects of Air Pollution," TP 139, 69pp (1970).

G. Popular and Children's Books

"Air Pollution Primer," National Tuberculosis and Respiratory Disease Assoc., New York, New York, vi + 104pp (1969).

"Controlling Air Pollution—A Primer on Stationary Source Control Techniques," American Lung Assoc., Washington, D.C., iv + 55pp (1974).

Bates, D. V., "A Citizen's Guide to Air Pollution—Environmental Damage and Control in Canada," McGill–Queen's University Press, Montreal, Quebec, xv + 140pp (1972).

Battan, L. J., "The Unclean Sky," Doubleday, Garden City, New York (1966).

Bregman, J., and Lenormand, S., "The Pollution Paradox," Books, New York, New York (1966).

Brodine, V., "Air Pollution," Harcourt, Brace, Jovanovich, New York, New York, xvi + 205pp (1973).

Carr, D. E., "The Breath of Life," Norton, New York, New York, 175pp (1965).

Connolly, C. H., "Air Pollution and Public Health," Dryden Press, New York, New York, x + 262pp (1972).

Edelson, E., and Warshofsky, F., "Poisons in the Air," Pocket Books, New York, New York, (1966).

Elliott, S. M., "Our Dirty Air," Julian Messner, New York, New York, 64pp (1971).

Esposito, J. C., and Silverman, L. J., "Vanishing Air, The Ralph Nader Study Group on Air Pollution," Grossman, New York, New York, 328pp (1970).

Hunter, D. C., "Air Pollution Experiments for Junior and Senior High School Science Classes" (2d ed.), Air Pollution Control Association, Pittsburgh, Pennsylvania (1972).

Herber, L., "Crisis in Our Cities," Prentice-Hall, Englewood Cliffs, New Jersey, 239pp (1965).

Laycock, G., "Air Pollution," Grosset and Dunlap, New York, New York (1972).

Lewis, A., "Clean the Air," McGraw-Hill, New York, New York, 96pp (1965).

Lewis, H. R., "With Every Breath You Take," Crown, New York, New York, 322pp (1965).

Marshall, J. A., "The Air We Live In," Coward-McCann, New York, New York, 96pp (1969).

Mills, C. A., "This Air We Breathe," Christopher Publ. House, Boston, Massachusetts, (1962).

Perry, J., "Our Polluted World," Franklin Watts, New York, New York, 214pp (1967).

Table V *(Cont.)*

Wise, W., "Killer Smog—The World's Worst Air Pollution Disaster," Rand-McNally, New York, New York, 181pp (1968).

H. Specific Pollutants

"Carbon Monoxide—Origin, Measurement and Air Quality Criterion," VDI-Verlag, Düsseldorf, Federal Republic of Germany, 127pp (1972).

"Data Concerning Environmental Standards for Nitrogen Oxides (Air Pollution Research Vol. 7, No. 3)"; "Chisso Sankabutsuto ni Kankyo Kijun ni Kansuru Shiryo (Taiki Osen Kenkyu, Dai 7 Kan, Dai 3 Go)," Taiki Osen Kenkyu Zenkoku Kyogikai, Tokyo, Japan, 123pp (1972).

Schroeder, H. A., "Chromium—Air Quality Monograph #70-15" 28pp; Nickel, #70-14, 24pp; Vanadium, #70-13, 32pp, American Petroleum Institute, Washington, D.C. (1970).

Vallee, B. L., "Phosphorus—Air Quality Monograph #73-19," American Petroleum Institute, Washington, D.C., 35pp (1973).

I. Vegetation Effects

"First European Congress on the Influence of Air Pollution on Plants and Animals—Proceedings," "Centrum Voor Land Bouw Publikaties en Land bouw Documentatie," Wageningen, The Netherlands, 415pp (1969).

"Impact of Air Pollution on Vegetation—Proceedings for a Specialty Conference," Ontario Section, Air Pollution Control Association, Pittsburgh, Pennsylvania, 132pp (1970).

"Recognition of Air Pollution Injury to Vegetation, a Pictorial Atlas." Informational Report No. 1, TR-7 Agricultural Committee (J. S. Jacob and A. C. Hill, eds.), Air Pollution Control Association, Pittsburgh, Pennsylvania, vii + 112pp (1970).

Denison, W. C., and Carpenter, S. M., "A Guide to Air Quality Monitoring with Lichens," Lichen Technology, Corvallis, Oregon, vii + 39pp (1973).

Ferry, B. W., Braddeley, M. S., and Hawksworth, D. L., "Air Pollution and Lichens," University of Toronto Press, Toronto, Ontario, 389pp (1973).

Krog, H., "Lav go luftforurensninger: Veilednig ved innsamling av prover med frageillustrasjoner; "Lichens and Air Pollution; Guide to Collecting Samples," with color illustrations, Norsk Institutt for Luftforskning, Oslo, Norway (1970).

MacSwan, I. C., Davison, A. D., and Fenwick, H. S. (eds.), "Proceedings: Workshop on Air Pollution and How it Affects Plants," Oregon State University, Federal Cooperative Extension Services, Corvallis, Oregon, vii + 207pp (1973).

Mansfield, T. A. (ed.), "Effects of Air Pollutants on Plants," Cambridge University, New York, New York, 209pp (1976).

Mudd, J. B., and Kozlawski, T. T., "Responses of Plants to Air Pollution," Academic Press, New York, New York, xii + 383pp (1975).

Naegele, J. A. (ed.), "Air Pollution Damage to Vegetation," "Advances in Chemistry Series 112," American Chemical Society, Washington, D.C., xiii + 137pp (1973).

Silver, I. H. (ed.), "Aerobiology—Proceedings of the First International Symposium," Academic Press, New York, New York, xvi + 278pp (1970).

Table VI Books on Air Pollution Control Technology

A. General Information

"Air and Gas Cleanup Equipment" (2nd ed.), Noyes Data Corp., Park Ridge, New Jersey, 554pp (1972).

"Proceedings of Air Pollution Control Seminar," The University of Arizona Engineering Experiment Station, Tucson, Arizona, 161pp (1970).

"The Air Pollution Abatement Market," Frost and Sullivan, Inc., New York, New York, 233pp (1974).

"Processes for Air Pollution Control," Chemical Rubber Co., Cleveland, Ohio (1972).

"Recent Advances in Air Pollution Control," American Institute of Chemical Engineering, New York, New York (1974).

Batel, W., "Dust Extraction Technology" (R. Hardbottle, trans.), Radley Communications, New York, New York, 272pp (1976).

Buonicore, A. J., and Theodore, L., "Industrial Control Equipment for Gaseous Pollutants," CRC Press, Cleveland, Ohio, Vol. 1, 225pp; Vol. 2, 158pp (1975).

Butt, J. B. and Coughlin, R. W. (eds.), "Important Chemical Reactions in Air Pollution Control," American Institute of Chemical Engineering, New York, New York, 92pp (1971).

Crawford, M., "Air Pollution Control Theory," McGraw-Hill, New York, New York, xv + 624pp (1976).

Cross, F. L., Jr., "Handbook on Air Pollution Control," Technomic Pub., Westport, Connecticut, 171pp (1973).

Dietz, V. (ed.), "Removal of Trace Contaminants from Air," American Chemical Society, Washington, D.C., 207pp (1975).

Kallard, T. (ed.), "Electret Devices for Air Pollution Control," Optosonic, New York, New York, 126pp (1962).

Klein, A., "Clean Air; Air Pollution and the Technical and Practical Possibilities of Regenerating Clean Air (Cold-Heat-Climate Current Topics, Vol. 5)," "Reine Luft; die Verschmutzung der Luft und die Technischen und Praktischen Moeglichkeiten zur Wiederherstellun Reiner Luft (Kaelte-Waerme-Klima Aktuell, Bd. 5)," C. F. Mueller, Karlsrune, Federal Republic of Germany, 165pp (1971).

Knob, W., Heller, A., and Lahmann, E., "Technology of Air Purity Maintenance" (2nd ed.), Technik der Luftreinhaltung, Krausskopf, Mainz, Federal Republic of Germany, 490pp (1972).

Kuznetsov, T. E., "Protection of the Atmospheric Air from Pollution," Zashchita Atmosfernago Vozdukha ot Zagryzneniy, Tavriya, Simferopol; USSR, 125pp (1973).

Marchello, J. M., "Control of Air Pollution Sources," Marcel Dekker, New York, New York, viii + 630pp (1976).

Marchello, J., and Kelly, J. J. (eds.), "Gas Cleaning for Air Quality," Marcel Dekker, New York, New York, 432pp (1975).

McDermott, H. J., "Handbook of Ventilation for Contaminant Control," Ann Arbor Science, Ann Arbor, Michigan, iv + 368pp (1976).

McIlvaine (Company, The), "Scrubber Manual," "Electrostatic Precipitator Manual," "Fabric Filter Manual," loose leaf, with monthly additions, Northbrook, Illinois (current in 1976).

Ledbetter, J. O., "Air Pollution—Part B. Prevention and Control," Marcel Dekker, New York, New York, xiii + 286pp (1974).

Table VI *(Cont.)*

Liptak, B., "Environmental Engineers Handbook, Vol. 2, Air Pollution," Chilton, Philadelphia, Pennsylvania, 1340pp (1974).

Moor, L. F., and Shargorodskii, Y. A., "Measures for Controlling Air Pollution in Cities due to Industrial Plant Emissions," Mery Bor'by s Zagryazneniem Vozdushnogo Basseina Gordov Vybrosami Pred-Privatii," Nauch. Infor. Stroitel 'stvu Arkhitckturc Cosstroya SSSR, Tsent. Inst., Moscow, USSR, 53pp (1970).

Noll, K. E., and Davis, W. T. (eds.), "Power Generation: Air Pollution Monitoring and Control," Ann Arbor Science, Ann Arbor, Michigan, ix + 555pp (1976).

Noll, K. E., and Duncan, J. R. (eds.), "Industrial Air Pollution Control," Ann Arbor Science, Ann Arbor, Michigan, vii + 343pp (1973).

Nonhebel, G. (ed.), "Gas Purification Processes," Chemical Rubber, Cleveland, Ohio, 894pp (1966).

Strauss, W., "Industrial Gas Cleaning: The Principles and Practice of the Control of Gaseous and Particulate Emissions," Pergamon, Long Island City, New York, 472pp (1966).

Vilesov, N. G., and Kostyukovskaya, A. A., "Cleaning of Waste Gases," Ochistka Vybrosnykh Gazov, Tekhnika, Kiev, Ukr. SSR, 194pp (1971).

B. Automobiles

"Effect of Automotive Emission Requirements on Gasoline Characteristics," American Society for Testing and Materials, Philadelphia, Pennsylvania, 165pp (1971).

Ayres, R. U., and McKenna, R. P., "Alternatives to the Internal Combustion Engine: Impacts on Environmental Quality," Johns Hopkins University Press, Baltimore, Maryland, 324pp (1972).

Chandler, R. H., "Automotive Emission Control, Vol. 1—Silencers (Mufflers) and Catalysts, A Review of Patents and Literature for 1972," R. H. Chandler, Braintree, England, 75pp (1973).

Crause, W. H., "Automotive Emission Control," McGraw-Hill, New York, New York, 136pp (1971).

Filiposyants, T. R., and Kratko, A. P., "Ways of Reducing the Smokiness and Toxicity of Exhaust Gases of Diesel Engines"; Puti Snizheniya Dymosti i Toksichnosti Otrabotavshikh Gazov Dizel 'nykh Dvigatelei," Nauch.-Issled. Inst. Inform. Avtomob. Prom, Moscow, USSR, 72pp (1973).

Hurn, R. W. (ed.), "Approaches to Automotive Emission Control, Symposium Series No. 1," American Chemical Society, Washington, D.C., 211pp (1974).

Patterson, D. J., and Henein, N. A., "Emissions from Combustion Engines and Their Control," Ann Arbor Science, Ann Arbor, Michigan, 360pp (1962).

Post, D., "Noncatalytic Auto Exhaust Reduction," Noyes Data Corp, Park Ridge, New Jersey, 278pp (1972).

Springer, G. S., and Patterson, D. J. (eds.), "Engine Emissions: Pollutant Formation and Measurement," Plenum, New York, New York, xii + 371pp (1973).

C. Odors

"State of the Art of Odor Control Technology—SP-5; Proceedings of a Specialty Conference of the Western Pennsylvania Section," Air Pollution Control Association, Pittsburgh, Pennsylvania, x + 230pp (1974).

Table VI (*Cont.*)

Cheremisinoff, P. N., and Young, R. A., "Industrial Odor Technology Assessment," Ann Arbor Science, Ann Arbor, Michigan, x + 509pp (1975).

Cox, J. P., "Odor Control and Olfaction," Sciences Publishing Co., Lyden, Washington, D.C., 500pp (1975).

Cross, F. L., "Air Pollution Odor Control Primer," Technomic, Westport, Connecticut (1973).

Summer, W., "Odour Pollution of Air: Causes and Control," Leonard Hill, London, England, 310pp (1971).

Summer, W., "Methods of Air Deodorization," Elsevier, Amsterdam, The Netherlands, ix + 397pp (1963).

D. Particulate Matter

"Design, Operation and Maintenance of High Efficiency Particulate Control Equipment," Proceedings of a Specialty Conference in St. Louis, Missouri, 1973, Air Pollution Control Association, Pittsburgh, Pennsylvania, 160pp (1973).

"First World Filtration Congress, Paris, France," Halsted Press, New York, New York (1974).

"The User and Fabric Filtration Equipment, SP-4," Proceedings of a Specialty Conference of the Niagara Frontier Section, Air Pollution Control Association, Pittsburgh, Pennsylvania, ix + 154pp (1974); Second Conference, ix + 189pp (1976).

Bamford, W. D., "Control of Airborne Dust," British Cast Iron Res. Assoc., Birmingham, England (1961).

Berlin, B. M., and Bunin, L. V., "Modern Scrubbers and Cyclones for Removing Dust from Gases in the USSR and Abroad," "Sovremennye Skrubbery i Tskilony dlya Pyleochistki Gasov SSR i za Rubezhom," Ekon. Issled. Khim, Neft. Mashinostr. Moscow, USSR, 62pp (1972).

Davies, C. N., "Air Filtration," Academic Press, New York, New York, 138pp (1973).

Englund, H. M., and Berry. W. T. (eds.) "Fine Particulate Control Technology," Air Pollution Control Association, Pittsburgh, Pennsylvania, vii + 214pp (1975).

Heikoff, J. M., "Management of Industrial Particulates; Corporate, Government, Citizen Action," Ann Arbor Science, Ann Arbor, Michigan, ix + 260pp (1975).

Jones, H. R., "Fine Dust Particulates Removal," Noyes Data Corp., Park Ridge, New Jersey, (1972).

Kurkin, V. P., Uzhov, V. N., and Urbakh, I. I., "Removal of Soot from Industrial Gases (Industrial and Sanitational Purification of Gases)," "Ochistaka Promsyshlennykh Gazov ot Sazhi (Promyshlennaya i Sanitarnaya Ochistka Gazov)," Nauch.-Issled. Inst. Tekh. Ekon. Issled. Nefteprererab. Neftekhim. Prom., Moscow, USSR, 126pp (1969).

White, H. J., "Industrial Electrostatic Precipitation," Addison-Welsey, Reading, Massachusetts, 376pp (1963).

E. Source Sampling

"Continuous Monitoring of Stationary Air Pollution Sources," Proceedings of a Specialty Conference, Air Pollution Control Association, Pittsburgh, Pennsylvania, ix + 211pp (1975).

Table VI (*Cont.*)

Brenchely, D. L., Turley, C. D., and Yarmac, R. F., "Industrial Source Sampling," Ann Arbor Science, Ann Arbor, Michigan, xxii + 484pp (1973).
Cooper, H. B. H., and Rossano, A. T., "Source Testing for Pollution Control," Environmental Sciences Services Division, Wilton, Connecticut, 227pp (1970).
Driscoll. J. N., "Flue Gas Monitoring Techniques," Ann Arbor Science, Ann Arbor, Michigan, 440pp (1974).
Driscoll, J. N., "Instrumentation for Process and Environmental Measurements," Ann Arbor Science, Ann Arbor, Michigan, (1975).
Howes, J. E., Jr., Pesut, R. N., and Foster, J. F., "The Average Velocity in a Duct Using ASTM Method D3154-72," American Society for Testing and Materials, Philadelphia, Pennsylvania, ii + 60pp (1974).
Howes, J. E., Jr., Pesut, R. N., and Foster, J. F., "The Relative Density of Black Smoke Using ASTM Method D3211-73T," American Society for Testing and Materials, Philadelphia, Pennsylvania, ii + 48pp (1974).
Hudson, A. E., and Mau, E. E. (eds.), "Introduction to Stack Sampling for Particulates," Lear Siegler, Englewood, California, vi + 43pp (1973).
Konopinski, V. J., Powals, R. J., and Zaner, L. J., "Stack Sampling Estimating Manual," Midwest Environmental Management, Maumee, Ohio, 68pp (1972).
Powals, R. J., and Zaner, L. J., "Sampling Time Estimating Manual," Technomic, Westport, Connecticut, vii + 48pp (1974).

F. Specific Pollutants

1. Hydrogen Sulfide

Ferguson, P. A., "Hydrogen Sulfide Removal from Gases, Air and Liquids," Noyes Data Corp., Park Ridge, New Jersey, x + 350pp (1975).
Stecher, P. G., "Hydrogen Sulfide Removal Processes," Noyes Data Corp., Park Ridge, New Jersey, 280pp (1971).

2. Nitrogen Oxides

"Oxides of Nitrogen Air Pollution Technology: Measurement and Control," Air Pollution Control Association, Pittsburgh, Pennsylvania, ii + 100pp (1973).
Lawrence, A. A., "Nitrogen Oxides Emission Control," Noyes Data Corp, Park Ridge, New Jersey, 212pp (1972).
Rozenfel'd E. I., "Results of Science and Technology, Heat-Engineering Characteristics of Fuel, Use of Gas and Mazut in Industry, Vol. 3: Gas Burners (with Reference to Nitrogen Oxide Pollution)," "Itogi Nauki i Tekhniki, Teplotekhnicheskie Kharakteristiki Topliva, Ispol'zovanie Gaza i Mazuta v Promyshlennosti, T. 3: Gazovye Gorelki," Viniti, Moscow, USSR, 124pp (1973).

3. Sulfur Dioxide

"Coal Utilization Symposium—Focus on SO_2 Emission Control," National Coal Assoc., Washington, D.C., 220pp (1974).

Table VI (*Cont.*)

"Complying with Sulfur Dioxide Regulations," Noyes Data Corp., Park Ridge, New Jersey, 166pp (1972).

Ando, J., "SO_x and NO_x Removal Technology in Japan—1976," Japan Management Assn., Tokyo, Japan, 75pp (1976).

Brodskii, Y. N., Balycheva, K. V., and Brodetskaya, R. N., "Modern Methods of Removing Sulfur from Flue Gases and Their Economy," "Sovremennye Metody Ochistiki Dymovykh Gazov ot Sernistogo Angidrida i Ikh Ekonomika," Tsent. Nauch.-Issled. Inst. Inform. Tekh.-Ekon. Issled. Nefteperab, Nefteknim. Prom., Moscow, USSR, 89pp (1973).

Pfeiffer, J. B., (ed.), "Sulfur Removal and Recovery from Industrial Processes," American Chemical Society, Washington, D.C., 221pp (1975).

Slack, A. V., "Sulfur Removal from Waste Gases," Noyes Data Corp., Park Ridge, New Jersey, 200pp (1971). Also [with Hollindan, G. A.—xii + 294pp (1975)].

G. Specific Processes

1. Agricultural Processes

"Control Technology for Agricultural Air Pollutants—SP-6: Proceedings of a Specialty Conference in Memphis, Tennessee, 1974," Air Pollution Control Association, Pittsburgh, Pennsylvania, 192pp (1974).

2. Fuel and Refuse Combustion

"Air Pollution from Fuel Combustion in Stationary Sources," Organization for Economic Cooperation and Development, Paris, France (1973).

"How Significant are Residential Combustion Emissions?" Proceedings of a Specialty Session, SP-8, Sponsored by the TS-2.3 Residence Committee of the Technical Council, Air Pollution Control Association, Pittsburgh, Pennsylvania, v + 105pp (1974).

"Recent Developments in Controlling Air Pollution from Stationary Combustion Sources," Air Pollution Control Association, Pittsburgh, Pennsylvania, 92pp (1973).

Cohen, A. S., Fishelson, G., and Gardner, J. L., "Residential Fuel Policy and the Environment," Ballinger, Cambridge, Massachusetts, xv + 175pp (1974).

Corey, R. C. (ed.), "Principles and Practices of Incineration," Wiley (Interscience), New York, New York, vii + 297pp (1969).

Edwards, J. B., "Combustion, Formation and Emission of Trace Species," Ann Arbor Science, Ann Arbor, Michigan, 256pp (1974).

Gordynya, R. C., and Voloshchenko, O. I., "Prevention of Atmospheric Pollution Due to Soot," "Profilaktika Zabrudnennya Atmosferi Sazhoyu," Zdorov'ya, Kiev, Ukr. SSR, 40pp (1974).

Noll, K. E., Davis, W. T., and Duncan, J. B. (eds.), "Air Pollution Control and Industrial Energy Production," Ann Arbor Science, Ann Arbor, Michigan, vi + 367pp (1975).

Table VI *(Cont.)*

Scott, D. L., "Pollution in the Electric Power Industry: Its Control and Costs," Heath, Lexington, Massachusetts, xvi + 104pp (1973).

Starkman, E. S., "Combustion-Generated Air Pollution," Plenum, New York, New York, 335pp (1971).

3. Graphic Arts

David, M. P. (ed.), "Proceedings of the 2nd Graphic Arts Technical Foundation Conference on Air Quality Control in the Printing Industry," Graphic Arts Technical Foundation, Pittsburgh, Pennsylvania, 152pp (1973).

4. Metallurgical Processes

"Proceedings of Symposium on Soviet Coke Dry Quenching, May 9–11 1973," Universal Technology Corp., Dayton, Ohio, 142pp (1973).

Andon'ev, S. M., and Filip'ev, O. V., "Dust and Gas Emissions of Ferrous-Metallurgy Plants," Pylegazovye Vybrosy Predpriyatii Chernoi Metallurgii," Metallurgiya, Moscow, USSR, 199pp (1973).

Englund, H. M. and Berry, H. M. (eds.), "Air Pollution Control in the Steel Industry," Air Pollution Control Association, Pittsburgh, Pennsylvania, v + 58pp (1976).

Jones, H. R., "Pollution Control in the Nonferrous Metals Industry," Noyes Data Corp, Park Ridge, New Jersey, 201pp (1972).

Lugarskii, S. I., and Andrianov, I. S., "Cleaning of Gases Leaving Cupolas and Electric Steel Melting Furnaces," "Ochistka Gazov, Otkhodyashchikh ot Vagranok i Elektrostaleplavil'nykh Pechei," Mashinostroenie, Moscow, USSR, 142pp (1972).

5. Petroleum Refining

Jones, H. R., "Pollution Control in the Petroleum Industry," Noyes Data Corp., Park Ridge, New Jersey, x + 322pp (1973).

Krasovitskaya, M. L., "Problems of Air Pollution in the Region of Petroleum Refineries and Petrochemical Plants," "Voprosy Gigieny Atmosfernogo Vozdukha v Raione Neftepererabatyvayushchikh i Neftekhimicheskikh Predpriyatiim," Meditsina, Moscow, USSR, 171pp (1972).

6. Pulp Manufacturing

Jones, H. R., "Pollution Control and Chemical Recovery in the Pulp and Paper Industry," Noyes Data Corp, Park Ridge, New Jersey, 337pp (1973).

Weiner, J., "Air Pollution in the Pulp and Paper Industry," Institute of Paper Chemistry, Appleton, Wisconsin, 101pp (1973).

7. Woodworking

Adams, D. G. (ed.), "Air Pollution Abatement in the Forest Products Industry," Texas Forest Products Laboratory, Lufkin, Texas, 66pp (1971).

environmental pollution, covering not only air pollution, but also pollution of water and land and their measurement, monitoring, effects, and control (Table IV). Table IV is divided into parts: (A) General Information; (B) Control Technology; and (C) Health Effects. Books on ecology have been excluded from this table.

Second, there are books exclusively on air pollution (Tables V and VI). Table V covers books on air pollution, its measurement, monitoring, and effects. It is divided into the following parts: (A) General Information; (B) Aerosols; (C) Ambient Air Measuring and Monitoring; (D) Atmospheric Chemistry and Radioactivity; (E) Health Effects; (F) Meteorology and Modeling; (G) Popular and Children's books; (H) Specific Pollutants; and (I) Vegetation Effects. Table VI covers books on air pollution control technology. It is divided into the following parts: (A) General Information; (B) Automobiles; (C) Odors; (D) Particulate Matter; (E) Source Sampling; (F) Specific Pollutants, which is subdivided into (1) Hydrogen Sulfide; (2) Nitrogen Oxides; and (3) Sulfur Dioxide; and (G) Specific Processes, which is subdivided into (1) Agricultural Processes; (2) Fuel and Refuse Combustion; (3) Graphic Arts; (4) Metallurgical Processes; (5) Petroleum Refining; (6) Pulp Manufacturing; and (7) Woodworking.

Tables IV, V and VI are not all-inclusive. Books specific to the chapters of this book will be found in the reference lists to each chapter and may or may not be repeated in these tables. A list of air pollution books published in Russian in the Soviet Union appears on pages 165–168 of "Profile Study of Air Pollution Control Activities in Foreign Countries" APTD 0601, published by National Air Pollution Control Administration, Public Health Service, U.S. Department of Health, Education, and Welfare, Research Triangle Park, North Carolina.

III. Secondary Resources

In order to facilitate access to the literature, a wide variety of secondary resources have been established and are accessible via searches or via a publication. Each of these resources may be characterized by type of service, its processing and delivery system, and its indexing system.

A. Information Retrieval Services

1. Searching Services

These services provide (1) searches for the client on their own information system, (2) direct access to their searching system for subscribers who may then do it themselves on-line, or (3) availability of tapes for sale or lease (Table VII).

2. Publishing Services

The service provided is a publication the reader may use to retrieve information. They include the common index and abstract services, such as *Index Medicus, Chemical Abstracts, Physics Abstracts, and Engineering Index,* plus specialized pollution abstracting services.

Table VII Search Services

Search[b] *service*	*Performs searches*	*National on-line access*	*Publication*
APTIC	Yes	Yes	*Air Pollution Abstracts*
BIOSIS	No	Yes	*Biological Abstracts*
CAIN (AGRICOLA)	No	Yes	No
CHEMCON	No	Yes	*Chemical Abstracts* (without the abstracts)
COMPENDEX	No	Yes	*Engineering Index*
ENVIROLINE	No	Yes	Environment Abstracts
GEOREF	No	Yes	*Bibliography and Index of Geology*
INFORM	No	Yes	No
INSPEC	No	Yes	*Electronics Abstracts; Computer & Control Abstracts; Physics Abstracts*
MARC	No	Yes	No
MEDLINE	No	Yes	*Index Medicus* (in part)
MGA	No	Yes	*Meteorological and Geoastrophysical Abstracts*
NIOSH	Yes	No	No
NTIS	No	Yes	Government Reports Announcements
PANDEX	No	Yes	No
POLLUTION ABS.	No	Yes	*Pollution Abstracts*
PTS	No	Yes	Several
SCISEARCH	No	Yes	No
SSIE	Yes	Yes	No
SWIRS	Yes	No	*Accession Bulletin*
TOXLINE	No	Yes	*Pesticides Abstract Bulletin; Index Medicus* (in part)

[a] The earliest year of input that is searchable varies widely among these services. If searches for earlier years are needed, these will require relatively expensive, supplemental manual searches. Many of these services overlap in their coverage; therefore, when the same search is run on two or more services, duplicate items must be manually deleted.

[b] Additional information may be sought in "Encyclopedia of Information Systems and Services" (A. T. Kruzas, ed.), Second International Edition, Edwards Brothers, Ann Arbor, Michigan (1974).

B. Processing and Delivery Systems (9)

1. Computerized Systems

In these systems, input is generally in the form of a master record containing the bibliographic description of the literature being added, assigned descriptor index terms (if used), and an abstract (if any). Although simple sequential access methods have been used for non-numeric information retrieval, they have been largely displaced by inverted file methods (indexed sequential access). The master record is processed by "inverting" it; i.e., the accession number for each document is added to a file for each index term that was selected for the document. The master records are maintained in a sequential file. When a search is made on any index term, the inverted file for that index file yields a list of the relevant accession numbers, and the master records (or specified parts) are retrieved from its sequential file. When a logical combination of index terms is used, the lists of accession numbers from the inverted file for each of the index terms are first matched with each other, and then the resulting hit list is used to retrieve the specified part(s) of the master records. In some systems, the assignment of index terms is omitted, and all words in the title and/or abstract (if any) are inverted to produce a kind of concordance. Searches may be made in the same way as with assigned index terms. In some systems, which use a tree structure, the master records may be processed in such a way that any element of them is stored only once, yet access to every record containing that element is maintained.

In computerized systems, information is usually stored on magnetic disks, magnetic drums, magnetic tape, or punched cards. Magnetic disks and drums are random access devices; i.e., information stored anywhere on them may be accessed without passing through the other information stored on them. Magnetic tape and punched cards are sequential access media; i.e., the Nth record may be accessed only by first passing through $N - 1$ records. When immediate access to the information can be waived, any of the storage media may be used, depending on the system design, cost, availability, and other criteria. However, when access must be available in real time, disks or drums must be used. In the APTIC system, the input records were on magnetic tape. These records were inverted, and the accession numbers (APTIC numbers) were posted to the file of each authorized term that was previously assigned by the indexer. The inverted files were stored on magnetic disk. Each bibliographic description was added to a sequential citation file, which was also stored on a disk. The processed master record, including the abstract, was maintained sequentially on magnetic tape. When real time access may be waived, the index term or combination of index terms to be used for each search was accumulated into a batch and submitted to the computerized information retrieval system at night. For each search, the lists of accession numbers from the inverted file for each of the index terms were matched to produce a hit list of accession numbers. Each hit list was merged with hit lists resulting from the other searches within this batch. The merged hit list was then used to retrieve the bibliographic description and the abstract for each accession number during a single pass of the sequential master file on magnetic tape. The retrieved information was then sorted in accordance with the hit list for each search and printed out. When real time access must be obtained, the index term or combination of index terms to be used for the search was entered on a terminal connected on-line to the computerized information retrieval system. The lists of accession numbers from the inverted file for each of the index terms were matched to produce a hit list of accession numbers. At the user's option, the hit list could then be used to retrieve the bibliographic descriptions from the citation file on disk and the descriptions were displayed on the terminal continuously or in groups of five.

2. Manual Systems

Manual systems continue to play an important role with relatively small collections (<10,000 accessions) such as personal information retrieval systems. There are some successful systems that use a manual system combined with a computerized system. Conventional filing systems (vertical files, visible files, rotary files, shelf files, etc.) are manual systems. The files are structured, arranged, or classified for direct searching according to the key that would be most frequently used to look up information (name, subject, etc.). Manual cross reference indexes may also be maintained within the same files or separately to help find information when users ask for it with a key that is different from any by which the file is structured. Because a single copy of the document can be filed under only one key at any given time, it can only be retrieved by that single key or its synonymous cross references. Such conventional filing systems are most successful where the information in the documents are clearly mutually exclusive and little or no doubt exists regarding the one "true" classification for each document.

Conventional systems have the following advantages: (a) design and operation are simple; (b) general office equipment may be used; (c) direct access and browsing are facilitated; and (d) input costs are relatively low. Their disadvantages are: (a) classification schemes can rarely satisfy every need; (b) classification schemes are difficult to revise; (c) classification schemes are difficult to apply to documents that have multiple subjects; and (d) cross references may be lost or misfiled.

Some other manual methods are edge-notched cards; multiple index cards, columnar cards, and optical coincidence cards. These other systems generally file the documents by a simple accession number rather than according to a structure or classification scheme designed for direct retrieval. Many independent keys can be assigned to each document, since they identify but do not predetermine its location in the file. Each document may be described with as much detail or specificity as desired. These other systems are most successful where the information in the documents is rarely mutually exclusive and often may be sought to match multiple criteria. These systems have two advantages: (a) they can be used to retrieve information with greater specificity and (b) they can retrieve information with greater speed. Their disadvantages are (a) specially qualified personnel are usually required to design and operate the system; (b) special equipment is usually required; (c) special procedures and techniques are often required to retrieve information; and (d) input costs are usually higher.

Edge-notched card systems employ cards with holes (usually round) along one or more of the edges. Usually, a single card is prepared for each document added to the system. Each of the holes in the card represents a key that may be assigned to the document. To assign a particular key to the document, the area of the card between the hole and the edge is punched out, or notched. To search the file of cards, a needle is passed through the appropriate hole in the deck of edge-notched cards and the needle is pulled toward the edge of the cards. The cards that are notched at that hole will remain behind, because the needle will be pulled through the notched out area. Unnotched cards will go with the needle. An additional pass of the needle through the remaining cards is usually required for each additional key in the search criteria. The cards are available in various sizes and formats, and the interior of each card may be used to type or otherwise record additional information about the document. Edge-notched cards have these advantages: (a) low cost; (b) simplicity; (c) browsability; (d) direct access to all information recorded on the card; and (e) frequent possibility of random filing. Their disadvantages are (a) the quantity of keys that may be used throughout the system is limited by the physical size of the card, or conversely, the perimeter of the card is proportional to the quantity of keys specified in the design of the system; (b) the duration of the search is proportional to the quantity of keys in the search criteria; and (c) the size of the file has a relatively low practical limit.

Multiple index card systems record the descriptive information for each document on a card and duplicate it; a sufficient number of duplicate cards is made so that one may be filed for each key assigned to each document. Multiple index card systems have these advantages: (a) simple equipment may be used; (b) design and operation are simple; (c) browsability; and (d) direct access to all information recorded on the card. The disadvantage of these multiple index card systems is that the bulkiness of the files is directly proportional to the total quantity of keys assigned to all of the documents.

Columnar card systems use a single card for each key. Each card is divided into ten columns, numbered 0 through 9, and the accession number of each document assigned any key is posted in the column matching the last digit of that number. The cards are searched by pulling the card for each key in the search criteria, and matching accession numbers column by column to locate hits. Columnar card systems have also been enhanced by so-called dual dictionary systems. Dual dictionaries provide two identical sets of cards, usually bound together side by side. Instead of pulling each card from a file, the searcher turns to one card in each dictionary and matches the accession numbers. A further enhancement of the dual dictionary system is a template with two openings separated by the distance between identically numbered columns and each with the same shape and size. The template may be moved along the width of the cards to expose only the two columns of numbers to be matched. Columnar card systems have these advantages: (a) relatively low costs for supplies and equipment; (b) parallel searching can be done using two keys at a time; (c) simplicity of design, maintenance, and use; and (d) enhanceability by addition of the dual dictionary and template. Their disadvantages are: (a) no direct access to all recorded descriptive information; (b) duration of the average search is proportional to the quantity of documents in the system; and (c) if the dual dictionary is used, the second set of cards makes the system relatively more costly. A columnar card system was used in one of the first major air pollution information retrieval systems, that of the nine-county Bay Area Air Pollution Control District of California.

Optical coincidence, or "peak-a-boo," systems employ cards with a fixed number of dedicated positions for making holes to represent the location of documents being indexed. For each key used in the system, a separate card is maintained. After each additional document has been assigned a four-digit accession number plus index keys, the cards for all of the assigned keys are pulled from the file, overlaid, and a hole is made in the cards at the position at which the rectangular coordinates correspond to the leading and trailing two digits of the accession number. To search, the card for each key in the search criteria is pulled from the file, overlaid on a light source, and the coordinates of the illuminated "optically coincident" holes are read with a graduated T square. The accession number is immediately reconstructed from the coordinates. The cards are usually made of a durable plastic and are approximately 9 inches (23 cm) square. The hole-making device and graduated T square accommodate one hundred positions (numbered 0 through 99) on each coordinate axis, representing a total of 10,000 documents per deck of cards. Prescored punched cards accommodating 480 documents have also been used. Optical coincidence systems have these advantages: (a) relatively rapid, parallel searching can be done using as many keys as necessary at one time; (b) relatively low cost for supplies and equipment; and (c) simplicity of design, maintenance, and use. Their disadvantages are (a) no direct access to all information recorded for the document; (b) relatively difficult to correct errors; and (c) relatively difficult to convert to a computerized system. An optical coincidence system was successfully used in the Air Pollution Technical Information Center for several years before conversion to a computerized system (*10*).

C. Indexing Methods and Formats

In conventional manual systems, the method traditionally used is either a hierarchical classification scheme that predetermines the organization and arrangement of the documents themselves, or a subject-heading scheme that assigns several broad stand-alone headings to cards for each document.

1. Coordinate Indexing

The foundation upon which most computerized systems and nonconventional manual systems are built is coordinate indexing. The information in each document is described without limitation to a single classification or to a broad subject heading or two; rather, it may be described with greater detail or specificity by using as many keys as warranted. This is possible because such systems primarily rely not on a structure imposed on the files of documents or their descriptive information, but rather on much greater specificity (and therefore quantity) of keys assigned to the documents, together with a system to efficiently manipulate those keys. The available keys are not intended to be used alone but rather to be logically combined or "coordinated" in conceptually describing the information in each document. In searching, the keys may also be freely coordinated in a great variety of ways. Coordinate indexing methods have these advantages: (a) greater specificity or depth in describing information; (b) greater adaptability to changing situations; and (c) greater manipulative possibilities in searching for information.

2. Key Word Methods

Although "key word" is often loosely used to refer to any subject index key, in the parlance of information retrieval, it is generally intended to mean words selected from the title of a document. The vocabulary of index keys is thus a by-product of the indexing process. The key words may be recorded by manual or machine (automatic) indexing methods. Indexes based on key words are also known as permuted indexes. Two of the better known types of permuted index are the Keyword-In-Context (KWIC) index and the Keyword-Out-of-Context (KWOC) index. In the KWIC index, the title is printed in such a way that each keyword is started in a column in the middle of the line, with the other words in the title in their normal relation to the key word. If the trailing part of the title exceeds the length of the line, it may be wrapped around and be printed at the other end of the line if the leading part of the title would not otherwise be printed there. In the KWOC index, the key word is extracted from the title and is printed at one of the margins, usually the left margin, and the title is printed verbatim next to it. Permuted indexes are sometimes enhanced by appending the author's name and/or descriptors (form controlled vocabularies) to the titles before processing. Permuted indexes have these advantages: (a) relatively low cost; (b) speed and ease of preparation; (c) ease of revisions; and (d) relative familiarity and browsability to the reader. Permuted indexes have the disadvantage that synonyms are not cross referenced; therefore, the reader must serve himself. The completeness of any search is limited by the searcher's personal recall of synonyms of words the authors used in their titles and by variations in the usage and meanings of the words searched. Permuted indexes have been used in the Pennsylvania State University Center for Air Environment Studies' "Air Pollution Titles" (*11*).

3. Descriptor Methods

Descriptors are index terms that have been specially prepared through a continuous or recurring process of analysis of the documents being indexed and the search questions being asked. Descriptors are formalized and controlled by means of a thesaurus. Analysts assign the authorized descriptors on a conceptual basis to each document. Descriptor methods have these advantages: (a) greater ease and accuracy in searching if properly used; (b) greater possibility of indexing concepts rather than merely words; and (c) better conservation of space, especially if the title does not appear in the index. Their disadvantages are (a) specially qualified staff is required; (b) method is limited in proportion to human inconsistency and error; and (c) more human effort is required for indexing.

IV. Tertiary Resources

There are many resource organizations (libraries, dissemination centers, etc.) that cannot be considered secondary resources, because they do not provide searches from their own information systems, nor direct access to a searching system, nor tapes for sale or lease. They function as middle men. They are intermediaries in two senses: (a) they provide searches from outside third-party information systems, and (b) they have an intermediate level of knowledge of the information systems, somewhere between that of the uninitiated requester and the persons responsible for producing the information system. In many cases, the requester needs to use these resources because the producers of the information systems will not provide searches, or direct access, or tapes directly to the ultimate user. If they will provide tapes, they may often be too expensive for the user. In other cases, the producers would supply the needed service, but the requestor relies on the intermediary's knowledge of the availability and suitability of information systems and their best efforts to retrieve and sometimes to review the information. Many of the information systems that will not provide either searches or tapes use key word automatic indexing methods, are directly accessable, and encourage searches by the ultimate user, who, of course, is most familiar with the terminology in his field and is thus best able to use these systems' capability of revising the search criteria and browsing on-line. With these systems, the searcher needs no special knowledge of information systems or the sets of descriptors that may be used by them.

Tertiary searching services have many limitations. Third-party searching systems often rely primarily on free text or automatic key word indexing. Even when descriptors are also indexed, tertiary services may not invest in the training needed to effectively use them for retrieval. The simple-to-use free text searching systems are usually not available to the user, because the service often insists on interposing itself as another layer between the ultimate user and the system, with increased probability of errors of transmission. Even when the tertiary service refers requests to the searching service, primary communication between the requester and the service has been interdicted. Systems that rely on free text key words generally consume more computer time and connect time than they would if they relied on a descriptor system with the same specificity; this is because more inverted files must be maintained (for synonyms, spelling variants, etc.) and the files for all of the synonymous key words must be searched. In free text systems there is no possibility of retrieving any information that does not contain the word searched; therefore, the results will, superficially, appear to be precisely relevant. However, because of variations in usage and meanings of words, that precision is often illusory. Also, because no one can be omniscient about synonyms, the ultimate user cannot be confident about the completeness of recall of the relevant information.

In general, the following procedure is recommended:

a. If producers will provide searches from their own information systems, it is best to deal directly with them, especially since they are most familiar with their systems, are very likely to have access to resources not available to an intermediary, and will afford communication without any errors of transmission by an intermediary.

b. If the producer will not provide searches from its own information system but will provide direct access to a key word searching system, make hands-on use of it yourself on-line.

c. If the producer will provide tape, analyze whether you would use it frequently enough to make it the most economical option.

Table VIII National Aeronautics and Space Administration (NASA) Industrial Applications Centers

Aerospace Research Applications Center (ARAC) Indiana University 400 East 7 Street Bloomington, Indiana 47401	North Carolina Science and Technology Research Center (NC/STRC) P. O. Box 12235 Research Triangle Park, North Carolina 27709
Knowledge Availability Systems Center (KASC) University of Pittsburgh Pittsburgh, Pennsylvania 15260	Technology Application Center (TAC) The University of New Mexico Albuquerque, New Mexico 87131
New England Research Application Center (NERAC) Mansfield Professional Park, Box U-41N The University of Connecticut Storrs, Connecticut 06268	Western Research Application Center (WESRAC) University of Southern California 809 West 34 Street Los Angeles, California 90007

d. If you cannot determine which information systems are most likely to be suitable, or if you wish to rely on an intermediary's special knowledge (if any) of various descriptor searching systems or on his efforts to review the retrieved information, use his services. For some users this fourth step may be the method of choice.

As examples of tertiary resources that provide more than just dissemination of printouts, Table VIII lists the NASA Industrial Applications Centers (formerly Regional Dissemination Centers). There are also some resources that are tertiary in the sense that they use the output of secondary resources as their primary source of input. One searching service of this type is the Canadian National Research Council's pollution information project (*12*); a publishing service of this type is "Pollution Look-Out" (Stockholm) (*13*).

V. Newsletters (See also Chapter 5, Vol. V)

Numerous organizations issue publications which are intended to keep the reader aware of recent developments; they are known as newsletters. State and local air pollution control agencies often are sources of newsletters. Other sources include universities, trade associations, private industries, and contract research organizations. Many newsletters also cover other fields in addition to air pollution. If the frequency of publication is much less than monthly, the term "newsletter" may be a misnomer.

VI. Translations

A. Into English

Many documents dealing with air pollution have been translated and are available from commercial and governmental sources (Table IX). A clearinghouse which disseminates translations is: National Translations Center; The John Crerar Library; 35 West 33 Street; Chicago, Illinois 60616. In addition, a cover-to-cover translation of *"Staub-Reinhaltung der Luft"* (German); is available as "Staub-in-English" from Foreign Resources Associates; P. O. Box 2352; Fort Collins, Colorado 80521, and *"Ochrona Powietrza"* (Polish) is available as "Air Conservation" from the United States National Technical Information Service (NTIS) (*14*).

Table IX Sources of Translations into English

BISITS (primarily regarding metals) 1 Carlton House Terrace London SW1Y 5DB, England	Associated Technical Services, Inc. P. O. Box 271 East Orange, New Jersey 07019
Henry Brutcher Technical Translations (primarily regarding metals) P. O. Box 157 Altadena, California 91001	Israel Program for Scientific Translations Kiriat Moshe P. O. Box 7145 Jerusalem, Israel
Find Company Limited (from Japanese) Togo Building 1-5-7, Yagumo, Meguro-ku Tokyo 152, Japan	National Technical Information Service (NTIS), U.S. Department of Commerce 5285 Port Royal Road Springfield, Virginia 22161
	British Library, Lending Division Boston Spa Yorkshire, England
Trans Chem., Inc. (chemistry) P. O. Box 6219 Knoxville, Tennessee 37914	Plenum Publishing Corporation 227 West 17 Street New York, New York 10011

Finally, the publication "Air Pollution Translations: A Bibliography with Abstracts" is available in five volumes (*15*).

B. Into French and German

A prime source of air pollution documentation in French is: Centre de Documentation; Centre Interprofessional Technique d'études de la Pollution Atmospherique; 3, Rue Henri-Heine, 75016 Paris, France. A prime source of air pollution documentation in German is: Verein Deutscher Ingenieure; VDI-Kommission Reinhaltung der Luft; 4, Dusseldorf 1, Postbox 1139, Federal Republic of Germany.

VII. Audiovisual Materials

There are numerous slides, motion pictures, audio and video recordings, and other audiovisual media available. One source of information about such media is the United States Environmental Protection Agency Office of Public Affairs, Washington, D.C., and counterpart offices in the agency's ten regional offices and four environmental research centers. The Washington office includes the agency's Documerica project for still photography and an audiovisual branch. Many films are distributed through the National Audiovisual Center.*

VIII. The Information Network

All efforts to establish a United States national environmental information network have foundered on the rocks of budgetary constraints. For the present, we must be content with a network of relationships within the information rather than among information systems (*16*).

IX. Bibliographies, Directories, and Guides

A. *Bibliographies*

The first comprehensive air pollution bibliography was published by the United States Bureau of Mines in 1954 and covered the literature back to the mid 1850s (*17*). This was followed by two Library of Congress air pollution bibliographies published in 1957 and 1959, respectively (*18, 19*). This was paralleled and followed by the publication by the Air Pollution Control Association of *APCA Abstracts* from 1955 to 1971 (*20*). *Air Pollution Titles* has been published bimonthly from 1965 to date, with each November–December issue being cumulative for the year (*11*). *Air Pollution Abstracts* (*7*) started publication in 1970 as *NAPCA* (National Air Pollution Control Administration, Public Health Service, U.S. Department of Health, Education and Welfare) *Abstract Bulletin* and changed its name to *Air Pollution Abstracts* in 1971. It ceased publication in 1976. An annual review of the air pollution literature of special interest to the paper and pulp industry has been published since 1964 (*21*). The Air Pollution Technical Information Center (APTIC) of the United States Environmental Protection Agency and its predecessors has published a number of specialized bibliographies (Table X).

B. *Directories and Guides*

Air pollution directories and guides are of two types. There are guides to equipment published in one issue each year of *Chemical Engineering*—"Equipment Buyer's Guide"; *Environmental Science and Technology*—"Pollution Control Directory"; *Journal of the Air Pollution Control Association*—"Product Guide"; and *Science*—"Guide to Scientific Instru-

* National Audiovisual Center, Washington, D.C.

Table X Bibliographies Published by Air Pollution Technical Information Center[a]

Air Pollution Aspects of Emission Sources: A Bibliography with Abstracts
Boilers, AP-105, v + 125pp (1972)
Cement Manufacturing, AP-94, vii + 44pp (1971)
Coke Ovens, EPA-450/1-74-002, iv + 55pp (1974)
Electric Power Production, AP-96, vii + 312pp (1971)
Ferrous Foundries, EPA-450/1-74-004, iv + 38pp (1974)
Iron and Steel Mills, AP-107, v + 84pp (1972)
Municipal Incineration, AP-92, vii + 95pp (1971)
Nitric Acid Manufacturing, AP-93, vii + 31pp (1971)
Petroleum Refineries, AP-110, v + 68pp (1972)
Primary Aluminum Production, AP-119, v + 57pp (1973)
Primary Copper Production, AP-125, v + 40pp (1973)
Primary Lead Production, AP-126, v + 29pp (1973)
Primary Zinc Production, EPA-450/1-74-003, iv + 30pp (1974)
Pulp and Paper Industry, AP-121, v + 166pp (1973)
Sulfuric Acid Manufacturing, AP-95, vii + 58pp (1971)
Surface Coatings—Their Production and Use, EPA-450/1-74-005, vi + 68pp (1974)
Asbestos and Air Pollution: An Annotated Bibliography, AP-82, iii + 101pp (1971)
Beryllium and Air Pollution: An Annotated Bibliography, AP-83, iii + 75pp (1971)
Carbon Monoxide: A Bibliography with Abstracts, Public Health Service Publication 1503, viii + 440pp (1966)
Chlorine and Air Pollution: An Annotated Bibliography, AP-99, iii + 113pp (1971)
Fluorine, Its Compounds, and Air Pollution: A Bibliography with Abstracts, EPA-450/1-76-003, v + 598pp (1976)
Hydrocarbons and Air Pollution: An Annotated Bibliography, AP-75, Part 1, iii + 582pp, Part 2, iii + 600pp (1970)
Hydrochloric Acid and Air Pollution: An Annotated Bibliography, AP-100, iii + 107pp (1970)
Lead and Air Pollution: A Bibliography with Abstracts, EPA-450/1-74-001, v + 431pp (1974)
Mercury and Air Pollution: A Bibliography with Abstracts, AP-114, v + 59pp (1972)
Nitrogen Oxides: An Annotated Bibliography, AP-72, iii + 633pp (1970)
Odors and Air Pollution: A Bibliography with Abstracts, AP-113, v + 257pp (1972)
Photochemical Oxidants and Air Pollution: An Annotated Bibliography, AP-88, Part 1, iii + 811pp; Part 2, iii + 714pp (1971)
Sulfur Oxides and Other Sulfur Compounds: A Bibliography with Abstracts, Public Health Service Publication 1093 (Bibliography Series 56), xi + 383pp (1965)

[a] U.S. Environmental Protection Agency (and its predecessors), Research Triangle Park, North Carolina

ments," and others. There is a "Directory of Governmental Air Pollution Agencies" published annually by the Air Pollution Control Association, Pittsburgh, Pennsylvania.

REFERENCES

1. W. Seidl, *in* "Proceedings of the International Clean Air Congress," Part I, pp. 287–290. Nat. Soc. Clean Air, London, England, 1966.

2. W. Seidl, *VDI* (*Ver. Deut. Ing.*) **149,** 359–366 (1970).
3. H. M. England, *in* "Air Pollution Control" (W. Strauss, ed.), Part II, pp. 255–289. Wiley (Interscience), New York, New York, 1972.
4. J. S. Nader, *in* "Air Pollution" (A. C. Stern, ed.), 2nd ed., Vol. 3, pp. 813–823. Academic Press, New York, New York, 1968.
5. American Chemical Society, "CAS Today: Facts and Figures About Chemical Abstracts Service." Chem. Abstr. Serv., Columbus, Ohio, 1974.
6. G. Anderla, "Information in 1985: A Forecasting Study of Information Needs and Resources." Organization for Economic Cooperation and Development, Paris, France, 1973.
7. "Air Pollution Abstracts." Air Pollution Technical Information Center, Research Triangle Park, North Carolina, 1970/1976.
8. F. Renner, P. Halpin, and B. E. Epstein, *Spec. Libr.* **62,** 421 (1971).
9. General Services Administration, "Information Retrieval." Gen. Serv. Admin., Washington, D.C., 1972.
10. S. A. Tancredi and O. D. Nichols, *Amer. Doc.* **19,** 66 (1968).
11. "Air Pollution Titles." Center for Air Environment Studies, Pennsylvania State University, University Park, Pennsylvania.
12. "Pollution Information Project." National Science Library of Canada, Ottawa, Ontario.
13. "Pollution Look-out." Karolinska Institute, Stockholm, Sweden.
14. "Air Conservation." National Technical Information Service, U.S. Department of Commerce, Springfield, Virginia.
15. "Air Pollution Translations: A Bibliography with Abstracts," Vol. 1 (AP-56) (1969); Vol. 2 (AP-69) (1970); Vol. 3 (AP-120) (1973); Vol. 4 (AP-122) (1973); Vol. 5 (EPA-220/2-74-001) (1974). United States Environmental Protection Agency, Research Triangle Park, North Carolina (and predecessor organizations), 1969–1974.
16. E. Garfield, *Science,* **178,** 471 (1972).
17. S. J. Davenport and G. G. Morgis, *U.S., Bur. Mines, Bull.* **537,** 1954.
18. J. R. Gibson, W. E. Culver, and M. E. Kurz, "The Air Pollution Bibliography," Vol. I. Library of Congress, Washington, D.C., 1957.
19. A. J. Jacobius, J. R. Gibson, V. S. Wright, W. E. Culver, and L. Kassianoff, "The Air Pollution Bibliography," Vol. II. Library of Congress, Washington, D.C., 1959.
20. *APCA Abstracts,* Vol. 1 (1955)/Vol. 16 (1970). Air Pollution Control Association, Pittsburgh, Pennsylvania.
21. "Atmospheric Quality Protection Literature Review," Atmospheric Quality Improvement Tech. Bull., NCASI Tech. Bull. No. 22 (1965); No. 33 (1967); No. 36 (1968); No. 40 (1969); No. 45 (1970); No. 53 (1971); No. 61 (1972); No. 65 (1973); No. 74 (1974); No. 77 (1975); National Council of The Paper Industry for Air and Stream Improvement, New York, New York, 1965–1975.

Subject Index

B

C

D

F

I

J

K

L

  ambient air level, 534, 536–537

P

Q

R

T

U

V

ENVIRONMENTAL SCIENCES

An Interdisciplinary Monograph Series

ARTHUR C. STERN, editor, AIR POLLUTION, Second Edition, Volumes I–III, 1968; Third Edition, Volumes I, III, 1976; Volumes II, IV, 1977; Volume V, in preparation

L. FISHBEIN, W. G. FLAMM, and H. L. FALK, CHEMICAL MUTAGENS: Environmental Effects on Biological Systems, 1970

DOUGLAS H. K. LEE and DAVID MINARD, editors, PHYSIOLOGY, ENVIRONMENT, AND MAN, 1970

KARL D. KRYTER, THE EFFECTS OF NOISE ON MAN, 1970

R. E. MUNN, BIOMETEOROLOGICAL METHODS, 1970

M. M. KEY, L. E. KERR, and M. BUNDY, PULMONARY REACTIONS TO COAL DUST: "A Review of U. S. Experience," 1971

DOUGLAS H. K. LEE, editor, METALLIC CONTAMINANTS AND HUMAN HEALTH, 1972

DOUGLAS H. K. LEE, editor, ENVIRONMENTAL FACTORS IN RESPIRATORY DISEASE, 1972

H. ELDON SUTTON and MAUREEN I. HARRIS, editors, MUTAGENIC EFFECTS OF ENVIRONMENTAL CONTAMINANTS, 1972

RAY T. OGLESBY, CLARENCE A. CARLSON, and JAMES A. McCANN, editors, RIVER ECOLOGY AND MAN, 1972

LESTER V. CRALLEY, LEWIS T. CRALLEY, GEORGE D. CLAYTON, and JOHN A. JURGIEL, editors, INDUSTRIAL ENVIRONMENTAL HEALTH: The Worker and the Community, 1972

MOHAMMED K. YOUSEF, STEVEN M. HORVATH, and ROBERT W. BULLARD, PHYSIOLOGICAL ADAPTATIONS: Desert and Mountain, 1972

DOUGLAS H. K. LEE and PAUL KOTIN, editors, MULTIPLE FACTORS IN THE CAUSATION OF ENVIRONMENTALLY INDUCED DISEASE, 1972

MERRIL EISENBUD, ENVIRONMENTAL RADIOACTIVITY, Second Edition, 1973

JAMES G. WILSON, ENVIRONMENT AND BIRTH DEFECTS, 1973

RAYMOND C. LOEHR, AGRICULTURAL WASTE MANAGEMENT: Problems, Processes, and Approaches, 1974

LESTER V. CRALLEY, PATRICK R. ATKINS, LEWIS J. CRALLEY, and GEORGE D. CLAYTON, editors, INDUSTRIAL ENVIRONMENTAL HEALTH: The Worker and the Community, Second Edition, 1975

A 7
B 8
C 9
D 0
E 1
F 2
G 3
H 4
I 5
J 6